ROUTLEDGE HANDBOOK OF THE RESOURCE NEXUS

In recent years the concept of the resource "nexus" has been both hotly debated and widely adopted in research and policy circles. It is a powerful new way to understand the myriad complex relationships between multiple resources, actors and their security concerns. Particular attention has been paid to water, energy and food interactions, but land and materials emerge as critical too. This comprehensive handbook presents a detailed review of current knowledge about resource nexus-related frameworks, methods and governance, including a broad set of inter-disciplinary perspectives.

Written by an international group of scholars and practitioners, the volume focuses on rigorous research, including tools, methods and modelling approaches to analyse resource use patterns across societies and scales from a "nexus perspective". It also provides numerous examples from political economy to demonstrate how resource nexus frameworks can illuminate issues such as land grabs, mining, renewable energy and the growing importance of economies such as China, as well as to propose lessons and outlooks for sound governance.

The volume seeks to serve as an essential reference text, source book and state-of-the-art, science-based assessment of this increasingly important topic – the resource nexus – and its utility in efforts to enhance sustainability of many kinds and implement the United Nations Sustainable Development Goals in an era of environmental and geopolitical change.

Raimund Bleischwitz is Chair in Sustainable Global Resources and Deputy Director at the UCL Institute for Sustainable Resources, University College London, UK.

Holger Hoff has a joint appointment as a Senior Researcher in the Resources and Development Group at the Stockholm Environment Institute, Sweden, and in the Research Domain "Earth System Analysis" at the Potsdam Institute for Climate Impact Research, Germany.

Catalina Spataru is a Lecturer on Energy Systems and Networks and Director of the MRes course Energy Demand Studies at UCL Energy Institute, University College London, UK.

Ester van der Voet is an Associate Professor of Industrial Ecology at the Institute of Environmental Sciences, Leiden University, The Netherlands. She is also a member of the UN International Resource Panel.

Stacy D. VanDeveer is a Professor of Global Governance and Human Security at the McCormack School of Policy and Global Studies at the University of Massachusetts Boston, USA.

"The language of the nexus highlights the need for interconnected thinking between the natural and social sciences, and between the research community and decision makers. By bringing together such an outstanding range of thinkers and perspectives from across the world, *Routledge Handbook of the Resource Nexus* looks set to become an indispensable volume for all those engaged in these debates. It sets out with admirable clarity the theoretical, empirical and research underpinnings of this field, and will provide nexus brokers and boundary-spanners with the intellectual tools they need to achieve tangible progress."

– James Wilsdon, *Professor of Research Policy, University of Sheffield, UK and Director, ESRC Nexus Network*

"No energy myopia, no water myopia and no food cry of alarm, but a great handbook addressing the mutual relations."

– Ernst von Weizsäcker, *Past Co-Chair, International Resource Panel*

"This handbook provides a qualified response, given by key experts, to the recurring question of what the Resource Nexus actually means. That nexus is presented as a useful heuristic for addressing the new challenges of the Anthropocene and for evidence-based support of a sustainability transition. Integrative methods and tools for systemic resource management and governance are illustrated, adaptable to a wide range of thematic and geographic contexts and scales."

– Professor Hans Joachim Schellnhuber, *Director of the Potsdam Institute for Climate Impact Research (PIK), Germany and Senior Research Fellow at the Stockholm Resilience Centre, Sweden*

ROUTLEDGE HANDBOOK OF THE RESOURCE NEXUS

Edited by Raimund Bleischwitz, Holger Hoff, Catalina Spataru, Ester van der Voet and Stacy D. VanDeveer

Routledge
Taylor & Francis Group
LONDON AND NEW YORK

earthscan
from Routledge

First published 2018 by Routledge

2 Park Square, Milton Park, Abingdon, Oxfordshire OX14 4RN
52 Vanderbilt Avenue, New York, NY 10017

Routledge is an imprint of the Taylor & Francis Group, an informa business

First issued in paperback 2020

British Library Cataloguing-in-Publication Data
A catalogue record for this book is available from the British Library

Library of Congress Cataloging-in-Publication Data
A catalog record for this book has been requested

ISBN: 978-1-138-67549-0 (hbk)
ISBN: 978-0-367-50528-8 (pbk)

Typeset in Bembo
by Apex CoVantage, LLC

CONTENTS

FIGURES

CONTRIBUTORS

Tundi Agardy is Director of the MARES Program at Forest Trends, Washington DC, USA.

Paolo Agnolucci is Senior Lecturer in Resource and Environmental Economics at the Institute for Sustainable Resources, University College London, UK.

Gerd Ahlert is Senior Researcher at the Institute of Economic Structures Research (GWS), Osnabrück, Germany.

Jeremy Allouche is a Research Fellow at the Institute for Development Studies (IDS) at the University of Sussex, UK.

Philip Andrews-Speed is a Principal Fellow at the Energy Studies Institute (ESI) of the National University of Singapore.

Jeremiah O. Asaka is a Doctoral Researcher in Global Governance and Human Security at University of Massachusetts Boston, USA.

Eva Barteková is a project manager at the Ellen MacArthur Foundation, UK, and an affiliated researcher at the United Nations University, Maastricht Economic and Social Research Institute on Innovation and Technology (UNU-MERIT), Netherlands.

Raimund Bleischwitz is Chair in Sustainable Global Resources at University College London (UCL), and Deputy Director at the UCL Institute for Sustainable Resources (UCL ISR), London, UK.

Tim Boersma is a Senior Research Scholar and Director of Global Natural Gas Markets at the Center on Global Energy Policy, Columbia University, New York, USA.

Martin Bruckner is a Research Associate at the Institute for Ecological Economics, Vienna University of Economics and Business (WU), Austria.

Benjamin Bryant is a Postdoctoral Scholar at the Stanford Woods Institute for the Environment, USA, and member of Water in the West and the Natural Capital Project.

Alvaro Calzadilla is Senior Lecturer in Macro-economic Modelling for a Green Economy at the Institute for Sustainable Resources, University College London, UK.

Rebecca Chaplin-Kramer is a Research Associate at the Stanford Woods Institute for the Environment, USA, and Lead Scientist at the Natural Capital Project.

Bassel Daher is a Research Associate at the Department of Biological and Agricultural Engineering at the Texas A&M University, USA.

Carole Dalin is a NERC Research Fellow at the Institute for Sustainable Resources, University College London, USA.

Vincenzo de Lipsis is Research Associate in Econometric Modelling for Energy and the Environment at the Institute for Sustainable Resources, University College London, UK.

Martin Distelkamp is Senior Researcher at the Institute of Economic Structures Research (GWS), Osnabrück, Germany.

Natalia Escobar-Pemberthy is an Assistant Professor at the Universidad EAFIT, Medellín, Colombia.

David Font Vivanco is a Marie Curie Fellow at the Institute for Sustainable Resources at University College London, UK, and formerly a Postdoctoral Researcher at the Center for Industrial Ecology at Yale University, New Haven, USA.

Stefan Giljum is Senior Researcher and Head of the Research Group "Sustainable Resource Use" at the Institute for Ecological Economics, Vienna University of Economics and Business (WU), Austria.

Jeroen B. Guinée is an Associate Professor at the Institute of Environmental Sciences (CML), Leiden University, Netherlands.

Perrine Hamel is a Postdoctoral Scholar at the Stanford Woods Institute for the Environment, USA, and the lead hydrologist at the Natural Capital Project.

Holger Hoff is Senior Scientist at the Stockholm Environment Institute (SEI) in Sweden and at the Potsdam Institute for Climate Impact Research (PIK) in Germany.

Joseph A. Houghton is a Doctoral Researcher at the Green Chemistry Centre of Excellence (GCCE) at the University of York, UK.

David Humphreys is an independent mining sector consultant, and served as Chief Economist of the London-based mining company Rio Tinto until 2004.

Maria Ivanova Associate Professor of Global Governance and Director, Center for Governance and Sustainability, John W. McCormack Graduate School of Policy and Global Studies, University of Massachusetts Boston, USA.

Corey Johnson is Associate Professor and Department Head of Geography at the University of North Carolina at Greensboro, USA.

Kaysara Khatun is a Marie Curie Fellow at the Environmental Change Institute, Oxford University Centre for the Environment, Oxford, UK.

Fred Kizito is a Senior Scientist at the International Center for Tropical Agriculture (CIAT), Kigali, Rwanda.

René Kleijn is an Associate Professor at Institute of Environmental Sciences (CML) at Leiden University in the Netherlands, and the Director of Education of the joint Leiden Delft master programme in Industrial Ecology.

Andreas Koch is Group Manager for Energy Planning and Geo-simulation at the European Institute for Energy Research (EIFER), Karlsruhe, Germany.

Enrique Kremers is a Researcher and Team Coordinator at the European Institute for Energy Research (EIFER), Karlsruhe, Germany.

Sanghyun Lee is a Visiting Scientist at Texas A&M University's Department of Biological and Agricultural Engineering, College Station, USA.

Stephan Lutter is a Postdoctoral Research Associate at the Institute for Ecological Economics, Vienna University of Economics and Business (WU), Austria.

Avtar Matharu is a Senior Lecturer and Deputy Director of the Green Chemistry Centre of Excellence (GCCE) at the University of York, UK.

Will McDowall is a Lecturer in Eco-innovation at the Institute for Sustainable Resources, University College London, UK.

Eduardo Melo is a Doctoral Researcher at the Green Chemistry Centre of Excellence (GCCE) at the University of York, UK.

Mark Meyer is Senior Researcher at the Institute of Economic Structures Research (GWS), Osnabrück, Germany.

Michał Miedziński is a Senior Research Associate at the Institute for Sustainable Resources, University College London, UK.

Rabi H. Mohtar is a TEES Endowed Professor at Texas A&M University, College Station, Texas.

Gavin M. Mudd is an Associate Professor in Environmental Engineering at RMIT University, Melbourne, Australia.

Marloes L. Mul is Senior Researcher in Hydrology and Water Resources at the International Water Management Institute (IWMI), Accra, Ghana.

Ramiro Parrado is a Senior Researcher of the Climate Change Economic Impacts and Adaptation Programme at the Fondazione Eni Enrico Mattei (FEEM); and also works at the Economic Analysis of Climate Impacts and Policy Division at the Fondazione Centro Euro-Mediterraneo sui Cambiamenti Climatici (CMCC), Venice, Italy.

Blas L. Pérez Henríquez founded and serves as Director of the California Global Energy, Water and Infrastructure Innovation Initiative at Stanford University, USA.

Darshini Ravindranath is a Doctoral Researcher at the Institute for Sustainable Resources, University College London, UK.

Jun Rentschler is a Doctoral Researcher at the Institute for Sustainable Resources (UCL), UK, and a Research Fellow at the Oxford Institute for Energy Studies, UK, and the Graduate Institute for Policy Studies in Tokyo, Japan.

Gloria Salmoral is a Research Fellow at the College of Engineering, Mathematics and Physical Sciences at the University of Exeter, UK.

Anke Schaffartzik is a Senior Researcher at the Institute of Social Ecology Vienna, Alpen-Adria University, Austria.

Shilpi Srivastava is a Postdoctoral Researcher at the Institute for Development Studies (IDS) at the University of Sussex, UK.

Catalina Spataru is a Lecturer in Energy Systems and Networks at the Energy Institute and Course Director MRes Energy Demand Studies, University College London, UK.

Julia Tomei is a Senior Research Associate at the Institute for Sustainable Resources, University College London, UK.

Arnold Tukker is the Professor of Industrial Ecology and Director of the Institute of Environmental Sciences (CML) at Leiden University, Netherlands, and Senior Researcher at the Netherlands Organisation for Applied Scientific Research TNO, the Netherlands.

Ester van der Voet is an Associate Professor at the Department Industrial Ecology of the Institute of Environmental Sciences at Leiden University, Netherlands, and a member of the UNEP International Resource Panel.

Stacy D. VanDeveer is Professor of Global Governance and Human Security, at the McCormack Graduate School of Policy and Global Studies at the University of Massachusetts Boston, USA.

Adrian Vogl is a Senior Scientist at the Natural Capital Project and Research Associate at the Stanford Woods Institute for the Environment, Stanford University, USA.

Jochen Wendel is a Research Scientist at the European Institute for Energy Research (EIFER), Karlsruhe, Germany.

Dominik Wiedenhofer is a Researcher at the Institute of Social Ecology, Alpen-Adria University, Austria.

Markku Wilenius is Professor of Future Studies at the Finland Futures Research Centre at the University of Turku, Finland.

Timothy O. Williams is Principal Researcher in Economics and Director for Africa at the International Water Management Institute (IWMI), Accra, Ghana.

PART I

Understanding the resource nexus

Setting scenes

1

THE RESOURCE NEXUS

Preface and introduction to the Routledge Handbook

Raimund Bleischwitz, Holger Hoff, Catalina Spataru, Ester van der Voet and Stacy D. VanDeveer

Preface

Demand for natural resources has grown rapidly for decades, and is expected to continue growing. These trends lead to repercussions, risks, and threats for humans and ecosystems at different scales. The challenges of sustainable resource management and governance are on numerous agendas, ranging from the G7 and G20 summits to UNEP's International Resource Panel, World Economic Forum, SDG implementation, and a growing community of international scholars. Research highlights the importance of accounting for the interdependencies of resource use and sustainability goals such as eliminating hunger, mitigating climate change, and expanding energy access. There is a need to understand interdependencies and the feasibility of more integrated approaches.

Debate is often framed in terms of a "nexus" between water, energy, and food (sometimes including other resources).[1] The main aim of this handbook is to come to grips with what the nexus[2] is about, provide a reference textbook with an overview, and a survey on emerging and cutting-edge research, and application of the concept.

This handbook is edited by five dedicated scholars, drawing on different schools of thought from different continents. Assembling a wide group of more than 50 authors across a host of disciplines and interdisciplinary fields, this volume rests on a thorough review of relevant literature and, in emerging with a distinct and original perspective, it conceptualizes the resource nexus as *a heuristic for understanding critical interlinkages between uses of different natural resources for systems of provision such as water, energy, and food.* The editors organized a symposium which took place in London in March 2015, debating various aspects of the resource nexus and refining the concept and defining the structure of the handbook. All chapters have been reviewed several times.

Many chapters seek to contribute to realization and implementation of the UN Sustainable Development Goals (SDGs, see Chapter 32, as well as Chapters 4, 19, 22, 26, 27, 28, 29), which endeavor to achieve greater opportunities and a better life for the world population as a whole and for our globe's poorest citizens, in particular while reducing environmental pressures. These goals – especially SDG2 (food), SDG6 (water), SDG7 (energy), SDG12 (sustainable consumption and production), and SDG15 (sustainable use of terrestrial ecosystems) – have extensive and enumerable links to natural resource use, underlining the need for integrative approaches. If implemented in ways that overlook critical interlinkages, the SDGs may well risk a further acceleration of natural resource demand and degradation, ensuing numerous knock-on effects on individuals, communities, businesses and societies – and the ecosystems on which all depend.

Similarly, we perceive this nexus handbook to be connected to topics such as *resource efficiency*, *circular economy*, and many others (Bleischwitz, Welfens, Zhang 2011) – all grappling with solutions aimed at more sustainable use of natural resources at different levels (micro, meso, and macro).

This handbook enables readers to understand (Part I), measure (Part II), assess and model (Part III), compare political economies (Part IV), learn from applications (Part V), and upscale solutions (Part VI). The handbook's six parts and 32 chapters are carefully organized around these aims. As a whole, the handbook seeks to combine analytical rigor with attempts to be transformative – i.e. shaping transformations towards sustainability – in realms of research and knowledge-making, as well as practice and implementation.

Introduction

What is the nexus? An integrated approach

In the past, resource governance mostly focused on single resource categories such as water or energy along a supply chain that ran from primary natural resource, through processing, distribution, and final consumption and disposal. Some of these supply chains are global in scope, such as those for oil, while others are more commonly national or local, such as those for coal and water. Graedel and van der Voet (2010) emphasized the need for a more integrative approach, signaling the existing linkages between the different resources. The nexus concept has been formulated and has become widely used in analytical and practitioners communities at least since the Bonn Conference 2011 (Hoff, 2011) and work at the Transatlantic Academy in 2011–2012 (Andrews-Speed et al., 2012), in response to the predominant single-resource "silo" thinking, emphasizing critical interlinkages across resources, particularly synergies and trade-offs, in a more integrated manner.

Authors in this handbook define the resource nexus as a set of context-specific critical interlinkages between two or more natural resources used as inputs into systems providing essential services to humans, such as water, energy, and food.[3] While a multitude of possible relationships, resources, and systems can be considered, we outline in this chapter a clearly defined five-node nexus for the systems of water, energy, food, land, and materials that seeks to provide consistency, focus, and adaptability to the respective scope and context of analysis and application. Advantages of this particular scope of a five-node nexus are detailed in the following section, with a focus on making robust choices across different contexts, linking with ecosystem services, and bringing in the dimensions of scale and the socio-economic metabolism.

Systems thinking is key. In general terms, natural resources serve as direct or functional inputs for socio-economic systems of provision, either for the production of another input, for general production and consumption purposes, or for the built environment. Figure 1.1 illustrates the *main resource interlinkages* between five essential resources and how these provide a basis for societies and sustainable development. Looking at those interlinkages, some may be more obvious to many readers than others, such as the bi-directional connections between energy and water. Others become more critical during periods of rapid increase in the use when sticking to typical silo approaches without assessing the availability of core inputs from other resources, such as the materials needed for energy production.

Figure 1.1 also introduces three *layers* in order to illustrate the value chains from nature to consumers for each resource. The first layer gives categories for the primary production of the respective natural resource; the second layer adds the socio-economic supply systems based

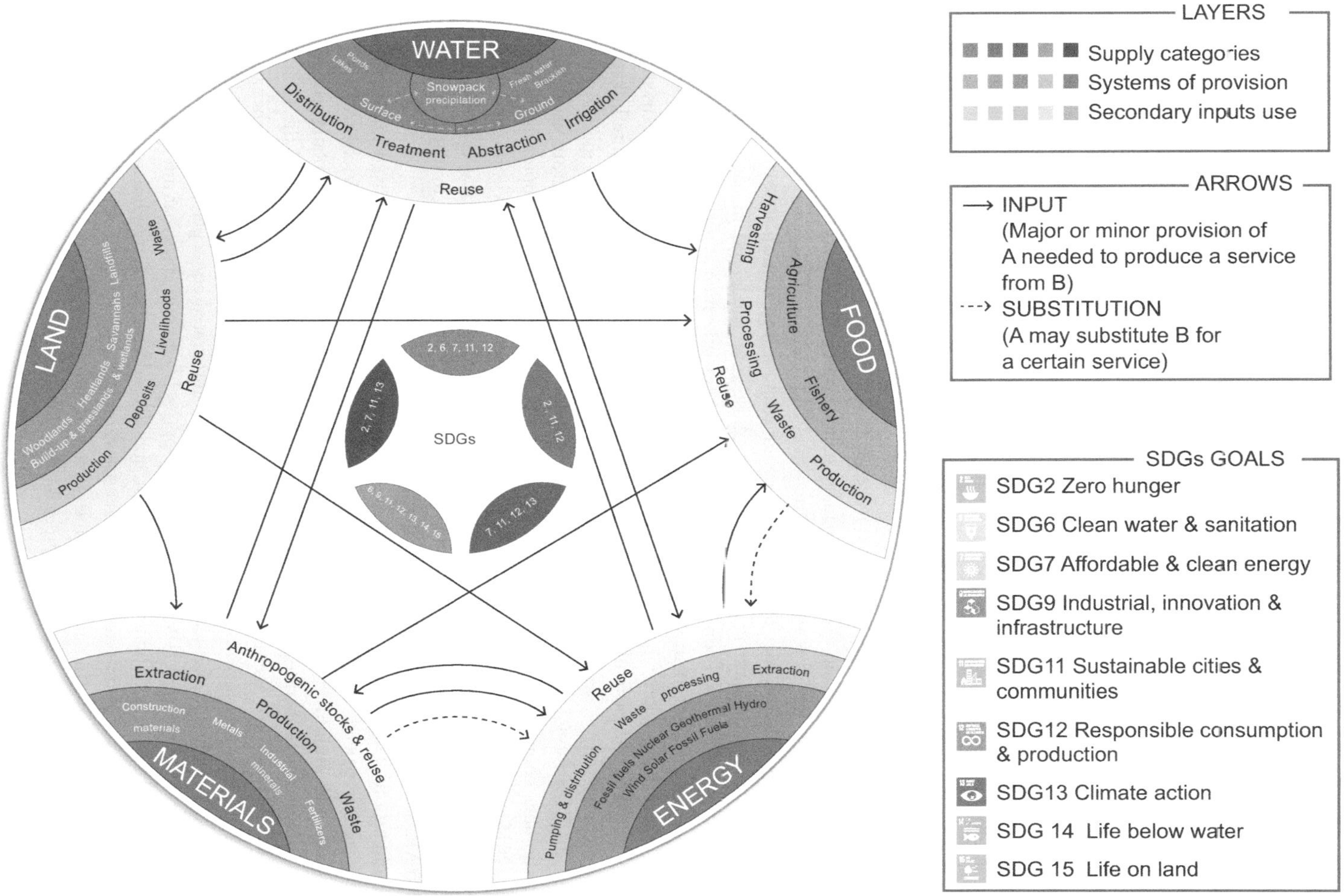

Figure 1.1 The resource nexus

Source: Author compilation (developed by Catalina Spataru with Raimund Bleischwitz) on such resources. The third layer adds the dimension of recycling and re-use and inputs from secondary resources – essential for enhanced resource use efficiencies, a circular economy, and a more sustainable use of resources in general. Critical interlinkages may occur between corresponding or different layers, as illustrated by energy needs for pumping water through distribution systems to end-users. We will discuss our proposed categories in the section on the scope of the nexus.

The figure illustrates main interlinkages in a generic manner based on the many existing studies. Certainly more such interlinkages exist, and the figure looks at them from the perspective of resource inputs being transformed and providing essential services for humans. Nexus research should be quite explicit about those layers, their scales, and how critical interlinkages can be identified. As this figure is applied throughout the entire handbook, we seek to demonstrate the usefulness of a nexus approach. Systems thinking, however, suggests research and practitioners should start from a broader nexus understanding but may well focus on certain critical interlinkages across selected layers. Accordingly, the handbook also entails chapters on cases related to water–energy–food in China (23), metals and energy (24), unconventional fuels and the nexus (25), and energy and water in California (31).

The importance of natural resources for development was repeatedly identified at Earth Summits in 1992, 2002, and 2012, and in SDGs such as (renewable) energy (SDG7), (sustainable) food production (SDG2), and sustainable management of water (SDG6). As such, Figure 1.1 also illustrates some of the links between natural resources and relevant SDGs.

The nexus approach stresses the need to generate relevant information about critical interlinkages that enable decision-makers to plan for robust governance and management across resources and spatial scales – consistent with planning frameworks such as national development plans, sustainable development strategies, or energy or agricultural transitions. Planners 'on the ground' are probably a key target group of a nexus approach. While many previous nexus studies are limited to one particular scale, e.g. the national or regional levels, analyzing interlinkages helps to connect across scales, from small, local places to international trade and global cycles (Hoff, 2017, see also Chapter 3 and Chapter 28 on urban governance). In the future we expect more findings connecting to global scales such as the planetary boundaries approach (Steffen et al., 2015) which emphasizes interlinkages between different large-scale environmental processes. Likewise, the nexus approach should enable more consistency as critical interlinkages become part of mission-oriented strategies such as resource efficiency or a circular economy (Ellen MacArthur Foundation and McKinsey, 2014).

Building on diverse strands of expertise, a nexus approach generates improved knowledge of cross-resource needs and impacts for decision-making and management. As Wichelns (2017) points out, such multifunctional management approaches have a long tradition of being embedded in UN principles and legal frameworks, especially in forestry management and the Dublin principles on water management. Yet, integrated approaches are still an exception rather than the rule. If practitioners or scholars in one sector attempt to reach out to others, feedback from other sectors or policy implementation remains weak in integrated concepts such as integrated water resource management (IWRM). The nexus approach acknowledges that integration adds complexity and hence is difficult to implement, and that addressing all interlinkages is impossible. Yet it rests on the assumptions that (i) identification and assessment of critical interlinkages is essential, and (ii) managing and governing such interlinkages is a key to achieve the SDGs, clearly superior to managing single resources in silos (Le Blanc, 2015; UNEP, 2015). The nexus framework enables actors to identify risks of overuse, to exploit synergies (ancillary benefits, co-benefits, more sustainable investments, cost reduction in joint policy implementation, etc.), and to manage trade-offs and contradictions.

Thus, a nexus approach seeks a more efficient resource management that addresses multiple targets in a more integrated manner. This is why the solution-oriented concepts of *resource efficiency* and a *circular economy* are closely linked with the nexus; Part VI of this handbook is devoted to such solutions. Note that such integrated approaches could also be applied when one resource (e.g. a forest) is governed for multiple and often conflicting goals, such as protecting biodiversity and water resources, community livelihoods, and timber production.

The scope of the nexus

Little agreement exists in the literature as to what natural resources are included in the nexus. The most widely acknowledged nexus approach covers *water–energy–food* (WEF) (Hoff 2011; Bazilian et al., 2011; Lawford et al., 2013; Green et al., 2016). Other studies and some chapters of this handbook focus on:

- The *water–energy* nexus (Ackerman and Fisher, 2013; Howells and Rogner, 2014; Talati et al., 2016; see also Chapters 23, 25, and 31) inspired by the huge amounts of energy needed for water pumping or desalination and vice versa: the large amount of water needed in the energy sector, as illustrated by the impact a drought might have on electricity production;
- *Water–energy–land* (European Commission, 2012; Ringler et al., 2013; Senger and Spataru, 2015; Sharmina et al., 2016; Obersteiner et al., 2016; see also Chapters 16, 22 and 26) all pointing to the manifold ecosystem services provided by land and the intersections with water, biomass, biodiversity, and energy;
- *Water–energy–mineral fertilizer* (Mo and Zhang, 2013; see also Chapter 26) highlights the potential depletion of non-renewable natural resources (minerals), their relevance for food security, their complex supply chains with recycling and recovery opportunities from e.g. wastewater or agricultural residues, and the potential environmental risks from accumulation of these minerals via eutrophication;
- *Water–energy–minerals* (Giurco et al., 2014; Kleijn et al., 2011; see also Chapters 18, 19, 21, 24) as illustrated by the increasing intensity of water and energy use in mineral extraction processes with declining ore grades. The reverse interlinkage is an increasing demand for minerals including metals for energy infrastructures, renewable energy such as photovoltaics or wind power, batteries, and unconventional fuels.

More policy-oriented studies published by Chatham House (Lee et al., 2012) and authors at the Transatlantic Academy (Andrews-Speed et al., 2012, 2015) share a wider recognition of natural resources as manifold inputs into socio-economic processes and metabolism in line with Figure 1.1. A similar approach is taken in the analysis by the McKinsey Global Institute (Dobbs et al., 2011), which focuses on opportunities of a 'resource revolution' for steel and related industrial sectors.

In order to provide a meaningful scope, this handbook proposes a five-node nexus to address:

- Water
- Energy
- Food
- Land
- Materials

While food – like water and energy – has been an essential category of the nexus debate from the very beginning, one may discuss consistency by pointing at the need to produce food rather than seeing it as a primary resource. Yet, our approach of including food points at its relevance as a system of provision (SDG2) with inputs needed from all other resources, manifold critical interlinkages, and resource-intensive value chains at all layers. Indeed, a life cycle approach is essential for systems thinking and for all categories of the nexus.

Inclusion of land in a resource nexus approach is necessary because of its many critical environmental functions, and as a prerequisite to relevant provisioning services and development. Figure 1.1 illustrates land as an input into all other categories, and its critical interlinkages with water.

This nexus handbook further includes materials in the resource nexus for at least four reasons. First, non-energy abiotic resources are essential for housing and shelter and account for about 50% of natural resource use in most industrialized countries measured in physical units according to material flow analysis (see Bringezu and Bleischwitz, 2009; Wiedmann et al., 2013). Second, base metals, critical minerals, and construction minerals have significant implications for energy production, storage, and distribution (SDG7); water provision and re-use (SDG6); and urbanization (SDGs9+11). Mineral fertilizers are also critical inputs for food production (SDG2). Third, costs associated with purchasing and processing materials in manufacturing industries are significant and have been estimated in the order of 40% of gross production costs throughout the 2000s (Wilting and Hanemaaijer, 2014). Lastly, base metals and nutrients cause particularly significant environmental impacts, including land and water resource degradation and GHG emissions (Hertwich et al., 2010). Nevertheless, intensifying agriculture with optimal fertilizer inputs can also prevent or slow the expansion of the agricultural frontier. Within materials one may have subcategories for metals and critical minerals, construction, and industrial minerals and mineral fertilizers. Like for 'food', our use of the term 'materials'[4] refers to a life cycle approach of the nexus rather than primary resources alone and is in line with the industrial ecology approach (see also Chapter 10).

Having those five categories is consistent but not necessarily exclusive. Biomass, for instance, should be considered relevant and has manifold interlinkages with land, food, energy, water, and materials. Those interlinkages across layers could be specifically addressed for biomass at the interface of ecosystems assessments and market analysis. Our Chapters 21 (related to land), 26 (feeding Africa), and 30 (green chemistry) seek to address this debate. As expressed in the section on the scope of the nexus, the five-node nexus is adaptable to specific contexts and should help to focus on the respective most relevant resource interlinkages across layers. Research also explores the interlinkages with biosphere integrity and climate change, the two core planetary boundaries (Dodds and Bartram, 2016; see also Chapters 2 and 4).

Thus, while we propose such a five-node nexus we also suggest to utilize it in a flexible manner. Having such nodes may add complexity compared to most previous studies that analyse a two-node or a three-node nexus. Like Liu et al. (2015) we however argue that it better captures the realities and complexities of human-environment systems and related environment-development goals as specified in the SDGs, and it should make research more relevant.

The nexus and the wider picture of environment, resource scarcities, and human securities

The rapid increase in global resource use along with an increase of the human influence on their global environment since the Industrial Revolution led some researchers to refer to the current era as the '*anthropocene*' (Steffen et al., 2015b): the era where humans are a geological force. While the exact implications of this situation and classification remain under debate, some of the ensuing implications and risks are apparent: global resource use has reached such a magnitude that it causes scarcities and degradation of natural resources at different scales, and on the capacity of ecosystems and the earth systems to regenerate, to provide the required services, and to absorb the resulting by-products and waste products such as CO_2 in a life-sustaining manner for humanity. In particular biosphere integrity and the world oceans are areas of concern from an environmental perspective. Climate change acts as a threat multiplier in this context.

Nexus research is about understanding, assessing, and predicting dynamic interlinkages across the spectrum of natural resources and the value chains by which essential services can be derived out of these resources. Focusing on inputs it also acknowledges wider environmental functions of water and land as supporting, regulating, and stabilizing the provision of inputs for other resources, such as biomass for food production. This links to research on the environment as natural capital[5] (see also Chapter 2). The resource nexus thus complements broader environmental sustainability research in regard to quantifying and assessing critical interlinkages in regard to natural resource management and governance. It provides a basis through which to reduce negative externalities of mono-sectoral approaches and increase system-wide resource efficiency, while looking at the bottom of a green economy in a consistent manner.

Another novel element of our proposed nexus scope is attention paid to inter-related *scarcities* and constraints on using different natural resources. While earlier discussions have often been on 'limits to growth' and limited availabilities of single resources, the geological surveys (USGS, BGS, and others) usually confirm the availability of a sufficient reserve base for mineral resource supplies in coming decades. Among the new constraints that nexus research can address are the following:

- Improve integrated knowledge of input-side resource constraints, e.g. due to decreasing ore grades and increasing energy and water intensity of extraction of minerals, or increasing water or land intensity of some renewable energies. Such evidence shall inform extractive industries and help to develop guidelines for planetary resource consumption, as suggested by Nickless (2016) and discussed in Chapters 18, 19, and 29.
- Constraints on global biomass production which are likely to compromise food security and the achievement of biofuel and bio-economy strategies, all requiring an integrated approach to manage the biosphere, land, water, and other natural resources.
- Apply research findings on inter-related constraints at the output side ('limited absorptive capacities of ecosystems and environment') to analyse and assess inter-related environmental impacts of resource use (e.g. due to potentially higher energy demand and CO_2 emissions of land- and water-smart agriculture).

Our handbook Parts II on metrics, III on modelling, and IV on applications are especially useful to study these nexus topics. Clearly, more research on relationships between natural resource use and inputs, socio-economic processes and metabolism and environmental degradation will be needed. Figure 1.1, with its five-node nexus, may help to orient such work. The figure highlights the potential of nexus research to grasp how water systems, energy systems, and other systems actually use natural resources via multiple and complex pathways and relationships. For example, UNEP's International Resource Panel proposes data and modelling approaches to quantify critical interlinkages between different resources, addressing questions such as: how will demand growth in one resource impact its use efficiency and demand for and use efficiency of other resources? Which (co-)constraints are likely to become more serious over time and how can a nexus approach help to mitigate these? How do the different footprints (water, land, carbon, etc.), as driven by consumption patterns and international commodity trade, affect each other? How can they be reduced jointly? How can a nexus approach help to avoid the transgression of planetary boundaries? Our handbook Part III on modelling touches those issues in more depth, see in particular Chapter 16 on a modelling perspective.

In addition, concern about the intersections of resources, environment, and *security* – in terms of both traditional national and inter-state security and a much broader human security agenda and livelihoods (see here Adger et al., 2014; Biggs et al., 2015) – resulted in a rapid proliferation of scholarly research and reports from think tanks, research institutes, national governments, and international organizations (see also Chapter 4). For example, related research and policymaking efforts have been seen within the UN system (including within both UNEP the Security Council), NATO, the G7, the World Bank, a host of national foreign policy and security think tanks and institutes, the US Department of Defense, Central Intelligence Agency, Department of State, and National Intelligence Council – to name only a few. A growing set of case studies, assessment and modelling exercises, and pilot programs have been produced, engendering a need for more systematic assessment of what is known and how this might be better integrated in the resource nexus framework. Certainly numerous national, international, and human security threats in and among wealthier countries and across the global south are identifiable.

Security-related research has long focused on resources such as water, energy, and minerals access and distribution. But here again, one sees mostly single-resource framings. Interest now turns towards the need to more systematically assess multi-resource linkages to security-related aspects, including those of climate change – its ecological and social impacts, as well as the resource implications and pre-conditions of mitigation and adaptation policies. The governance challenges at the intersection of the resource nexus, climate change, and traditional and human security are legion, occurring at and across multiple scales, including provision of resources needed to survive and thrive, and identifiable risks of local, national, and transnational violence. While research has begun to focus on the ways such connections raise the risks of conflict, violence, and loss of human security and health, these challenges also connect to research on criminality, peacebuilding, and post-conflict reconstruction and a host of vexing justice and equity-related concerns embedded within natural resources, climate change mitigation and adaptation, and Sustainable Development Goals and governance. Traditional and human security concerns are thus central to SDG implementation, and many aspects of climate change and resource nexus governance. We consider it a strategic importance of nexus research to address those issues and, therefore, be able to connect to important areas of policies and decision-making. Special contributions to this security angle can be studied especially in Chapter 4, as well as in Chapter 3 on different scales, Chapter 9 on criticality, Chapter 17 on a non-equilibrium perspective, Chapter 20 on international trade, and Chapter 32 on the UN.

What should a nexus perspective help to accomplish?

Nexus terminology is increasingly popular, and possibly at risk of becoming a 'buzzword' as a recent editorial in *Nature* suggests (Nature, 2016; Cairns and Krzywoszynska, 2016). Acknowledging this debate, the authors of this handbook clearly see a value in a nexus approach that is well defined, conceptualized, and applied. Nexus research should be able to improve understanding of resource interdependency by complementing and adding value to existing strands of research – rather than replacing them; relevant contributions can be expected in particular to these areas:

- Closing gaps between environmental sustainability research by adding analysis of resource degradation, exploitation, and use across scales and value-chain systems;

- Industrial ecology research on material flows, stocks, and footprints tracked through economies, by analysing interlinkages across resources and adding socio-economic dimensions;
- Ambitions towards integration coming from one sector (such as integrated water resource management, integrated land-use planning, integrated resource planning as known from energy), by looking more specifically at interactions and feedbacks with other resources, critical thresholds, and use patterns;
- Economic and social science research on commodities, conflicts, and security issues, by adding analysis on natural resource inputs and critical thresholds of resource use;
- Informing resource-related governance and decision-making by public, private, and civil society sector actors.

Clearly, nexus research comes with an inter- and transdisciplinary agenda (Stirling, 2015). The years ahead will require such approaches if states, local communities, societies, and transnational actors are to deliver on the goals and promises made in the SDGs and the Paris Agreement on climate change, for example, as well as for long-term visions of low-carbon, circular economies around a globe in which the worst manifestations of poverty and human deprivation are alleviated.

Organizing the nexus handbook

Part I: Understanding the resource nexus: setting scenes

Part I sets the scene by providing a conceptual overview (Chapter 1), clarifying the relationship of the nexus to environmental sustainability research (Chapter 2), outlining the importance of scale (Chapter 3), and emphasizing the inescapable relevance of climate change and security concerns (Chapter 4).

Part II: Analysing the resource nexus: tools and metrics

Part II responds to the need for quantification, especially on metrics accounting for resource interlinkages. This part introduces a number of existing methods from the field of industrial ecology. These methods each describe, in some manner, stocks and flows of materials and energy through society. As such, they link different resources from a functional point of view being useful for analysing the interdependencies across resources in fulfilling society's needs for energy, food, shelter, infrastructures, and products. The following methods are included in Part II:

- Life Cycle Analysis (LCA), capturing the environmental impacts especially of products (Chapter 5);
- Material Flow Analysis (MFA), tracking the material use along value chains and across countries (Chapter 6);
- The various footprints on water, carbon, and materials that have been developed (Chapter 7);
- Input–output analysis as a worldwide data tool and method needed for other tools (Chapter 8);
- Criticality assessments and underlying methodologies as introduced within a number of countries responding to demand for critical materials (such as rare earth elements) (Chapter 9);
- A concluding chapter on such tools with a nexus perspective (Chapter 10).

Part III: Resource nexus modelling: practices and future transformations

Part III introduces key modelling tools cutting across different approaches. Modelling is a valuable tool for assessing system implications and dynamics for existing interlinkages and for future strategies, as well as building knowledge about these interdependencies. Accordingly, chapters in this part focus on modelling sustainability transformations with both environmental and social impacts (Chapter 11), challenges of integrating different scales (Chapter 12), the long-term dynamics of K-Waves (Chapter 13), foresight and scenarios (Chapter 14), and macro-economic modelling (Chapter 15). In addition to collecting contributions on existing modelling approaches with relevance for the nexus, Part III concludes with a chapter (Chapter 16) on future nexus modelling, proposing a next generation modelling approach of integrated methods to analyse trade offs and interlinkages of the five-node resource nexus dynamics.

Part IV: International political economy and the resource nexus

Beyond better metrics and modelling approaches, comparative research on governance will be needed to better connect nexus analysis to worlds of business transactions, public and civil society practice, and cautioning against mono-sectoral optimization efforts inherent in scenario and modelling processes. This part of the handbook intends to complement and strengthen other research on the anthropocene, on 'planetary boundaries', and on 'earth systems governance'. Accordingly, chapters address mining, commodity trade, and rare earth elements, the latter being a case for critical materials needed for transitions to low-carbon economies. The chapters also attend political economy in the global South, with chapters on equity in the real world of uncertainties, on land governance, and the risks of new resource curses.

Part V: Applying the resource nexus: regional and global scale

A significant number of regional nexus studies have been conducted. They assess the resource interlinkages most relevant in respective regions. Such case studies demonstrate the ways regional contexts often determine the selection of relevant resources and critical interlinkages. In addition, many published papers contribute to understanding resource interlinkages without explicitly using the term 'nexus'. As in Part I, the scale dimension is essential for the nexus. While we propose a five-node nexus, we also encourage research addressing more focused approaches as appropriate. This part of the handbook allows readers to zoom in, with chapters on China, on the energy demand for metals needed for future low-carbon economies, and the nexus dimension of new energy fuels. Another chapter sheds light on the nexus dimension to hunger in Africa and opportunities for improving food security. Finally, this part also addresses oceans covering large portions of our planet and relevant for all nexus dimensions.

Part VI: Governing the nexus: emerging responses

While the previous parts intend to analyse nexus challenges and seek to enable actors to apply metrics and modelling as well as to compare cases, this part addresses a few illustrative, emerging response strategies. The array of such strategies has grown as many actors on the ground, in business and among policymakers, deal with nexus risks and seek to turn them into opportunities. One thing is already clear: Overcoming silo-type of thinking and planning is easier said than done. This part of the handbook addresses cities and the UN as the two angles of the political spectrum beyond states. It also offers chapters on eco-innovation and decoupling, on green chemistry as a key enabler for the supply of sustainable resources, and on improved infrastructure planning, illustrated by the state of California. The illustrative chapters in this section seek to inspire future research and future practice, as analysts and policymakers alike grapple with the complexities, challenges, and opportunities presented by the resource nexus.

Notes

1 See e.g. http://futureearth.org/future-earth-water-energy-food-nexus; www.thenexusnetwork.org; www.water-energy-food.org, www.youtube.com/watch?v=T-eATjiXTOQ.
2 "Nexus" is the Latin word for interlinkages.
3 One may note this definition resembles the term 'systems of provision' originating from research on consumption and work done by Ben Fine and Ellen Leopold in the early nineties of last century; while this strand of research puts consumption in relation to processes of production, distribution, and retail, and into a historical and socio-cultural context, our definition of the nexus focuses on the natural resource base of such systems, the value chains from nature to consumers across such systems, and the interlinkages across them.
4 The term 'materials' refers here to non-energetic raw materials; note some glossaries define materials rather broad including energy and other resources, which we consider in other categories of the nexus approach.
5 Natural capital can be defined as "the elements of nature that directly and indirectly produce value or benefits to people, including ecosystems, species, freshwater, land, minerals, the air and oceans, as well as natural processes and functions", a broad definition with roots in environmental research.

References

Ackerman, F., and Fisher, J. (2013). Is there a water–energy nexus in electricity generation? Long-term scenarios for the western United States. *Energy Policy*, 59, 235–241.

Adger, W. N., Pulhin, J. M., Barnett, J., Dabelko, G. D., Hovelsrud, G. K., Levy, M., Oswald Spring, U., and Vogel, C. H. (2014). Human security. In *Climate change 2014: Impacts, adaptation, and vulnerability. Part A: Global and sectoral aspects. Contribution of working group II to the fifth assessment report of the intergovernmental panel on climate change*, Cambridge, United Kingdom and New York, pp. 755–791.

Andrews-Speed, P., Bleischwitz, R., Boersma, T., Johnson, C., Kemp, G., and VanDeveer, S. (2012). *The global resource nexus: The struggles for land, energy, food, water, and minerals*. Washington, DC: Transatlantic Academy.

Andrews-Speed, P., Bleischwitz, R., Boersma, T., Johnson, C., Kemp, G., and VanDeveer, S. (2015). *Want, waste or war? The global resource nexus and the struggle for land, energy, food, water and minerals*. London: Routledge/Earthscan.

Bazilian, M., Rogner, H., Howells, M., et al. (2011). Considering the energy, water and food nexus: Towards an integrated modelling approach. *Energy Policy*, 39(12), 7896–7906.

Biggs, E. M., Bruce, E., Boruff, B., et al. (2015). Sustainable development and the water–energy–food nexus: A perspective on livelihoods. *Environmental Science & Policy*, 54, 389–397.

Bleischwitz, R., Welfens, P., and Zhang, Z. X. (Eds.). (2011). *International economics of resource efficiency: Eco-innovation policies for a green economy*. Berlin/Heidelberg: Springer Publisher.

Bringezu, S., and Bleischwitz, R. (Eds.). (2009). *Sustainable resource management: Global trends, visions and policies*. Sheffield: Greenleaf Publishing.

Cairns, R., and Krzywoszynska, A. (2016). Anatomy of a buzzword: The emergence of 'the water-energy-food nexus' in UK natural resource debates. *Environmental Science & Policy*, 64, 164–170.

Dobbs, R., Oppenheim, J., Thompson, F., Brinkman, M., and Zornes, M. (2011). *Resource Revolution: Meeting the world's energy, materials, food, and water needs*. New York: McKinsey Global Institute.

Dodds, F., and Bartram, J. (Eds.). (2016). *The water, food, energy and climate nexus: Challenges and an agenda for action*. London and New York: Earthscan/Routledge.

Ellen MacArthur Foundation and McKinsey & Co. (2014). *Towards the circular economy: Accelerating the scale-up across global supply chains*. Geneva: World Economic Forum.

European Commission. (2012). *Confronting scarcity: Managing water, energy and land for inclusive and sustainable growth*. Brussels: European Union. https://ec.europa.eu/europeaid/sites/devco/files/erd-consca-report-20110101_en_0.pdf.

Giurco, D., McLellan, B., Franks, D. M. T., et al. (2014). Responsible mineral and energy futures: Views at the nexus. *Journal of Cleaner Production*, 84, 322–338.

Graedel, T. E., and Voet, E. v. d. (2010). *Linkages of sustainability*. Cambridge, MA: MIT Press.

Green, J. M. H., Cranston, G. R., Sutherland, W. J., et al. (2016). Research priorities for managing the impacts and dependencies of business upon food, energy, water and the environment. *Sustainability Science*, 1–13. DOI 10.1007/s11625-016-0402-4

Hertwich, E., Voet, E. v. d., Suh, S., and Tukker, A. (2010). *Assessing the environmental impacts of consumption and production: Priority products and materials*. Nairobi: UNEP.

Hoff, H. (2011). *Understanding the Nexus*. Background paper for Bonn 2011 conference: The water, energy and food security nexus, Stockholm Environment Institute.

Hoff, H. (2017). *Integrated SDG implementation – how a vertical (cross-scale and cross-regional) nexus approach can complement horizontal (cross-sectoral) integration*. Background paper for the Dresden Nexus Conference 2017.

Howells, M., and Rogner, H-H. (2014). Water-energy nexus: Assessing integrated systems. *Nature Climate Change*, 4(4), 246–247.

Kleijn, R., Van der Voet, E., Kramer, G. J., Van Oers, L., and Van der Giesen, C. (2011). Metal requirements of low-carbon power generation. *Energy*, 36(9), 5640–5648.

Lawford, R., Bogardi, J., Marx, S., et al. (2013). Basin perspectives on the water – energy – food security nexus. *Current Opinion in Environmental Sustainability*, 5(6), 607–616.

Le Blanc, D. (2015). Towards integration at last? The sustainable development goals as a network of targets. *Sustainable Development*, 23(3), 176–187.

Lee, B., Preston, F., Kooroshy, J., Bailey, R., and Lahn, G. (2012). *Resources futures*. London: Chatham House.

Liu, J., Mooney, H., Hull, V., et al. (2015). Systems integration for global sustainability. *Science*, 347(6225), 1258832: 1–9.

Mo, W., and Zhang, Q. (2013). Energy–nutrients–water nexus: Integrated resource recovery in municipal wastewater treatment plants. *Journal of Environmental Management*, 127, 255–267.

Nature. (2016). Editorial: Scientific buzzwords obscure meaning: 'Nexus' is enjoying new-found popularity. But what does it actually mean? *Nature*, 538, 7624.

Nickless, E. (2016). Resourcing future generations: A global effort to meet the world's future needs head-on. *European Geologist European Geologist*, 42, 46–50.

Obersteiner, M., Walsh, B., Frank, S., et al. (2016). Assessing the land resource – food price nexus of the sustainable development goals. *Science Advances*, 2(9), 16 September, 11 pages.

Ringler, C., Bhaduri, A., and Lawford, R. (2013). The nexus across water, energy, land and food (WELF): Potential for improved resource use efficiency? *Current Opinion in Environmental Sustainability*, 5(6), 617–624.

Senger, M., and Spataru, C. (2015). *Water-energy-land nexus- modelling long-term scenarios for Brazil*, Published in Modelling Symposium (EMS), 2015 IEEE European, IEEE. doi:10.1109/EMS.2015.47

Sharmina, M., Hoolohana, C., Bows-Larkina, A., et al. (2016). A nexus perspective on competing land demands: Wider lessons from a UK policy case study. *Environmental Science & Policy*, 59, 74–84.

Steffen, W., Richardson, K., Rockstroem, J., et al. (2015). Planetary boundaries: Guiding human development on a changing planet. *Science*, 347(6223), 1259855, 10 pages.

Steffen, W., Broadgate, W., Deutsch, L., et al. (2015b). The trajectory of the anthropocene: The great acceleration. *The Anthropocene Review*, 2(1), 81–98.

Stirling, A. (2015). *Developing 'Nexus Capabilities': Towards transdisciplinary methodologies*. Brighton: University of Sussex.

Talati, S., Zhai, H., Page Kyle, G., et al. (2016). Consumptive water use from electricity generation in the Southwest under alternative climate, technology, and policy futures. *Environmental Science & Technology*, 50(22), 12095–12104.

UNEP. (2015). *Policy coherence of the sustainable development goals: A natural resource perspective*. Paris: International Resource Panel.

Wichelns, D. (2017). The water-energy-food nexus: Is the increasing attention warranted, from either a research or policy perspective? *Environmental Science & Policy*, 69, 113–123.

Wiedmann, T. O., Schandl, H., Lenzen, T., et al. (2013). The material footprint of nations. *Proceedings of the National Academy of Sciences*. doi:10.1073/pnas.1220362110.

Wilting, H. and Hanemaaijer, A. (2014). Share of raw material costs in total production costs. The Hague: PBL Netherlands Environmental Assessment Agency.

2

APPRECIATING WIDER ENVIRONMENTAL ANGLES

Kaysara Khatun and Gloria Salmoral

Introduction: natural capital and ecosystems services: towards the maintenance of sustainable societies

In this chapter, we will introduce and discuss the concepts of natural capital, ecosystem services (ES), the resilience and complexity issues that surround their management, as well as the role natural capital and ES can play in the resource nexus. From an ecosystem management perspective, diverse natural capital (e.g. tropical forests) and the services they provide require deliberate maintenance and optimal utilisation as unsustainable patterns of use will leave many regions with scarce resources, food-insecure populations and poorer economies (Avishek et al., 2012). Moreover, in light of the short timelines available to mitigate climate change, large gaps exist between the immediate objectives of food, energy and water security and the long-term goals of conservation initiatives of natural capital that are simultaneously taking place in many countries especially in the global south. These themes are connected at landscape level yet remain divided by different purposes, stakeholder interests and conflicting policies at local, national and international scales. This presents a set of scientific, policy and management issues that are critical for ecosystem sustainability and which need to address the complex interrelationships within natural capital and ecosystem service research and policy.

Natural capital can be defined as the world's stocks of natural assets which include all living things and is a multidimensional meta-concept for a plurality of interrelated and heterogeneous stocks that perform various functions and services for human society (Natural Capital Committee, 2014). Natural capital is widely used to signify a variety of components (e.g. water resources, geology, biodiversity, soil, ozone layer) and properties (e.g. ecological resilience, ecosystem health, integrity). It is also used by economists to convert the aggregate natural resources that are used or can be used for producing various goods into monetary value (de Groot, 2006). The availability of all the Earth's natural resources is limited either due to insufficient reserves, complicated extraction or transportation and/or inadequate technology of extraction and processing. The overexploitation of natural capital can be the cause of catastrophic loss, not just in terms of biodiversity, but also for humans as ecosystem productivity and resilience decline over time. As a result, some regions, especially many developing nations will become more prone to extreme events such as floods and droughts (MEA, 2005). Drawing down stocks of natural capital without allowing or encouraging the natural environment to recover runs the risk of local, regional or even global ecosystem collapse as well as ecological debt which needs to be paid back, for example by replanting clear-cut forests, or allowing aquifers to replenish themselves after we have abstracted water (Dumont et al., 2013). This can lead to many communities being unable to sustain themselves particularly in already stressed ecosystems, potentially leading to starvation, conflict over resource scarcity and displacement of populations.

The term 'natural capital', while being praised for its terminological strengths, has been criticised for its reductionism, in that nature is simplified to the status of a commodity to be marketed at its exchange value (Daly, 2014). Moreover, it is still controversial, what types of capital – i.e. natural capital (e.g. ecosystems, air, water), cultivated natural capital (e.g. salmon farms, wineries), man-made capital (e.g. infrastructure), social capital (e.g. political institutions), human capital (e.g. skills, education) and knowledge capital – ought to be preserved in favour of current and future generations. In this vein, the concept of 'critical' natural capital signifies the part of natural capital that performs important and irreplaceable environmental functions, such as ES for sustaining both human and natural systems that cannot be substituted by other types of capital. Thus 'critical natural capital' represents the part of the natural environment that ought to be maintained in any circumstances. When the demand for certain natural capital exceeds that of their availability (in terms of access to extraction) poor management of such resources becomes not only an ecological liability, but a social and economic liability too. It is this importance; for quality of life, development and the survival of humans that makes critical natural capital such an essential component of sustainability and thus well-founded criteria is required to assess the specific criticality of natural capital stocks. Yet, confusion of the term critical natural capital is immense and it is unclear what is important and irreplaceable, and therefore critical due to the multidimensional character of natural capital. Moreover, numerous scientific disciplines and societal groups bring their own perspective in valuing natural capital, and what is considered critical, adding further levels of complexity to the understanding of ecosystems, and their resilience as well as the management of the natural capital and the services that they provide, which will be discussed in the next sections. The resource nexus approach can eventually address these weaknesses by understanding how the five systems (water, food, energy, soils and minerals) interact to further equity and decision-making, as well as to integrate segments of society and various scientific disciplines.

Understanding complexity, resilience and critical thresholds

Complex systems such as natural ecosystems can be decomposed into structure and underlying processes, which are generally scale specific and operate at distinct spatiotemporal scales (Allen et al., 2014). For instance, in a typical boreal forest, fresh needles cycle yearly; the crown of the foliage cycles with a decadal period; and trees and stands cycle at a period of about a century or more (Franklin and Downing, 2004). Complexity does not refer much to the properties of the individual components within a system, but rather to their relationships and the complex behaviour with each other.

The resilience of a system is achieved due to its adaptive capacity and is a measure of the vulnerability to unexpected disturbances and surprises (Holling, 2001). Under more resilient social-ecological systems larger changes can be absorbed without modifying the structure and functioning, since they are able to deal, adjust or restructure without renouncing the actual provision of ES (Folke et al., 2002). Resilience is often enhanced by diversity (e.g. of species, of economic options), based on the premise that diversity helps a system return more quickly to its pre-disturbance conditions, supporting and promoting both adaptation and social learning (Gunderson, 2010).

When a system's resilience threshold (or also the so-called critical threshold) is overpassed, regime changes are forced to take place and the processes responsible for a system's structure and functioning change, creating new self-organised structures (Allen et al., 2014). At this point, regime changes can have negative consequences on ES such as reducing freshwater or food provisioning. Although ecological regime shifts can be understood retrospectively, it is difficult to predict them in advance and to develop assessment programs that adapt as fast as thresholds are overpassed (Folke et al., 2002). While managers can anticipate some of the types of impacts that are associated with different disturbances, such as habitat loss, climate change and invasions of exotic species, many of these impacts cannot be known, foreseen or predicted (Gunderson, 2010). Moreover, disasters that are yet out of human control such as weather extremes, fires and pest outbreak are also contributing to the resilience of a system (Khatun, 2011).

Ecosystem services

It is from natural capital that humans derive a wide range of life-sustaining services, often referred to as ecosystem services (ES), and defined as processes that produce or support the production of ecosystem goods and benefits (MEA, 2005) (see Figure 2.1). The most obvious of these include the food we eat, the water we drink and the plant materials we use for fuel, building materials and medicines.

Figure 2.2 shows the main grouping of ES classified and popularised by the Millennium Ecosystem Assessment (MEA) in 2005. These are namely: cultural, provisioning, and regulatory and supporting services. Supporting services (e.g. such as nutrient cycles and crop pollination) are often overlooked in ES accounting as these are difficult to monitor. Natural capital is the stock of resources that generate ES, which in turn are referred to the flows of benefits which

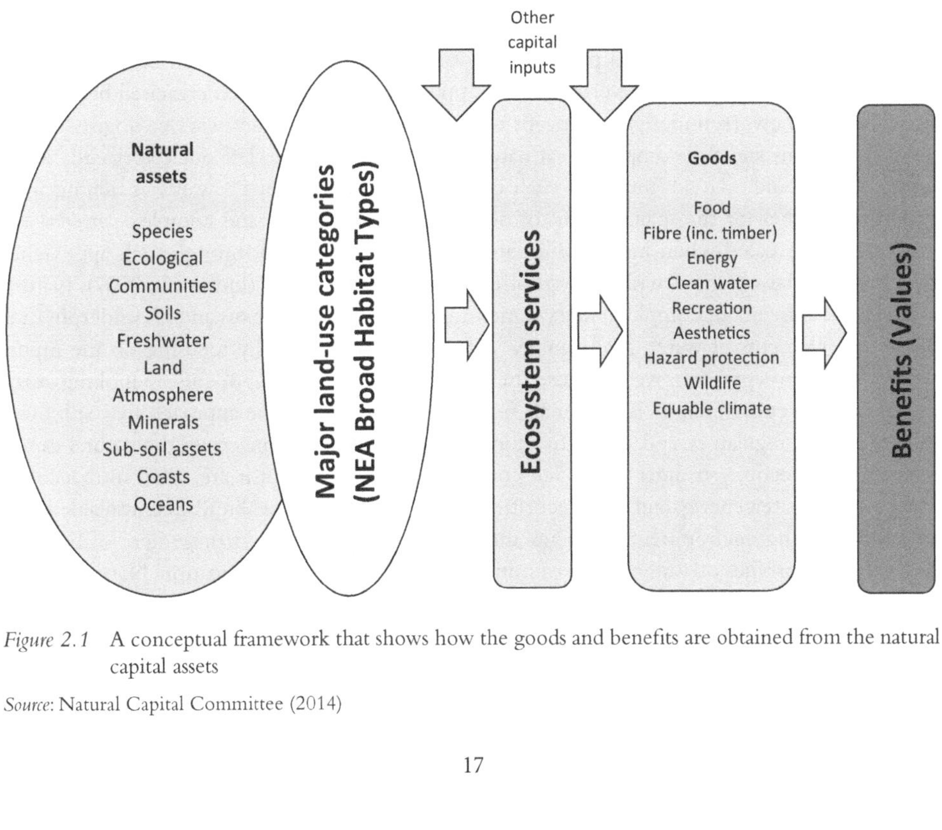

Figure 2.1 A conceptual framework that shows how the goods and benefits are obtained from the natural capital assets

Source: Natural Capital Committee (2014)

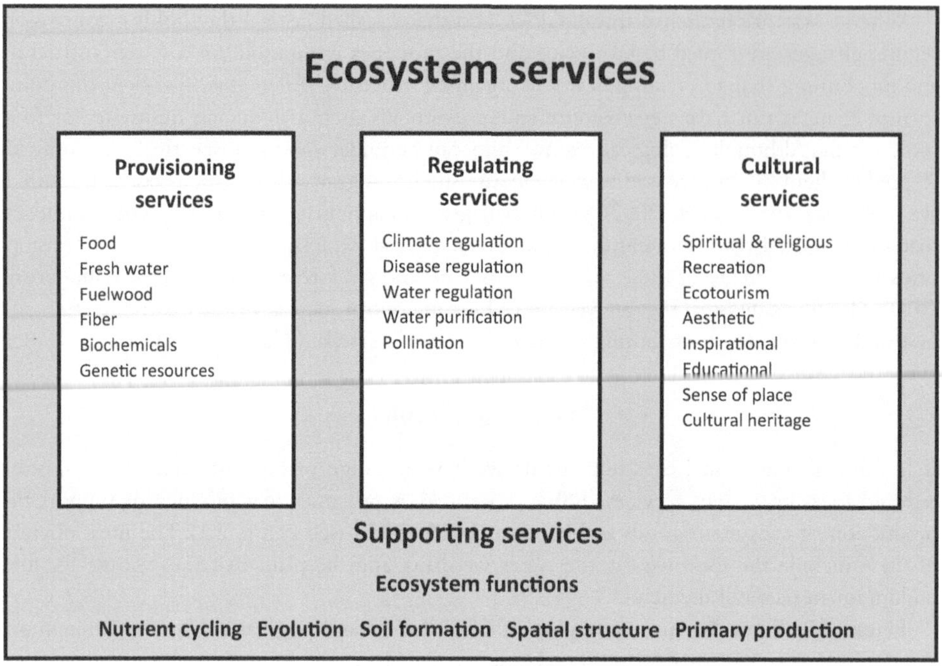

Figure 2.2 Ecosystem services defined under the MEA (2005)

people gain from natural ecosystems (Jarret, 2012). The crucial link between natural capital and ES is that when some classes of ES are appropriated by humanity at an unsustainable rate, the stocks of natural capital which provide them may be depleted, as with timber and forests.

The concept of ES provides a powerful way of examining the interaction between ecosystems and human well-being. Nevertheless, the application of ES has been criticised because the stock-flow framework underlying concept only uses one of the many ways ecologists understand ecosystems and the complexity of natural-human systems are left out (Norgaard, 2010). Instead, we should take advantage of such complexity in ecosystems towards enriching integrated approaches for the management of our natural capital. New and complex patterns and processes are revealed when human and natural systems are studied through such approaches, which cannot be obtained with ecological or social research alone (Liu et al., 2007). Natural capital is the unseen but complementary dimension of the resource nexus and ES underpin each of the five elements of water, food, energy, soil and minerals. Thus by focusing on the inputs, within the nexus approach, we can close the gap between these components by looking at the interface between natural, social and man-made capital. However, the approach may still overlook the wider regulatory and supportive function of ecosystems; conversely, the natural capital concept may become stronger if all such components derived from it are taken into account. For example, water, energy and food securities can be affected by the abundance, available infrastructure, existing environmental policies and capacity of sustainable management of fisheries, flows to turn turbines or water stored to mitigate scarcity and supply irrigation. Natural capital, when not depleted – through overexploitation (e.g. of mineral, timber) – also acts as a buffer that can mediate the nexus links, for example by storing, moving, cleaning and buffering flows of water, making drought and floods less severe as well as food and energy production more reliable (MEA, 2005).

Another development along with the ES terminology in recent years has been the amalgamation of biodiversity with ES into one concept: 'BES', an acronym referring to 'biodiversity and ecosystem services' (Jarret, 2012). Biodiversity loss is one of the most important challenges facing the international community. The rapid disruption of these ecosystems probably imperils global biodiversity more than any other contemporary phenomenon and in many parts of the world forests are increasingly becoming final refuges for threatened species and natural ecosystem processes (Laurance et al., 2012). Furthermore, according to the UN Environment Programme's report on *The Economics of Ecosystems and Biodiversity* (TEEB, 2008), business risks and opportunities associated with biodiversity and ES are growing, with increasing pressure on and more restricted access to natural resources in the future. The nexus approach can also be useful in this context, for instance for better biofuel management, or a water management minimising energy use, etc., towards integrating a more holistic approach to resource efficiency.

However, it needs to be noted that references to BES are problematic because flows of ES and stocks of biodiversity interact in very complex ways, often in a contrary nature. The concept (as with ES) fails to account for the inseparability and the multiple relationships within and between nature and people that are regional and context specific. The flows of ES (examples are shown in Figure 2.3) appropriated by humans from natural ecosystems, are *responsible* for depletion of biodiversity and natural capital. For example, in the case of the natural capital that are tropical forests, which provide ES ranging from less visible services such as climate regulation and carbon storage, flood defence, the pollination of crops, the inspiration/enjoyment we take from the natural environment (cultural services) to visible ES such as food, medicine and timber

Figure 2.3 Natural capital, ecosystem services and biodiversity

Source: Authors' own

Figure 2.3 Continued

among others. The unsustainable appropriation of timber, some foods or fibres, competes with other ES and creates losses in biodiversity, carbon storage and water services (Khatun, 2011), and thus requires an integrated assessment, as with the nexus approach. Notwithstanding, the fundamental lack of understanding of the challenges of such complexity and interconnections, amalgamating the terms biodiversity and ES together in one acronym as if the relationship is always symbiotic, makes little sense (Jarret, 2012). Steffen et al. (2015) coin a more integrated term 'biosphere integrity' as a new concept, which describes the rate of biodiversity loss.

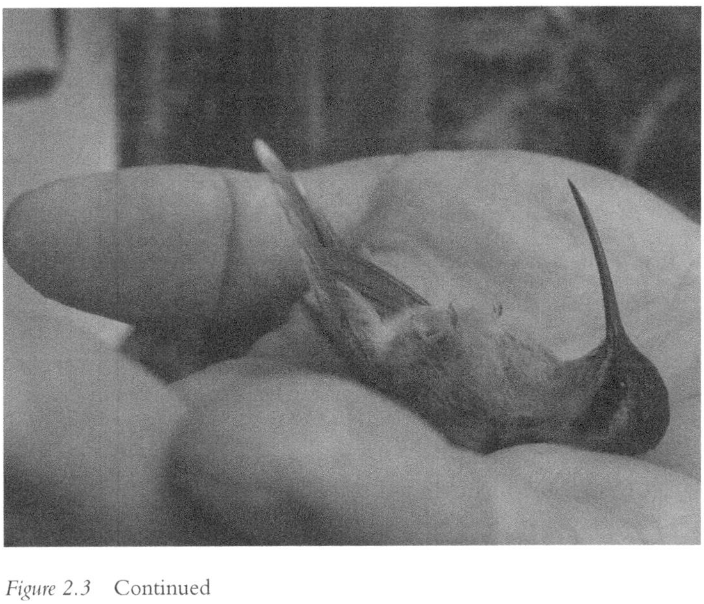

Figure 2.3 Continued

Clarification of terminology is crucial as stakeholders including businesses, governments, and NGOs are attempting to build the frameworks and policies that are needed to maintain natural capital stocks, ensure sustainable flows of ES and conserve biodiversity.

Payments for ecosystem services (PES)

Payments for ecosystem services (PES) have emerged as a relatively new paradigm for natural capital management such as for forests and their governance. By relating the provision of financial incentives to conservation and the reduction of land-use emissions in a given country, it is indirectly imposing a view of how land and forest should be managed in developing nations. PES will potentially take many forms and involve a combination of positive incentives, policy programs (e.g. sustainable forest management), and various sub-national activities including private-led conservation projects (Corbera and Schroeder, 2011). Such initiatives will not operate in 'empty landscapes', as the rural poor and those without land require resilient sustainable livelihoods that are flexible in the short term, whereas conservation efforts such as the carbon sinks approach to reduce emissions, promote sustainable land use or preserve biodiversity will require a longer perspective. This long-term requirement may conflict with the short-term development needs of the poor, while simultaneously being at odds with other actors involved in the land-use sector. Moreover, the conditions under which PES initiatives are likely to succeed are taking a long time to come to fruition and remain very much in the development phase. PES has been advocated by global institutions such as the World Bank as a mean to secure the provision of these services and ameliorate wide-ranging problems, from protecting biodiversity and habitat loss, carbon fixation and mitigating climate change and poverty alleviation. Furthermore, in the emerging field of PES, as power relations among stakeholders are unequal, it is often the resource-strong stakeholders who are able to define the terms of ES trade. For this reason, it has been argued that forest conservation programs should consider equity and legitimacy issues, as well as environmental effectiveness and economic efficiency (Corbera et al., 2007). Thus if real life complexities, including people's livelihoods, are ignored, simplified and/or assessed partially, negative consequences are unavoidable. Frameworks such as the resource nexus need to explicitly or adequately incorporate such elements given that livelihoods are key to achieving sustainable development. The following section focuses one the most well-known examples of PES, namely reducing emissions from deforestation and forest degradation (REDD+), and looks at its ability to meet some of these objectives.

Reducing emissions from deforestation and forest degradation (REDD+)

As well as regulating climate, forests support livelihoods of 90% of 1.2 billion people classified as living in extreme poverty, they are home to 90% of terrestrial biodiversity and they provide life-sustaining ES (FAO, 2014). Moreover, land-use change is the biggest threat to terrestrial ecosystems, and livelihoods 'resilience' for many of the world's poor. The main drivers of deforestation vary from region to region. In Latin America, a majority of deforestation is the result of industrial activities, e.g. cattle ranching in the Amazon, and in Southeast Asia, drivers are large-scale agriculture and intensive logging. Subsistence agriculture, often using slash-and-burn/fallow techniques, is the most important driver of forest loss in Africa (FAO, 2014).

The role of forestry in the stabilisation of Greenhouse Gas (GHG) emissions was officially recognised at the Thirteenth Conference of the Parties (COP13) in Bali to the UN Convention on Climate Change (UNFCCC) in December 2007 as a vital issue. The five elements of

REDD+ were laid out in the Bali Action Plan (2007), with the core being reducing emissions from deforestation and forest degradation (REDD) 'plus' the three supplementary elements – conservation, sustainable management of forests and enhancement of forest carbon stocks (Parker et al., 2009). However, the REDD text that emerged in the Copenhagen Accord at the Conference of the Parties (COP15) in December 2009 followed by the Cancún negotiations in 2010 included all five elements on an equal basis. Thus the term 'REDD' has been replaced in all key texts and discussions by 'REDD+'. These enhancement activities are not linked to emissions reductions; rather, it is a call for investment in tropical forests for the other services they provide. REDD+ has the potential to deliver a range of 'co-benefits', or non-carbon benefits e.g. biodiversity, indigenous peoples' rights, human rights, land rights, livelihoods, transparency and accountability (Lang, 2012). The recent COP (21) in Paris in 2015 did little to move REDD+ forward. The Paris Agreement encouraged countries to 'take action' but remains legally non-binding, with the co-benefits aspects already covered in REDD safeguards which remain weak, where countries 'should' provide a compliance report every two years, with the least developed countries being exempt. Co-benefits in theory allow for a system of practices for stewardship and use of forest land aimed at fulfilling relevant ecological, economic and social functions of the forest in a sustainable manner. In the past, the rationale for forest conservation was simply to sustain the forests' productive role for the timber industry; it is now acknowledged that forests provide a range of services, timber production is just one (Khatun, 2011).

The study of land use and ES, such as those under REDD+, is a complex interrelated system, and is closely related to the resource nexus. It is characterised by feedbacks and lies at the interface in the study of human-environment interactions. Any strategy to increase conservation or promote sustainable land use will need to take into account all the underlying causes of deforestation and degradation, symptomatic of governance, legislative issues, institutional and market weaknesses, among others (Humphreys, 2006). This challenge is far from trivial as there are highly complex interactions among different levels of organisations (local versus national capacity), human decisions (e.g. land-use change influenced by economic pressures) and climatic changes-based factors to name a few.

Analysis by Nordhaus (2009) indicates that if society follows an 'optimal' carbon abatement policy through forestry, it could accomplish roughly 30% of total abatement over the century. Around 42% of this would arise from avoided deforestation, with the rest roughly equally split between afforestation and forest management options. Many land-use activities can be carried out immediately, appear to present a relatively cost-effective emission reduction component of international climate policy, and may generate environmental co-benefits. However, uncertainties exist in the form of 'permanence' as sustainable land-use (or forest) management does not represent a permanent change in the cumulative flux of carbon dioxide. Nilsson and Persson (2012) also emphasised the need to integrate environmental policies such as those designed to reduce greenhouse gas emissions (e.g. REDD+) and those that govern land and water management, which affect the resource nexus interplay. One example is how to strengthen concordance in the case of hydropower generation at the same time that freshwater bodies are maintained under good ecological conditions i.e. meeting the environmental flows, in that the production of bioenergy is required to reinforce the policies for energy, land use and agriculture (*ibid*). Moreover, mitigation of land degradation can save water and energy for human use by increasing soil water storage and groundwater recharge, as well as diminishing the application of energy-intensive fertiliser (Rockström et al., 2014). In this context, the nexus approach can aid towards an innovative climate policy by supporting climate change mitigation and adaptation while matching research findings with national socio-economic and technology interests and trends in countries such as India and China (Bleischwitz et al., 2017).

There is a need to highlight the co-benefits that ecosystems such as forests provide reflecting multiple ecological and socio-economic functions. Revenues under the REDD+ option would be higher if co-benefits such as those from watershed protection, soil erosion control, recreation, and other non-carbon ES are included in the estimates. The question of whether or not protection of existing forests through REDD+ poses any sort of biodiversity 'negative' also needs to be explored. Thus by implication REDD+ should not focus on forests that have high carbon values at the expense of those with lesser carbon content but rich biodiversity, processing of any materials or energy derived out of these projects as well as cultural values for conservation efforts (Khatun, 2011).

Natural capital and the resource nexus

At the international level, there has been increasing interest and discussions regarding how to govern interactions between land (or food), water, and energy systems, in what has become the 'nexus perspective' as shown in Figure 2.4. In fact, natural capital is required to support actual and future resource nexus for the development of human activities (e.g. agriculture, tourism and

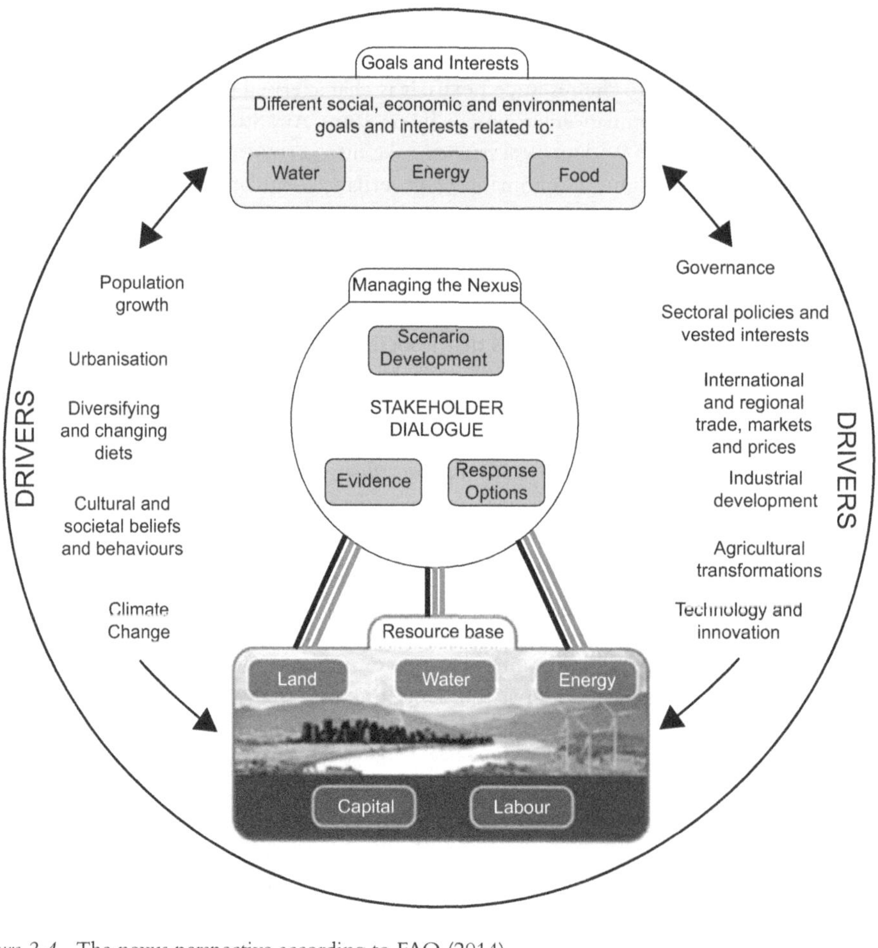

Figure 2.4 The nexus perspective according to FAO (2014)

industry) and basic needs (e.g. drinking water and food). For instance, water, energy and food are inextricably linked with the *water for energy*, *water for food*, *energy for water*, *energy for food* and *food for energy* relationships (Hardy and Garrido, 2012; Mayor et al., 2015).

The resource nexus can and 'should' play a key role in maintaining natural capital. For example, water plays a fundamental function on maintaining landscape productivity and biomass growth, essential also for sustaining a growing world population (Rockström et al., 2014). *Water for food* faces several options and challenges towards increasing food production under sustainable boundaries for aquatic and terrestrial environments. When we think of how to perform a more efficient and sustainable *water for energy* nexus in the future, we may thus arrive at the conclusion that minimising the use of water for energy generation, is with the implementation of high-efficient and low-water consumer technologies during thermal processes.

The role of nature and hence natural capital in the nexus can help understand what investment is needed to ensure water, food and energy security. Nature and ES then become part of the investment focus for the nexus and for economic and social development and a keystone of a future green economy. The results from such investment should be accounted for in terms of economic returns and water, food and energy security, as well as, critically, social equity and resilience. There are a wide variety of footprints for the water, land, carbon and material resources that can be measured as part of nexus approaches. These tools along with methods addressing other metrics are discussed in the second part of this book. The footprint analysis can apply several approaches – e.g. life cycle analysis, input–output analysis, national footprint accounting and the scale of analysis – range from a product/food/energy through organisation/sector/trade/building up to region/nation/world. However, the most challenging part is to create alternatives that are able to embrace the nexus with a broader and inclusive approach for each resource towards maintaining (critical) ES. The modernisation of irrigated systems in Spain provides a good example of the failure in the integration of the resource nexus and the implications on existing ES, where the consideration of the local conditions are required in order to avoid unexpected changes (see Box 2.1).

Box 2.1 The resource nexus and unexpected consequences: the modernisation of irrigation systems in Spain

The state-driven modernisation of irrigated systems in Spain during the period from 2002 to 2008 is an example of the relevance of having a previous and detailed analysis of possible consequences in the nexus when public measures are implemented. The plan has been possibly the most important in terms of surface area and investment in the whole of Europe and one of the largest programs in the world. The modernisation was put into action to increase water efficiency for irrigation purposes, to avoid overexploitation of aquifers and to generate water savings at plot and basin level with the aim of reducing water stress during drought periods (López-Gunn et al., 2012). This measure was to provide farmers enough water to produce food, but in contrast larger energy requirements were needed for the new pressurised pipes systems with an increase from 0.02–0.15 kWh/m^3 for flood irrigation to 0.28–0.68 kWh/m^3 for sprinkler and drip irrigation (Hardy and Garrido, 2012). On one hand, the investment in the irrigation modernisation contributed to water purification. A reduction of water dose at field level led to a reduction in diffuse pollution due to the lower volume of return flows, and as result of pollutant loads. This positive outcome has been particularly important with the onset of fertirrigation, since it helps farmers to apply water and nutrients at the same time and more efficiently reducing the nutrients losses (López-Gunn et al., 2012). On the other hand, some evidence envisages that planned water savings were not met. In fact, it identified an opposite effect or the so-called rebound effect, where improvements in efficiency led to higher consumptive use (Rodríguez-Díaz et al., 2011). For instance, at field level in Guadalquivir basin (South Spain) water savings took place (e.g. in the amount of water that is diverted and applied on the field), but as a result the consumptive use increased. The reason was that additional areas were converted into irrigation and/or crop patterns changed to more water-consuming crops (*ibid*). Although water saving might not be met at basin scale, it is true that with more efficient systems there is more margin for water allocation among activities (e.g. industry and agriculture and among crops, such as low and high water-demanding crops). Additionally, an unexpected change was found in 2008 with a process of energy liberalisation that removed the preferential binomial tariffs and led to a rise of electricity price, and particularly of contracted capacity, between 30 to 70% (Ederra and Murugarren, 2010). Before modernisation the water costs were approximately 100 € ha^{-1}, whereas after modernisation they rose up to 400 € ha^{-1} (Rodríguez-Díaz et al., 2011). In regions that mostly depend on groundwater source such as in the high part of Guadalquivir basin, higher energy costs constrained farmers from irrigating the amounts they needed, particularly in deep groundwater wells (Salmoral and Chico, 2012). An additional measure to control the water deficit in the basin was the gradual introduction of decreasing water allowances particularly for olive orchards, which changed from approximately 2520 m^3/ha (in 1997) to 1500 m^3/ha (after 2002). However, in the end, in the Guadalquivir basin, absolute water savings have not been met due to higher irrigation efficiency and reduction of water allowances or restriction of irrigation by water costs because of the opposite effect of conversion to irrigated conditions. In fact, the gross water demand[1] increased. According to the 2002 and 2015 Hydrological Plans of the Guadalquivir River basin, the irrigated crops in 2002 comprised 647,570 ha and 2,996 hm^3 gross water demand (i.e. 4630 m^3/ha) and in 2015 the area reached up to 856,430 ha and 3,357 hm^3 gross water demand (i.e. 3920 m^3/ha).

Criticism and weaknesses of current approaches

There are significant information gaps in our knowledge on biodiversity, carbon flux, species prevalence, identification and the quantification of ES, among other environmental and socio-economic factors. Thus before we can 'value' the services that ecosystems provide, we need to understand what it is we are losing when complexity issues are not addressed. The current wave of ecology-commodity creation under PES requires the rejection of prior values, practices, access or use rights so that land itself becomes capital (Sullivan, 2011). McAfee (2012) gets to the core through the simple statement 'selling nature to save it', but as Sullivan (2011) states, the ensuing payments for the services do not return to 'nature' as provider, but to those who are able to annex them successfully. The current new initiative is the 'green economy' highlighted at the UN Conference on Sustainable Development (Rio+20) in June 2012. Under the green economy, developing countries are particularly likely to benefit, by harnessing new trade opportunities, including, for example, the export of certified commodities in the fisheries, forests or agriculture sectors (UNEP, 2011). Commodities rely on the simple premise that they can be traded, thus if we look at where PES sits in the international policy and trade context, large inconsistencies remain with agricultural objectives and conservation policies. Inefficient food production and harmful agriculture subsidies are also contributing to deforestation, water shortages and pollution which can exacerbate impacts of climate change and have far-reaching consequences for food security and the development priorities of many nations (Khatun, 2012) and can have negative consequences for the natural capital maintenance and for the resource nexus. An example is the European Union's (EU) long-established Common Agricultural Policy (CAP), where as far back as 1962 the decision was taken for the CAP to exclude animal feed from any subsidies. Subsequently, the EU abandoned its own production of plant proteins, and cheaply imports animal feed from Latin America, including soy beans, thus encouraging nations like Brazil and Argentina to start large-scale soy beans production at the expense of food crops and pasture land, thereby exacerbating deforestation rates (Soares-Filho et al., 2006). EU support to its own agricultural producers has been at the expense of developing nations, and has kept commodities such as beef artificially low. The recently approved CAP reform 2014–2020 aims at stimulating viable food production, sustainable natural resources management (e.g. water and land) and adaptation to climate change with the application of the compulsory cross-compliance and greening measures by farmers. However, factors such as scarcity, market fluctuations, or profit by speculations will impact the access to food products and the 'value' of ES, and even the risk of liquidation. Therefore, it can be argued that there is a need to monitor subsidies carefully and phase out environmentally harmful subsidies. A resource nexus approach in this case would require the inclusion of interlinkages, e.g. energy, agricultural and water subsidies in the food chain.

In the global south today, safeguards against the 'new colonisation' has become linked to preservation of forests and ancestral lands. Losing rights to these commons has led to the separation and expatriation of people from their land and subsistence, with the loser's found to be traditional communities. Numerous struggles against large-scale extraction of resources, and enclosures and other forms of privatisation have accelerated and resulted in the eviction and loss of traditional rights for many indigenous societies. These range from the Maoist movement in India, created among other factors due to lack of access to resources, and land reforms, their goal to create 'a mighty mass movement against imperialism' (Gupta, 2006) to Africa where Chomsky (2012) writes that, 'the conversion of public goods into private property through the privatisation of our otherwise commonly held natural environment is one way neoliberal institutions remove the fragile threads that hold African nations together'. Poor, yet resource-rich, Bolivia, made the strongest stand in Cancún, in 2011. Evo Morales rejected 'the process of mercantilism of environmental functions, including the REDD+' (Lang, 2012). These examples highlight the need to go beyond market opportunities and to find pathways that can conserve natural capital, which are inclusive of the people who reside in many ecosystems, and rely on these for their livelihoods and food, water and energy needs.

Knowledge gaps in natural capital and ecosystem service research

Since the seminal works by some economists in the 1990s, research on ES has grown significantly, with the MEA (2005), the Intergovernmental Panel on Biodiversity and Ecosystem Services (IPBES) and the UK's Natural Capital Committee bringing the concepts into broader environmental planning and natural resource management. While several disciplines study natural capital and ES – each stressing particular aspects of the social, technical, economic or political – there remains a lack of consensus in the levels of natural capital degradation. Despite progress in understanding the links between ES, their functioning and the socio-economic benefits that depend on the flow of ES, research and application are also still largely experimental and exploratory as multi-scale studies across global sites show little comparability. Integrated modelling approaches exist and are improving, yet the valuation issue is as critical due to the system boundary about what elements of the nexus ought to be included (Bleischwitz et al., 2017). Further conceptual and empirical work is needed to understand how these concepts can be utilised for integrating ES into natural capital management and decision-making overall. There is also a need to understand how the concepts can be embedded into existing practices or used to transform current resource management and policy approaches relating to agriculture, forestry and other efforts to address resource efficiency. Holistic perspective can only be obtained through a multidisciplinary perspective; thus research objectives with regional collaborators should seek to identify and address the gaps between science, policy and 'on the ground' realities. The 'state of the art' literature shows that integrated empirical research on any part of world where food security and land-use issues/conflicts exist is sorely needed. This is particularly relevant in regions where biodiversity and forest cover are high, yet agriculture or fisheries are the primary and often the only source of income for poor rural people. The focus on the economic valuation of ES, tends to apply narrow approaches of monetary values with a bias towards provisioning, cultural or regulatory functions where a willingness to pay can be assumed. Additional opportunities towards sustainable development might be detected if the nexus is addressed holistically highlighting the interrelationships between the biosphere and the development needs of society (Folke et al., 2002) through multi-level governance, from local to global level.

Conclusions

To continue to obtain life-sustaining and other ES, preservation and the sustainable use of natural capital is essential. Humankind has enjoyed benefits from nature, including food, clothing, energy and in many places even shelter (FAO, 2014). However, the availability of water, energy or food for societies not only depends on the existing climatic conditions, their spatial temporal relationships, soil productivity or abundance of primary energy sources (i.e. natural biophysical boundaries), but also on the social technical means (e.g. infrastructure for water distribution and storage, crop intensification or capacity for energy generation) and the existing management measures (e.g. water allocation among and within economic activities, soil conservation practices or stability for energy generation). In addition, the identification and management of critical natural capital – to ensure important environmental functions for human development continue – and urgent context-specific research are needed for a sound approach to environmental policy. Thus the nexus might support natural capital maintenance by creating opportunities for policy and technological innovation aimed at building socio-ecological resilience. The coordination and integration of policies would represent a key role while reducing risks, allowing for solutions 'on the ground' and avoiding the 'panacea' problem. Science and policy efforts will have to adapt under changing and uncertain conditions, as well as to get benefit undertaking the complexity of resource systems with the development and integration of diverse approaches and scientific disciplines. From this premise, research is required which extends existing knowledge on practices that act as obstacles as well as those that offer opportunities, but also forge new links with other fields (and collaborators) that are currently understudied. Water, food and energy security as well as climate change mitigation/adaptation are urgent issues now and thus long-term ecosystem resilience is required. Policy decisions based on informed socio-ecological research/data is needed in conjunction with discussions on food security, resource management, climate mitigation, which in turn can contribute to evolving policies towards social transformation and sustainable futures.

Many initiatives around conservation (REDD+, PES) are works in progress and there is a consistent shifting of key concerns and positions among countries, but for many nations sustainable development remains the ultimate aspiration. Moving towards this goal presents tremendous challenges. Therefore, the nexus approach can aid towards the management of natural capital and ES by interlinking the different components, and accounting for co-benefits, in the face of potential shocks to water, energy and food, land and minerals stocks across global value chains. Moreover, there is a need to incorporate decisions and planning that allow for better holistic decision-making where approaches such as ES are viewed by nations as more than just money-making tools. This in turn may provide interconnectedness from local to global scale between the system governance and the dynamics of the biosphere, internalising environmental degradation into the global economy and promoting institutional reforms that would support the adaptive capacity of the complex behaviour of social-ecological systems, as well as encouraging ecosystem management processes that link communities, natural capital practitioners and scientists.

Acknowledgements

Kaysara Khatun would like to acknowledge the support of the PROMOTEO project of the Secretariat of Higher Education, Science, Technology and Innovations of Ecuador (SENESCYT).

Note

1 Total water withdrawals, taking into account the efficiency of transport (i.e. channels), distribution (i.e. ditches) and application system (i.e. drip, sprinkler and gravitatory).

References

Allen, C. R., Angeler, D. G., Garmestani, A. S., Gunderson, L. H., and Holling, C. S. (2014). Panarchy: Theory and application. *Ecosystems*, 17, 578–589.

Avishek, K., Liu, J., and Yu, X. (2012). *Challenges and opportunities for management of terrestrial ecosystems in Asia*. UNEP policy series, Policy Brief, 9.

Bleischwitz, R., Hoff, H., and Spataru, C (Ed.). (2017). *Modelling approaches and future scenarios on the nexus, part III of. handbook on the resource nexus*. London: Routledge.

Chomsky, N. (2012). How the Magna Carta became a minor carta, part 1. *The Guardian*.

Corbera, E., Brown, K., and Adger, W. N. (2007). The equity and legitimacy of markets for ecosystem services. *Development and Change*, 38, 587–613.

Corbera, E., and Schroeder, H. (2011). Governing and implementing PES. *Environmental Science and Policy*, 14, 89–99.

Daly, H. (2014). *Use and abuse of the "Natural Capital" concept*. Published by reslienece.org.

de Groot, R. (2006). Function analysis and valuation as a tool to assess land use conflicts in planning for sustainable, multi-functional landscapes. *Landscape and Urban Planning*, 75, 175–186.

Dumont, A., Salmoral, G., and Llamas, M. R. (2013). The water footprint of a river basin with a special focus on groundwater: The case of Guadalquivir basin (Spain). *Water Resources and Industry*, 1–2, 60–76.

Ederra, I., and Murugarren, N. (2010). La nueva tarifa eléctrica: la escalada de precios del agua de riego. *Navarra agraria* 179.

FAO (2014). *The water-energy-food nexus – a new approach in support of food security and sustainable agriculture*. Rome: Food and Agricuture Organization of the United Nations.

Folke, C., Carpenter, S., Elmqvist, T., Gunderson, L., Holling, C. S., and Walker, B. (2002). Resilience and sustainable development: Building adaptive capacity in a world of transformations. *Ambio*, 31, 437–440.

Franklin, S., and Downing, T. (2004). *Resilience and vulnerability*. Poverty and Vulnerability Programme GECAFS Project, Stockholm Environment Institute.

Gunderson, L. (2010). Ecological and human community resilience in response to natural disasters. *Ecology and Society*, 15, 29.

Gupta, T. D. (2006). Maoism in India: Ideology, programme and armed struggle. *Economic and Political Weekly*, 41(29), 3172–3176.
Hardy, L., and Garrido, A. (2012). Challenges and opportunities related to the Spanish water-energy nexus. In L. De Stefano and M. Llamas (Eds.) *Water, agriculture and the environment in Spain: Can we square the circle?* pp. 177–190.
Holling, C. S. (2001). Understanding the complexity of economic, ecological, and social systems. *Ecosystems*, 4, 390–405.
Humphreys, D. (2006). *Logjam: Deforestation and the crisis of global governance*. London: Earthscan.
Jarret, D. (2012). *Biodiversity, ecosystem services and natural capital: Terms matter*. Retrieved January 2016, from http://ecometrica.com/article/biodiversity-ecosystem-services-and-natural-capital-terms-matter
Khatun, K. (2011). Reconciling timber provision with carbon sequestration opportunities in the tropical forests of Central America. *Environmental Science & Policy*, 14, 1091–1102.
Khatun, K. (2012). Reform or reversal; implications of the Common Agricultural Policy (CAP) on Land Use, Land Use Change and Forestry (LULUCF) in developing countries. *Conservation Letters*, 5, 99–106.
Lang, C. (2012). *Bolivia's joint mitigation and adaptation mechanism: An alternative to REDD?* Retrieved from www.redd-monitor.org/2012/10/11/bolivias-joint-mitigation-and-adaptation mechanism-an alternative-to-redd/
Laurance, W. F., Useche, D. C., Rendeiro, J., Kalka, M., Bradshaw, C. J. A., Sloan, S. P., Laurance, S. G., Campbell, M., et al. (2012). Averting biodiversity collapse in tropical forest protected areas. *Nature*, 489, 290–294.
Liu, J., Dietz, T., Carpenter, S. R., Alberti, M., Folke, C., Moran, E., Pell, A. N., Deadman, P., Kratz, T., Lubchenco, J., Ostrom, E., Ouyang, Z., Provencher, W., Redman, C. L., Schneider, S. H., and Taylor, W. W. (2007). Complexity of coupled human and natural systems. *Science* (New York), 317, 1513–1516.
López-Gunn, E., Mayor, B., and Dumont, A. (2012). Implications of the modernization of irrigation. In L. De Stefano and M. Llamas (Eds.), *Water, agriculture and the environment in Spain: Can we square the circle?* Abingdon, Oxon: Taylor & Francis, pp. 241–256.
Mayor, B., López-Gunn, E., Villarroya, F. I., and Montero, E. (2015). Application of a water–energy food nexus framework for the Duero river basin in Spain. *Water International*, 40(5), 1–18.
McAfee, K. (2012). The contradictory logic of global ecosystem services markets. *Development and Change*, 42(1), 105–131.
Millennium Ecosystem Assessment (MEA). (2005). *Ecosystems and human well-being*. Washington, DC: Synthesis Island Press.
Natural Capital Committee. (2014). *The state of natural capital: Restoring our natural assets*. Second report to the Economic Affairs Committee. Retrieved from www.naturalcapitalinitiative.org.uk/wp-content/uploads/2014/06/1f06ffa9eaf0af134d7022af36fe17cf.pdf.
Nilsson, M., and Persson, Å. (2012). Reprint of "Can earth system interactions be governed? Governance functions for linking climate change mitigation with land use, freshwater and biodiversity protection." *Ecological Economics*, 81, 10–20.
Nordhaus, W. (2009). *A question of balance*. New Haven, CT: Yale University Press.
Norgaard, R. B. (2010). Ecosystem services: From eye-opening metaphor to complexity blinder. *Ecological Economics*, 69, 1219–1227.
Parker, C., Mitchell, A., Trivedi, M., and Mardas, N. (2009). The little REDD book. Oxford: Global Canopy Program.
Rockström, J., Falkenmark, M., Allan, T., Folke, C., Gordon, L., A. Jägerskog, Kummu, M., Lannerstad, M., Meybeck, M., Molden, D., Postel, S., Savenije, H. H. G., Svedin, U., Turton, A., and Varis, O. (2014). The unfolding water drama in the Anthropocene: Towards a resilience based perspective on water for global sustainability. *Ecohydrology*, 1261, 1249–1261.
Rodríguez-Díaz, J. A., Pérez-Urrestarazu, L., Camacho-Poyato, E., and Montesinos, P. (2011). The paradox of irrigation scheme modernization: More efficient water use linked to higher energy demand. *Spanish Journal of Agricultural Research*, 9, 1000–1008.
Salmoral, G., and Chico, D. (2012). Lessons learnt from analyses of the water footprint of tomatoes and olive oil in Spain. In L. De Stefano and M. R. Llamas (Eds.), *Water, agriculture and the environment in Spain: Can we square the circle?* Abingdon, Oxon: Taylor & Francis, pp. 123–136.
Soares-Filho, B. S., Nepstad, D. C., Curran, L. M., Cerqueira, G. C., Garcia, R. A., Ramos, C. A., Voll, E., McDonald, A., Lefebvre P., and Schlesinger, P. (2006). Modeling conservation in the Amazon basin. *Nature*, 440, 520–523.

Steffen, W., Richardson, K., Rockström, J., Cornell, S. E., Fetzer, I., Bennett, E. M., Biggs, R., Carpenter, S. R., de Vries, W., and de Wit, C. A. (2015). Planetary boundaries: Guiding human development on a changing planet. *Science*, 347(6223). DOI: 10.1126/science.1259855

Sullivan, S. (2011). *Banking nature? The financialisation of environmental conservation*. Working Paper Series 8, OAC Press. ISSN 2045-5763.

TEEB (2008) the economics of ecosystems and biodiversity: An Interim Report. European Communities. Retrieved from http://ec.europa.eu/environment/nature/biodiversity/economics/pdf/teeb_report.pdf

UNEP (2011). *The green economy report*. United Nations Environment Programme. Retrieved from www.unep.org/greeneconomy/greeneconomyreport/tabid/29846/default.aspx

3
SCALE AND THE RESOURCE NEXUS

Corey Johnson and Stacy D. VanDeveer

Introduction

The thrust of nexus thinking is that understanding the causes and implications of resource extraction and consumption require thinking across boundaries of resource categories. The potential pitfalls of expanding the methodological ambit in such ways, however, are significant. Among these pitfalls is that collapsing boundaries renders the questions about resources and environments impossibly complicated and therefore difficult to model, visualize, and communicate. Of course such challenges are not new in scientific inquiry and governance, and there are many possible ways to constrain inquiries so that problems can be semi-isolated for the purposes of understanding them or attempting to govern them. The concept of scale is inherent to nexus thinking and necessary for any fruitful nexus-based analysis. Each nexus dimension – from water to energy to food – crosses scales.

Yet scale is understood and operationalized differently in different fields (and even within single fields). We therefore attempt here to give an overview of scalar thinking in all its heterodoxy as well as offering possibilities for incorporating scalar thinking into the nexus. As the chapter makes clear, contrary to the impression of everything being "global" in scale and scope (e.g., international value chains, global virtual networks), when it comes to resources the interlinkages across scales – from micro to macro – are a central feature of the resource nexus. So, for example, it is well known that "global" economic forces may shape "local" water, food, land, and energy usage and availability in many places around the globe. Such economic forces can impact different countries, cities, or rural areas – or different economic classes – very differently. But less often appreciated is that "global" economic forces cross scales, both in their impacts and in their causes. One cannot explain or govern the impacts of global minerals markets, for example, without attention to the hundreds of millions of consumer decisions and the state, local and firm policies that shape them. Thinking about scale means interrogating and reflecting on concepts framed almost exclusively at one scale, such as "earth governance," "global public goods," or individual purchasing decisions. Similarly, concepts that seem to ignore scale, such as "value-chain management," may hide more than they reveal about the complexity of nexus interactions and challenges without some attention to scale.

The chapter first explores the meaning and utility of the concept of scale, and scalar thinking and analysis, by discussing its use and development in disciplines such as geography and ecology, where it is of central intellectual importance and where it has been engaged in resource-related analysis for some time. Next, the chapter seeks to connect the concept more explicitly to the resource nexus, and to analysis in politics, policy and governance. Latter sections explore the connections between scale and resource and climate governance and draw out a few conclusions.

Scalar concepts and analysis

In disciplines such as geography and ecology, scale is a foundational concept that, in spite of a lack of common definitional agreement, provides a sort of unifying vocabulary for thinking through the spatiality of phenomena, events, etc. (Silver, 2008). Sayre (2009) provides a number of examples of prominent scientists and journalists situating scalar thinking at the heart of their inquiry, from Princeton ecologist Simon Levin who said in his Robert H. MacArthur Award lecture in 1989 that relating phenomena across scales is "the fundamental conceptual problem in ecology, if not all of science" (Levin, 1992, 1944) to journalists Elizabeth Kolbert and Thomas Friedman, who have made similar such claims. The term is used frequently in both colloquial and academic contexts, which gives rise to a considerable lack of clarity of what the concept means and how it can best be used.

In common usage, scale is often used interchangeably with "level," and it suggests a nesting of hierarchical levels from the micro to macro or the inverse (e.g., from local to global, with various intermediary steps). In cocktail party usage, one can imagine a retort to someone highlighting having seen a seemingly one-off dead tree in the nearby forest with "The scale of the problem is actually much greater," typically meaning that the localized event of one dead tree is actually situated in a wider – more geographically dispersed – set of occurrences as a result of some systemic set of environmental phenomena.

In the discipline of ecology, scale has both spatial and temporal connotations, and when one considers observing patterns in the environment related to, for example, a species of organism, being explicit about scale makes sense.

> When we observe the environment, we necessarily do so on only a limited range of scales; therefore, our perception of events provides us with only a low-dimensional slice through a high-dimensional cake. In some cases, the scales of observation may be chosen deliberately to elucidate key features of the natural system; more often, the scales are imposed on us by our perceptual capabilities, or by technological or logistical constraints.
>
> (Levin, 1992, 1945)

Levin emphasizes that there is no "correct" scale at which to observe ecosystems, and in fact that what ecologists refer to as a species "community" or "ecosystem" is in fact "an arbitrary subdivision of a continuous gradation of local species assemblages" (Levin, 1992, 1960). The important takeaway from this for the purposes of this chapter is that like natural resources, ecosystems are not closed systems, but rather are defined by porous borders; the use of scale is nevertheless useful in constraining inquiry and quantifying variability in time and space.

Scale has a long conceptual lineage in geography, first being deployed by cartographers to express the relationship between something represented on a map to space in the world. The fractions associated with "cartographic scale" are usually displayed graphically to relate the map to the world in such a way: 1 unit (cm, inch, etc.) on the map is equivalent to y units in the world being represented on the map. While primarily a *representational* matter (Sayre, 2009), cartographic scalar relationships are implicitly asking maps users to think across and make connections among spatial scales, relating what appears on paper or one's smartphone to the environment to facilitate, for example, successful navigation. This relational aspect will be returned to, as it helps to understand what the analytical distinction is between scale and level, two terms that are often use interchangeably in common parlance.

Physical geographers tend to reflect ecologists in treating scale as "the temporal and/or spatial range and magnitude of a process or observation" (Silver, 2008, 922). Human geographers typically use scale methodologically in order to define the resolution of data collection or constrain analysis to fit a particular set of issues (Gregory et al., 2009). In that way, scale is more an *observational* matter than a representational one (Sayre, 2005). As an example, when a human geographer or other social scientist wishes to study the issue of food availability and hunger, the scale of analysis chosen to take on the question is not simply a matter of location but of granularity of the lens being used to ask the question. A relatively prosperous community in North America, for example, may compare favorably at the aggregate to other communities in terms of percent of the population that is food insecure, but if one focuses on neighborhood scale, pockets of food deserts and high rates of hunger may be found. Typically, then, human geographers attempt to think through the linkages across scales of analysis. In the example of hunger, a change in food availability at the neighborhood scale might hypothetically be tied to changes in global trade patterns that cause price hikes to various commodities. Such a linkage would not be immediately evident if thinking only about neighborhoods in a city or globally about food productions, but the ties between the two phenomena occurring at different spatial scales are nevertheless very important.

Geographers and others involve scale deliberatively in epistemological decisions about approaching research questions and methods, but "scalar thinking" is something that most of us do unwittingly as a cognitive means of making sense of complexities in the environments around us. When it rains, for example, it would be very strange indeed to think about the effects of the rainfall only in terms of some global impact of the precipitation on global water balances or a regional drought. Rather, most observers would cognitively organize the event at multiple scales: the micro-scale effects of the rain on my eyeglasses, to the large puddle forming between me and where I want to walk, to community- and regional-scale impacts on a persistent drought. This scalar thinking is so much a part of daily life that it requires little effort or reflection.

A significant body of literature in recent decades has examined the scalar aspects of water governance. Water is, as Giordano (2003) puts it, a *fugitive resource*, whose flows and availability varies over time and space as a consequence of human and environmental factors (Feitelson and Fischhendler, 2009). The rights regimes that overlay the resource are also rarely perfectly coincident, and as a result, especially in areas of the world where water is scarce, there is potential for conflict and a need for governance regimes that allocate the common resource according to norms of equitability, legal frameworks, etc. (Giordano, 2003).

Feitelson and Fischhendler (2009) provide an overview of the historical evolution of water resource allocation in various contexts, making the case that "The spatial scale determines the areas to which a governance regime pertains and thus the issues and parties with which it has to content" (p728). In water-rich or sparsely populated areas, often water management historically occurred at the household scale, with each household having its own well or open access to a stream. In the case of the American West, household-scale water access came to be managed as an "individual asset," and was married to individual property rights. As the West was developing, though, making water available to encourage widespread settlement and agricultural productivity was also framed as part of a national-scale political imperative, and this gave rise to massive "reclamation" projects by state and, above all, the federal government. These projects sought to impound, divert, and direct water across the largely arid region, but the access regime that focused on individual rights was left largely unchanged (Reisner, 1993). Over time, as demands on the limited water resources in the West grew and an uncooperative climate brought lengthy droughts to the region, water governance grew increasingly contentious. One could argue that the scalar politics of the household – the socially constructed nature of rugged individualism and rights – thwarted attempts to more effectively manage water usage of key sources such as the Colorado River or California Central Valley aquifers, and it is this mismatch that helps to explain why the Colorado is among the most over-allocated freshwater sources in the world, pitting in various contexts farmer against farmer, city against city, urban against agricultural interests, states against states, and even neighboring countries (US and Mexico) against each other.

Yet the scalar aspects of water governance also provide opportunities for cooperative management. In areas where water is not abundant, water governance is typically at the community or higher scale. The case of Israel and its neighbors offers a good example of the scalar dynamics of water governance, and it also provides some insight into the material as well as the socially constructed elements of scale as it pertains to water resources and their management. Prior to the 1930s, water management for Jews and Arabs in the Levant was largely a community affair, but with the arrival of larger numbers of Jews and "the state-building development ethos espoused by the Zionist movement from the outset" (Feitelson and Fischhendler, 2009, 733), basin- and interbasin-scale projects were developed, such as the Kishon project started in 1937 (ibid.). With the creation of the State of Israel in 1948, a national-scale effort at water management was introduced, one that sought to encourage development in water-poor parts of the country through large-scale interbasin transfers such as the National Water Carrier that moves water from the Upper Jordan River basin (especially the Sea of Galilee) to communities in the dryer Negev in the south of the country. During the 1980s, a growing environmental movement in Israel successfully lobbied for the construction of wastewater facilities, giving rise to a water governance system wherein freshwater supply is still largely national in scope, while wastewater and drainage is managed at the basin scale.

In Israel, there is of course also a geopolitical and international aspect to water. As part of the peace process of the 1990s, a number of agreements were reached between Israel and its neighbors around issues of water access. That Palestinians and Jordanians experienced greater precariousness of their water supplies than Israel was well known and contentious; how this would be governed in the context of a negotiated peace was touchy on all sides. Feitelson and Fischhendler (2009) point out that the supranational-scale solutions to water supply between Palestinian territories, Israel, Jordan, and other neighboring countries were partial and largely reflected national positions rather than more nuanced, multi-scalar water management regimes that might have served to diminish the importance of international borders (see also Cohen and Frank, 2009).

Scale figures in the work of nearly all human geographers in a methodological sense, but for a subset of political economy focused geographers, scale becomes the object of inquiry in the sense of thinking through the so-called politics of scale. For example, contrast the flow resource of water described earlier in this section with the another resource nexus-related case, that of the fixed assets of minerals. Scale has been at the center of defining the ways in which inequalities are produced and reproduced. Huber and Emel (2009), for example, use the 1872 Mining Law in the US – and recent efforts to reform the law – to illustrate a scalar politics around mineral deposits and efforts to capitalize on them. The issue with the law for some was that it was written at a time when small, private mining companies were incentivized to develop mineral resources by offering virtually unlimited rights to explore and extract on publicly owned lands with no requirement for compensating the public for those resources. Over the next century, however, the amount of mining (and the environmental consequences of it) increased dramatically, and the mining companies also were increasingly large, multinational corporations. In the debate over reforming the law to require the payment of royalties, debates involved arguments and counterarguments about implications of such changes locally, nationally, and globally. Mining concerns threatened that their capital would go elsewhere globally to avoid paying such royalties, presenting a tension between the fixed geography of mineral deposits and the highly mobile, increasingly globalized capital seeking to exploit those deposits.

Scale and resource interlinkages

The two previous examples of minerals and water are instructive of how scalar analysis can enrich the understanding of resource issues, but they also underscore the tendency in resource scholarship to place individual resource categories in silos. The story of mineral extraction on public lands in the US is as much a story of the environmental consequences of mining on freshwater and the lands where the mines were located. In 2015, a US Environmental Protection Agency team, while working on a remediation project on the Gold King Mine near Silverton, Colorado, triggered a mine blowout that resulted in the release of 3 million gallons of acid and arsenic-laden mine water into the Animas River (a tributary of the Colorado River), turning the river bright orange and carrying the environmental legacy of a mine that had closed in 1923 downstream to Lake Powell and beyond (U.S. Department of the Interior, 2015). The millions of tons of ore extracted from that mine enriched its owners, but Gold King Mine and the more than 100,000 abandoned mines in the western US, are a considerable liability for other resources, not limited to water, timber, agricultural lands, and fisheries, hanging, as an article in the *Atlantic* magazine put it, like "chemical swords of Damocles" over the rivers below them (Parker, 2015). Similarly, the governance of water resources in the western US or the Middle East is part of a broader, interlinked resource issues around agriculture, energy, mineral extraction, etc.

This chapter argues then that, as the resource nexus is developed conceptually and methodologically, issues of scale should be integrated much as they have been in the research on individual resources cited in the section "Scalar concepts and analysis." In some circles this is already occurring. Geographers have recently taken up nexus thinking, and scale has played a key role. In her presidential address to the Royal Geographical Society meeting in 2013, Judith Rees argued

> The type and importance of the linkages between the resources in the nexus will vary spatially as will the feasibility and acceptability of measures to reduce external costs and improve overall sustainability and resource use efficiency. Moreover, it will often be the case that changes at multiple spatial scales (community to international) will be needed to reduce resource insecurity, even if the problems manifest themselves at a relatively local level.
>
> (Rees, 2013, 280)

One of the great challenges in integrating scalar thinking into any host of resource nexus issues is that of factoring negative externalities into analysis, a theme that figures prominently in this handbook (see e.g. Part II). Externalities are often shifted across geographic spaces, time, and scales (Andrews-Speed et al., 2015). Among the best-known examples of this can be found in US carbon emissions, the costs of which will be borne differentially around the world for decades to come. But even beyond fossil-fuel consumption, the massive resource through-put of the increasingly globalized high-consumptive lifestyles of (largely) urban upper- and upper-middle classes impose myriad externalities around the world and in the future. Who pays the ecological, social, and political costs associated with illicit markets in coltan, diamonds, drugs, weapons, or endangered species, for example? But this is hardly a problem exclusive to illicit markets. Some peoples' beef, oil, or minerals consumption clearly imposes costs on others, whether these costs be toxic or biological pollution, higher food prices, government corruption, or ecological and cultural destruction. In terms of resource interlinkages, the challenge of "governing oil" or "governing beef" are not exclusive to the resource silo in which each one sits. Both can impact water quality and supply across localities and continents, and both can alter the prices, supply, and distribution of water, land, and many foodstuffs.

Scalar concepts in politics, policy, and governance

Scholars in the overlapping fields of political science, international relations, and public policy have usually avoided using (and learning from) the concept of "scale," but they have spent a great deal of time and ink grappling with related concepts. Generally speaking, such fields deploy comparatively state-centric theories and concepts, and thinking more in terms of "levels," which as noted in the section "Scalar concepts and analysis" is a simpler and less nuanced conceptualization. So, for example, for decades nearly every introductory text for international relations has taught students to organize causal factors driving international relations events and outcomes within "levels of analysis." These levels – the international system, the state/society sphere, and the individual – are posited as a way to organize and separate significant causal variables for outcomes (like war) (Nye Jr and Welch, 2013; Rourke, 2008; Singer, 1961). Since the 1960s, students and scholars have been cautioned against making uncritical choices regarding which level to examine most closely for a given analysis, the need to think about trade-offs between the levels, and indeed whether seeing the world through these levels contributes more to our collective understanding than it restricts. International relations scholarship has often taken these "levels" for granted and treated them as quite static (unlike the "scalar thinking" discussed earlier in the chapter), with debate often revolving around the relative importance of structural and agent-based theories for understanding how actors and institutions shape outcomes at either the system or the state levels (O'Neill et al., 2004). So, for example, one might seek to understand how global markets shape incentives for rational actors (at any level of scale) around the world, or whether domestic or international system-level factors were the most significant causes of a particular violent conflict, or seek to understand how the foreign policies of particular states and the preferences or identities of their polities shape resource consumption.

Recent scholarship, particularly related to understanding contemporary European politics, and politics and governance related to energy and climate change at multiple scales has often deployed the concept of "multilevel governance" (Hooghe and Marks, 2001; Selin and Van-Deveer, 2012). In short, multilevel governance scholarship emerges as politics scholars seek to understand the relationships of institutions, actors, and issues operation across governance scales and jurisdictions. The concept helped scholars understand how urban scale actors interact with each and build networks for common purposes related to climate change mitigation, for example, even as they engage national, transnational, and inter-governmental organizations and institutions (Betsill and Bulkeley, 2006). As such, climate-concerned actors might interact or build links horizontally (with other actors or institutions at a similar "level"), or do so vertically with actors and institutions at levels above or below their own. So, for example, state-level officials in Massachusetts and California might work horizontally with those from other states and provinces in the US, Mexico, and Canada even as they seek to engage joint actions with officials or non-state actors working in local, national, and international organizations (Selin and VanDeveer, 2009). Political science, international relations, and public policy scholars often focus on the implications of these multilevel interactions, the opportunities and challenges for policymaking and institution building, and so on. More "scalar thinking" might also seek to understand the changing nature of legitimacy and authority within these interactions, and the ways that individuals', communities', and whole polities' relationships to each other might be changing as a result. Because some multilevel governance scholarship explores the ways that different "levels" constitute each other, and the actors and processes within and across levels, one can see more influence of scalar thinking in some multilevel governance scholarship than is commonly found in political science, international relations, and public policy literatures.

Until recently, the very old traditions of studying the roles and implications of federalism(s) remained largely unconnected to multilevel governance research (Selin and VanDeveer, 2012) and the politics of scale. Comparative federalism scholars often demonstrate the federalist institutions change over time, distributing and redistributing authority, legitimacy, resources, and so on. The structure and operation of federalist structures can have substantial influence on policy outcomes across a host of issue areas. Certainly it is difficult to understand US environmental or climate change politics, movements, policy outcomes, and other dynamics without attention to American federalism (Kamieniecki and Kraft, 2013; Rabe, 2010), and the same might well be said of many environmental, energy, and climate change-related processes and policies in Germany, Canada, Brazil, or India. While federalist scholars have generally ignored many of the more dynamic and nuanced aspects of the scale concept and scalar thinking – focusing instead on how levels interact or incentive actors differently, for example – it is clear that some of them grapple with similar concerns and processes as those explicitly using the concept. Also, federalism scholars have mostly ignored global and international interactional, and the many transnational links across levels of authority (and the implications of these) (Selin and VanDeveer, 2012), which greater attention to the concept scale and scalar thinking might addresses.

State centrism, geopolitics, and nexus analysis

Two final points about the state centrism and strategic logic of much political science, international relations, and geoeconomics are worth making. The first regards non-governmental organizations (NGOs) and conceptualizations of civil society, where this large and complex realm of social organization is defined primarily as "not the state." Critics of such state centrism, from a host of disciplines and theoretical perspectives, often note that the exact boundaries of the state – in theory and in empirical practice – tend to be much more difficult to identify than generally assumed. One often sees similar and related challenges in economics, as well. Where does the state end or begin, in a world in which statist institutions (ideas, norms, laws, rules, and so on) can shape fundamental aspects of individual and collective identity, family and community life, sexual and reproductive health, and notions of human rights? Where exactly does one draw lines between interactions that are "in the market" and those shaped primarily by states?

A second concern related to state centrism, with significant implications for resource nexus analysis and institutions, is captured in both the concept of geopolitics and its critiques. Geopolitics emerged as a field of inquiry and a mode of understanding international politics at the turn of the 20th century in Europe. Unsurprisingly it was deeply tied up with European empires' attempts to scientifically rationalize and better understand the impacts of factors such as physical space, resource distributions, technological innovations (especially in the realms of transport and military), and cultural phenomena on the exercise of international affairs. As such, the spatial ambit of geopolitics, conceptually as well as in praxis, was the global scale, but with the clear understanding that local events and phenomena could in fact impact global ambitions.

Although the term was reclaimed after World War II by scholars of international relations as a broader set of questions about strategic statecraft (especially by the postwar superpowers), the term cannot be simply divorced from its imperial origins, with all that those origins portended for things like securing access for the powerful to territories and natural resources they deemed necessary. Lately policy circles have become enamored by the arguably no less problematic term "geoeconomics," a term that seems to suggest that market access, market manipulation, and the relationship between political-cum-military power and economy, has superseded raw power politics as the primary focus of states and corporations in the international context. Yet, the two concepts must be viewed critically as two sides of the same coin (Blackwill and Harris, 2016).

The resource nexus, then, if it is to have analytical value, must critically acknowledge and grapple with geopolitical-geoeconomic framings of resource issues. Geopolitics and geoeconomics suggest a particular worldview – and expectation that resources serve and are controlled by (or, *should* serve and be controlled by) the interests of those with more material power and wealth. To think about and analyze the resource nexus in these types of "strategic" terms suggests little room for community or household level resource concerns, and precious little attention to equity, justice, or human rights frameworks, for example. If water and food are treated as primarily strategic resources – as is common in scholarship and policymaking about oil, uranium, or a particularly desirable rare earth mineral – does nexus analysis assume that those with less material power and wealth get none? Does it assume they will either have to become the highest bidder for scarce resources, or resort to taking what they need by force in order survive? If the answers to these questions are "no," than geo-strategic views (among others) of resource competition and governance must address how and why resource governance institutions can (or should) embody values, worldviews, rules or other social institutions beyond strategic logics based on power and wealth.

Linking scale to resource and climate governance research

A spate of recent work in various fields has sought to incorporate scalar thinking into questions of environmental and resource governance. Along similar lines, a number of scholars in the field of global environmental change have called for attending to definitional and methodological shortcomings in how scales of analysis are conceptualized and operationalized. An important intervention from over 25 years ago is still relevant today:

> To address adequately the human dimensions of global environmental change poses a formidable challenge. A reconciliation of two competing trajectories in the social sciences will be required – of those tending towards understanding through broad, macro scale social forces that affect nature-society relationships with those tending towards understanding through the complexity of these relationships in their local space-time contexts.
>
> (Turner et al., 1990)

A similar point can be made about nexus thinking and governance. Even if scientists were able to agree on the appropriate scale of governance for a particular issue, such a consensus inevitably hits up against the inflexibility of the scales of governance available to address problems. In no area has this been more apparent in recent decades than in debates over climate change, but it extends to a host of other issues: the territorialization of political decision-making in cities, regions, and above all, sovereign nation-states, make the scalar politics of governance far from simply an academic question. Water, again, provides an interesting case study. As the European Union (EU) progressively rescaled more and more governance areas from the national level to the supranational level over many decades, water management emerged as an area where national management of a flow resource that by definition crosses borders without deference to political lines and provided a key test case of how to supranationalize environmental issues. The solution, which emerged in the early 2000s, was the EU Water Framework Directive (WFD) (Johnson, 2012). It mandated that EU member states work with neighboring states where there was a shared river basin in order to manage the cross-border flows of water for purposes of monitoring and improving water quality. In this instance, there was a clear instance where the territorialization of decision-making was sorely lacking, and the EU provided a political venue through which to re-negotiate the inadequacies of political borders that emerged historically for many reasons other than which way water happened to run off and sought to provide a new "scalar fix" to the issue. The EU's WFD has not been without its critics or challenges in implementation, but there it represents a contemporary case of where scale the scale of a governance challenge is tackled in the policymaking realm.

Geographer Harriet Bulkeley has critiqued the taken-for-granted use of scale and space in governance circles as nested, hierarchical, and biased towards thinking that the macro necessarily is more important than the micro (Bulkeley, 2005). Much of her subsequent work, then, has sought to escape what has been referred to as the "territorial trap" of much international relations and political science literature (Agnew, 1994) by examining non-hierarchical, networked forms of governance around environmental challenges such as climate change (e.g., Bulkeley, 2012) and has spawned a host of interdisciplinary work challenging the assumption that a truly global problem such as climate change (albeit one with profoundly localized consequences) can only be governed at the global scale and within the context of traditional sovereign states acting in tandem with one another (Bulkeley et al., 2014). Examining these issues through scalar lenses, considering the polycentricity of decision-making related to the environment (Osofsky, 2013), is a profoundly helpful corrective. Bulkeley and her colleagues argue that while paying more attention to various levels of scale in climate governance – i.e., getting beyond the attachment to the "global" – enriches our understanding of governance processes, actors, institutions and outcomes, it must not tacitly assume that "different levels of governance are rather neatly separated from one another in distinct jurisdictional levels" (p184). When one sees discrete and different levels, one often focuses on how these levels interact. Yet, attention to governance operating at various levels and across various levels must also attend to the ways that such governance can shape or reshape the levels and/or the governance itself. So, for example, the emergence of multiple and varied carbon markets at various levels of scale – global, national, and subnational, for example, but also within and among firms and civil society organizations as well as within a particular public sector jurisdiction or between such jurisdictions – helps drive the creation of varies processes like carbon trading and carbon accounting across scales. Such dynamics change actors, behaviors, and institutions in the public, private, and civil society sectors – which in turn create various constituencies and norms across and within scales. These changes help implications for governance that may extend well beyond the functioning of a particular regulated carbon market.

Political science and international relations scholars Andonova and Mitchell (2010) catalogued the struggles with, and growing acceptance of, the ongoing "rescaling of global environmental politics." They state:

> An international sphere dominated by interaction among nation-states has been replaced by one in which international organizations, substate governments, scientists, nongovernmental organizations (NGOs), and multinational corporations play major roles. (p256)

Their work treats global environmental politics and increasingly complex and interconnected, rather than a set of statist processes and actors operating primarily at a global or international level. Such work, like that of Bulkeley and her colleagues, has much to teach analysts of resource-related dynamics about the ongoing "vertical and horizontal rescaling across political arenas, actors and issues" (p273). So, for example, Ken Conca's *Governing Water* (2006) explains the challenges to building effective international water-related governance institutions by exploring the substantial differences in norms and values that are institutionalized across different transnational and local organizations and communities. Furthermore, Ostrom and her colleagues' work (Brondizio et al., 2009; Dolšak and Ostrom, 2003; Ostrom, 2009) demonstrates that governance is also rescaled through of proliferation of centers of authority and coordination, leading them to examine this increasing common phenomenon via a "polycentric approach." As such, the resource nexus approach not only challenges us to understand governance in similar ways, but to focus greater attention of the interactions, dynamics, challenges, and opportunities in the spaces between one or more resources. While energy politics and governance – and water or food politics and governance – might be understood has having multiple centers and critical levels, examining the interactions of these resource governance silos might well "rescale" understanding. Perhaps, for example, premier centers of research or of rulemaking regarding the food–energy–water nexus are differently identifiably located than key processes or actors in any one of the resources. Long-standing national and international energy organizations like the International Energy Agency (IEA) or the US Energy Information Agency (EIA) might well not be the places producing the highest quality data and analysis about the food–energy–water nexus, for example. Furthermore, this nexus might look much different at local and urban scales – or in more or less arid regions – than it does in globally framed quantitative assessment models or international trade datasets.

One might also focus on the ways in which mining or minerals governance has been rescaled over the last 20 years. Thousands – perhaps tens of thousands – of mining-related protests; activist and media attention to myriad corruption and environmental scandals; a proliferation of activist campaigns to "clean up" gold, copper, and iron-ore mining; and a growing set of certification schemes, codes of conduct, and transnational transparency and accountability campaigns have changed the way mining, manufacturing, and retailing firms, states, and international organizations collect and analyze mining-related information and how they seek to govern mining operations and minerals trade and use (Andrews-Speed et al., 2015; Bleischwitz, 2014; VanDeveer, 2013, 2015). Few would say that the industry, or its governance at local mining sites, international markets, or transnational organizations, is unchanged from 20 years ago. Some activist tactics or particular pieces of information in one location, or in one governance forum like the Extractive Industries Transparency Initiative or the industry-supported International Council on Mining and Metals, can alter ideas, practices, or expectations in other locations or fora. Furthermore, governance rules and expectations change, sometimes becoming more restrictive, illicit markets in some minerals can expand to become an important fraction of international trade (Bleischwitz et al., 2015), posing additional governance challenges. As more attention is paid to the roles of minerals and mining within a resource nexus framework – as these related to water, food or land use, quality and prices, for example – one might well expect more changes in governance practices from artisanal mining areas or giant corporate mines to national and transnational and global venues such as the Brazilian or EU law making or the World Bank. Meanwhile, global commitments such as those embodied in the 2015 Paris Agreement on climate change may well shape changes across some areas of energy policy-related coal consumption and mining in some countries and localities, while in other regions and states coal mining and consumption may continue to grow rapidly, given differing political, social, and economic dynamics (Boersma and VanDeveer, 2016).

Returning to questions of resource nexus governance, scalar thinking asks us to understand governance as much more than the tracking and regulation of a particular resource (like water) across different jurisdictions and levels of authority. So, for example, Andrews-Speed et al. (2015) identified a host of gaps and asymmetries in basic data about resource use, flows, and interactions at local, national, and global scales. What might data monitoring and analysis institutions look like and how would they operate, if we wanted them to produce much more robust local, national, and global scale data? If we take the resource nexus approach seriously, we cannot rely (exclusively or primarily) on international market values and institutions to often determine what data is available about the food eaten by the globe's two billion poorest people, or the amounts of water used in energy or agricultural production or contaminated by minerals mining and processing and other industrial uses. Taking nexus connections and challenges seriously requires not just more data – which indeed is needed – but also different data and different analytical goals and institutions if various needs related to human security and sustainability are to be prioritized with economic growth and state security. Furthermore, how might politics and governance change if local communities, residents of large urban areas, and national citizenries had ready access to more high-quality information about resource use and resource nexus interactions, as well as their current and future ramifications?

Data, knowledge, and information also have scalar dynamics, of course. They always have organizational and institutional contexts, and their links to different actors, organizations, and processes shape their creation, use, and interpretation. Information and knowledge about resources – such as that produced by intergovernmental mentor organizations like the IEA, the World Bank, or the UN Environment Program – is inevitably shaped by the institutional structures in which it is produced. Scalar thinking suggests that both the international organizations and the knowledge they produce is more than mere products of the interaction among states (Andonova and Mitchell, 2010). Actors, institutions, and political processes across a host of issues, disciplines, and actor types shape the processes by which "global" data is produced and how it is understood. Naturally, such influences also help to explain what data is not gathered or assessed.

Conclusion

Scalar thinking about the resource nexus does not merely ask us to consider connections between local actions and global processes. Among other things, it asks to think about how different many aspects of governance would be if we take seriously the growing challenges identified and analyzed via the nexus framework. What governance institutions, for example, would value knowledge about the food eaten by the "bottom billion" as highly as knowledge about the most valuable of the grains traded internationally? What governance institutions would facilitate the inclusion of substantially more of externalities associated with fossil-fuel extraction and use and/or minerals mining, processes, transportation, use, and disposal? What governance institutions – in which kinds of places – help to balance the high monetary value of oil, gas, or minerals extraction with the general much cheaper prices of food and water? Questions such as these bring resource nexus analysis and governance into direct contact with conceptions of equity, justice, ecological value, and/or human rights. Furthermore, if policymakers or activists are successful in some places, and balancing resource extraction, externalities, individual and community needs, and conceptions of fairness, what governance institutions protect other people and ecosystems bearing the burden of human rights and environmental protection elsewhere? The fifth assessment report of the Intergovernmental Panel on Climate Change (IPCC) reviewed vast bodies of research relating to the overlapping challenges of climate change, human security, sustainable development, and equity (Adger et al., 2014; Fleurbaey et al., 2014). If analysis or policymaking related to resource nexus challenges – in the public, private, and civil society sectors – is to improve governance or the knowledge intended to inform governance decisions, it cannot ignore scalar thinking or the many complexities surrounding conceptions of human security, sustainable development, and equity across scales.

This chapter argues that scale matters more than in often acknowledged. Analytically, it involves more than simply listing a few "levels" but rather requires a more systematic incorporating of scalar thinking into a host of resource-related questions. Just as global environmental politics and governance have been "rescaled" (Andonova and Mitchell, 2010), so too must resource nexus analysis and governance.

References

Adger, W. N., Pulhin, J. M., Barnett, J., Dabelko, G. D., Hovelsrud, G. K., Levy, M., Spring, Ú. O., and Vogel, C. H. (2014). Human security. In C. B. Field, V. R. Barros, D. J. Dokken, K. J. Mach, M. D. Mastrandrea, T. E. Bilir, M. Chatterjee, E. L. Ebi, Y. O. Estrada, R. C. Genova, B. Girma, E. S. Kissel, A. N. Levy, S. MacCracken, P. R. Mastrandrea, and L. L. White (Eds.), *Climate change 2014: Impacts, adaptation, and vulnerability. Part A: Global and sectoral aspects. Contribution of working group II to the fifth assessment report of the intergovernmental panel on climate change*. Cambridge: Cambridge University Press.

Agnew, J. A. (1994). The territorial trap: The geographical assumptions of international relations theory. *Review of International Political Economy*, 1(1), 53–80.

Andonova, L. B., and Mitchell, R. B. (2010). The rescaling of global environmental politics. *Annual Review of Environment and Resources*, 35, 255–282.

Andrews-Speed, P., Bleischwitz, R., Boersma, T., Johnson, C., Kemp, G., and VanDeveer, S. D. (2015). *Want, waste or war? The global resource nexus and the struggle for land, energy, food, water and minerals*. London and New York: Routledge.

Betsill, M. M., and Bulkeley, H. (2006). Cities and the multilevel governance of global climate change. *Global Governance: A Review of Multilateralism and International Organizations*, 12(2), 141–159.

Blackwill, R. D., and Harris, J. M. (2016). *War by other means: Geoeconomics and statecraft*. Cambridge, MA: The Belknap Press of Harvard University Press.

Bleischwitz, R. (2014). Transparency in the extractive industries: Time to ask for more. *Global Environmental Politics*, 14(4), 1–9.

Bleischwitz, R., Dittrich, M., and Pierdicca, C. (2015). Illicit trade with coltan and implications for certification. In S. Hartard and W. Liebert (Eds.), *Competition and conflicts on resource use*. Berlin: Springer.

Boersma, T., and VanDeveer, S. D. (2016). Coal after the Paris agreement. *Foreign Affairs*. Retrieved from www.foreignaffairs.com/articles/2016-06-06/coal-after-paris-agreement

Brondizio, E. S., Ostrom, E., and Young, O. R. (2009). Connectivity and the governance of multilevel social-ecological systems: The role of social capital. *Annual Review of Environment and Resources*, 34, 253–278.

Bulkeley, H. (2005). Reconfiguring environmental governance: Towards a politics of scales and networks. *Political Geography*, 24(8), 875–902.

Bulkeley, H. (2012). Governance and the geography of authority: Modalities of authorisation and the transnational governing of climate change. *Environment and Planning A*, 44(10), 2428–2444.

Bulkeley, H., Andonova, L., Betsill, M., Compagnon, D., Hale, T., Hoffman, M., Newell, P., Paterson, M., Roger, C., and VanDeveer, S. D. (2014). *Transnational climate change governance*. Cambridge: Cambridge University Press.

Cohen, S., and Frank, D. (2009). Innovative approaches to territorial disputes: Using principles of riparian conflict management. *Annals of the Association of American Geographers*, 99(5), 948–955.

Conca, K. (2006). *Governing water: Contentious transnational politics and global institution building*. Cambridge, MA: MIT Press.

Dolšak, N., and Ostrom, E. (2003). *The commons in the new millennium: Challenges and adaptation*. Cambridge, MA: MIT Press.

Feitelson, E., and Fischhendler, I. (2009). Spaces of water governance: The case of Israel and its neighbors. *Annals of the Association of American Geographers*, 99(4), 728–745.

Fleurbaey, M., Kartha, S., Bolwig, S., Chee, Y. L., Chen, Y., Corbera, E., Lecocq, F., Lutz, W., Muylaert, M. S., Norgaard, R. B., Okereke, C., and Sagar, A. D. (2014). Sustainable development and equity. In O. Edenhofer, R. Pichs-Madruga, Y. Sokona, E. Farahani, S. Kadner, K. Seyboth, A. Adler, I. Baum, S. Brunner, P. Eickemeier, B. Kriemann, J. Savolainen, S. Schlömer, C. v. Stechow, T. Zwickel, and J. C. Minx (Eds.), *Climate change 2014: Mitigation of climate change. Contribution of working group III to the fifth assessment report of the intergovernmental panel on climate change*. Cambridge: Cambridge University Press.

Giordano, M. (2003). The geography of the commons: The role of scale and space. *Annals of the Association of American Geographers*, 93(2), 365–375.

Gregory, D., Johnston, R., Pratt, G., Watts, M., and Whatmore, S. J. (Eds.). (2009). *The dictionary of human geography*. Malden, MA: Blackwell.

Hooghe, L., and Marks, G. (2001). *Multi-level governance and European integration*. Lanham, MD: Rowman & Littlefield.

Huber, M. T., and Emel, J. (2009). Fixed minerals, scalar politics: The weight of scale in conflicts over the '1872 Mining Law' in the United States. *Environment and Planning A*, 41(2), 371–388.

Johnson, C. (2012). Toward post-sovereign environmental governance? Politics, scale, and EU water framework directive. *Water Alternatives*, 5(1), 83–97.

Kamieniecki, S., and Kraft, M. (2013). *The oxford handbook of US environmental policy*. Oxford: Oxford University Press.

Levin, S. A. (1992). The problem of pattern and scale in ecology: The Robert H. MacArthur award lecture. *Ecology*, 73(6), 1943–1967.

Nye Jr, J. S., and Welch, D. A. (2013). *Understanding global conflict and cooperation: An introduction to theory and history* (10th ed.). Hoboken, NJ: Pearson.

O'Neill, K., Balsiger, J., and VanDeveer, S. D. (2004). Actors, norms, and impact: Recent international cooperation theory and the influence of the agent-structure debate. *Annual Review of Political Science*, 7, 149–175.

Osofsky, H. M. (2013). Geography of solving global environmental problems: Reflections on polycentric efforts to address climate change. *The New York Law School Law Review*, 58(4), 777–828.

Ostrom, E. (2009). *A polycentric approach for coping with climate change*. Policy Research Working Paper, The World Bank, Washington, DC.

Parker, R. (2015). A river runs yellow. *The Atlantic*, 21 August. Retrieved from https://www.theatlantic.com/national/archive/2015/08/a-river-runs-yellow/401966/

Rabe, B. G. (2010). *Greenhouse governance: Addressing climate change in America*. Washington, DC: Brookings Institution Press

Rees, J. (2013). Geography and the nexus: Presidential address and record of the Royal Geographical Society (with IBG) AGM 2013. *The Geographical Journal*, 179(3), 279–282.

Reisner, M. (1993). *Cadillac desert: The American West and its disappearing water*. New York: Penguin.

Rourke, J. T. (2008). *International politics on the world stage* (12th ed.). New York: McGraw-Hill.

Sayre, N. F. (2005). Ecological and geographical scale: Parallels and potential for integration. *Progress in Human Geography*, 29(3), 276–290.

Sayre, N. F. (2009). Scale. In N. Castree, D. Demeritt, D. Liverman, and B. Rhoads (Eds.), *A companion to environmental geography*. Malden, MA: Wiley-Blackwell.

Selin, H., and VanDeveer, S. D. (2009). *Changing climates in North American politics: Institutions, policymaking, and multilevel governance*. Cambridge, MA: MIT Press.

Selin, H., and VanDeveer, S. D. (2012). Federalism, multilevel governance and climate change politics across the Atlantic. In P. F. Steinberg and S. D. VanDeveer (Eds.), *Comparative environmental politics: Theory, practice, and prospects*. Cambridge, MA: MIT Press.

Silver, J. J. (2008). Weighing in on scale: Synthesizing disciplinary approaches to scale in the context of building interdisciplinary resource management. *Society & Natural Resources*, 21(10), 921–929.

Singer, J. D. (1961). The level-of-analysis problem in international relations. *World Politics*, 14(1), 77–92.

Turner, B. L., Kasperson, R. E., Meyer, W. B., Dow, K. M., Golding, D., Kasperson, J. X., Mitchell, R. C., and Ratick, S. J. (1990). Two types of global environmental change. *Global Environmental Change*, 1(1), 14–22.

U. S. Department of the Interior. (2015). *Technical evaluation of the gold king mine incident*. Bureau of Reclamation, Denver, Colorado. www.usbr.gov/docs/goldkingminereport.pdf

VanDeveer, S. D. (2013). *Still digging: Extractive industries, resource curses, and transnational governance in the anthropocene*. Washington, DC: Transatlantic Academy.

VanDeveer, S. D. (2015). Consumption, commodity chains, and global and local environments. In R. S. Axelrod and S. D. VanDeveer (Eds.), *The global environment: Institutions, law, and policy* (4th ed.). Thousand Oaks, CA: CQ Press.

4

SECURITY, CLIMATE CHANGE, AND THE RESOURCE NEXUS

Bassel Daher, Sanghyun Lee, Rabi H. Mohtar, Jeremiah O. Asaka and Stacy D. VanDeveer

Almost every January we observe a nearly annual ritual of running and reading stories in major media outlets, government reports, and science blogs declaring the previous year to be either the hottest on record, globally, or among the warmest. Add to this common story those on topics such as weather volatility, record high local temperatures, Arctic sea ice levels, glacial and Antarctic icecap retreat, and a picture of popular coverage of our changing climate begins to emerge. Less covered on page one of major newspapers or the nightly news is the steadily growing concern among security analysts and scientists from many disciplines about the myriad links between climate change and national, international, local, and human security. To such concerns we must now add – and integrate – the resource nexus approach and the governance challenges it presents.

"Security" is often called an essentially contested concept. Its various meanings and their implications cannot be resolved into a single conceptualization – rather like "sustainable development," in fact. Instead, the term security is often preceded by one of many modifiers, including "national," "international," "global," or "human." More traditional conceptualizations are generally said to relate to national and state security, while broader and newer conceptualizations often focus either on international or global scale framings, on the one hand, or on individuals and communities, on the other hand. In recent years, human security has emerged as a commonly used term to denote the broad inclusion of individuals' and communities' physical, economic, and social well-being within a security framework. The term now often encompasses issues denoted by popular concepts like food security, water security, and health security, and it often attends to gendered aspects of these issues, as well.

This chapter briefly presents major nodes of research and analysis around climate change and security linkages, around more recent attempts to link security to the resource nexus, and the need to position equity and justice-related concerns at the center of these complex perspectives. The chapter highlights a broad view of security at national, regional, and global scales, which span the traditional geopolitical definition to include perspectives on water, energy, and food securities as key elements of economic and physical security. It suggests that better and more integrative resource planning might alleviate some of these related stresses – stresses that traditional security analysis and research are also grappling with as geopolitics and resource conflicts receive increased attention. The chapter explores the role of climate change in increasing the challenges of responding to water and food security challenges, and considers the roles of trade and localization in reducing those stresses. After the sections focused on the water–energy–food nexus across three scales – national, regional, and global – attention turns to an additional illustrative case focused on climate-security and resource nexus connections in northwestern Kenya, where climate variability and scarce land and grassland resources increase the risks of lethal, inter-group violence. The Kenyan case illustrates what Andrews-Speed et al. (2015) called the "nexus on the ground."

Climate change and security

The 1990s saw an acceleration of research and analysis – and a good deal of hyperbole – about many possible connections between environmental change and security (Dabelko, 2008). Such work often focused attention on the connection between environmental change and violent conflict within and between states – often, it must be said, via a research focus on places where war was the known outcome. As scientific consensus about the growing threats posed by climate change stabilized and grew more certain, such research tended to focus, in particular, on the security impacts of climate change. In recent years, concern about the intersections of climate change and security – in terms of both traditional national and interstate security and a much broader human security agenda – resulted in a rapid proliferation of scholarly research and reports from think tanks, research institutes, national governments, and international organizations (Campbell et al., 2007; CNA, 2007; Andrews-Speed et al., 2015; Lee et al., 2012; Adger et al., 2014; Nordkvelle et al., 2017; Nett and Ruttinger, 2016). For example, related research and policymaking efforts have been seen within the UN system (including within both UNEP the Security Council), NATO, the G7, the World Bank, a host of national foreign policy and security think tanks and institutes around the world, the US Department of Defense, Central Intelligence Agency, Department of State, and National Intelligence Council – to name only a few. A growing set of case studies, assessment and modeling exercises, and pilot programs have been produced, engendering a need for more systematic assessment of what is known and how this might be better integrated in the resource nexus framework. Certainly climate change threatens aspects of national, international, and human security in and among both wealthier, OECD countries and across nations and communities in the global south.

That climate change and security connections have achieved global, high-level scientific and policymaker attention is confirmed – and perhaps enhanced – by the inclusion of a "Human Security" chapter in the IPCC 5th Assessment report (Adger et al., 2014) and by the incorporation of climate change as a central theme in the 2014 US Department of Defense Quadrennial Defense Review (QDR) (USDOD, 2014). Together, these two documents illustrate that a diverse set of security threats, challenges, and frameworks are now connected to climate change trends, impacts, and implications. While the two documents are framed differently – with the QDR pitched primarily in terms of US national security and US interests around the world and the IPCC chapter more concerned with a broad human security agenda – both include themes related to national, international, and human security and both rely on largely overlapping set of expertise and research. Many high-level US admirals and generals are now conversant in climate change impacts and their security implications, as are most NATO defense ministries, with some positioning climate change among the most important threats to international security in the coming years.

In fact, in 2016, a group of over 40 experts called "The Climate and Security Working Group", including former US defense and foreign policymakers and many academics and think tank analysts, issues a report urging concrete action on over 75 enumerated items for the US White House, defense, foreign policy, energy, intelligence, and homeland security agencies

(CSAG, 2016). Even if the Trump administration ignores the recommendations, at least initially, the extensive list of suggested actions illustrates that this broad area of research and analysis has extensive policymaking implications. Certainly the US government is far from the only state to begin planning for and worrying about the dangerous implications of changing global, regional, and local climates. Attention to many states' security planning processes, military spending patterns, border securitization efforts, energy policies and planning, and emergency and disaster response planning and investments suggest that threats posed by climate change are on the political agenda of many large and smaller states (Andrews-Speed et al., 2015).

As the IPCC "Human Security" chapter (Adger et al., 2014) makes clear, the causal pathways through which climate change connects to national, interstate, and human security concerns are myriad and often complex. While this research continues, it is clear through extensive, focused case study research and large-n quantitative analysis – stemming from a host of disciplinary backgrounds – that climate change poses a number of worrisome threats. The chapter's review and assessment of existing research pays particular attention to climate change as a driver of human displacement, as a risk or threat multiplier of violent conflict (usually civil or inter-group violence) in part because it exacerbates poverty and economic volatility, and as an increasing threat to the integrity of critical state/public infrastructures and territorial integrity.

Security and the resource nexus

National and international security analysts have long been concerned about resource access, flows, and physical control of strategic resources such as coal, oil, food and arable land, minerals, uranium, rivers, seas, and so on (Osisanya, 2015). Indeed, just about any resources considered essential for national or state survival or well-being can be "securitized" by analysts and/or policymakers. Not only is such securitized thinking about resources not new, it seems to date back at least as far as written records. Security analysis over the past decade or more revived concern about the role of resources scarcities and resource competition at multiple levels of scale, from local and national civil conflicts to increased prospects or great power resource wars (Klare, 2002, 2008, 2012). This work has often focused on the politics around oil, gas, and water and the particular territories (on land and at sea) in which these resources rest.

What is newer is the resource nexus perspective and the growing linkages between climate change and security. Like so much previous resource-related analysis, security planners, analysts, and practitioners tended to focus on security-related threats associated with one or two resources, at any given time. Growing human demand on resource flows and governance and accelerating climate change, however, reframe these security concerns (Andrews-Speed et al., 2015; VanDeveer, 2015). For example, a country or community dependent on some combination of fossil fuels and hydropower for electricity most be concerned about oil, gas, and coal extraction and supply; water availability and rainfall volatility; and a host of related minerals-, land-, food-, and water-quality issues. Traditionally, policymakers and analysts worried about governance aspects of particular resources in isolation, but increasing interlinkages between resource use, consumption, and demand calls this single-resource approach to governance and security analysis into question. So, for example, while energy geopolitics has a long history among security researchers and practitioners, they have paid little attention to the nexus between energy and other resources.

Security-related research has focused on resources such as water, energy, and minerals access and distribution. Interest now turns towards the need to more systematically assess resource nexus linkages to security-related aspects of climate change – its ecological and social impacts, as well as the implications of mitigation and adaptation policies (Adger et al., 2014; Dabelko et al.,

2013). The governance challenges at the intersection of the resource nexus, climate change, and traditional and human security are legion, occurring at and across multiple scales, including provision of resources needed to survive and thrive, and identifiable risks of local, national, and transnational violence. While research has focused on the ways such connections raise the risks of violence and human degradation, these challenges also connect to research on criminality (Ranson, 2014), climate change and public health (Lockwood, 2016), water, peacebuilding and post-conflict reconstruction (Weinthal et al., 2013), and a host of vexing justice- and equity-related concerns embedded within climate change mitigation and adaptation. as well as Sustainable Development Goals (SDGs) and governance (Fleurbaey and Kartha, 2014; Ciplet et al., 2016). Traditional and human security concerns are thus central to SDG implementation, as well as many aspects of climate change and resource nexus governance.

Traditional conceptualizations of the elusive concept of security (Baldwin, 1997) can be largely attributed to the World War II and Cold-War eras, in which particular focus was given to military power and the ability to ensure physical safety and territorial integrity. Certainly secure access to oil, uranium, and other strategic commodities was critically important. However, the pressures and grand challenges of today's world, the nature of their tight interconnectedness, and their effects on modern economies and societies demands a broader, more holistic definition of the boundaries of security. Growing competition over access to resources, such as water, minerals, energy, and food carry local, regional, and global economic and social implications with the potential to create or aggravate geopolitical and civil unrest. Since 2011, Global Risks Reports consistently raise interconnected environment related risks that include water, food, energy, climate, and natural disasters at the top of global risks and potential impact (WEF, 2016, 2017). For example, trends point towards water stress as a threat multiplier, in which societal, economic, energy, and agriculture stability and productivity depend on the very availability of water.

Equity challenges across scales

Put simply, the world is a very unequal place. For decisions and institutions related to governance – and for needs and goals associated with more sustainable development, as well as the climate change causes, impacts, and responses – equity is an unavoidable topic. Resource endowments are very unevenly distributed – geophysically, within societies, and across nations and world regions. Furthermore, across contexts, some actors' ability to control or purchase access to resources always exceeds that of many other actors. This reality is clearly described in the IPCC (2007) 4th Assessment "Sustainable Development and Equity" chapter and, like all IPCC chapters, backed by substantial research (Fleurbaey and Kartha, 2014).

Equity and issues related to justice emerge across generations, and within and between nations and communities across scales. As is well known, high consumption societies and communities contribute more to climate change than do poorer, lower consumption communities. And they consume substantially more of the globe's resources. Furthermore, the impacts of climate change are likely to be more severe and more threatening on those who have contributed least to the changing global climate. Similar dynamics can be seen around the extraction, processing, trade, and consumption of high-value energy, mineral, and agricultural resources (including land). The often extensive ecological and social costs of resource exploitation are often borne more heavily by poorer individuals and communities along commodities consumption and production chains (VanDeveer, 2015). For SDG6 (water and sanitation) alone, 2.4 billion people lack adequate sanitation facilities, with nearly 1 billion having none at all and relying only on open defecation. For drinking water, depending on the minimal treatment standard needed, between 600 million and 1.8 billion still lack reliable access to safely treated water.[1] Unsafe drinking water and inadequate sanitation negatively impact individual and public health, lifespan, nutrition intake, and child development. Meanwhile, the global trade in food and grains drives alarming aquifer depletion in India, Pakistan, and the US (Dalin et al., 2017).

Among the most vexing aspects of equity concerns for policymaker and analysts alike is found at the intersection of equity with the design and implementation of policy responses to address identifiable climate change and/or resource nexus challenges. So, for example, economists often extol the efficiency-related virtues of extremely high carbon taxes as a needed – indeed as the preferable – policy response to the clear need to reduce carbon emissions. But substantially raising the costs of heating homes in winter and cooling them in summer also has regressive and dangerous implications – as growing concerns about "energy poverty" in many wealthier societies illustrate. Similarly, privatizing water resources is often advised by international organizations and private sector consultancies as a way to incentivize better water management and more water conservation. But such strategies can undermine some people's livelihoods and directly contradict the notion that sufficient, clean water is a human right. Furthermore, water privatization and the reduction of energy subsidies can raise the risks of social unrest, and even social violence. These examples should not be taken to mean that price incentives cannot or should not be used in policy. Rather, they illustrate that equity dimensions emerge around most proposed or enacted policy responses. They are usually inescapable in our unequal world. Furthermore, most states and a host of global institutions have repeatedly asserted that equity-related goals such as poverty reduction, improved public health, women's equality – and a host of other social goals – are global priorities. The Sustainable Development Goals (SDGs), like their forerunners the Millennium Development Goals, demonstrate that the equity and justice-related goals of global actors and institutions are more than mere rhetoric.

Global equity-related efforts like the SDGs are directly and indirectly resource related. For example, some SDGs are explicitly related to food, water, energy, climate change, and sustainable consumption and production, while others are more indirect implications for land, minerals, and biodiversity (Obersteiner et al., 2016).[2] Throughout SDG debates, lack of access to sufficient resources is framed in part as an equity issue. But the reality is still more complex, because any actor or institutional intervention designed to meet such ecological and social goals will also have equity implications. Bruce et al. (2015) argue that resource nexus research and analysis has often failed to incorporate sustainable livelihoods perspectives. The posit the need for frameworks to measure and assess "environmental livelihood security," seeking to balance natural resource supplies and human demands on the environment in ways that promote more sustainable development. They see such approaches as critical for both sustainable development target implementation and equity promotion. Similarly, Shi et al. (2016) demonstrate that most climate change adaptation strategies also have significant procedural and distributive justice implications. In short, equity and justice concerns are inescapable in the politics of the resource nexus, climate change, and security.

Redefining primary resource security: multiple scales and a water–energy–food nexus[3]

Debate in academia, policy, business, and civil society often centers on economic security, wherein water, energy, and food securities constitute the primary pillars, together with threats to physical security such as terrorism and disease (Axworthy and Adeel, 2014). Discourse on the correlation of resource scarcity with conflict and war is ongoing. Lagi et al. (2011) argue that the spike in food prices played a major role in the revolutions of Tunisia and Egypt, and led to the social and political unrest of both countries. Similar views regarding the roles that water shortage and drought may have played in escalating the early stages of the Syrian conflict (van der Heijden and Maddocks, 2015) abound: the intense droughts between 2006 and 2010 contributed to the heavy displacement of agricultural workers and drove them away from their rural farms and into cities before violent conflict erupted (Jaafar et al., 2015; IDMC, 2016). The recent Russian annexation of Crimea and the internal conflict in Ukraine, a strategically located transit state for natural gas and petroleum export from Russia to Europe, are often attributed to the Russian state's inclination to use of energy security as a political tool in the region (NATO, 2015). While debate rages about these example among analysts, each illustrates that resource politics are once again "high politics."

Such conflicts have implications for resource securities: insurgency and war are usually harmful to the political, economic, and social frameworks of communities. Agriculture and food production face significant losses due, in part, to increased food prices (Obersteiner et al., 2016), damaged assets and infrastructure, and disrupted supply chains (Flowers, K., 2015). Indeed, "food insecure populations are likely to express frustration with troubled regimes, perpetuating a cycle of political instability and further undermining long-term economic development" (Flowers, K., 2015). The overall impact of armed conflict on food production and food availability is significant in terms of public health risks and economic losses. The FAO estimated an average of $4.3 billion in loss of agricultural output over a span of 27 years, due to war and conflict (FAO, 2002). The post-conflict dimension of resource security is often overlooked: food shortages tied to conflict often place countries in an extended state of food emergency long after the war or conflict has ended (Allouche, J., 2011; FAO, 2008). Also, food and water scarcity may well increase the risks of minerals exports disruption (Bleischwitz, Johnson and Dozler, 2013). The technological advancement and increased labor productivity in agriculture over the last two centuries, argues Allouche (2011), has made humankind notably resilient to water and food conflicts. In fact, datasets monitoring conflict reveal that only seven minor conflicts in the last century were linked to water resources (Böhmelt et al., 2014). Water and food insecurities are generally assumed to be issues of overconsumption and competition. However, the insecurity of these resources are more often a reflection of the politics of allocation and inequality (Allouche, J., 2011).

Stresses and levers for WEF security at multiple scales: national, regional, and global

This section identifies major stresses facing water, energy, and food securities at national, regional, and global scales. It highlights some of the main levers for alleviating those stresses and moves towards increased resilience at each scale (Figure 4.1). Climate change is a stressor across all scales. It affects water availability and agricultural yields, and extends to major shifts beyond country borders and regions. Consumption by the world's growing population also puts pressures on global resource systems and planetary boundaries. Global agreements such as the Paris Climate Change agreement (UNFCC, 2015) and the Sustainable Development Goals (United Nations, 2015) play important roles in realigning national strategies towards identified global goals.

Encouraging technological innovation and investment through cutting-edge research across scales will also contribute to reducing some of the stresses on resource systems securities. An additional effective lever lies in the adoption of balanced trade strategies that help meet market needs without jeopardizing the sustainability of local resources or the resource security through increased dependability on other markets.

While many of these pressures are global and regional in scope, decisions and solutions are often offered at the national scale. When addressing WEF security challenges at a national scale, a

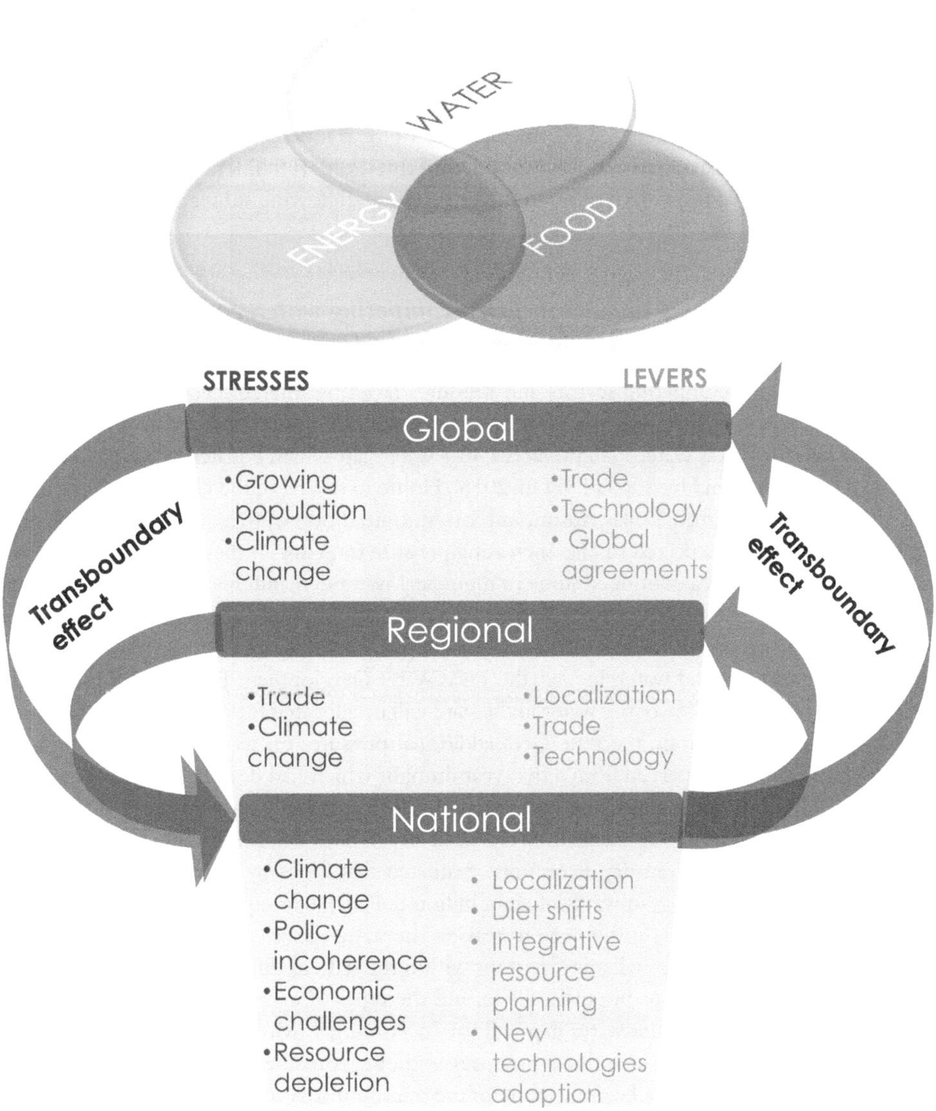

Figure 4.1 Multiscale stress and levers of WEF nexus and feedback loops across scales

main challenge lies in the lack of policy coherence across decision-making entities, and across scales. Lack of coordination and conflicting national goals could lead to unsound allocation practices with the potential to harm resource sustainability. Levers at the national scale could be driven by policy, through better integrative planning across sectors and scales, by industry, through improved efficiencies and management practices across supply chains, or by civil society, through altered habits of resource consumption.

An important aspect of the interaction of those securities across scales is the effect of one decision on another, particularly between scales.[4] We saw examples of this effect during the 2008 food crisis, where national energy security policies related to increases in biofuel costs resulted in unintended consequences in global food processes and food insecurities worldwide. Other critical questions that capture examples of feedbacks and transboundary effects include: What would be the state of global food security if China, the world's top producer of phosphate, essential for fertilizer production, decides to decrease phosphate production? How would decisions by OPEC countries to reduce energy production affect global food security? How might climate change effect drought patterns and food production thresholds in vulnerable nations? And what potential does each of these have to result in conflict? The following subsections highlight three cases, one at each scale.

National or state security: competing water, energy, and food securities in Texas

This case highlights competing sectors and pressures faced by interconnected resource systems in Texas, and potential economic security and transboundary implications of possible pathways forward. The 2016 Texas Water Plan predicts a 40% water gap, which equates to a supply–demand deficit of 8.24 billion m3 by 2060 (TWDB, 2016). Home to several of the fastest growing cities in the US, including Houston, Dallas, Austin, and San Antonio (Forbes, 2015), Texas' resource systems and infrastructure are expected to face increasing pressure to maintain these centers of economic activity, which indicates a continued surge in municipal water demand. Texas is home to the Eagle Ford Shale play, which provides major support for its economy, while adding to the demand for water for energy production. Texas is also a major US producer of dairy, cotton, and cattle (USDA, 2016), which also demand water; indeed, the Texas Water Development Board Water Plan (2016) projects that more than 70% of the water in the state will be allocated for irrigation by 2020. Prone to frequent drought periods, the state faces additional pressures on its interconnected resource systems. In 2015, Texas emerged from a five-year drought which had devastating effects on different sectors. This drought resulted in a series of wildfires that destroyed more than 1300 homes in Bastrop in 2011. Over 1000 residents in Spicewood Beach have been relying on tankers for their water since 2012 as wells ran dry. Economist estimated that losses by farmers and industries in drought-affected regions was upwards of eight billion dollars (StateImpact, 2017).

These projections highlight the need to reassess the status quo trends across different sectors. The projected statewide gap will affect regions within the state in different ways due to the high variability in ecology, climate, population size, and the types of economic activities. Alternative pathways towards bridging the water gap will vary accordingly (Mohtar et al., 2015). The question, then, lies in how to reduce those pressures without jeopardizing the sustainability of the state's resources and economy, keeping aware of the transboundary affect that decisions regarding each of these sectors might cause to other states (e.g. New Mexico, Arizona, Oklahoma, Louisiana), or countries (e.g. Mexico). With regard to agriculture, a major trade-off lies in incentivizing *local* agriculture, rather than importing it. The question becomes whether Texas has sufficient resources to maintain local agricultural activity and the livelihoods connected to it, given the increased competition from other sectors. Would a change in farming practices, technologies, or choice of products alleviate those pressures? Are there alternative sources for dairy and cotton to meet market demand if their local production was de-incentivized? How would that impact other states that depend on Texas as a source for these products?

Similar trade-offs exist for energy production. As a producer of energy, nationally and globally, the Texas energy sector provides major support for its economy. To alleviate stress on water and environmental systems, alternative energy sources with lower water footprints (Chapagain and Hoekstra, 2004), wind for example, could be encouraged and phased into the state's energy portfolio, while ensuring an economically sustainable transition. Moreover, as a border state that potentially shares 36 transboundary aquifers with Mexico (Sanchez et al., 2015), there are further limitations in terms of maintaining the equity assured through bilateral treaties and international water agreements. Sustaining economic security and its interconnected constituents requires integrated, coherent policy incentives that ensure no single sector infringes on the success of the others. Industry also has a role to play through choices within supply chains controlled by them. Civil society likewise has a role through addressing those behavioral changes that might contribute increased security through the release of resource stresses by changing demand.

Regional security: WEF nexus in the Arab region

The internal and external challenges besetting the management, sustainability, and security of water, energy, and food resources in the Arab region are multifold. Decision-makers are challenged with a region characterized by arid and semi-arid climates, recurring water stresses, high population growth rates, frequent political unrest, and climate change. The high variability of challenges across the region calls for localized strategies of food production and farming systems; adaptation plans are needed to insure increased resilience against the region's key stressors: population growth, climate change, and shifting socio-economic and political dynamics.

The state of the Arab region under the influence of climate change is highly variable. Average climatic scenarios predict through multiple Global Climate Models that climate change will be responsible for about 20% of the gap in supply–demand by 2050 (Mohtar et al., 2016; Immerzeel et al., 2011; Droogers et al., 2012). Climate change is further expected to accelerate the hydrological cycle (Oki, 2005), driving intense rainfall in some areas of the Arabian Peninsula, while a concurrent increase in temperatures is likely to increase evapotranspiration rates and lead to an overall rise in water demand primarily for the agricultural sector. This persistent uncertainty in extreme weather conditions (flooding and desertification), carries substantial implications for water and food security in the region. Beyond the physical security of water and food resources, extreme weather events have the potential to create or aggravate socio-economic and political frictions in the region (Mohtar et al., 2016).

Agriculture is essential to the national identity and security fabric of Arab societies whose most vulnerable populations depend on this sector for its economic stability. Yet, the majority of the countries in the Arab region depend on food imports, and consequent virtual water trade, to sustain food supply and demand. This virtual water is anticipated to provide over 60% of the food basket by 2030 (World Bank, 2009), and while virtual water trade alleviates the gap between water and food supply for the region's expanding population, it carries long-term repercussions, such as degradation of arable soil, socio-economic impact on farmers, and vulnerability to global markets (Mohtar et al., 2016). The potential for improving levels of local food production is overlooked or yet to be developed, but include the use of green water sources, defined as "rainwater stored in soil as soil moisture," and the application of saline/brackish water in agriculture, as piloted by the Sahara Forest Project.

Going forward, farmers and rural inhabitants of the region will play key roles in increasing the resilience of their agricultural system (Verner et al., 2013) and consequently, their water–food security, against the variable effects of climate change. Due to the enormous variability in resource availability, external factors, and range of involved stakeholders across the Arab region, water–food security strategy cannot be developed through an "umbrella approach." Holistic yet localized adaptation strategies are needed to increase resilience and sustainable levels of water–food security.

Global security: the role of virtual water trade

Food trade is regarded as a significant lever for reducing global water stresses and managing regional water resources in water-poor regions. Trading in food commodities can be converted to virtual water, that water embedded in food products (Hoekstra, 2003). The virtual water concept was developed to assess water footprints and it is divided into green and blue water. Green water indicates available water in soil supplied by rainfall, and blue water indicates irrigation water supplied by technological facilities. Virtual water trade allows sharing water resources indirectly within and between countries or regions. Approximately 13% of the water used for worldwide crop production was traded internationally between 1995–1999 (Hoekstra and Hung, 2005), and 16% of the global water resources were used in production and export (Chapagain and Hoekstra, 2008). However, virtual water trade not only generates water savings for importing countries, but also causes water "losses" for the exporting countries (Chapagain et al., 2006). The water exported through commodities into the global market contributes significantly to changes in regional water systems of major exporters (Chapagain and Hoekstra, 2008). On the other hand, pushing to increase food self-sufficiency in traditionally food-importing countries may lead to depletion of local resources and competition with other sectors. Thus, understanding the impacts of food trade on water and food security can become a major driver towards more sustainable resource management.

Asian countries are major food suppliers, thus food trade within and outside of Asia can be an important factor for integrated water management in certain regions of Asia. Lee et al. (2016c) analyzed the contribution of virtual water from main exporters in Asia via crop trade between 2000 and 2012. Based on virtual water trade, approximately 46.9% and 40.9% of the total green and blue water exports in Asia were discharged from Asia to non-Asian countries. In the European Union (EU), the rate of virtual water export to non-EU countries was 30.2% (green water) and 25.2% (blue water). Importing food can be also be viewed as allowing local resource saving in water, land, and energy. For example, Korea is a representative country, importing food and feed crops, with a volume of 96.2 Gm^3 of virtual water imported via 41 crops traded between 2006 and 2010 (Lee et al., 2016b). A significant increase in domestic food production would be required for food security. However, significant water and energy requirement increases would happen as a result.

Changes in global trade can be the overriding variable for managing global water and food sustainably; Lee et al. (2016a) constructed a trade network of main food crops (primarily wheat, barley, and rice), and analyzed the characteristics of global virtual water trade, such as vulnerable and influential traders, using degree and eigenvector centrality. For example, the vulnerable importers are identified by large amount of virtual water import with low connectivity to exporters which could be a low-resilience structure. Mexico, Iran, the Philippines, and Japan were identified as vulnerable importers. The US, Russian Federation, Thailand, and Canada were identified as influential traders in global virtual water trade of green water, while western Asia, Pakistan, and India were identified as influential traders in global virtual water trade of blue water, an indicator of irrigation water supplied from an artificial facility.

Climate change also plays a major role in affecting availability of water supplies. It is necessary to tie in the climate change influences on the water–energy–food nexus boundary and further assess the holistic impacts of climate change in relation to food trade, food security in importers, and water security in exporters. Looking at global trade through a water–energy–food nexus lens can provide a basis for assessing holistic impacts of different international trade schemes on national and regional resource management.

The climate–security–resource nexus in northwestern Kenya[5]

About 80% of Kenya's landmass is considered arid and semi-arid lands (ASALs). Northwestern Kenya is part of the country's drylands, and pastoralism and wildlife conservation are the dominant land-use systems in the region. The region is infamous for intermittent, long-standing inter-ethnic violent conflict, which dates back decades. Most human security studies in this region adopt a narrow focus concentrating either on the interplay between climate change and conflict, or resource scarcity and conflict (Adano, Witsenburg and Dietz, 2009; Campbell et al., 2009), but such narrow frameworks fail to capture the complex nature of the human insecurity problem in the region. Moreover, past studies of the same phenomenon mostly relied on a narrow state-centric conceptualization of security.

But unlike the state-centric security analytical framework, a human security analytical framework compels us to pay more attention to security of people – as individuals and/or communities. Furthermore, it broadens the range of security threats from a single focus on violent conflict (including war) to include, among others, non-traditional threats such as food insecurity, water insecurity, disease and even the state in certain cases (Owen, 2004). By focusing on the complex interplay between climate variability, resources, and inter-ethnic violent conflict, this illustrative case analysis makes the case for a five-node climate–security–land–energy–water nexus as a better alternative for analyzing and understanding human security than the commonly known water–energy–food nexus, particularly in a northwestern Kenya context.

The region

Northwestern Kenya is comprised of seven arid and semi-arid counties: Samburu, Laikipia, Baringo, Isiolo, West Pokot, Turkana, and Marsabit (see Figure 4.2). The region forms the Kenyan portion of the Karamoja cluster – a transboundary, conflict-prone area that sits astride the borders of Kenya, Uganda, and South Sudan (Simonse, 2011). The main economic activity for majority of the residents of the region is pastoralism – subsistence agriculture centered on the raising of various types of livestock. Other economic activities in the region include tourism (wildlife and cultural), small-scale trading, and cattle ranching among others.

For the most part, land in northwestern Kenya is owned communally under a system known as group ranch (Wayumba, 2004). Nomadic pastoralism practiced by the region's pastoral communities dictates that they move livestock frequently and relatively freely across the vast landscape (Asaka and Smucker, 2016). In that sense, a group ranch system of land ownership serves the pastoralists' interest better than a private property (in this case, land) ownership system would, because of mobility restrictions associated with the latter. In fact, private land ownership – an increasing reality in northwestern Kenya – elicits mixed reactions with some residents in favor and others against. The privatization trend in the region is informed in part by a strong privatization wave that began in the Maasai lands of southern Kenya, which has had significant implications (both positive and negative) for livelihood security and therefore human security of Maasai pastoralists (Wayumba, 2015).

Moreover, northwestern Kenya is experiencing considerable development activity, and therefore change. Discovery of oil and water reservoirs in Turkana, establishment of a large wind farm in Marsabit, construction of an international airport in Isiolo, and the building of the Kenyan section of the Nairobi-Addis Ababa highway are just a few of the ongoing development projects that may transform the region considerably. Also, community-based conservation continues to expand considerably in the region with implications for pastoralism both as a production system and as a source of livelihood for a majority of the region's residents.

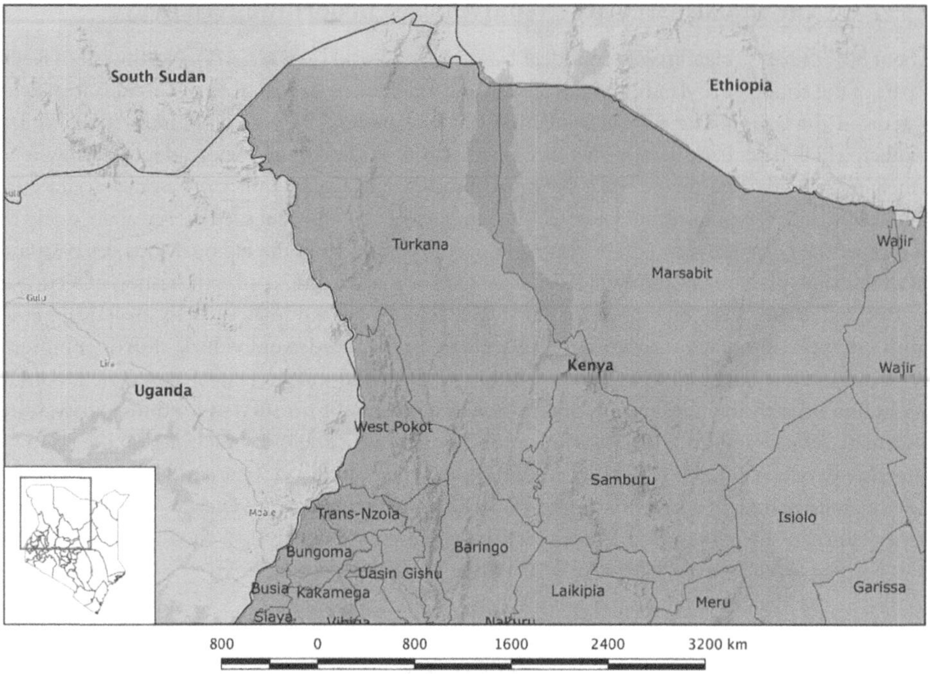

Figure 4.2 Map of northwestern Kenya

Source: Authors

Factors contributing to inter-ethnic violent conflict in northwestern Kenya

Inter-ethnic violent conflict is a dominant character of northwestern Kenya. A variety of factors contribute to inter-ethnic violent conflict in the region, including cattle rustling, proliferation of small arms, and competition over limited natural resources (mainly water and pasture) among others. For illustrative purposes the following discussion is focused on only the three mentioned contributing factors (for more in-depth discussion on violent conflict in the region see, for example, Njiru, 2012; Opiyo et al., 2012; Schilling et al., 2012; Mkutu, 2007; Mkutu, 2002).

Cattle rustling: One of the most common contributors of inter-ethnic violent conflict in the region is cattle rustling. Originally a small-scale, rarely deadly traditional practice where a group from one ethnic group would invade another ethnic group and forcefully take away livestock primarily for restocking following a bad drought, cattle rustling today is a complex, large-scale, and deadly endeavor. In recent years, it has been commercialized (Schilling et al., 2012) and made worse by proliferation of small arms "from a variety of sources, including conflict areas in Sudan, Northern Uganda and elsewhere in the Horn *of Africa*, as well as from insecure official weapons stockpiles" (Mkutu, 2002 italics added).

Proliferation of small arms: Proliferation of illegal firearms complicates the security situation in northwestern Kenya. Due to inter-ethnic mistrust and the culture of cattle rustling, rival ethnic communities in the region have taken to stockpiling illegal firearms for their own protection

(Mkutu, 2007). Over time, weaponry gets more sophisticated and conflicts become more deadly (Schilling et al., 2012; Mkutu, 2007). The result is a state of perpetual intermittent inter-ethnic violent conflict, usually involving deaths. Disarmament attempts by the Kenyan State have been generally unsuccessful, in part because pastoral communities, having endured a long period of systemic marginalization imposed by consecutive government regimes (British colonial regime, the Kenyatta I regime, and Moi regime), are suspicious (or less trusting) of any government. In addition, porous borders in the Karamoja cluster make it easy to acquire guns from black market in the relatively lawless trans-border region.

Competition over scarce and limited resources: Perhaps the most significant contributor to inter-ethnic violent conflict in northwestern Kenya is competition for limited pasture and water particularly during dry periods. Since the majority of the region's population depend on pastoralism for their livelihood, availability of pasture and water for livestock is central to people's survival. It is therefore in the interest of each of the region's ethnic groups to secure grazing and watering rights for their livestock all year round. When water and pasture availability begins to dwindle, in-group cooperation and out-group competition becomes almost inevitable – given the lack of other options. Severe pasture and/or water scarcity can make competition degenerate quickly into conflict, usually along ethnic lines. In fact, research suggests that most conflict in the region emanates from the escalation of untamed inter-ethnic resource competition situations complicated by proliferation of small arms as well as simmering land and political issues among other factors (Njiru, 2012; Opiyo et al., 2012).

Regional and international dimensions

Resource-related inter-ethnic violent conflict in northwestern Kenya is not confined only to rival ethnic groups who inhabit this part of the country. Cross border inter-ethnic violent conflicts also do occur. In fact, a key defining character of the Karamoja cluster is cross border inter-ethnic violent conflict with strong links to cattle rustling. Importantly here, the term "border" is used loosely to denote either a regional or international boundary. In the context of this illustrative case, an inter-ethnic violent conflict is considered regional if it is between an ethnic group from northwestern Kenya and another ethnic group from a different region of the country, but international if it occurs between a Kenyan ethnic group and non-Kenyan ethnic groups from Uganda and/or South Sudan. Examples of inter-ethnic violent conflicts with an international dimension include long-standing conflicts between: Toposa ethnic group of South Sudan and Turkana ethnic group of Kenya across the Nadapal-South Sudan border; Pokot ethnic group of Kenya and Matheniko (Karimojong) ethnic group of Uganda across the Kenya-Uganda border; and Turkana ethnic group of Kenya and Matheniko (Karimojong) ethnic group of Uganda.

Moreover, the emergence of "new" resources with both conflict and cooperation potential is likely to further complicate issues in the region. These new resources include recent discoveries of oil and underground water reservoir in Turkana County (Fong, 2015; Johannes et al., 2015), as well as the rise of clean energy alternatives, notably the Lake Turkana Wind Power project located in Loiyangalani, Marsabit County. Importantly, it remains unclear how the entry of these new resources will shape the human security landscape in northwestern Kenya and the Karamoja cluster, and across the greater Horn of Africa region. Energy and water are already emerging as frontiers for potential conflict, primarily at the sub-national level (e.g., the case of Turkana oil exploration project), and cooperation, primarily at the international level (e.g., the energy sharing pact between Kenya and Ethiopia) (Fong, 2015).

Making the case for a climate–security–resource nexus-based analytical framework

While all the seven counties that make up northwestern Kenya are categorized as arid and semi-arid lands (ASALs), they do not experience a uniform weather pattern. In fact, Samburu County, for example, typically experiences dry weather in one part (the low lands) and wet weather in another (the highlands) concurrently. In essence, climate variability has created an alternating pattern of resource abundance/scarcity – regarding both water and pasture – in northwestern Kenya. In the past, such climate variability was largely predictable and a nomadic lifestyle enabled northwestern Kenya's pastoral communities to cope with recurrent droughts, storms, and flash floods. Basically, during dry periods livestock would be moved from the low lands to the highlands and then back to the low lands during wet season. Galvin et al. (2001) point out that, "Climate variability by affecting ecosystem structure and function, has always influenced *northwestern Kenya's* livestock production and rainfed agriculture production" (p. 163 italics added).

But in the past few decades, climate variability has become increasingly unpredictable, thus making coping with drought, for example, more challenging for most of the region's pastoralists and agro-pastoralists. Irregular droughts, longer than normal droughts, irregular rainfall patterns, abrupt storms, and flash flooding are increasingly the norm in northwestern Kenya today. In part, the result is disruption of traditional livestock mobility patterns and increased competition for limited pasture and water, particularly during long droughts, often resulting in conflict. For instance, *The Guardian* recently reported that, "thousands of heavily-armed herders are invading conservancies, private properties and smallholdings in Laikipia, one of Kenya's most important wildlife areas, as they search for pasture for their cattle" (Cruise and van der Zee, 2017). Also, more rain in the highlands of northwestern Kenya (e.g. parts of Samburu County) has encouraged the spread of agriculture, which creates significant mobility challenges for pastoralists, raising the risks of conflict between them and their agro-pastoral counterparts.

Climate change seems likely to make these regional dynamics more complicated (Barnett and Adger, 2007). The "human security" chapter in the IPCC 5th Assessment Report described climate change as a human security threat multiplier, arguing that "human security will be progressively threatened as the Climate changes" (Adger et al., 2014). Considering the strong connection between inter-ethnic violent conflict and scarcity/accessibility of natural resources (mainly, water and pasture) in northwestern Kenya on one hand, and clear links between climate variability and the success or failure of the region's main source of livelihood (pastoralism) on the other hand, there is a real need to rethink the water–energy–food (WEF) nexus as a framework for understanding human security in the context of northwestern Kenya.

The three-node WEF nexus analytical framework offers useful insight(s) into the northwestern Kenya regional human security dynamics. For instance, through a WEF nexus framework it is possible to simultaneously consider the complex interplay between water, energy, and food in order to understand their role, or lack thereof, of in resource-related inter-ethnic violent conflicts in the region. However for the case of northwestern Kenya the WEF nexus framework leaves out two critical aspects that complete the inter-ethnic violent conflict puzzle in the region: climate and land. The complex nature of inter-ethnic violent conflict in the region requires nuanced understanding and a framework that accounts not only water, energy, and food dimensions, but also the climate and land dimensions. On this basis that a climate-security-resources nexus is a better alternative in the context of northwestern Kenya.

This five-node nexus framework consists of climate–security–land–energy–water (CSLEW). Thus, this case suggests flexible application and context-specific modification of the five-node nexus around which this volume is organized (see Chapter 1). With a CSLEW nexus framework it is possible to improve understanding of the complex interplay between climate change, human security, and natural resources in the inter-ethnic violent conflict-prone region. Finally, this illustrative case makes two additional contributions to the dynamic and growing body of nexus research. First, it succeeds in showing that context matters in nexus research. Whereas a WEF nexus framework might be sufficient to understand human security in certain contexts, it is evidently not well suited to provide a nuanced understanding of the same in the context of northwestern Kenya. Second, it highlights the significance of attention to scale in nexus research.[6] Particularly, the case describes how at one scale the emergence of new resources seems to enable cooperation while at another scale the same phenomena seem to generate conflict.

Conclusion

Climate change and resource nexus challenges and analysis are increasingly linked to traditional and human security concerns across scales. There is little doubt that accelerating climate change impacts and their social, political, and economic implications add stress at and across multiple scales, and therefore must be accounted for in nexus research, foresight analysis, and governance (see e.g. Bleischwitz, Johnson and Dozler, 2013). Resource gaps and governance challenges occur across scales, resulting in pressures that potentially perpetuate insecurities and increase the risks of violence and social unrest. Such threat multipliers may manifest in local communities or within a region of a single country, or spill over borders or involve whole societies. This is not a simple claim that resources or climate change cause conflict on their own. Rather, such factors may increase the risks of violence in particular contexts. Nexus research is not simply about modeling (Braudeau and Mohtar, 2014) or "managing resources." It also intersects with long-standing social and political conflicts, some of which are – or could be – quite violent.

While resource security challenges are experienced in different scales, actions and implementation may also occur at and across scales. In an effort to improve levels of water, energy, and food securities, integrated and coherent policies and strategies may be needed at a national scale, including horizontal coherence across sectors and vertical coherence across more granular scales within the national context. Yet measures enacted by local actors and institutions – as well as international and global actors – will also be needed. Options for shifting dependence of nations' or communities' resource securities to increased imports or localization must be evaluated and the trade-offs for the relative choices in either direction should guide planning and policy action. In the case of food security, for example, such an assessment would span beyond the national borders to account for the resilience of exporting countries to climate change, their political stability, and overall vulnerability to global market fluctuations. Such analysis – and related policy actions – will likely need to grapple with the host of resources (water, energy, etc.) embedded in food production and the food and agricultural trade. Such interlinkages are likely to reveal that risks across countries and world regions are considerably more interconnected than they appear at first glance. The same is likely true for many climate change impacts. In the northwestern Kenyan case, some combination of local and community engagement with national and international actors may well be necessary, if violent conflicts are to be reduced in the future rather than multiplied. Policymakers may be able to leverage existing global and regional agreements to improve the alignment of global targets and resource securities. There is a role for policy, industry, and civil society to play in alleviating those pressures and reducing the risks of instability and insecurity across the different scales, while being aware to the feedback and transboundary effects. Finally, various resource nexus approaches to both analysis and decision-making in practice will increasingly have to grapple with climate change and security interactions. Too often, policymakers, activists, experts, and combatants are likely overlook the important resource and climate-related drivers of social and political conflict. Nexus approaches seek to reveal and anticipate such connections – and inform actions intended to reduce conflict risks and costs.

Notes

1 See the SDGs Goal 6 webpage: https://sustainabledevelopment.un.org/sdg6
2 For more on the nexus and the SDGs, see Ivanova and Escobar-Pemberthy, this volume.
3 This section draws from the ongoing work of Rabi H. Mohtar, Bassel Daher, and Sanghyun Lee.
4 For more on scale and the nexus, see Johnson and VanDeveer, this volume.
5 Jeremiah O. Asaka's doctoral research examines transformations in conservation governance and implications for human security in northwestern Kenya, where he conducted fieldwork in 2013 and 2016.
6 On scale and the resource nexus, see also Johnson and VanDeveer, this volume.

References

Adano, W. R., Witsenburg, K., and Dietz, T. (2009). Scarcity of natural resources and pastoral conflicts in northern Kenya: An inquiry. *Horn of Africa Bulletin*, 1(21), 1–5.

Adger, W. N., Pulhin, J. M., Barnett, J., Dabelko, G. D., Hovelsrud, G. K., Levy, M., Spring, U. O., Vogel, C., Field, C. B., Baros, V. R., Dokken, D. J., Mach, K. J., Mastrandrea, M. D., Bilir, T. E., Chatterjee, M., Ebi, K. L., Estrada, Y. O., Genova, R. C., Girma, B., Kissel, E. S., Levy, A. N., MacCracken, S., Mastrandrea, P. R., and . . . White, L. L. (2014). Human security. In *Climate change 2014: Impacts, adaptation, and vulnerability. Part A: Global and sectoral aspects. Contribution of working group II to the Fifth assessment report of the intergovernmental panel on climate change*. New York: Cambridge University Press, pp. 755–791.

Allouche, J. (2011). The sustainability and resilience of global water and food systems: Political analysis of the interplay between security, resource scarcity, political systems and global trade. *Food Policy*, 36, 53–58. Retrieved from https://sustainability.water.ca.gov/documents/18/3334111/Water%2Band%2BFood%2BSecurity.pdf

Andrews-Speed, P., Bleischwitz, R., Boersma, T., Johnson, C., Kemp, G., VanDeveer, S. D. (2015). *Want, waste or war? The global resource nexus and the struggle for land, energy, food, water and minerals*. London: Routledge/Earthscan.

Asaka, J. O., and Smucker, T. A. (2016). Assessing the role of mobile phone communication in drought-related mobility patterns of Samburu pastoralists. *Journal of Arid Environments*, 128, 12–16. http://doi.org/10.1016/j.jaridenv.2015.12.001

Axworthy, T. S., and Adeel, Z. (2014). *Water, energy and the Arab awakening*. N. p.: InterAction Council. Retrieved from www.interactioncouncil.org/book-launch water-energy-and-arabawakening

Baldwin, D. A. (1997). The concept of security. *Review of International Studies*, 23, 5–26. Retrieved from www.cambridge.org/core/journals/review-of-international-studies/article/div-classtitlethe-concept-of-securitydiv/67188B6038200A97C0B0A370FDC9D6B8

Barnett, J., and Adger, W. N. (2007). Climate change, human security and violent conflict. *Political Geography*, 26(6), 639–655. http://doi.org/10.1016/j.polgeo.2007.03.003

Bruce, E., Boruffc, B., Duncana, J. M. A., Horsley, J., Pauli, N., McNeill, K., Neef, A., Van Ogtrop, F., Curnow, J., Haworth, B., Duce, S., and Imanarig, Y. (2015). Sustainable development and the water–energy–food nexus: A perspective on livelihoods. *Environmental Science & Policy*, 54, 389–397.

Bleischwitz, R., Johnson, C., and Dozler, M. (2013). Re-Assessing resource dependency and criticality. *European Journal of Futures Research*, 15, 34.

Böhmelt, T., Bernauer, T., Buhaug, H., Gleditsch, N. P., Tribaldos, T., and Wischnath, G. (2014). Demand, supply, and restraint: Determinants of domestic water conflict and cooperation. *Global Environmental Change*, 29, 337–348. doi:10.1016/j.gloenvcha.2013.11.018

Braudeau, E., and Mohtar, R. H. (2014). Integrative environmental modeling. In *Encyclopedia of agricultural, food, and biological engineering* (2nd ed.). Taylor and Francis: New York, Published online: 07 May 2014; 1–6. doi:10.1081/E-EAFE2-120049111.

Campbell, I., Dalrymple, S., Craig, R., and Crawford, A. (2009). *Climate change and conflict: Lessons from community conservancies in northern Kenya*. The Conservation Development Centre (CDC), the International Institute for Sustainable Development (IISD) and Saferworld.

Campbell, K. M., Gullidge, J., McNeill, J. R., Podesta, J., Ogden, P., Fuerth, L., Woolsey, R. J., Lennon, A. T., Smith, J., and Weitz, R. (2007). *The age of consequences: The foreign policy and national security implications of global climate change.* Washington, DC: CSIS.

Chapagain, A. K., and Hoekstra, A. Y. (2004). *Water footprints of nations.* Value of Water Research Report Series No. 16, UNESCO-IHE.

Chapagain, A. K., and Hoekstra, A. Y. (2008). The global component of freshwater demand and supply: An assessment of virtual water flows between nations as a result of trade in agricultural and industrial products. *Water International*, 33(1), 19–32.

Chapagain, A. K., Hoekstra, A. Y., and Savenije, H. H. G. (2006). Water saving through international trade of agricultural products. *Hydrology and Earth System Sciences*, 10, 455–468.

Ciplet, D., Roberts, J. T., and Khan, M. R. (2016). *Power in a warming world.* Cambridge, MA: MIT Press.

Climate and Security Advisory Group (CSAG). (2016). *Briefing book for a new administration.* Washington, DC: Center for Climate and Security.

CNA. (2007). *National security and the threat climate change.* Alexandria, VA: CNA.

Cruise, A., and van der Zee, B. (2017). Armed herders invade Kenya's most important wildlife conservancy. *The Guardian.*

Dabelko, G. (2008). An uncommon peace: Environment, development and the global security agenda. *Environment*, 50(3), 32–45.

Dabelko, G. D., Herzer, L., Null, S., Parker, M., and Sticklor, R. (2013). Backdraft: The conflict potential of climate change adaptation and mitigation. Washington, DC: Woodrow Wilson International Center for Scholars.

Dalin, C., Wade, Y., Kastner, T., and Puma, M. J. (2017). Groundwater depletion embedded in international food trade. *Nature*, 543, 700–704.

Droogers, P., Immerzeel, W. W., Terink, W., Hoogeveen, J., Bierkens, M. F. P., Van Beek, L. P. H., and Debele, B. (2012). Water resources trends in Middle East and North Africa towards 2050. *Hydrology and Earth System Sciences*, 16(9), 3101–3114. doi:10.5194/hess-163-101-2012.

FAO. (2002). *Food, security, justice and peace.* World Food Summit 2002. Retrieved from www.fao.org/world-foodsummit/msd/Y6808e.htm

FAO. (2008). *Crop prospects and food situation*, April, No. 2. Electronic Document. Retrieved from www.fao.org/docrep/010/ai465e/ai465e03.htm

Fleurbaey, M. and Kartha, S. (2014). Sustainable development and equity. In O. Edenhofer, R. Pichs-Madruga, Y. Sokona, E. Farahani, S. Kadner, K. Seyboth, A. Adler, I. Baum, S. Brunner, P. Eickemeier, et al. (Eds.), *Climate change 2014: Mitigation of climate change. Contribution of working group III to the fifth assessment report of the intergovernmental panel on climate change.* Cambridge: Cambridge University Press, pp. 283–350.

Flowers, K. (2015). *The conflict, displacement, and food security nexus.* Retrieved from www.csis.org/analysis/conflict-displacement-and-food-security-nexus

Fong, C. (2015). *Water, land and oil: Drivers of major upheaval in the lower Omo Valley and Turkana.* International Rivers.

Forbes. (2015). America's fastest-growing cities 2015. Released on: January 27, 2015. Retrieved from www.forbes.com/sites/erincarlyle/2015/01/27/americas-fastest-growing-cities-2015/#3ab5c5b76158

Galvin, K. A., Smith, N. M., Boone, R. B., and Lynn, S. J. (2001). Impacts of climate variability on East African pastoralists: Linking social science and remote sensing. *Climate Research*, 19, 161–172.

Hoekstra, A. Y. (2003). *Virtual water trade: Proceedings of the international expert meeting on virtual water trade.* Value of Water Research Report Series No. 12, UNESCO-IHE.

Hoekstra, A. Y., and Hung, P. Q. (2005). Globalisation of water resources: International virtual water flows in relation to crop trade. *Global Environmental Change*, 15(1), 45–56.

Immerzeel, W., van Beek, L. P. H., Konz, M., Shrestha, A. B., and Bierkens, M. F. P. (2011). Hydrological response to climate change in a glacierized catchment in the Himalayas. Climatic Change 2012(110), 721–736. DOI 10.1007/s10584-011-0143-4 Retrieved from: https://link.springer.com/content/pdf/10.1007%2Fs10584-011-0143-4.pdf

Intergovernmental Panel on Climate Change (IPCC). (2007). *Climate change impacts, adaption and vulnerability.* Cambridge: Cambridge University Press.

Internal Displacement Monitoring Centre (IDMC). (2016). *Global report on internal displacement.* Retrieved from www.internal-displacement.org/assets/publications/2016/2016-global-report-internal-displacement-IDMC.pdf

Jaafar, H. H., Zuraykb, R., King, C., Ahmadd, F., and Al-Outa, R. (2015). "Impact of the Syrian conflict on irrigated agriculture in the Orontes Basin. International Journal of Water Resources Development, 31(3), 436–449.

Johannes, E. M., Zulu, L. C., and Kalipeni, E. (2015). Oil discovery in Turkana County, Kenya: A source of conflict or development? *African Geographical Review*, 34(2), 142–164.

Klare, M. (2002). *Resource wars: The new landscape of global conflict*. New York: Owl Books.

Klare, M. (2008). *Rising powers, shrinking planet: The global geopolitics of energy*. New York: Henry Holt and Company.

Klare, M. (2012). *The race for what's left: The global scramble for the world's last resources*. New York: Metropolitan Books.

Lagi, M., Bertrand, K. Z., and Bar-Yam, Y. (2011). *The food crises and political instability in North Africa and the Middle East*, August 15. Retrieved from SSRN: https://ssrn.com/abstract=1910031 or http://dx.doi.org/10.2139/ssrn.1910031

Lee, B., Preston, F., Kooroshy, J., Bailey, R., and Lahn, G. (2012). *Resources futures*. London: Chatham House

Lee, S. H., Mohtar, R. H., Choi, J. Y., and Yoo, S. H. (2016a). Analysis of the global agricultural virtual water trade network using degree and eigenvector centrality. *Hydrology and Earth System Sciences*, 20, 4223–4235.

Lee, S. H., Yoo, S. H., Choi, J. Y., and Mohtar, R. H. (2016c). *Evaluation of external virtual water export and dependency through crop trade: An Asian case study*. Paddy and Water Environment (PAWE) (Accepted).

Lee, S. H., Yoo, S. H., Shin, A., and Choi, J. Y. (2016b). Evaluation of the dependency and intensity of virtual water trade in Korea. *Irrigation and Drainage*, 65, 48–56.

Lockwood, A. H. (2016). *Heat advisory: Protecting health on a warming planet*. Cambridge, MA: MIT Press.

Mkutu, K. (2002). *Pastoralism and conflict in the Horn of Africa*.

Mkutu, K. A. (2007). Small arms and light weapons among pastoral groups in the Kenya-Uganda border area. *African Affairs*, 106(422), 47–70.

Mohtar, R. H., Assi, A. T., and Daher, B. T. (2016). Current Water for Food Situational Analysis in the Arab Region and Expected Changes Due to Dynamic Externalities. *The Water, Energy, and Food Security Nexus in the Arab Region*, 193–208. DOI: 10.1007/978-3-319-48408-2_10

Mohtar, R. H., Assi, A. T., and Daher, B. T. (2015). *Bridging the water and food gap: The role of the water-energy-food nexus*. UNUFLORES Working Paper Series 5 (H. Hettiarachchi, Ed.). Dresden: United Nations University Institute for Integrated Management of Material Fluxes and of Resources (UNU-FLORES).

NATO. (2015). *Russian-Ukrainian-EU gas conflict: Who stands to lose most?* Retrieved from www.nato.int/docu/review/2014/nato-energy-security-running-on-empty/Ukrainian-conflict-Russia-annexation-of-Crimea/EN/index.htm

Nett, K., and Ruttinger, L. (2016). *Insurgency, terrorism and organizaed crime in a warming world*. Berlin: Adelphi.

Njiru, B. N. (2012). Climate change, resource competition and conflict amongst pastoral communties of northern Kenya. In J. Scheffran, M. Brzoska, G. H. Brauch, and J. Schilling (Eds.), *Climate change, human security and violent conflict*. Springer, Berlin, Heidelberg, pp. 513–527.

Nordkville, J., Rustad, S., and Salmivalli, M. (2017). Identifying the effect of climate variablity on communal conflict through randomization. *Climatic Change*, 14, 627–639.

Obersteiner, M., Walsh, B., Frank, S., Havlík, P., Cantele, M., Liu, J., Palazzo, A., Herrero, M., Lu, Y., Mosnier, A., Valin, H., Riah, K., Kraxner, F., Fritz, S., and van Vuuren, D. (2016). Assessing the land resource – food price nexus of the sustainable development goals. *Science Advances*, 2(9). e1501499 DOI: 10.1126/sciadv.1501499.

Oki, T. (2005). The hydrologic cycles and global circulation. In M. G. Anderson (Ed.), *Encyclopaedia of hydrological sciences*. Chichester: John Wiley and Sons.

Opiyo, F. E., Wasonga, O. V, Schilling, J., and Mureithi, S. M. (2012). Resource-based conflicts in drought prone Northwestern Kenya: The drivers and mitigation mechanisms. *Wudpecker Journal of Agricultural Research*, 1(11), 442–453.

Osisanya, S. (2015). National security versus global security. *UN Chronicle*. Retrieved from https://unchronicle.un.org/article/national-security-versus-global-security

Owen, T. (2004). Human security – conflict, critique and consensus: Colloquium remarks and a proposal for a threshold-based definition. *Security Dialogue*, 35(3), 373–387. http://doi.org/10.1177/0967010604047555

Ranson, M. (2014). Crime, weather, and climate change. *Journal of Environmental Economics and Management*, 67(3), 274–302.

Sanchez, R., Lopez, V., and Eckstein, G. (2015). Identifying and characterizing transboundary aquifers along the Mexico – US border: An initial assessment. *Journal of Hydrology*, 535, 101–119, April 2016. Retrieved from http://dx.doi.org/10.1016/j.jhydrol.2016.01.070

Schilling, J., Opiyo, F. E., and Scheffran, J. (2012). Raiding pastoral livelihoods: Motives and effects of violent conflict in north-western Kenya. *Pastoralism: Research, Policy and Practice*, 2–25. Retrieved from https://doi.org/10.1186/2041-7136-2-25.

Shi, L., Chu, E., Anguelovski, I., Aylett, A., Debats, J., Goh, K., Schenk, T., Seto, K. C., Dodman, D., Roberts, D., Roberts, J. T., VanDeveer, S. D. (2016). Roadmap toward justice in urban climate adaptation research. *Nature Climate Change*, 6, 131–137.

Simonse, S. (2011). *Human security in the borderlands of Sudan, Uganda and Kenya: Key advocacy issues from the perspective of a grassroots peace building programme for youth warriors.*

StateImpact. (2017). *Everything you need to know about the texas drought.* Energy and Environment Reporting for Texas. Retrieved from https://stateimpact.npr.org/texas/tag/drought/

TWDB. (2016). *2017 state water plan.* Retrieved from www.twdb.texas.gov/waterplanning/swp/2017/

UNFCC. (2015). *Paris agreement.* Retrieved from http://unfccc.int/paris_agreement/items/9485.php

United Nations. (2015). *Sustainable development goals.* Retrieved from http://sustainabledevelopment.un.org/?menu=1300

USDA. (2016). *State fact sheets: Texas.* Retrieved from www.ers.usda.gov/data-products/state-fact-sheets/state-data.aspx?StateFIPS=48&StateName=Texas

US Department of Defense (DOD) Quadrennial Defense Review 2014. (2014). Washington, DC: DOD.

Van der Heijden, K., and Maddocks, A. (2015, November 3). *Beyond conflict, water stress contributed to Europe's migration crisis.* Retrieved from www.wri.org/blog/2015/11/beyond-conflict-water-stress-contributed-europe%E2%80%99s-migration-crisis

VanDeveer, S. (2015). Consumption, commodity chains and global and local environments. In R. Axerod and S. D. VanDeveer (Eds.), *The global environment.* New York: CQ Press.

Verner, D., Lee, D. R., Ashwill, M., and Wilby, R. (2013). *Increasing resilience to climate change in the agricultural sector of the Middle East: The cases of Jordan and Lebanon.* Washington, DC. World Bank. doi:10.1596/978-0-8213-9844-9

Wayumba, G. (2004). *A review of special land tenure issues in Kenya.* Nairobi, Kenya. Retrieved from www.fig.net/resources/proceedings/2004/nairobi_2004_comm7/papers/ts_06_3_wayumba.pdf

Wayumba, G. (2015). The challenge of sustaining pastoralism land tenure system for ecological conservation of the Maasai Mara. *International Journal of Scientific Research and Engineering Studies*, 2(7), 98–103.

Weinthal, E., Troell, J., and Nakayama, M. (2013). *Water & post-conflict peacebuilding.* Abingdon: Routledge.

World Bank. (2009). *Improving food security in Arab countries.* A joint report by FAO, IFAD and the World Bank, The World Bank, Washington, DC.

World Economic Forum. (2016). *The global risks report 2016* (11th ed.). Geneva. Retrieved from http://www3.weforum.org/docs/Media/TheGlobalRisksReport2016.pdf

World Economic Forum. (2017). *The top 5 global risks in terms of likelihood.* Retrieved from http://reports.weforum.org/global-risks-2017/the-matrix-of-top-5-risks-from-2007-to-2017/

PART II

Analysing the resource nexus

Tools and metrics

5

LIFE CYCLE ASSESSMENT FOR RESOURCE NEXUS ANALYSIS

Ester van der Voet and Jeroen B. Guinée

Introduction

Life Cycle Assessment (LCA) is a method to assess environmental impacts of cradle-to-grave chains connected to products or services. It compiles and evaluates the inputs, outputs and the potential environmental impacts of a product system throughout its life cycle (ISO, 1996). The environmental inputs and outputs include emissions, but also the use of resources such as land, water and minerals. Traditionally, this analysis is done at the micro-level and uses a static, steady-state approach. Recently, the LCA field has started to broaden its horizon. Life cycle thinking is now also used at a higher scale level, for forward-looking analyses and for assessing social and economic impacts. This broadened field of science is nowadays indicated as Life Cycle Sustainability Assessment, which is rather a framework than a method in itself. The development of quantitative methods for LSCA is now in full progress. This may open up possibilities to account for nexus issues in a more dynamic and integrated way. In this chapter, we briefly describe the LCA method and outline the potential of this method to address resource nexus issues. Then we will explore the possibilities of LCSA to include more of these aspects. We conclude by a summing up of the strengths and limitations of the LCA approach in addressing the resource nexus.

The LCA method

Environmental Life Cycle Assessment (LCA) of products and services is a method that has been developed over nearly 50 years. The first studies that are now recognized as (partial) LCAs date from the late 1960s and early 1970s (Assies, 1992; Guinée et al., 2011). Since then, an ISO standard has been published (ISO, 1996; ISO, 2006), and several LCA guides and handbooks have been produced (ILV et al., 1991; Lindfors, 1992; Grieshammer et al., 1991; Heijungs et al., 1992; Vigon et al., 1993; Lindfors et al., 1995; Curran, 1996; Hauschild and Wenzel, 1998; Guinée et al., 2002; Baumann and Tillman, 2004; EC, 2010; Curran, 2012; Klöpffer and Grahl, 2014) providing general guidance and specific guidelines for practical LCA studies.

According to the ISO Standard, the method consists of four phases (see Figure 5.1).

In the goal and scope definition, the purpose of the LCA study is stated and a choice is made for the functional unit: the basis for comparing different product or service systems fulfilling the

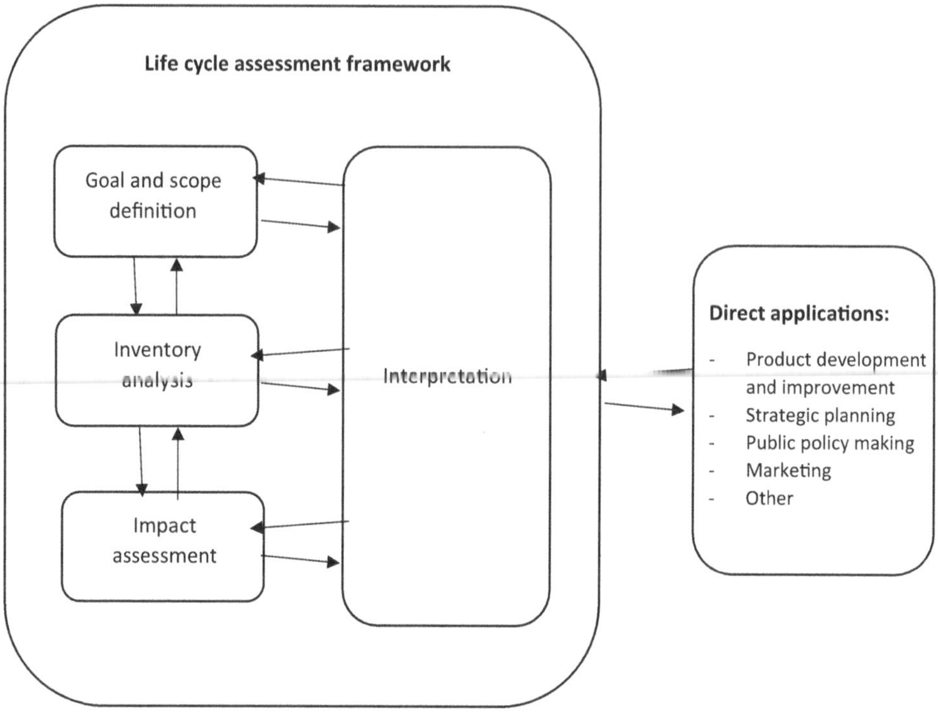

Figure 5.1 The general methodological framework for LCA (ISO, 1996)

same function. These choices, among others, determine the further work: system boundaries, data collection, impact assessment, etc.

The next phase is the Life Cycle Inventory (LCI) analysis. In this phase, the product system is specified. In LCA the system is typically cradle-to-grave: it includes all processes related to the production, use and waste management of the product or service. In the LCI phase, the system boundaries are established and economic and environmental flows of all unit processes[1] in the chain leading up to the functional unit are quantified. The LCI can be considered as the most demanding part of the LCA since representative data need to be collected for all unit processes of the product system. Unit processes are specified according to environmental inputs (resources), economic inputs (products, services, energy and materials), environmental outputs (emissions) and economic outputs (products, goods, services and waste to be processed). To make up the flow chart of the product system, these unit processes are linked up into chains. Several LCI databases are available (see e.g. www.rivm.nl/Onderwerpen/L/Life_Cycle_Assessment_LCA/Databases). Usually, specific "foreground" data are added to tailor the LCA to the needs of the user. One of the methodologically most influential steps within the LCI phase refers to allocation: the treatment of multifunctional processes. It has been established that different ways of allocation can lead to widely different outcomes (see for example van der Voet et al., 2010). The choice for allocation is therefore a critical one. Unfortunately, the ISO standard is not unambiguous here: ISO offers a hierarchy of different methods to solve multifunctionality but the hierarchy can be interpreted in different ways in practice. We will show examples of how the choice for allocation can be relevant for resource nexus issues as well. The result of the life cycle inventory phase is the inventory table showing all the environmental interventions associated with a product system (Guinée et al., 2002).

The next phase is the Life Cycle Impact Assessment (LCIA). In this phase, all extractions from and emissions to the environment are evaluated and aggregated into potential contributions to environmental impact categories. There are many ways of doing this. A general distinction is made into "midpoint" methods and "endpoint" methods. The midpoint methods translate the extractions and emissions into a potential contribution to a number of environmental impact categories, such as global warming, resource depletion, human toxicity and ecotoxicity. Multiplyers are used, in LCA called characterization factors, expressing a particular environmental intervention in terms of a common unit, for example, Global Warming Potentials (GWPs) expressing global warming gases in CO_2-equivalents representing the midpoint infrared radiative forcing. The endpoint methods try to translate all emissions and extractions into impacts on human and ecosystem health, and on natural resources. To some extent, midpoint and endpoint methods can be translated into each other as, for example, shown by the ReCiPe (Goedkoop et al., 2013) and IMPACT2002+ (Jolliet et al., 2003) methods. A very controversial step in the LCIA is weighting: by giving weights to the different impact categories, a single score can be obtained for the product or service system's environmental performance, which can be used for comparison with other systems. The ISO standard objects to the use of weighting in comparative assertions.

Present developments in LCA go in several directions:

- Finetuning the LCIA by adding new impact categories or refining existing ones;
- Developing LCA into a forward-looking tool by using consequential LCA or adding scenario techniques to attributional LCA (Spielmann et al., 2005; Hertwich et al., 2014);
- Adding context, for example via hybrid LCA or IPAT LCA: encompassing the LCA system into a model of the total economy to discover side-effects via broader economic mechanisms;
- Attempting to include more spatial and temporal differentiation, both in the LCI and the LCIA;
- Developing uncertainty and sensitivity analyses to comment on the robustness of the LCA outcomes, including data uncertainty as well as variability due to methodological choices;
- Broadening LCA into LCSA: Life Cycle Sustainability Analysis. This development is treated in a little more detail below.

Life Cycle Sustainability Analysis

LCSA is a new and potentially very relevant research field, which may in the end encompass many other methods of analysis. A generally accepted definition of LCSA is not yet available, but nowadays there is a small body of studies claiming to be LSCA studies. An up-to-date overview of the literature is provided by Guinée (2016). This publication also contains a discussion on LCSA definitions. LCSA is perceived as a framework and not as a specific method (Guinée et al., 2011). The idea of cradle-to-grave chains is maintained, but applied in a broader way than just to environmental impacts related to micro-level product or service systems. The whole development stems from the need for a life cycle approach to apply to other, broader types of questions than can be answered with the classic LCA.

Generally, the following aspects are included in the notion of LCSA:

- Broadening the scope by including social and economic impacts: adding Life Cycle Costing and social LCA to the environmental LCA to cover the three pillars of sustainability in one framework;

- Broadening the spatial scale by life cycle thinking to larger systems such as sectors or national economies that may change society's metabolism as a whole;
- Broadening the temporal scale by forward-looking analysis: Life Cycle Scenario Analysis uses life cycle thinking in forecasting or backcasting scenario analysis to overcome the usually narrow focus and cover a broader array of relevant changes;
- Deepening the analysis by including economic and behavioural relations within the system besides technical relations.

To this end, other more macro-scale models are used, for example environmentally extended input–output models (see Chapter 9). Aspects such as the rebound effect, indirect effects of changes in technology via income effects, also require the use of macro-level models in addition to LCA models (Hertwich, 2005; Hofstetter et al., 2006; Thiesen et al., 2008; Girod et al., 2011; Druckman et al., 2011; Font Vivanco and van der Voet, 2014). A more elaborate treatment can be found in Guinée (2016). This whole area of research might become very relevant for nexus issues as well.

LCA and the resource nexus

In LCA, resource extraction and energy use is specified over the full life cycle of the product or service. We therefore know the complete resource requirements of certain products/services. This information can also be used to calculate "footprints" (see Chapter 8). At the micro-level, such footprints are essentially LCA studies with a single focus, such as water or land use, energy use, or GHG emissions in the case of the carbon footprint (ISO, 2012; BSI, 2008), related to the product or service.

This means that LCA can be used to specify the resource intensity of products/services. This is relevant information to quantify the resource nexus. By appointing 1 kg of a certain resource as the functional unit, it is possible to calculate resource and energy requirements to produce 1 kg of this resource. The LCA is then used in the cradle-to-gate mode, ignoring the application and the end-of-life phase of the resource. Kleijn et al. (2010, 2011) take such an LCA approach in specifying the linkages between energy and metals. This is elaborated in more detail in Chapter 28. Similarly, linkages between energy and water and energy and materials are explored by Graedel et al. (2015) (see Figure 5.2) and by Eckelman et al. (2013). Linkages between water and materials are also investigated on an LCA basis, a.o. by Norgate and Haque (2010) (see Figure 5.3). They specify their results in terms of MJ per kg produced material, or kg copper per MJ produced: the energy intensity of material production, or the material intensity of energy production.

This is a first step in the analysis of the resource nexus. A next step is then to identify places where the linkages are mutual or reciprocal. For example, to produce copper we need electricity, and to produce electricity we need copper. Thus, feedback loops are identified and to some extent quantified. The concept of the Energy Return on Investment (EROI) is used as an indicator for such a feedback loop, which can be calculated based on LCA outcomes (Fthenakis et al., 2011). The EROI specifies how much MJ of energy is required to produce 1 MJ of energy. The EROI is used mostly in the assessment of alternative fossil resources such as shale oil or gas from fracking (Hall et al., 2014), but also in the assessment of renewable energy technologies (see Figure 5.4). The same can also be done, for example, with environmental technologies: to reduce GHG emissions implies having extra GHG emissions related to building and applying the equipment (Singh et al., 2011).

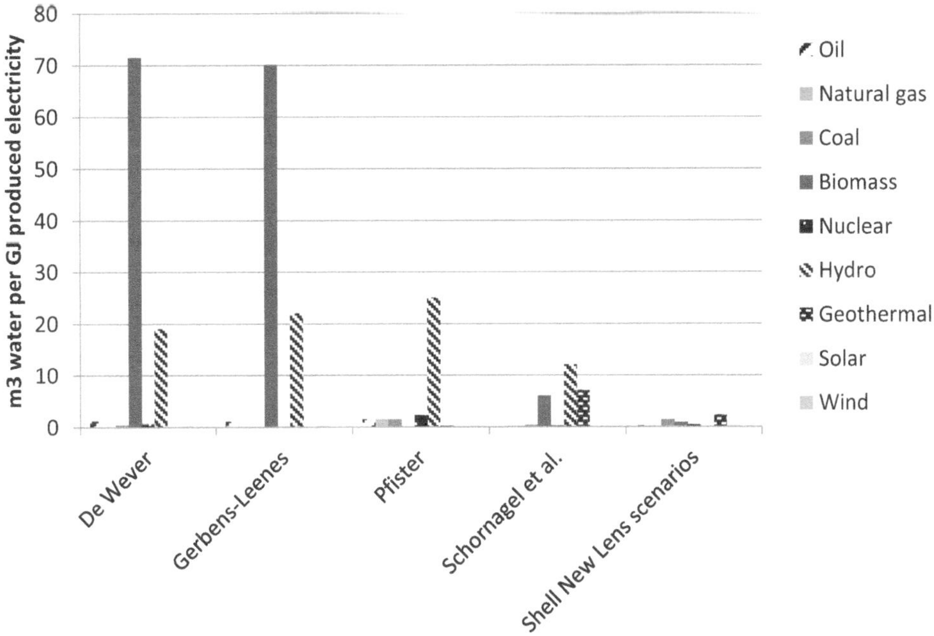

Figure 5.2 Water requirement of energy technologies according to different authors

Source: Graedel et al. (2015)

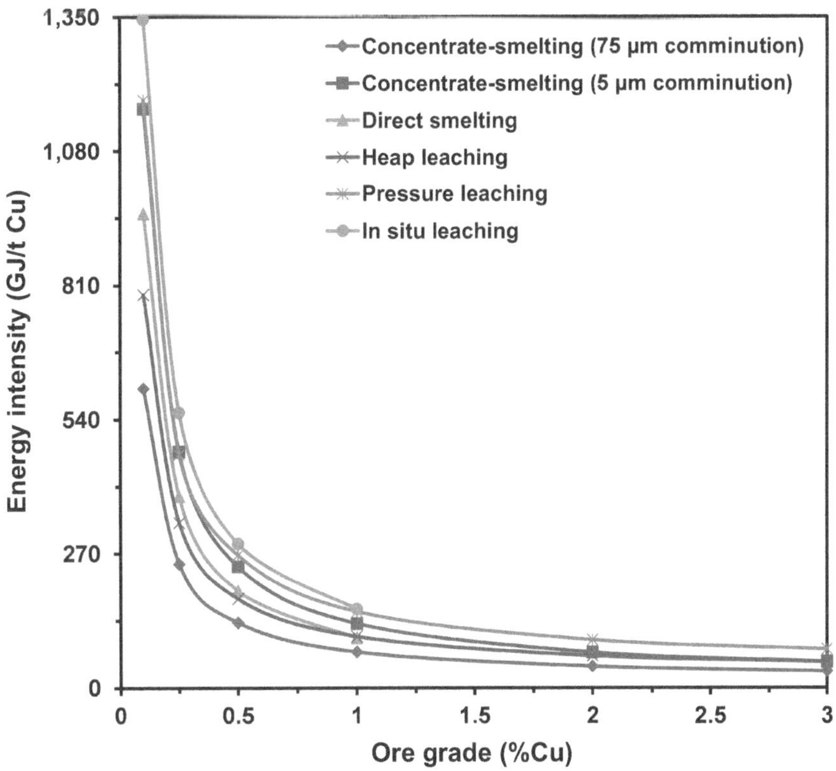

Figure 5.3 Energy requirement of metal production as a function of ore grades

Source: Norgate and Haque (2010)

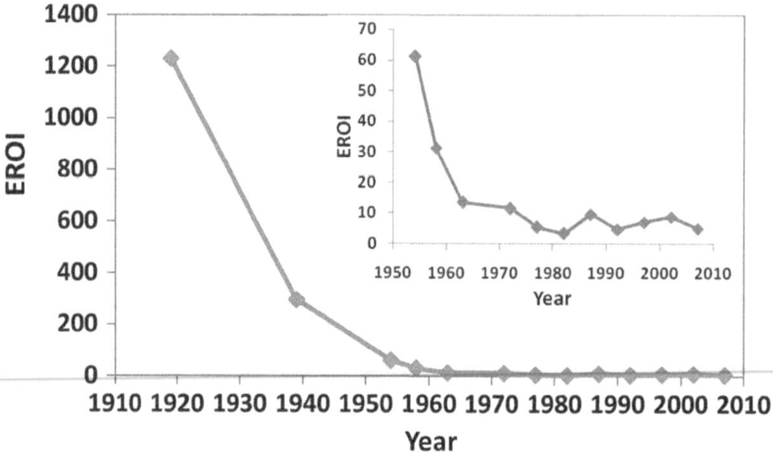

Figure 5.4 EROI for discoveries for the US oil and gas industry. The inset is the same data plotted on a different scale.

Source: Guilford et al. (2011)

With LCA, therefore, we can map the resource intensity of and the linkages between resource usage for a particular product or service system as a steady state (steel needed for electricity production, electricity needed for steel production). This is without doubt very useful information in the analysis of resource nexus phenomena. What cannot be done with LCA is to map the intensity and linkages under changing circumstances, for example as a consequence of degrading ore grades. Therefore, the dynamic nature of the linkages and the influence of feedback loops on those linkages is probably better investigated by other assessment methods. One of the characteristics of LCA, in section 5.2 is the need for allocation in case of multifunctional processes. This need for allocation is also present in case of circular processes: waste recycling and secondary production. Although allocation is essential to having model outcomes in LCA, and probably in some other methods as well, it is not helpful for insight in the dynamics of the resource nexus. It is also a major source of variability in LCA outcomes.

Apart from unit process data (Henriksson et al., 2015), uncertainty and variability in calculating resource intensities arises mainly from allocation choices (Mendoza et al., 2015), allocation being a methodological step that is always required in cases of transforming waste into resources again. Examples can be found in the areas of bio-based energy and of metal production. In the case of metal production, mining delivers ores which usually contain many different metals. Not in all cases is it clear which metal is the "main" product and which are co-products or by-products. Some metals are even produced completely as by-products, or originate mainly from waste-treatment processes. In some cases, metals could be co-produced but are not for economic reasons, and then may be recovered later. In such cases, how to allocate the extractions and emissions related to the production processes? Using partitioning based on mass means the by-products have very little impacts allocated to them. Resource and energy intensity of by-products then would always be low. Using economic proceeds does justice to the driving forces but means the allocation changes whenever (relative) market prices change. Such difficulties arise, and need to be solved somehow, when calculating resource and energy intensities of metals.

In the case of bio-energy, allocation occurs especially when waste streams or by-products such as corn stover, flax shives or wheat stems are used to make fuel or generate electricity. This can be 'solved' by declaring such by-products to be waste. In that case, the resource and energy intensity is zero. Allocating on a mass basis usually means that very large parts of the burden are allocated to the bio-energy, usually rendering it useless because it is worse than fossil fuels. Economic allocation makes more sense, but is again unstable with fluctuating (relative) prices. A method often used here is 'avoided burden': by producing bio-energy in addition to food crops, the production of fossil energy is 'avoided' and the hypothetical emissions of fossil energy production may be subtracted from the emissions of the bio-energy production. As it leaves the path of thinking in supply chains, this method is not applicable for calculating resource intensities in the framework of a resource nexus analysis. Allocation thus is a complicating factor, but such issues invariably arise when dealing with complex chains and networks of supply, even while aiming to focus on one resource or product only. They will show up in the analysis of the resource nexus as well.

Thus, we can conclude that resource and energy intensities of resource production, both single and reciprocal, can be calculated using the traditional, steady-state LCA. Allocation may be a complicating factor and some allocation methods lead to meaningless results in the analysis of the resource nexus. The need for allocation however will arise in all cases we address a system within a wider context, and therefore has to be dealt with.

However the dynamics of the linkages and the way they influence each other cannot be specified with LCA. Another issue is the upscaling of the information to encompass larger systems than just the micro-level functional unit. LCA is typically not suitable for that. It may be that the broader LCSA can capture these aspects. This will be explored in the next section. Figure 5.5 shows an example of the influence of allocation on LCA outcomes for the case of biofuels.

These are just first insights, following from a brief overview. To identify the contribution that can or cannot be made by LCA in studying the resource nexus in more detail and with

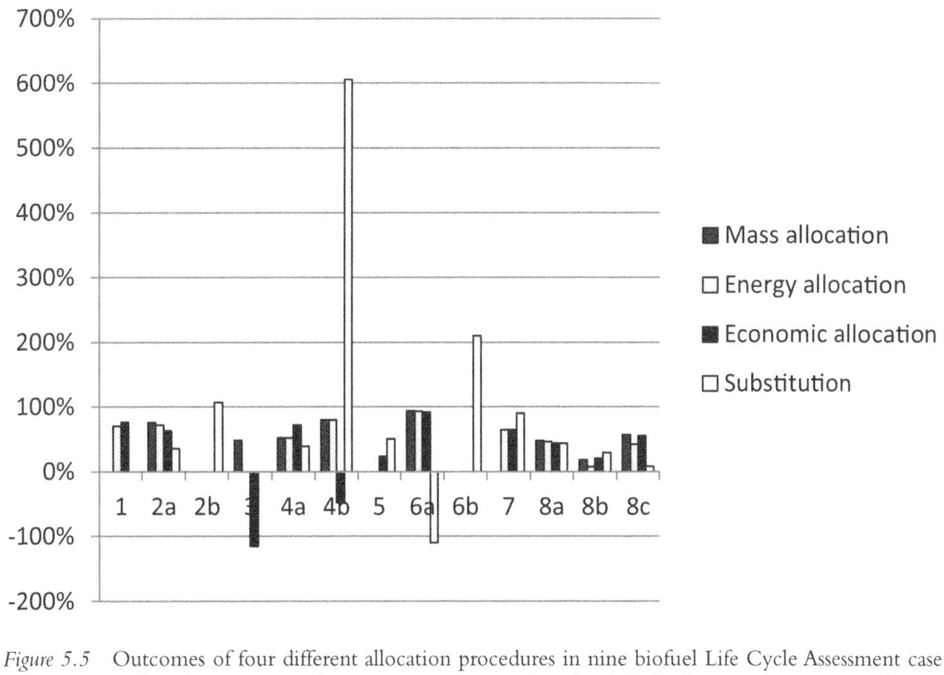

Figure 5.5 Outcomes of four different allocation procedures in nine biofuel Life Cycle Assessment case studies, in percentage improvement compared to fossil fuels, for greenhouse gas emissions (cases 1–8) and energy requirement (case 9)

Source: van der Voet et al. (2010)

more rigour, it is important that the developing field research into the resource nexus should first develop clear definitions of the resource nexus itself, the consequences and impacts related to it, and related indicators measuring these impacts. LCA (and other assessment methods) may then supply these indicators (e.g., as part of the LCIA phase) as long as the indicators don't ask for input information that LCA cannot provide, or require too much spatial and/or temporal details. If the information requirements are too high for LCA, LCA developers may decide to adapt those indicators again to the constraints of LCA, but will then never be able to assess the resource nexus problem comprehensively. This is very similar to the relation between LCA and human and ecosystem health risk assessments, studies of biodiversity or climate science. They are different fields of science: LCA doesn't develop risk assessment, climate change or biodiversity models, but it will try to apply models and indicators developed by scientists and experts of these fields as far as possible and sensible, and possibly involving some adaptations.

LCSA and the resource nexus

Extending LCA to include social and economic impacts

The resource nexus is about connections between resources. In LCA, resource extraction is specified, as well as emissions to the environment. In the LCIA, resource extraction is translated into rather straightforward characterisation factors related to resource depletion for individual resources. This does not really add to insight in the dynamic nature of the linkages between resources. The field is developing, however, and there are some ideas that are interesting from a resource nexus point of view (see van Oers and Guinée (2016) for an overview). Vieira et al. (2012) for example propose to use ore grade degradation as an indicator for scarcity.

Expanding LCA to include economic and social impacts may allow for a more sophisticated treatment of resource extraction. There are some attempts to include resource criticality into the LCA framework (Drielsma, Russel-Vacari et al., 2016; Schneider et al., 2015). Resource criticality is typically a topic that is unrelated to environmental issues, but is very relevant when LCA is expanded to include economic impacts. These developments are quite new and have not yet crystallised out into generally accepted impact factors. One pitfall to be avoided is to introduce economic driving forces as environmental impacts. That would turn the train of thought around in an unfortunate direction. An example is introducing GDP as an indicator for environmental pressure. No doubt GDP correlates heavily with environmental pressure, but the information is meaningless in a context where society aims at reducing environmental impacts while maintaining economic growth. To use price information as an indicator in LCA may lead to similar results (Guinée et al., 2011; Drielsma, Allington et al., 2016). It is imaginable that resource nexus issues have social aspects that could be included in an LCSA framework. We are not aware of any attempts so far.

This direction of development may open up possibilities to include a larger array of impacts in the analysis of the resource nexus. That in any case could be useful. It is too early to be able to conclude which specific additions would be applicable and would make sense.

Increasing the spatial scale of application

A second aspect of LCSA is the scale level: from the micro-level of the individual product or service to a relevant system at the regional, national, continental or global scale. Enlarging the scale level is certainly relevant for the resource nexus. At the micro-level there is never a problem; therefore, much of the relevance of the resource nexus operates at the regional or global scale. Analysing the resource nexus at such a larger scale level with LCSA means preserving the life cycle idea, but extending or complementing the 'classic' LCA tool. Some attempts have been made to quantify the resource nexus at the global level. Mostly this has been done by upscaling micro-level intensities: for example, multiply the MJ/kg data of the energy intensity of material production with the number of kilograms produced in one year worldwide. In UNEP (2013) the conclusion is drawn that at the global level 7–8% of total global energy consumption is used for metals production, the majority of that for iron and steel. Graedel et al. (2015) conclude that the energy–water linkage is asymmetric: "about 1% of the total global energy production is needed for water, whereas tens of percents of global water use are involved in energy production". Such explorations show the importance of the resource nexus and put it into a global level perspective, adding to the already existing global sustainability agendas. This upscaling is justified for past situations, although data needs are considerable and the uncertainties high. For future situations and emerging technologies this is not straightforward.

At the regional level, many attempts have been made to explore nexus issues. These are discussed in Chapters 26–31. However, a life cycle perspective is not often used in these studies.

So far, there has been no standard use of models for quantifying the nexus at larger scale levels. Models such as multiregional-environmentally extended input–output analysis or Material Flow Analysis could be used to highlight aspects of the resource nexus at the higher scale level. These are discussed in Chapters 7 and 9 of this volume. Both allow for a life cycle perspective, although this is not automatically the case. MR-EE-IOA models are used, for example, to calculate water or carbon footprints at the larger scale level of national economies (see Chapter 8). Material Flow Accounting is also used at the level of national economies, but falls short on the life cycle perspective. The more versatile Material Flow Analysis can include the life cycle perspective but has a very narrow view in other respects (see Chapter 7).

Forward-looking analysis: including the time dimension

The resource nexus is not static, but has many dynamic aspects. Even the rather straightforward resource or energy intensities change over time as a result of improving efficiencies or otherwise changing technologies. The whole modus of resource use, including all interdependencies, changes over time. Not just technology, but also consumption patterns and industrial production change, thereby leading to modified or even completely new relations between different types of resources.

A relevant question is whether improved/extended LCA models could address the dynamics of the nexus. There is an increased attention for forward-looking analysis in the LC(S)A community (Spielmann et al., 2005; Hertwich et al., 2014; Koning et al., 2015). Relevant aspects include, among others:

- Changes in background processes for the linked resources. An example of that is the question, posed by German federal governments, whether the 'Energiewende' has consequences for the usefulness of improvements in the built environment aimed at a reduction of energy use. Measures such as thermal insulation do reduce energy use, but the potential to reduce GHG emissions is very limited when the energy system is renewable. Such dynamics can be captured by LCA time series with changing 'background data' for the energy system. This could be called a quasi-dynamic analysis: a series of (static) LCAs of subsequent years, together providing a simulated time series.
- Changes or assumed changes in production processes of the resources. Such changes can be captured adequately with LCA time series, and will lead to changed resource or energy efficiencies. Here we deal with uncertainties related to economies of scale, and sometimes with uncertainties related to the conversion from lab and pilot data to real-world data. Such

uncertainties are inherent to explorations of future situations and must be dealt with in one way or another anyway. They are not limited to LC(S)A.

- Changes in resource scarcity or criticality. Resource availability may change in the future. It may grow less, as more and more resources are 'used up'. It may also grow better, in case substitutes are found for some specific resources and when the share of secondary production increases, reducing the demand for virgin resources. To assess such dynamics, LCA is not a suitable tool. LC(S)A may not be much better in that respect. The other way around, it means that any scarcity or criticality related characterization factors used in LC(S)A need to be updated frequently to remain relevant.

Conclusions

Life Cycle Assessment is a relevant tool for analysing the resource nexus. As is shown in this chapter, it is excellently suited to calculate resource or energy intensities, which are essential information for the resource nexus. These intensities can be used to calculate footprints at the micro-level.

Life Cycle Sustainability Analysis offers options to upscale the information on intensities or footprints. LCSA is a framework rather than a method. It therefore does not offer a recipe for analysis, but it does allow both for a more forward-looking analysis and for an exploration of change in the interrelations between resources. It also allows including the economic and social dimension of sustainability, which may be relevant for analysing the resource nexus. Methods to do so are in development, and are partly described in other chapters of this book.

Although LC(S)A can include resource scarcity in its array of impact categories, it is not suitable to analyse the phenomenon of resource scarcity or criticality itself. LSCA can add certain aspects of dynamics, if strictly defined and bounded. The main advantage, even at the macro-level, is the technological detail involved linked to the idea of the life cycle. With the LCA framework, however, it is not possible to assess feedback loops, so the dynamics of the linkages themselves remain unaddressed. Other tools are better suited to do that.

In all, LCA is able to have a limited but essential contribution to the analysis of the resource nexus. The broader framework of LC(S)A has more potential. The life cycle approach should be an integral, although in itself not sufficient, part of the analysis of any resource nexus.

Note

1 According to ISO 14044, a unit process represents the smallest portion of a product system for which data are collected in an LCA.

References

Assies, J. A. (1992). *Introduction paper to SETAC-Europe workshop on environmental life cycle analysis of products.* In Life-Cycle Assessment, Proceedings of a SETAC-Europe workshop on Environmental Life Cycle Assessment of Products, December 2–3, 1991, Leiden; SETAC-Europe: Brussels, Belgium.

Baumann, H., and Tillman, A-M. (2004). *The hitchhiker's guide to LCA: An orientation in life cycle assessment methodology and application.* Lund: Studentlitteratur.

Beltran, M., Angelica, J. G., Heijungs, R., and Tukker, A. (2015). A pseudo-statistical approach to treat choice uncertainty: The example of partitioning allocation methods. *International Journal of Life Cycle Assessment.* doi:10.1007/s11367-015-0994-4.

BSI. (2008). *PAS 2050 – specification for the assessment of the life cycle greenhouse gas emissions of goods and services.*

Curran, M. A. (1996). *Environmental life-cycle assessment.* New York: McGraw-Hill.

Curran, M. A. (Ed.). (2012). *Life cycle assessment handbook: A guide for environmentally sustainable products.* Beverly: Scrivener Publishing.

Drielsma, J. A., Russell-Vaccari, A. J., Drnek, T., Brady, T., Weihed, P., Mistry, M., et al. (2016). Mineral resources in life cycle impact assessment – defining the path forward. *International Journal Life Cycle Assessment*, 21, 85–105.

Drielsma, J. A., Allington, R., Brady, T., Guinée, J. B., Hammarstrom, J., Hummen, J., Russell-Vaccari, A., Schneider, L., Sonnemann, G., and Weihed, P. (2016). Abiotic raw-materials in life cycle impact assessments: An emerging consensus across disciplines. *Resources*, 5(1), 12. doi:10.3390/resources5010012.

Druckman, A., Chitnis, M., Sorrell, S., and Jackson, T. (2011). Missing carbon reductions? Exploring rebound and backfire effects in UK households. *Energy Policy*, 39, 3572–3581.

Eckelman, M. J. (2013). Life cycle assessment in support of sustainable Transportation. Environ. Res. Lett. 8 021004.

European Commission – Joint Research Centre – Institute for Environment and Sustainability. (2010). *International Reference Life Cycle Data System (ILCD) Handbook: General guide for life cycle assessment*. Luxembourg: Publications Office of the European Union.

Font Vivanco, D., and Voet, E. van der. (2014). The rebound effect through industrial ecology's eyes: A review of LCA-based studies. *International Journal of Life Cycle Assessment*, 19(12), 1933–1947.

Fthenakis, R. F., Raugei, M., Kim, H. C., Alsema, E., Held, M., and de Wild-Scholten, M. (2011). *Methodology guidelines on life-cycle assessment of photovoltaic electricity*. Report IEA-PVPS T12-03:2011.

Girod, B., de Haan, P., and Scholz, R. W. (2011). Consumption-as-usual instead of ceteris paribus assumption for demand: Integration of potential rebound effects into LCA. *International Journal of Life Cycle Assessment*, 16(1), 3–11.

Goedkoop, M., Heijungs, R., Huijbregts, M., De Schryver, A., Struijs, J., and van Zelm, R. (2013). *ReCiPe 2008: A life cycle impact assessment method which comprises harmonised category indicators at the midpoint and the endpoint level* (1st ed.) (vesrion 1.08). Report I: Characterisation.

Graedel, T., Elshkaki, A., and van der Voet, E. (2015, December). Entangled circles – energy and its resource connections. In G. J. Kramer and B. Vermeer (Eds.), *The colours of energy, essays on the future of our energy system*. Amsterdam, the Netherlands: Shell International BV. ISBN: 9789082398403

Grieshammer, R., Schmincke, E., Fendler, R., Geiler, N., and Lütge, E. (1991). *Entwicklung eines Verfahrens zur ökologischen Beurteilung und zum Vergleich verschiedener Wasch- und Reinigungsmittel; Band 1 und 2*. Berlin: Umweltbundesamt.

Guilford, M. C., Hall, C. A. S., O'Connor, P., and Cleveland, C. J. (2011). A new long term assessment of Energy Return on Investment (EROI) for U. S. oil and gas discovery and production. *Sustainability*, 3(10), 1866–1887. doi:10.3390/su3101866.

Guinée, J. B. (2016). Life cycle sustainability assessment: What is it and what are its challenges? Chapter 3 in R. Clift and A. Druckman (Eds.), *Taking stock of industrial ecology*. New York: Springer. ISBN: 978-3-319-20570-0 and 978-3-319-20571-7 (eBook).

Guinée, J. B., Gorrée, M., Heijungs, R., Huppes, G., Kleijn, R., Koning, A. de, Oers, L. van, Wegener Sleeswijk, A., Suh, S., Udo de Haes, H. A., Bruijn, J. A. de, Duin, R. van, and Huijbregts, M. A. J. (2002). *Handbook on life cycle assessment: Operational guide to the ISO standards*. Series: Eco-Efficiency in Industry and Science, Vol. 7. Dordrecht, The Netherlands: Springer.

Guinée, J. B., Heijungs, R., Huppes, G., Zamagni, A., Masoni, P., Buonamici, R., Ekvall, T., and Rydberg, T. (2011). Life cycle assessment: Past, present and future. *Environmental Science & Technology*, 45(1), 90–96. doi:10.1021/es101316v.

Hall, C. A. S., Lambert, J. G., and Balogh, S. B. (2014). EROI of different fuels and the implications for society. *Energy Policy*, 64, 141–152.

Hauschild, M., and Wenzel, H. (1998). *Environmental Assessment of products. Volume 1: Methodology, tools and case studies in product development – Volume 2: Scientific background*. London: Chapman & Hall.

Heijungs, R., Guinée, J. B., Huppes, G., Lankreijer, R. M., Udo de Haes, H. A., Wegener Sleeswijk, A., Ansems, A. M. M., Eggels, P. G., Duin, R. van, and Goede, H. P. de. (1992). *Environmental life cycle assessment of products: Guide & backgrounds – October 1992*. Leiden: Leiden University, Centre of Environmental Science.

Henriksson, P. J. G., Heijungs, R., Dao, H. M., Phan, L. T., Snoo, G. R. de, and Guinée, J. B. (2015). Product carbon footprints and their uncertainties in comparative decision contexts. *PLoS One*. doi:10.1371/journal.pone.0121221.

Hertwich, E. G. (2005). Consumption and the rebound effect: An industrial ecology perspective. *Journal of Industrial Ecology*, 9(1–2), 85–98.

Hertwich, E. G., Gibon, T., Bouman, E. A., Arvesen, A., Suh, S., Heath, G. A., Bergesen, J. D., Ramirez, A., Vega, M. I., and Shi, L. (2014). *Integrated life-cycle assessment of electricity-supply scenarios confirms global environmental benefit of low-carbon technologies*. "Proceedings of the National Academy of Sciences of the

United States of America." Proc Natl Acad Sci U S A. 2015 May 19;112(20):6277-82. doi: 10.1073/pnas.1312753111. Epub 2014 Oct.

Hofstetter, P., Madjar, M., and Ozawa, T. (2006). Happiness and sustainable consumption: Psychological and physical rebound effects at work in a tool for sustainable design. *International Journal of Life Cycle Assessment*, 11(SI1), 105–115.

ILV, GVM, and IFEU. (1991). *Umweltprofile von Packstoffen und Packmitteln, Methode (Entwurf)*. München, Wiesbaden and Heidelberg: Fraunhofer-Institut für Lebensmitteltechnologie und Verpackung (ILV), Gesellschaft für Verpackungsmarktforschung Wiesbaden (GVM) and Institut für Energie- und Umweltforschung Heidelberg (IFEU).

ISO 14067. (2012). *Greenhouse gases – carbon footprint of products – requirements and guidelines for quantification and communication*. Geneva, Switzerland. Retrieved from www.iso.org/iso/catalogue_detail?csnumber=59521.

ISO International Standard 14040. (1996). *Environmental management – life cycle assessment – principles and framework*. Geneva: International Organisation for Standardisation (ISO).

ISO International Standard 14044. (2006). *Environmental management – life cycle assessment – requirements and guidelines*. Geneva: International Organization for Standardization.

Jolliet, O., Margni, M., Charles, R., Humbert, S., Payet, J., Rebitzer, G., and Rosenbaum, R. (2003). IMPACT 2002+: A new life cycle impact assessment methodology. *International Journal of Life Cycle Assessment*, 8, 324–330.

Kleijn, R., and van der Voet, E. (2010). Resource constraints in a hydrogen economy based on renewable energy sources: An exploration. *Renewable and Sustainable Energy Reviews* 14 (2010) 2784–2795.

Kleijn, R., E. van der Voet, G.J. Kramer, L. van Oers, and C. van der Giesen, 2011. Metal requirements of low-carbon power generation. *Energy* 36 (2011) 5640–5648.

Klöpffer, W., and Grahl, B. (2014). *Life Cycle Assessment (LCA) – a guide to best practice*. Weinheim: Wiley-VCH.

Koning, A. de, Huppes, G., Deetman, S., and Tukker, A. (2015). Scenarios for a 2 °C world: A trade-linked input-output model with high sector detail. *Climate Policy*. dooi:10.1080/14693062.2014.999224.

Lindfors, L-G. (Ed.). (1992). *Product life cycle assessment – principles and methodology*. Nord, 1992:9. Copenhagen: Nordic Council of Ministers.

Lindfors, L-G., Christiansen, K., Hoffman, L., Virtanen, Y., Juntilla, V., Hanssen, O. J., Rønning, A., Ekvall, T., and Finnveden, G. (1995). *Nordic guidelines on life-cycle assessment*. Nord 1995:20. Copenhagen: Nordic Council of Ministers.

Norgate, T., and Haque, N. (2010). Energy and greenhouse gas impacts of mining and mineral processing operations. *Journal of Cleaner Production*, 18(3), 226–274.

Oers, L. Van, and Guinée, J. B. (2016). The abiotic depletion potential: Background, updates, and future. *Resources*, 5, 16. doi:10.3390/resources5010016.

Schneider, L., Berger, M., and Finkbeiner, M. (2015). Abiotic resource depletion in LCA – background and update of the anthropogenic stock extended abiotic depletion potential (AADP) model. *International Journal of Life Cycle Assessment*, 20, 709–721.

Singh, B., Strømman, A. H., and Hertwich, E. G. (2011). Comparative life cycle environmental assessment of CCS technologies. *International Journal of Greenhouse Gas Control*, 5, 911–921.

Spielmann, M., Scholz, R.W., Tietje, O., and Haan, P. de. (2005). Scenario modelling in prospective LCA of transport systems: Application of formative scenario analysis. *International Journal of Life Cycle Assessment*, 10(5), 325–335.

Thiesen, J., Christensen, T. S., Kristensen, T. G., Andersen, R. D., Brunoe, B., Gregersen, T. K., Thrane, M., and Weidema, B. P. (2008). Rebound effects of price differences. *International Journal of Life Cycle Assessment*, 13(2), 104–114.

UNEP, Voet, E. van der, Salminen, R., Eckelman, M., Mudd, G., Norgate, T. and Hischier, R. (2013). *Environmental risks and challenges of anthropogenic metals flows and cycles*, A Report of the Working Group on the Global Metal Flows to the International Resource Panel. UNEP. ISBN: 978-92-807-3266-5.

Vieira, M. D. M., Goedkoop, M. J., Storm, P., and Huijbregts, M. A. J. (2012). Ore grade decrease as life cycle impact indicator for metal scarcity: The case of copper. *Environmental Science and Technology*, 46(23), 12772–12778.

Vigon, B. W., Tolle, D. A., Cornaby, B. W., Latham, H. C., Harrison, C. L., Boguski, T. L., Hunt, R. G., and Sellers, J. D. (1993). *Life-cycle assessment: Inventory guidelines and principles*. EPA/600/R-92/245. Washington, DC: Environmental Protection Agency.

Voet, E. van der, Lifset, R., and Luo, L. (2010). Life-cycle assessment of biofuels, convergence and divergence. *Biofuels*, 1(3), 435–449.

6

LINKING SOCIETY AND NATURE

Material flows and the resource nexus

Anke Schaffartzik and Dominik Wiedenhofer

Society's metabolism: the link between social organization and resource use

In an analogy to the metabolism of a living organism, the concept of social metabolism proposes that a society requires material and energy inputs which it processes and assimilates for the purposes of societal reproduction and then discharges as wastes or emissions (Ayres and Simonis, 1994; Fischer-Kowalski, 1998; Martinez-Alier, 1987). The magnitude and composition of the biophysical flows and stocks, or the metabolic profile, which characterize a socio-economic system strongly depend on how the society in question is organized (Sieferle, 2003). Functional links between different types of material and energy strongly influence metabolic profiles. Societies of hunters and gatherers passively use solar energy by collecting wild-growing plants and hunting wild animals. This metabolic mode supports low population densities and a nomadic life which in turn prohibits the accumulation of material possessions. Biotic materials 'fuel' this form of societal organization and the use of other materials is only marginally possible and necessary. By actively harnessing solar energy and cultivating plants and domesticating animals, agrarian societies enable higher population density in a sedentary way of life with a significant impact on material use for buildings, tools, and other durable possessions. The dramatic shift in society's resource basis is accompanied by comparable shifts in organization, including the ability of agrarian societies to support labor that is not directly required for food provisioning as well as higher fertility rates. The agricultural surplus which can be produced also allows for some degree of urbanization. The use of fossil energy in industrial societies is linked to a completely new form of social organization in which the majority of the population must no longer work in the provisioning of food. Fossil energy both requires and enables the use of materials (such as metals, for example) which were only marginally used in agrarian societies. With high shares of construction minerals, fossil energy carriers, and metals, the metabolic profile of industrial societies is not only characterized by much higher per capita values for material use but also by a composition completely distinct from that of hunters and gatherers or agrarian societies in which biomass was dominant. It has been suggested that the increase of per capita energy availability in industrial societies may be linked to the occurrence of social revolutions (Fischer-Kowalski et al., 2014), possibly related to the degree to which the use of fossil energy allowed human societies to defy what was previously seen as the 'natural order' of things and therefore unchangeable: to light up the dark, to develop machines with superhuman strength and endurance, to alter the temperature of the environment (Lord, 2014).

By recognizing the need to consider biophysical societal structures as well as socio-cultural forms of societal organization in order to understand resource use, the social metabolism concept answers a central requirement of resource nexus research. One of the most important applications of the social metabolism concept is the analysis of material or substance flows. While economy-wide material flow accounting and analysis is a top-down approach which covers all materials crossing the border between a socio-economic system and its natural environment or other socio-economic systems, material or substance flow analysis (SFA) traces the inputs, transformations, and outputs of specific substances or materials in a bottom-up manner. At the national level, economy-wide material flow accounting (ew-MFA) currently distinguishes 46 types of extracted material and 49 types of traded material (Krausmann et al., 2015). To this detailed information on extracted and traded materials, the analysis of individual substances or materials adds an understanding of the stages from mining and extraction to further processing, use, and waste treatment or disposal within the socio-economic system (Brunner and Rechberger, 2003). The flows of individual materials are primarily tracked within an accounting framework which can be used to combine statistical data into comprehensive indicators. While substances can be accounted for in a comparable manner, the use of models is increasingly popular in SFA. Dynamic models, in particular, afford the advantage of considering stock-flow relationships and their driving forces so that they may be used for forecasting (e.g., Elshkaki et al., 2016; Krausmann et al., 2017).

Accounting for and modelling of substance and material flows are central tools in the study of social metabolism. An important conceptual achievement of such research is the consistent definition and application of the system boundary which a flow must cross in order to be considered as a societal input or output. In ew-MFA, the socio-economic system is commonly defined to include humans, their artefacts (infrastructure and durable goods), and livestock. In SFA, the system boundary definition differs between studies that may or may not include agriculture and/or landfills within the boundaries of the socio-economic system, for example. SFA may also trace flows into, out of, and within the environment which are not accounted for in the ew-MFA framework. Apart from these differences, both ew-MFA and SFA consider as inflows those material flows which enter the socio-economic system by domestic extraction or import and are used to build or maintain the societal. Conversely, outflows are those flows which exit the socio-economic system either as exports to other systems or as discharges (waste or emissions) to the environment (Figure 6.1; also see Fischer-Kowalski and Haberl, 2015). Ew-MFA focuses on the flows into and out of the socio-economic system, treating their transformation or integration within the system as balancing items. SFA focuses on individual substances (chemicals or chemical compounds) and their successive production and use within the socio-economic system in detail. The two tools are thus highly complementary with ew-MFA affording the advantage of complete coverage and SFA providing insights on use patterns of individual substances.

Through a tremendous international effort, the ew-MFA methods and standards have been harmonized (Fischer-Kowalski et al., 2011), resulting in the integration of material flow data into the regular statistical accounting practices of many countries (e.g., Eurostat, 2015; OECD, 2014) and, within the framework of the UN Environmental Programme (UNEP, 2016), in an international database reaching as far back as 1970 and covering the material extraction and trade of over 200 countries. The wealth of material flow data available for research enables very detailed comparisons of the material use patterns of different socio-economic systems and their development pathways (Steinberger et al., 2010, 2013), an important prerequisite to

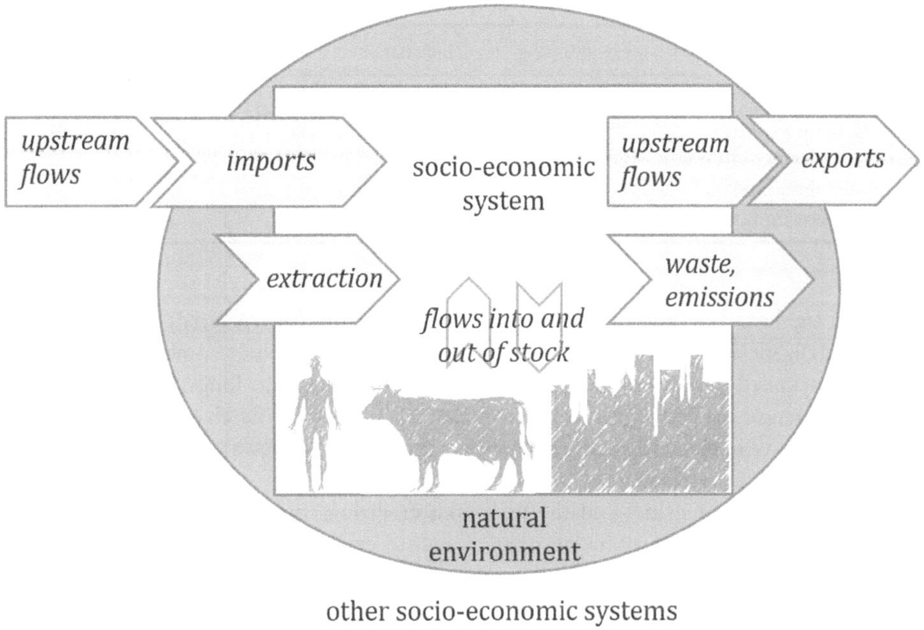

Figure 6.1 Sketch of material flows into and out of a socio-economic system. The scaling of the arrows does not represent the relative size of the flows. Please see Table 6.1 for the material flow indicators corresponding to these flows.

understanding the resource nexus as it applies to material resources. Material flow accounts can be compiled for systems at very different levels of scale. Studies spanning from the global (e.g., Krausmann et al., 2009) to world regions (e.g., Schaffartzik et al., 2014) to countries (e.g., Manrique et al., 2013) and subnational units such as cities (e.g., Rosado et al., 2014) allow for investigations of the material use nexus across levels of spatial scale.

The standardized ew-MFA indicators (Table 6.1) are calculated from the flows sketched out in Figure 6.1. SFA indicators are not standardized, and in practice, many different indicators are used to assess flows and stocks (van der Voet, 1996). Here, we provide only a short introduction to the standardized ew-MFA indicators referred to in the research presented in this chapter. Extraction from the domestic environment is referred to as *domestic extraction* (DE) and encompasses all those materials extracted by agriculture, forestry, fisheries and hunting (biomass), and by mining operations (minerals and fossil energy carriers). DE is accounted for if it is subject to further societal use; material which is moved at the site of extraction but not further used is referred to as "unused extraction". An important example of unused extraction with far-reaching socio-ecological impacts is the overburden removed in open-pit mining. The balance between biophysical imports and exports is the *physical trade balance* (PTB). Money and materials flow in opposite directions (exports represent a physical output from the socio-economic system but also a form of monetary income) and while the monetary trade balance is expressed as net exports, the PTB is expressed as net imports. DE plus PTB yields *domestic material consumption* (DMC), the main indicator of material use. DMC is a measure of apparent use, i.e., it includes all materials which are used within a system, regardless of whether this use occurs in the production for export or for domestic final consumption. Because all material used by a society eventually turns into waste and emissions, DMC can also be understood as an indicator of waste potential.

Table 6.1 Overview of central indicators in material flow accounting

Flow(s)	*Indicator*	*Abbreviation*
Extraction	Domestic extraction	DE
Imports minus exports	Physical trade balance	PTB
Extraction plus imports minus exports	Domestic material consumption	DMC
Extraction plus upstream flows and imports minus upstream flows and exports	Raw material consumption	RMC

The time lag, however, differs widely depending on whether the materials are, e.g., through-flow materials such as biomass used for human or animal nutrition or combusted fossil energy carriers, stockpiled (e.g., fossil energy carriers), or integrated into long-lasting infrastructure (e.g., construction materials) (Krausmann et al., 2017). The specific characteristics of societal stocks are decisive in this relationship and stocks can similarly be seen as an indicator of potential waste or as a secondary resource.

As globalization intensifies and the early industrializing countries increasingly meet their material demands via imports rather than domestic extraction, understanding the material dynamics of trade patterns becomes a prerequisite to understanding the dynamics and implications of material use (Schaffartzik et al., 2014; Steinberger et al., 2013). As imports play a greater role for some countries, others increasingly focus on exports. Along the supply chain, products become increasingly 'lighter' (e.g., from the extraction of copper ore via refining and processing into copper wires). Much of the material 'burden' lies in the countries of extraction (even more so if unused extraction is considered). If the country exports these materials, however, this material use is related to final demand elsewhere. In order to capture the material implications of such patterns of extraction, production, and trade, the ew-MFA framework is currently being expanded to account for the upstream flows associated with the production of traded goods and services which capture the material extraction, no matter where in the world it occurs, that was necessary in order to produce a specific trade flow. Based on this information, the *raw material consumption* (RMC), also referred to as the material footprint, can be calculated (Wiedmann et al., 2015; also see chapter on footprints, this volume).

The sketch of the socio-economic system presented in Figure 6.1 is highly stylized and filling this conceptual understanding with a greater amount of detail is a strong focus in the material flow accounting community. The human population of a socio-economic system is an important determinant for material availability and requirements. The mutual influence between social structures and forms of organization and material demand has been conceptually explored (Fischer-Kowalski and Weisz, 1999) and historically described for the different socio-metabolic modes of hunters and gatherers, agrarian societies, and industrial societies (Sieferle, 2003). Empirical detail is added by research which finds, for example, strong interrelations between the use of fossil fuels, metals, and construction minerals; links between these materials and GDP; as well as a strong relation between biomass use and population size (Steinberger et al., 2010, 2013) and which explores the role of energy use for human development (Steinberger et al., 2012). The topic of material extraction and (potential for) environmental conflict (Martinez-Alier et al., 2010) has been explored in conjunction with the resource nexus (e.g., Bleischwitz, 2014) and conflicts often arise out of competing claims to the same resource for different uses (e.g., for land: Haberl, 2015).

As the ew-MFA approach becomes increasingly consolidated, it is also constantly expanded upon in order to enable analyses of some of the most pressing sustainability issues. Researchers in the field continue to seek meaningful ways to represent the complexity of factors governing material availability and use patterns. Important links exist between use and extraction of different types of materials: fossil energy carrier inputs into agriculture may enable higher rates of biomass extraction per area (yields), use of construction minerals may predetermine fossil energy carrier requirements (e.g., if these construction minerals were used on the construction of roads, if buildings were insulated or not). Highly relevant linkages also exist between materials and other resources such as biomass and land, material extraction, and water (Andrews-Speed et al., 2012). The following section of this chapter is devoted to turning the spotlight on four current research topics in material flow analysis and their contribution to exploring the resource nexus.

Material flows and the resource nexus: spotlights

Synergies between ew-MFA studies and resource nexus research can often be traced back to the common interest in these two approaches in analytically bridging the perceived disconnect between societies and their resource base. The following four spotlights are work-in-progress, some of it more conceptual and some rather data-driven, relating to this theme. We use the topics of material productivity, material use and human development, material stocks and flows, and material extractivism to discuss potential synergies with the resource nexus concept rather than to present definitive analyses in any one of these areas.

Material productivity: the resource–economy nexus

Material productivity as the indication of how much economic wealth can be generated per unit of material use is commonly expressed as GDP per DMC or RMC (see Table 6.1) and is rapidly becoming an indicator of choice in evaluating environmental performance (UNEP, 2011). In this spotlight, we will illustrate how, by situating this indicator within a resource nexus framework, we simultaneously learn more about the indicator and about options for decoupling economic growth and resource use.

For the purposes of illustration, we will consider in generalized terms the role which policy measures to stimulate economic growth by increasing capital spending on infrastructure may play in determining how the material productivity indicator performs (Figure 6.2). Such policy measures are motivated by the observation or expectation of low economic growth. If successful, they would increase construction activity alongside related employment and wages and, by extension, final consumption on the one hand and intermediate consumption on the other hand. Both 'positive' loops would cause growth in GDP, as the theory goes.

The material productivity indicator requires us to examine not only the economic performance of an economy but also the associated material use. If the stimulus to infrastructure spending is successful, it commonly also causes (depending on which specific infrastructure is targeted) an increase in built-up land which may pose at least a potential conflict with other forms of land use including biomass extraction and nature protection. Either directly through deforestation or via the consumption of fossil energy carriers in the use-phase of buildings or infrastructure, the increase in built-up land may be related to changing levels of emissions, notably of greenhouse gases. Growth in final as well as intermediate consumption – via which the stimulus to the construction sector also promotes GDP growth – may be associated with increased material use, especially of non-metallic (construction) minerals, metals, and fossil energy carriers in the case of the intermediate consumption related to construction activity. In this way, the economic stimulus to construction activity leads to increased material use. Considering material productivity as the relation of GDP to RMC would allow us to additionally

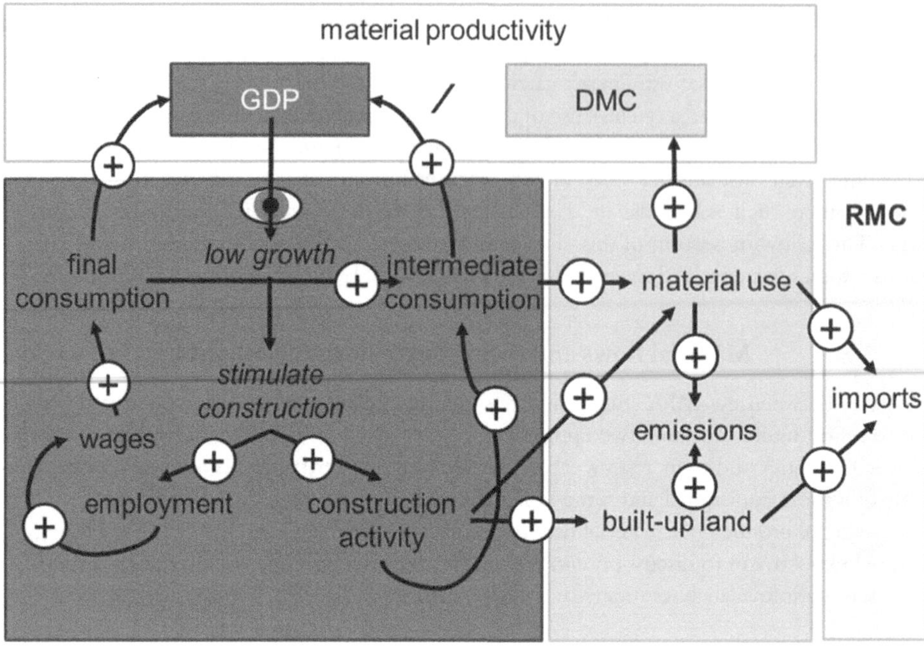

Figure 6.2 Simplified conceptual sketch of material productivity (measured as GDP per domestic material consumption (DMC)) and the involved interlinkages between political and economic processes on the left and different types of resource use on the right

capture the indirect effects of increasing requirements for imports which may associated with the rise in material use. How capital is spent on infrastructure will additionally strongly influence performance of the material productivity indicator in the future: Construction of new roads, possibly combined with failure to maintain existing roads, could lead to higher future demand for fossil energy carriers; where new buildings are situated, could, depending on their function, have an influence on average commute times and again impact energy requirements; if the stimulus creates long-term jobs, an impact could be made on future final consumption. Even in this most cursory of manners, the example of increased capital spending on infrastructure and its relation to the material productivity indicator illustrates that what may appear to be an economic policy must be understood with regard to its biophysical impact. This is only possible if the resource nexus is understood.

Not only in our conceptual example does the construction sector play a potentially pivotal role in shaping material productivity. In 2013, non-metallic minerals – which mainly consist of bulk construction materials such as sand and gravel – made up almost half of the DMC of the 28 member states of the European Union (EU-28) (Eurostat, 2016). Between 2002 and 2013, material productivity in the EU-28 increased by approximately 30% from 1432 €/t to 1812 €/t for GDP/DMC; with very similar trends for GDP/RMC (Figure 6.3). During this time period, 93% of the variance in GDP/DMC can be explained by the productivity of the non-metallic minerals.

If GDP and DMC grow at the same rate, the material productivity indicator remains unchanged; it improves if GDP grows faster than DMC and declines if DMC should grow more quickly than DMC. When we look beneath the surface of this indicator's performance, we find that we must consider what drives the components and invariably focus on interlinkages between political and economic processes and different forms of resource use.

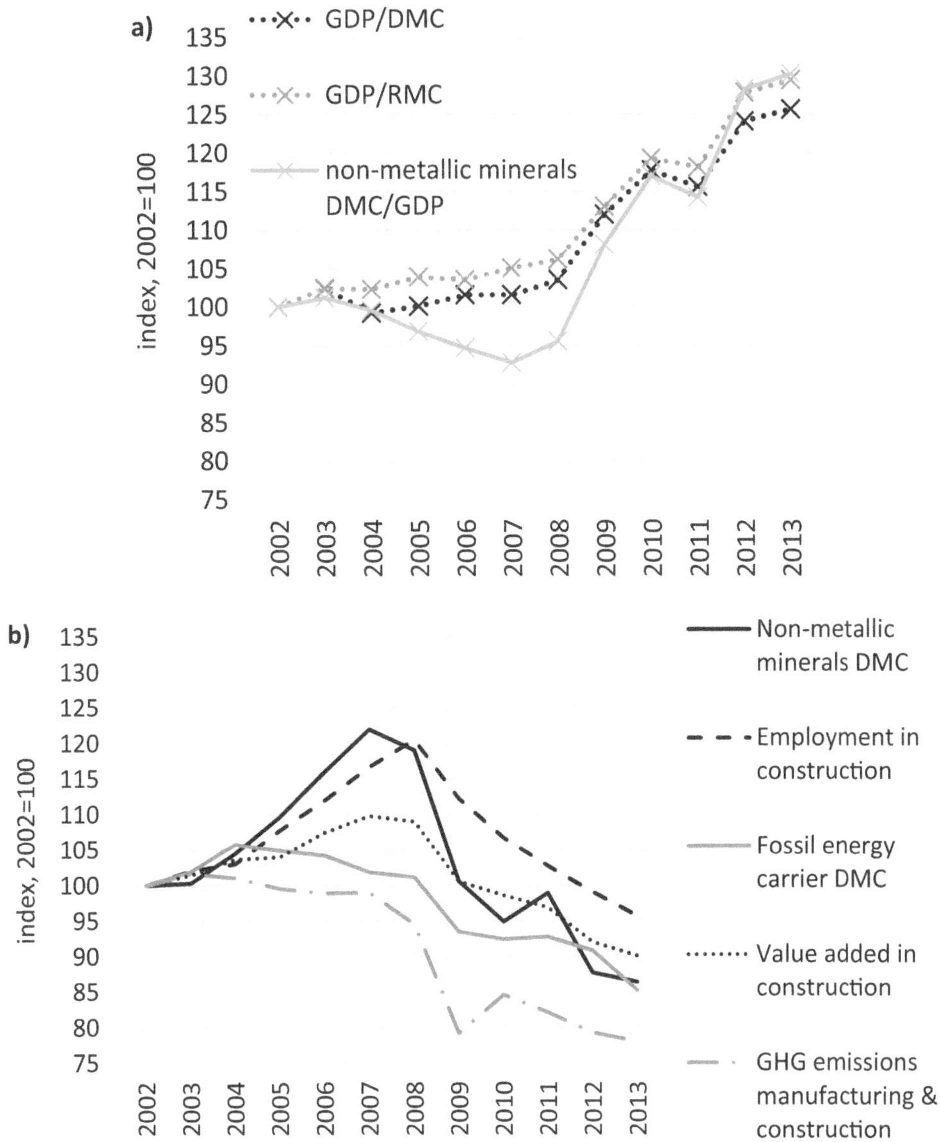

Figure 6.3a and 6.3b Trends in material productivity in the EU-28 based on domestic material consumption (DMC), raw material consumption (RMC), and consumption of non-metallic minerals between 2002 and 2013 calculated with GDP in 2005 chain-linked volumes (top); trends in DMC by material categories, employment, and value added in construction and greenhouse gas emissions in the EU-28 between 2002 and 2013 (bottom)

Source: Eurostat (2016)

If we turn back to the example sketched out in Figure 6.2 and specifically examine the resource nexus linked to the construction sector, we find that during the period of the European financial crisis, the positive feedback loops take effect in reverse with declining value added and employment in construction as reflected in the economic indicators and declining use of non-metallic minerals and fossil energy carriers as well as GHG emissions reflected in the biophysical indicators. Data on built-up land is unfortunately too sparsely available to allow inclusion in this depiction. By using material flow data along with other indicators, specific interlinkages of the resource nexus as it applies to the material productivity indicator can be examined.

Material use and human development: the resource–society nexus

While the concept of material productivity suggests that the functional use of resources (in this case materials) may be assessed by determining how much GDP can be generated per unit of material use, one might also argue that material use has (or ought to have) a function for human societies that goes beyond enabling GDP growth. In this spotlight on material use and human development, we examine the relationship between the human development index (HDI) and DMC. Of course, one of the components of the HDI is per capita GDP but it additionally includes measures of life expectancy and education. We use this example to shed light on the role of material use for human development and what role the link between the two plays in understanding the resource nexus.

The development of the resource nexus concept is motivated by the desire to gain a holistic understanding of resource use, including potential trade-offs between resource use and development. Based on material flow data and the composite HDI, we can assess whether a certain level of material wealth is a prerequisite to 'high' development levels (in the logic of the HDI) and how this relationship has developed over time. We examine a sample of 114 countries which span all continents and all development levels. The average material use threshold for a high HDI of 0.72 (2010 definition) or above lay at 26.7 t/cap/a in 1980 and had decreased significantly to 11.8 t/cap/a by 2010 (Figure 6.4). Those countries which used between 11.5 and 12.0 t/cap of materials in 2010 exhibited HDIs ranging from 0.45 (indicative of low human development) to 0.88 (just below the threshold for very high human development). In general, the logarithmic relationship between per capita DMC and HDI indicates that while at low levels of DMC per capita more material use is related to strong improvements in HDI, there is a level of material use at which this curve flattens and more material use no longer results in relevant HDI improvements.

To consider the societal 'value' of given levels of resource use, these results suggest that it is important not only to consider whether resource use is increasing or decreasing (on which there's a significant focus in the material productivity indicator) but also the level of resource use. This analysis further directs our attention to countries obtaining high levels of HDI with comparatively low levels of material use which may potentially yield best-practice examples. If we can understand the interlinkages shaping the resource nexus in these countries, we may advance an important step to considering policy recommendations on important aspects to consider in governing resource use.

Material stocks: past–present–future nexus

From a socio-ecological perspective, resource use is functionally related to the reproduction of societal stocks of people, livestock, buildings, and infrastructure (Pauliuk and Müller, 2014; Weisz et al., 2015) which are therefore decisive drivers of annual flows of materials and energy. Examples are biomass harvested to ultimately feed the population (which is a stock of people; also see Figure 6.1), energy use required to operate transport infrastructure and

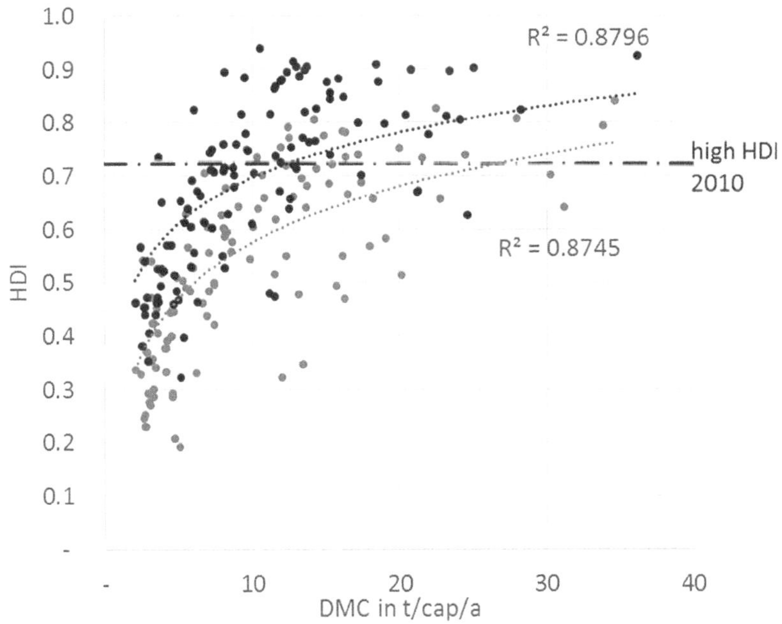

Figure 6.4 Domestic material consumption (DMC) in metric tons per capita and year (t/cap/a) plotted against the human development index (HDI) for a sample of 114 countries in 1980 and 2010. The logarithmic curves have been fitted to population-weighted data (allowing countries with large populations to have a greater impact on the curve than countries with small populations). The goodness-of-fit is depicted for the population-weighted least squares. The horizontal line indicates the threshold for a high HDI (0.72) in 2010.

Source: Schaffartzik et al. (2014) and UNDP (2014)

buildings, the specific energy carriers and materials required to operate a national power grid (coal dominant versus renewable, for example), or large amounts of construction minerals such as asphalt or gravel required to maintain road networks. Therefore, the specific quantity and quality of stocks strongly influences the amounts and types of resources needed and the types and timings of waste flows arising from the social metabolism. In terms of energy use, emissions usually occur in conjunction with use, with no significant time lag, while the buildings and infrastructure which require this energy to be useful tend to be in service for many decades before they are demolished and materials are recycled or landfilled. On a more general level, this means that decisions of the past shape present stocks and resource use and that decisions made today with regard to stocks can determine the resource nexus many decades into the future.

During the 20th century, global resource use increased tenfold, from approximately 7 to 78 Gigatons per year (Gt/a) between 1900 and 2010 (Figure 6.5). A large fraction of these resources provide energy, be it fossil fuels or biomass for food and feed. An increasingly large share of materials is accumulated in infrastructure and buildings, rising from approximately 14% of global resource use in 1900 to 55% in 2010 (Figure 6.5). The few existing studies on economy-wide material stocks suggest that globally approximately 115 tons of stock per capita are currently in use and that this figure may be as high as 300–360 tons per capita in industrial economies like

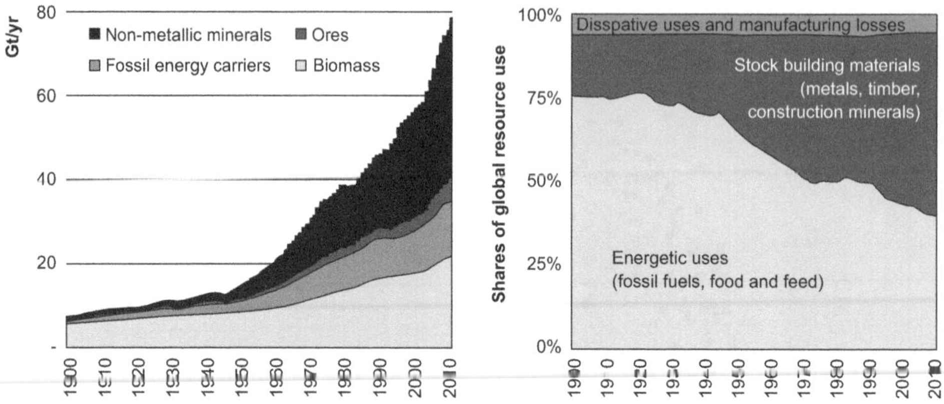

Figure 6.5 Global resource use and its main uses from 1900–2010. Gigatons (Gt) are 10^9 metric tons.

Source: adapted from Schaffartzik et al. (2014) and Krausmann et al. (2017)

the US or Japan (Fishman et al., 2014; Krausmann et al., 2017). The long-service lifetimes of these stocks translate into current waste flows coming from stocks constructed a few decades ago. In 2005, waste and emissions amounted to approximately 41 Gt globally, of which only 4 Gt or 10% were recycled (Haas et al., 2015), mainly because the majority of global resource use ends up as emissions from energy use (Figure 6.5).

As research on the mutual links between material stocks and flows progresses, dynamic models are being developed with the aim of predicting how the development of stocks over time will shape future material flows (e.g., Pauliuk and Müller, 2014). Because the dynamics differ considerably according to the material studied, such models require detailed input data on individual materials and their specific use patterns. The analysis of consumption and/or policy scenarios is an important application for such models. A prerequisite to this type of modelling is detailed data on the composition of societal stock including the age and expected lifetime of in-use stocks and the material input requirements as well as the generation of waste and emissions by age of the stock. These models are often rooted in SFA rather than in economy-wide MFA approaches because of the detailed information they require on each substance contained in stock. By analysing the stock dynamics of the global aluminum cycle, for example, Liu and colleagues (2013) have found that convergence of the aluminum stocks in developing countries to the levels in industrial countries would significantly undermine climate change mitigation efforts due to the greenhouse gas emissions associated with aluminum production. In analysing the resource nexus, this type of modelling can be used to dynamically explore the mutual links between policy, long-term stock dynamics, and environmental pressures for different forms of resource use.

Resource nexus thinking leads us to consider improved recycling, the management of existing stocks, and the reductions of future stock requirements as key ingredients in strategies towards a circular economy and alleviated pressure on the resource base and the environment. A prerequisite to such strategies is the better understanding of the long-term relationships between energy and resource use, wastes and emissions, and stock dynamics (Haas et al., 2015; Liu et al., 2013; Pauliuk and Müller, 2014; Krausmann et al., 2017). These topics are currently on the frontier of socio-metabolic research and are one of the specific contributions of this field to the wider field of sustainability science (Weisz et al., 2015) and to identifying potentials for intervention into the resource nexus.

Material extractivism: resource-development nexus

With regard to the sustainability transition, the situation of extractive economies (i.e., of those countries highly dependent on their primary sectors and the export of unprocessed goods in terms of their economic performance) poses a particular challenge, especially when these extractive economies generate low levels of per capita income, possibly coupled with high income inequality.

If we consider the resource nexus as an interlinkage between resources which a socio-economic system requires as inputs, then extractive economies provide a very interesting case: On the one hand, the extraction of primary materials is associated with pressures on other resources and the environment, often requiring significant inputs of fossil energy carriers and vast amounts of land and water. On the other hand, the focus on the extractive industries may be associated with an import dependency for more highly processed goods (paradoxically even for those goods made in other countries from the raw materials exported by the extractive economy), so that we can also identify a linkage between domestic extraction (as an input flow), exports and wastes/emissions (as output flows), and imports (as an input flow).

Whether the endowment with resources is a 'curse' or a 'blessing' is widely debated and seems to differ substantially among countries. In a first approach to the issue, we examine only two indicators for a small sample of 68 countries: the PTB of primary products in t/cap/a and gross national income (GNI) in US$/cap (Figure 6.6). What is striking even based on this small amount of data is that only four out of the 68 countries (6% of the sample) are high-income net-exporting countries: Saudi Arabia, Australia, Canada and New Zealand. The other 22 net-exporting countries (32% of the sample) fall below the 2010 threshold for high income at 12 276 US$/cap. Conversely, out of the 28 countries in this sample which are high-income countries, 24 (86%) are net importers of primary products.

Based on an understanding of the resource nexus as it particularly effects extractive economies and of the factors that allow countries to prosper as extractors of raw materials and

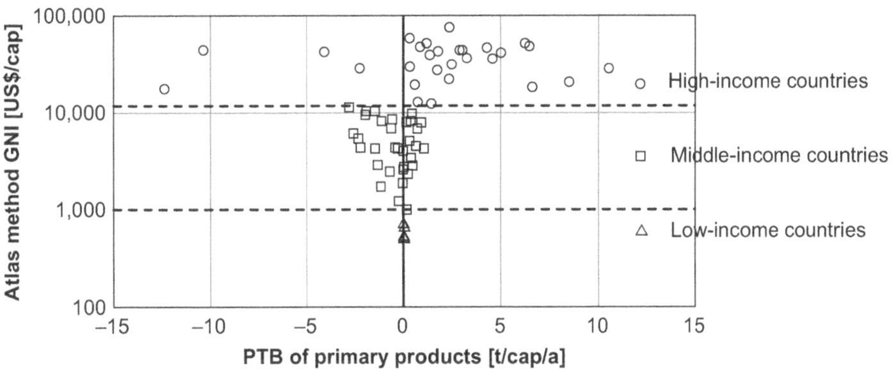

Figure 6.6 Plot of per capita gross national income (GNI) and physical trade balance (PTB = imports - exports) for primary products for a sample of 68 countries in 2010. The horizontal lines indicate the threshold for high-income (12 276 US$/cap) and low-income (1 005 US$/cap) countries in 2010; the vertical line separates the net exporters from the net importers (at 0 t/cap). Please note the logarithmic scale of the y-axis.

Source: Schaffartzik et al. (2014) and The World Bank (2013)

exporters of primary products (whether this prosperity is best investigated by considering GNI is, of course, very much up for debate), policy suggestions might be made which could have positive impacts on both the environmental as well as the social sustainability of a country's development trajectory.

We can know what we cannot feel: the case for analytically bridging disconnects

From the championing of accounting for economies in biophysical terms at a time of complete dominance of monetary indicators (Ayres and Kneese, 1969) to the currently emerging focus on the resource nexus, interdisciplinary sustainability research often follows an implicit or explicit agenda of re-embedding societies within their natural environment, i.e. to bridge what, especially in the 'modern' industrial societies, has become disconnected: social processes and their resource base. Even the current buzzword that is "planetary boundaries" (Steffen et al., 2015) serves as a reminder that we are fast approaching the limits to growth even if, especially in well-off population segments of richer countries, these limits are not yet palpable.

In our ability to develop effective responses to the environmental crisis, indicators of resource use play a decisive role because the information they contain can allow us to mentally connect actions to their impact. The further development of indicators which live up to this purpose is therefore an important research frontier. The current focus in MFA on developing indicators which reflect upstream material use; i.e., material use no matter where in the world it occurs associated with a given level of final consumption can be understood as an attempt to analytically bridge a disconnect.

The concept of the resource nexus draws our attention to the many ways in which social, political, and economic processes directly and indirectly govern the use of resources, while the access to resources and the feasibility of their use simultaneously shapes these societal processes. By more strongly considering material flows in this frame of mind, our attention returns to the origin of MFA in the concept of societal metabolism (i.e., in the inter-dependence between societal organization and its biophysical input and output requirements). This nudges research towards those topics which are currently at the frontier of research into societal metabolism and which include but are not limited to: policy messages which can be derived from MFA (e.g., Should countries aim to improve their material productivity? How much material use for a 'good life'?), the relationship between stocks and flows, and the role of resource use regimes (including the extractive economies discussed in this chapter).

From the evolution of clothing and shelter to the 'metabolic rift' between cities and rural areas and the distinction of entire economies as global providers of natural resources, much of human development has historically served to create a greater buffer between society and its natural environment. This, however, is not a functional disconnect – we still rely on ultimately limited biophysical resources, even if they are extracted (and/or their wastes disposed) at a considerable distance. Many of the most important emerging concepts in the sustainability sciences aim to re-connect analytically what seems disconnected in day-to-day life. The ability to do this is one of the great appeals of both material flow analysis and the concept of the resource nexus. If the application of this concept can generate compelling narratives of how fundamentally the workings of our human societies continue to depend on our limited resource base, then this may serve as the basis for developing policies which can address sustainability issues across space, time, and resources.

References

Data sources

Eurostat. (2016). *Database*. Luxembourg: Eurostat.
UNDP. (2014). *Human development index trends, 1980–2013*, Human Development Reports. Paris, France: United Nations Development Programme.
The World Bank. (2013). *World development indicators 2013*. Washington, DC: World Bank Publications.

Literature

Andrews-Speed, P., Bleischwitz, R., Boersma, T., Johnson, C., Kemp, G., and VanDeveer, S. D. (2012). *The global resource nexus: The struggles for land, energy, food, water, and minerals*. Washington DC, USA: Transatlantic Academy.
Ayres, R. U., and Kneese, A. V. (1969). Production, consumption, and externalities. *The American Economic Review*, 59, 282–297.
Ayres, R. U., and Simonis, U. E. (1994). *Industrial metabolism: Restructuring for sustainable development*. New York: United Nations University Press.
Bleischwitz, R. (2014). Der Ressourcen-Nexus als Frühwarnsystem für zukünftige zwischenstaatliche Konflikte [The resource nexus as a risk radar for future interstate conflicts]. *Z Außen Sicherheitspolit*, 8, 9–22. doi:10.1007/s12399-014-0435-7
Brunner, P. H., and Rechberger, H. (2003). *Practical handbook of material flow analysis, advanced methods in resource & waste management*. Boca Raton: CRC Press.
Elshkaki, A., Graedel, T. E., Ciacci, L., and Reck, B. K. (2016). Copper demand, supply, and associated energy use to 2050. *Global Environmental Change*, 39, 305–315. doi:10.1016/j.gloenvcha.2016.06.006
Eurostat. (2015). *Material flow accounts, Material flows and resource productivity*. Luxembourg: Eurostat.
Fischer-Kowalski, M. (1998). Society's metabolism: The intellectual history of material flow analysis, Part I: 1860–1970. *Journal of Industrial Ecology*, 2, 61–78. doi:10.1162/jiec.1998.2.1.61
Fischer-Kowalski, M., and Haberl, H. (2015). Social metabolism: A metric for biophysical growth and degrowth. In J. Martinez-Alier (Ed.), *Handbook of ecological economics*. Cheltenham, UK/Northampton, MA, USA: Edward Elgar Publishing, pp. 100–138.
Fischer-Kowalski, M., Krausmann, F., Giljum, S., Lutter, S., Mayer, A., Bringezu, S., Moriguchi, Y., Schütz, H., Schandl, H., and Weisz, H. (2011). Methodology and indicators of economy-wide material flow accounting. *Journal of Industrial Ecology*, 15, 855–876. doi:10.1111/j.1530-9290.2011.00366.x
Fischer-Kowalski, M., Krausmann, F., Pallua, I., and Heinz, M. (2014). The role of social revolutions in major historical energy transitions. In M. Fischer-Kowalski and D. Hausknost (Eds.), *Large scale societal transitions in the past. The role of social revolutions and the 1970s syndrome, social ecology working paper*. Vienna: Institute of Social Ecology.
Fischer-Kowalski, M., and Weisz, H. (1999). Society as hybrid between material and symbolic realms: Toward a theoretical framework of society-nature interaction. *Advances in Human Ecology*, 8, 215–252.
Fishman, T., Schandl, H., Tanikawa, H., Walker, P., and Krausmann, F. (2014). Accounting for the material stock of nations: Accounting for the material stock of nations. *Journal of Industrial Ecology*, 18, 407–420. doi:10.1111/jiec.12114
Haas, W., Krausmann, F., Wiedenhofer, D., and Heinz, M. (2015). How circular is the global economy? An assessment of material flows, waste production, and recycling in the European Union and the world in 2005: How circular is the global economy? *Journal of Industrial Ecology*, 19, 765–777. doi:10.1111/jiec.12244
Haberl, H. (2015). Competition for land: A sociometabolic perspective. *Ecological Economics*, 119, 424–431. doi:10.1016/j.ecolecon.2014.10.002
Krausmann, F., Gingrich, S., Eisenmenger, N., Erb, K-H., Haberl, H., and Fischer-Kowalski, M. (2009). Growth in global materials use, GDP and population during the 20th century. *Ecological Economics*, 68, 2696–2705. doi:10.1016/j.ecolecon.2009.05.007
Krausmann, F., Weisz, H., Eisenmenger, N., Schütz, H., Haas, W., and Schaffartzik, A. (2015). *Economy-wide material flow accounting introduction and guide (No. 151)*, Social Ecology Working Papers, Institute of Social Ecology, Vienna.
Krausmann, F., Wiedenhofer, D., Lauk, C., Haas, W., Tanikawa, H., Fishman, T., Miatto, A., Schandl, H., and Haberl, H. (2017). Global socioeconomic material stocks rise 23-fold over the 20th century and require half of annual resource use. *Proceedings of the National Academy of Sciences*.

Liu, G., Bangs, C. E., and Muller, D. B. (2013). Stock dynamics and emission pathways of the global aluminium cycle. *Nature Climate Change*, 3, 338–342. doi:10.1038/nclimate1698

Lord, B. (2014). *Art & energy: How culture changes*. Arlington, TX: AAM Press.

Manrique, P. L. P., Brun, J., González-Martínez, A. C., Walter, M., and Martínez-Alier, J. (2013). The biophysical performance of Argentina (1970–2009). *Journal of Industrial Ecology*, 17, 590–604. doi:10.1111/jiec.12027

Martinez-Alier, J. (1987). *Ecological economics: Energy, environment and society*. Oxford: Basil Blackwell.

Martinez-Alier, J., Kallis, G., Veuthey, S., Walter, M., and Temper, L. (2010). Social metabolism, ecological distribution conflicts, and valuation languages. *Ecological Economics*, Special Section: *Ecological Distribution Conflicts*, 70, 153–158. doi:10.1016/j.ecolecon.2010.09.024

OECD. (2014). *Material resources*. Paris: Organisation for Economic Co-Operation and Development.

Pauliuk, S., and Müller, D. B. (2014). The role of in-use stocks in the social metabolism and in climate change mitigation. *Global Environmental Change*, 24, 132–142. doi:10.1016/j.gloenvcha.2013.11.006

Rosado, L., Niza, S., and Ferrão, P. (2014). A material flow accounting case study of the lisbon metropolitan area using the urban metabolism analyst model. *Journal of Industrial Ecology*, 18, 84–101. doi:10.1111/jiec.12083

Schaffartzik, A., Mayer, A., Gingrich, S., Eisenmenger, N., Loy, C., and Krausmann, F. (2014). The global metabolic transition: Regional patterns and trends of global material flows, 1950–2010. *Global Environmental Change*, 26, 87–97. doi:10.1016/j.gloenvcha.2014.03.013

Sieferle, R. P. (2003). Sustainability in a world history perspective. In B. Benzig (Ed.), *Exploitation and overexploitation in societies past and present*. Münster: LIT Publishing House, pp. 123–142.

Steffen, W., Richardson, K., Rockström, J., Cornell, S. E., Fetzer, I., Bennett, E. M., Biggs, R., Carpenter, S. R., Vries, W. de, Wit, C. A. de, Folke, C., Gerten, D., Heinke, J., Mace, G. M., Persson, L. M., Ramanathan, V., Reyers, B., and Sörlin, S. (2015). Planetary boundaries: Guiding human development on a changing planet. *Science*, 1259855. doi:10.1126/science.1259855

Steinberger, J. K., Krausmann, F., and Eisenmenger, N. (2010). Global patterns of materials use: A socioeconomic and geophysical analysis. *Ecological Economics*, 69, 1148–1158. doi:10.1016/j.ecolecon.2009.12.009

Steinberger, J. K., Roberts, J. T., Peters, G. P., and Baiocchi, G. (2012). Pathways of human development and carbon emissions embodied in trade. *Nature Climate Change*, 2, 81–85.

Steinberger, J. K., Krausmann, F., Getzner, M., Schandl, H., and West, J. (2013). Development and dematerialization: An international study. *PLoS One*, 8, e70385. doi:10.1371/journal.pone.0070385

UNEP. (2011). *Decoupling natural resource use and environmental impacts from economic growth*, A Report of the Working Group on Decoupling to the International Resource Panel (M. Fischer-Kowalski, Swilling, M., von Weizsäcker, E. U., Ren, Y., Moriguchi, Y., Crane, W., Krausmann, F., Eisenmenger, N., Giljum, S., Hennicke, P., Romero Lankao, P., Siriban Manalang, A., and Sewerin, S.). Nairobi: United Nations Environment Programme.

UNEP. (2016). *Global material flows and resource productivity* (H. Schandl, M. Fischer-Kowalski, J. West, S. Giljum, M. Dittrich, N. Eisenmenger, A. Geschke, M. Lieber, H. Wieland, A. Schaffartzik, F. Krausmann, S. Gierlinger, K. Hosking, M. Lenzen, H. Tanikawa, A. Miatto, and T. Fishman). Nairobi: United Nations Environment Programme, International Resource Panel.

Voet, E. van der. (1996). *Substances from cradle to grave, development of a methodology for the analysis of substance flows through the economy and the environment of a region*. PhD thesis Leiden University, defended May 28, 1996.

Weisz, H., Suh, S., and Graedel, T. E. (2015). Industrial ecology: The role of manufactured capital in sustainability. *Proceedings of the National Academy of Sciences*, 112, 6260–6264. doi:10.1073/pnas.1506532112

Wiedmann, T. O., Schandl, H., Lenzen, M., Moran, D., Suh, S., West, J., and Kanemoto, K. (2015). The material footprint of nations. *PNAS*, 112, 6271–6276. doi:10.1073/pnas.1220362110

7
RESOURCE FOOTPRINTS

Stefan Giljum, Martin Bruckner and Stephan Lutter

Introduction

In the era of globalisation, supply chains are increasingly organised on the international level, thus disconnecting the location of production from final consumption (Liu et al., 2013). Consumption therefore has developed into a major, geographically distant driver of various local environmental and social impacts in – often developing – countries, which extract and process raw materials or manufacture products that are then consumed in rich countries, e.g. in Europe or North America.

While global resource consumption is rapidly increasing (Giljum et al., 2014), resources embodied in international trade augmented even faster over the past 15 years. Today, already more than a third of globally extracted materials serve as direct and indirect inputs for the production of internationally traded products (Wiedmann et al., 2015). This implies that traditional production-oriented, national perspectives to investigate the resource nexus might no longer be sufficient, as important drivers for the local situation are being missed. Furthermore, territorial indicators of a country's environmental performance are not robust against the displacement of environmental burden through outsourcing of resource- and pollution-intensive production stages via international trade. In contrast, consumption-based – or "footprint" – indicators consider the environmental impacts embodied in internationally traded products (Wiedmann, 2016) and are thus independent from the location of production. The design of policy responses to tackle these "teleconnections" between production and consumption therefore also require a focus on sustainable product supply chains and the reduction of global environmental and social impacts of consumption.

The need to consider a "footprint" perspective is also addressed in a number of policy initiatives on the European and international level. For example, the EU Resource Efficiency Roadmap demands the further development and application of indicators that take into account the global aspects of EU consumption in areas such as water, land and material use as well as regarding climate change and other environmental impacts (European Commission, 2011). The OECD is pursuing an initiative to develop demand-based indicators for carbon and material flows as part of its work on green growth indicators (OECD, 2014). Also the Group of 7 (G7) in their final declaration of the 2015 summit declared that, given their prominent role in shaping globalisation processes, the G7 countries need to promote environmental protection and decent working conditions in global supply chains, in order to reduce the negative social and environmental effects of products (G7, 2015).

This chapter aims at introducing the footprint concept as a tool for assessing the global environmental consequences related to consumption and at discussing the role of footprints in analyses of the resource nexus. First, we explain the basic concept of footprints and describes which types of footprint indicators can be calculated. Afterwards, we elaborate on the various methodological approaches that have been developed over the past 15 years to calculate footprint-type indicators. Finally, we illustrate the role of resource footprints in assessments of the resource nexus referring to examples from the literature. In the concluding section, we discuss the future prospects of footprint-type indicators in the context of the global resource nexus.

Footprints: basic concepts

Environmental assessments generally apply a territorial – or production-based – perspective to analyse environmental pressures and impacts that occur within the borders of a country or region. Monitoring of current environmental policies mostly relies on indicators applying this perspective. Examples include the accounting of greenhouse gas emissions in the context of the UNFCCC climate treaties, assessments of water abstraction in comparison to the available water or changes of land cover and related impacts on ecosystems and biodiversity (for an assessment in the European environmental policy context, see EEA, 2015).

However, in a globalised economy, supply chains are becoming increasingly international, often involving a large number of countries along a product's life cycle from the extraction of raw materials, via processing and manufacturing to the delivery of a product to the final consumer. Production-oriented indicators cannot account for the actual environmental consequences induced by the consumption of certain products, as impacts can be located in other world regions. Driven by the need to better understand these "teleconnections" between distant places of production and consumption, a new type of indicator has become increasingly popular, which focuses on the global, life cycle-wide consequences related to consumption, i.e. footprint indicators.

The term footprint was first coined by Mathis Wackernagel and William Rees when introducing their concept of the Ecological Footprint in the early 1990s (Wackernagel and Rees, 1996). Their intention was to create a measure which illustrates the amount of nature required to sustain the consumption of goods and services by an individual person, a city or a country. The traditional concept of carrying capacity as applied in ecology addresses the question of the maximum size of an animal population that can be sustained on a given territory. With the Ecological Footprint concept, Wackernagel and Rees turned this question around by inquiring how much global bio-productive land a given population requires to maintain certain consumption levels (Wackernagel et al., 1999).

Following the Ecological Footprint concept, footprint-type indicators have been developed for a large number of other environmental issues over the past 15 years (for overviews, see Čuček et al., 2012; Giljum et al., 2013; Hoekstra and Wiedmann, 2014; Wiedmann, 2016). The following Table 7.1 provides an overview of existing environmental footprint indicators. The footprint approaches can be clustered into three main categories: (1) input-oriented footprints, which focus on natural resources serving as inputs to economic processes; (2) output-oriented footprints, which address issues related to pollution; and (3) impact-oriented footprints, which illustrate environmental impacts related to certain input or output flows.

Table 7.1 illustrates that a large number of footprint indicators have already been developed and tested in all three footprint categories. Input-oriented footprints have been widely applied

Table 7.1 Overview of existing footprint indicators

Category	*Name*	*Specification of covered environmental issues*	*Unit of measurement*
INPUT-ORIENTED FOOTPRINTS	Material footprint	Extraction of abiotic raw materials (fossil fuels, metals and non-metallic minerals) and biotic raw materials from agriculture, forestry, fishery	Tonnes
	Water footprint	Water abstraction or water consumption of green/blue/grey water	Cubic meters
	Land footprint	Actual land appropriation by land type, e.g. agricultural, forestry or built-up land	Hectares
OUTPUT-ORIENTED FOOTPRINTS	Carbon footprint	Emissions of greenhouse gases, particularly CO_2, CH_4, N_20	Tonnes CO_2 equivalents
	Air pollution footprint	Emissions of air pollutants, such as CO, NH_3, NO_x or SO_x	Tonnes
	Water pollution footprint	Emissions of water pollutants, such as nitrogen	Tonnes
	Waste footprint	Production of waste along supply chains, by waste type	Tonnes
IMPACT-ORIENTED FOOTPRINTS	Ecological Footprint	Biologically productive land and water areas (actually used areas plus theoretically required carbon sequestration areas)	Global hectares (ha with average global productivity)
	Biodiversity footprint	Species threatened or lost through socio-economic activities	Number of threatened/lost species
	Deforestation footprint	Deforestation related to the production of agricultural products	Hectares
	Embodied HANPP	Human appropriation of net primary production embodied in traded products	Tonnes of carbon
	Water scarcity footprint	Water withdrawals as percentage of local renewable freshwater resources	Scarcity-weighted cubic meters
	Product/organisation environmental footprint	14 impact categories derived from life cycle assessments, including global warming, ozone depletion, acidification and nitrification	Depending on impact category

for all major types of natural resources, i.e. materials, water and land. These indicators have particular policy relevance, as both the EU and the OECD intend to include resource footprint indicators – alongside carbon footprints – in their future policy monitoring. Regarding outputs of waste and emissions, the carbon footprint is the best known and most widely applied indicator on both the national and the product level. However, also air and water pollution as well as waste generation have been examined from a footprint perspective. In the category of impact-oriented indicators, the Ecological Footprint with its focus on biological productivity has the longest tradition (Table 7.1). However, in recent years, also footprint indicators on other environmental impacts, such as biodiversity loss, deforestation or water scarcity have been developed.

With the intention to facilitate widespread application of product life cycle assessments (LCA), the European Commission has developed the product environmental footprint, a simplified and standardised product LCA illustrating the environmental performance of products and services as well as of organisations (JRC, 2013).

Methodologies to calculate footprints

While the life cycle assessment (LCA) technique was continuously advanced over the past four decades in order to increase comprehensiveness in the assessment of the various environmental impacts, more recently emerging footprinting methods aim at a reduction of complexity. In most cases (see Table 7.1), footprint methodologies narrow their focus to a single type of impact in order to foster wide applicability and facilitate communication with the broader public. The rapid expansion of footprint indicators was not only driven by the research community, but has been particularly promoted by private initiatives from NGOs and companies.

The methodology for the calculation of the Ecological Footprint, the first widely known footprint indicator (Table 7.1), was developed independently of previous achievements in LCA, using a form of economy-wide inventory analysis by accounting for a country's product in- and outflows and converting them into corresponding land areas. More recently, the carbon footprint dominated the scientific and public footprint discourse, which is more in line with the tradition of LCA and is sometimes even considered equal to the global warming potential (GWP) indicator in LCA (Weidema et al., 2008). Therefore, in the beginning of carbon footprinting, well-established process data and bottom-up methods from the field of LCA were used, particularly of products and corporations. However, these approaches were quickly followed by studies applying input–output analysis (IOA), in particular multi-regional input–output (MRIO) analysis, with the purpose of expanding the scope of footprint indicators to the economy-wide, national and international levels (Lifset, 2014).

While narrowing the scope of considered impacts, the development of economy-wide footprinting methods also termed consumption-based accounting, opened up previously inconceivable areas of applications, most remarkably in the international political debate (Peters and Hertwich, 2008). Although starting from a single-impact perspective, developments in the area of footprinting increasingly intend to provide a more comprehensive picture of the worldwide connections between production and consumption and argue for the integration of various footprints into a family of footprint indicators (Ridoutt and Pfister, 2013; Tukker et al., 2014). Ongoing research also investigates what would be a minimum set of footprint indicators that reproduces the overall result of a comprehensive LCA while still being easy to calculate and to communicate.

This is also the line of thought behind the European Commission's efforts in developing a harmonised methodology for the calculation and communication of the environmental performance of products and organisations, the Product/Organisation Environmental Footprint (PEF/OEF), which is closely related to established LCA methodologies but preselects a set of impact categories depending on the product group (JRC, 2013).

In the following sections, this chapter primarily focuses on economy-wide footprinting methods operating on the level of geographical or political entities, particularly countries. For economy-wide applications, it is crucial to account for trade flows and, to the extent possible, to consider regional differences in technologies and environmental intensities. Therefore, data on international trade in commodities and services are combined with information on production processes and related environmental effects (i.e. resource inputs, waste and emission outputs, or environmental impacts). Three variants of trade and footprint accounting models can be distinguished: bottom-up approaches, top-down approaches and mixed approaches. The presentation of the main features of these three variants is followed by a discussion on the use of monetary versus physical units in footprinting methods.

Bottom-up approaches

The National Footprint Accounts used to calculate the Ecological Footprint were the first established bottom-up footprint accounting system. They comprise detailed data on bilateral trade and domestic production expressed in quantities (e.g. tonnes) and derive the 'apparent consumption' of a country by calculating production plus imports minus exports. This form of economy-wide inventory analysis was later also applied for other footprints. The quantities of each product consumed in a country are eventually characterised by multiplying them with coefficients reflecting the related upstream environmental effects. For example in the case of the material footprints, a coefficient provides information on the quantities and composition of materials required along a product supply chain. These coefficients are primarily obtained from LCAs. While LCA studies are technically detailed, they are often limited in their spatial detail and representativeness on the aggregate national level. Therefore, if applied on the global level, the sum of all footprints from bottom-up calculations will inevitably differ from the sum of all resource inputs and emission outputs.

Due to the important role of regional climatic and soil conditions, among other factors, regional differences are particularly relevant in the case of land and water footprints. Therefore, current bottom-up methods for these indicators consider country-specific characterization factors. The availability of detailed global agricultural statistics as well as spatially explicit climatic and agro-ecological information allows for the estimation of water and land footprints on a high level of spatial detail (Hoekstra and Mekonnen, 2012; Pfister and Bayer, 2014).

Top-down approaches

In contrast to bottom-up methods, top-down approaches use input–output analysis (IOA), which builds on an accounting framework focusing on the economic structure of a country in the form of matrices that depict inter-industry flows, i.e. input–output tables (for more details see Chapter 7). Each vector of an IO table can be interpreted as a set of production recipes or, in analogy with LCA, an inventory of production inputs, while the environmental extensions of an IO table can be considered an inventory of environmental inputs and outputs such as raw materials and carbon emissions. A sector's structure of intermediary inputs as well as environmental inputs and outputs per production unit may differ widely between countries, among other things, reflecting technological differences.

While single-region IO models assume domestic technology for imported products, in the case of MRIO models, country IO tables are linked together via bilateral trade data (Wiedmann et al., 2011). MRIO analysis allows tracking product value chains and related environmental effects along all life cycle stages of all products and services from resource extraction to final consumption, considering specific environmental intensities across countries (Tukker et al., 2013). Environmental MRIO analysis, thus, simply redistributes global environmental inputs and outputs from the point of occurrence (an economic sector) to the point of consumption (e.g. final demand by households and government). The sum of all footprints hence is always equal to the sum of all resource inputs and emission outputs.

There are several variations of the MRIO method, including studies considering only unidirectional trade between a central country and its major trade partners (for example, Nijdam et al.,

2005), and, more recently, a method called SNAC-MRIO (single-country national accounts consistent MRIO), which uses an available MRIO table and replaces the IO table and trade data for the country of interest with data from national statistics (Edens et al., 2015). This approach allows statisticians to generate footprint results fully relying on available national expert knowledge and data (particularly information on imports and exports properly considering transit trade and re-exports) and thus improving consistency and robustness. Finally, also the emissions-embodied-in-bilateral-trade (EEBT) method (Peters and Hertwich, 2008) and its analogous applications to water and land (Hubacek and Feng, 2016) need to be mentioned. EEBT uses a MRIO table covering domestic supply chains only and calculates emissions embodied in bilateral trade directly by multiplying the Leontief inverse with a bilateral trade matrix.

A major disadvantage of IOA is the limited commodity and regional detail determined by the sector and region definitions of an IO model. The assumption of homogenous environmental characteristics of all products within a product group leads to inaccurate results, for example, when products with widely diverging mass-value-ratios are aggregated into one product group (de Koning et al., 2015).

Mixed approaches

Mixed or hybrid approaches combine elements from IOA with physical trade accounts or process-based coefficients and aim at exploiting the advantages of all available methods and data. Hybrid methods apply a differentiated approach for the calculation of footprint indicators for different products, depending on the processing stage and data quality and availability. Typically, they use physical trade accounts and environmental characterisation factors for raw materials and products with a low level of processing, thereby considering detailed environmental properties for those products where intragroup variation is particularly high. Processed commodities and finished goods with more complex production chains are treated with IOA, which allows considering all indirect effects and thus all upstream resource requirements (Ewing et al., 2012).

Calculations based on physical and/or monetary units

The operating unit, particularly whether trade flows are accounted for in monetary or physical values, are a further major distinguishing characteristic of footprinting methods. While all available bottom-up approaches are based on physical trade flows, top-down methods mostly apply monetary data. If all flows of the product groups specified in an IO model within and between countries were fully homogenous in terms of prices, the use of monetary or physical trade flows for tracking embodied impacts could be considered equal from a conceptual point of view, disregarding differences in data availability. However, current data availability is still limited. Quality and resulting price differences in product flows result in differences of value-based and mass-based allocation methods. For example, Bruckner et al. (2015) showed that the rice footprint of wealthier countries increases when flows are allocated according to monetary values as opposed to a mass-based allocation, while for low-income countries footprints are lower when using monetary trade flows. Quality and related price differences between rice imports to high-income and low-income countries can be assumed to cause these differences.

There are different ways of integrating physical supply chain information into IOA. One option is to set up physical satellite accounts to model commodity flows from raw material extraction to the first use stage and track the remaining parts of the supply chains within monetary IO models (Ewing et al., 2012; Weinzettel et al., 2013). An alternative option is to integrate detailed use structures of selected products in mass units into monetary IO tables, thereby creating mixed-unit IO tables (Schoer et al., 2012).

Besides trade accounting, the operating unit also matters for the allocation of environmental effects in the case of joint and by-products. While LCAs draw on a variety of different allocation methods, for example, in relation to the products' economic value, mass or energy content, allocation in IOA is determined by the unit of the IO table. Thus, currently applied MRIO models use economic value-based allocation. It is often argued that monetary values would represent the economic incentives, and thus the drivers behind production activities and ultimately environmental impacts. Therefore, value-based allocation is suggested as the basic rule also for LCA.

From a conceptual viewpoint, all economy-wide footprinting methodologies are equal as they aim at a full life cycle-wide coverage of upstream effects, account for them in the form of tree or matrix structures, which eventually are equivalent and interchangeable, and consider trade relations with other countries in addition to domestic production. In an ideal world of perfect data availability, footprinting methods based on bottom-up and top-down methods would therefore generate equal results – under the condition of identical allocation rules. Considering the steadily growing amounts of data collected and processed, a convergence of methods may be expected in the future. However, current data situation is still limited; therefore the selection of suitable methodologies depends on the level of application.

On the micro level of products and services and the meso level of economic sectors, footprint studies usually apply a simplified LCA technique, whether purely based on unit process data (process-LCA) or in combination with economic input–output data (IO-LCA). More details on the LCA methodology can be found in Chapter 4.

Footprint indicators and the resource nexus

Possibilities and limits of footprint indicators for nexus assessments

Footprint indicators illustrate the global, life cycle-wide impacts of consumption, thereby elucidating to what extent environmental pressures and impacts occur along the supply chains of products and services, often far away from the location where they are consumed. Hence, in the context of the resource nexus, the main contribution of footprint indicators is that, in contrast to territorial analyses of resource use, they add information on the upstream requirements of resources. Footprint indicators thereby provide a more comprehensive picture of how consumption of products in one country or region interacts with the use and management of resources in producing countries. This additional dimension allows analysing resource nexus issues from a global perspective, identifying to what extent the extraction and use of different types of resources in a specific country is driven by local requirements and decisions versus by developments on international markets and trade relationships.

A prominent example in the literature is the changing global demand for water and land, i.e. water and land footprints, related to increased consumption of biofuels. Figure 7.1 provides an illustrative example, how a substantial increase in the production of biofuels could impact on the domestic and foreign parts of various footprint indicators.

In this notional example, a country decides to increase the amount of biofuels in its national energy mix. A footprint analysis of the nexus effects will reveal both synergies and trade-offs between emissions and resources, and regional relocation effects. Such a shift, for instance, will decrease the country's carbon footprint from fossil-fuel combustion, but increase land-use

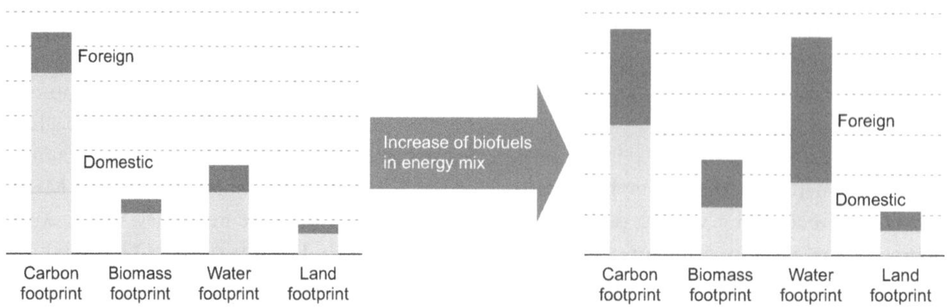

Figure 7.1 Illustrative example of a comparative footprint assessment related to changes in the energy mix

change-related carbon emissions from biofuel production (for example, Plevin et al., 2010). Moreover, the increased use of biomass for energetic purposes, illustrated by a growing biomass footprint, will have impacts on other resource footprints, in particular on water and land (see, for example, Dale et al., 2011; Elena and Esther, 2010). In case biofuels or energy crops are imported, increases in footprints will occur in other countries, resulting in significantly larger relocation effects of environmental burden.

However, footprint indicators also have clear limitations with regard to nexus-related assessments. These hold true both for LCA, as the best available tool to measure footprints at the micro level of products and organisations, as well as for MRIO-based methods for analyses on the meso- and macro-levels, i.e. of industries and countries. In input–output analysis, just like in LCA, input–output relations are static for any level of production and consumption. Footprint indicators are therefore incapable of capturing the dynamic nature of the interdependencies of various resources over time.

Hence, the main limitation of footprint indicators for investigating the resource nexus is that the underlying methodologies do not explain causalities and thus have mainly a descriptive value. Linkages can only be observed by contrasting the development of two or more footprint indicators. This will reveal parallel or reverse developments in the past, but does not help understanding the complex dynamics of the resource nexus. Fully addressing these interlinkages and their underlying causal relationships requires additional models, which describe the complexity of the involved systems and thereby allow the analysis of how changes in stocks or flows for different resources impact on each other.

Examples for footprint-type assessments in the context of the resource nexus

Footprint analyses can play an important role in the evaluation of current and future options to decrease pressures and impacts brought about by resource consumption. If used in parallel, they can illustrate that supply chain-wide improvements in one category can actually lead to negative trends in other categories. Examples of existing studies encompass, for instance, an analysis of biofuel production in general and its effects on carbon, waste, land and water footprints (Hammond and Li, 2015). In this study, the authors illustrate that by 2050, bio-productive land will likely constitute the largest component in the overall environmental footprint of biofuels (a 48% share in 2050), followed by the carbon footprint (23%), energy footprint (16%) and the water footprint (9%). Another study on the impacts of food production in the Mediterranean showed that, while contribution of vegetal- and animal-based food-product groups to food supply is variable among the Mediterranean countries, it has important implications in terms of water and other resource requirements. Hence, a change in diets will also result in changing footprints (Lacirignola et al., 2014).

Especially in the context of development assistance, first case studies for specific countries or regions which analyse resource use in the context of specific (inter-sectoral) development programmes from a nexus perspective have recently been carried out. For instance, for the case of Ethiopia, a study found that government policies, strategies and flagship programmes currently promote integrated approaches for planning and implementation of development efforts. However, "nexus" as an approach taking the different types of resources into planning account does not necessarily mean that all water–food–energy-related problems have to be solved by one programme or project (Skat, 2014).

Also on the macro level, comparative analyses of various types of resource footprints have been carried out for a large number of countries. These studies used MRIO frameworks, hence top-down approaches, as these approaches offer one consistent modelling framework enabling footprint analyses for different types of natural resources and other environmental categories (see Table 7.1). For example, in a study for the EU-27 and its 15 main trading partner countries and regions for the year 2007, the "EXIOBASE" MRIO model was used by Tukker and colleagues (2014) to carry out footprint analyses for materials, land, water and carbon emissions. They showed that the EU as a whole is a net importer from other world regions with regard to all four different types of resources. When looking at the specific countries in detail, a number of countries were identified following this EU-27 trend of being import dependent in all categories. Apart from a large number of European countries, South Korea and Mexico also showed this specific footprint pattern. Other countries were identified as being net importers in some categories and net exporters in others. For example, the US are a net exporter of embodied water, but net importer of the other resource categories, while in the case of South Africa the opposite pattern is observed. These numbers reflect both the national resource endowments as well as the level of economic development and specialisation. The US have developed a highly automatised agricultural system with considerable nutrient inputs (fertilizers), very high yields and high production, which is exported to other countries. On the other hand, with regard to other types of raw materials such as metals, the lack of national sources to sustain industrial production is being compensated with imports from other countries. In comparison, South Africa has very high resource endowments in terms of metals and minerals but imports high quantities of land and water via biomass-based products.

When analysing resource footprints and the nexus between them, it is of relevance not only to identify net importer or net exporters but to investigate also absolute quantities. While different MRIO models still diverge in the resulting absolute footprint numbers, they coincide with regard to the order of magnitude indicating which countries consume above or below the global average (see Table 7.2).

Tukker et al. (2014) found that, from a consumption perspective, especially countries such as Australia, Luxembourg and the US have per capita footprints far above the global average with regard to all the four different types of resources. But there are also examples such as South Africa which are below the global resource footprint average in some categories (carbon and water), and close to it (materials) or above it (land) in others. On the other end of the spectrum, countries such as Indonesia are always located below the global average. The same holds true for India, with the exception of water where India's water footprint is slightly above the global average due to its considerable agricultural production.

Table 7.2 Overview of selected countries with high versus low resource footprints per capita, year 2007

Country	***Carbon footprint t/cap***	***Country***	***Water footprint m³/cap***	***Country***	***Land footprint km²/cap***	***Country***	***Material footprint t/cap***
Australia	27.0	Australia	701	Australia	0.208	Ireland	59.0
Luxembourg	26.8	Luxembourg	697	Canada	0.075	Australia	48.1
USA	24.8	USA	664	Russia	0.066	Luxembourg	44.8
Global average	*5.7*	*Global average*	*250*	*Global average*	*0.013*	*Global average*	*9.9*
Brazil	3.0	Poland	134	Cyprus	0.007	South Africa	7.4
Indonesia	1.9	Cyprus	113	Indonesia	0.005	Indonesia	4.1
India	1.4	Indonesia	88	India	0.002	India	3.7

Source: Tukker et al. (2014)

Conclusions and outlook

This chapter provided an introduction to footprint-type indicators as one of the available tools to perform nexus assessments across various categories of resources. In a global economy with a growing degree of division of labour and production processes between different countries and world regions, consumption increasingly constitutes a geographically distant driver of local and regional environmental and social problems. With their global, life cycle-wide perspective, footprint-type indicators are therefore an important tool to analyse, which international supply chains and final products contribute to increasing resource use.

Footprint indicators received wide attention in the academic and policy fields in recent years, initially triggered by the developments of the Ecological Footprint and later the carbon footprint. Some policy initiatives, such as the European Resource Efficiency Roadmap, already acknowledge the need to complement the traditional production-based perspective with consumption-based indicators. To the present, around a dozen of different footprint indicators have been calculated for all countries worldwide using a range of methodological approaches, from bottom-up approaches based on physical trade accounts to top-down methodologies based on economic input–output data.

Given the increasing demand for a growing number of natural resources, such as energy, water or biomass, from various parts of society, it can be expected that the importance of footprint analyses will further increase in the future. Current research efforts in the area of footprint methodologies will facilitate this development. On the one hand, data availability is increasing, as MRIO databases are continuously advanced to increase their product and country detail. Also, the improved availability of data for single products will allow linking footprint assessments closer with policy initiatives, which are often targeted towards particular sectors and supply chains. On the other hand, an essential focus currently lies in prolonging time coverage towards "now-casting" and "forecasting" future trends. This will allow not only performing ex-post footprint analyses but also evaluating future options for sustainable resource management from an international perspective.

However, such an evaluation requires integrating footprint-type indicators with other assessment models, which are capable of reflecting the dynamic relations between and within socio-economic and biophysical systems. This integration would allow, for example, evaluating low-carbon solutions such as various forms of renewable energy currently discussed in Europe and other developed countries with regard to their interlinked global requirements for materials, land and water. Such an integrated assessment is required, in order to identify those technological options, which have the best performance across a range of resource categories. Furthermore, scenarios on changes in consumption patterns based on such integrated models can illustrate the global impacts on the resource nexus of different lifestyles, such as those related to food or mobility.

References

Bruckner, M., Fischer, G., Tramberend, S., and Giljum, S. (2015). Measuring telecouplings in the global land system: A review and comparative evaluation of land footprint accounting methods. *Ecological Economics*, 114, 11–21.

Čuček, L., Klemeš, J. J., and Kravanja, Z. (2012). A review of footprint analysis tools for monitoring impacts on sustainability. *Journal of Cleaner Production*, 34, 9–20.

Dale, V. H., Efroymson, R. A., and Kline, K. L. (2011). The land use–climate change–energy nexus. *Landscape Ecology*, 26, 755–773.

de Koning, A., Bruckner, M., Lutter, S., Wood, R., Stadler, K., and Tukker, A. (2015). Effect of aggregation and disaggregation on embodied material use of products in input–output analysis. *Ecological Economics*, 116, 289–299.

Edens, B., Hoekstra, R., Zult, D., Lemmers, O., Wilting, H., and Wu, R. (2015). A method to create carbon footprint estimates consistent with national accounts. *Economic Systems Research*, 27, 440–457.

EEA. (2015). *The European environment: State and Outlook 2015*. Synthesis report, European Environment Agency, Copenhagen.

Elena, G.-d.-C., and Esther, V. (2010). From water to energy: The virtual water content and water footprint of biofuel consumption in Spain. *Energy Policy*, 38, 1345–1352.

European Commission. (2011). *Roadmap to a resource efficient Europe*. Brussels: European Commission.

Ewing, B. R., Hawkins, T. R., Wiedmann, T. O., Galli, A., Ertug Ercin, A., Weinzettel, J., and Steen-Olsen, K. (2012). Integrating ecological and water footprint accounting in a multi-regional input – output framework. *Ecological Indicators*, 23, 1–8.

G7. (2015). *Think ahead. Act together. Leaders' declaration*. G7 Summit, June 7–8, 2015, Elmau, Germany.

Giljum, S., Dittrich, M., Lieber, M., and Lutter, S. (2014). Global patterns of material flows and their socio-economic and environmental implications: A MFA study on all countries world-wide from 1980 to 2009. *Resources*, 3, 319–339.

Giljum, S., Lutter, S., Bruckner, M., and Aparcana, S. (2013). *State of play of national consumption-based indicators: A review and evaluation of available methods and data to calculate footprint-type (consumption-based) indicators for materials, water, land and carbon*. Report for DG Environment of the European Commission. Sustainable Europe Research Institute (SERI), Vienna.

Hammond, G. P., and Li, B. (2015). Environmental and resource burdens associated with world biofuel production out to 2050: Footprint components from carbon emissions and land use to waste arisings and water consumption. *GCB Bioenergy*, 18, 894–908.

Hoekstra, A. Y., and Mekonnen, M. M. (2012). The water footprint of humanity. *Proceedings of the National Academy of Sciences*, 109, 3232–3237.

Hoekstra, A. Y., and Wiedmann, T. O. (2014). Humanity's unsustainable environmental footprint. *Science*, 344, 1114–1117.

Hubacek, K., and Feng, K. (2016). Comparing apples and oranges: Some confusion about using and interpreting physical trade matrices versus multi-regional input–output analysis. *Land Use Policy*, 50, 194–201.

JRC. (2013). Product Environmental Footprint (PEF) guide. Ispra, Italy: JRC IES.

Lacirignola, C., Capone, R., Debs, P., El Bilali, H., and Bottalico, F. (2014). Natural resources–food nexus: Food-related environmental footprints in the Mediterranean countries. *Frontiers in Nutrition*, 1.

Lifset, R. (2014). Frontiers in footprinting. *Journal of Industrial Ecology*, 18, 1–3.

Liu, J., Hull, V., Batistella, M., DeFries, R., Dietz, T., Fu, F., Hertel, T. W., Izaurralde, R. C., Lambin, E. F., and Li, S. (2013). Framing sustainability in a telecoupled world. *Ecology and Society*, 18.

Nijdam, D. S., Wilting, H. C., Goedkoop, M. J., and Madsen, J. (2005). Environmental load from Dutch private consumption: How much damage takes place abroad? *Journal of Industrial Ecology*, 9, 147–168.

OECD. (2014). *Green growth indicators 2014*. Paris: Organization for Economic Cooperation and Development.

Peters, G., and Hertwich, E. (2008). CO2 embodied in international trade with implications for global climate policy. *Environmental Science and Technology*, 42, 1401–1407.

Pfister, S., and Bayer, P. (2014). Monthly water stress: Spatially and temporally explicit consumptive water footprint of global crop production. *Journal of Cleaner Production*, 73, 52–62.

Plevin, R. J., Jones, A. D., Torn, M. S., and Gibbs, H. K. (2010). Greenhouse gas emissions from biofuels' indirect land use change are uncertain but may be much greater than previously estimated. *Environmental Science & Technology*, 44, 8015–8021.

Ridoutt, B. G., and Pfister, S. (2013). Towards an integrated family of footprint indicators. *Journal of Industrial Ecology*, 17, 337–339.

Schoer, K., Weinzettel, J., Kovanda, J., Giegrich, J., and Lauwigi, C. (2012). Raw material consumption of the European Union – concept, calculation method, and results. *Environmental Science & Technology*, 46, 8903–8909.

Skat. (2014). *Nexus case study ethiopia*. Sankt Gallen, Switzerland: Swiss Resource Centre and Consultancies for Development.

Tukker, A., Bulavskaya, T., Giljum, S., de Koning, A., Lutter, S., Simas, M., Stadler, K., and Wood, R. (2014). *The global resource footprint of nations: Carbon, water, land and materials embodied in trade and final consumption*. The Netherlands Organisation for Applied Scientific Research – NL- 2628 XE Delft; Leiden University, NL-2300 RA Leiden; Vienna University of Economics and Business, A-1020 Vienna, Norwegian University of Science and Technology, NO-7491 Trondheim.

Tukker, A., de Koning, A., Wood, R., Hawkins, T., Lutter, S., Acosta, J., Rueda Cantuche, J. M., Bouwmeester, M., Oosterhaven, J., and Drosdowski, T. (2013). EXIOPOL – development and illustratvie analyses of detailed global MR EE SUT/IOT. *Economic Systems Research*, 25, 50–70.

Wackernagel, M., Onisto, L., Bello, P., Linares, A. C., Falfan, I. S. L., Garcia, J. M., Guerrero, A. I. S., and Guerrero, M. G. S. (1999). National natural capital accounting with the ecological footprint concept. *Ecological Economics*, 29, 375–390.

Wackernagel, M., and Rees, W. E. (1996). *Our ecological footprint – reducing human impact on the earth*. Gabriola Island, BC, Canada: New Society Publishers.

Weidema, B. P., Thrane, M., Christensen, P., Schmidt, J., and Løkke, S. (2008). Carbon footprint. *Journal of Industrial Ecology*, 12, 3–6.

Weinzettel, J., Hertwich, E. G., Peters, G. P., Steen-Olsen, K., and Galli, A. (2013). Affluence drives the global displacement of land use. *Global Environmental Change*, 23, 433–438.

Wiedmann, T. (2016). *Impacts embodied in global trade flows, taking stock of industrial ecology*. New York: Springer, pp. 159–180.

Wiedmann, T., Schandl, H., Lenzen, M., Moran, D., Suh, S., West, J., and Kanemoto, K. (2015). The material footprint of nations. *Proceedings of the National Academy of Sciences (PNAS)*, 112, 6271–6276.

Wiedmann, T., Wilting, H. C., Lenzen, M., Lutter, S., and Palm, V. (2011). Quo Vadis MRIO? Methodological, data and institutional requirements for multi-region input–output analysis. *Ecological Economics*, 70, 1937–1945.

8

INPUT–OUTPUT ANALYSIS AND RESOURCE NEXUS ASSESSMENT

Arnold Tukker and David Font Vivanco

Introduction

The premise of this book is that problems related to climate, energy, water, land use, biodiversity and materials use cannot be seen any more in isolation. Water shortage will have impact on biodiversity and productivity of land, and so on.

This chapter looks primarily at the nexus problem from the perspective of the economic system. It is after all the economic system that creates pressures on the energy, water, land, geological and climate systems simultaneously. It is however likely that specific production and consumption chains will differ in their relative contribution to pressures of the different nexus elements. One could argue that economic processes that create pressures with regard to energy, water, land, materials and climate simultaneously are more problematic as economic processes that have just limited contribution to pressures on nexus elements, or to just one or two nexus elements. In fact, according to Liu et al. (2015), about 80% of all nexus studies have analysed only two-node nexuses, being water–energy, and water–food and energy–food being popular alternatives. This evidence highlights the need for a comprehensive scope in which multiple nexus nodes are studied in combination. The risks of a limited scope gained importance during the biofuels debate, as the eventual consideration of the GHG emissions from land-use change proved to be key in their overall contribution to climate change mitigation strategies (Searchinger et al., 2008).

The chapter of Giljum et al. in this volume already indicated there are various ways to calculate the life cycle pressures of final consumption on nexus elements such as climate change, energy extraction, water use and land use. As will be explained later on, input–output economics (IOE), and particularly the method of Multi Regional Environmentally Extended Input–Output analysis (MR EE IO) has the advantage that it puts all information with regard to economic relations, (local/territorial) pressures on nexus elements and how final consumption relates to these pressures in a consistent framework. Against this background, this chapter sets out to discuss the following:

- The relation between the nexus framework and IOE – what is the state of the art? What is MR EE IO, what does it cover, and via what perspectives can it analyse nexus problems?

- How can we measure stress on nexus elements by producers and final consumption with MR EE IO, and via what kind of approaches can we weight this?
- What empirical results can we provide in terms of potential country rankings?

This chapter makes in part use of the work of the EXIOBASE consortium, which has built one of the most extensive global MR EE IO databases available (e.g. Tukker et al., 2009, 2013; Wood et al., 2014, 2015). The EXIOBASE consortium produced a booklet reflecting pressures by consumption on nexus elements by country for the year 2007, which is summarised here (Tukker et al., 2014). The proposal for a weighting method was developed as an experimental approach for the World Resources Forum (Tukker et al., 2015).

Input–output economics and the nexus framework

This section describes the current state of the art of the study of nexus issues through IOE. Furthermore, it briefly introduces the concept of MR EE IO and relates it to an analysis of what a proper tool for analysing nexus problems should capture. We then end with conclusions about what type of insights in nexus problems IOE and MR EE IO can contribute.

State of the art

The interplay between IOE and the nexus framework has been limited so far, yet a handful of studies provide a glimpse of its potential. One of the first studies was carried out by Marsh (2008), which studied water–energy nexus issues in Australia using various input–output (IO) techniques. The author claimed that a diversity of techniques within IOE could serve different purposes for the study of nexus issues, and focused on four: linkage analysis, dependency analysis, multiplier analysis and scenario analysis. Linkage analysis describes how much economic agents (economic sectors and final consumers) rely on each other, and the implications in terms of overall economic activity. Dependency analysis focuses on the interdependencies between economic agents, while multiplier analysis is used to measure the effects of changes in final demand on economic indicators such as income and employment. Lastly, scenario analysis is concerned with possible future scenarios under plausible assumptions. An important finding of this study is the value of understanding the role of key economic sectors in nexus issues, that is, those sectors with stronger linkages with other sectors and final consumers will propagate to a larger extent issues arising from the nexus, such as resource scarcity.

China has been at the spotlight of the study of nexus issues through IOE, especially the water–energy nexus. Kahrl and Roland-Holst (2008) studied the water–energy nexus in China and defined three relevant metrics using IOA: physical (the embodied energy and water intensity of water and electricity provision, respectively), monetary (the links between energy and water prices) and distributive (the energy implications of water allocation). These metrics led the authors to a better understanding of the relative impact of the nexus on total resource use as well as the repercussions of price changes and changes in economic structure due to the nexus. Li and colleagues (2012) used a hybrid model that combined IOA with life cycle assessment (LCA) of electricity production technologies, also to assess the water–energy nexus in China. The increased technology detail of the model allowed the authors to confirm the potential of wind energy to mitigate the detrimental impacts of the nexus, albeit with a lack of detail of water quality and importance at the local level. This limitation was dealt with by Feng and colleagues (2014), which used a hybrid model in combination with the water stress index developed by Pfister et al. (2009) to identify water scarcity from the water–energy nexus at China's provincial level. This approach showed that key sectors in terms of water use were largely located in areas with limited water supply. Okadera et al. (2015) also considered regional differences in energy supply systems and water availability in the Liaoning Province, and found that the limited availability of water resources could jeopardise the energy security of the region. Gu et al. (2014) focused on analysing the water-saving effects from energy conservation policies in China using IO interdependencies, and identified key resource-intensive sectors. Outside of China, and focusing on the energy–water nexus instead, Mo et al. (2014) applied a hybrid IO-LCA model to study water supply scenarios in the US. The authors concluded that water reclamation would minimise the effects of the nexus from future water demand in some coastal areas.

A MR EE IO model (see the following section for a description) was used by Guo and Shen (2015) to assess the land–water nexus in China, which allowed the development of indicators of resources embodied in interregional trade, specifically related to imports, exports and trade balance. The results highlighted the imbalanced distribution of physical land and water resources among provinces, which resulted in uneven resource pressures. A combination of MR EE IO and ecological network analysis (ENA) was applied by Wang and Chen (2016) to assess the energy–water nexus of urban agglomeration in China. The authors explored structural properties and sectorial interactions of the system, and highlighted the effects of the nexus on those regions which rely importantly on imported resources as well as on certain 'dominant' sectors, which propagate nexus issues throughout the economy. A similar approach was applied by Duan and Chen (2017) for the study of the energy–water nexus of China's international trade. The results revealed the risks of resource depletion in some regions due to exploitation relationships from international trade and the use of resources as means to control other regions through trade relationships.

This section offers a brief overview of existing studies that applied IOE under the nexus framework. While they largely focus on the interdependencies between economic agents via de Leontief model (Leontief, 1970), a variety of other IO techniques have been applied, such as linkage, scenario and network analysis. Each technique fits a different purpose to study nexus issues, showing the value of the framework. For instance, dependency analyses are useful to approach the absolute magnitude of nexus issues, whereas linkages and structural properties of economic agents can give insights into how nexus issues propagate throughout the economy. These studies also demonstrate that known limitations of IOE can be overcome by means of additional modelling and/or data. For instance, the increase in technology detail via LCA and the improvement of resource indicators via additional data (e.g. water scarcity). While the study of water–energy nexus predominates, IO techniques can be potentially applied to other types of resources, such as land use and materials, with the help of environmental satellite accounts. Lastly, the inclusion of trade data in the form of multi-regional databases help to understand the global effects of nexus issues through international trade.

Multi-regional environmental extended input–output analysis

To get an impression of environmental impacts caused by a country, it has been custom to monitor resource extraction as well as emissions due to production and consumption processes within the own territory. But in the last decades, due to ongoing liberation of trade and economic specialisation, growth in international trade has outpaced growth in global GDP. Impacts related to consumption in one country hence increasingly take place abroad (e.g. Peters et al., 2011; Wiedman et al., 2010). Consumption in one country hence drives production in value chains spanning many countries, creating a complex, global web of activities impacting the environment in multi-faceted ways (cf. Tukker and Dietzenbacher, 2013).

The so-called Multi-regional Environmentally Extended Supply and Use/Input–Output Tables (MR EE SUT/IOT) are now widely seen as a powerful tool for the calculation of both production and consumption-based indicators reflecting pressure on nexus elements (or 'footprints'), particularly if one wants to calculate different footprints side-by-side with one consistent dataset (see Figure 8.1).

In short, a SUT/IOT shows a country's total economy, with production divided in a few dozen industry sectors, and consumption divided into a few dozen product (and service) groups. The tables show how much of these specific products (output), for instance cars, each industry sector produces, expressed in Euros. The tables also show for each industry sector how much of other products they need to realise this production (input) – e.g. the amount of steel, glass, plastics, electricity and electronics the car industry in that country needs to produce the output of cars. Furthermore, for each industry one can identify the primary resource use and emissions ('environmental extensions') – for instance land use by the agricultural sector, or CO_2 emissions by the electricity production sector. By these means, at country level, it now becomes possible to analyse how the economy is interconnected. For instance, for the final use of cars by consumers, it becomes possible to analyse how much production value was contributed by the car industry, the glass production industry, the steel industry and so on. But since we also know for each industry the emissions and primary resource extraction per Euro, we now can estimate the total primary resource extraction and life cycle emissions for the total consumption of cars in that country (cf. Miller and Blair, 2009). This example is for one country only, and we know that imports and exports in the current global economy are substantial. So, one also must understand the emissions and primary resource use involved in imports. For that, one

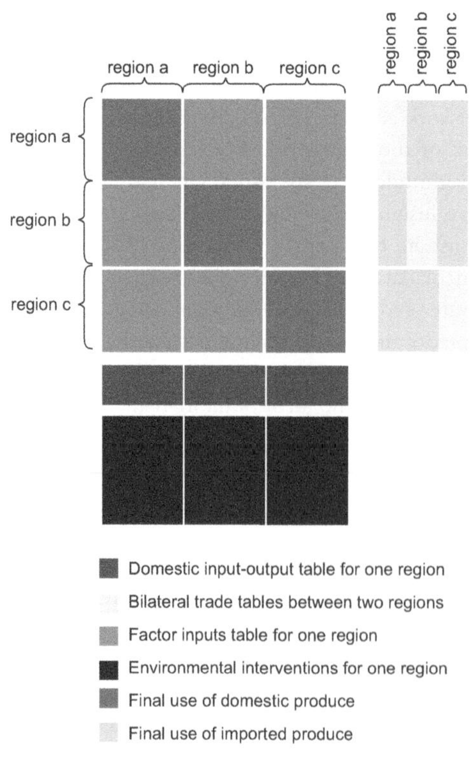

Figure 8.1 Stylised MR EE IOT for three countries

needs to create EE SUT/IOTs for the most important economies of the world and identify the trade flows between the specific sectors of all the countries (Tukker et al., 2013; Peters et al., 2011). This exercise results in the aforementioned MR EE SUT, which paints a detailed picture of all linkages between production and consumption in the global economy (see Figure 8.1 for a three-region example). One of the great strengths of this MR EE SUT/IOT approach is that it is inherently consistent. All direct emissions of GHG and primary extraction/use of water, land and materials, next to number of jobs and added value created by industries are, by definition, directly related to the final consumption of products – they cannot be 'lost' in the calculations.

Nexus problems

This book already has given an introduction to nexus problems, but to make clear what the MR EE IO method can cover and what not, a summary and our own insights into how nexus elements are related will be given here, particularly via Figure 8.2. Usually, various environmental or natural subsystems are identified making up the nexus. It concerns the climate system, the water system, the land system, biodiversity resources, and energy and material resources. These all can be seen as part of the 'natural system' from which the global economic system extracts resources and to which it expels emissions and waste.

So, on the one hand, the concept of a 'nexus' can be used to analyse how these environmental subsystems interact. Figure 8.2 shows in this context six reciprocal relations between four subsystems for land, water, energy and climate. Each of these subsystems influences the other subsystems in specific ways, and for each of these interactions strategies can be thought of to

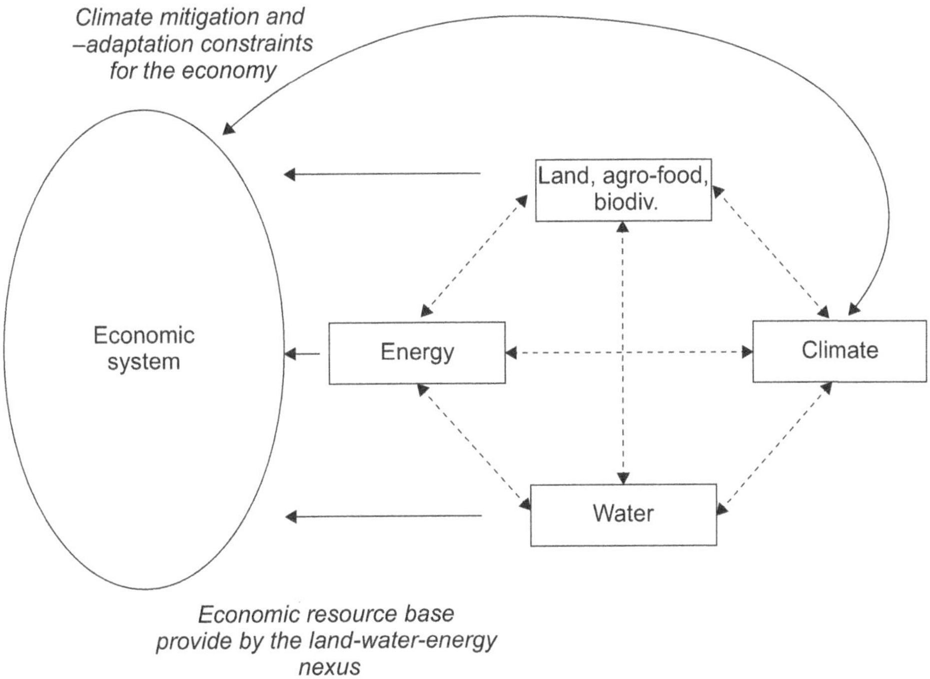

Figure 8.2 The land–water–energy–climate nexus and its relation with economic development, specified in more detail in the boxes below the figure

reduce negative interactions and feedback loops. For instance, the quantity and quality of land and water available will be affected by the nature and extent of climate change. The quantity of land and water required will depend on the size of population and various lifestyle factors such as diet, the rural/urban split and the population densities of urban areas. Land will also be required for the generation of different forms of biomass (fodder, food, fiber), each of which has its own requirements for water, and each of them will be differentially affected by the climate change occurring in those regions. Land is also required for forests, not just for the generation of timber and non-timber forest products, but also as carbon sinks, both in the timber and the soil underneath. The boxes below Figure 8.2 give a fairly comprehensive review of such relations. Such an analysis clearly puts the environmental system central, rather than the economic system.

But on the other hand, the concept of a 'nexus' can also be used what boundary conditions nexus elements set to economic development, and more specifically, what trade-offs the economic system may create – i.e. reducing the pressure on one nexus element, but then at cost of enhancing pressures on another nexus element. For instance, the climate element of the nexus may work out as follows. Climate impacts may result in reduced agricultural productivity or the need for climate adaptation measures. Climate adaptation measures may in turn lead to a higher need for resources or energy use. GHG emission reduction targets may lead to investment in low-carbon technologies that usually are more material intensive, or outright limit production. Such an analysis clearly puts the economic system at the center stage. Again, the boxes below Figure 8.2 give a fairly comprehensive review of such relations.

Land/agro-food/biodiversity -> Energy	**Energy -> Land/agro-food/biodiversity**
• Potential for energy generation via 1st, 2nd, 3rd and 4th generation biomass. . . • . . . in relation to biomass needs for diets, food waste, and agricultural and land productivity)	• Direct energy requirements for food production • Indirect energy requirements for food production (e.g. for production of N and P fertilizers) • Biomass generation versus food production and biodiversity trade-offs

Land/agro-food/biodiversity -> Water	**Water -> Land/agro-food/biodiversity**
• Pollution of (blue) water by N and P loss, pesticides • Level of green and blue water use, changes in evapotranspiration patterns and possibly precipitation patterns by choices in food consumption and related agriculture and land use	• Available green and blue water by region by crop • Disruptions in agriculture/food production due to extreme events and droughts • Impacts of local water quality and quantity on biodiversity

Water -> Energy	**Energy -> Water**
• Availability of cooling and process water for thermal power plants • Availability of water used by processes in the energy supply chain (biomass, other renewables (mining and refining of metals used in low-carbon energy technologies); unconventional energy sources (shale gas, tar sand extraction) • Availability of water for hydropower • Disruptions due to extreme events	• Energy requirements for water extraction • Energy requirements for water desalination

Climate -> Energy	Energy -> Climate
• GHG emission budgets imposed on the energy system (review of RCPs and SSPs) • Energy costs of climate adaptation • Indirect impacts of climate change via, for example, water and agriculture on energy (e.g. availability of biomass)	• Direct GHG emissions by the energy system • Indirect GHG emissions in supply chains for energy (e.g. land-use change for biomass and hydropower, materials extraction and refining for, for example, solar and wind, higher energy requirements to refine lower ore grades)

Climate -> Water	Water -> Climate

• Change in temperature patterns
• By this, changes in the water cycle
  ▪ Precipitation patterns
  ▪ River systems
  ▪ Evapotranspiration (combined effect of land use and temperature)
	• Change in albedo effect due to change in ice- and snow-covered surface

Climate -> Land/agro-food/biodiversity	Land/agro-food/biodiversity -> climate

• Change in temperature patterns
By this changes in
  ▪ Available agricultural land
  ▪ Agricultural productivity per crop
	• GHG emissions due to land-use cover change
	• Change in albedo effect due to changes in land-use cover
	• Carbon sequestration by crop and biomass production

Relation between MR EE IO and nexus problems

From the description of the MR EE IO approach it is clear that MR EE IO is not suitable to analyse the full nexus problem. MR EE IO is in essence a tool that captures the relations in the global economic system. It is hence very well suited to analyse trade-offs created by the economic system when dealing with nexus pressures – e.g. if a reduction in e.g. water use in one country will not have trade-offs in terms of higher land use and water use elsewhere. For that purpose, the tool is extremely powerful, although also there it has some limitations. MR EE IO can do an excellent job to analyse the carbon pressure, water requirements, land-use requirements and bulk-material requirements of different configurations of our economic system, as we will illustrate next. Limitations are however the following. Given a typical resolution of product groups and sectors of 100–150 in MR EE IO tables, more fine-grained tools like substance flow analysis (SFA) or life cycle assessment (LCA) are needed to assess the requirement of specific minor metals and materials of different configuration of our economic system (compare e.g. Hertwich et al., 2015; de Koning et al., 2016; see also Chapter 10 by Mudd). Further, MR EE IO, like SFA or LCA, can only be used as a scenario tool when developments are imposed exogenously. Unlike for instance dynamic Computable General Equilibrium (CGE) or econometric models such as GTAP, Fidelio and E3ME, there is no explicit assessment if the levels of economic savings allow for economic investments in new infrastructure, or how dynamics in price developments will influence demand for specific products, and how investments lead to a higher GDP and available income for final consumption in future. In this sense, increasing efforts are being made to integrate both frameworks in the form of dynamic economic-environment models. An example of such model is EXIOMOD, a CGE model with a high level of environmental and economic completeness thanks to using EXIOBASE as the underlying database (Ivanova, 2014).

MR EE IO further does not include the environmental system. Hence, for analysing for instance the relation between global warming, the water cycle, and agricultural and land productivity, a totally different set of models is required that is outside the scope of this chapter. Examples include models such as Water GAP or LPJmL that focus on water, land use and agricultural productivity issues, climate circulation models, etc.

Ideally, a comprehensive set of a models covering the environment, combined with a dynamic economic model covering the economy in even more detail as MR EE IO databases do, would be required to cover the full nexus both at the environmental as the economic side. The closest to this are integrated assessment models such as GCAM and IMAGE, but these cannot realise the economic detail and global consistency of MR EE IO databases and the detailed representation of environmental relations in specialised water and land-use change models.

Materials and methods: analysis of stress on nexus elements by MR EE IO and illustrative weighting

Introduction

This section analyses the global pressures on nexus elements both from a production and consumption perspective, and it further gives an illustration of how these pressures could be weighted. It is not the aim to present a formal ranking based on a tried-and-tested methodology. So all findings presented have to be seen against this background, and be seen merely as an example of possible presentations of results once a tried-and-tested methodology is available. In the following sections, we provide the data inventory and methods to calculate nexus pressures, present the weighting approach, and present the results.

Data inventory and methods

We explained in earlier that the environmental pressures of consumption of a country can take place outside that country itself. But once a global MR EE IO such as EXIOBASE is available, there exists in essence an input–output matrix representing the global economy and reflects all economic transactions between all economic sectors and all countries. It now becomes possible for each final consumption category in a country – say 1 Mio Euro of bread – to track back which sector in which country contributed how much added value to this 1 Mio Euro of bread (e.g. say 300.000 Euro by the retail sector, 200.000 Euro by the bakery sector, 200.000 Euro by the wheat production sector/farmers, next to all kind of smaller contributions of e.g. the electricity production sector and the fertilizer sector). But since the MR EE IOT gives also for each sector in each country the emissions and total turnover, it is possible to calculate per sector the emissions per unit turnover. This together allows to estimate how the emissions and resource uses per sector and country (the production-based approach) can be re-distributed over the final consumption categories in the different countries (which is the consumption-based approach).

Mathematically, this approach was first proposed by Leontief, who later won the Noble prize for this work. Using this now well-known standard calculation approach based on the so-called Leontief inverse, a MR EE IO allows calculation of the emissions and primary resource extraction related to a final demand by category per country as follows (c.f. Miller and Blair, 2009) (equation 1):

$$e_i = s_i x = s_i L y = s_i (I - A)^{-1} y \quad (1)$$

Where y is a given final demand vector, A is the matrix of direct input coefficients, I is an identity matrix, s is a direct impact coefficient vector for a given pressure i, L is the Leontief matrix, x is the total output (intermediate output plus final demand) required to satisfy f and e is the total requirement of environmental factors. Alternatively, direct resource extraction and emissions (f) can be calculated as (equation 2):

$$f_i = s_i \quad (2)$$

Equation 1 thus provides information on the direct plus indirect emissions and resource use induced by a given final demand (e.g. final consumption from a given country) (consumption perspective), whereas equation 2 can be used to calculate direct resource use and emissions by country and sector. Both approaches provide essentially different yet complementary insights into nexus issues.

For the practical illustrative calculations of which the results are shown in this volume, we took the following, pragmatic approach. This pragmatic approach was mainly driven by the fact that the EXIOBASE team had already performed analysis of the water, land, material and carbon footprint of final consumption of nations, which we use here as input (Tukker et al., 2014 and 2016).

a) Data inventory methods: We pragmatically used EXIOBASE, a multi-regional input–output database that contains data for 43 countries and five continents, and 163 economic sectors by country including a dozen different agricultural sectors as a basis. For each economic sector and country, the database includes its green and blue water extraction, some 80 extracted resources including energy resources, and some 40 emissions including all relevant GHG emissions.
b) Choice of pressures related to nexus elements: We pragmatically used the carbon, water, land and material footprint as the four main pressures to be covered. These relate quite well to the nexus elements of land use, water extraction, climate change, and other resource extractions and footprint calculations are already available from Tukker et al. (2014, 2016).
c) Aggregation methods within resource categories again were pragmatically based on what had been calculated already by Tukker et al. (2014), i.e. for carbon: GWP; for water: m3 blue water extraction (using the convention applied by the water footprint network rather than scarcity indices as suggested by Pfister et al., 2009); for land: land-use occupation in m^2 (without any correction for land quality or productivity); and for materials: material use in tons, in line with the conventions of the MFA community.
d) Weighting methods: To aggregate carbon, water, land and materials into one single score, we used the approach developed by van Oers and Tukker (2016). In short, they developed four weighting sets – a) one with equal weights, b) a second based on 'distance to target' between current pressures and maximum desirable pressures; c) a third one using weights derived from expert panel studies; and d) a fourth one based on damage costs. The weighting sets are summarised in Table 8.1.

Table 8.1 Weighting sets used in this study

Weighting set name	*Carbon*	*Materials*	*Water*	*Land*
Equal weight	25%	25%	25%	25%
Weighted panels	51%	16%	11%	22%
DtT	42,9%	29,7%	7,5%	19,9%
Damage costs	8,7%	0,0%	15,3%	76,0%

*All weights to be applied to scores normalised on global total pressures. The shadow prices for calculating damage costs weights are 25 Euro/ton CO_2-eq., 0 Euro/ton material extraction, 1 Euro/m3 water extraction and 940 Euro/ha/yr land-use occupation

Some analysis and results: direct nexus pressures and nexus footprints of final consumption

This section presents quantitative results of direct nexus pressures (production perspective) and nexus footprints of final consumption (consumption perspective) respectively.

Direct resource use and emissions

We first approach nexus issues from a production perspective, and present the global direct resource use and carbon emissions by country and sector. Figures 8.3, 8.4, 8.5 and 8.6 present heat maps for direct pressures related to carbon emissions as well as water, land and material use, respectively, as coefficients with respect to country totals (where one means a given sector is responsible for all the emissions/resource use of a given country). These results therefore help to understand which sectors are directly responsible for a larger amount of the selected pressures as well as country differences. To facilitate both interpretation and presentation of the results, the 163 industry categories from EXIOBASE have been aggregated to NACE rev 1.1 statistical classification at the 60 sector level (Koning et al., 2013). Furthermore, only sectors that contribute to more than 0.05% for all indicators and countries with a total output larger than 0.01% from the maximum are presented. The results show a largely unequal distribution of carbon emissions and resource use among economic sectors.

GHG emissions concentrate on sectors related to food production, manufacturing (e.g. chemicals and metal products), electricity and gas generation as well as sewage and waste management. Blue water consumption generally takes place in agricultural sectors and, in some countries, also to a considerable extent in the manufacture of chemicals. Land use focuses on activities related to agriculture and forestry, whereas material extraction concentrates on mining activities (mostly metals and minerals) and agriculture (mostly biomass). Under the nexus framework, however, nodes (emissions and/or resources) are analysed as a whole rather than in isolation (Liu et al., 2015). To assess the extent to which nexus nodes are emitted/used in conjunction throughout the economy, the four different weighting methods described in Table 8.1 have been applied to calculate a single coefficient value for each country-sector. As an illustration, Figure 8.7 presents the aggregated results according to the equal weighting scheme. The aggregated results are relatively consistent across weighting methods, with combined nexuses focusing mostly on agriculture and forestry activities and, to a lesser degree, mining activities, manufacture of metal products, electricity generation as well as sewage and waste management sectors. The method 'equal weighting' highlights the agriculture industries as the strongest combined nexus, mostly driven by water and land use. The methods 'weighted panels' and

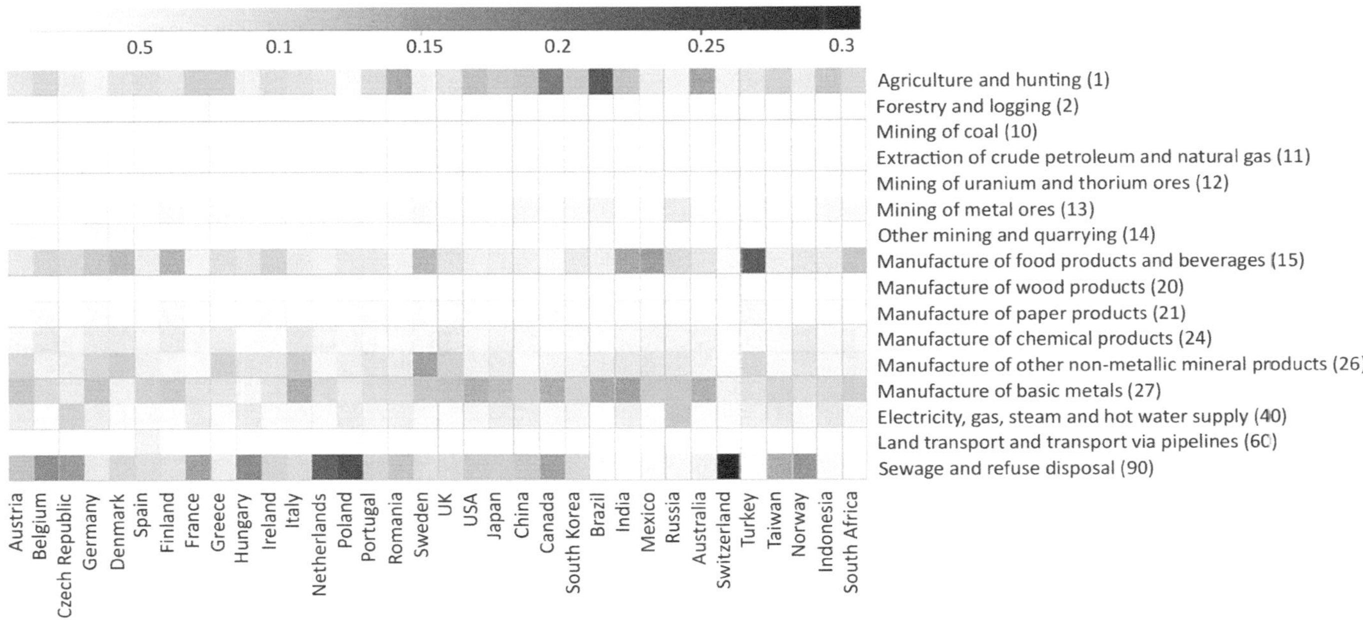

Figure 8.3 Direct GHG emissions per country and sector (as % from country total). Within brackets, the corresponding NACE 1.1 code

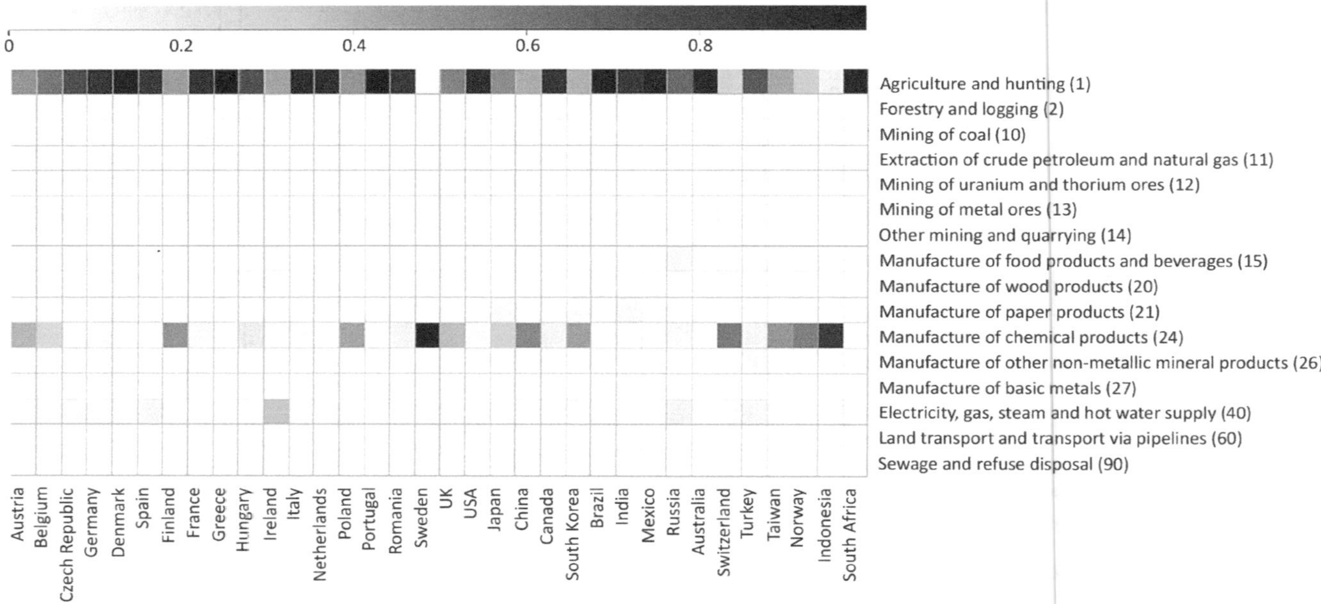

Figure 8.4 Direct consumption of blue water per country and sector (as % from country total). Within brackets, the corresponding NACE 1.1 code

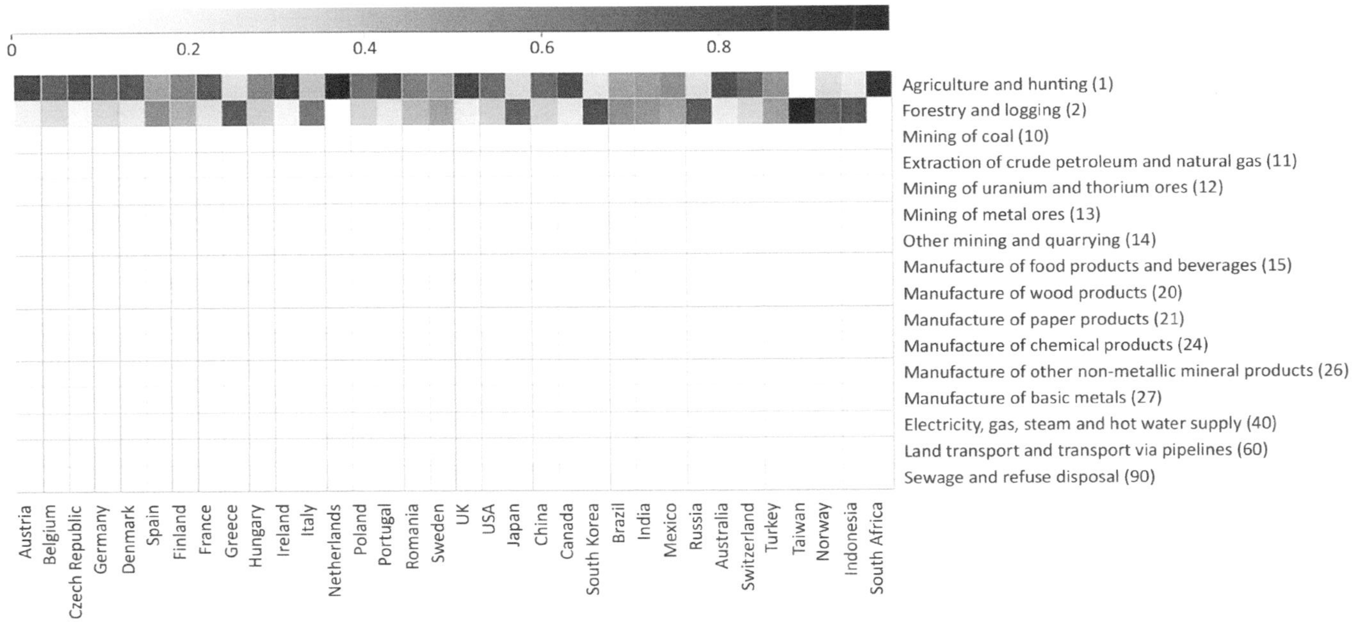

Figure 8.5 Direct land use per country and sector (as % from country total). Within brackets, the corresponding NACE 1.1 code

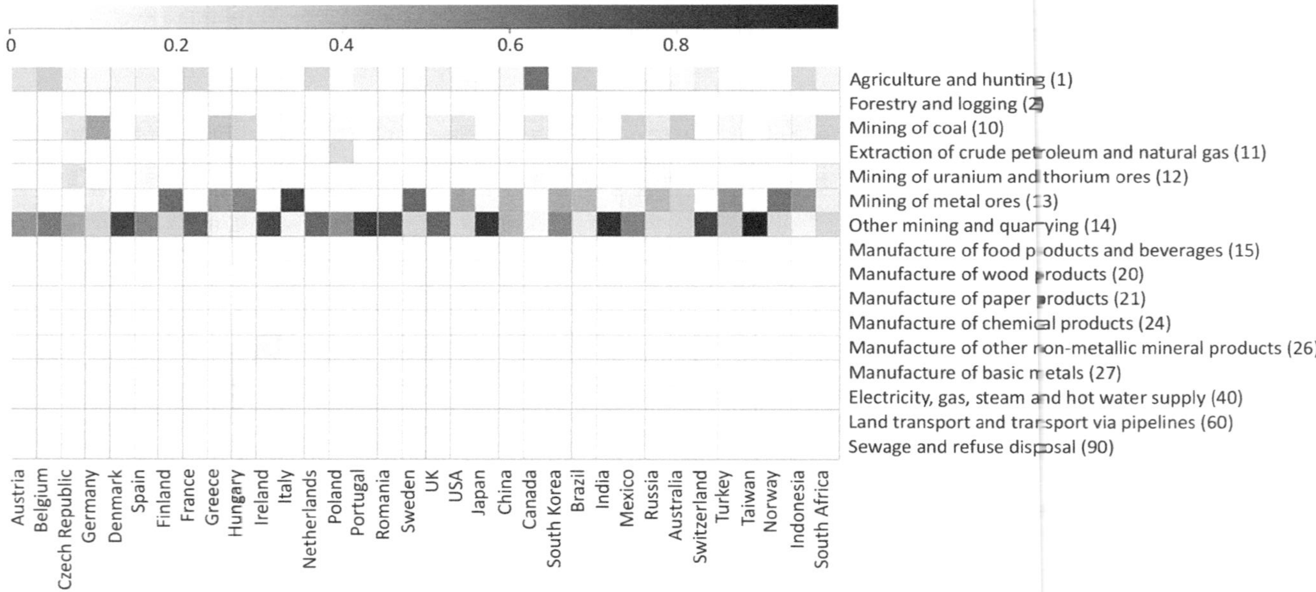

Figure 8.6 Direct extraction of materials per country and sector (as % from country total). Within brackets, the corresponding NACE 1.1 code

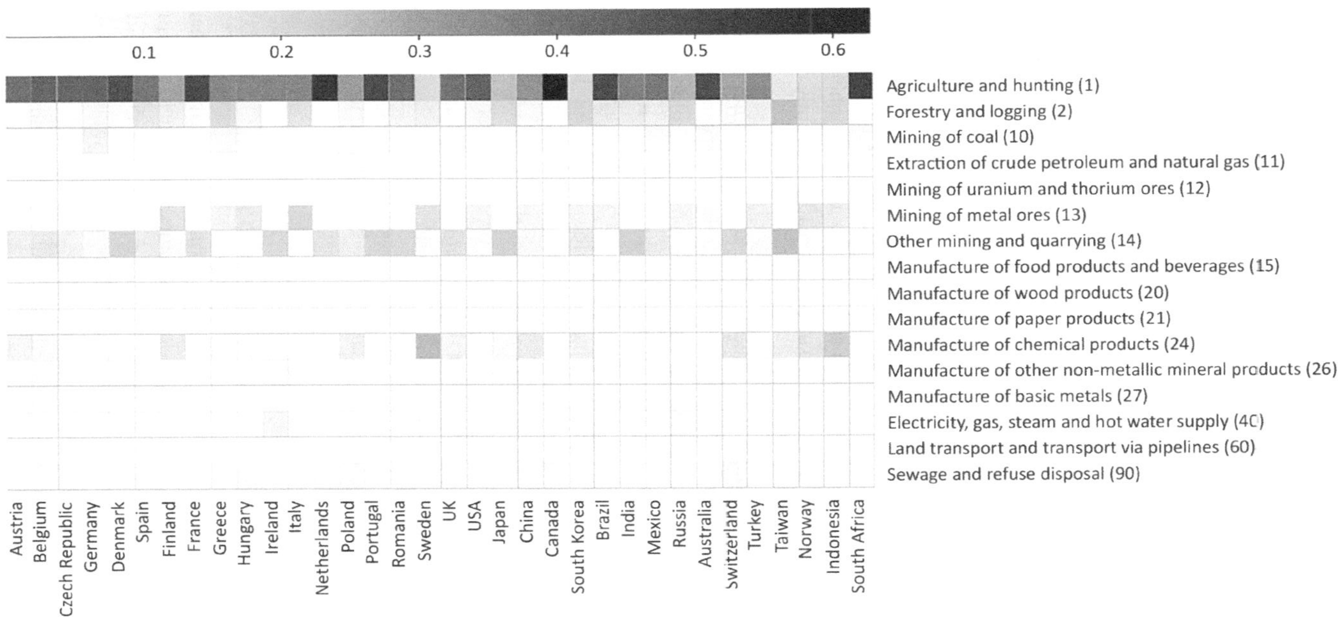

Figure 8.7 Aggregated direct emissions and resource use according to equal weighting. Within brackets, the corresponding NACE 1.1 code

'distance-to-target' increase mostly the scores of those sectors that are GHG-intensive, keeping agricultural sectors on top but giving more notoriety to sectors related to manufacture of metal products, electricity production as well as sewage and waste management. Furthermore, the method 'distance-to-target' also highlights mining sectors, mostly driven by increased weight of material extraction. Lastly, the method 'shadow prices' highlights land-intensive sectors, describing the agriculture and forestry sectors as the strongest combined nexus.

Footprint indicators

This section presents the footprints per capita that were the basis for the calculations in Figure 8.8. This figure has been taken from Tukker et al. (2014, 2016). The figure shows that the different resource footprints vary significantly among countries, particularly if one looks at per capita figures, regardless if it concerns the carbon, water, land or material footprint. Consistently, rich, developed countries such as Australia, the US, Luxembourg and some other EU countries have the highest carbon, water, land and material footprints per capita. Less-developed countries such as Indonesia and India tend, in general, to have the lowest footprints per capita. This result is well in line with the earlier, separate footprint analyses (e.g. Hoekstra and Chapagain, 2007; Davis and Caldeira, 2010; Bruckner et al., 2012). From the developed nations, the EU27 belongs to the most resource efficient. The US, Canada and Australia all have higher footprints per capita, particularly for carbon, materials and land. Japan and Korea have slightly lower water, land and material footprints as the EU27.

Figure 8.9 further applies as illustration the Distance to Target weighting set to the carbon, water, land and materials footprints. We used the dimensionless normalised total carbon, water, land and material footprint per country as a starting point (the total global normalised carbon, water, land and material footprint being 1). With all weighting factors together being 100%, this resulted in a dimensionless 'Integrated use index' per country, which again adds up to 1 for the world. As reflected by Table 8.1, weighting using the shadow costs we took from literature, gives a very high emphasis on land use, neglects materials, and does not emphasise carbon or water. It appeared that all weighting sets lead to a similar ranking as the DtT weighting set in Figure 8.9. We again see that rich countries can create per capita resource pressures which are factors higher as of the global average citizen. It is likely that the weighting method applied is less relevant for the outcome as the rather simple approaches to calculate the materials, land and water footprints themselves, all based on simply adding tons of materials, ha of land, and m3 of water, without taking into account impact of the materials, quality of the land, or scarcity of water in the region where it is extracted.

Conclusions and outlook

This chapter explored the potential of input–output economics to study human-nature nexus issues. While this field of study presents well-known limitations that have hampered to some extent its application under the nexus framework (e.g. technology resolution and human-nature dynamics), its value to study nexus issues is evident. Some valuable applications relate, for example, to understanding the role of economic relationships, trade and the economic structure in the creation and propagation of nexus issues throughout the economy. Furthermore, a quantitative analysis was performed from both a production and a consumption-based perspective, regarding the pressures related to carbon emissions, land use, materials extraction and water use, which are among the most important nexus nodes identified in the literature (Liu et al., 2015).

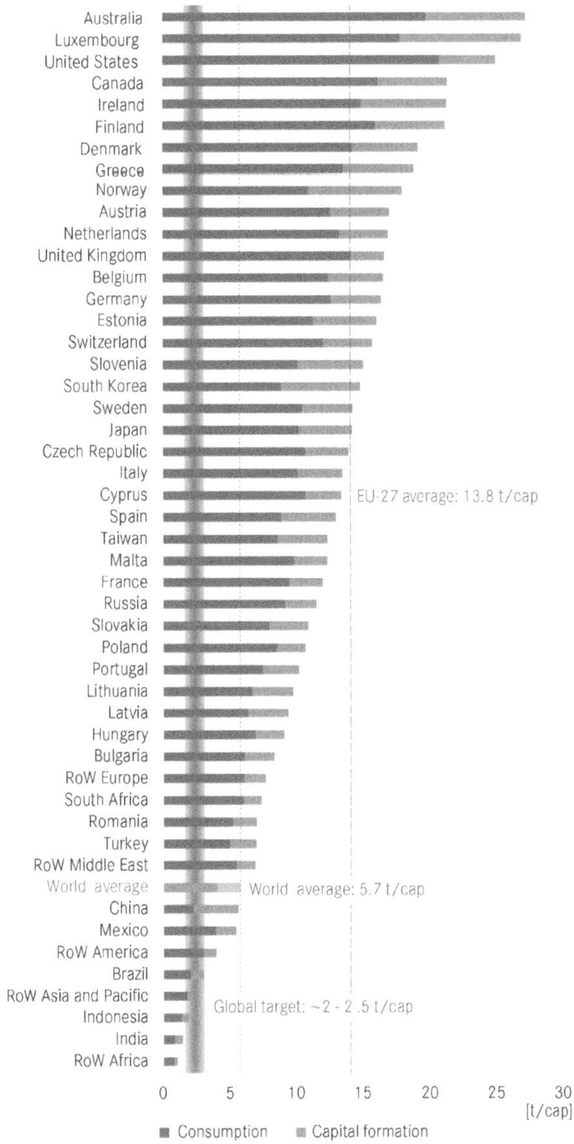

Figure 8.8 Carbon, water, land and materials footprint per capita per country

Source: calculated with EXIOBASE v2.0, data for 2007; Tukker et al. (2014, 2016)

The main results suggest that a few economic sectors such as agriculture, mining, manufacture as well as sewage and waste management are central in the nexus debate, as they are responsible for multiple direct pressures simultaneously. While this holds true in general terms, other sectors can also be important in specific regions. This highlights the value of comprehensive analyses which consider the whole economy, such as the one presented here. The results also show the importance of material resources as nexus nodes, albeit these have received little to no attention in the literature (Graedel and van der Voet, 2010).

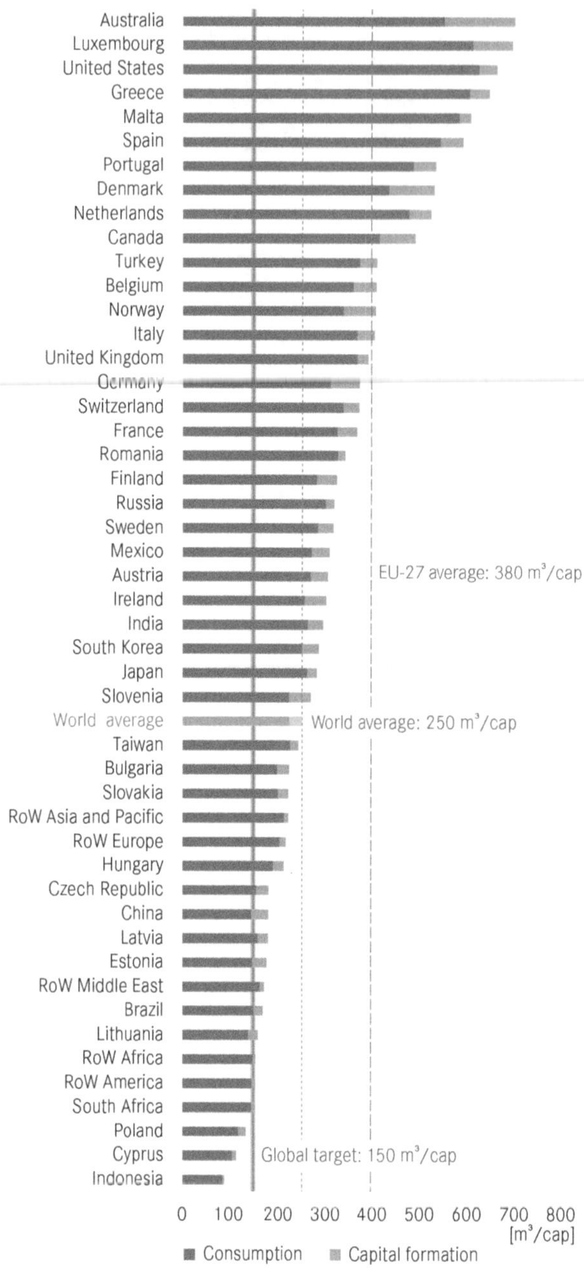

Figure 8.8 Continued

This chapter also did a relatively pragmatic analysis based on the earlier work of Tukker et al. (2014), showing in essence that the individual pressures almost invariably are determined by the level of final consumption per capita in countries. We further applied quite diverging weighting methods, based on external costs, panel methods, and the distance-to-target approach, next to an approach based on equal weights. The ranking in weighted scores is hardly differed between weighting methods.

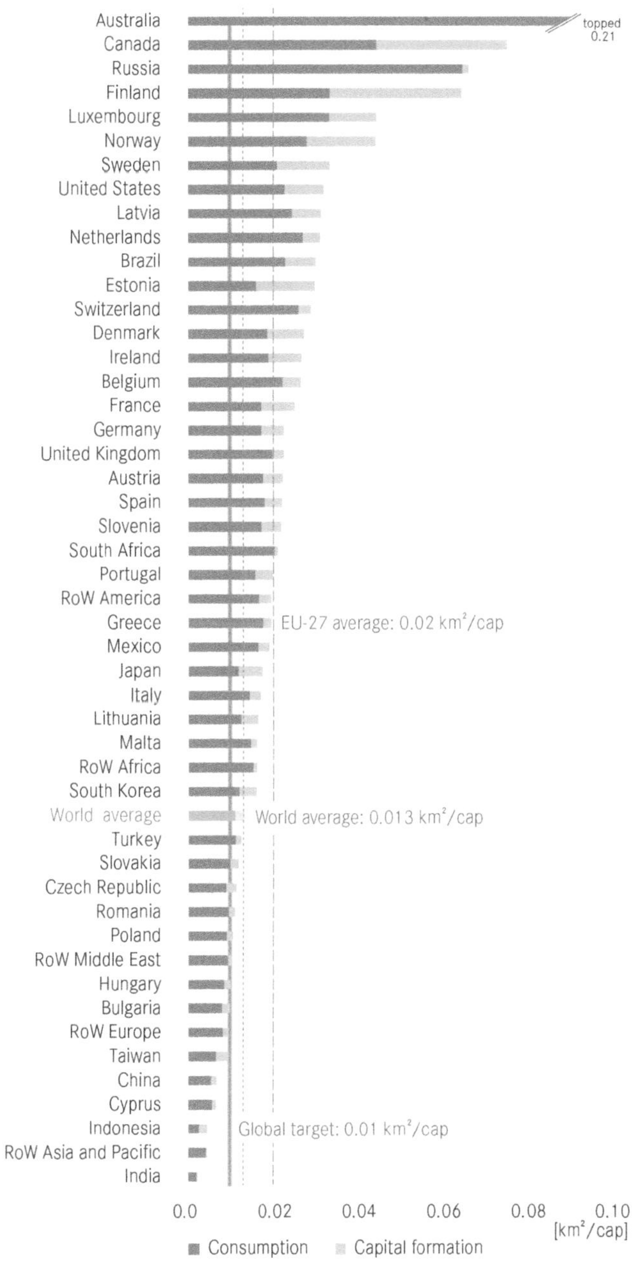

Figure 8.8 Continued

The analysis shows that the combined pressures on the nexus of climate, energy, water, land use and biodiversity are driven by final consumption in the richest countries. This paper did not do a more specific analysis by consumption category, but it is pretty obvious that consumption of food and biomass drives water and land use, whereas energy using processes (heating, cooling, appliances, transport) drive energy and climate problems. Nexus problems hence must be solved

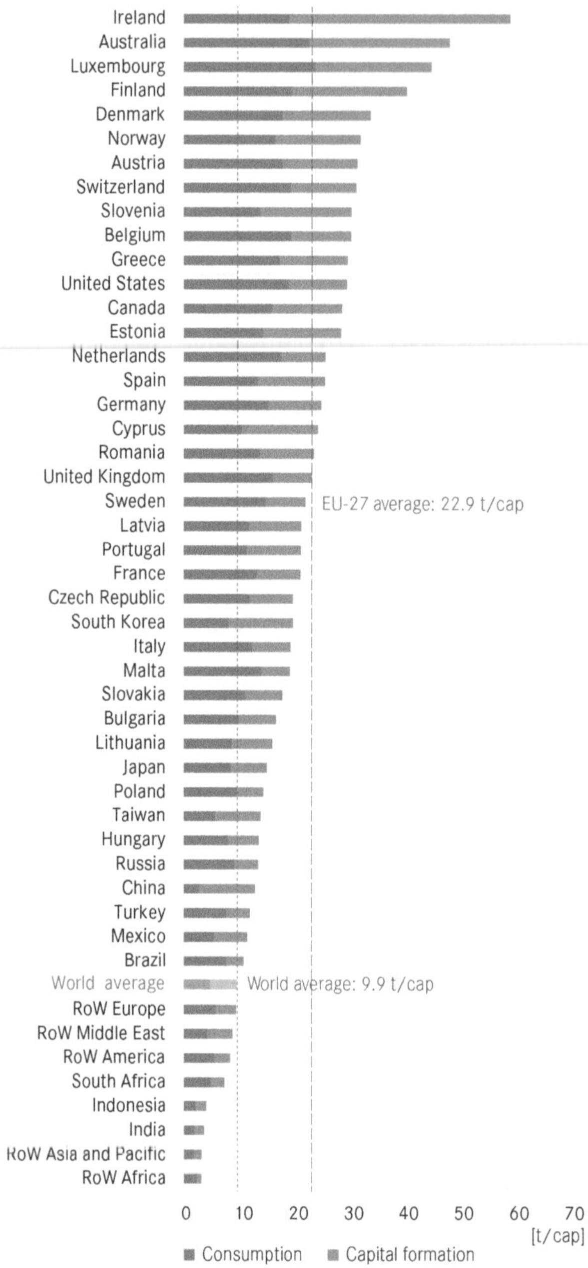

Figure 8.8 Continued

by reducing impacts of final consumption of the well-known 'big four': food, housing, appliances and transport (cf. Tukker and Jansen, 2006).

Despite being in the present and future agenda of the input–output community (Dietzenbacher et al., 2013), the interplay between IOE and the nexus framework is still relatively

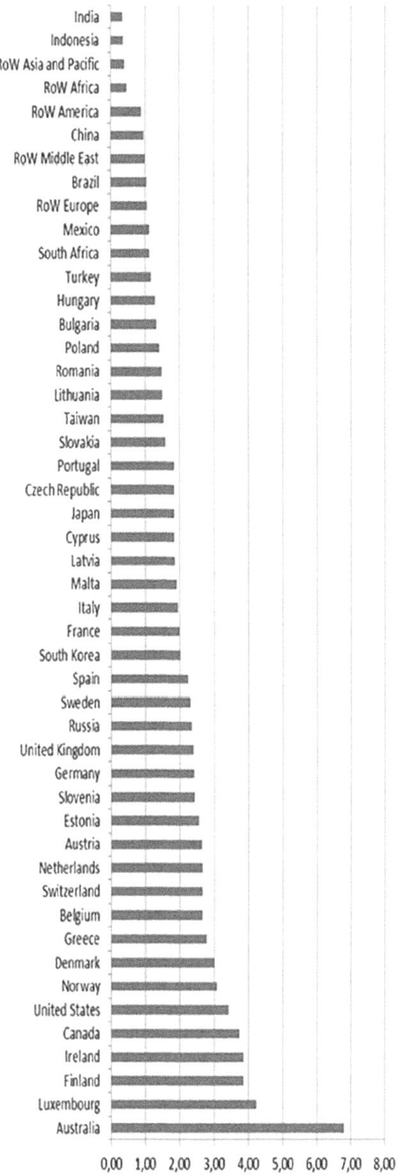

Figure 8.9 Resource index of nations – weighted resource pressures by country as % of global total compared with share as % of the global population of that country. The population in a country scoring 1 would have an average contribution per person to weighted resource pressures. Example for the distance to target method

Source: calculated with EXIOBASE v2.0, 2007

weak. The section beginning on page 106 shows that such interplay has been gaining momentum during the past decade, yet many venues for further research are still unexplored. For example, while we have attempted to study nexus nodes in conjunction by means weighting schemes, more elaborated techniques can be applied to deal with multi-dimensional issues and the well-known 'environmental incommensurability' (Funtowicz et al., 1999). To this end, multi-criteria decision analysis (MCDA) tools can be valuable to identify and rank the most challenging nexus issues (Font Vivanco et al., 2017), and their application to IOE is relatively straightforward (Miller and Blair, 2009). Furthermore, while the aforementioned section depicts a considerable diversity of IO techniques being applied to study nexus issues, other techniques still remain unexplored. For example, linkage analysis could be applied to study how key sectors are connected to other economic actors, allowing to study how nexus issues and potential resource scarcities could propagate across the economy, among other applications. Also, scenario analysis can be applied to study the repercussions of changes in the economic structure (e.g. increase in renewable energies) in terms of nexus issues. Other potential venues of research include the use of hybrid LCA models to increase the technology detail, the study of the role of feedback loops from inter-industry relationships on nexus issues and the relationship between nexus issues and macro-economic indicators such as productivity, employment and income. Input–output economics thus presents extensive uncovered potential to study the global magnitude of nexus issues, and the extent to which these can constrain global economic growth and welfare.

References

Bruckner, M., Giljum, S., Lutz, C., Wiebe, K.S. 2012. Materials embodied in international trade – Global material extraction and consumption between 1995 and 2005. *Global Environmental Change* 22(3), 568–576.

Davis, S., and Caldeira, K. (2010). Consumption-based accounting of CO2 emissions. *PNAS*, 107(12), 5687–5892.

Dietzenbacher, E., Lenzen, M., Los, B., Guan, D., Lahr, M. L., Sancho, F., Suh, S., and Yang, C. (2013). Input–output analysis: The next 25 years. *Economic Systems Research*, 25, 369–389.

Duan, C., and Chen, B. (2017). Energy–water nexus of international energy trade of China. *Applied Energy*, 194, 725–734.

Feng, K., Hubacek, K., Siu, Y. L., and Li, X. (2014). The energy and water nexus in Chinese electricity production: A hybrid life cycle analysis. *Renewable and Sustainable Energy Reviews*, 39, 342–355.

Font Vivanco, D., Wang, R., and Hertwich, E. (2017). Nexus strength: A novel metric for assessing the global resource nexus. *Journal of Industrial Ecology* (accepted for publication).

Funtowicz, S. O., Alier, J. M., Munda, G., and Ravetz, J. R. (1999). *Information tools for environmental policy under conditions of complexity*. Luxembourg: Office for Official Publications of the European Communities.

Graedel, T. E., and van der Voet, E. (2010). *Linkages of sustainability*. Cambridge, MA: MT Press.

Gu, A., Teng, F., and Wang, Y. (2014). China energy-water nexus: Assessing the water-saving synergy effects of energy-saving policies during the eleventh five-year plan. *Energy Conversion and Management*, 85, 630–637.

Guo, S., and Shen, G. Q. (2015). Multiregional input – output model for China's farm land and water use. *Environmental Science & Technology*, 49, 403–414.

Hertwich, E. G, Gibon, T., Bouman, E. A., Arvesen, A., Suh, S., Heath, G. A., Bergesen, J. D., Ramirez, A., Vega, M. I., and Shi, L. (2015). Integrated life-cycle assessment of electricity-supply scenarios confirms global environmental benefit of low-carbon technologies. *PNAS*, 112(20), 6277–6282. doi:10.1073/pnas.1312753111

Hoekstra, A.Y., and Chapagain, A. K. (2007). Water footprints of nations: Water use by people as a function of their consumption pattern. *Water Resources Management*, 21(1), 35–48.

Ivanova, O. (2014). Modelling inter-regional freight demand with input-output, gravity and SCGE methodologies. In L. A. Travasszy and G. de Jong (Eds.), *Modelling freight transport*. London: Elsevier, p. 13.

Kahrl, F., and Roland-Holst, D. (2008). China's water–energy nexus. *Water Policy*, 10, 51–65.

Koning, A. d., Eisenmenger, N., and Van der Voet, E. (2013). Deliverable 1 of Project ENV.F.1/ETU/2011/0044. Assessment of Scenarios and Options towards a Resource Efficient Europe. *Topical*

Paper 1: Resource-efficiency in the built environment – a broad-brushed, top-down assessment of priorities E. C. D.-G. Environment (Ed.). Delft, the Netherlands: TNO
Koning, A. de, Kleijn, R., Huppes, G., Sprecher, B., van Engelen, G., and Tukker, A. (2016). Challenges for future metal supply not fundamentally enhanced by a transition to low carbon energy technologies.
Leontief, W. (1970). Environmental repercussions and the economic structure: An input-output approach. *The Review of Economics and Statistics*, 52, 262–271.
Li, X., Feng, K., Siu, Y. L., and Hubacek, K. (2012). Energy-water nexus of wind power in China: The balancing act between CO2 emissions and water consumption. *Energy Policy*, 45, 440–448.
Liu, J., Mooney, H., Hull, V., Davis, S. J., Gaskell, J., Hertel, T., Lubchenco, J., Seto, K. C., Gleick, P., Kremen, C., et al. (2015). Systems integration for global sustainability. *Science*, 347.
Marsh, D. M. (2008). *The water-energy nexus: A comprehensive analysis in the context of New South Wales*. Sydney Australia: Faculty of Engineering and Information Technology, Sydney University of Technology.
Miller, R., and Blair, P. R. (2009). *Input-output analysis: Foundations and extensions* (2nd ed.). Cambridge: Cambridge University Press.
Mo, W., Wang, R., and Zimmerman, J. B. (2014). Energy – water nexus analysis of enhanced water supply scenarios: A regional comparison of Tampa Bay, Florida, and San Diego, California. *Environmental Science & Technology*, 48, 5883–5891.
Oers, L. v., and Tukker, A. (2016). *Development of a resource efficiency index of nations*. St. Gallen, Switzerland: World Resources Association.
Okadera, T., Geng, Y., Fujita, T., Dong, H., Liu, Z., Yoshida, N., and Kanazawa, T. (2015). Evaluating the water footprint of the energy supply of Liaoning Province, China: A regional input–output analysis approach. *Energy Policy*, 78, 148–157.
Peters, G. P., Minx, J. C., Weber, C. L., and Edenhofer, O. (2011). Growth in emission transfers via international trade from 1990 to 2008. *Proceedings of the National Academy of Sciences*, 108, 8903–8908.
Pfister, S., Koehler, A., and Hellweg, S. (2009). Assessing the environmental impacts of freshwater consumption in LCA. *Environmental Science & Technology*, 43, 4098–4104.
Searchinger, T., Heimlich, R., Houghton, R. A., Dong, F., Elobeid, A., Fabiosa, J., Tokgoz, S., Hayes, D., and Yu, T-H. (2008). Use of U. S. Croplands for biofuels increases Greenhouse gases through emissions from land-use change. *Science*, 319, 1238–1240.
Tukker, A., Bulavskaya, T., Giljum, S., de Koning, A., Lutter, S., Simas, M., Stadler, K., and Wood, R. (2014). *The global resource footprint of nations: Carbon, water, land and materials embodied in trade and final consumption*. Retrieved December 5, 2014, from http://creea.eu/index.php/7-project/8-creea-booklet
Tukker, A., Bulavskaya, T., Giljum, S., de Koning, A., Lutter, S., Simas, M., Stadler, K., and Wood, R. (2016). Environmental and resource footprints in a global context: Europe's structural deficit in resource endowments. *Global Environmental Change*, 40, 171–181.
Tukker, A., de Koning, A., Wood, R., Hawkins, T., Lutter, S., Acosta, J., Rueda Cantuche, J. M., Bouwmeester, M., Oosterhaven, J., Drosdowski, T., and Kuenen, J. (2013), EXIOPOL – development and illustrative analysis of a detailed global MR EE SUT/IOT. *Economic Systems Research*, 25(1), 50–70.
Tukker, A., and Dietzenbacher, E. (2013). Global multi-regional input output frameworks: An introduction and outlook. *Economic Systems Research*, 25(1), 1–19.
Tukker, A., Guinée, J., van Oers, L., and van der Voet, E. (2015). *Towards a resource efficiency index of nations*. Discussion paper. Leiden University, CML, Leiden and World Resources Forum, St Gallen.
Tukker, A., and Jansen, B. (2006). Environmental impacts of products: A detailed review of studies. *Journal of Industrial Ecology*, 10(3), 159–182.
Tukker, A., Poliakov, E., Heijungs, R., Hawkins, T., Neuwahl, F., Rueda-Cantuche, J. M., Giljum, S., Moll, S., Oosterhaven, J., and Bouwmeester, M. (2009). Towards a global multi-regional environmentally extended input-output database. *Ecological Economics*, 68(7), 1928–1937.
Wang, S., and Chen, B. (2016). Energy–water nexus of urban agglomeration based on multiregional input–output tables and ecological network analysis: A case study of the Beijing–Tianjin–Hebei region. *Applied Energy*, 178, 773–783.
Wiedman, T., Wood, R., Minx, J. C., Lenzen, M., Guan, D., and Harris, R. (2010). A carbon footprint time series of the UK – results from a multi-regional input-output model. *Economic Systems Research*, 22(1), 19–42.

Wood, R., Hawkins, T. R., Hertwich, E. G., and Tukker, A. (2014). Harmonising national input–output tables for consumption-based accounting – experiences from exiopol. *Economic Systems Research*, 26(4), 387–409.

Wood, R., Stadler, K., Bulavskaya, T., Lutter, S., Giljum, S., de Koning, A., Kuenen, J., Schütz, H., Acosta-Fernández, J., Usubiaga, A., Simas, M., Ivanova, O., Weinzettel, J., Schmidt, J. H., Merciai, S., and Tukker, A. (2015). Global sustainability accounting-developing EXIOBASE for multi-regional footprint analysis. *Sustainability* (Switzerland), 7(1), 138–163.

9

MATERIAL CRITICALITY ASSESSMENT AND RESOURCE NEXUS ANALYSIS

Gavin M. Mudd

Introduction and brief background

There has always been basic human needs for reliable water, food, shelter and community – but the past few centuries has seen substantial growth in demands for a range of metals and minerals to meet the demands of new technologies, infrastructure and modern lifestyles, including both civilian and military needs and desires (e.g. Rankin, 2011; von Gleich et al., 2006). Throughout the 20th century the common metals of copper, iron, lead, zinc, nickel, tin, silver and gold found increasingly widespread use in a range of contexts, such as buildings and infrastructure, piping systems for gas, water and sewerage, telecommunications, power transmission systems and electrical wiring, military equipment, munitions and technologies and consumer appliances, among others (e.g. Graedel et al., 2015; Spitz and Trudinger, 2008). The past few decades, however, has seen a major transformation with modern demands now including a considerably wider range of metals for specialty alloys, high-performance electronics (especially mobile phones, flat screens and computers), chemical catalysts, renewable energy technologies (especially wind turbines, solar photovoltaic panels), hybrid electric-petrol or fully electric transport vehicles and military technologies (e.g. NRC, 2008; Skirrow et al., 2013). These new 'technology' metals include indium, molybdenum, gallium, germanium, selenium, tellurium, cobalt, rhenium, the six platinum group elements, the family of rare earth elements (REEs) and perhaps a few others. Given that the demands for technology metals include uses such as consumer electronics and renewable energy, uses which are expected to grow considerably in coming decades, these metals are often considered vital to meet the reasonable technological, social and environmental needs of our 21st century global society – hence they are often labelled as 'critical' due to their fundamental importance.

These 'technology' metals, however, are generally not mined in their own right but are mostly (if not entirely) derived as by-products incidental to the mining of the primary metals – meaning that supply of a particular technology metal is entirely dependent on the host primary metal, a situation often referred to as 'companion metals'. At present, many technology metals are produced by a small number of countries or companies (or less), meaning any disruption to supplies could have significant consequences for uses such as consumer electronics or renewable energy. For example, rhenium is dominantly sourced from Chile and China has a near monopoly on rare earths – meaning any changes to export policies or supply potential from these countries would significantly and negatively affect all the technological uses of such metals.

Over the past decade there have been a variety of formal assessments of which metals in particular are critical (e.g. BGS, 2015; EC, 2014; NRC, 2008; Skirrow et al., 2013) – along with important developments and improvements in the methodologies adopted to assess the extent to which an individual element can be regarded as critical. This chapter examines the primary concepts involved in mineral resources, mining and supply, examines the factors used to assess criticality, and discusses the wide variety of complex and often inter-related factors which affect the criticality of numerous metals which are increasingly in demand in the modern world. Finally, these issues are discussed in content of the resource nexus – the relationships between metals supply and other environmental, social and economic factors which will underpin our ongoing ability to meet potential supply scenarios for numerous technology metals.

Theoretical approaches to assessing criticality

Over the past decade there has been widespread interest in determining which metals are critical and in improving the methodology used to assess criticality. The most common groups interested in such assessments are government agencies responsible for policy (e.g. US Department of Energy), national or international scientific agencies (e.g. US or British Geological Surveys, US National Academy of Sciences, European Commission), companies (suppliers or consumers) and the industrial ecology research community. In general, there are a range of factors which can be incorporated into criticality assessment, but the fundamental concerns involve the degree of supply risk and the impact of supply disruption to particular uses.

In the US, the National Research Council (NRC, 2008) used a two-axis approach to assessing criticality, involving the importance of use and availability of a given metal or mineral, as shown in Figure 9.1. The first component is focussed on the impacts of possible supply

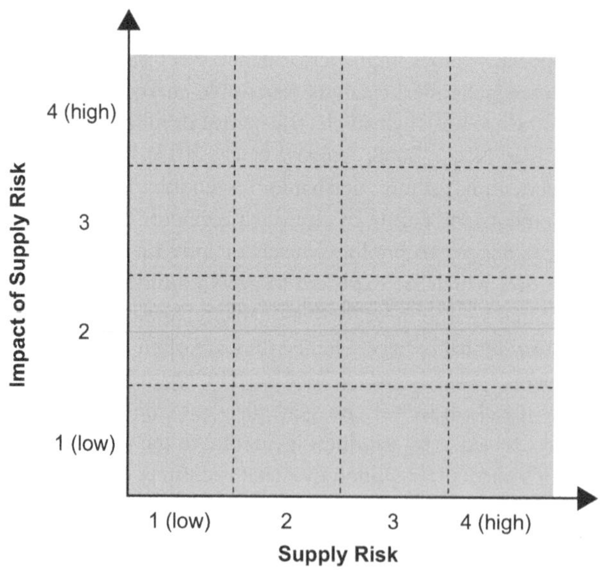

Figure 9.1 US National Research Council framework for assessing criticality

Source: NRC (2008)

disruptions and considers the various uses of a particular metal, the ease and relative cost effectiveness of substitution and the scale of its use. The second component addresses where supply is currently sourced from, including geologic or mineable reserves, current production, the ease and degree of recycling, environmental, economic and social constraints on production (especially political stability), as well as the spread of production across countries or regions. Similarly, the European Commission (EC, 2014) uses dimensions of economic importance and supply risk in their assessments of critical metals and minerals, with their most recent assessment shown in Figure 9.2.

More recently, the Centre for Industrial Ecology at Yale University developed a more comprehensive methodology (see Graedel et al., 2012) for quantifying criticality, keeping supply risk and disruption as key elements but also adding a third dimension of the environmental implications. An example of the matrix approach used to quantify supply risk is shown in Figure 9.3, with a similar matrix for the vulnerability to supply risk. The environmental implications are derived from life cycle impact assessment values for human health and ecosystem impacts. By varying the input values and their relative weighting, different perspectives can be adopted (such as corporate, national or international) as well as the sensitivity of criticality to particular factors (also shown in Figure 9.3).

The major challenge with any assessment of the criticality of a particular metal or mineral is the extent and quality of data which underpins it. For some aspects, such as current primary supply, there is reasonably reliable annual production data for many major metals (e.g. copper, iron, gold, zinc, etc.) but not for some of the technology metals (e.g. gallium, scandium) – but for reserves in particular, there are often major gaps in knowledge of global reserves for many technology metals (in contrast to regular assessments for major metals). Furthermore, understanding key issues such as substitution and vulnerability to a supply disruption require extensive

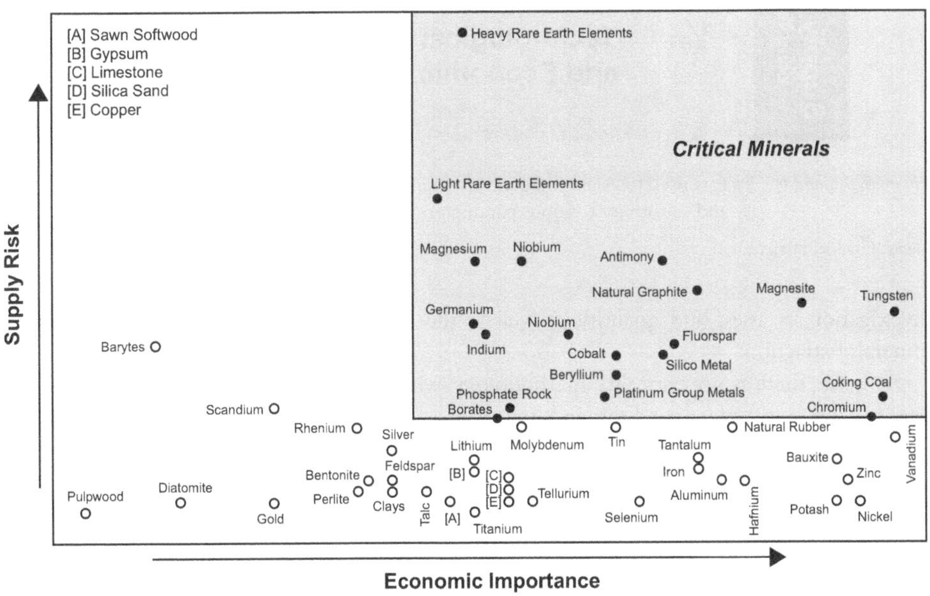

Figure 9.2 European Commission assessment of critical minerals and metals for 2014

Source: redrawn from EC (2014)

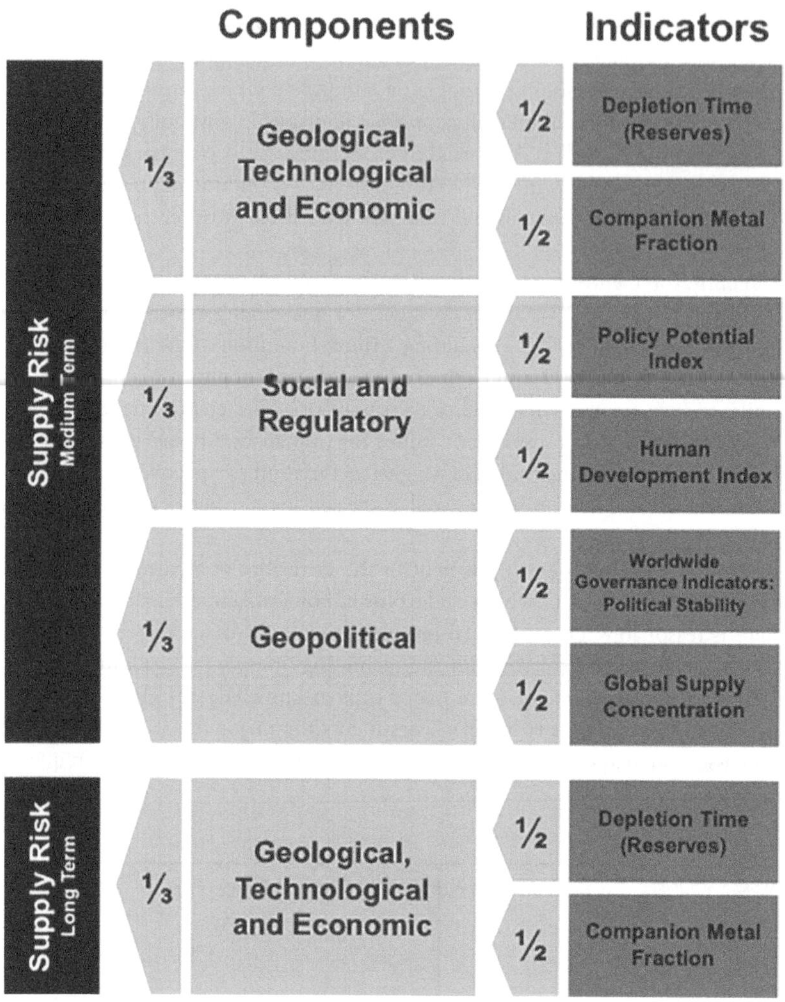

Figure 9.3 a and b Factors used to assess and quantify supply risk (left) and visual representation of criticality and sensitivity to input parameters (right)

Source: Graedel et al. (2012)

investigation to assess and quantify, as well as the environmental implications of metals and minerals extraction.

At heart, there is no criticality methodology which explicitly includes the nexus between energy, food, water, metals and minerals (or even social issues). To an extent they may be able to be included indirectly in some of the factors shown in Figure 9.3 (e.g. higher energy costs as ore grades decline could be included under technological in supply risk), but this leaves such assessments as somewhat subjective. Furthermore, the environmental implications do not directly address the complex relationship between energy and metals and growing impacts as ore grades decline – although they could be indirectly assessed, or values varied for a sensitivity analysis, this is not a satisfactory or comprehensive approach. Recycling can be a key component of current supply (e.g. lead), and it is difficult to address future trends in metals recycling in criticality assessments. Finally, it must be remembered that criticality assessments are a snapshot in time

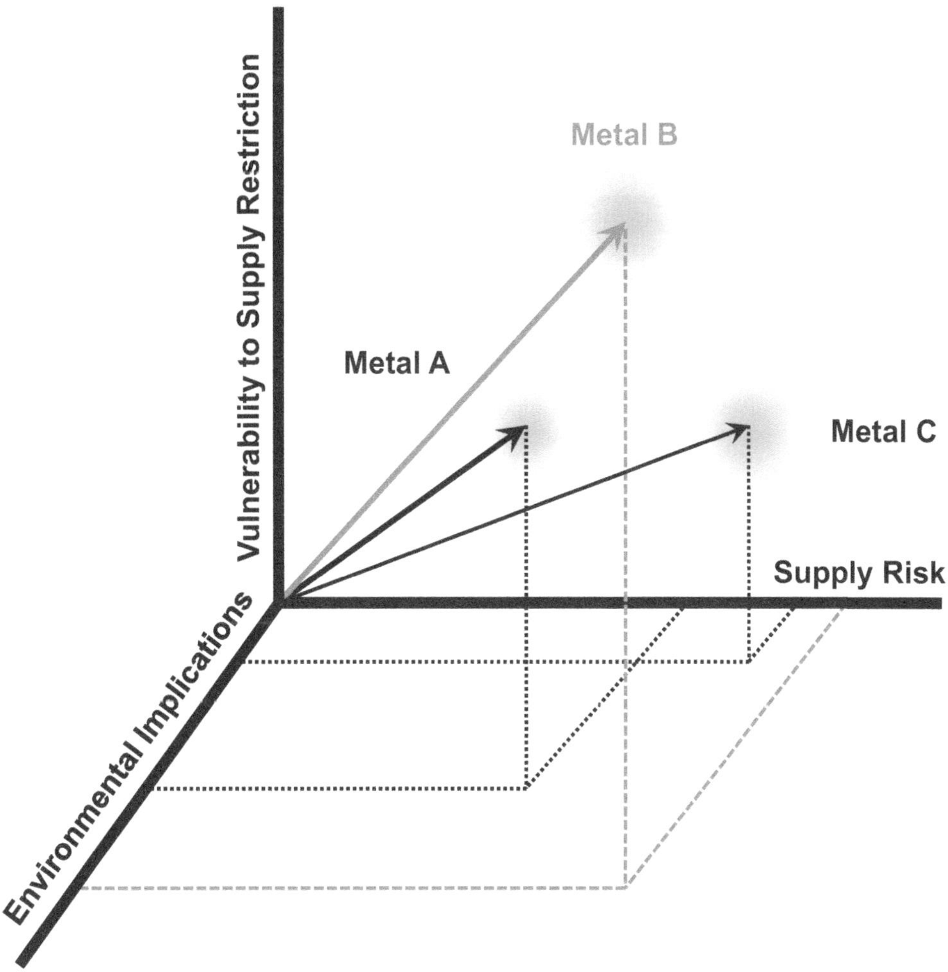

Figure 9.3 a and b Continued

and are not intended to cover temporal changes in metals supply, vulnerability to supply or the growing environmental implications of mining (e.g. more mine wastes, greater energy and water needs, etc.).

In the remaining parts of this chapter, the focus will be on mineral resources and mine production, as these are generally this most crucial aspects in understanding the criticality of any metal or mineral – that is, a review of the key trends and factors which affect metals supply.

Essential concepts and issues in metal mining

Mineral resources

In essence, to make a metal worth extracting from the earth's crust there needs to a sizeable body of enriched rock – that is, rich concentrations in rock from which the metals of interest can be extracted, with typical enrichment factors ranging from tens to thousands of times crustal abundance. Over the past century, and especially the past few decades, the global mining industry has developed standardised methodologies to conduct such assessments, often embodied in codes, guidelines or standards which are required by industry associations or statutory legislation. For example, Australia uses the JORC Code, Canada uses National Instrument 43–101, South Africa has the SAMREC Code, Europe has PERC and so on. The common elements in these and similar approaches are the concepts of ore reserves and mineral resources, which are crucial in understanding the criticality of all metals (or other resources such as oil, phosphate, etc.). In simple terms, ore reserves are based upon detailed drilling, geological, metallurgical, mine planning and numerous other studies to demonstrate that mining has a high likelihood to be profitable, whereas mineral resources are less studied and certain than ore reserves but are similar in nature and there are reasonable prospects for eventual economic extraction. Ore reserves consist of proved and probable reserves, while mineral resources include measured, indicated and inferred resources, with the key components being increasing geological confidence in a mineral deposit and the range of modifying factors which can affect the potential to mine a deposit, such as environmental, economic, legal, metallurgical and other factors which facilitate or hinder development, as shown in Figure 9.4.

For many primary metals, it is possible to compile a detailed assessment of reported mineral resources (or ore reserves), to synthesise a regional, national or even global assessment of a particular metal of interest – arriving at an estimate of tonnes of metal 'X', which can be compared to current global annual production rates to derive a crude estimate of the years of supply currently estimated to be remaining.

While the primary metals enjoy wide reporting, often due to their widespread nature and large volume and economic importance, almost all technological metals are not the main target during primary mining and there is minimal to negligible data reported for such metals in ore reserves or mineral resources – often related to the lack of value attributed to such metals during primary mining. This paucity of reserves/resources data makes it very difficult to assess the ability to meet potential future demand scenarios – which can often be (easily) confused with a lack of reserves/resources. In other words, although we don't have extensive data on the reserves-resources of technology metals does not mean we are potentially constrained by geological resources in the ability to meet growing future demands.

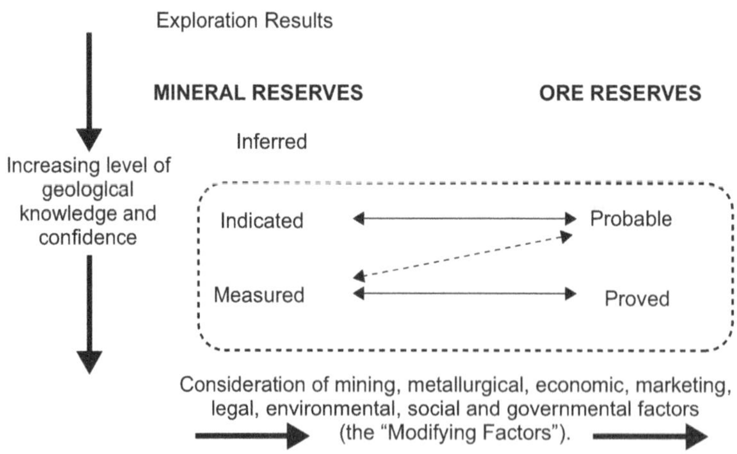

Figure 9.4 Typical conceptual relationship between ore reserves and mineral resources

Source: AusIMM et al. (2012)

Current trends in reserves-resources and future potential

The main group which compiles estimates of global reserves of metals and minerals is the US Geological Survey, published annually in their Mineral Commodity Summaries (USGS, 2016) – although they have data for copper, gold, iron, lead, zinc and many other primary metals, they have no current estimates for gallium, germanium, hafnium, indium, scandium and thallium (and incomplete estimates for others such as tantalum, tellurium).

Overall, there is good evidence that global reserves-resources of most major metals continues to grow – as shown for copper (Mudd et al., 2013a), nickel (Mudd and Jowitt, 2014), platinum group elements (Mudd, 2012), cobalt (Mudd et al., 2013b), lithium (Mohr et al., 2012) and more recently indium (Werner et al., 2017). Although the same has yet to be demonstrated for the various technology metals and minerals (excluding the recent indium study), this is largely a function of the fact that these sectors are still relatively minor in value for the global mining industry and the extent of reporting to date for numerous mineral deposits remains minimal (albeit increasing in recent years). That is, based on experience in the major metals sectors, it is reasonable to expect that similar trends will eventually be discernible for the various technology metals – especially since most technology metals depend on primary metal resources in the first place. The reported reserves or resources of copper are shown in Figure 9.5 over time, including both global and selected countries.

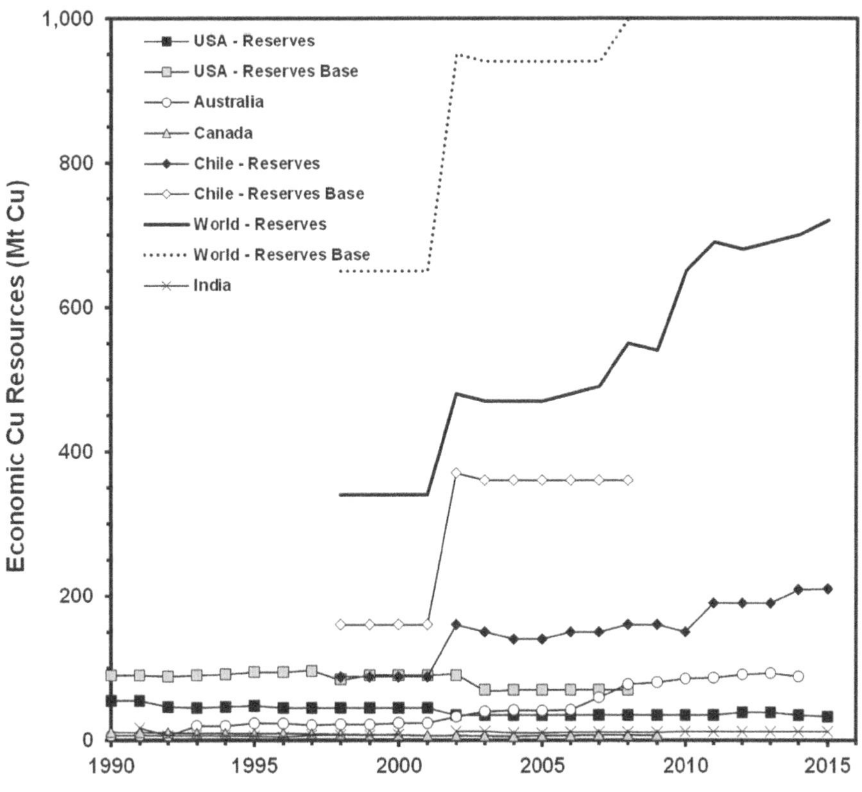

Figure 9.5 Historical trends in copper reserves or resources by global or selected countries

Source: data updated from Mudd et al. (2013a)

Table 9.1 Current global production, reserves and/or resources of various primary and technological metals by world or country

Primary metals	*Copper*	*Lead*	*Zinc*	*Nickel*	*Iron ore*	*Gold*
Global production (2015) [1]	18.7 Mt Cu	4.71 Mt Pb	13.4 Mt Zn	2.53 Mt Ni	3,320 Mt Fe ore	3,000 t Au
Global **reserves** (2015) [1]	720 Mt Cu	89 Mt Pb	200 Mt Zn	79 Mt Ni	186,000 Mt Fe ore	56,000 t Au
Years remaining	~38.5 years	~18.9 years	~14.9 years	~31.2 years	~56.0 years	~18.7 years
Australia – *resources (2014)* [2]	141.0 Mt Cu	50.42 Mt Pb	92.73 Mt Zn	43.1 Mt Ni	139,875 Mt Fe ore	14,013 t Au
Canada – **reserves (2010)** [3]	10.75 Mt Cu	0.4 Mt Pb	4.133 Mt Zn	3.07 Mt Ni	***no data***	1,473 t Au
South Africa – **reserves (2013)** [4]	11 Mt Cu	0.3 Mt Pb	14 Mt Zn	3.7 Mt Ni	650 Mt Fe ore	6,000 t Au
Recent detailed global studies	1,861 Mt Cu [5]	179.0 Mt Pb [6]	536 Mt Zn [6]	299.6 Mt Ni [7]	-	188,680 t Au [8]

Technology metals	*Cobalt*	*Indium*	*Gallium*	*Germanium*	*Lithium*	*Platinum*	*Palladium*	*Rhenium*	*Selenium*	*Tellurium*	*Rare earths*
Global production (2015) [1]	124 kt Co	755 t In	~435 t Ga	~165 t Ge	~32.5 kt Li	178 t Pt	208 t Pd	46 t Re	***no data***	***no data***	124 kt REOs
Global **reserves** (2015) [1]	7,100 kt Co	***no data***	***no data***	***no data***	14 Mt t Li	66,000 **t PGEs**#		2,500 t Re	120 kt Se	25 kt Te	130 Mt REOs
Years remaining	~57.3 years	***no data***	***no data***	***no data***	~430 years	~150 years		~54.[illegible] years	***no data***	***no data***	~1,048 years
Australia – *resources (2014)* [2]	2,610 kt Co	***no data***	***no data***	***no data***	1.712 Mt Li	343 **t PGEs**#		***no data***	***no data***	***no data***	58.1 Mt REOs
Canada – **reserves (2010)** [3]	***no data***	***no data***	***no data***	***no data***	***no data***	***no data***	***no data***	***no data***	***no data***	***no data***	***no data***
South Africa – **reserves (2013)** [4]	***no data***	***no data***	***no data***	***no data***	***no data***	66,000 **t PGEs**#		***no data***	***no data***	***no data***	***no data***
Recent detailed global studies	26,793 kt Co [9]	-	-	-	~23.6 Mt Li [10]	90,783 t PGEs [11]		-	-	-	619.5 Mt REOs [12]
Common Source Primary Metals [1]	Ni, Cu	Zn	Al, Zn	Zn	-	Ni-Cu		Cu-Mo	Cu	Cu, Pb	-

Sources: [1] USGS (2016); [2] GA (2014); [3] Drake (2012); [4] CMSA (2014); [5] Mudd et al. (2013a); [6] Mudd et al. (2017); [7] Mudd and Jowitt (2014); [8] data updated from Jowitt and Mudd (2014); [9] Mudd et al. (2013b); [10] Mohr et al. (2012); [11] Mudd (2012); [12] Weng et al. (2015).

A compilation of various sources of global production and reserves-resources data is shown in Table 9.1, along with some estimates at a national level. The first critical aspect to note is whether the estimate is reserves or resources. For example, Canada reports reserves of 10.75 million tonnes (Mt) of copper and 3.07 Mt of nickel – yet detailed assessments of resources shows 54.12 Mt Cu (Mudd et al., 2013a) and 21.92 Mt Ni (Mudd and Jowitt, 2014). Similarly, South Africa's reserves of PGEs are reported as 63,000 tonnes of PGEs (CMSA, 2014; USGS, 2016) yet when analysing project by project as strictly reported by PGE companies in South Africa, Mudd (2012) estimates reserves of 9,884 tonnes of PGEs with an additional 62,455 t PGEs in resources (i.e. total ~72,339 t PGEs, or more than the global reserves of 66,000 t PGEs from USGS (2016). Although many of the metals display ~20–30 years remaining, this is based on reserves only, and this aligns with the shorter term timeframes for mine planning and not the larger mineral resources – and the value of ~20–30 years of remaining reserves has been maintained for several decades despite growing global production.

Given ongoing mineral exploration, new discoveries, resource expansions, technological improvements in mining and ore processing, favourable economics (especially demand and prices versus input costs), it is reasonable to justify an optimistic view of meeting resource needs for some decades – across both primary metals as well as technology metals. There are several caveats to this, discussed later, as these are crucial in understanding the criticality of different metals.

Current trends in global metals production

In general, almost all metals have continued to display long-term growth in annual production globally. The trends are shown for primary and technological metals in Figure 9.6, using production

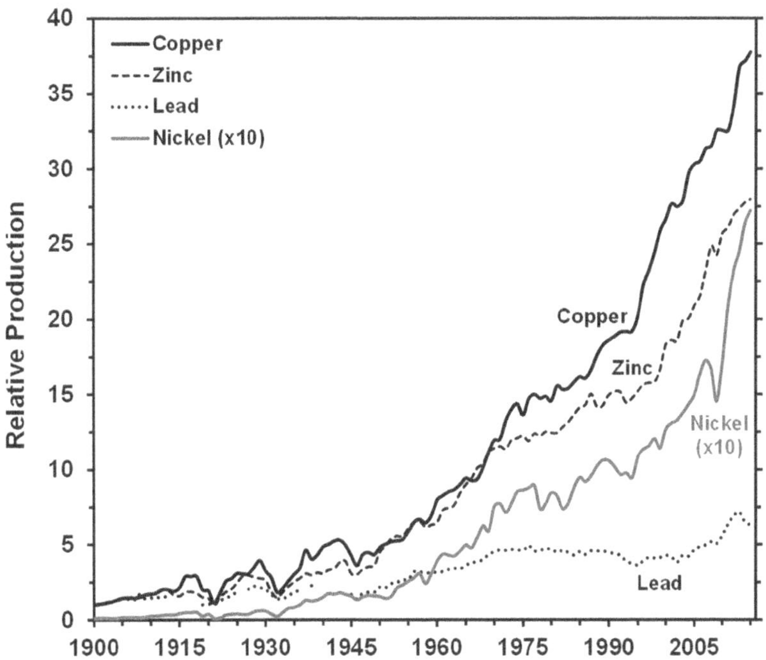

Figure 9.6 Long-term relative trends in global production trends in base and precious metals (top) and several technology metals (bottom) (data relative to 1900 or earliest year of production data)

Source: data from Kelly et al. (2016) and OCE (2014)

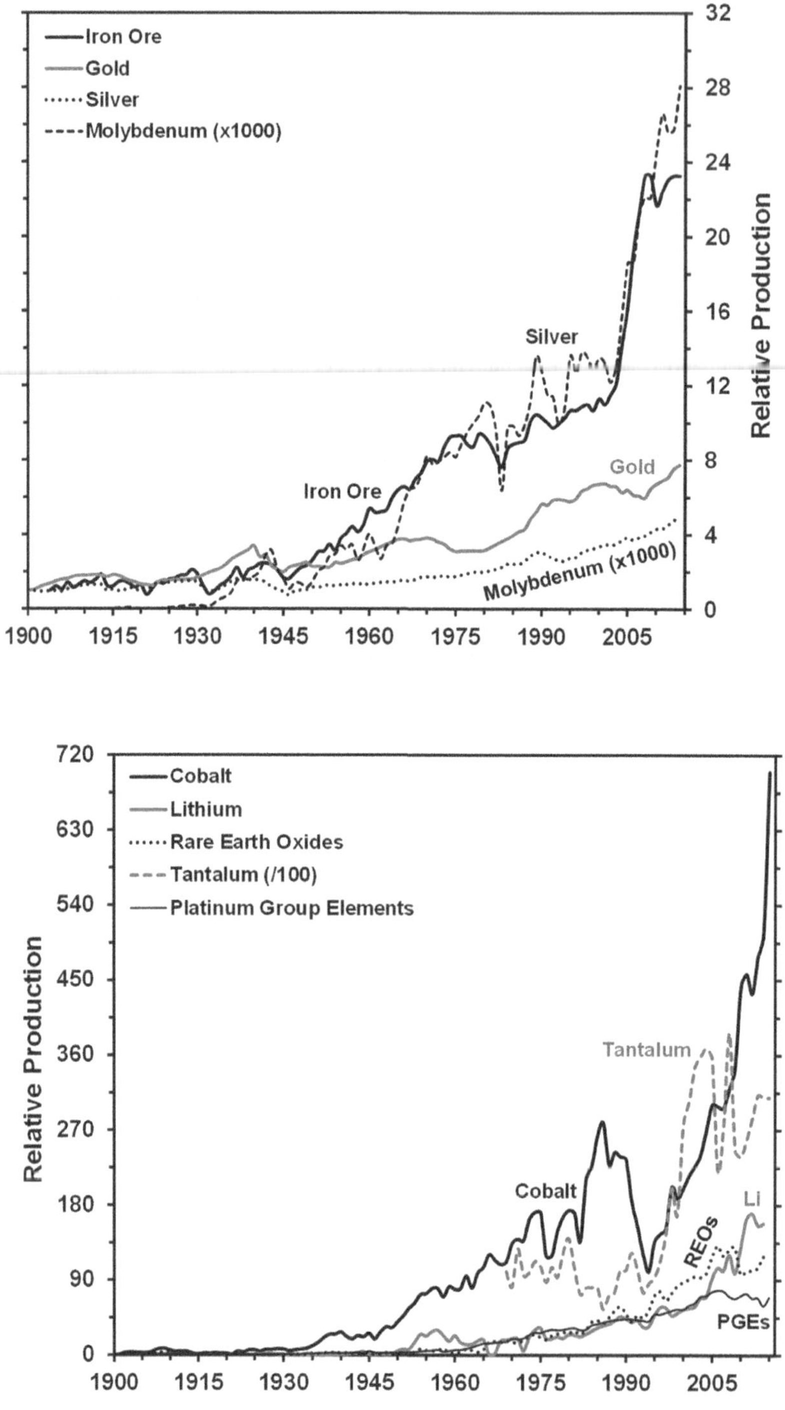

Figure 9.6 Continued

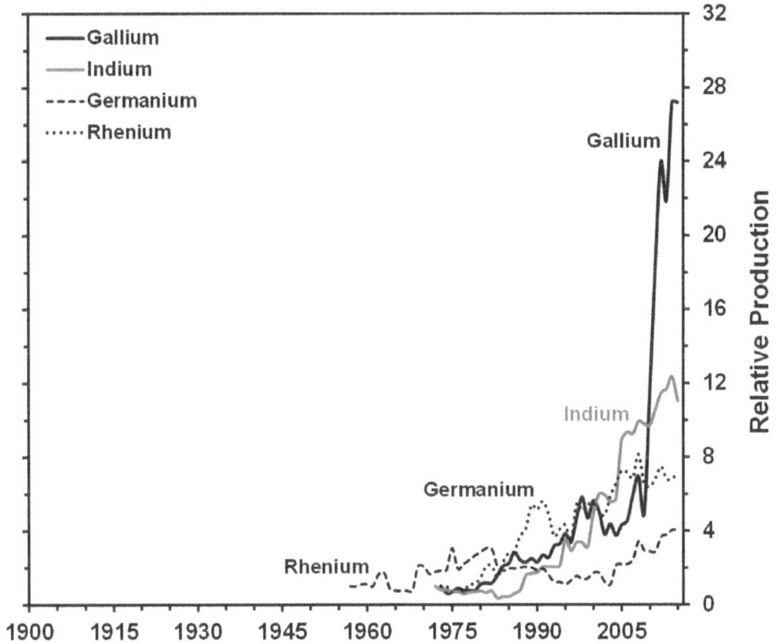

Figure 9.6 Continued

Table 9.2 Global production in 2015, main supplier countries and relative production to 1900 or 2000

Metal	*Global production 2015*	*Largest national producer (fraction)*	*Global production relative*	
			to 1900#	*to 2000*#
Copper	18.7 Mt Cu	Chile (30.5%)	37.8	1.42
Lead	4.71 Mt Pb	China (48.8%)	6.29	1.47
Zinc	13.4 Mt Zn	China (36.6%)	28.0	1.53
Nickel	2.53 Mt Ni	Philippines (20.9%)	272	2.15
Gold	3,000 t Au	China (16.3%)	7.77	1.16
Silver	27,300 t Ag	Mexico (19.8%)	5.06	1.51
Iron ore	3,320 Mt Fe ore	China (41.6%)	23.2[a]	2.06[a]
Molybdenum	267 kt Mo	China (37.8%)	26,700	1.98
Cobalt	124 kt Co	DRC (50.8%)	507[a],[b]	2.54[a]
Gallium	435 t Ga	China	27.2[c]	4.83
Germanium	165 t Ge	China (72.7%)	4.04[d]	2.36
Indium	755 t In	China (49.0%)	11.0[e]	2.25
Lithium[f]	590 kt Li[a,f]	Australia (41.2%)	158[g]	2.89[a]
Platinum group elements	~460 t PGEs	South Africa (51.3%)	69.4	1.26
Rare earth elements	124 kt REOs	China (84.7%)	119	1.36
Rhenium	46 t Re	Chile (56.5%)	7.04[c]	1.28
Tantalum	1,200 t Ta	Rwanda (50.0%)	3.09[h]	1.12

Source: USGS (2016)

Notes: [a]data for 2014; [b]relative to 1901; [c]relative to 1973; [d]relative to 1957; [e]relative to 1972; [f]data remains inconsistent for lithium production, with some sources appearing to be spodumene concentrate (e.g. 590 kt in 2015) and others containing lithium (e.g. 32.5 kt in 2015); [g]relative to 1925; [h]relative to 1969.

relative to 1900 (or the earliest year of production data). Overall, most metals show strong growth, but some display exceptional growth, such as nickel, molybdenum and cobalt – all alloying metals – or lithium and the rare earths, which are both closely aligned with the rise of modern electronics and allied uses. A summary of 2014 global production and the major supplier countries is shown in Table 9.2, showing the clear dominance of China in particular for numerous metals but also Chile for copper and rhenium, Australia for lithium, Rwanda for tantalum and South Africa for PGEs.

A key aspect in assessing criticality is the supply risk – or in other words, whether a supply is dominated by particular countries (or country) or a small number of supplier companies. As shown by Table 9.2, there is a strong concentration of almost all of the important technology metals in just a few countries (as well as several primary metals such as copper) – meaning any changes to export policies could significantly affect the ability to continue to produce the range of technologies which rely on such metals (e.g. renewable energy, consumer electronics). Another important point to note is that some metals in particular are derived from regions which are perceived to be socially unstable or perhaps somewhat volatile (e.g. Rwanda, DRC, South Africa), highlighting that the rest of the world has a clear economic (let alone humane) interest in seeing real progress on human development in these countries.

Finally, given that many technology metals are extracted as opportunistic by-products from metal refineries, it is important to understand these trends over time. For example, indium is dominantly extracted from zinc refineries, rhenium from molybdenum refineries or tellurium from copper refineries – with the ratio of global refined technology metal to host primary metal for these examples shown in Figure 9.7. Curiously, tellurium shows a gradual decline in apparent

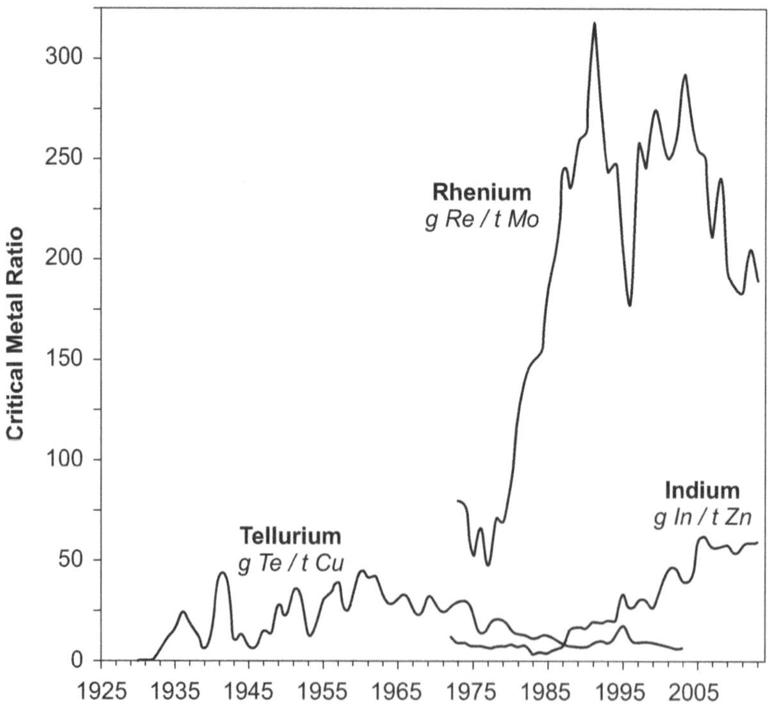

Figure 9.7 Trends in the ratio of selected refined technology metals to dominant host primary metal
Source: data from Kelly et al. (2016)

refinery extraction efficiency, while rhenium appears to have plateaued but indium continues to increase. While these collectively may be a reflection of refinery processing technology, especially the growing dominance of solvent extraction-electrowinning technology in copper refineries (meaning tellurium recovery becomes less economic), it is unclear whether this is also a function of declining ore grades in or concentrate grades from source mining projects. Given that current process technology and recovery rates in primary metal refineries are commonly geared towards the primary metal and not the by-product technology metal, this would suggest that it is possible to increase technology metal production (assuming the normal economic factors of demands, prices and costs are aligned).

Potential constraints to future supply

The depletion of the world's endowment of non-renewable mineral, metal and energy resources often focusses on 'how much is left' or 'how many years left' – yet in reality there are many complex factors at play which will influence how much and at what rate various resources are made available for human use. This is a large area of literature and scholarly debate, and the most important factors are listed and briefly noted herein:

- ***Declining ore grades*** – As documented in the various global resource studies noted earlier, ore grades of primary mining continue to decline for all metals – meaning that more ore has to be mined and processed and generating more tailings per unit metal, increasing the environmental footprint exponentially. At some point in the distant future (i.e. latter 21st century perhaps), it may be more economic and preferable to mine secondary stocks and wastes for their metal content rather than primary mining;
- ***Waste rock (overburden)*** – As mining has moved to increasingly large open-cut mine techniques, this has exponentially increased the quantities of waste rock involved, which increases costs during mining as well as rehabilitation needs and long-term environmental liabilities;
- ***Life cycle assessment (LCA) studies*** – LCA studies show that as ore grades decline, energy, greenhouse gas emissions and other aspects increase exponentially, increasing the environmental footprint of mining substantially;
- ***Value proposition for technology/by-product metals*** – At present, most of the technology metals with strong growth prospects are all produced as a by-product of a primary metal at a refinery (see Table 9.1) and mines receive negligible to zero value for such metals – thus the economic incentives are not clear and direct for increasing production, especially since technology metals remain dependent on their primary host metal in the first place (i.e. there appears to be a need to increase primary metal production to increase technology metal production);
- ***Environmental issues*** – A variety of environmental risks and impacts need to be managed pro-actively during mining and refining of all metals, and many communities remain concerned about such risks, especially mine waste management, acid and metalliferous drainage (i.e. acid mine drainage or AMD), tailings management (especially the potential for catastrophic failures), rehabilitation effectiveness and post-mining land use, carbon price and responses to climate change, dust management, ecological and biodiversity changes, surface water and groundwater resources impacts (especially long-term changes to local or regional hydrologic regimes), etc.
- ***Role of recycling*** – Although metals are in theory recycling, at present many uses are diffuse or in small quantities, making recycling technologically or economically difficult, despite recycling commonly having a lower environmental footprint in the supply of a metal;

- ***Social issues*** – Concerns over social impacts or risks, especially economic benefit sharing, can lead to significant community opposition to new mining projects.

There are many examples globally where deeply held concerns about existing or proposed mining projects has led to some deposits becoming inaccessible to any development. For example, concerns over AMD and other environmental risks led to the exclusion of any development of the large Windy Craggy deposit in British Columbia, Canada (the area is now a world heritage-listed national park); concerns over extensive environmental and social impacts from the (now-former) substantial Panguna copper-gold-silver mine on Bougainville Island in Papua New Guinea led to its shutdown – the ensuing civil war between locals and the national government led to an estimated death toll of thousands of Bougainvillean people; political, security and social concerns has delayed the development of the sizeable Reko Diq project in Pakistan; deeply held environmental and social concerns appears to have effectively delayed any development of the super-giant Pebble project in Alaska, US. This list could continue, but it is sufficient to show that there are numerous and often complex but well-justified reasons to opposition to new mining projects or the expansions of existing mines.

Case study: the complex family of the rare earth elements

Uses of rare earth elements

The rare earths family of elements, strictly the lanthanide family of elements plus scandium and yttrium (i.e. 17 elements), are finding increasing uses in a range of modern technologies due to their unique physical and chemical characteristics. These uses include permanent magnets in renewable energy, medical equipment, military technology, consumer electronics, phosphors, alloys and chemical catalysts – almost all of which are expected to grow in demand substantially in coming decades. For example, the development of high-performance neodymium-iron-boron permanent magnets in the 1980s has allowed the rapid evolution and deployment of various consumer electronics using such magnets (especially computers) as well as improving the performance of turbines for electricity generation (especially wind turbines). Overall, there is a strong growth trajectory expected for rare earth elements to meet the demands of an increasingly affluent global population and its increasingly complex technological mix.

The generation of electricity from wind energy has been the fastest growing source of electricity over the past two decades, growing from a cumulative installed capacity of 7.6 GW in 1997 to 369.6 GW in 2014 (GWEC, 2015). Over time, wind turbines have become larger and more efficient, with current designs for onshore and offshore turbines increasingly using rare earths to improve economics and productivity. For example, recent designs by wind turbine manufacturer Vestas include ~0.5 t REOs to reduce the need for steel, aluminium, polymers and other materials in their V112 wind turbine model, leading to an ~15% reduction in lifetime greenhouse gas emissions intensity from these turbines compared to those turbine models which do not use rare earths (D'Souza et al., 2011; Garrett and Rønde, 2014).

Other strong sources of demand growth for rare earths include electric vehicles, solar photovoltaic (PV) panels, batteries for energy storage (especially in consumer electronics), among others.

The primary problem, however, is that the uses with the best prospects for demand growth are known as the heavy rare earths (or HREEs) (Jowitt et al., 2013; Zepf, 2016), which constitute only a minor fraction of total rare earths production. There are consistent concerns that to meet future HREE demands, especially from secure and stable suppliers, that total REE production will need to substantially exceed current total production of REEs just to meet an individual REE need – this is known as the 'balance' problem. To solve the balance problem is difficult – it implies the need for massive over-production of all REEs just to meet a single REE demand, or the need to find a substitute material with similar properties or effectiveness which hitherto has not been discovered. But from a mining-refining perspective, the total costs and revenues are not driven by a single element but the whole REE family, as well as potential co/by-products such as phosphate, uranium, niobium, iron, zircon and hafnium – making the economics of REE projects difficult to assess and predict.

Mineral resources of rare earths

A recent study by Weng et al. (2015) examined in detail the currently known global mineral resources of rare earths, which are found in a wide range of mineral deposit types. Despite the name 'rare', rare earth deposits are certainly widespread. Although current production is dominated by China due to the super-giant Bayan Obo iron-REE mining complex in Inner Mongolia and the HREE-dominant mines in the Seven Southern Provinces, future production could be sourced from a wide variety of possible projects. Weng et al. (2015) showed that there is at least 619.5 Mt of total rare earths (including REO+Y) known in reported and estimated resources – and given 2015 global production of ~124 kt REOs (see Table 9.1 or 9.2), this means of the order of thousands of years of potential future supply.

Therefore, there is clearly an abundance of already known mineral resources (let alone other potential sources of REOs in poorly quantified mineral resources such as heavy mineral sands, see Mudd and Jowitt, 2016) which can meet optimistic long-term demand scenarios – so the primary issue remains short-to-medium term production and demands and especially the fraction of HREEs.

In the near term, potential REE projects outside of China may need to develop new ore processing technologies, focus on HREE over LREE products, or focus on developing HREE-dominant projects. This is already happening, as the recently approved Dubbo-Toongi project in Australia recently has chosen to stockpile LREE from their proposed process configuration to focus on sales of HREE and other co-products of zirconium, hafnium and ferro-niobium. Of further note is that despite the recent crisis in the rare earths sector (i.e. China's 2010 export restrictions and the massive spike in prices over 2010–2011), only one new REE mine was developed globally (Mount Weld, Australia with a refinery in Malaysia) – highlighting the particular difficulty in developing REE mines. Overall, there needs to be clarity in understanding the difference between known resources and current and future production, especially the HREE-LREE split for projects (although detailed statistics by individual REE are rarely published, especially on a global basis).

Criticality of rare earths

In reality, despite extensive global mineral resources and the potential for production outside of China, the rare earths will remain labelled as critical due to the ongoing supply dominance of China, the fact that there remain little substitutes for the REE in most of their uses and the balance problem makes the economics of REE production difficult to predict.

Discussion: criticality assessments of metals and the resource nexus

The application of criticality methodologies are intended to help in understanding the dynamics of a particular metal (or mineral, or any other resource) – to explore where bottlenecks may exist (e.g. technology metals dependent on primary metals), where opportunities are (e.g. new sources of supply, new process technology), or to investigate policy options (e.g. conflict minerals). At heart, criticality allows the assessment and management of risk (e.g. prices, policy changes, social unrest, etc.). With respect to the resources nexus, however, a variety of aspects remain which current criticality methodology does not address directly (or at least could be addressed indirectly but not clearly), and these are discussed below in sections 5.1 to 5.4.

Water resources and impacts from mining

As mines grow in scale and ore grades decline (e.g. Mudd, 2007; Mudd et al., 2013a), so too does the need for water to supply mines as well as their footprint in catchments (see Northey et al., 2013). That is, bigger mines consume more water but can also present significant risks to water resources (both quantity and quality), as evidenced by recent major tailings dam failures (e.g. Mount Polley, Canada; Samarco, Brazil) or ongoing acid mine drainage (e.g. the Tinto region of southern Spain, Witwatersrand Basin, South Africa), among others. Although the mining industry recognises water as a strategic sustainability issue (e.g. IIED and WBCSD, 2002), with considerable work done by many companies, water-related risks will grow continue to grow in importance and drive a greater nexus between metals supply and water issues. For example, concerns over potential impacts to water resources and associated biodiversity and ecosystems can drive significant social opposition to mining projects (e.g. Windy Craggy, Canada, now stopped; Pebble, Alaska, US, where development is delayed). Although a recent study (Sonderegger et al., 2015) adapted criticality methodology for water, it was intended for water quantity only and does not account for water quality. Therefore, at present, criticality assessments do not address the metals–water nexus.

Energy sources and consumption

Similarly to water, as mines grow in scale and ore grades decline (or if the ore becomes more refractory and difficult to process), they require exponentially more energy (both diesel for machinery and electricity). This represents not only a direct operating cost to mines, but also magnifies the environmental impacts due to the additional impacts from this higher energy consumption (especially if the energy sources are carbon based, such as diesel and coal, which give rise to significant greenhouse gas emissions linked to climate change risks). In theory, increasing energy costs could be quantified using LCA studies and included in the 'geological, technological and economic' component of criticality (see Figure 9.3), but this is currently not done and it would require a comprehensive assessment of all existing mines and mineral resources for their energy intensity – clearly outside the scope of a standard criticality assessment.

The energy needs for the mining of metals could be met with renewable energy sources, such as biodiesel (for machinery), wind turbines, solar photovoltaic panels or solar thermal with heat storage (or a myriad of other options which be present locally), and this could also accommodate greater energy needs – it depends on economics, local renewable energy resources, and ultimately whether the focus is on energy security, capital and operating costs or climate change adaptation.

A related issue is that as the energy intensity of metals increases (assuming no improvements in technology and/or energy efficiency during mining), this will increase the embodied energy of any product using that metal. For example, a wind turbine uses a range of metals (e.g. copper, iron, neodymium, etc.) and as the energy costs increase for these metals, the embodied energy of the wind turbine increases, meaning the energy return on energy invested, declines also. This leads to a greater need for energy productivity in the wind turbine, which given the rapid evolution of wind technology in recent decades, shows that there is certainly room for optimism on this front. There are similar arguments which could also be presented for other technologies used in energy systems, such as solar panels (especially the wide array of possible metals which can be used in various designs), lithium-based batteries or even geothermal or solar thermal systems, but there is a strong need for considerable research in this area.

In the end, although criticality assessments may conduct sensitivity analyses to address the energy issues for metals supply (and use for energy supply), they cannot explicitly capture the complex relationships and issues which link the supply of metals and energy.

Metals required to supply metals

A rarely examined issue is the metals required to mine and produce metals (e.g. t Cu input per t Cu supply). As ore grades decline, the size of mineral deposits often increases (see Mudd et al., 2017; Mudd et al., 2013a), meaning that larger project scales are possible and economic and they can therefore produce greater amounts of metals over a mine's life. The larger processing plants and bigger excavators and haul trucks clearly require more metals to manufacture, but whether this is offset by the larger size of the project and greater amounts of metals produced is unclear. While it can reasonably be expected not to be a significant issue, especially in comparison to energy and water nexus issues, it is one which remains poorly addressed.

Social and economic aspects

Social and economic issues, perhaps unsurprisingly, can be especially difficult to address well in criticality assessments (as noted with some examples previously).

The current Yale-led methodology includes components of social, regulatory and geopolitical factors (see Figure 9.3), using data from regular surveys such as the Fraser Institute's annual survey of the mining industry's perception of investment in a particular country or jurisdiction or the United Nation's Human Development Index (UNHDI). While these are useful, they are surveys conducted external to communities and do not necessarily represent a local community's views towards mining – and therefore supply risk may be incorrect and often overly optimistic (e.g. numerous mining-related social conflicts across the breadth of the Americas; e.g. Tambogrande, Pebble, Famatina, Pascua Lama, Mount Tolman, Crandon, Marlin, etc.). The Fraser Institute is an industry body and certainly does not represent community views. To address this, community surveys could be undertaken or literature analysed to explore sources of conflict (e.g. Davis and Franks, 2014), but this requires extensive work – although it may deliver very different results to those suggested by the Fraser Institute survey or the UNHDI on their own. The common themes in community concerns with respect to mining projects include environmental risks (e.g. water depletion, water quality impacts from AMD, landscape change), economic impacts (e.g. benefits go to multinational companies and not locally compared to the level of negative impacts received, such as higher costs of living, no investment in education by government, etc.) or cultural issues and sacred sites (i.e. sites fundamental to local culture and need protection; e.g. former Coronation Hill Au-Pd-Pt-U project in Australia). Another area of great social complexity is the issue of conflict minerals in central Africa – although the US included sections in the Dodd-Frank Act to address this issue, with some of the most critical metals emanating from this region, it remains controversial as to whether policy intervention in this way has helped or hindered the negative impacts on the ground. In reality, the incorporation of social aspects into criticality assessments will always be constrained by the methodology adopted, and especially victim to the sources of data used (and their independence and relevance to the communities involved), but it remains fundamental that such issues be acknowledged and considered.

Economic (or in reality, more socio-economic) issues are another area of great difficulty for criticality assessments. While standards measures for different regions have readily available metrics, such as income per capita or level of education, these are not necessarily indicative of social concerns with respect to mining and therefore supply risk. To a limited extent, they can be included in the Fraser Institute survey results, as opinions by mining company management on social risks are implied in their results, but this is far from a satisfactory basis upon which to explore such factors which could affect supply risk. In addition, costs of labour, energy, chemicals, infrastructure and taxation regimes may also affect economic risks of mining projects, especially in some countries perceived to be more volatile politically.

Overall, it is possible to conceive of a variety of ways to add several axes for criticality assessment to the current three-axis approach (perhaps leading to a matrix-based assessment) – such as social, economic, energy, water and so on. In the end, the level of detail is up to the user of such assessments, and in particular their purpose – for example, a corporation sourcing metals, a government managing high-technology industry sectors, military defence procurement strategies, individuals concerned about the supply of the metals used in goods and services they consume, or similar.

Summary: insights into the criticality assessments of metals and the resource nexus

Criticality assessments provide a methodology or framework to understand the current supply and uses of metals and minerals (or other materials or resources), and allow deeper insights into the dynamics of a particular metal and its applications. A wide variety of metals, especially important technology metals, are currently labelled as critical due to the supply dominance of a few countries (or less) and the lack of their substitutability for crucial applications. However, the labelling of such metals as critical has often led to confusion that we are 'running out' – despite strong evidence that there are still abundant mineral resources to meet current and future demands for some decades. Furthermore, there is little recognition that such criticality assessments are based on a wide array of data and assumptions, and that often there is a major lack of data to underpin the detail required for a thorough criticality assessment. There is clearly a need to consider a variety of nexus-related issues in criticality, as they can indeed be important and even affect supply – the extent of analysis can be widened, deepened and expanded in scope to include a wide array of factors and issues (e.g. social, economic, energy, water), but this also makes the assessment more difficult to complete. Ultimately, it depends on your purpose as to what level of detail one works towards – it remains fundamental to acknowledge that the criticality of a particular metal is dependent on a considerably more complex array of factors than those currently addressed in criticality assessments.

Acknowledgements

This work has benefited from strong funding support from Australia's CSIRO over several years, especially through the Collaboration Clusters on Mineral Futures (2009–2012) and Wealth From Waste (2013–2016), as well as funding of PhD students (especially Zhehan Weng and Stephen Northey). Many community groups around Australia and the world also continue to help me understand the modern mining industry and its environmental issues. I also acknowledge the generous collaboration with Prof. Thomas Graedel and his industrial ecology group at Yale University and Dr Simon Jowitt at the University of Nevada Las Vegas (especially during his former time at Monash University).

References

AusIMM, MCA, and AIG. (2012). *Australasian code for reporting of exploration results, mineral resources and ore reserves: The JORC code*. Parkville, VIC: Joint Ore Reserves Committee (JORC) of The Australasian Institute of Mining and Metallurgy (AusIMM), Minerals Council of Australia (MCA) and Australian Institute of Geoscientists (AIG), December, 44 p.

BGS. (2015). *Risk list 2015 – current supply risk for chemical elements or element groups which are of economic value*. British Geological Survey (BGS), 11 p.

CMSA. (2014). *Facts and figures 2013–2014*. Johannesburg, South Africa: Chamber of Mines of South Africa (CMSA), 44 p. Retrieved from www.bullion.org.za.

Davis, R., and Franks, D. M. (2014). *Costs of company-community conflict in the extractive sector*. Corporate Social Responsibility Initiative Report No. 66. Cambridge, MA: Harvard Kennedy School, Harvard University, 56 p.

Drake, A. (2012). *Canadian reserves of selected major metals and recent production decisions*. Ottawa, Ontario: Natural Resources Canada (NRC). Retrieved from www.nrcan.gc.ca/mining-materials/exploration/8294.

D'Souza, N., Gbegbaje-Das, E., and Shonfield, P. (2011). *Life cycle assessment of electricity production from a V112 Turbine Wind Park*. Randers, Denmark: Vestas Wind Systems, 87 p.

EC. (2014, May). *Critical raw materials for the EU – report of the Ad-Hoc working group on defining critical raw materials*. Brussels, Belgium: European Commission (EC), 41 p. Retrieved September 25, 2015, from http://ec.europa.eu/enterprise/policies/raw-materials/files/docs/report-b_en.pdf

GA. (2014). *Australia's identified mineral resources 2014 – summary*. Canberra, ACT: Geoscience Australia (GA), 16 p. Retrieved from www.ga.gov.au.

Garrett, P., and Rønde, K. (2014). *Life cycle assessment of electricity production from an onshore V126–3.3 MW wind plant*. Aaehus, Denmark: Vestas Wind Systems, 116 p.

Graedel, T. E., Barr, R., Chandler, C., Chase, T., Choi, J., Christoffersen, L., Friedlander, E., Henly, C., Jun, C., Nassar, N. T., Schechner, D., Warren, S., Yang, M-Y., and Zhu, C. (2012). Methodology of metal criticality determination. *Environmental Science & Technology*, 46(2), 1063–1070.

Graedel, T. E., Harper, E. M., Nassar, N. T., and Reck, B. K. (2015). On the Materials Basis of Modern Society. *Proceedings of the National Academy of Sciences (PNAS)*, 112(20), 6295–6300.

GWEC. (2015, March). *Global wind report – annual market update 2014*. Brussels, Belgium: Global Wind Energy Council (GWEC), 80 p.

IIED, and WBCSD. (2002). *Breaking new ground: Mining, minerals and sustainable development*. London: Earthscan for International Institute for Environment and Development (IIED) and World Business Council for Sustainable Development (WBCSD).

Jowitt, S. M., and Mudd, G. M. (2014). *Global gold: Current resources, dominant deposit types and implications*. Proc. "Gold14 @ Kalgoorlie", Kalgoorlie, Western Australia, pp. 56–58.

Jowitt, S. M., Weng, Z., and Mudd, G. M. (2013, June). Rare Earth Elements: Deposits, Uncertainties and Wasted Opportunities. *Materials World*, 21(6), 22–24.

Kelly, T. D., Matos, G. R., Buckingham, D. A., DiFrancesco, C. A., Porter, K. E., Berry, C., Crane, M., Goonan, T., and Sznopek, J. (2016). *Historical statistics for mineral and material commodities in the United States*. US Geological Survey (USGS), Reston, Virginia, USA, Data Series 140 (Supersedes Open-File Report 01–006), Version 2014 (Online Only). Retrieved February 6, 2017 and December 22, 2016, from minerals.usgs.gov/minerals/pubs/historical-statistics/

Mohr, S. H., Mudd, G. M., and Giurco, D. (2012). Lithium resources, production: Critical assessment and global projections. *Minerals*, 2(1), 65–84.

Mudd, G. M. (2007). An analysis of historic production trends in Australian base metal mining. *Ore Geology Reviews*, 32(1–2), 227–261.

Mudd, G. M. (2012). Key trends in the resource sustainability of platinum group elements. *Ore Geology Reviews*, 46(1), 106–117.

Mudd, G. M., and Jowitt, S. M. (2014). A detailed assessment of global nickel resource trends and endowments. *Economic Geology*, 109(7), 1813–1841.

Mudd, G. M., and Jowitt, S. M. (2016). Rare earth elements from heavy mineral sands: Assessing the potential of a forgotten resource. *Applied Earth Science* (Trans. Inst. Min. Metall. B), 125(3), 107–113.
Mudd, G. M., Jowitt, S. M., and Werner, T. T. (2017). The world's lead-zinc mineral resources: Scarcity, data, issues and opportunities. *Ore Geology Reviews*, 80, 1160–1190.
Mudd, G. M., Weng, Z., and Jowitt, S. M. (2013a). A detailed assessment of global Cu resource trends and endowments. *Economic Geology*, 108(5), 1163–1183.
Mudd, G. M., Weng, Z., Jowitt, S. M., Turnbull, I. D., and Graedel, T. E. (2013b). Quantifying the recoverable resources of by-product metals: The case of Cobalt. *Ore Geology Reviews*, 55, 87–98.
Northey, S.A., Haque, N., and Mudd, G. M. (2013). Using sustainability reporting to assess the environmental footprint of copper mining. *Journal of Cleaner Production*, 40, 118–128.
NRC. (2008), *Minerals, critical minerals and the US economy*. Washington, DC: Committee on Critical Mineral Impacts on the U. S. Economy & Committee on Earth Resources, National Research Council (NRC) of the National Academies, 263 p.
OCE. (2014). *Resources and energy statistics 2014*. Canberra, ACT: Office of the Chief Economist (OCE, formerly Bureau of Resources & Energy Economics, BREE), Department of Industry, Australian Government, 174 p.
Rankin, W. J. (2011). *Minerals, metals and sustainability – meeting future material needs*. Melbourne, Australia: CSIRO Publishing and CRC Press.
Skirrow, R. G., Huston, D. L, Mernagh, T P., Thorne, J. P., Dulfer, H., and Senior, A. B. (2013). *Critical commodities for a high-tech world: Australia's potential to supply global demand*. Canberra, Australia: Geoscience Australia (GA), 128 p.
Sonderegger, T., Pfister, S., and Hellweg, S. (2015). Criticality of Water: Aligning Water and Mineral Resources Assessment. *Environmental Science & Technology*, 49, 12315–12323.
Spitz, K., and Trudinger, J. (2008). *Mining and the environment – from ore to metal*. Boca Raton, FL: CRC Press, Taylor & Francis Group. 891 p.
USGS. (2016). *Mineral commodity summaries 2016*. Reston, VA: US Geological Survey (USGS), 205 p. Retrieved from minerals.usgs.gov/minerals/pubs/mcs/.
von Gleich, A., Ayres, R. U., and Gößling-Reisemann, S. (Eds.). (2006). *Sustainable metals management: Securing our future – steps towards a closed loop economy*. Dordrecht, The Netherlands: Springer, 607 p.
Weng, Z., Jowitt, S. M., Mudd, G. M., and Haque, N. (2015). A detailed assessment of global rare earth resources: Opportunities and challenges. *Economic Geology*, 110(8), 1925–1952.
Werner, T. T., Mudd, G. M., and Jowitt, S. M. (2017). The world's by-product and critical metal resources Part III: Indium. *Ore Geology Reviews*. doi:10.1016/j.oregeorev.2017.01.015.
Zepf, V. (2016). An overview of the usefulness and strategic value of rare earth metals. In I. B. De Lima and W. L. Filho (Eds.), *Rare earths industry – technological, economic, and environmental implications*. Amsterdam, The Netherlands: Elsevier, 485 p.

10

INDUSTRIAL ECOLOGY METHODS AND THE RESOURCE NEXUS

Ester van der Voet

Introduction

In the previous chapters, several methods have been discussed from the field of industrial ecology through the lens of the resource nexus. The industrial ecology object of study is the economy, or even broader, society, in terms of the flows and stocks of materials and energy. This is referred to as society's metabolism, in analogy to the metabolism of an organism ingesting food and energy to build up its biomass, and exhausting or excreting waste in the process. Other terms used are the technosphere (as a separate physical "sphere" in addition to atmosphere, biosphere and geosphere), and the physical economy (to set it apart from the monetary economy). It acknowledges the physical nature of our society, its commodities and its processes, and helps us realise that the laws of nature apply in our economy, in addition to the laws of our societies. Over the years, it has become apparent that the study of society's metabolism generates valuable information to help understand the complex relation between society and economy on the one hand, and the environment on the other hand. Figure 10.1 depicts this.

In this figure, the connection between the symbolic aspects of society and the physical environment is depicted as the (still symbolic) economy and its physical counterpart the technosphere. It shows that the impact of society and economy on the environment is never direct, but in all cases operates via the technosphere. Economy, politics and (cultural) behaviour serve as driving forces and sometimes as correcting mechanisms, but ultimately these must work out in the technosphere to have any impact on the environment. Therefore, the technosphere is an essential element in studying resource nexus issues, and consequently methods modelling the technosphere are essential for an analysis of resource nexus issues.

The previous chapters address methods that in one way or another describe society in physical terms. Each of these methods has some relevance for nexus issues. In this chapter, we attempt a summing up. What do the various methods tell us, is a combination of methods worthwhile, and can we move towards an integrative framework? These questions will be, as far as possible, answered in this chapter.

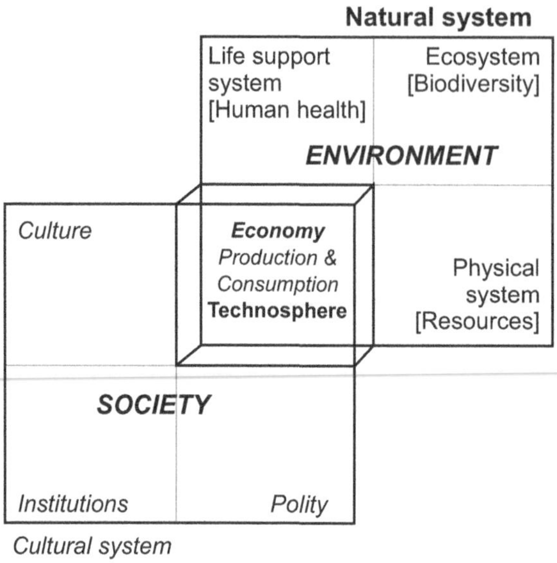

Figure 10.1 The industrial ecology system

Source: after Huppes and Ishikawa (2009)

Industrial ecology methods and the resource nexus

In the previous chapters, we have described the following methods and their relevance for the resource nexus:

- Life Cycle Assessment and Life Cycle Sustainability Assessment (LCA and LCSA) (van der Voet and Guinée, Chapter 4);
- Material Flow Analysis and Accounting (MFA) (Schaffartzik et al., Chapter 5);
- Footprint analysis (Giljum et al., Chapter 6);
- Environmentally extended input–output analysis (EE-IOA) (Tukker and Font Vivanco, Chapter 8);
- Criticality assessment (Mudd, Chapter 9).

These methods can be classified into two main families:

- The LCA family: methods that have functional production-consumption systems and specify supply chains: LCA, footprinting and MR-EE-IOA;
- The MFA family: methods addressing resources and resource flows with geographically bounded systems: MFA and criticality assessment.

Note that IOA, and also the EE-IOA variant relevant for resource analysis, in fact also have geographically bounded systems. The relevance for resources and the resource nexus, however, is derived from the multi-regional variant covering the whole world, as is described in Chapter 7. In the MR-EE-IOA system, it is possible to start from consumption and trace back (global) supply chains until the point of resource extractions.

The LCA family allows to follow supply chains from the cradle to the grave, which enables for a comprehensive analysis from a functional point of view, including all steps in the chain wherever they may occur. With LCA, the resource needs of products (including other resources) can be assessed at the micro-level. LCA is suitable to calculate the resource intensity of producing certain resources at a detailed level. Multiplyers of energy for metals, metals for energy, water for energy, energy for water and many other such linkages can be derived, covering the complete suite of five nexus dimensions. This corresponds to the notion of micro-level resource footprints. MR-EE-IOA can do similar things at the macro-level. The footprints of energy, water, land and a limited selection of material resources can be covered at present. The information is not based on technology but on monetary relations, linked to the environmental extensions that represent resource extraction. Although this hardly offers options to address nexus issues, it does put them in a larger scale perspective: by combining the national or global level extractions with information on the availability of resources, we can conclude whether certain developments may be problematic.

Both LCA and MR-EE-IOA work with a system that links (production) processes into chains, which is essential to follow all steps in the supply chain back to the extraction of resources. Both offer opportunities of linking resource use with other aspects: LCA with technology, technology development and innovation, and IOA with macro-economic variables such as trade, national income, employment and labour productivity. Compared to LCA, EE-IOA lacks the technological detail which is essential to really understand these linkages. On the other hand, it covers global supply chains and can be comprehensive as well as location specific. These features are important to assess the importance and consequences of such linkages.

Both LCA and MR-EE-IOA are, next to more specifically targeted methods, used to calculate different types of footprints (energy, land, water, materials, carbon). LCA calculated those for micro-level products or services, IOA does so for macro-level goods and services or consumption categories. Such footprints are really valuable to understand the linkages between resources as well: they, in fact, give us the resource intensity of producing resources.

The MFA family differs from those. They have a geographic system boundary, usually corresponding to administrative units. Material Flow Accounting is used at the level of national economies. Substance Flow Analysis is used at different scale levels, from global to local (municipalities or industrial parks).

SFA's main drawback is that it typically covers only one resource, thereby excluding linkages between resources by definition. Such linkages can be brought in by combining the analysis of different substances into one, as is done sometimes for co-produced metals (refs) and recently for metals and energy (refs). It doesn't come naturally to the method, though, and will never be encompassing.

SFA also has some strong points that make it suitable for resource analysis. Its flexibility in scale level allows for combining detailed technical information with global level assessments. The main advantage of SFA is the dynamic nature of the tool, allowing to include stock dynamics and making forecasts that include such dynamics. No other tool has these valuable characteristics. These features are presently hardly used in the analysis of the resource nexus, but might be more important in future, as the need arises for more quantitative forecasting and exploring of options for mitigation. In criticality analysis we can see a start: co-produced metals and the environmental dimension of criticality indices show a direction for that.

Material Flow Accounting is a statistical activity that in principle covers all material resources, and for which time-series data are now available covering the world and several decades. It cannot be used as a model and does not have causal relationships built in. However it can be used to derive resource-related indicators to link to other statistical outputs, such as the GDP. This allows to calculate indicators on resource efficiency and resource productivity at an aggregate level. There is a global coverage now and a significant time series, starting in 1970, which allows to follow these indicators through time. Like MR-EE-IOA, it specifies trade relations, implying that also changes over time in the consumption as well as production patterns of resource use can be detected. A recent trend is the establishment of a link between MR-EE-IOA and MFA, enabling to calculate material or resource footprints.

Integration: a framework of methods for analysis of the resource nexus

All of the above, it can be concluded that different methods cover different aspects of the resource nexus. That implies that it would be helpful to apply more than one method, or even to try and integrate methods to cover a larger share of the relevant aspects.

Table 10.1 summarises the different aspects of the methods covered in this chapter. It shows the strengths and weaknesses of the different methods in assessing the resource nexus. The features that are particularly interesting for the analysis of resource nexus issues are highlighted in bold.

The LCA family in essence have more relevant features than the MFA family, because of the functional systems. However the MFA family adds some relevant characteristics that are also very useful, even essential, in an analysis of the resource nexus. Possibly the way forward lies in the integration of different methods to cover more relevant aspects.

Integration of methods to some extent has been attempted already (Pauliuk et al., 2015). Integration of LCA and IOA is already quite established as hybrid LCA. It is regarded as a variant of LCA where the IO system provides the background, so no cut-off is needed and side-effects are to some extent visible (Suh and Huppes, 2002). Integrating LCA in IOA linked to economic models is done to assess rebound effects (Font Vivanco et al., 2015). For the macro-level footprint analysis, MFA and IOA are integrated to some extent: the Domestic Extraction indicator of the MFA system is used as environmental extension in the IOA system, both at the aggregate and the specific level (Wiedmann et al., 2015). This enables to calculate material or resource footprints at the macro-level. Attempts to integrate IOA and SFA so far has not been very successful – not because the models or systems cannot be reconciled, but because the highly specific information related to individual substances cannot be linked to the much lower resolution of the IOA system. It is possible to make a "soft" link using identical systems and sector classifications, but bringing in much more detail (Deetman et al., 2017). MFA and LCA are linked in recent attempts to specify resource scenarios and their environmental impacts, thus providing a link between materials use and energy requirements or emissions to the environment (refs).

The common element of all methods seems to be commodities. Products are the object of LCA. In IOA they are aggregated to consumption categories, containing a class of products. MFA is about materials but in the accounting as well as modelling information is needed on commodities: the composition of products to identify flows of materials, and the lifespan of products as an essential element in dynamic modelling. Demand for commodities is an important driving force for material flows, and also has linkages to the economy in terms of prices. It is therefore also essential information for criticality assessments.

Commodities as a starting point allows for detailed, technology-specific information. This is essential for understanding linkages between resources, and therefore for generating options to generate options to mitigate any unwanted effects. Commodities are also a good starting point

Table 10.1 Summary of different aspects of methods

Methods	***Features***								
	Scale level	*Time horizon*	*System*	*Functional/ territorial*	*Coverage of resources*	*Coverage of linkages*	*Insight in mechanism*	*(Global) reference*	*Linkable to socio-economic aspects*
LCA	Micro	N/A	**Product/service**	**f**	**All**	**All**	**y**	n	**y**
MFA	Macro	Past (since 1970), present	National economies	r	**Minerals, metals, fossil fuels, biomass**	None	n	**Potentially**	**y**
SFA	**Any**	**Past, present, future**	Geographical at any scale level	r	Any material, but one at a time	None or limited	**y**	**y**	n
MR-EE-IOA	Macro	Present	National economies	**f**	**A selected number**	**All within scope**	**y**	**Potentially**	**y**
Footprinting	**Any**	Past, present	**Service, national economy, lifestyle, company**	**f**	Any but one at a time	None	n	**y**	n
Criticality assessment	Micro	N/A	Global resource	r	Any but one at a time	None	n	**y**	**y**

for a bottom-up analysis at larger scale levels. The larger scale is essential to assess whether the physical linkages between resources actually are problematic in some way, and to assess policy measures on their potential effectiveness as well as side-effects. For a quantified analysis of resource nexus issues, an integration of methods would have to be attempted bottom-up, with commodities at the core, and aiming at linking all these commodities in one comprehensive system.

However, integrating these methods into one super-tool would be a rather grandiose affair, which would take many years of interdisciplinary work to develop. A more modest and therefore feasible road to travel could be to define a methodological framework. Since life cycle information is crucial to understanding resource nexus issues and their dynamics, the best possibility for a resource nexus analytic framework therefore seems to be the Life Cycle Sustainability Analysis (LCSA) framework. This framework is not prescriptive in the methods used, but has the life cycle approach as its starting point, and intends to expand the narrowly and strictly defined micro-level LCA to encompass larger scale levels and wider time horizons, as well as linking the purely physical analysis to economic and social aspects. MR-EE-IOA as well as various brands of MFA could find a place in such a framework. Which methods to use would then depend on the questions at hand. The underlying database that would be most useful is one containing information on commodities: their composition and lifespan, but also time series of their demand and supply, that can be used to derive driving force variables for forecasts. Data on demand and supply are to some extent available in statistics such as PRODCOM or COMTRADE. There is presently no comprehensive database on composition or lifespan of products. The body of knowledge in that area is growing via a number of projects on urban mining and the circular economy.

Concluding remarks

Industrial ecology is an essential research field for analyzing the resource nexus. Its object is the technosphere, also referred to as the physical economy or as society's metabolism: a description and modelling of society in terms of flows and stocks of materials and energy. Any analysis of resource nexus issues starts there, at the issues themselves, before it can be expanded to cover socio-economical aspects and policy. Therefore, industrial ecology methods are essential methods to use in an analysis of resource nexus issues. The various methods covered in Part II each have their own way of approaching these issues, showing relevant features but also blind spots. Taken as a group, the relevant aspects of the physical economy are covered. To gain analytic power, an integration of methods could be attempted. A core shared by all these methods is commodities: flows and stocks of resources, materials and products throughout society. Commodities therefore provide a good starting point to connect the methods. An integrated method should cover different scale levels, dynamics in time and explicit linkages. A less ambitious but probably also effective and more flexible option could be to define a framework of methods under the LCSA umbrella, allowing the user to pick and choose methods to combine, specifically for the topic at hand.

Industrial ecology methods allow for analyzing resource nexus issues in physical terms. The analysis offers openings for mitigation in a technical sense. Exploration of the future by scenario analysis is an emerging activity, and a promising new direction for development of industrial ecology methods that is also relevant for resource nexus issues. Developing policy scenarios cannot be done by these methods alone. Other types of information, from economics, sociology, politicology and governance, is then necessary as well, as is shown in other chapters of this volume. Industrial ecology methods however should be included in all attempts to investigate resource nexus issues.

References

Deetman, S., van Oers, L., van der Voet, E., and Tukker, A. (2017). Deriving European tantalum flows using trade and production statistics. *Journal of Industrial Ecology*. Doi:10.1111/jiec.12533.

Font Vivanco, D., Kemp, R., and Voet, E. van der. (2015). The relativity of eco-innovation: Environmental rebound effects from past transport innovations in Europe. *Journal of Cleaner Production*, 101, 71–85.

Giljum, S., Lutter, S., and Bruckner, M. (2017). Resource footprints. Chapter 6 of this book.

Huppes, G., and Ishikawa, M. (2009). Eco-efficiency guiding micro-level actions towards sustainability: Ten basic steps for analysis. *Ecological Economics*, 68, 1687–1700.

Mudd, G. M. (2017). Criticality assessments. Chapter 8 of this book.

Pauliuk, S., Majeau-Bettez, G., Mutel, C. L., Steubing, B., and Stadler, K. (2015, December). Lifting industrial ecology modeling to a new level of quality and transparency: A call for more transparent publications and a collaborative open source software framework. *Journal of Industrial Ecology*, 19(6), 937–949.

Schaffartzik, A., and Wiedenhofer, D. (2017). Linking society and nature: Material flows and the resource nexus. Chapter 5 of this book.

Suh, S., and Huppes, G. (2002). Missing inventory estimation tool using extended input–output analysis. International Journal of Life Cycle Assessment, 7(3), 134–140. http://dx.doi.org/10.1007/BF02994047

Tukker, A., and Font Vivanco, D. (2017). The role of environmentally extended input–output analysis in assessing nexus problems. Chapter 7 of this book.

Voet, E. van der, and Guinée, J. B. (2017). Life cycle assessment, life cycle sustainability analysis and the resource nexus. Chapter 4 of this book.

Wiedmann, T., Schandl, H., Lenzen, M., Moran, D., Suh, S., West, J., and Kanemoto, K. (2015). The material footprint of nations. *Proceedings of the National Academy of Sciences (PNAS)*, 112, 6271–6276.

PART III

Resource nexus modelling

Practices and future transformations

11

INTEGRATING ENVIRONMENTAL AND SOCIAL IMPACTS WITH ECOSYSTEM SERVICES ANALYSIS

Perrine Hamel, Benjamin Bryant, Rebecca Chaplin-Kramer and Adrian Vogl

Highlights: modeling the transformation of the food–water–energy nexus requires an understanding of i) the impact of land management on current and future natural ecosystems; and ii) the socio-economic context and distribution of benefits gained from natural capital. Ecosystem services assessments may be used as a tool to inform decisions that affect this transformation.

Introduction

Population growth and the evolution of the modern diet have led to a sharp increase in food production over the past century. The resulting increase in land brought under cultivation, together with the application of fertilizers and pesticides, is responsible for major environmental impacts through both direct and indirect pathways. For example, deforestation attributed to soy production in the Amazon has been associated with major changes in water quality and quantity, biodiversity reduction, as well as carbon emissions that affect global climate (Macedo et al., 2012). Food production also impacts marine environments: overfishing and unsustainable practices have led to the depletion of 90% of important species over the past century (Lotze et al., 2006).

The social impacts of modern food production are both positive and negative. On one hand, of course, agricultural production helps respond to our basic need for food. On the other hand, it affects the well-being of many around of the globe through its impact on livelihoods, health and environmental amenities. A number of benefits that humans derive from natural ecosystems, also called ecosystem services, are indeed affected by agricultural expansion (Millennium Ecosystem Assessment, 2005). Access to clean and sufficient water can be compromised by the increase in soil loss or fertilizer application due to agricultural expansion (Filoso et al., 2006). Air quality can be affected by forest burning for agricultural expansion, a phenomenon that has raised great concern in Indonesia and its neighbors (Gaveau et al., 2014).

Not only are the possible impacts of agricultural expansion very broad but they also occur at different scales. Given the complexity of the food–water–energy nexus, actions that are overall beneficial at one scale or place may be harmful when examined at a different scale or place. For example, agricultural expansion or intensification may provide net benefits by improving food security and raise incomes at many points along the supply chain within a country, but also impact the water quality or flooding risk of communities living close to agricultural lands (Power, 2010). Conversely, locally beneficial activities like forest clearing with wildfire may benefit locals in the short term while the associated pollution causes net negative impacts regionally.

Understanding the feedbacks between water, energy, and food production (WEF) is key to addressing the major environmental and social issues we are facing. Importantly, understanding the *transformation* of the WEF nexus under future drivers, planned and unplanned, is necessary to take action. Modeling offers a means to address systems' complexity and integrate the multiple dimensions of the WEF nexus. It allows analysts to develop system's thinking; that is, to understand and assess the interactions between resource issues not in isolation but as part of a system.

Models are developed for a variety of purposes, such as exploring possible futures, designing policies, raising awareness, or forecasting to prevent undesirable consequences for the resource nexus. Irrespective of the particular objective, a robust and efficient modeling approach is needed, one that decision-makers and all stakeholders find credible and legitimate. These conditions are necessary for science to effectively influence policy and action (Cash et al., 2014). Therefore, scientists not only need to understand individual processes that shape the nexus, but represent them through approaches that are co-developed with decision-makers. Recent scientific advances in fields as diverse as natural sciences, social sciences, and integrated modeling contribute to developing these modeling approaches.

In this chapter, we briefly explore options and challenges to modeling the transformation of the WEF nexus, which requires the representation of both the biophysical and social systems. We propose that modeling the nexus and its transformation requires an understanding of i) the impact of current and future land management on natural ecosystems; and ii) the socio-economic context and distribution of benefits gained from natural capital. We illustrate these points with two examples of recent socio-ecological assessments: one focusing on integrated watershed management for food and energy production, the other assessing strategies for sustainable commodity sourcing. Throughout the chapter, we rely on the framework of ecosystem services, which has been successfully applied to shape decisions in this realm (Ruckelshaus et al., 2015); we recognize, however, that other environmental and social assessment frameworks may equally be used as a tool to understand the WEF nexus transformation.

Modeling changes in the biophysical and social systems

Principles of integrated modeling

Practical tools are needed to translate the WEF nexus from a conceptual framework to policy and action. While the design of these tools depends on the application (e.g., exploratory analyses, environmental or social impact assessments), they have in common the integration of two areas: the biophysical environment and human society. Interactions between the two can be considered in terms of impacts and dependencies, with the main focus determined by the objective and scope of the analysis. We note that the terminology to describe this conceptualization varies depending on the field of research: for example, in the ecosystem services framework it can be described as the interaction between "biophysical" or "ecological" systems with "social" systems, or the interaction of the Earth and human systems in integrated environmental assessments (US Department of Energy, 2014).

The integration and coupling of biophysical and socio-economic models has been the focus of many researchers, under the umbrella term "integrated assessment" (Jakeman et al., 2008). As such, WEF nexus modeling can be seen as a particular case of this field, which, in the words of Jakeman et al., aims at "integrating knowledge from different disciplines [. . .] to understand

and solve complex societal problems, that arise from the interaction between humans and the environment." Such integrated assessments already adopt the systems thinking that is required to represent the WEF nexus.

Since the WEF nexus comprises many well-studied interactions, numerous models already exist to represent various aspects of the biophysical and human systems of interest. Biggs et al. (2015) provide examples of these interactions in a matrix (Figure 11.1), which suggests that understanding of WEF biophysical processes will require knowledge from geographical, hydraulic, atmospheric, chemistry and engineering sciences, to cite a few examples. Popular biophysical modeling approaches include physics-based or statistical techniques (e.g., using landscape indices or proxies to represent ecosystem functions, as analyzed in Eigenbrod et al., 2010); their outputs comprise hydrological, ecological, energy use and climate metrics that are used to assess the current and future states of the biophysical environment (see Bagstad et al., 2013 for a review

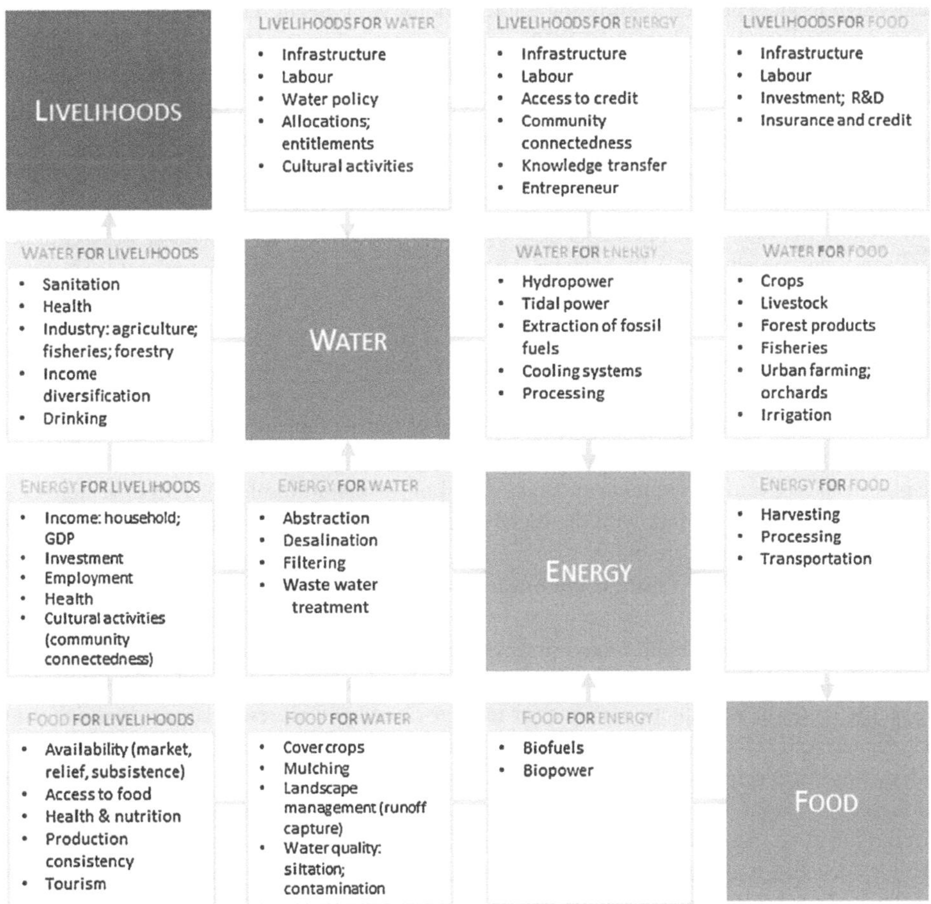

Figure 11.1 Examples of livelihoods, water, energy and food system interactions. Modeling the WEF nexus requires modeling dynamics within the biophysical and the socio-economic systems and their interactions. For each interaction, a number of modeling approaches are available, drawing from the literature on natural and socio-economic sciences.

Source: modified from Biggs et al. (2015)

of generalist ecosystem services models; and Vigerstol and Aukema, 2011 for a review of hydrological models).

A number of models are also available to represent the socio-economic systems, drawing from the literature in social science and, in particular, economics and policy analysis. These models are often conceptual or statistically based, as required by the objectives of the analyses and available data. Vulnerability and livelihood assessment approaches can be used if the analyses focus on social equity (Turner et al., 2003), for example in rural development policies (Brocklesby and Fisher, 2003). Technology-specific models, incorporating social impacts and adaptation options are of interest when assessing the effect of particular technology on the WEF (see examples in the review by the US Department of Energy, 2014, Ch. 6). Again, this list is not exhaustive given the large number of models used in the disciplines cited in this subsection.

Examples of integrated assessments in the current literature

Ecosystem services provide a promising framework for WEF nexus modeling, as evidenced by a growing literature that utilizes them within integrated assessments. Hurford and Harou (2014), provide a recent example, wherein they present an integrated assessment of potential trade-offs between hydropower production and food security. Related work examines how landscape interventions can benefit multiple groups of stakeholders in a watershed. In a different context, Goldstein et al. (2012) provide insights into the trade-offs between food production, carbon storage and water quality. While ecosystem assessments are not always concerned with food production, the tools and approaches developed in this field are useful to integrate biophysical and social components to WEF analyses. The reader is referred to Bagstad et al. (2013) for an overview of ecosystem services models.

Existing practices in the corporate sustainability literature, such as life cycle analysis (LCA) or value-chain accounting, also have the capacity to integrate multiple sectors and levels of information. LCAs typically start by defining the product, scope and goal of the analyses. The next two steps, the inventory and impact assessment, aim to quantify impacts and dependencies of production on the environment, through one or more models of the biophysical system (e.g., calculating tons of sediment in streams, carbon in the atmosphere). The final step, LCA interpretation, makes the link with the social system. In particular, characterization factors serve as simple models to translate biophysical metrics into benefits, or avoided negative impacts, for society. In that sense, LCAs share commonalities with ecosystem services assessment and benefit from advances in this field.

While these approaches illustrate options to model the WEF nexus, they are not applicable across all contexts. The methods may be limited in their scope (e.g., life cycle assessments have a specific focus on a product or service) or in their capacity to represent large systems' dynamics; for example, focusing on one particular element of the nexus and neglecting important interactions between other elements. As noted by Bazilian et al. (2011), integrated assessments models often "focus on one resource and ignore the interconnections with other resources; have overly simplified spatial representations; are grand policy 'research' rather that short term applied 'policy'/decision support models, or analyse scenarios which are impractically long term."

These limitations have led some researchers to develop new models that rely more strictly on the WEF nexus framework and cover a broad range of contexts. These include CLEWS (IAEA, 2009), MuSIASEM (Giampietro et al., 2013), and frameworks developed by the Stockholm Environmental Institute (Hoff, 2011) or the FAO (FAO, 2014). Additional integrated assessment models were developed with a holistic approach and to address broad applications, which makes them good candidates for WEF nexus: among the most common ones are GCAM and IGSM

(US Department of Energy, 2014). Often, these models represent large-scale systems and may be limited in their representation of the social systems, for example the consideration of livelihoods (Biggs et al., 2015). In general, modeling approaches need to be adapted for the particular objective and available data for the study.

Challenges for integrated modeling

Successfully integrating diverse biophysical and social processes into WEF nexus modeling poses several challenges related to the model's scope and specific objectives, and to the spatial and temporal scale of the study. In the next paragraph, we provide a brief overview of these challenges, which we further detail at the end of each case study.

Scope and purpose. Model scope may vary broadly, ranging from a simple interaction of a few nexus processes or conversely aim for a more holistic representation of the system of interest. The matrix proposed by Biggs et al. (2015, cf. Figure 11.1) hints at the multiplicity of possible combinations and foci of interest for a given study. This means that there is no one-size-fits-all approach, and it is difficult to provide a general assessment of the suitability of a particular tool for WEF modeling. Models of the WEF nexus may also serve several purposes: raise awareness, improve system understanding, or predict the complex interactions characterizing WEF systems. Depending on the purpose and the audience, different levels of complexity and certainty will likely be needed, thereby constraining the range of modeling approaches for the analyses.

Scale. Since the WEF nexus concept is not fundamentally defined at particular spatial and temporal scales, nexus models may be applied to a broad range of both, as dictated by the problem at hand and available resources. Once appropriate scales are identified, however, this choice further constrains the types of models that can be used for nexus modeling. For example, daily flow predictions may be needed for run-of-river hydropower production, while monthly to annual estimates of water yields may be sufficient for global assessment of irrigation water use. Spatial dynamics are also important to include, as they drive how and where the biophysical and social systems are linked.

Uncertainty assessment. A cross-cutting issue in these challenges is that of modeling uncertainty. While different applications may require different levels of confidence, they all require thorough consideration of underlying assumptions and model interpretation to produce credible results. However, because of model complexity and feedbacks, it is difficult to track and a fortiori to quantify the various sources of uncertainty. This challenge is well known from the field of integrated assessment (Hamel and Bryant, 2017), as explicated in the next section.

Case studies: simulating changes to the nexus to inform decisions

Designing watershed interventions for food production, water supply and hydropower in the Upper Tana watershed, Kenya

Highlights:

- Programs for integrated watershed management can have net positive social impacts on all aspects of the food, water and energy nexus;
- Integrating biophysical and economic models highlights impacts and dependencies among food, water and energy production;

- Involving stakeholders in project scoping, data collection and model development resulted in increased buy-in to results; however, it may also constrain exploration of potentially relevant nexus linkages or scenarios.

Context: design of an investment in watershed services program

Investments in watershed services are increasingly recognized as effective policy mechanisms that can diversify the portfolio of strategies for communities to address water security challenges (Goldman-Benner et al., 2012; Palmer et al., 2015). These programs aim to link downstream beneficiaries of clean water to upstream land managers, through financial transfers or through technical assistance (Bremer et al., 2016). In multiple-use watersheds, such programs present opportunities for managing trade-offs and synergies between land management and the resulting ecosystem services that provide food, water and energy.

The Upper Tana watershed, located northeast of Nairobi, Kenya (Figure 11.2), provides critical water supply and electricity to the city of Nairobi (population 3.1 million), subsistence and export agriculture to the surrounding population, reservoir storage that supports agriculture in the lower Tana and hydro-electricity production in a series of seven dams beginning with the Masinga Reservoir. Water is especially valuable during the dry season when all the beneficial uses are most in competition.

Following earlier work that sought to show the value of green and blue water for sustaining watershed health (Hunink et al., 2012; Kauffman et al., 2014), The Nature Conservancy, along with other major stakeholders, NGOs, and development agencies, formed the Upper Tana-Nairobi Water Fund. The Fund seeks to engage stakeholders who benefit from clean, reliable water provided by the watershed, and to leverage investments from these actors to manage and maintain service provision. It invests in activities such as maintaining and restoring riparian buffers, implementing terracing and cross-slope grass strips in croplands, protecting natural forests from agricultural encroachment into protected areas, and mitigating sediment erosion from unpaved roads.

A stakeholder steering committee represents the diverse interests of both downstream beneficiaries and upstream service 'providers' in the Water Fund: Nairobi City Water and Sewerage Company (NCWSC), KenGen (hydropower operator), Water Resources Management Authority (WRMA), the Tana and Athi Rivers Development Authority (TARDA), agricultural producers, government agencies, as well as local and international development NGOs. The stakeholder steering committee, in consultation with the Nature Conservancy and The Natural Capital Project, recognized the trade-offs that might exist when managing the landscape for potentially competing goals of agricultural production, water quality, hydropower and livelihoods. The study's purpose was to determine if investments in watershed services in this area would have a net positive social return on investment (to leverage public and international investment for ongoing activities of the Water Fund), as well as positive financial returns to major stakeholders (to leverage private investment in the scheme). The scope of the study and the modeling approach taken was co-developed with these major stakeholders; thus the services and outputs considered reflect the priorities and goals of the Water Fund and its supporters.

Simulating future intervention scenarios

To understand the potential benefits and trade-offs that come with managing the Upper Tana, two of us (Bryant and Vogl, as part of a larger study team) examined scenarios of watershed

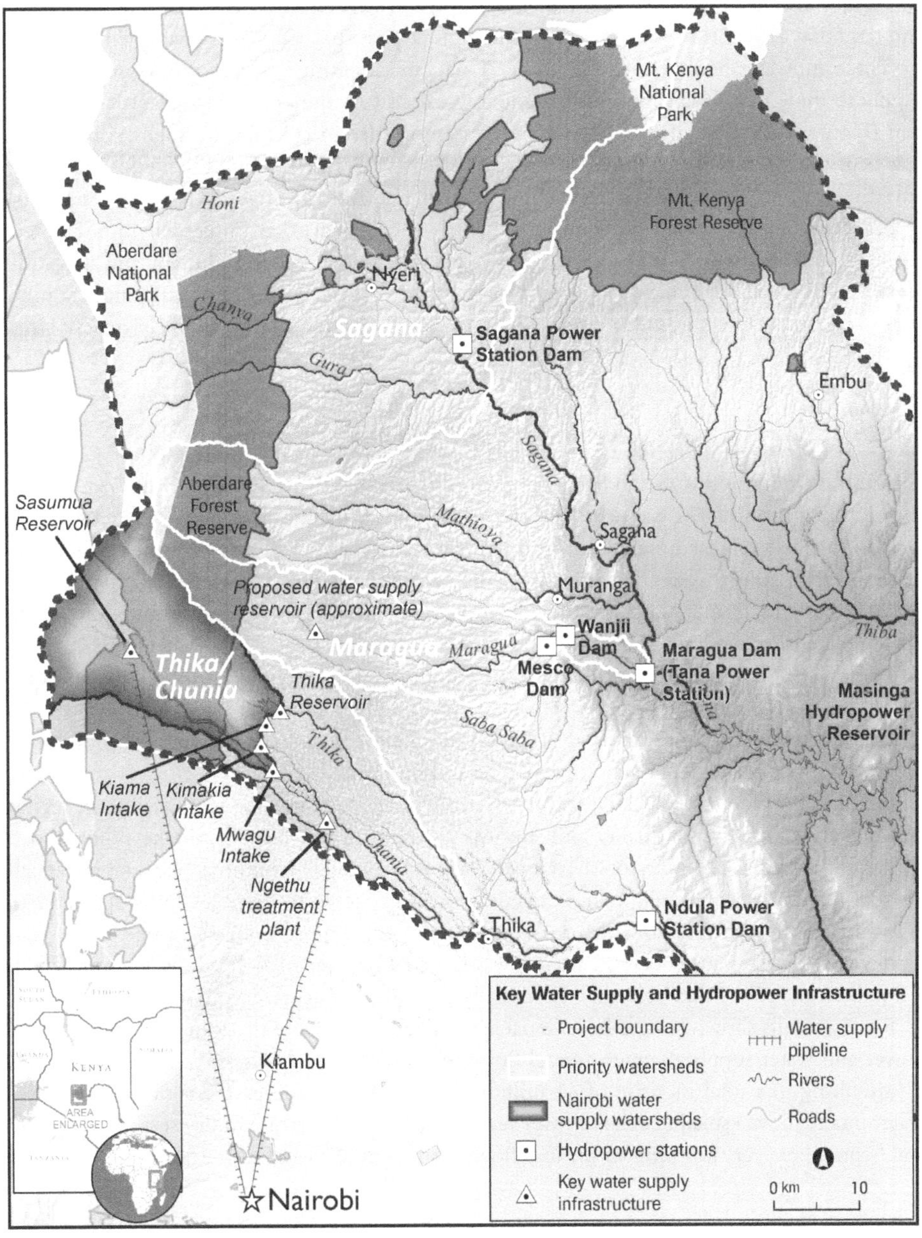

Figure 11.2 The priority watersheds and key infrastructure in the Upper Tana

Source: TNC (2015)

interventions at different levels of investment and quantified the impacts on sediment, water yield and agricultural production. We also examined benefits to beneficiaries within the watershed, in terms of financial impacts where possible, and other relevant metrics. Stakeholder input was solicited in all stages of the analysis – including study objectives and design, biophysical and economic data inputs, reviewing results, and ensuring that outputs are feasible and useful to meet two major goals of the Water Fund: quantifying return on investment for the Fund's activities and targeting activities to locations with the highest potential return on investment.

Three subwatersheds within the Upper Tana were identified as critical for providing water supply to the City of Nairobi, and for reducing excessive sedimentation to the Masinga Reservoir (Figure 11.2). Our modeling analysis focused on these three subwatersheds, and combined i) a heuristic optimization model to identify places where Water Fund activities are likely to have the greatest impact on baseflow and sediment retention objectives (RIOS;Vogl et al., 2015); ii) process-based watershed modeling to quantify the impacts that implementation of recommended activities would have on the ecosystem services of water, sediment and agricultural yields (SWAT; Arnold et al., 1998); and iii) an estimate of economic benefits using a combination of quantitative econometric modeling and qualitative description of likely benefit streams and impacts.

Assessing impacts on the biophysical system

The results from the RIOS and SWAT models showed that Water Fund activities can influence the biophysical system in ways that directly impact food, water and energy production: reducing soil loss, reducing sedimentation into streams and reservoirs, and improvements in seasonal flow and total annual water yield. First, model predictions suggested a major reduction in soil loss that benefits farmers and food production. Reductions in soil loss throughout the three subwatersheds were considerable, with the greatest total reductions in degraded lands, coffee and other croplands. Mitigation of sediment lost from unpaved roads also contributed significantly to overall reduced sediment delivery to service points in all scenarios. These reductions in soil loss translate to improved soil fertility, agricultural production and increased yields through implementing terracing and other activities to keep soil on farms.

Results also predicted a major reduction in sediment load in streams as well as in reservoir sedimentation. For water quality, our analysis focused on the Mwagu intake point, the single largest withdrawal point for the city of Nairobi's Ngethu treatment plant (approximately 400,000 m^3 day^{-1}), where sediment concentration is expected to be reduced by approximately 50% during the rainy seasons. More broadly, sediment export from the three study watersheds over the 10-year simulation period was reduced by over 35%, resulting in approximately 1.2 million m^3 of avoided volume loss annually for the Masinga reservoir.

Finally, results show changes in seasonal and annual water yield that can benefit both hydropower and water supply. Average annual flows into the Masinga Reservoir increased by nearly 4%, resulting in a total increase of 41 million m^3 into Masinga reservoir. Flows at the Mwagu water intake show a small decrease in wet seasons and a small increase in dry seasons. July, August and September – critical months for low flows – show increases of over 15%.

Assessing impacts on the social system

Following the biophysical modeling, we quantified expected benefits for three major stakeholders, most with multiple benefit streams. Table 11.1 itemizes the quantified benefit streams, with 5% discounting over a 30-year time horizon, in a way that also accounts for the time it takes for implementation and for benefits to materialize. Additional benefit streams that had less firm grounding were quantified as part of the sensitivity analysis, while still other benefit streams were addressed qualitatively (Table 11.2). Overall, the project was found to have positive net return on investment, expected to generate over US$21 million in benefits for total costs of

Table 11.1 Benefits assessed for monetized and non-monetized benefit streams. Parentheses indicate negative values

Stakeholder	*Benefit/cost*	*USD*
Water fund	Investment cost	(7,110,000)
Ag producers	Net additional cost (maintenance, etc.)	(8,520,000)
Ag producers	Upstream farmers	12,000,000
NCWSC	Avoided flocculant costs	394,000
NCWSC	Avoided electricity costs	36,700
NCWSC	Net revenue from saved process water	2,090,000
NCWSC	***Total NCWSC benefits with scale-up***	***3,390,000***
KenGen	Avoided interruptions	281,000
KenGen	Increased generation	5,870,000
KenGen	***Total KenGen benefits***	***6,160,000***
	Present value of benefits	21,500,000
	Present value of costs	(15,600,000)
	Net present value	**5,900,000**

Table 11.2 Non-monetized benefits from improved water yield and reduced sediment

Stakeholder	*Non-monetized benefits*
NCWSC	Reduction in wet sludge disposal
NCWSC	Avoided service interruptions
NCWSC	Increased dry season flows
Other water suppliers	Lowered sediment levels
Municipal water processors	More reliable water supply
KenGen	Reduction in reservoir sedimentation
KenGen	Avoided turbine intake maintenance costs
Upstream farmers	Increased fodder for livestock
Upstream famers	Additional income and employment opportunities
Urban private sector processors	Improved water supply
Local communities	Cleaner drinking water
General ES	More habitat for pollinators
General ES	Increased carbon storage in new trees planted

under US$16 million. (The total costs are higher than the investment costs due to maintenance and reinvestment necessary to generate benefits over the long term.)

Comparison across Tables 11.1 and 11.2 reveals multiple benefit streams for each stakeholder group, and that some stakeholder groups have no quantified benefit streams. The diversity of stakeholders and benefit streams required a diversity of approaches, which were developed based on data availability, system understanding and modeling resources. For example, for NCWSC treatment cost data, we had access to some detailed data that allowed us to build statistical cost functions, though we could not represent all processes expected to change, such as saved process water. In-depth assessment of the benefits to farmers could also be performed: in our work, a simple approach based on soil depth relationships and water productivity was used to assess the impacts of reduced erosion on agricultural production, although with more time and data, sophisticated crop growth models and economic models could provide better insights into this benefit stream.

Challenges and lessons for decision-making

The modeling approach employed here demonstrates the integrated modeling framework for nexus modeling: connecting land management and planning to the biophysical system (by linking RIOS to SWAT), connecting biophysical changes to socio-economic environments (by linking SWAT outputs to monetary and non-monetary metrics) and engaging stakeholders to ensure legitimacy and credibility of results. At the same time, it illustrated the challenges related to scale, scope, purpose and uncertainty:

Scope. The goal of establishing the net social return on investment due to program activities was challenged by the fact that some potentially significant benefit streams were difficult to quantify, either through lack of data, uncertainty around system dynamics or as a consequence of decisions made early in the process to limit the scope of analysis. For example, connecting Water Fund investments to improved service reliability for end users in the Nairobi water supply system was infeasible with the available data. Essentially, the system boundaries were drawn at the treatment plant during the initial project scoping, so benefits further downstream could not be quantified. For hydropower, we were able to overcome the fact that we had high-quality data only for the Masinga reservoir (but not the remainder of the hydropower cascade) by making a variety of bounding assumptions to consider the impact of other generating stations in the system.

Spatial scale. Even though each phase of the analysis (scenario generation, biophysical modeling and benefits estimation) occurred over the same spatial extent, the analysis required integrating inputs and outputs (and consequent uncertain assumptions) between models operating on different computational units (pixels for RIOS; hydrologic response units for SWAT). In the second stage, impacts on agricultural yield could be evaluated spatially, but the third-stage (economic analysis) required focusing on key points of interest, even though users along waterways were potentially broadly impacted (e.g., by access to clean water).

Time scale: SWAT provided daily outputs which would in theory be useful for water treatment cost estimates, but ultimately we were only able to consider most cost functions on annual or monthly time steps due to data limitations at other points in the results chain. In this particular case, a SWAT model had already been largely developed, but in a different situation, this may have constituted unnecessary modeling effort if a simpler model could have provided the same outputs with approximately equivalent fidelity. This highlights the value of an iterative process of stakeholder engagement, data collection and modeling work.

Uncertainty assessment. Ultimately, uncertainties and assumptions were less critical than they might otherwise have been, because the purpose of this particular analysis was to demonstrate that total benefits were likely to exceed the cost of intervention. To the extent we could justify the estimates as lower bounds or generally conservative yet still producing a positive social return on investment, we were able to meet the study objectives despite omission of major benefit streams. Later work that seeks to better target interventions or to more precisely align payments with benefits received would require greater precision, or at minimum, greater confidence in the ranges of plausible outcomes so that parties can account for this in deciding on terms of payment structures. These can be achieved through alternative modeling approaches and additional data collection, including ongoing monitoring of early interventions as part of the project.

Overall, this study demonstrates the importance of stakeholder engagement to achieve the kind of integration of biophysical and social impacts called for in our nexus approach. As others have shown, analytical approaches relating to dynamics of these systems must be perceived as credible and legitimate in order to significantly influence resource management (McKenzie et al., 2014; Posner et al., 2016). The analysis done for the Upper Tana Water Fund, by engaging stakeholders in an iterative and participatory modeling process, is more likely to be accepted and used in guiding future activities by the Fund.

Modeling the impact of sourcing decisions on land-use change and ecosystem services

Highlights:

- Scenario comparison is a useful tool to inform sourcing strategies and mitigate environmental impacts;
- Spatially explicit environmental models provide a better understanding of the impact of land-use change on ecosystem services;
- Estimating future land-use change as a result of changing demand is an important challenge for assessing impacts of sourcing decisions.

Context: enhancing life cycle analysis with spatial ecosystem services modeling

Corporations are a powerful force for change in commodity production, with many companies generating revenues on par with the GDP of some nations; as such, an individual company may substantially drive demand for whole sectors of production (Kareiva et al., 2015; O'Rourke, 2014). At the same time, increasing attention is being paid to corporate sustainability and more and more commitments being made to reduce environmental impacts (e.g., Zero Net Deforestation,[1] Carbon Disclosure Project,[2] Natural Capital Protocol,[3] etc.). Certification is one lever for corporations to promote more sustainable production, and indeed is featured as an element of many companies' sustainability strategy in terms of committing to buy a certain proportion (or all) of key commodities from certified sources, such as McDonald's and the Marine Stewardship Council (Chaplin-Kramer et al., 2015a). However, certification accounts for less than 10% of the market for most commodities (Potts et al., 2014), and therefore does not provide a full-scale solution for the majority of supply chain decisions, at least in the near term.

Many companies are developing and applying increasingly sophisticated methods to assess the environmental impacts of their procurement and product innovation decisions. As noted earlier, life cycle assessments (LCA) are commonly used for this purpose. For agricultural commodities, the inventory conducted in the LCA identifies the impacts of agricultural land on the environment, among other things. Next, the characterization factors modulate these impacts to account for local characteristics; that is, weighting the biophysical metrics as a function of the negative impacts in the particular environment in which they occur.

Recent advances in LCA methodology have enabled the incorporation of impacts of land-use transformation (not only occupation) and its impacts on ecosystem services (not only emissions or pollutants) (Hellweg and Milà i Canals, 2014). However, even the most advanced methods are typically based on static characterization factors related to the area transformed or occupied, which may be regionally specific, but not sensitive to the landscape configuration. In reality, impacts of agricultural expansion or intensification depend on the biophysical environment and land-use patterns: for example, impacts near the stream will affect water quality more than due to the buffering effect of riparian areas (Liu et al., 2008; Poeppl et al., 2012). Therefore, to inform supply chain decisions about commodity sourcing regions, one needs to understand how much spatial patterns of land transformation matter to different ecosystem services. As illustrated in the following sections, this has implications for modeling, since more sophisticated models may be needed to represent spatially explicit processes, rather than assuming a fixed relationship with amount of land cover types.

Assessing impact of land transformation on ecosystem services

In collaboration with corporate partners (Unilever), we conducted exploratory analyses to assess the impact of land transformation on ecosystem services, in particular for carbon sequestration, biodiversity and water purification (Chaplin-Kramer et al., 2016, 2015c). Our studies focused on four different regions in the US, Brazil, and China where commodities such as soy and corn are sourced from. We used the InVEST models (Sharp et al., 2015) to assess the biophysical impacts of land-use conversion on ecosystem services.

The results have shown that the pattern of agricultural expansion and the spatial context of remaining natural vegetation can determine vastly different outcomes for ecosystem services. For instance, forest fragmentation can double losses of carbon relative to deforestation from the forest edge, for the same amount of total forest area converted (Chaplin-Kramer et al., 2015c). This phenomenon of lower carbon storage at forest edges is seen across tropics, and is stronger in regions associated with human influences, suggesting that there are both physical and anthropogenic mechanisms for carbon degradation at forest edges (Chaplin-Kramer et al., 2015b). This has major implications for zero deforestation commitments, many of which are based on the principle of zero *net* deforestation. If restoration or afforestation in small fragments are being used to offset losses in previously unfragmented forest, we cannot assume that zero net deforestation means zero carbon loss.

The proximity of land transformation to streams can also lead to strong nonlinearities with respect to hydrologic ecosystem services. Through theoretical land-use change scenarios, recent analyses showed that loss of vegetation nearer to streams can account for an order of magnitude difference in sediment export for the same total area of land converted (Chaplin-Kramer et al., 2016), overturning vast differences in potential soil loss due to climate and soil in different regions (Figure 11.3). Riparian buffers are important but not enough to fully mitigate the impacts from habitat conversion in the broader watershed; the pattern of land-use change beyond the buffer matters. This suggests that using geographically distinct indicators such as erosion potential based on physical factors alone as a surrogate for water quality impacts in different sourcing regions could be misleading.

Assessing impacts on the socio-economic system

In addition to the land-use change dependency, the magnitude of ecosystem services impacts will be determined by the resources, preferences, and location of the affected beneficiaries. In the examples presented earlier in this section, several groups can be identified. The first one is

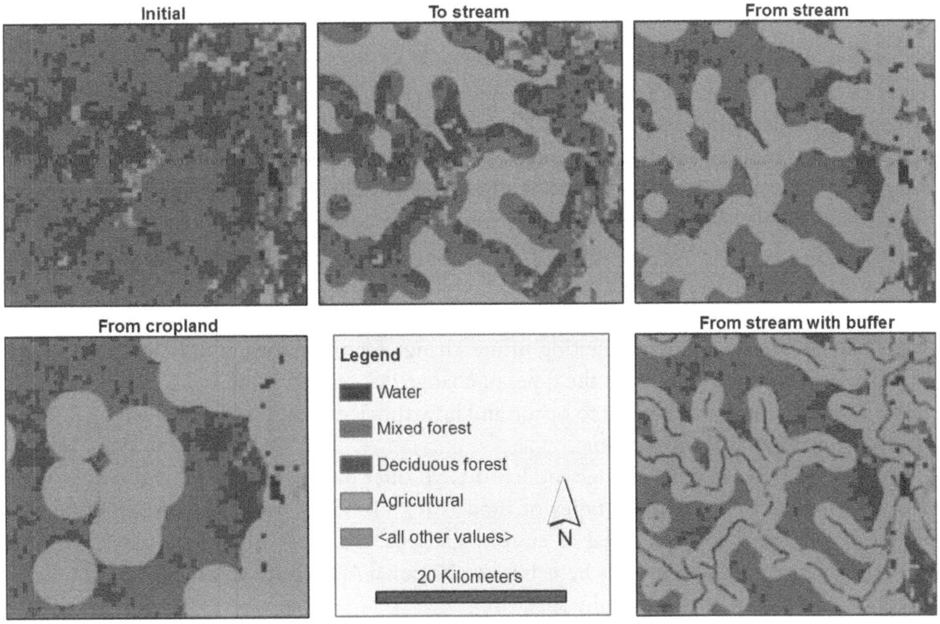

Figure 11.3 Hypothetical scenarios of agricultural expansion in a watershed in China. Initial landscape is shown in the top left hand panel; the four other maps represent land conversion to agriculture based on different expansion rules, and are used to illustrate the effect of land-use configuration on the provision of ecosystem services

Source: Chaplin-Kramer et al. (2016)

the global population, since carbon sequestration arguably benefits all humans through the many direct and indirect effects of climate change. The second group is the subset benefiting from clean water due to sediment retention. As a first approximation, this beneficiary group can be considered as the population living in the watershed. However, some people may directly drink from the stream, while others draw water from other sources, thus being less affected by a change in sediment retention. Furthermore, some beneficiaries are better able to cope with change than others; in the case of water quality, the capacity to treat the water will determine how vulnerable the beneficiaries are to changes in service provided and the value they ascribe to it. We provided further details on common benefit streams associated with the sediment retention service above. Here, we simply highlight that the extra step of considering beneficiaries is often left out of LCAs (and environmental impact assessments in general), since these data require understanding of the regional or local socio-economic context, and therefore may give only a partial picture of the environmental and social impacts of production.

Overall, these analyses have important ramifications for the use of LCAs in business decisions, since misrepresentation or lack of accounting for important biophysical processes and social factors may result in misleading results and recommendations. The spatially explicit scenario analysis illustrated how different land-use expansion patterns may lead to contrasted changes in ecosystem services. The possible range can be usefully restricted by generating more plausible trajectories along which land transformation may proceed, enabling businesses to make more informed sourcing decisions.

Simulating plausible land-use change

Generating such plausible future land-use trajectories is the domain of land change modeling, a broad category of techniques that encompasses many approaches for both testing theories about drivers of past land-use change as well as simulating future landscapes. A recent US National Academies review (Committee on Needs and Research Requirements for Land Change Modeling, 2014) describes modeling approaches that vary widely in how they represent the *processes* behind land change: At one end of the spectrum are "machine learning" and "cellular" models that tend to rely on spatial information from past patterns or change without representing human decision processes, while at the other end are "econometric" or "agent-based" models with structures that reflect assumptions about human behavior (e.g., profit maximization). Fundamentally, all approaches to generating future change scenarios specify how parcels on a map may change by imposing rules for the types of change that are allowable, what factors influences propensity for different transitions to occur, and how those are constrained (if at all) by historical information about change over time.

Practically, some land-use change models (LCM) offer interfaces or platforms within which one can construct one or more types of models (e.g. TerrSet,[4] LuccME,[5] and SITE),[6] while others are effectively implemented as custom statistical scripts. We do not attempt to discuss the wide range of existing models here, but the National Academies report provides numerous examples within each class along the spectrum.

For a sustainable sourcing application to be scalable in multiple corporate contexts, specific criteria apply to the modeling approach: it should i) be applicable anywhere using globally available data; ii) not require proprietary software; and iii) be applicable without detailed land change modeling expertise. Models that generate suitability maps using regressions on land use or environmental information meet these criteria, provided time series exist to allow validation. Suitability maps (sometimes called "transition potential maps") assign rankings to each unit on the landscape so that the score provides a *relative* indication of a pixel's likelihood of being converted. Multiple approaches can be used to generate a suitability map (Mas et al., 2014), the simplest of which is a regression where the dependent variable is determined by assigning existing or recently converted agriculture a value of one, and other all other landscape values a zero. Independent (or "driver") variables are based on available data that has some theoretical relationship to determining land conversion. To model the actual land change from time step to time step, these maps are combined with rules for choosing transitions for each period, the simplest of which is to assume that pixels with the highest suitability value are converted. While attractive in their simplicity, suitability maps are specific to one type of transition, so in applications that require predicting multiple types of land-use change, the LCM model must specify additional details for how to aggregate suitability values and choose the transition that "actually" occurs.

Challenges and lessons for decision-making

These analyses illustrated how decision support for sustainable commodity sourcing can benefit from more detailed biophysical and social modeling. We return here to the challenges associated with integrated modeling and how they were addressed in this application.

Spatial and temporal scales. In the analyses presented, we used the InVEST models for the assessment of carbon sequestration and sediment export (Sharp et al., 2015). These models have the advantage of being process based, representing spatially explicit landscape processes, but using input data that are globally available. One challenge, however, is to reconcile datasets that have been developed at different times, for different purposes, and ensure consistency in the analyses.

For example, land-use and climate data may come from datasets that were compiled 10 years apart, which makes model verification challenging (e.g., it is not possible to use observations in stream gauge since they do not represent the scenario actually represented in the analyses). In addition, data on beneficiaries (e.g., the social system) are often difficult to obtain globally: in our case study, we could use data on actual drinking sources for the local population, but these estimates are not available globally. There is a trade-off between the complexity of the modeling approach, and the objective to assess actual benefits to humans, instead of environmental impacts.

Uncertainty assessment. Our work has looked in particular at the uncertainties associated with the land-use change assumptions. Other sources of uncertainty related to the InVEST model parameters or structure add to this context. Regardless of the LCM approach, it is important to assess the performance of the model itself, *and* to assess the relationship of land change model performance as it relates to ecosystem services impacts. Many approaches exist for assessing LCM performance though more work is needed on this front (Brown et al., 2013), and such methods often require careful interpretation (Pontius and Millones, 2011). Depending on the ecosystem service, imperfect land change predictions may be magnified or attenuated by the structure of the ES model. One common example of this is when the ES model is not sensitive to small-scale variations in pixel locations – in this case, pixel level accuracy of the overall LCM may be low, but it may have a minimal effect on output. Essentially "where matters", but "exactly where" does not. In general, both land change and ES models exist for many spatial scales, but harmonizing to find the most appropriate scales for joint modeling of the two remains an important and underexplored issue. Multiple resolution analysis is a promising avenue for achieving this goal (Pontius et al., 2011).

Summary and outlook

Our two case studies demonstrated the practical steps involved in modeling the transformation of social and environmental systems: simulating plausible land-use changes, assessing the impacts of land-use change on the biophysical, and translating these impacts on the social systems (i.e. evaluating impacts on the ecosystem processes and services and beneficiaries).

In the first study, we illustrated how integrated watershed management, informed by frequent stakeholder inputs, can be used to design interventions that have net positive impacts. Land-use change modeling was done through a heuristic optimization approach, which allowed refining and capturing stakeholders' values and expectations. Impact assessment was done through the SWAT hydrological model and benefit stream analyses. The second study looked at the impact of supply chain decisions on ecosystem services. Indeed, recently completed work (Chaplin-Kramer et al., 2017) suggests that tools currently used by businesses may overlook important ecological and social processes and be misleading to support decision-making.

Both case studies point to a number of challenges for modeling transformation of the nexus and inform decision-making. First, the scope and purpose of such modeling exercises ranges broadly. Therefore, there is no one-size-fits-all approach or model approach. The concepts and steps highlighted here, however, can be used to guide the modeling approach. Second, the various spatial and temporal scales considered in integrated modeling call for close consideration of the modeling assumptions and interpretation. Finally, assessing uncertainties in the integrated models used for simulating the nexus process is challenging. Recognizing this challenge, modeling the nexus to inform decision-making should be guided by the three principles underlying "effective knowledge systems": credibility, salience and transparency (Cash et al., 2014). Such robust decision frameworks have the potential to address the challenges posed by competing demands at the WEF nexus.

Notes

1. www.theconsumergoodsforum.com/sustainability-strategic-focus/sustainability-resolutions/deforestation-resolution
2. www.cdp.net/en-US/Pages/HomePage.aspx
3. www.naturalcapitalcoalition.org/natural-capital-protocol.html
4. https://clarklabs.org/terrset/
5. http://luccme.ccst.inpe.br/conteudo_en/index.html
6. www.ufz.de/index.php?en=37508

References

Arnold, J. G., Srinivasan, R., Muttiah, R. S., and Williams, J. R. (1998). Large area hydrologic modeling and assessment part I: Model development. *Journal of the American Water Resource Association*, 34, 73–89.

Bagstad, K. J., Semmens, D. J., Waage, S., and Winthrop, R. (2013). A comparative assessment of decision-support tools for ecosystem services quantification and valuation. *Ecosystem Services*, 5, 27–39.

Bazilian, M., Rogner, H., Howells, M., Hermann, S., Arent, D., Gielen, D., Steduto, P., Mueller, A., Komor, P., Tol, R. S. J., and Yumkella, K. K. (2011). Considering the energy, water and food nexus: Towards an integrated modelling approach. *Energy Policy*, 39, 7896–7906.

Biggs, E. M., Bruce, E., Boruff, B., Duncan, J. M. A., Horsley, J., Pauli, N., McNeill, K., Neef, A., Van Ogtrop, F., Curnow, J., Haworth, B., Duce, S., and Imanari, Y. (2015). Sustainable development and the water-energy-food nexus: A perspective on livelihoods. *Environmental Science & Policy*, 54, 389–397.

Bremer, L. L., Auerbach, D. A., Goldstein, J. H., Vogl, A. L., Shemie, D., Kroeger, T., Nelson, J. L., Benítez, S. P., Calvache, A., Guimarães, J., Herron, C., Higgins, J., Klemz, C., León, J., Sebastián Lozano, J., Moreno, P. H., Nuñez, F., Veiga, F., and Tiepolo, G. (2016). One size does not fit all: Natural infrastructure investments within the Latin American Water Funds Partnership. *Ecosystem Services*, 17, 217–236.

Brocklesby, M. a., and Fisher, E. (2003). Community development in sustainable livelihoods approaches – an introduction. *Community Development Journal*, 38, 185–198.

Brown, D. G., Verburg, P. H., Pontius, R. G., and Lange, M. D. (2013). Opportunities to improve impact, integration, and evaluation of land change models. *Current Opinion in Environmental Sustainability*, 5, 452–457.

Cash, D. W., Clark, W. C., Alcock, F., Dickson, N. M., Eckley, N., Guston, D. H., Jäger, J., Mitchell, R. B., Cash, D. W., Clark, W. C., Alcock, F., Dickson, N. M., Eckley, N., Guston, D. H., Jager, J., and Mitchelll, R. B. (2014). Knowledge systems for sustainable development. Proceedings of the National Acadamy of Science of United States of America, 100, 8086–8091.

Chaplin-Kramer, R., Jonell, M., Guerry, A., Lambin, E. F., Morgan, A. J., Pennington, D., Smith, N., Franch, J. A., and Polasky, S. (2015a). Ecosystem service information to benefit sustainability standards for commodity supply chains. Annals of the New York Academy of Sciences, 1355, 77–97.

Chaplin-Kramer, R., Ramler, I., Sharp, R., Haddad, N. M., Gerber, J. S., West, P. C., Mandle, L., Engstrom, P., Baccini, A., Sim, S., Mueller, C., and King, H. (2015b). Degradation in carbon stocks near tropical forest edges. *Nature Communications*, 6, 10158.

Chaplin-Kramer, R., Sharp, R. P., Mandle, L., Sim, S., Johnson, J., Butnar, I., Milà i Canals, L., Eichelberger, B. A., Ramler, I., Mueller, C., McLachlan, N., Yousefi, A., King, H., and Kareiva, P. M. (2015c). Spatial patterns of agricultural expansion determine impacts on biodiversity and carbon storage. *Proceedings of the National Acadamy of Sciences*, 112, 201406485.

Chaplin-Kramer, R., Sim, S., Hamel, P., Bryant, B., Noe, R., Mueller, C., Rigarlsford, G., Kulak, M., Kowal, V., Sharp, R., Clavreul, J., Price, E., Polasky, S., Ruckelshaus, M., and Daily, G. (2017). Life Cycle Assessment Needs Predictive, Spatial Modelling for Biodiversity and Ecosystem Services. *Nature Communications*, 8(15065). Retrieved from https://doi.org/10.1038/ncomms15065.

Chaplin-Kramer, B., Hamel, P., Sharp, R., Kowal, V., Wolny, S., Sim, S., and Mueller, C. (2016). Landscape configuration is the primary driver of impacts on water quality associated with agricultural expansion. *Environmental Research Letters*. Retrieved from http://iopscience.iop.org/article/10.1088/1748-9326/11/7/074012/meta

Committee on Needs and Research Requirements for Land Change Modeling. (2014). *Advancing land change modeling: Opportunities and research requirements*.

Eigenbrod, F., Armsworth, P. R., Anderson, B. J., Heinemeyer, A., Gillings, S., Roy, D. B., Thomas, C. D., and Gaston, K. J. (2010). The impact of proxy-based methods on mapping the distribution of ecosystem services. *Journal of Applied Ecology*, 47, 377–385.

FAO. (2014). The water – energy – food nexus: A new approach in support of food security and sustainable agriculture.

Filoso, S., Martinelli, L. A., Howarth, R. W., Boyer, E. W., and Dentener, F. (2006). Human activities changing the nitrogen cycle in Brazil. *Biogeochemistry*, 79, 61–89.

Gaveau, D. L. A., Salim, M. A., Hergoualc'h, K., Locatelli, B., Sloan, S., Wooster, M., Marlier, M. E., Molidena, E., Yaen, H., DeFries, R., Verchot, L., Murdiyarso, D., Nasi, R., Holmgren, P., and Sheil, D. (2014). Major atmospheric emissions from peat fires in Southeast Asia during non-drought years: Evidence from the 2013 Sumatran fires. *Science Report*, 4, 6112.

Giampietro, M., Aspinall, R. J., Bukkens, S. G. F., Cadillo Benalcazar, J., Diaz-Maurin, F., Flammini, A., Gomiero, T., Kovacic, Z., Madrid, C., Ramos-Martin, J., and Serrano-Tovar, T. (2013). *An innovative accounting framework for the food – energy – water nexus – application of the MuSIASEM approach to three case studies*. Environment and Natural Resources Working Paper No. 56. Rome: Food and Agriculture Organisation of the United Nations.

Goldman-Benner, R. L., Benitez, S., Boucher, T., Calvache, A., Daily, G., Kareiva, P., Kroeger, T., and Ramos, A. (2012). Water funds and payments for ecosystem services: Practice learns from theory and theory can learn from practice. *Oryx*, 46, 55–63.

Goldstein, J. H., Caldarone, G., Duarte, T. K., Ennaanay, D., Hannahs, N., Mendoza, G., Polasky, S., Wolny, S., and Daily, G. C. (2012). Integrating ecosystem-service tradeoffs into land-use decisions. *Proceedings of the National Acadamy Science of United States of America*, 109, 7565–7570.

Hamel, P., and Bryant, B. P. (2017). Uncertainty assessment in ecosystem services analyses: Seven challenges and practical responses. *Ecosystem Services*, 24, 1–15.

Hellweg, S., and Milà i Canals, L. (2014). Emerging approaches, challenges and opportunities in life cycle assessment. *Science*, 344, 1109–1113.

Hoff, H. (2011). *Understanding the nexus: Background paper for the Bonn 2011 conference: The water, energy and food security nexus*. Stockholm: Stockholm Environment Institute.

Hunink, J. E., Droogers, P., Kauffman, S., Mwaniki, B. M., and Bouma, J. (2012). Quantitative simulation tools to analyze up- and downstream interactions of soil and water conservation measures: Supporting policy making in the Green Water Credits program of Kenya. *Journal of Environmental Management*, 111, 187–194.

Hurford, A. P., and Harou, J. J. (2014). Balancing ecosystem services with energy and food security – assessing trade-offs from reservoir operation and irrigation investments in Kenya's Tana Basin. *Hydrology and Earth System Sciences*, 18, 3259–3277.

IAEA. (2009). AnnexVI: Seeking sustainable climate land energy and water (CLEW) strategies. *Nuclear Technology Review*.

Jakeman, A. J., Chen, S. H., Rizzoli, A. E., and Voinovc, A. A. (2008). Modelling and Software as Instruments for Advancing Sustainability pp. 1–13, *In* Environmental modelling, software and decision support. Edited by Jakeman, A. J., Voinov, A. A., Rizzoli, A. E., and Chen, S. H. Elsevier, Series.

Kareiva, P., McNally, B., McCormick, S., Miller, T., and Ruckelshaus, M. (2015). Improving global environmental management with standard corporate reporting. *Proceedings of the National Acadamy of Sciences*, 112, 7375–7382.

Kauffman, S., Droogers, P., Hunink, J., Mwaniki, B., Muchena, F., Gicheru, P., Bindraban, P., Onduru, D., Cleveringa, R., and Bouma, J. (2014). Green water credits – exploring its potential to enhance ecosystem services by reducing soil erosion in the Upper Tana basin, Kenya. *International Journal of Biodiversity Science, Ecosystem Services & Management*, 10, 133–143.

Liu, X., Zhang, X., and Zhang, M. (2008). Major factors influencing the efficacy of vegetated buffers on sediment trapping: A review and analysis. *Journal of Environmental Quality*, 37, 1667.

Lotze, H. K., Lenihan, H. S., Bourque, B. J., Bradbury, R. H., Cooke, R. G., Kay, M. C., Kidwell, S. M., Kirby, M. X., Peterson, C. H., and Jackson, J. B. C. (2006). Depletion, degradation, and recovery potential of estuaries and coastal seas. *Science* (80-.), 312, 1806–1809.

Macedo, M. N., DeFries, R. S., Morton, D. C., Stickler, C. M., Galford, G. L., and Shimabukuro, Y. E. (2012). Decoupling of deforestation and soy production in the southern Amazon during the late 2000s. *Proceedings of the National Acadamy of Sciences*, 109, 1341–1346.

Mas, J.-F., Kolb, M., Paegelow, M., Camacho Olmedo, M. T., and Houet, T. (2014). Inductive pattern-based land use/cover change models: A comparison of four software packages. *Environmental Modelling & Software*, 51, 94–111.

McKenzie, E., Posner, S., Tillmann, P., Bernhardt, J. R., Howard, K., and Rosenthal, A. (2014). Understanding the use of ecosystem service knowledge in decision making: Lessons from international experiences of spatial planning. *Environment and Planning C: Government and Policy Journal*, 32, 320–340.

Millennium Ecosystem Assessment. (2005). *Ecosystems and human well-being: Synthesis*. Washington, DC: World Resources Institute, Island Press.

O'Rourke, D. (2014). The science of sustainable supply chains. *Science*, 344, 1124–1127.

Palmer, M. A., Liu, J., Matthews, J. H., Mumba, M., and D'Odorico, P. (2015). Manage water in a green way. *Science* (80-.), 349, 584–585.

Poeppl, R. E., Keiler, M., von Elverfeldt, K., Zweimueller, I., and Glade, T. (2012). The influence of Riparian vegetation cover on diffuse lateral sediment connectivity and biogeomorphic processes in a medium-sized agricultural catchment, Austria. *Geografiska Annaler Series: A Physical Geography*, 94, 511–529.

Pontius, R. G., and Millones, M. (2011). Death to Kappa: Birth of quantity disagreement and allocation disagreement for accuracy assessment. *International Jorunal of Remote Sensing*, 32, 4407–4429.

Pontius, R. G. J., Peethambaram, S., and Castella, J. C. (2011). Comparison of three maps at multiple resolutions: A case study of land change simulation in Cho Don District, Vietnam. *Annals of the American Association of Geographers*, 101(1), 45–62.

Posner, S. M., McKenzie, E., and Ricketts, T. H. (2016). Policy impacts of ecosystem services knowledge. *Proceedings of the National Acadamy of Sciences*, 113(7), 1760–1765, doi:10.1073/pnas.1502452113.

Potts, J., Lynch, M., Wilkings, A., Huppe, G., Cunningham, M., and Voora., V. (2014). *The state of sustainability initiatives review 2014*. London: International Institute for Sustainable Development, Winnipeg and International Institute for Environment and Development.

Power, A. G. (2010). Ecosystem services and agriculture: Tradeoffs and synergies. *Philosophical Transactions of the Royal Society of London. Series B: Biological Sciences*, 365, 2959–2971.

Ruckelshaus, M., McKenzie, E., Tallis, H., Guerry, A., Daily, G., Kareiva, P., Polasky, S., Ricketts, T., Bhagabati, N., Wood, S. A., and Bernhardt, J. (2015). Notes from the field: Lessons learned from using ecosystem service approaches to inform real-world decisions. *Ecological Economics*, 115, 11–21.

Sharp, R., Tallis, H. T., Ricketts, T., Guerry, A. D., Wood, S. A., Nelson, E., Ennaanay, D., Wolny, S., Olwero, N., Vigerstol, K., Pennington, D., Mendoza, G., Aukema, J., Foster, J., Forrest, J., Cameron, D., Arkema, K., Lonsdorf, E., Kennedy, C., Verutes, G., Kim, C. K., Guannel, G., Papenfus, M., Toft, J., Marsik, M., Bernhardt, J., Griffin, R., Glowinski, K., Chaumont, N., Perelman, A., Lacayo, M., Mandle, L., Hamel, P., and Chaplin-Kramer, R. (2015). *InVEST 3.0 user's guide*. Retrieved June 2016, from http://ncp-dev.stanford.edu/~dataportal/invest-releases/documentation/current_release/

TNC. (2015). *Upper Tana-Nairobi water fund: A business case. Version 2*. Nairobi, Kenya: https://www.nature.org/ourinitiatives/regions/africa/upper-tana-nairobi-water-fund-business-case.pdf.

Turner, B. L., Kasperson, R. E., Matson, P. A., McCarthy, J. J., Corell, R. W., Christensen, L., Eckley, N., Kasperson, J. X., Luers, A., Martello, M. L., Polsky, C., Pulsipher, A., and Schiller, A. (2003). A framework for vulnerability analysis in sustainability science. Proceedings of the National Acadamy of Sciences, 100, 8074–8079.

US Department of Energy. (2014). *The water energy nexus: Challenges and opportunities*. Report 238.

Vigerstol, K. L., and Aukema, J. E. (2011). A comparison of tools for modeling freshwater ecosystem services. *Journal of Environmental Management*, 92, 2403–2409.

Vogl, A., Tallis, H., Douglass, J., Sharp, R., Veiga, F., Benitez, S., León, J., Game, E., Petry, P., Guimerães, J., and Lozano, J. S. (2015). *Resource investment optimization system: Introduction & theoretical documentation*. The Natural Capital Project. Retrieved 29 Oct 2017, from https://www.naturalcapitalproject.org/software/#rios.

12

MODELLING PRACTICES FROM LOCAL TO GLOBAL

Enrique Kremers, Andreas Koch and Jochen Wendel

Introduction

Man-made systems use and transform resources in all sectors. As human activities surpass the available resources the sum of our activities put the basis for future generations and other species at stake. The results of such unsustainable growth are system-wide crisis such as climate change as well as sectorial failures of the environmental capacity to sustain human activities such as depletion of resources (e.g. peak oil). When regarding our planet as such it seems obvious that limited natural resources are provided and thus use of this global reservoir is highly interconnected. In practice this is often ignored by relying on sectorial or national assessments of the economy and its environmental impact. Multi-scale modelling can support decision-making at all levels to identify local targets that are in line with global sustainable development.

In all sectors systems are currently undergoing dramatic changes as the global need is recognised to reduce the immediate environmental impact of human activities and global greenhouse gas emissions in order to maintain functioning ecosystems. These developments concern the use of resources globally and have varying impact on the individual sectors. While actions have to be addressed at the scale of individual actors or subsystems the impacts are effecting systems globally. As a result a discrepancy between the investment in local mitigation measures and the (locally) perceived effects can be identified often blocking ambitious engagement which is less the case for adaptation measures.

This chapter argues that in order to capture the full breadth of such system-wide effects the measures and impacts have to be assessed at multiple scales. As such a multi-scale approach allows for a consistent description of the problem statement and solutions at different system levels.

As concrete example, in the energy sector three main strategic options can be identified (source) to transform today's energy systems: reducing the overall needs (sufficiency), satisfying needs with less resources (efficiency) and finally using renewable resources to satisfy the needs (consistency). While efficiency and energy savings are often identified as first measures to take (source) action is taken to reduce the CO_2 content of energy supply technologies. In the latter context renewable energy sources play a key role to achieve the often ambitious targets at international or national level. The employed renewable energy sources such as wind and solar power are fluctuating by nature and are integrated at different levels of the energy grid. As the electrical grid balances demand and supply at each instant over the whole network, it provides for a good example for local measures (e.g. solar PV installation) that impact overall system characteristics (e.g. CO_2 content of the electricity in the grid).

In order to analyze such systems or make prediction on their future evolution often models are applied that predict specific characteristics relevant for specific questions of analysis. This perspective is usually specific for the question at hand but also corresponds to the resolution in which the model is designed which strongly depends on the scale of system evaluation. From this need to investigate man-made systems at different scales emerges a strong requirement to represent phenomena at different scales. This clearly shows the relevance and need to conduct multi-scale modelling. Improvements in calculation power make simulations of complex systems possible. The advances in computing power make available facilities which allow to scale up and extrapolate models in dimensions which were not even imaginable some time ago. This opens new doors to benefit from this additional performance, in which the modeller has not continuously take care about resource-efficient modelling that usually constraints the expressivity of the simulations, especially in the case of complex systems.

General characterisation and purpose of multi-scale modelling

Models reflect simplified representation of reality, used to explain specific aspects of a, mostly closed, observed system. The concept of representing reality in the most reduced way that still allows for a meaningful assessment is known as Occams Razor (Young, Parkinson et al., 1996). According to this concept the simplest model possible should be selected to describe a system. While this allows for a high efficiency in modelling, such models are not suitable to describe phenomena beyond their initial purpose.

System-of-systems and scalability

The term system-of-systems (SoS) is a collection of task-oriented or dedicated systems that pool their resources and capabilities together to achieve a new, more complex metasystem. This metasystem itself offers more functionality and performance than simply the sum of the individual systems from which it is constituted (Markarian, Kölle et al., 2011).

The concept of systems-of-systems point of view is promoted that acknowledges the effect individual objects or subsystems have at a system level. For example, complex behaviour can be observed at a higher level that emerges from individual behaviour. This emergent behaviour cannot always be fully inferred by the individual behaviours. The higher level behaviour sum is more than the pure sum of the individuals and can only be explained by the dynamics and interactions that take place at the different levels of the system. This requires a better understanding of, for example, the cross-scale relations among systems that might exist and which determine the relations between different scales.

According to the definition of Maier (1998), a SoS can be characterised by the following criteria:

- Operational independence of the elements
- Managerial independence of the elements
- Evolutionary development
- Emergent behaviour
- Geographic distribution

As an example, the energy–water nexus is subdivided into many subsystems, such as power and water utilities, which typically are operationally and managerially independent of each other. Even further, the liberalisation and subsequent unbundling processes clearly describe an independence of different stakeholders in each field. Network operators are considered natural monopolies, and thus regulated accordingly. On the trading and generation side, however, a market approach has been chosen. An evolutionary development is given, as the systems are not in a static state but evolve adapting to the environmental and boundary conditions. So, for example, the energy sector is facing a large evolution due to CO_2 reduction policies, and the water sector is impacted by demand increase and resource shortage. To describe an emergent behaviour, the nexus must be considered as a whole.

Increasing energy needs especially in regions developing fast towards advanced industry economies, such as China and India, will thus be main driver for water scarcity. Interestingly, renewable energy resources such as wind and solar require much less water. The increasing introduction of these resources will have an impact on the energy–water nexus, which can be considered as a complex emergent phenomenon, as both energy and water sectors will depend on it. Currently, water shortage is not a global issue, but it could be an additional driver towards alternative energies in future, if scarcity drives prices up and makes conventional power plants less profitable.

Representation of different scales in a model

Multi-scale modelling refers to a way of modelling in which multiple models at different scales are used simultaneously to describe a system. It involves combining models from different scales already in its early stages of development. The purpose, besides quantitative modelling, should aim at better understanding and exploring a complex system.

Concerning the modelling of systems considering different scales, we can differ between:

- **Modelling of systems at different scales**: Modelling of the system or subsystem at different scales. For example, for a part of the system which is to be described more in detail, a more accurate model is set up and separately run. The results can then be compared, but no direct exchange between the models exist. Analytically, combining the different outputs aims to give a better view of the system modelled, by modelling subsystemic effects through separate models.
- **Multi-scale modelling**: Multi-scale modelling refers to a style of modelling in which multiple models at different scales are used *simultaneously* to describe a system. They are related to each other, so more accurate models or subsystem parts are put in relation with a more global model. Exchanges of information exist primarily in within the scales. Between the scales, models are linked manually.
- **Cross-scale modelling**: Cross-scale models are multi-scale models that include the ability to interact dynamically among different scales. They can be seen as a more advanced form of multi-scale models, in which the interactions represented are more copious and can cross over different scales. Exchanges of information across scales are modelled endogenously.

Finer scale processes often govern processes seen at coarser scales and vice versa.

The complexity of the models increases from separate modelling up to cross-scale modelling. However, to represent complex cause-effect relationships among scales, more complex models can be useful as they can handle feedback loops that are relevant especially in large-scale

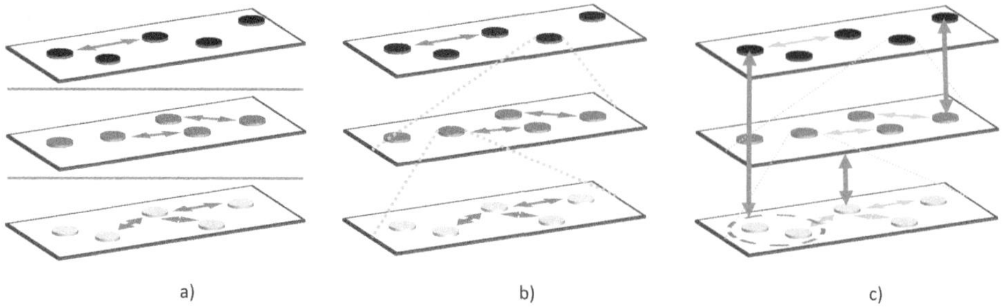

Figure 12.1 a) Different scale models non-related to each other, b) multi-scale model with manual or weak interactions, c) cross-scale model with dynamic interactions across scales

systems with high number of entities and relationships among them. This is typically the case for socio-technical, socio-economical or socio-techno-economical systems. The water–energy–land–food–minerals nexus shows all these relevant aspects.

Concerning the setup of multi-scale models, two main approaches used to tackle multiple scale questions, which often are opposed to each other, are:

- Top-down approaches: Relate global phenomena to local systems and thus allow to investigate possible causes of system-wide effects in greater detail.

  A top-down methodology neglects details of the system, but in return it enables to have an overview representation of the system. This overview can then be divided into subsystems that subsequently are broken down to further understand the impact at lower system scales. The top-down idea corresponds to a deductive reasoning, in which premises are generally applied and narrowed down until the conclusions are left.

  The water–energy–food nexus can be considered from a top-down perspective: different sectors are considered inextricably linked with each other – the implication that the actions in one area typically affect one or more other areas. To tackle the impact of the effects in each area, however, a breakdown from this top model to straight down has to be made.

  Top-down approaches typically aggregate individual effects and use statistical or averaged values at system level to quantify systemic effects. For this purpose, often black-box models are used, which may in some cases over-simplify relevant system behaviour, which can be a drawback of the methodology.

  Top-down approaches typically offer a reduced variability in comparison to bottom-up approaches, due to the assumptions made to characterise the system at the chosen modelling scale. At strategical level, top down approaches are commonly used as they allow for much simpler ways (and therefore avoiding additional effort) to represent systems.

  The main purpose of this approach is to plan and understand globally a system and its transversal relationships.

  A top-down modelling approach example would be the System Dynamics methodology, which allows to qualitatively and quantitatively describe the dynamics of large complex systems at a global scale, taking into account their non-linear relations such as feedback loops.
- Bottom-up approaches: They allow to quantify global impact as the sum of local processes. A bottom-up methodology targets first the individual components of the system, which then are put in relation with each other. The overall model aggregates the different parts and allows in this way to draw conclusions at more abstract system scales.

The information acquisition and processing at the individual entities of the system is based on a local perception of information of their environment. Bottom-up is related to inductive reasoning, where conclusions are achieved by generalising from the more specific considerations towards the global picture.

The lack of accurate disaggregated data can be a drawback for this methodology.

Bottom-up approaches offer their full potential if large, low system level entity datasets are available, as they enhance the detailed modelling. This is fortunately increasingly the case due to the spreading of (mobile) sensing devices, such as smartphones or distributed sensors (e.g. smart meters).

Bottom-up is well suited for the extrapolation of system effects and impacts to larger scale (upscaling), due to its modular nature. Bottom-up approaches are usually more "close to reality" in terms of their modelling (lower abstraction level) than top-down approaches – with all the implications that this has: if the purpose of modelling is to simplify a problem which can be easily quantified at system level, a more realistic viewpoint might not be needed. However, if the problem to be tackled is of complex nature and, for example, causal relationship chains are not fully understood, a more detailed approach is useful to be able to explore, better understand and quantify them. A main challenge in describing the nexus lies in harmonising the subsystems structure to describe phenomena at the system level.

Bottom-up models are typically object-oriented models or agent-based models, which are able to capture individual entities of the system and aggregated them towards a higher system scale. Due to their nature, they allow to capture emergent phenomena; that is, that higher level effects can be captured through the interaction of lower level components and their aggregation.

To describe and better understand processes in complex systems, such as a sector-transversal system like the nexus, both the principles of top-down and bottom-up modelling can be used to generate an appropriate manner to represent the system. It has however to be noted that models, as simplified representations of reality, never can depict a complex system in its full integrity. For this reason, the model-related uncertainty has to be taken into account.

As we have seen, multi- and cross-scale models somehow overcome the strict separation of bottom-up and top-down approaches. By linking different scales, there is no more a clearly unidirectional aggregation or disaggregation of the system behaviour, but we can rather observe cyclic feedback interactions between the macro, meso and micro scales. Only in this way can complex phenomena be better understood, as real systems as well are not only pure bottom-up or top-down systems, but rather include these aspects in a more or less-developed manner.

Going further than the conceptual or structural scale modelling of a system, which was just discussed, multi-scale models have to deal with different spatial and time resolutions, which have to be adapted to each scale. When modelling cross-scale interactions, different representations have to be considered. Geographical information systems (GIS) can be used to represent different spatial resolutions and the location of the entities. They can also help to relate these entities to each other and serve as aggregation boundaries.

Lessons learned from other fields: an excursus to biomedicine research

In biomedicine, multi-scale modelling is already used and quite advanced in comparison to other fields. For example, multi-scale modelling is applied for research on cancer, which is characterised by out-of-control cell growth and tissue invasion (Deisboeck, Wang et al., 2011). The cancer initiation and progression process needs a more systemic understanding, as it is composed of the emerging behaviour of individual cells in interaction with each other.

Modelling in the cancer research field occurs over different temporal and spatial scales (e.g. from nanoseconds up to days or years) and over the different spatial scales from atomic, molecular, microscopic, tissue/multicellular) to macroscopic (organ) level (see Figure 12.2). In the context of biology and physiology at large, a model is considered to be multi-scale if it spans two or more different spatial scales and/or includes processes that occur at two or more temporal scales. Recent models link lower-scale modules with those at higher scales, with the purpose of developing more realistic and predictive models for cancer diseases.

Different modelling methods are used at each scale, in accordance to the phenomena represented, which are ODE (Ordinary Differential Equations), PDE (Partial Differential Equations) or ABM (Agent-Based Models). The methods used are of varying nature, so for example ODE/PDE are continuous approaches which are used for larger scale volumetric tumour growth dynamics, at a lower computational effort. Discrete approaches such as DEM (Discrete Event Models) or ABM include rules based on data or data-driven assumptions at cell or cell cluster level, defining their behaviour and cell-to-cell interactions. Hybrid modelling approaches attempt to integrate and draw on the strengths of both continuum and discrete representations. Figure 12.3 shows such an example, in which a tumour growth process is approached with a combined continuum-discrete model. The dead cells dynamic is aggregated and modelled continuously, whereas the alive cells are models as agents using discrete rules at individual level. This makes the overall model more flexible and lightweight, as the agent-based modelling would have little interest on the necrotic core, which can be well characterised by an equation model.

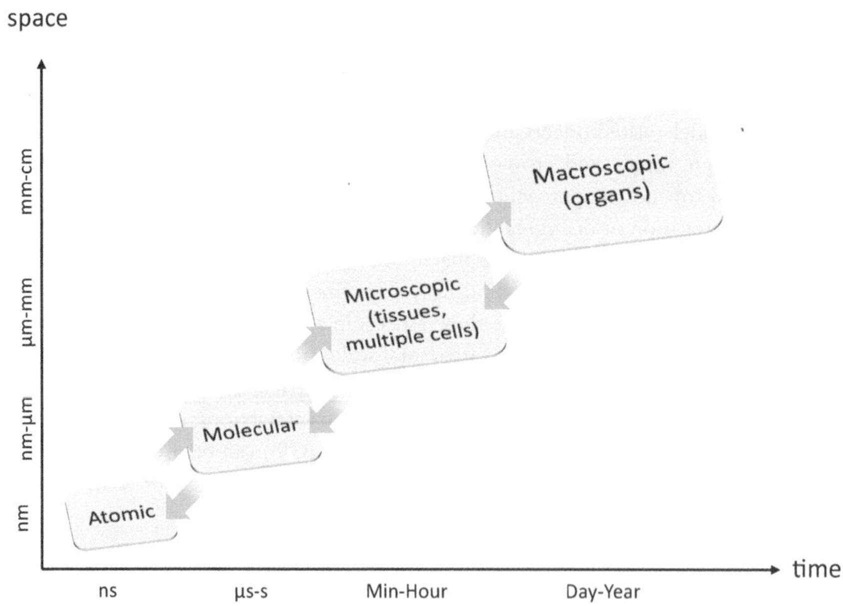

Figure 12.2 Multi-scale modelling in biomedicine cancer applications. Different time and space scales and modelling methods used

Source: based on Deisboeck, Wang et al. (2011)

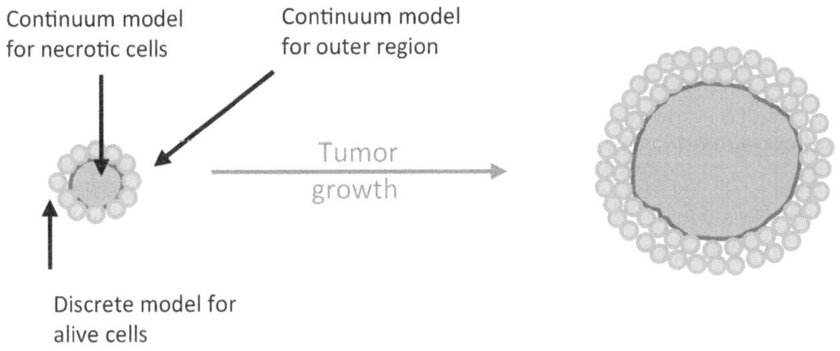

Figure 12.3 Hybrid modelling of tumour growth: living cells are represented with a discrete model, in which every tumour cell is modelled individually. The cluster of cell growths following biophysical rules coded in each cell. Once the cells die, they are integrated into a continuous model of the necrotic region, and only the shape is considered. As only a few cells at the outer region are modelled individually, greater detail can be included. The alive proliferating cells interact through a mechanical force-based model and are later added to the continuous model using geometrical models and a least square projection. The outer gel region's mechanic response is also modelled using a continuous model for their viscoelastic material properties

Both models are however dynamically interconnected with each other, reflecting the mechanical interactions between the different types of cells.

This kind of modelling can be applied for personalised diagnosis, in which a model can be calibrated with personal data from a patient to make more accurate previsions of its cancer evolution (ibid).

The following lessons learned can be drawn from today's outcomes of multi-scale cancer modelling:

- Parameter estimation: Whereas more and more data is currently available, not always the correct parameters can be estimated or quantified at the needed scales. Two techniques are mainly used for parameter estimation: local optimisation (e.g. direct-search and gradient-based methods) and global optimisation (e.g. simulated annealing, branch and bound, and evolutionary algorithms)
- Computational power: Discrete or discrete-based hybrid models are typically computationally more intensive as continuous ones, because they can be too detailed in some time periods of the simulation. Adaptive hybrid modelling (with a dynamically changing granularity of the model to avoid these too-detailed periods) are currently in an early stage of development.
- Data sharing and model reusability: Data sharing of public-funded works has to be made better available to the research community. To avoid duplication and waste of resources a standardisation of both data and models is needed. In biochemistry, several XML-based standards (e.g. SBML, CellML or FieldML) are already defined and recognised for different purposes. However they do not cover yet all fields necessary for a more fluent exchange, so work is still ongoing.

These lessons extracted from a distant, but currently very active and relevant field, are of interest to anticipate the modelling challenges arising for the nexus. Parameter estimation in a microscopic system has been identified as an important question. Laboratory experiments, such as the ones carried out for cancer research, can provide some reliable quantitative data. This data has still to be fitted to be conform to the parameters that the designed models use. A similar topic arises when dealing with nexus models, however, in macroscopic systems laboratory experiments that would, for example, represent a nexus or parts of it, are not at all realisable within a realistic scope. Real data and analysis has to be acquired from the field and from estimations. Furthermore, *softer* data is typically needed, when dealing with non-fully quantifiable systems, including especially social subsystems (e.g. socio-technical, socio-economical). This is an important difference from more quantitative sciences, where *hard* data is measurable and can be more easily transferred to models. Virtual simulation environments in which the system is represented *in-silico* might help to overcome these questions.

Examples of multi-scale modelling

As mentioned in the previous section, we can differ between representation of different scales in separated models, multi-scale models with punctual interactions and dynamic cross-scale models. Concerning space and time, the multi-scalability aspect can, though, be further declined. This is the purpose of this section which will exemplarily show how multiple scales are treated in both space and time, and then discuss their relevance and application of these concepts to the nexus.

Spatial abstraction levels of data for multi-scale modelling

As described in the previous sections, many application examples in the multi-scale modelling domain are conducted at different abstraction levels. In order for data to be used in multi-scale modelling data has to be stored at different abstraction levels. From a geographical perspective, according to Montello (1993), scale can be divided into geographic, attribute and temporal scale. In multi-scale modelling all three types of scale for data can be applied.

On the geographic abstraction level the unit of analysis can range from a single building entity to whole geographic regions. Data is thereby spatially aggregated or disaggregated to geographic units or boundaries also referred to micro-, meso- and macroscale. The domain or phenomena that the data represents is scale dependant and defines in part the level of abstraction. Various geographic abstraction levels can be nested or emerge from a smaller or larger scale, also referred to as "hierarchy of scales". For example, national or international development incentives can have highly different outcomes on regional or local administration levels. Similarly, a global increase in the average temperature could have high differential impact on the global scale (Kolaczyk and Huang, 2001). Figure 12.4 depicts an example of geographic raster data of average wind speed in northern Germany at a resolution of 1 km (left) and at an aggregated scale of 50 km (right). Conceptually, the aggregated scale is derived via resampling to a coarser resolution based on statistical aggregation strategies such as sum, minimum, maximum, mean or median values.

Application cases of geographic abstraction levels in context to the energy–water nexus for example can be found in the estimation of drainage networks that can be incorporated into the modelling of water runoff and water network estimates for potential or existing hydropower stations (Stuebe and Johnston, 1990; Vieux, 2001). Thereby, beside hydrographic networks, Digital Elevation Models (DEM) are required for these estimations for identification of catchment areas

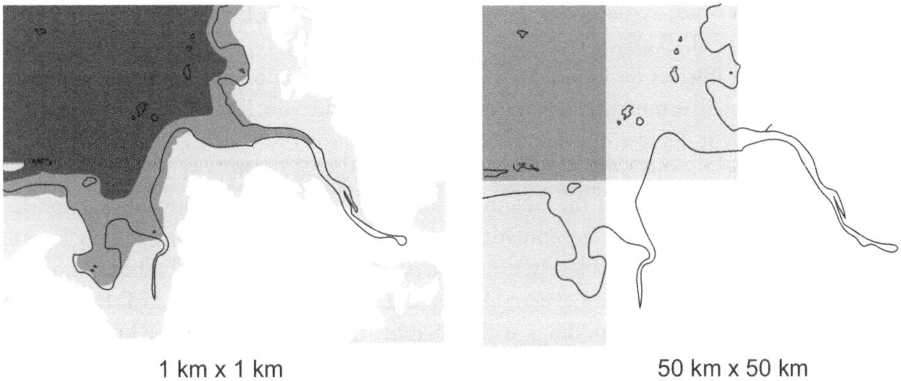

Figure 12.4 Spatial aggregation of raster cells by statistical means

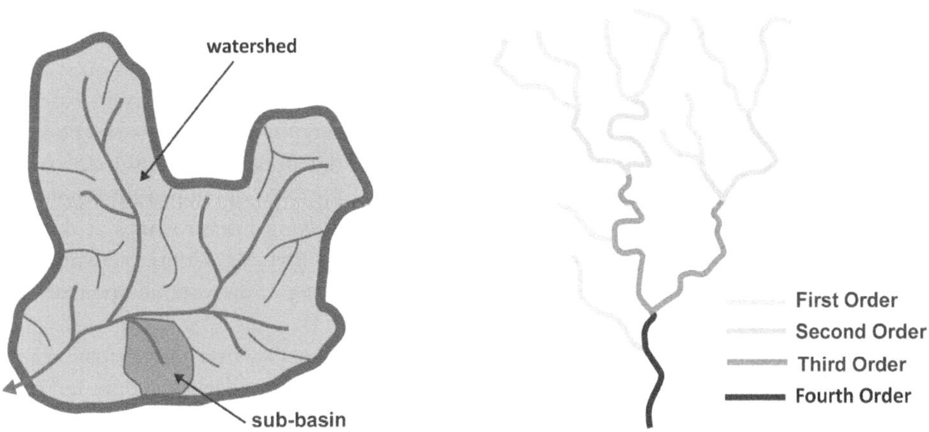

Figure 12.5 Left) Multi-scale watershed diagram consisting of watersheds, drainage basins and sub-basins. Right) Strahler stream classification according to Strahler (1957)

that are the basis for water runoff estimates (Mantilla and Gupta, 2005). Watersheds or catchment areas and water networks are thereby grouped into nested hierarchies that can be modelled at different scales (Figure 12.5).

The division of watersheds usually follows a hierarchical order defined by geographic features such as mountain ridges and hill tops. Watersheds are composed of many smaller drainage basins where interactions occur between and across different sub-basins. Multi-scale modelling can be applied at multiple scales following statistical aggregation schemes for water runoff estimations, in contrast to stream networks that are represented as hierarchical networks of channels at fine scales. A first-order stream is a channel with no tributaries, usually represented as a stream at its upper reaches and near its source. A second-order stream is a channel fed by at least two first-order tributaries. The joining of two-second order streams forms a third-order stream. Stream ranking continues in this manner until the highest-ordered channel is reached. First- and second-order streams are located in the headwater areas of watersheds and typically convey small volumes of water. This inherent hierarchical structure of stream networks is important in hydropower application as responses such as hydrological and thermal properties as well as interactions with other systems can be modelled (Singh, 1997; Buttle, 2006; Caissie, 2006; Peterson, Theobald et al., 2007).

On the attribute abstraction level for multi-scale modelling an object can be represented in different level of details simultaneously, enabling analysis and visualisation of the same object with regard to different degrees of resolution. Thereby the level of detail for analysis and visualisation do not have to be the same. For example, analysis can be conducted at a course abstraction level but the visualisation of the outcomes can be shown on a very low abstraction level depending on the scope and the computational complexity of the modelling task. A concrete example for attribute abstraction levels is the CityGML format used for building modelling applications consisting of different Level of Details (LoD) ranging for a coarse LoD1 model only consisting of simple prismatic buildings with flat roofs to LoD4 representing high resolution structural shapes and interior structures of a building. CityGML enables the storage of semantic information not only of each surface but also of each element of a building or object. Furthermore, object modelling specifications can be stored and displayed at different LoDs, specifically relevant of multi-scale modelling applications (Kolbe, Gröger et al., 2005).

Multiple domain specific extensions are currently in development for the CityGML standard. Relevant ADEs (Application Domain Extensions) for the energy nexus are for example the Energy ADE and the Utility Network ADE. The CityGML Energy ADE focuses on standard features necessary to perform energy simulation and to store corresponding results based on buildings (Nouvel, Bahu et al., 2015). However, building energy representations can be interfaced with utility networks that can represent energy relevant installations outside the building envelope through the CityGML Utility Network ADE that was developed for topological network model representations of utility networks and supplying infrastructures at different abstraction levels (Becker, Nagel et al., 2011). Included in the CityGML Utility Network ADE are, among others, network hierarchies of arbitrary depth, nesting of network components, and modelling of multi-modal networks (Becker, Nagel et al., 2011).

Spatio-temporal scale describes the temporal summarisation of data over a geographic region. Spatio-temporal aspects in datasets are fundamental in modelling any changes of time and space such as monitoring and prediction (Van Oosterom and Stoter, 2010). Thereby, the temporal abstraction level differs on the phenomena and technology (e.g. sensors) that is represented. Within multi-scale modelling the temporal scale can vary depending on the computational complexity of the processes that are being modelled. Figure 12.6 depicts an example of the

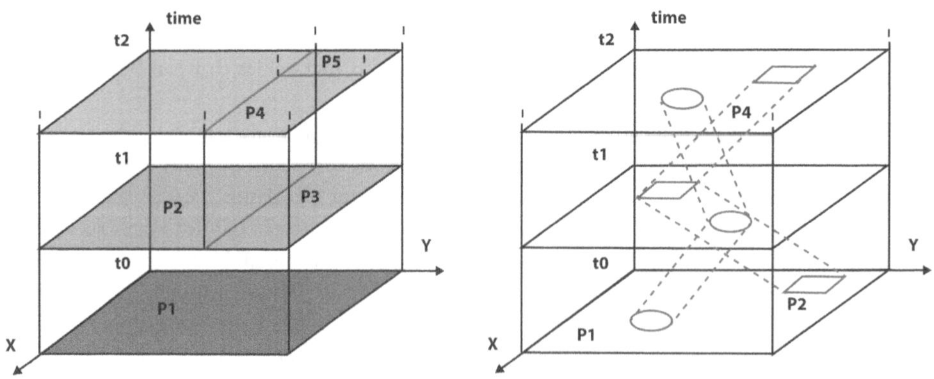

Figure 12.6 Time as a third dimension in spatio-temporal modelling of aggregated objects (left) and single moving objects (right)

Source: Van Oosterom, Ploeger et al. (2006)

status change of a set of related features (left) over time and abstracted spatio-temporal changes of single moving objects (right).

The literature provides numerous spatio-temporal data models (Hornsby and Egenhofer, 2000; Raper and Livingstone, 2001; Peuquet, 2002). According to Van Oosterom and Stoter (2010), the semantics of each model can be categorised by its temporal granularity, temporal operations, modelling foundation of time, system, valid time, lifespan and the representation of time.

Current and future research on the topic of multi-scale modelling is focusing on the storage and visualisation of information such as 5D modelling approaches where 2D, 3D space, time and scale aspects of geo-information are combined; and the interaction of geographic multi-scale models with models from other domains such as environmental, economic or health models.

Time scale aspects in multi-scale simulation

Electricity systems as part of the energy nexus are complex systems, composed of different subsystems (e.g. individual power plants which are quite complicated technical systems in themselves, connected to each other) and demonstrating a vast geographical spread, including many elements which are all interconnected to each other. In order to understand cross-scale effects on those systems, a disaggregated and bottom-up approach is used, which aggregates the effects of the systems at different scales. Examples of geographical, temporal and structural multi-scale modelling using an agent-based approach are shown through different case studies. Furthermore, due to their characteristics, electricity systems can be considered system-of-systems (Kremers, Viejo et al., 2010).

The background for the development of models vary: Some of them originate from a macroeconomic context and were provided with extensions towards the energy field. Others arose from some subareas (power plants, energy distribution) and evolved towards a representation of the entire energy sector. Still others are made from an assortment of sectorial sub-models (Koch, Harnisch et al., 2003).

However, many challenges are still open: Modelling of the right points of interactions for cross-scale relations, validation of multi-scale models, complexity and calculation power demand, as well as adaptive approaches will be discussed.

In this case, the multiple time aspect will be exemplarily explained by exploring the needs of different time resolutions for a wind power simulation system. When trying to represent a wind farm for different purposes, such as daily estimations and yearly evaluations, the authors identified different models for that purpose, but it was not evident to find an approach or the appropriate use of the different models together. Therefore, a modular multi-scale simulation model was created to cover a large range of time resolutions.

The model is described in (Kremers, Lewald et al., 2009; Kremers, Viejo et al., 2011) and provides a toolbox to generate simulated wind speeds at different time scales. It is composed of different modules for hourly up to minutely level. Three output scales are possible: hourly, quarter-hourly and minutely (see Figure 12.7). The hourly module is based on an ARMA (autoregressive-moving average) time-series model, and can be parameterised by its Auto-Regressive/Moving-Average coefficients. The interpolation module allows to interpolate these hourly values to obtain further lower resolution such as approximate minutely or quarter-hourly values.

In the following, the agent-based modelling of a wind turbine farm is described.

A bottom-up strategy is used to generate the total output of wind power of a wind farm. Stochastic behaviour such as failure and maintenance of the turbine is included. Further, variability between the turbines due to slight technical differences on the generation machines,

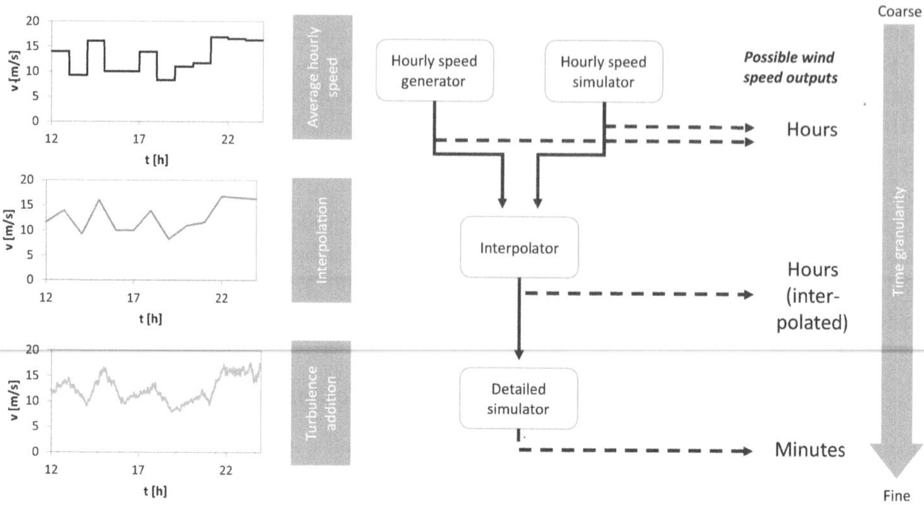

Figure 12.7 Structure of a multi-scale model for wind speeds and wind power generation

which are never 100% identical, as well as on the wind speeds at different locations are included. This provides a more realistic simulation result at global (= wind farm level).

In Figure 12.8 we can clearly observe the differences at different scales. In the upper plot the generated wind speed outputs of the system are shown, where the turbulences added by the minutely module can be recognised, in comparison to the hourly and interpolated hourly simulations. The effect at the generated power for the wind farm can be seen in the lower plot. The turbulences make the wind speed vary in between two hours, which in some cases can cause a so-called cut-off of the turbine. When wind speed is too high, turbines automatically stop working for security reasons and only start again at a cut-back-in wind speed, typically lower than the cut-off value. This can be seen at hour 4 of the simulation, where the minutely model shows a complete cut off of the farm. A gradual cut-back-in can be observed after, were it should be noted that the turbines do not start working back at a sudden, but gradually due to slightly different local conditions. Between hour 8 and 20, the hourly model almost shows no wind generation, due to averagely low wind speeds, however the minutely output shows times at which some or even mostly of the turbines are producing electricity. The turbulences captured by the minutely model overpass the so-called cut-in wind speed, at which the turbines start working. We can see that it is impossible to capture these with a coarse model.

Further it should be noted that in the given simulation a difference in the total energy produced (integration of the power curve). This indicates that the coarse simulation gives an approximate value, but has an error rate which can be quantified by the stochastic properties of the turbulences. This fact opens the question of validation of models at different time scales.

Energy systems cover phenomena at different time scales, which have to be correctly handled if representing the energy system from a sector perspective for the Nexus. However, while modelling effects within the energy system, different time scales are quite relevant. For example, the recent trend which aims at coupling the heat and the power sector faces quite interesting questions at a time scale level. Thermal effects typically take place in larger time scales as do electrical, so for example the time constants of thermal inertia are rarely described below a minutely resolution, whereas transient effects in electrical systems need for consideration of

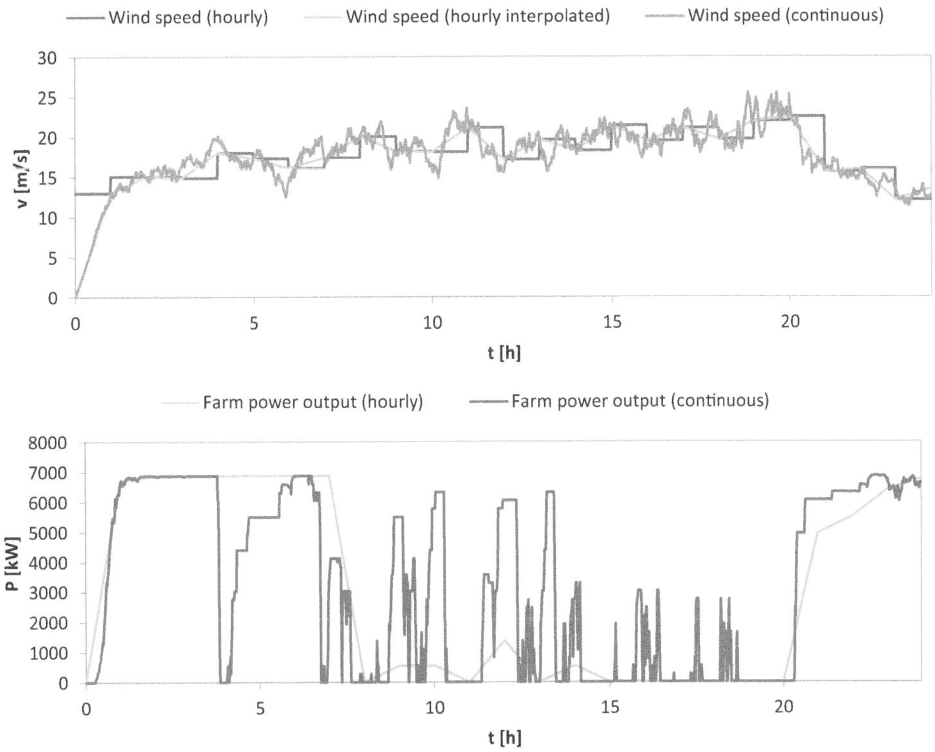

Figure 12.8 Simulation of wind speed and electric power generation of a wind farm at multiple time scales

milliseconds or even lower scales to capture for example the moment in which a high voltage line is connected or disconnected from the power grid. In contrary to the biomedicine example, where cancer propagation is modelled and a strong correlation between space and time scales could be observed (small phenomena require small time steps to be described), the power system can show quite short-term variations over large geographical scales. This is the case when large imbalances can propagate within a few milliseconds through the grid frequency across the whole power grid, leading to cascade effects and large blackouts. In this case we can argue that, for the nexus model, time and space scales would not be always correlated in a linear form.

Climate models

A prominent example of a multi-scale model approach that addresses the water–energy nexus is the assessment of thermal power plants' efficiency in relation to changing climate conditions. The challenge posed by this subject is bringing together engineering models of plant efficiency including the cooling water cycle which is impacted by climate change, which is represented at a macro level in dedicated models. Hoffmann, Häfele et al. (2013) propose an approach based on a system dynamics model representing the performance of German thermal power plants coupled to results of regional climate models. In Germany the majority of thermal power plants use river water for cooling purposes. In addition to extreme events in which a scarcity of cooling water can be observed especially the temperature level of the river water is identified as an important impacting factor for the plants' performance. On the one hand the inlet temperature for cooling cycles directly impacts the plant's efficiency, on the other hand the outlet temperature of cooling cycles underlies environmental regulation and thus the amount of cooling energy is limited depending on the overall water temperature (ibid).

Sieber (2013) further highlights the aspect of vulnerability of power infrastructure climate change and extreme weather events in particular. Among changes of climatic conditions, the impact on the cooling systems can be seen as mayor link of the two systems. However, in the case of climate change in addition to the multi-scale issue also asynchronous timelines are posing a major obstacle of a holistic depiction in a single model. In contrast extreme weather events are directly linked to failure of electric infrastructures described by Sieber (2013) in a GIS-based approach. In descending order of importance hail, storm and flood events are identified as causes for outages of power generation and transmission infrastructure (ibid).

Conclusions and outlook

By definition models imply a simplification of reality in order to describe observed phenomena. Insofar the recognition of the sensitivity of scale is not a recent phenomenon. Yet the availability of models at diverse scales and the ambition to understand systems by coupling models of different domains with different scales necessitates a better understanding of the multi-scale problematic. As discussed among other examples in this chapter the energy–water nexus is such a system for which the explanatory power of models can be increased by applying a multi-scale perspective. While in material science or biomedicine multi-scale models are well developed there is relatively little sound application at the macro level covered in this chapter. However, it has been identified that the correlations between spatial and temporal scales, which in cancer modelling seem increasingly correlated (the pertinent scales seem strongly related for space and time), can be more complex in the nexus. For example, large-scale phenomena can sometimes only be considered at very high small time scales, such as the stability of power grids.

Different forms for representing multi-scale systems have been discussed: from simple separate modelling of different scales, over multi-scale models with manual weak interactions up to dynamic cross-scale interaction models. Other fields have already had an experience with modelling systems at different scales, and techniques like hybrid modelling are state of the art. However, some open questions remain, concerning data accuracy, parameter estimation, validation and harmonisation of the models. Concerning validation, components can be validated individually, but systemic behaviour can only be validated considering the conjunction of the components. Validation at different scales needs coherency between the scales. This encompasses the challenge to find coherent data at different scales for a reference situation of the system at the same time.

Introducing different spatial scales has been showing to be a useful means as a common denominator for structuring multi-scale models. Time-relevant aspects have been addressed through the use of continuous models, coupled to regressive time-series models in a hybrid and modular approach. An application in the climate context describes the importance of such models, on the question of thermal power plants efficiency decrease related to global warming, which could be enhanced considering the previous aspects.

Further points to be tackled consider the introduction of hybrid or multi-method approaches, in which different modelling paradigms can be used to unlock the full potential of cross-scale models. This would allow model practitioners to use an adapted paradigm to their scale, while at the same time the connections to other scales would be possible. Adaptive hybrid modelling, in which the models adapt automatically with the optimal model paradigm (e.g. discrete or continuous) are still in early stage of development, but have proven their relevance. As shown in the example cancer modelling, these approaches are possible and useful, and could successfully be transferred to the nexus.

References

Becker, T., Nagel, C., and Kolbe, T. H. (2011). Integrated 3D modeling of multi-utility networks and their interdependencies for critical infrastructure analysis. In Kolbe, T. H, Nagel, C., and Koenig (eds.), *Advances in 3D geo-information sciences*. Berlin: Springer, pp. 1–20.

Buttle, J. (2006). Mapping first-order controls on streamflow from drainage basins: The T3 template. *Hydrological Processes*, 20(15), 3415–3422.

Caissie, D. (2006). The thermal regime of rivers: A review. *Freshwater Biology*, 51(8), 1389–1406.

Deisboeck, T. S., Wang, Z, Macklin, P., and Cristini, V. (2011). Multiscale cancer modeling. *Annual Review of Biomedical Engineering*, 13, 127–155.

Hoffmann, B., Häfele, S., and Karl, U. (2013). Analysis of performance losses of thermal power plants in Germany – a system dynamics model approach using data from regional climate modelling. *Energy*, 49, 193–203.

Hornsby, K., and Egenhofer, M. J. (2000). Identity-based change: A foundation for spatio-temporal knowledge representation. *International Journal of Geographical Information Science*, 14(3), 207–224.

Kolaczyk, E. D., and Huang, H. (2001). Multiscale statistical models for hierarchical spatial aggregation. *Geographical Analysis*, 33(2), 95–118.

Kolbe, T. H., Gröger, G., and Plümer, L. (2005). CityGML: Interoperable access to 3D city models. In *Geo-Information for Disaster Management*. Berlin: Springer, pp. 883–899. Retrieved from https://link.springer.com/chapter/10.1007%2F3-540-27468-5_63

Kremers, E., Lewald, N., Barambones, O., and González de Durana, J. M. (2009). *An agent-based multi-scale wind generation model.* The Ninth IASTED European Conference on Power and Engineering Systems, EuroPES 2009, Palma de Mallorca, Spain, Acta Press.

Kremers, E., Viejo, P., Barambones, O., and Gonzalez de Durana, J. M. (2010). *A complex systems modelling approach for decentralised simulation of electrical microgrids.* Engineering of Complex Computer Systems (ICECCS), 2010 15th IEEE International Conference on.

Kremers, E., Viejo, P., Lewald, N., González de Durana, J. M., and Barambones, O. (2011). Agent-based simulation of wind farm generation at multiple time scales. *Wind Farm – Impact in Power System and Alternatives to Improve the Integration.* G. O. Suvire, Intech. **2**.

Koch, M., J. Harnisch, Harnisch, K. Blok (2003). Systematische Analyse der Eigenschaft von Energiemodellen im Hinblick auf ihre Eignung für möglichst praktische Politik-Beratung zur Fortentwicklung der Klimaschutzstrategie. Climate Change. Umweltbundesamt. Berlin.

Koch, M., J. Harnisch, Harnisch, K. Blok (2003). Systematische Analyse der Eigenschaft von Energiemodellen im Hinblick auf ihre Eignung für möglichst praktische Politik-Beratung zur Fortentwicklung der Klimaschutzstrategie. Climate Change. Umweltbundesamt. Berlin.

M. W. Maier. (1998). Architecting principles for systems-of-systems. *Systems Engineering*, 1(4), 267–284.

Mantilla, R., and Gupta, V. K. (2005). A GIS numerical framework to study the process basis of scaling statistics in river networks. *IEEE Geoscience and Remote Sensing Letters*, 2(4), 404–408.

Markarian, G., Kölle, R., and Tarter, A. (2011). *Aviation security engineering: A holistic approach.* London: Artech House Publishers.

Montello, D. R. (1993). *Scale and multiple psychologies of space.* European Conference on Spatial Information Theory, Springer.

Nouvel, R., Bahu, J. M., Kaden, R., Kaempf, J., Cipriano, P., Lauster, M., Haefele, K-H., Munoz, E., Tournaire, O., and Casper, E. (2015). Development of the CityGML application domain extension energy for urban energy simulation. *Proceedings of Building Simulation 2015*, 14th International Conference of the international building performance simulation association, December 7–9, Hyderabad, IND, 559–565.

Peterson, E. E., Theobald, D. M., and Ver Hoef, J. M. (2007). Geostatistical modelling on stream networks: Developing valid covariance matrices based on hydrologic distance and stream flow. *Freshwater Biology*, 52(2), 267–279.

Peuquet, D. J. (2002). *Representations of space and time.* New York: Guilford Press.

Raper, J., and Livingstone, D. (2001). Let's get real: Spatio-temporal identity and geographic entities. *Transactions of the Institute of British Geographers*, 26(2), 237–242.

Sieber, J. (2013). Impacts of, and adaptation options to, extreme weather events and climate change concerning thermal power plants. *Climatic Change*, 121(1),55–66.

Singh, V. (1997). Effect of spatial and temporal variability in rainfall and watershed characteristics on stream flow hydrograph. *Hydrological Processes*, 11(12), 1649–1669.

Strahler, A. N. (1957). Quantitative analysis of watershed geomorphology, *Eos Trans. AGU*, 38(6), 913–920. doi: 10.1029/TR038i006p00913.

Stuebe, M. M., and Johnston, D. M. (1990). *Runoff volume estimation using GIS techniques*. Wiley Online Library.

Van Oosterom, P., Ploeger, H., Stoter, J., Thompson, R., and Lemmen, C. (2006). *Aspects of a 4D cadastre: A first exploration*. In Peoceedings of Shaping the Change, XXIII international FIG congress, Citeseer.

Van Oosterom, P., and Stoter, J. (2010). *5D data modelling: Full integration of 2D/3D space, time and scale dimensions*. International Conference on Geographic Information Science, Springer.

Vieux, B. E. (2001). *Distributed Hydrologic Modeling Using GIS*. Berlin: Springer, pp. 1–17.

Young, P., Parkinson, S., and Lees, M. (1996). Simplicity out of complexity in environmental modelling: Occam's razor revisited. *Journal of Applied Statistics*, 23(2–3), 165–210.

13
GLOBAL CHANGE AND K-WAVES
Exploring future nexus patterns

Markku Wilenius

Introduction

In the early 1970s, a group of leading system analysts at the Massachusetts Institute of Technology (MIT) built the world's first simulation model, which used a set of parameters[1] to evaluate global development over the course of the next 130 years. The model was commissioned by the Club of Rome (founded by Italian industrialist and scholar Aurelio Peccei), which had taken upon itself the task of examining key issues affecting the long-term survival of humanity. The result of the MIT simulation was the book *Limits to Growth* (1972), which upon its publication rocked the industrial world. (Meadows, 1972)

The book takes the form of a "scenario analysis," which seeks to answer one question: "What will happen over the next 130 years if humanity decides to adhere to certain policies?" It serves not as a forecast but as a mapping of possible futures. It describes how societal resources might be used to accelerate technological solutions to obvious problems of the time: population growth, food and resource shortages, and environment damage. The model has been used to assess the impacts of alternative policies on global emissions and other parameters. The approach is systemic; family size, for instance, can only be regulated with sufficient levels of health care and education.

Back in 1972, *The Limits to Growth* was met with an outcry, particularly by the industrial sector, largely because the report effectively challenged the prevalent notion at the time – and no less today – regarding the blessings of unfettered growth of materially intensive consumption. The book's conclusion is that we are at risk of overstepping the limits of physical growth unless profound changes are enacted.

Forty years since the book's release, we can now see that although alarms were being sounded, nothing has since been done to significantly redress the situation. For starters, in spite of international treaties, global greenhouse gas emissions continue to increase. Natural resources are being depleted, and most ecosystems have degraded seriously, as reported by the United Nation *Millennium Ecosystem Assessment*. (Millennium Assessment, 2005) The World Wildlife Fund's *Living Planet Report* similarly indicates that the population of vertebrates has declined by more than one-half from since 1972, when the Club of Rome report was published. (WWF, 2014)

If a company or state were to deteriorate at this rate, it would be facing bankruptcy and insolvency. And we are indeed driving nature into bankruptcy, as Anders Wijkman and Johan

Rockström observe in their Club of Rome report from 2012. (Wijkman and Rockström, 2012) Just like with Greece in its decade-long borrowing scheme that brought huge debt upon the country, we have refused to face the facts: Instead of trying to repay the debt in some way, we are borrowing more.

We need to be humble about the future. Most long-term forecasts or projections are wildly off the mark. The overshoot-scenarios that the Club of Rome produced back in 1972 have come true with alarming accuracy: We have failed to adopt global policies that will significantly alter the projected course of the business-as-usual scenario. (Meadows, 1972) In spite of all the international treaties, the situation has only worsened over the decades.

Excess and energy

It is obvious that energy consumption is at the heart of global problems. In his book *2052*, Jörgen Randers (an internationally acclaimed Norwegian business school professor and a member of the original MIT research team) examines the course that energy consumption will follow into the future. (Randers, 2012) In 2013, 81.4% of all energy was produced using the three fossil fuels (coal, oil and gas), while nuclear accounted for 5% and renewables covered the remaining. (IEA, 2015) If we wish to stop using fossil energy sources, there is a lot to do!

So what will the future be like? According to Randers, we will continue to be plagued by the inherent short-sightedness of politics and the structures of the capitalist system for 40 years to come. Based on this assumption, he has devised a scenario that suggests that the share of renewable energy sources will increase to 37% by the mid-century. That is not much, considering what we need to achieve. According to the World Wildlife Fund, we will need to produce 95% of our energy from renewables by 2050. (WWF, 2014) At the other end of the spectrum is the International Energy Agency, whose basic scenario some years back suggested an increase in the share of renewables by up to a mere 14% by 2050. (IEA, 2010)

It's important to remember that in the early stages of technological revolutions, very few people believe that a breakthrough will occur. Electronics pioneer Ken Olsen, who in the 1970s was director of Digital Equipment Corporation, famously observed in 1977, "There is no reason anyone would want a computer in their home." (Digital Trends, 2012) On the other hand, Jorma Ollila, the former CEO of Nokia, was the butt of jokes in the early 1990s when he announced his vision of putting a mobile phone in everyone's pocket. Human thinking follows the generally safe logic that things will go on just as they have before. The reality, however, is different, with real potential for both leaps or discontinuities and either exponential growth or collapse.

The world of energy production is currently dominated by large institutions. Researchers talk about a "lock-in." (Foxon, 2002) The predominant system of energy production has a built-in interest – backed by enormous power and resources – to bleed the market of all conceivable profit before "the sky falls on our heads" and before the consequences of fossil energy use become unbearable. Yet, there is reason to think that the current paradigm is changing in fundamental ways.

First, whatever our interpretations of the current situation and the future are, no one seems to be denying that the challenges facing us today are greater and more critical than any other time in human history. Climate change, land degradation, water and energy scarcity, waste management, fish stock depletion, and rain forest and biodiversity loss are only some of the most serious current environmental concerns. It is clear, however, that the gap between resources and their use will only widen in the future. Although growth has slowed, world population will still increase by more than 2 billion people. It is expected that the global population of 9.7 billion people will be reached in 2050. (UN, 2015) At the same time, however, millions of people around the world are rising from poverty towards a better quality of life.

Second, the factors discussed above seem to be incontrovertibly pointing to the conclusion that the expansion of the human race and the limited resources of the planet are on a collision course. We have simply consumed far too much of the planet's limited resources, and there is no longer enough left for the impending increase in population. The same processes that in the past 200 years have produced unprecedented wealth are now threatening to take us to extinction.

Third, awareness of the above-mentioned trends is inexorably spreading among a growing number of people. The concerns and values relating to nature and its exploitation have already changed, and the world's capacity for innovation is increasingly being focused on finding solutions to humanity's greatest problems. Another factor pushing development in this direction is awareness of the biosphere's imminent and irreversible tipping point.

Real change may sound like a utopian dream. We know how difficult it is to change the behavior of an individual, even if the positive consequences of change are obvious, while the history of international climate negotiations show us how difficult it is to achieve change on a collective scale.

Yet, there have been situations in the history of the human race where entire societies have successfully accomplished radical changes in just a few decades. Such changes are most likely to happen when the pressures for change become compelling.

Turning up the heat

In November 2012, the Potsdam Institute for Climate Impact Research (one of the world's most respected research institute on global warming) came up with new and rather dramatic results concerning the so-called 4° Celsius world, or a world in which the average global temperature surpasses preindustrial levels by 4° – which will happen unless much more radical steps are taken to curb greenhouse gas emissions. (The World Bank, 2012) And then, in June 2013, a sequence to the report was published that dealt with climate extreme and the regional impacts climate change will have.

In those reports commissioned by the World Bank, the major point is this: It is impossible to believe that we might possibly preserve decent living conditions for all people on this planet if this scenario should become actuality. Using various simulations, they convincingly demonstrate what will happen to coastal cities when sea level rises (a substantial share of big cities are located on the coast), and particularly what will happen to some of the least developed countries when they are hit by drought and heat waves. All this has been supported in the recent Intergovernmental Panel on Climate Change's *Fifth Assessment* report. (IPCC 2014) In a very short time, the world in those corners will become a rather unpleasant place to live. And for the rest of us living in the northern latitudes, indirect consequences, such as the pressures resulting from massive human migration, would be devastating as well.

These reports make it very clear that this decade is our last chance to start a more radical phase in global warming management. It is our last chance because of the necessary time delays between the various phases – from resolutions to binding targets to political decisions to measures to emissions to concentrations. The December 2015 UN Climate Change Conference (COP21) in Paris provided some hope that governments are slowly realizing the real task ahead, which is to actually reverse climate change.

All of our efforts must be geared towards building more sustainable economies and lifestyles. This is the ethical challenge of the coming decades, and we simply cannot afford to fail in it.

In essence, this entails the dematerialization of our economies on a massive scale, as well as the dematerialization of our lifestyles, but it also requires commitment from the international community to utilize the various means at our disposal for imposing the necessary legal structures.

In that "turn down the heat" report, there is a telling foreword by the chairman of the World Bank, Jim Yong King, in which he writes, "It is my hope that this report shocks us into action." Indeed, as we learn from the report, there has been a growing number of extreme climatic events in the last 10 years or so. Some of them have been caused by climate change and some by environmental degradation, such as deforestation. The massive scale of environmental destruction, so well observed in the UN-led Millennium Ecosystem Assessment, has already brought our biosphere into an increasingly fragile state. (Millennium Assessment, 2005) As the assessment and its follow-up studies have shown, virtually all of Earth's ecosystems have been significantly transformed through human action.

The power of long waves

In the 1930s, Soviet economist Nikolai Kondratiev developed a theory of socio-technological cycles that he considered to be an integral aspect of the economic structure (Kondratiev, 1984). According to his theory, modern economies change radically in cycles of 40–60 years. In my more holistic reinterpretation of Kondratiev waves, technological innovations percolate through economic and social structures, productivity increases constantly and the result is an economic upswing. Such a development is always accompanied by new value systems, new forms of social activity and new cultures of organization. At some point, however, investments in the networks of new technology and the profits they yield will inevitably begin to decline. This in turn leads to stagnating credit demand, which pushes interest rates ever closer to zero, thus resulting in economic recession. Sooner or later, some new form of technology again jump-starts the upward cycle. Major economic crises in the past 200 years – including the most recent one – have all followed this pattern (see Figure 13.1).

Kondratiev waves may be understood as an inherent, dynamic component of economic development, one which occurs with certain regularity. At the same time, however, modern societies have become ever-more complex networks of policy, technology and economy. The financial crisis that began in 2008 was the turning point upon which the fifth Kondratiev wave reached its demise and the new socio-economic cycle began. The emergence of information technologies, oil, electricity and the automobile industry were economic drivers of the preceding waves; the sixth wave will see the emergence of new drivers that will determine both the economy and the society of the future.

Technological advancement opens up whole new horizons for the comprehensive transformation of societies. One way of thinking about this is to envision the preceding wave, which was massively influenced by the development of information and communications technology (ICT), as the foundation for a new kind of planetary nervous system. This nervous system, known as the "Internet of Things," is one in which every piece of equipment has its own IP address and which will be part of a global network – and thereby more accessible and usable with greater ease than ever before. Some of the answers to the critical challenges of the world may come from increasing ICT capacities and the intelligent infrastructures spawned by them, which may even have a decisive impact in solving the current energy problem. (Diamandis and Kotler, 2012) The new technologies are already here, but we have only just begun to apply them.

Significant technological changes often go hand-in-hand with social change. Transformations of technological paradigms and mindsets will also alter the predominant social patterns of behavior. Such changes have an impact not only on structures but also on the values on which

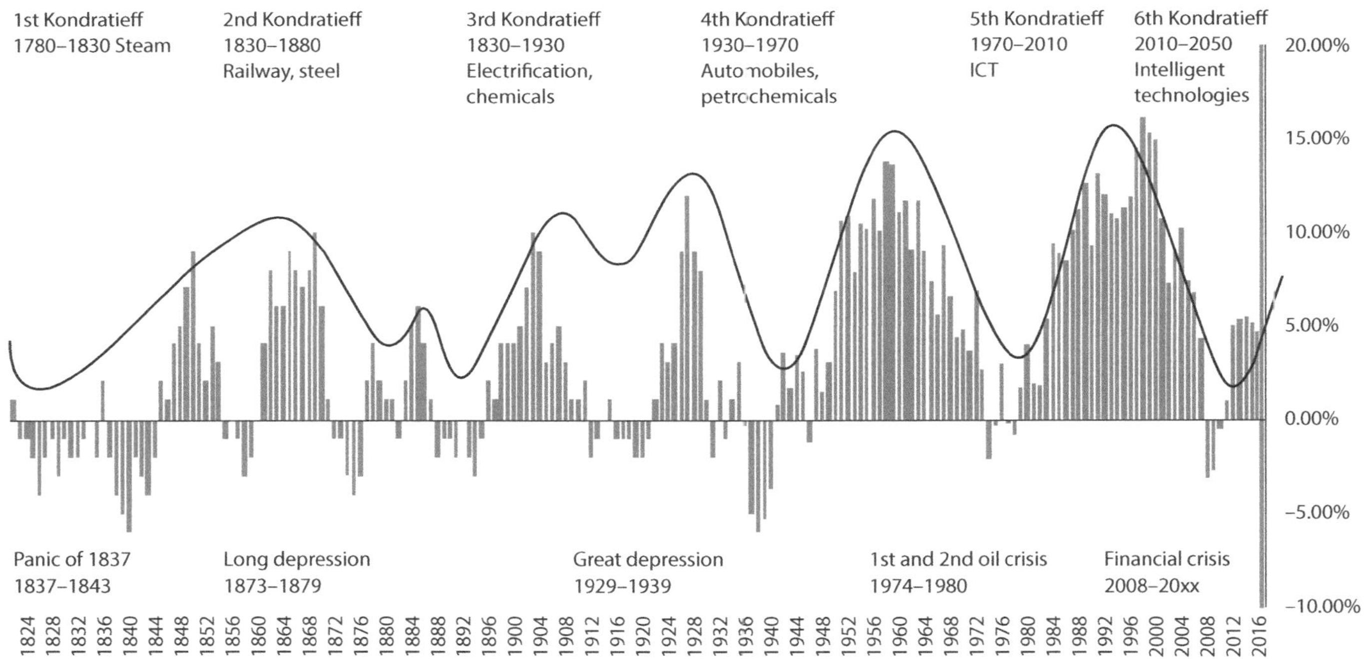

Figure 13.1 Pattern of social waves of advancement since the birth of industrialization

society rests. Current societal challenges and the solutions presented to them will form a new foundation for social and political action. Old demarcations will no longer apply. For example, the notion of industry and services as separate sectors of the economy is a division that is no longer tenable. Industry, in terms of its mode of operation, will increasingly come to resemble the service sector.

Change is possible, because humanity possesses the capacity for innovation – the capacity to create the future, as we have seen so often in the past. What we need is a new wave of creativity for producing technological and political innovations that can help us solve the world's critical resource problems. Such innovations can represent entirely new ways of thinking and utilize new inventions, but they can also be combinations of new and old solutions, such as industrial ecology, green nanotechnology or new energy technologies. The current progress of humanity requires even more intelligent solutions for infrastructure and urban design.

For the first time in history, the imminent wave is truly global. On the threshold of the sixth wave, competition for the next determinants of technological development, economic growth, social cohesion and even cultural activity will escalate. Awareness of the finiteness of the planet will increase both within the individual consciousness and among social groups. The key theme of innovation in the sixth wave will be a radically more efficient way to use natural resources. The impact of new innovations will be made greater by the fact that they are connected to the web-like technological and social information infrastructure that emerged in the preceding cycle.

Although the sixth wave will bring about systemic changes, the repercussions of which cannot be foretold, it is possible to pinpoint at least some basic aspects of the transformation. The focus in the sixth wave will be on key technologies and innovations that improve the efficiency of resource use. While the fifth wave saw a race towards more flexible technologies and lower costs, competition will now be focused on less material-intensive products and processes. Resource-intensive and non-recyclable solutions will be replaced by new intelligent technologies (many of which have yet to be invented), as well as by natural fiber-based products. All in all, we shall see that resource nexus will become the centerfold of political debates in the coming decades.

The world is undergoing a deep transformation from the fifth to the sixth wave. The crux of the matter is whether certain countries and regions hope to be successful in the next wave – a wave in which technological development will focus on improving material and energy efficiency by (among other things) employing digital technology and by focusing on mitigating the damage to the environment caused by human activity.

Thus, the following will be among the key areas of development in the coming decades:

- radically improved energy efficiency, particularly in end-use sectors;
- massive deployment of new energy technologies, including as-yet marginal wave and geothermal energy;
- advances in intelligent energy and power grids;
- new ways to sequester and store carbon.

In the big picture, almost every country is clearly struggling in gearing its resource use towards more sustainable ways. If the next energy transition – large-scale adoption of renewable energy sources – had occurred 20–30 years ago, we would be facing fewer problems now. If we are to avoid the escalation of climate disturbances, the rate of CO_2 emissions must be reversed to an annual decrease of 5–10% by 2020. This cannot be achieved without a massive investment in renewable energy and improvements in energy efficiency. In the latter case, we are talking about a minimum tenfold increase in efficiency in the long term. For instance, the energy efficiency of business properties has been largely ignored in all countries because there has been no incentive to do so. Along with residential properties, such properties form upwards of 40% of total energy consumption.

To understand the scale, this projected transition challenges our natural resource management and its linkages to the socio-economic system. If all this development were to occur, we need a fundamentally different assessment approach and control system to safeguard the sustainability of these transitions and investments. For this purpose, we should deploy a resource nexus approach that comprises the interlinkages between five key natural resources (water, energy carriers, biomass, land, minerals) in terms of extraction and consumption (see Bleischwitz et al., this volume). As we allow ourselves to take this more systemic view, it gives us the privilege to understand, among other things, the conflicting patterns of resource use. The well-known case is the use of agricultural land for growing biofuel material, replacing thus some food production. The global biofuel production grew over sevenfold between 2000–2012 and EU's goal in promoting the use of biofuels in transport is related to EU target to reduce greenhouse gas emissions (Bourguignon, 2015).[2] However, there are lots of critical voices arguing that growing crops for biofuel consumption takes away from food production. In FAO scenario towards 2050, much more arable land is needed to cover the need for global food and fibre production (FAO, 2009).

Thus, in keeping the target to improve radically our material and energy efficiency, all policies should evaluated on systemic basis (Schmidt-Bleek et al., 2014). This calls for a more integrated, interdisciplinary, nexus-like approach. For instance, the energy efficiency level of PV panels used in northern latitudes, like in Finland, do not necessarily suffice that technology to be major source of energy, due to low efficiency rate. Instead, the use of geothermal energy, wind and bioenergy could have much bigger share in our energy mix. In policies, we need to account for prohibiting factors as a part of Nexus approach. For instance, in Brazil, hard drought heat waves in the years 2014–2016 have wrecked the performance of hydroelectric dams in southeast Brazil, causing extensive damage to society and economy. Since there is robust evidence about human influence on the frequency of extreme weathers, a more integrated approach would connect and assess all the major factors and give then a more reliable policy option scenario for taking care of national energy security. In the complex human-environment systems, the nexus resource approach is ever more needed to provide the sufficient analysis of the existing situation as well as providing related environment-development related goals, expressed in more generic terms in Sustainable Development Goals, as adopted by the UN.[3]

The demographic factor

The rapid growth of Asia has already led to a swiftly growing consumption of raw materials, and there is no end in sight. The growth in China in particular is unprecedented: In the course of the next 10 years, it might even overtake the US and the EU (in terms of GDP per capita), as Euromonitor has estimated. (Euromonitor, 2012)

The population in many countries, especially in European ones, is aging, and as the baby boomer generation retires, the dependency ratio will change rapidly. By the year 2030, nearly one-third of the population in some European countries will be over 65 years old, which will be a radically different situation than the present (see Figure 13.2). In addition to slowing down economic growth, this will also lead to a considerable change in demand. Besides adding pressure on the public sector for basic services, the graying of Europe will also lead to the

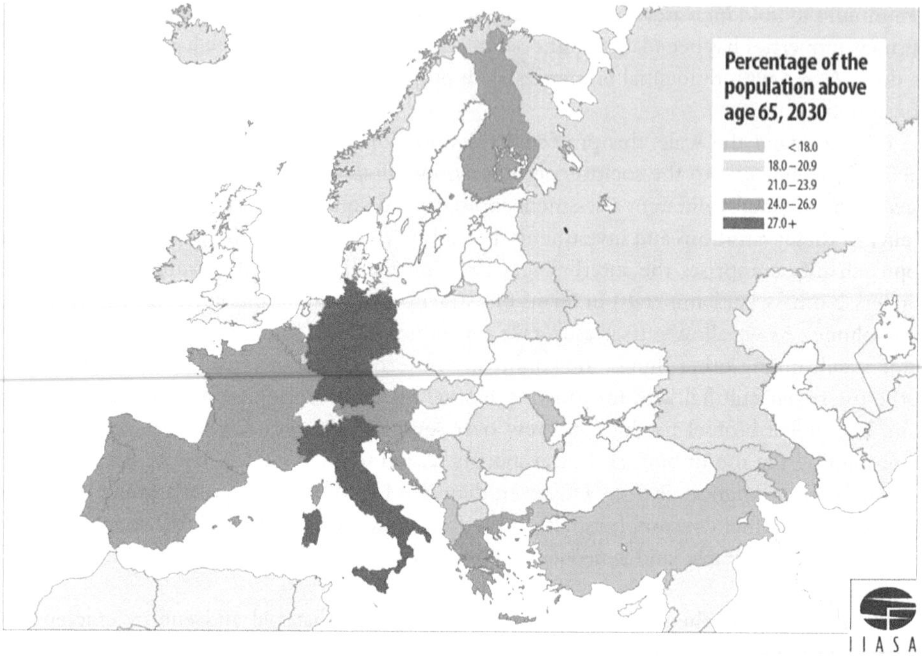

Figure 13.2 Aging population in Europe, 2010–2030

Source: Prof. Wolfgang Lutz/Global Education Trend scenario, World Population Program, IIASA 2009

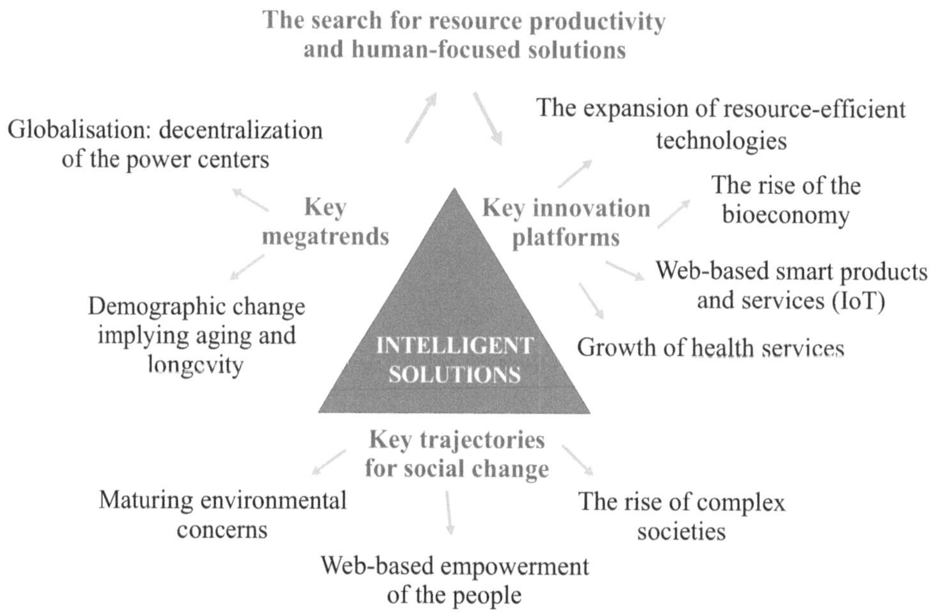

Figure 13.3 Key parameters in the sixth Kondratiev wave, 2010–2050

breakdown of the old industrial model, wherein the length of active working life was determined in detail and the structure of the welfare state was based on a sharp division between the public and the private sector.

While aging will take its toll, population growth continues in many other parts of the world, particularly in sub-Saharan Africa and Asia, which seems to lead, as stated earlier, to close to 10 billion inhabitants on Earth by 2050. The two megatrends of globalization and demographic change are together creating a world in which the key challenge will be how to increase resource productivity. At the same time, technology will advance in leaps and bounds: New, efficient energy technologies will gain ground and gradually supersede the current fossil fuel-based energy economy. Thanks to new developments in biotechnology, synthetic products can be replaced with organic materials. Nanotechnology already allows us to modify the properties of products so as to make them more durable, and it is also leading to the development of innovative new methods for diagnoses in the health care industry.

The transformational impact of the sixth Kondratiev wave on the development of our societies should not be underestimated (see Figure 13.3). Highly industrialized countries will experience a remarkable shift from manufacturing to service.

Take Finland as an example. While service currently accounts for about 65% of Finland's GDP (a relatively low percentage compared to other European countries), in 2030 that figure will in all likelihood be well over 80%. The growing prevalence of services will alter the economic and professional structure of society. Along with other countries, Finland will gradually shift towards a production structure that is less reliant on energy and raw materials and more towards livelihoods that are based on high-value products and services, uniqueness, and locality. In fact, dematerialization has been underway in Finland since the early 1970s, where production has increased in efficiency, with fewer resources being used per unit of manufacture. At the same time, however, the benefits of improved efficiency have been offset by strong economic growth so that total consumption of energy, for example, has in fact increased. From 2006 (before the financial crisis began), however, total energy consumption in Finland started to decline. (Statistics Finland, 2015)

Entering the resource revolution through waves

In our research framework based on Kondratiev cycle analysis, we have postulated that the key driver for the impending wave we are entering is the search for resource productivity. This is largely because economic globalization has developed to the point at which raw materials and commodities have become, or are becoming, increasingly scarce. This development has and will have a particular price indication: There has been a fundamental shift from falling commodity prices into rising ones (see Figure 13.4).

It is remarkable that in the course of the 20th century, the price of commodities dropped almost by one-half. (McKinsey Global Institute, 2011, 2013) This happened at the same time that economic output went through the roof, increasing 20-fold. The reason why resource prices fell was quite obvious: New supplies were discovered in abundance, and new techniques that increased productivity were developed. Moreover, prices did not always reflect the full cost of materials. The last 10 years have been dramatically different, as prices have trend-wise been on the rise. This is particularly due to China, whose huge demand has driven prices up. When the financial crisis hit the global economy in 2008, the price surge of raw materials stopped for a while, only to continue after the most difficult phase of the crisis. From 2011 onward, the prices have been dropping or leveling out, and it is all too early to call off the rising trend.

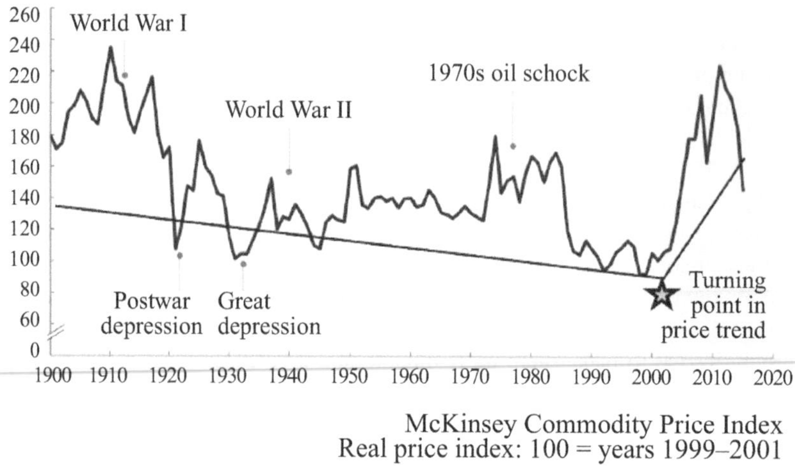

Figure 13.4 Resource prices have increased significantly since the turn of the century

Because our industries are still quite heavily material intensive, we are indeed running into trouble, and as a result the new battle for resources between global powers has started in earnest. Excessive use of both materials and energy sparked development in previous waves, coupled with inadequate technology for restricting the pollution that was largely caused by the use of petrochemicals. This development has now determined the agenda for the oncoming sixth Kondratiev wave, which will dominate the next 40–60 years of global progress. Increases in efficiency will be sought through the reduced use of materials – non-renewables in particular, such as metals and minerals – and from decreasing energy intensity in both production processes and in everyday life.

This quest for greater resource productivity will steer our businesses and societies and therefore set the tone for innovation. This is particularly important, because we have learned that to continue on the same growth trajectory while using resources as we do now will result in three times the use of resources by 2050. (UNEP, 2011) We have ample evidence that this is simply not possible, which means that the growth pattern will fundamentally have to change. Traditional industries must become much less materially intensive, while whole new industries will emerge, including some of which do not yet even have a name. Additionally, new professions will arise in order to facilitate these developments.

In the research and analysis side, there is clearly a need for a resource nexus approach, where interlinkages between several resources needs to be better understood in the socio-economic-ecological systems of provision. The Sustainable Development Goals adopted by UN acknowledge the fundamental role of five key resources defined by resource nexus thinking: water, energy, biomass, land and materials. Understanding the interlinkages of these key resources form the essence of any policies within the realm of these 17 SDGs.[4]

Above all, materials are the basis of our economy. Historically, the extraction and consumption of materials have strongly correlated with economic growth Interestingly, however, over the course of the last decades there has been a growing decoupling[5] of these two developments (Krausmann et al., 2009). On the other hand, the sheer rise in the amount of extraction has been massive: Between 1980 and 2009, the world's domestic extraction almost doubled (Giljum et al., 2014). The global production of concrete in 2012 was more than 3.5 billion tons, of which China had a share of one billion tons. Today, the total amount of rock and sand being extracted globally equals two times the volume of Mount Fuji (Bardi, 2014).

Obviously, this development has had a huge environmental impact. It has led to climate change, overexploitation of key natural resources, pollution and many other detrimental impacts. The result is that today a fairly large consensus believes that a change is needed; that is, that we need to strive for an absolute reduction in resource use in order put ourselves on the path to sustainability.

Simultaneously, our dwindling reserves have become increasingly subject to international political maneuvering, as they are not uniformly distributed across the globe. For example, 60% of the iron ore in the world comes from Australia, Brazil and China, while 83% of the world's exploitable phosphate rock is located in Morocco, China and South Africa (Vaccari, 2009). Phosphorus is a typical example of a raw material that is becoming increasingly scarce, thus creating ample opportunities for new innovations to take over the marketplace.

As implied, the key drivers of innovation for the next Kondratiev wave will be material scarcity stemming from the anticipated shortage of metal and mineral resources in the next few decades. Simply put, material scarcity is controlled by two factors: the supply of the material versus its demand. With a growing world population and growing consumption, the demand for material resources already far exceeds the supply. In the past century, the exploration of new locations and technological innovations in mining and extraction kept material reserves in balance with the demand, helping the price levels stay moderate. However, present knowledge indicates that this balance will not continue into the 21st century, as the production of a number of material resources has already peaked. (UNEP, 2013)

The most precious resources on our planet are food and water. In the decades to come, there will be a tremendous challenge concerning the sufficiency of these resources. In the food markets, we are already seeing the results of scarcity as food prices are rising. FAO food price index has risen 61% from 2004.[6] Global food production is under severe constraints, for reasons varying from the overuse of fertilizers and other harmful farming practices to the growing needs of the global middle class. One thing seems certain: The crisis will severely affect those countries that are ill-equipped to manage the situation.

According to Professors Harald Sverdrup and Vala Ragnarsdottir, experts in the study and modeling of the cycles of minerals and other natural resources, every year we are using 3,700 million tons of cement; 1,300 million tons of wood; 2,200 million tons of iron and steel, of which 1,500 tons comes from mines; and we produce 110 million tons of aluminum, 50 million of it from mines. The annual global output of copper, zinc and lead is 35 million tons, of which 16 million comes from within the earth. The other metals account for 50 million tons, 35 million of which are extracted from mines. (Sverdrup and Ragnarsdottir, 2014)

After examining the circulation of raw materials and assessing the use of natural resources in the future, they have arrived at the following conclusions. First, there will not be any serious shortage of copper, aluminum or iron for a long time. Iron production will peak around 2030, with serious scarcity emerging around 2080, which will in turn lead to a serious economic crisis, as iron is such a vital part of all infrastructure.

Second, there will be a serious crisis with respect to phosphorus – a crucial ingredient in modern fertilizers – but it will be due to erosion. In Sverdrup's and Ragnarsdottir's estimation, soil erosion due to phosphorus overuse will probably be the greatest threat to civilization in the long run. Without healthy soil, we have no hope of feeding the growing population on Earth. According to the Food and Agriculture Organization of the UN, we already saw the peak of arable soil around the year 2000.

Third, problems will escalate and eventually blow up in our faces with regard to the use of fossil fuels, the inefficiency and costs of our systems of production, and the scarcity of nickel and other metals needed for high-tech products. We must adapt our economies to these radically new circumstances. Our taxation system must shift from taxing the work towards taxing the use of scarce and polluting materials. There is promise of a new economic system in which the principles of three R's will reign – reduce, reuse and recycle – called circular economy. Now being adopted by EU, circular economy simply holds that the massive plundering of the earth's precious resources should stop. (European Commission, 2016) In the sixth wave, we simply must be much more intelligent in how we run our economies.

As new types of policies are gaining ground, a nexus approach to manage resources will be ever-more needed in the future since it emphasizes synergies in natural resource governance and emphasizes resource-related stress multipliers (see Bleischwitz et al., this volume). Moreover, to understand the challenges in the emerging sixth wave, we need such modeling that provides us with new business-as-usual scenarios that are based on better understanding of interactions between bio-physical and socio-economic systems. Only then can we expect fully informed policies facing the real challenges of the next decades.

Conclusion

We are entering a new socio-economic cycle – the sixth wave. This is an era for which we truly need a balance sheet for the planet. (Wijkman and Skånberg, 2014) It is a phase in our societies and economies (extending to the middle the century) in which we need to focus on becoming aware of both looming deficits and potential assets. Any serious attempt to make forecasts for the middle of the century cannot avoid asking what resources we might be living on in that time. Jörgen Randers determined that we most likely will experience a world where substantially less is being consumed by the typical Western citizen. The path to the future will be tumultuous and full of conflict. (Randers, 2012) However, as the Sustainable Development Goals should guide the agenda for international and national policies, the resource nexus approach, comprising critical natural resources and analyzing human-environment relationship, should provide us a more realistic and comprehensive research and policy agenda.

On country level, the real question is: Which country has the capacity to become a trendsetter? We can all probably contribute to making that happen. The most important thing, however, is that we must ourselves understand what it takes to promote the achievement of such a goal. We must all become responsible consumers and promote lifestyles and solutions that further this idea. What an enormous opportunity this is for any country, particularly when considering that the change in direction is inevitable in the next few decades. We are still looking for the country or group of countries that will start to redefine their legislation, taxation and overall policies something we have earlier referred to as "system policies." (Schmidt-Bleek et al., 2014) This should serve as a model for how the social regime of the country will be transformed during the sixth Kondratiev cycle in order to both reflect and respond to the great challenges of our age.

Notes

1 The model, called World3, utilized the following parameters: investments, population, pollution, natural resources and food.

2 See: www.europarl.europa.eu/RegData/etudes/BRIE/2015/548993/EPRS_BRI(2015)548993_REV1_EN.pdf

3 See: https://sustainabledevelopment.un.org/sdgs

4 See: www.un.org/sustainabledevelopment/sustainable-development-goals/

5 UNEP's International Resource Panel (IRP) has applied the concept of 'decoupling' to this challenge. While the term has been applied to everything from electronics to physical cosmology to linear algebra, in the sense used here decoupling means using less resources per unit of economic output and reducing the environmental impact of any resources that are used or economic activities that are undertaken.
6 World Food Situation, FAO Food Price Index. www.fao.org/worldfoodsituation/foodpricesindex/en/

References

Bardi, U. (2014). *Extracted: How the quest for mineral wealth is plundering the planet: A report to the club of Rome.* Vermont: Chelsea Green Publishing.

Bourguignon, D. (2015). *EU biofuels policy: Dealing with indirect land use change.* European Parliament. Briefing paper 15. Retrieved August 20, 2016, from www.europarl.europa.eu/RegData/etudes/BRIE/2015/548993/EPRS_BRI(2015)548993_REV1_EN.pdf

Diamandis, P., and Kotler, S. (2012). *Abundance: The future is better than you think.* New York: Free Press.

Digital Trends. (2012). *Top 10 bad tech predictions.* Retrieved March 31, 2016, from www.digitaltrends.com/features/top-10-bad-tech-predictions/7/.

Euromonitor. (2012). *Top 10 largest economies in 2020.* Retrieved March 31, 2016, from http://blog.euromonitor.com/2010/07/special-report-top-10-largest-economies-in-2020.html

European Commission. (2016). *Closing the loop: An EU action plan for the circular economy.* Retrieved March 8, 2016 from http://ec.europa.eu/environment/circular-economy.

Food and Agriculture Organization of the United Nations (FAO). (2009). Livestock in the balance. Retrieved from http://www.fao.org/publications/sofa/2009/en/

Foxon, T. J. (2002). *Technological and institutional 'lock-in' as a barrier to sustainable innovation.* Retrieved March 31, 2016, from http://www3.imperial.ac.uk/pls/portallive/docs/1/7294726.PDF.

Fridolin, K., Simone, G., Nina, E., Karl-Heinz, E., Helmut, H., and Marina, F.-K. (2009). Growth in global materials use, GDP and population during the 20th century. *Ecological Economics*, 68(10), 2696–2705.

Friedrich, S-B., Wilenius, M., and Lehmann, H. (2014). The challenge of the whole: Creating system policies to tackle sustainability. In M. Angrick et al. (Eds.), *Factor X: Policy, strategies and instruments in the sustainable resource use.* New York: Springer Verlag.

Giljum, S., Dittrich, M., Lieber, M., and Lutter, S. L. (2014). Global patterns of material flows and their socio-economic and environmental implications: A MFA study on all countries world-wide from 1980 to 2009. *Resources*, 3, 319–339. doi:10.3390/resources3010319

IEA, International Energy Agency. (2010). *Energy technology perspectives.* Scenarios & Strategies for 2050. Paris, France. Retrieved from www.iea.org/publications/freepublications/publication/etp2010.pdf

Intergovernmental Panel on Climate Change (IPCC). (2014). Retrieved March 8, 2016, from www.ipcc.ch/report/ar5/.

International Energy Agency (IAE). (2015). *Key world energy statistics.* Paris: OECD. Retrieved from www.iea.org/publications/freepublications/publication/KeyWorld_Statistics_2015.pdf.

Kondratiev, N. (1984). *The long wave cycle* (G. Daniels, Trans.). New York: Richardson & Snyder.

McKinsey Global Institute. (2011). *Resource revolution: Meeting the world's energy, materials, food, and water needs.* Retrieved March 31, 2016, from file:///Users/user/Desktop/MGI_Resource_revolution_full_report.pdf

McKinsey Global Institute. (2013). *Resource revolution: Tracking global commodity markets.* Retrieved March 31, 2016, from file:///Users/user/Desktop/MGI_Resources_survey_Full_report_Sep2013.pdf.

Meadows, D. (1972). *The limits to growth: A report of the club of Rome's project on the predicament of mankind.* New York: Universe Books.

Millennium Assessment. (2005). *Guide to millenium assessment reports.* Retrieved March 8, 2016, from www.millenniumassessment.org/en/index.html.

Randers, J. (2012). *2052: Global forecast for the next 40 years.* White River Junction: Chelsea Green Publishing.

Statistics Finland. (2015). Total energy consumption. Retrieved March 31, 2016, from www.stat.fi/tup/suoluk/suoluk_energia_en.html.

Sverdrup, H., and Ragnarsdottir, V. (2014, October). Natural resources in a planetary perspective. *Geoghemical Perspectives*, 3(2). Also: Lecture in Club of Rome Conference 2015.

Think Progress. (2012). *Jeremy Grantham on 'Welcome to Dystopia': We are 'entering a long-term and politically dangerous food crisis.* Retrieved March 8, 2016, from http://thinkprogress.org/climate/2012/08/16/681571/jeremy-grantham-on-welcome-to-dystopia-we-are-entering-a-long-term-and-politically-dangerous-food-crisis/.

UN. (2015). *World population prospects: The 2015 revision*. Retrieved March 31, 2016, from www.un.org/en/development/desa/news/population/2015-report.html

The United Nations Environment Programme (UNEP). (2011). *Decoupling natural resource use and environmental impacts from economic growth*. Retrieved March 31, 2016, from www.unep.org/resourcepanel/Publications/Decoupling/tabid/56048/Default.aspx.

The United Nations Environment Programme (UNEP). (2013). *Metal recycling: Opportunities, limits, infrastructure*. Retrieved March 31, 2016, from www.unep.org/resourcepanel/Publications/MetalRecycling/tabid/106143/Default.aspx.

Vaccari, D. (2009, June). Phosphorus: A looming crisis. *Scientific American*, pp. 54–59.

Wijkman, A., and Rockström, J. (2012). *Bankrupting nature: Denying our planetary boundaries*. Stockholm: Routledge.

Wijkman, A., and Skånberg, K. (2014). *The circular economy and benefits for society: Jobs and climate clear winners in an economy based on renewable energy and resource efficiency*. Rome: Club of Rome.

The World Bank. (2012). *Climate change report warns of dramatically warmer world this century*. Retrieved March 8, 2016, from www.worldbank.org/en/news/feature/2012/11/18/Climate-change-report-warns-dramatically-warmer-world-this-century.

World Wildlife Fund (WWF). (2014). *Living planet report 2014*. Retrieved March 8, 2016, from http://wwf.fi/mediabank/6426.pdf.

14
FORESIGHT AND SCENARIOS
Modelling practices and resource nexus assessment

Gerd Ahlert, Martin Distelkamp and Mark Meyer

Economic environmental models as modules in integrated assessment analysis

The dependency of emissions and resource extractions from economic development can only be understood in a deep disaggregation of the economy which allows identifying the environmentally important consumption, investment and production activities in their relation to the entire economy. So input–output modelling is essential for environmental studies. But the open static Leontief model is too simple. A system is needed that has all this and in addition endogenizes final demand, factor demand for labor and capital, substitution in final demand and factor demand and of course endogenous prices. Such a system is a macro model with a fully integrated flexible input–output system.

Based on the general equilibrium theory of Arrow and Debreu (1954), Johansen (1960) started the long tradition of Computable General Equilibrium (GCE) modelling. He developed a model in which households maximize their utility subject to their budget constraint. Industries decide on inputs minimizing costs subject to production function constraints. Investors allocate capital to industries to maximize their profits. In the 1970s the theory of "new classical" macroeconomics was widely recognized whereas the then-applied econometric methods came under criticism by means of the Lucas critique (Lucas, 1976). Therefore, the parameters of these models have (in most cases) not been estimated econometrically. Instead, assumptions about price elasticities for demand and supply equations were implemented and the rest of parameters have been calibrated by the information of one data point. With this procedure, the empirical test was abandoned and such models became pure theoretical systems based on so-called deep structural parameters. Because of its long-standing and diverse application GTAP (Global Trade Analysis Project) is the most successful CGE type of model, which first was developed for the analysis of global trade (Hertel, 1997) and has later been used in many variations as well as in economic environmental applications. Böhringer and Löschel (2006) give an overview of the state of the art and its prospects.

The macro-econometric (ME) models as a class of model are not different from CGE models in structure concerning the degree of disaggregation and the degree of endogenization. But these models standing in the Keynesian tradition do not assume optimization on perfect markets. Instead bounded rationality of agents acting on imperfect markets is assumed. The modelling of bounded rationality and imperfect markets is much more open than that of perfect competition. Therefore the parameters of these models are estimated econometrically, which gives the whole model empirical evaluation. Almon (1967) was the first to integrate the input–output approach in a macro-econometric (ME) framework based on Keynesian theory.

An example for a deeply disaggregated macro-econometric model with an extended environmental module is the model GINFORS (Meyer et al., 2013). The GINFORS model incorporates not only an endogenous mapping of detailed economic structures and global economic interrelationships due to globalized trade patterns. It also considers the embedment of these processes in the global earth system: To deal with environmental questions the economic system is consistently interlinked with subsystems that explain energy use, electricity production, air emissions and material extractions in physical terms. As regards the later, GINFORS is able to map quantitative indicators of material extractions embedded in regional consumption activities over the global supply chain. Current GINFORS applications thus provide medium-term projections (up to the year 2050) of national material footprint indicators like Raw Material Consumption (RMC) or the broader Raw Material Input (RMI) indicator.

Historical developments and selected drivers of national material footprint indicators have been analyzed in detail by different researchers over the last years. See, for example, Steger and Bleischwitz (2011), Wiebe et al. (2012), Weinzettel et al. (2013), Wiedmann et al. (2015) or Wood et al. (2015) for some selective recent applications in this regard.[1] However, although it is well recognized in the literature that a combination of modelling approaches with global MRIO datasets advances the identification of drivers of environmental impacts, provides guidance for scenario developments and enables ex-ante assessments of alternative resource-efficiency strategies (Wiedmann and Barrett, 2013), only a few research teams proofed their abilities to integrate dynamic input–output modelling expertise to this research agenda until now (but, see, for example, European Commission (2014) and UNEP (2017), for a recent example in this regard).

Accordingly, a common perception of necessary criteria for a sufficient mapping of global causal chains of international resource use does not seem to have been thoroughly established until now. But, at least three drivers seem to collude in a notable way in this regard: Technological progress and international supply chain developments, regional differences in prosperity developments and their implied effects on related lifestyle patterns and, finally, the economic feedback mechanisms initiating from resulting international trade developments. Apparently, a cohesive global modelling framework is needed in order to satisfy a self-contained mapping of the mutually interdependent interplay of these drivers.

Concerning this matter, we introduce our readers to the latest version of the global simulation model GINFORS, a dynamic input–output model whose early applications in resource-efficiency scenario analyses have already been published in a seminal contribution by Giljum et al. (2008). The current GINFORS version, which is based on the Multi Region Input-output WIOD dataset (see Timmer et al., 2015), has been re-engineered over the last couple of years and facilitates simulation studies until the year 2050.

Each GINFORS simulation is soundly rooted on an endogenous mapping of detailed national economic structures and international economic interrelationships due to globalized trade patterns. Income effects resulting from (i.a.) diversified investment expenditures, induced efficiency improvements or sustained shifts in consumption patterns are explicitly modelled. Hence, each simulation run accounts for potential macroeconomic rebound effects (Sorrell and Dimitropoulos, 2008; Sorell et al., 2009) in a variety of ways.

Concerning the modelling of the global resource nexus, the modelling of the economic system is consistently interlinked with subsystems that explain energy use, electricity production, air emissions and material extractions in physical terms (Ahlert et al., 2017). Hence, for 39 regions used, extraction is projected for the following material categories: coal, gas, crude oil, ores, construction minerals, industrial minerals. Furthermore, the physical imports of these raw materials are determined by their monetary imports in constant prices. Resources embedded in the imports of products can be calculated via the bilateral trade relations and the technologies used in the different producing countries, which both are endogenously explained in the whole system, insofar as it is possible to estimate for every country and every type of raw material a material footprint indicator.

In practical model applications comparisons of the simulation results from CGE models and macro-econometric models do often indicate significant deviations. Admittedly, it cannot be precluded that alternative economic foundations imply contradictory policy recommendations. The neoclassical paradigm is clear: All agents act on perfect markets, producers maximize their profits facing a substitutional production function and consumers maximize their utility constrained by income. But what is the theoretical base of modern Keynesian modelling and how can it be implemented in deeply disaggregated macro-econometric models? In the second section of this chapter we try to give an answer and meet the literature that Dornbusch at al. (2008) qualified as the "Neokeynesian Counter Revolution."

Furthermore, we discuss in the third section the role of the two different approaches in the construction of the reference. One approach is to produce a plausible forecast on the basis of the underlying model that excludes the policy that is analyzed in the alternative scenarios. The comparison of the alternative scenario with the reference gives then for all variables the direct and indirect effects of the policy instrument(s) in question. This procedure is generally unproblematic for Neokeynesian models since they are econometric, which means that their ability to forecast has been tested positively. In the case of CGE models the construction of the reference may turn out to be difficult if their parameterization is based on assumptions concerning some elasticities and a calibration for the rest of the parameters with one data point. In such a case a baseline is taken from an exogenous source and the model is calibrated to it. It may turn out that this baseline is unrelated to a specific "business-as-usual" policy that is implemented in the model.

In the fourth section the effects of the different methods of parameterization on policy results are discussed. Of course, this can only be done in a concrete model application, but then all three causes are responsible for differences in results that come into consideration simultaneously. We choose the POLFREE project where the two large-scale models EXIOMOD (CGE) and GINFORS (Keynesian) have been applied in an economic–environmental analysis. Four simulations have been produced – a reference and three alternative scenarios with different environmental targets to be reached, different governance assumptions and policy mixes. We will list the most striking variations in the simulation results of the two models in the light of the three mentioned causes.

Some conclusions in the final section close the chapter.

How to model market imperfections in the macro-econometric context

Keynes criticized classical theory indicating that the labor market and the capital market are imperfect. The wage rate has not the flexibility to equilibrate the labor market and the incomplete information of lenders and borrowers of capital does not guarantee that savings determine investment as neoclassical theory claims. With these corrections, Keynes was able to show that equilibrium in the circular flow of income could coexist together with permanent unemployment.

Modern Keynesian theory uses micro foundations for these two imperfections and adds stickiness of goods prices so that, in this approach, all markets fail to work as neoclassical theory presumes it.

First, let us look to the labor market. To have a model solution that allows for unemployment, it is not necessary to have a sticky wage. A theory is needed that shows that the real wage rate can be permanently above the market clearing rate. A simple explanation for such a situation could be a strong power of unions in the sector and country in question. In the wage bargaining process the expected productivity development of the industry in question and the inflation rate are the arguments to find the nominal wage rate for the contract, which holds for a certain period. This is the Phillips Curve approach in the interpretation of Samuelson and Solow (1960). The resulting real wage rate is not related to the market clearing rate.

Of course, strong unions are not to be found in all industries and countries. What pushes producers to pay voluntarily more than the market clearing rate? The answer is that it must pay for the producers. The efficiency wage theory (Stiglitz, 1976) shows that the higher wage rate raises the efforts of employees and reduces labor turnover. Each firm hires employees until the real wage rate equals the marginal productivity of labor. Yellen (1984) gives an overview of several variants of this theory.

In a second step, let us look to the capital market. The neoclassical equilibrium in the labor market determines the real wage rate and fixes labor demand to the level of labor supply. The capital stock is given from the investment of preceding periods and depreciation. With the inputs of labor and capital the output is given by the production function and cannot be influenced from the demand side of the economy. This means that in the neoclassical framework we are not free in modelling demand. Sen (1963) concluded that so-called closure rules have to be mentioned, if inconsistences between supply and demand shall be avoided. In a closed economy, a rise of consumption has to be accompanied by a fall of investment, which totally compensates the effect on total demand. Total GDP is defined as the sum of consumption and investment. Subtracting from both sides consumption gives the equality of savings and investment. The inconsistency problem can be solved either if investment is determined by savings or if savings are determined by investment. In the first case Sen (1963) speaks of a "neoclassical closure," in the other case of a "Johansen closure." In the second case we are still in a neoclassical surrounding with cleared labor market and insofar a fixed supply. Here a rise of investment induces a rise of savings, which means a reduction of consumption, which leaves total demand unchanged.

In the case of the neoclassical closure a rise of savings means a reduction of consumption and induces a rise of investment compensating the fall of consumption. It is used in growth theory and in by far most CGE models. It can be interpreted as the reduced form of a perfect capital market, where all savings are supplied and transformed into investment. The interest rate equilibrates the market: A rise of savings reduces the interest rate and raises investment. In an open economy the current account of international trade has to be taken into account additionally, which complicates the analysis requiring further closure rules in relation to exchange rate regimes (Taylor and Lysy, 1979; Dewatripont and Michel, 1987).

Sen (1963) defined the "Keynesian closure" allowing for an independent investment function, in combination with an imperfect labor market with unemployment, where labor demand is explained by a flexible real wage rate. In other words: In a Keynesian model with unemployment the neoclassical property of a fixed production in the short run does not exist. The nominal wage rate is fixed in the short run by contracts, but a change of the GDP price moves the real wage rate and this pushes employment and production. Insofar as the neoclassical consistency problem for supply and demand of GDP does not exist, and investment can be independent from savings, even though it has to match savings in the solution, which is of course the result of the interplay of all variables. So the term "Keynesian closure" is misleading since a Keynesian model does not require a closure rule because supply and demand are fully flexible and are determined interdependently in the equilibrium of the circular flow of income.

These considerations show that neoclassical models have the well-known property of the dominance of the supply side of the economy, whereas fully developed Keynesian models depict a well-balanced interdependency of both market sides. Therefore, it is not true that in Keynesian models the demand side dominates. Demand side models are only the simple multiplier models, which we better should call "Incomplete Keynesian Models."

In a final third step, let us look to the good markets. The Neokeynesian discussion on the stickiness of goods markets prices is focusing the question whether the neoclassical property that a fall in demand induces a reduction of the market price holds also under conditions of imperfect competition. This question is important, because there is much evidence that most goods markets have imperfect structures (Hall, 1986).

Akerloff and Yellen (1985) discuss the reaction of a firm supplying under monopolistic competition, which faces the reduction of demand of the total industry moving the demand function of the firm to the left. Before this movement, the firm maximized its profits choosing a price that equalizes marginal revenues and marginal costs. After the movement of demand at the actual price and the new quantity, marginal revenue is bigger than marginal cost. Therefore, profit maximization would require a reduction of the price until the marginal revenues and marginal costs again equilibrate. But it has to be mentioned that the incentive to reduce the price and to realize the additional profit depends on the steepness of the marginal cost curve and that of the marginal revenue curve. The more marginal costs fall when output declines the greater is the incentive of the firm to lower its price. The more the marginal revenue curve shifts to the left the smaller is the incentive for the firm to lower its price. In any case, the losses are very small so that already small costs of the price change itself (menu costs) can outweigh the advantage of the price change. Romer (1993) discusses four effects that influence this incentive. Blinder (1994) presents 12 different micro foundations for sticky prices and discusses the result of an empirical study testing their evidence.

Stiglitz (1984, 355) concludes in a paper with microeconomic arguments favoring price stickiness:

> I suspect that some of these theories may provide a better description of some markets, others of other markets; there is no compelling reason to believe a single theory provides the explanation of price rigidities in all markets. In these theories firms behave all "rationally"; an alternative "explanation" of seeming price rigidities is that firm's managers act according to certain rules of thumb, for example, those that entail a mark-up over average costs. Though such "theories" – if they can be called that – fail to explain how or when mark-ups change, as they undoubtedly do, they may provide as good a description of the short run behavior of the firm as our more sophisticated theories.

The recommendation of Stiglitz is very helpful, since the macro-econometric models applied in economic environmental research are typically multi-country/multi-industry models with thousands of prices that have to be explained, insofar as it would be a huge task to identify which of the more sophisticated approaches can be identified in which industry and country. Further, it is clear that the variables needed to estimate more sophisticated approaches are not available in such a context.

A further aspect that argues for mark-ups over average costs is the information that firms have about their marginal revenue function. It can be doubted whether this information is solid enough to base optimization on it.

If price rigidity is given with mark-up pricing then prices are not equilibrium prices and production is determined by the prices set and demand reacting to them. A determination of production via a production function is then inconsistent. In other words: The determination of output by a production function assumes that firms realize permanently an optimum: Optimal factor demand as input in the technology description of the production function creates optimal supply and prices are determined equalizing demand and supply. Either the price is determined by market clearing as well as supply is determined by the production function (neoclassical case), or the price is set by the supplier and production is given by the clearing of the market (Neokeynesian case).

From this follows that in the Neokeynesian case the specification of the factor demand system is more than deriving optimal factor inputs based on a production function. There are some degrees of freedom that can only be filled by empirical analysis. But of course the neoclassical approach is a starting point for the econometric estimation of the system of factor demand functions.

But the Neokeynesian case of price setting by the suppliers does not mean that prices are fixed. The iterative solutions of the models activate further channels that allow price flexibility. Let us discuss this in a model context with mark-up pricing. If demand rises this will raise production in the first iteration. This means that labor productivity will grow inducing higher wage rates. Since the elasticity of the real wage rate on labor demand is normally lower than 1, the labor costs will rise, which pushes the price of the good. In the next iteration the process goes into the opposite direction, because the rise of the price will reduce demand. If the model is stable the process converges, and in the end a rise of demand will also have raised the price in the same period.

The described differences between neoclassical and modern Keynesian modelling of factor inputs are very important for the discussion of environmental policy. There is a large base of literature[2] that gives theoretical arguments and empirical evidence for an inefficient use of resources and especially of materials. CGE models are not able to depict this because they assume a permanent optimality of all factor inputs. This means that the important class of information instruments focusing on the behavior of producers cannot be analyzed by CGE models.

Discrepancies between the reference scenarios: the problem of exogenous growth

In assessing the implications of specific policy measures a reference solution of the model is compared with an alternative solution, in which one or more variables have been shocked against their values in the reference. The deviations between both simulations give all direct and indirect effects of the shock.

Such simulations can be done for the past or for future periods. In the first case the reference would be a historical simulation with historic values only for the exogenous variables and the policy parameters. A so-called counterfactual simulation gives then the alternative with different parameters of the policy variables. Meyer and Meyer (2013) give an actual example. Especially in economic environmental simulations, the challenges for policy formulation are given for the future. Here the reference is a forecast with the model with policy parameters that exclude the policy in question in terms of business as usual. Econometric models are tested for the ability to produce a plausible forecast.

Most CGE models have not econometric underpinnings. Their parameters have been taken from the literature (elasticities) and the rest of them have been calibrated on the basis of one data point. It is rather unlikely that such a system produces a plausible long-run forecast. Therefore, pathways either for GDP or for factor productivities are taken from exogenous sources and the model is calibrated to that pathway. If GDP is exogenous, factor productivity can be calculated endogenously by the CGE model; in the other case GDP is estimated endogenously by the CGE model. An extensive discussion of the pros and cons of the different approaches is found in Fontagne et al. (2013).

We want to address a more fundamental point. This procedure seems to be acceptable, if the analysis is done to evaluate single policy instruments. The typical questions in economic environmental applications are: What is the effect on extractions and emissions and what is the effect of the instrument on socio-economic variables? Typically, these effects are measured as percentage deviations from the reference. Therefore, the levels of the variables of the reference are of minor interest. The argument is: If the reference is wrong, this failure is also contained in the alternative scenario. Since policy evaluation is interested only in the differences between both scenarios the failure will in most cases be eliminated.

But if the user of the model is interested to compare a situation without a certain policy with an alternative situation in which this policy is implemented in terms of levels of the central variables, we have a different situation. If for example the question is which consequences a business-as-usual policy has for the economy and the environment compared with a situation in which an engaged policy mix will be implemented during the next 35 years, the levels of the variables are of interest and not only the deviations from the reference. In this case, the exogenous construction of a reference is very problematic because the policy implications of business as usual do not enter the forecast. Dixon and Rimmer (2012) argue that these applications become more important and they recommend improving the forecasting qualities of CGE models.

When comparing model results of CGE and of modern Keynesian models one has to note that the levels of all economic variables, emissions of pollutants and extractions of resources may already be different in the business-as-usual simulation. It is consequently very difficult to understand to which extent the endogenous model relations and the calibration to an exogenous path is responsible for the differences. Further, it is clear that such differences in the business-as-usual simulation are mirrored in the alternative simulations: If a lower emission path is projected in the business-as-usual simulation, it will be easier to meet a concrete emission target for the future than in the opposite case.

An exemplifying comparison of simulation results for alternative modelling approaches to the resource nexus

This section complements our previous theoretical annotations with an outlook on exemplary simulations results from the POLFREE project.[3] The central research question of the POLFREE project was how to reach the following key environmental targets for Europe by 2050:

- CO_2 emissions reduction by 80% compared with 1990;
- Reduction of the cropland footprint by 30% compared with 2005;
- Raw material consumption 5 tons per capita;
- Water exploitation index below 20% in all EU countries.

Extensive integrated assessment studies have been carried out in order to analyze these nexus interrelationships together with their simultaneous impacts on the socio-economic system.[4]

In doing so, two different economic environmental models (GINFORS, Meyer et al., 2015 as well as EXIOMOD, Hu et al., 2015) had been independently linked with the global vegetation model LPJmL (Beringer et al., 2011). Thus, the resulting findings offer a (rare) opportunity to classify the impacts of alternative economic modelling approaches on the overall outcomes of applied simulation studies.

The reference for the GINFORS simulations was a business-as-usual projection assuming for the EU that the already implemented climate policy is maintained and that no activities to reduce material extractions are taken, and that in the Non-EU countries no environmental policy action at all will be implemented. World population was assumed to rise by one-third until 2050.

Given these scenario assumptions global CO_2 emissions were projected to rise by 50% (compared to actual levels) until 2050. Accordingly, exogenous prices for fossil fuels were taken from the (strongly rising) IEA ETP 6 degrees scenario and also exogenous metal price specifications were adjusted to a dynamic trajectory. The link with the vegetation model LPJmL facilitated the projections of productivities of land for different crops. Therefore, GINFORS was able to calculate price developments for crops endogenously; they rose even stronger than the fossil-fuel prices. Since price elasticities for food are low, nominal consumption shares for food increase in this scenario, which reduces the demand for other products.

Overall, increasing scarcity of resources and induced budget restrictions dampen the economic development and reduce GDP growth rates and employment. Notwithstanding the demographically induced reduction of labor supply in the EU, slight increases of unemployment levels were projected.

The reference for the EXIOMOD model was calibrated on the CEPII EconMap v2.2 data. Therefore, the macroeconomic development of the reference was exogenous for the model, insofar as it is not a business-as-usual forecast of the model itself and the results cannot be interpreted as such.

This example shows that both references cannot be compared, as the GINFORS reference represents an endogenous projection under specific policy assumptions whereas the EXIOMOD reference is rather independent from the modelling exercise itself. In fact it reflects a theory-rooted future that is consistent with standard supply side modelling approaches.

For three alternative scenarios a set of about 30 different policy instruments (information, economic and regulation instruments) have been allocated to three alternative scenarios with different governance and international cooperation assumptions (Distelkamp et al., 2015). A large set of preparatory simulations runs were executed with EXIOMOD and GINFORS (both independently linked with LPJmL) in order to parametrize policy mixes that meet the environmental targets mentioned at the beginning of this section.

In the case of GINFORS it was possible to derive policy mixes that met (more or less) the objectives. See Figure 14.1 for an illustrative representation of selected GDP, RMC abiotic and CO_2-emissions projections in the reference case (upper row) and two alternative scenarios Global Cooperation (middle row) and EU Goes Ahead (lower row). In case of EXIOMOD very strong rebound effects as well as emerging conflicts between the targets hampered an achievement of these objectives (Hu et al., 2015, 24).

A detailed comparison of the economic projections from both models points to the fact that alternative economic modelling approaches imply significant restrictions with regard to the represented causality chains (Distelkamp et al., 2015).

In the scenarios Global Cooperation and EU Goes Ahead, structural change towards a resource-efficient economy induces strong investment needs. In case of GINFORS these (at least partly) additional investment expenditures drive GDP and employment. Sectoral investment is

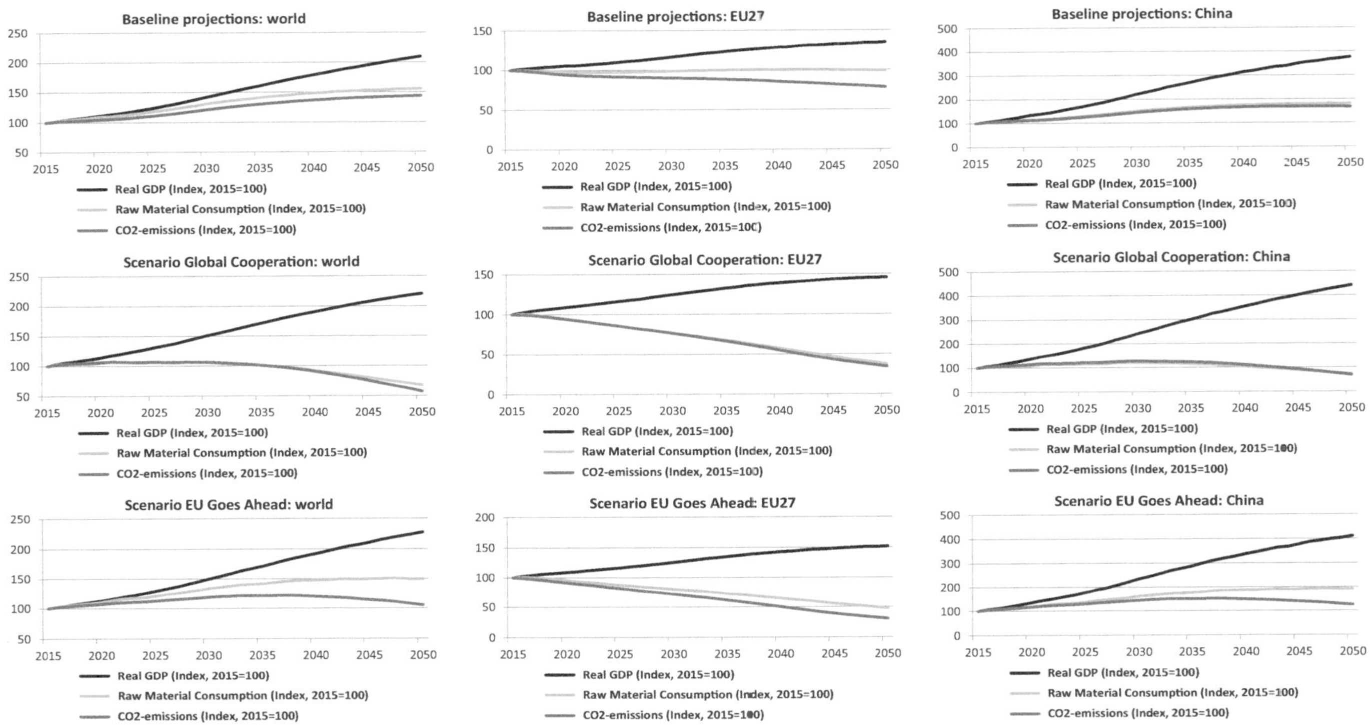

Figure 14.1 Annual GINFORS projections of real GDP, abiotic raw material consumption and CO_2-emissions developments in alternative scenario setting until 2050 for selected world regions

Source: authors

determined by the needed capital input of sectoral production, which is derived from econometrically estimated sectoral investment functions. Of course, investment in the old technologies is reduced, but the total effect is positive. EXIOMOD has a so-called neoclassical closure, which means that macroeconomic investment is determined by savings. Therefore, a rise in investment in new technologies is by assumption accompanied by a fall of other investments, which totally compensates the first effect. Since the labor market is cleared, there are also no effects on employment.

Further differences between both sets of simulation results are caused by diverse parameterization routines. GINFORS represents an empirically evaluated structure as all price and income elasticities in its demand and supply functions have been estimated econometrically. Compared to this, the related elasticities within EXIOMOD have been set exogenously, especially the assumptions about the price elasticities for intermediate demand, for consumption and for imports seem to be responsible for the result.

Price elasticities for intermediate demand are zero in EXIOMOD, which means that taxation of resource-intensive intermediate goods is not able to change the structure of production directly.

On the other side information instruments induce improvements in resource productivity, which over all stages of production shifts supply functions to the right. In the case of consumption goods the price elasticity of demand is with -1 rather high in absolute terms. Here the shift of the supply function to the right generates a new equilibrium with a lower price and a higher demand and production, which raises finally the input of resources. This rebound effect is strengthened by a strong reaction of imports, which have a price elasticity of -5. So the reduction of the consumer price will strongly reduce imports and push production further.

In GINFORS intermediate demand is price dependent with econometrically estimated specific elasticities for the different product groups. Further the econometrically estimated price elasticities of consumption and of imports are significantly lower than those of EXIOMOD, which reduces rebound effects.

To get near to the targets it was necessary for EXIOMOD to assume exogenously given ad hoc improvements in the energy and material efficiency. Consequently, the presented results are only partly induced by the policy mixes. The strong rebound effects of EXIOMOD are an artifact.

Conclusions

Summarizing we see that the GINFORS modelling approach, rooted in modern Keynesian thinking and based on econometrically specified elasticities, creates totally other properties than the EXIOMOD approach. All markets are imperfect and firms are price setters and not price takers. Permanent unemployment is possible in the Neokeynesian case, whereas in the neoclassical model full employment is given.

In GINFORS prices of industries react to unit costs or incorporate adjustment costs inducing stickiness of goods prices instead of being determined by the market equilibrium.

The different modelling of the capital market implies in the case of neoclassical model a macroeconomic closure that creates the dominance of the supply side in macroeconomic terms whereas our modelling allows for a balanced influence of macroeconomic supply and demand.

The labor market plays a central role: If a perfect market is modelled we get the problem of inconsistency between supply and demand; if not a closure rule is given, which guarantees that total macroeconomic demand equals supply. In most CGE models this rule determines investment by savings with variations coming from international trade. As can be seen in comparisons of simulations with the models EXIOMOD and GINFORS this closure rule is responsible for the strongest differences in results. Especially in environmental modelling technological change plays a central role. Investment is the economic activity that is driving the story. In neoclassical modelling investment in new technologies may happen, but with a neoclassical closure rule it will always reduce investment in other activities. This excludes by assumption that an investment in new technologies can push GDP and employment. In modern Keynesian modelling investment in new technologies will rise and boost the investment of the industries concerned and endogenously the investment of the industries with "old" technologies will be reduced. The total effect on investment is open; it can be positive or negative depending on the capital/output ratios of the technologies and other parameters.

The neoclassical model EXIOMOD describes the development and reactions of a perfect economic world, whereas GINFORS tries to identify the deviations from the perfect world testing the assumptions of the theory using the method of econometric estimation for the parameterization. This difference is very important since there is strong evidence that the inputs of resources and especially that of materials are not used efficiently.

It is no question that both modelling approaches have their justification and problems: The neoclassical approach can serve as a reference showing how the economy would develop, if a general equilibrium would exist, but to use it for concrete policy evaluation should be called into question. The modern Keynesian approach claims that it accomplishes this, but this ability depends on the empirical validity of the underlying models.

Notes

1 Hoekstra and Wiedmann (2014) provide for a recent overview of various environmental footprint assessments.
2 Wilts et al. (2014) give an overview.
3 This project has received funding from the EU's Seventh Programme for research, technological development and demonstration under grant agreement No 308371. Additional references to the POLFREE project as well as detailed reports about underlying methodological frameworks and comprehensive representations of the resulting simulation outcomes may be inferred from www.polfree.eu/.
4 These modelling works were embedded in a scenario framework (Jäger and Schanes, 2014) which defined EU targets and plausible international cooperation scenarios. Furthermore, suitable policy mixes (Wilts et al., 2014) had been allocated to different assumptions on governance (O'Keeffe et al., 2014).

References

Akerloff, G. A., and Yellen, J. L. (1985). A near rational model of the business cycle, with wage and price inertia. *The Quarterly Journal of Economics*, 100, 823–838.

Almon, C. (1967). *The american economy to 1975*. New York: Harper & Rowe.

Beringer, T., Lucht, W., and Schaphoff, S. (2011). Bioenergy production potential of global biomass plantations under environmental and agricultural constraints. *GCB Bioenergy*, 3(4), 299–312.

Blinder, A. S. (1994). On sticky prices: Academic theories meet the real world. In N. G. Mankiw (Ed.), *Monetary policy*. Chicago: The University of Chicago Press, 117–154.

Böhringer, C., and Löschel, A. (2006). Computable general equilibrium models for sustainability impact assessment: Status quo and prospects. *Ecological Economics*, 60(1), 49–64.

Dewatripont, M., and Michel, G. (1987). On closure rules, homogeneity, and dynamics in applied general equilibrium models. *Journal of Development Economics*, 26(1), 65–76.

Distelkamp, M., Meyer, B., and Moghayer, S. (2015). *Report about integrated scenario interpretation: Comparison of results*. Deliverable 3.7c POLFREE project.

Dixon, P. B., and Rimmer, M. T. (2012). Validation in computable general equilibrium modeling, p. 888. In P. B. Dixon and D. Jorgenson (Eds.), *Handbook of computable general equilibrium modeling*. North Holland: Elsevier.

Ahlert, G., Meyer, M., and Distelkamp, M. (2017): Structures of a welfare-oriented sustainable impact assessment model and practical implementation experiences. in: Diemer A. (Ed.), *Europe and Sustainable Developement, Challenges and Prospects*, Editions Oeconomia, 23–44.
Dornbusch, R., Fischer, S., and Startz, R. (2008). *Macroeconomics*. McGraw Hill.
European Comission. (2014). *Study on modelling of the economic and environmental impacts of raw material consumption*. Technical report 2014–2478. Retrieved from http://ec.europa.eu/environment/enveco/studies_modelling/index.htm.
Fontagne, L., Gourdon, J., and Jean, S., (2013). Transatlantic Trade: Whither Partnership, Which Economic Consequences? Policy Brief CEPII. No.1(September). Retrieved from http://www.cepii.fr/pdf_pub/pb/2013/pb2013-01.pdf
Giljum, S., Behrens, A., Hinterberger, F., Lutz, C., and Meyer, B. (2008). Modelling scenarios towards a sustainable use of natural resources in Europe. *Environmental Science & Policy*, 11, 204–216.
Hall, R. E. (1986). *Market structure and macroeconomic fluctuations*. Brookings Papers on Economic Activity 2.
Hertel, T. W. *Global Trade Analysis Modeling and Applications*. Cambridge, MA: Press Syndicate of the University of Cambridge, Cambridge University Press. Retrieved from https://www.gtap.agecon.purdue.edu/resources/download/7684.pdf
Hoekstra, A., and Wiedmann, T. (2014). Humanity's unsustainable environmental footprint. *Science*, 344, 1114–1117.
Hu, J., Moghayer, S., and Reynes, F. (2015). *Report about integrated scenario interpretation: EXIOMOD/LPJmL results*. Deliverable 3.7b POLFREE project.
Jäger, J., and Schanes, K. (2014). *Report on scenario formulation. Deliverable 3.5*. POLFREE project.
Johansen, L. (1960). *A multisectoral study of economic growth: Contributions to economic analysis 21*. Amsterdam: North-Holland Publishing Company, x+177.
Lucas, R. E. (1976). Econometric policy evaluation: A critique. *Carnegie-Rochester Conference Series on Public Policy*, 1, 19–46.
Meyer, B., Distelkamp, M., and Beringer, T. (2015). *Report about integrated scenario interpretation: GINFORS/LPJmL results*. Deliverable 3.7a POLFREE project.
Meyer, B., and Meyer, M. (2013). *Impact of the current economic instruments on economic activity*. CECILIA2050 WP 2. Deliverable 2.6.
O'Keeffe, M., Jäger, J., Hartwig, F., Armeni, C., and Bleischwitz, R. (2014). *Report on global governance for resource-efficient economies*. Deliverable 2.5 POLFREE project.
Romer, D. (1993). The new Keynesian synthesis. *The Journal of Economic Perspectives*, 7(1), 5–22.
Samuelson, P. A., and Solow, R. M. (1960). Analytical aspects of anti-inflation policy. *American Economic Review*, 50(2), 177–194.
Sen, A. (1963). Neo-classical and Neo-Keynesian theories of distribution. *Economic Record*, 39(85), 53–64.
Sorrell, S., and Dimitropoulos, J. (2008). The rebound effect: Microeconomic definitions, limitations and extensions. *Ecological Economics*, 65, 636–649.
Sorrell, S., Dimitropoulos, J., and Sommerville, M. (2009). Empirical estimates of the direct rebound effect: A review. *Energy Policy*, 37, 1356–1371.
Steger, S., and Bleischwitz, R. (2011). Drivers for the use of materials across countries. *Journal of Cleaner Production*, 19, 816–826.
Stiglitz, J. E. (1976). The efficiency wage hypothesis, surplus labour, and the distribution of income in L. C. D.'s. *Oxford Economic Papers, New Series*, 28(2), 185–207.
Stiglitz, J. E. (1984). Recent developments in the theory of industrial organization: Some macroeconomic implications. Price rigidities and market structure. *AEA Papers and Proceedings*, 74(2), 350–355.
Taylor, L., and Lysy, F. (1979). Vanishing income redistributions. Keynesian clues about model surprises in the short run. *Journal of Development Economics*, 6(1), 11–29.
Timmer, M. P., Dietzenbacher, E., Los, B., Stehrer, R., and de Vries, G. J. (2015). An illustrated user guide to the world input-output database: The case of global automotive production. *Review of International Economics*, 23, 575–605.
UNEP (2017) Resource Efficiency: Potential and Economic Implications. *A report of the International Resource Panel*. International overview on state of the art macroeconomic modelling activities by Ekins, P., Hughes, N., et al., 99–110.
Weinzettel, J., Hertwich, E., Peters, G., Steen-Olsen, K., Galli, A. (2013). Affluence drives the global displacement of land use. *Global Environmental Change-Human and Policy Dimensions*, 23, 433–438.

Wiebe, K. S., Bruckner, M., Giljum, S., Lutz, C., and Polzin, C. (2012b). Carbon and materials embodied in the international trade of emerging economies a multiregional input-output assessment of trends between 1995 and 2005. *Journal of Industrial Ecology*, 16, 636–646.

Wiedmann, T., and Barrett, J. (2013). Policy-relevant applications of environmentally extended MRIO databases – experiences from the UK. *Economic Systems Research*, 25, 143–156.

Wiedmann, T. O., Schandl, H., Lenzen, M., Moran, D., Suh, S., West, J., and Kanemoto, K. (2015). The material footprint of nations. *Proceedings of the National Academy of Sciences of the United States of America*, 112, 6271–6276.

Wilts, H., v. Gries, N., Bahn-Walkowiak, B., O'Brien, M., Busemann, J., Domenech, T., Bleischwitz, R., and Dijk, M. (2014). *Policy mixes for resource efficiency*. Deliverable 2.3 POLFREE project.

Wood, R., Stadler, K., Bulavskaya, T., Lutter, S., Giljum, S., de Koning, A., Kuenen, J., Schuetz, H., Acosta-Fernandez, J., Usubiaga, A., Simas, M., Ivanova, O., Weinzettel, J., Schmidt, J. H., Merciai, S., and Tukker, A. (2015). Global sustainability accounting-developing EXIOBASE for multi-regional footprint analysis. *Sustainability*, 7, 138–163.

Yellen, J. L. (1984). Efficiency wage models of unemployment. *American Economic Review*, 74(2), 200–205.

15

EXTENDING MACRO-ECONOMIC MODELLING INTO THE RESOURCE NEXUS

Alvaro Calzadilla and Ramiro Parrado

Introduction

Any assessment performed with a modelling tool should start from a comprehensive overview of the problem taking into account the single elements as well as their interactions. This preliminary evaluation is the first step before choosing the degree of detail for a modelling effort. Therefore, modelling the different nexus linkages from a macro-economic standpoint is a challenge that should carefully consider the issue from different angles. This chapter describes the recent modelling efforts tackling the nexus with different approaches and from several viewpoints that include direct and implicit links with the resource nexus as well as with a wider context of the nexus used in those frameworks. It is not intended as an extensive literature review but as a systematic review of the diverse methods used in combination with the main elements of the nexus susceptible of being modelled in a macro-economic approach. These elements constitute a subset of the broad nexus definition including interlinkages presented in Chapter 1 of Part I of the handbook. Therefore, the review takes into consideration the broad model categories which have been used for the modelling task in the current literature.

Main macro-economic modelling frameworks suited for a nexus analysis

Among the diverse modelling approaches that can be used to analyse the connotations of the resource nexus there are at least five types of models that can address the issue in a macro-economic context: *Partial Equilibrium*, *Econometric*, *Input–Output*, *Computable General Equilibrium*, and *Integrated Assessment*. These main categories have been selected based on a review of papers dealing directly or indirectly with the resource nexus. Establishing a clear-cut categorization of these models is difficult because they can tackle the resource nexus from different angles. Therefore, they are organized following an order starting first describing bottom-up methodologies that model with more detail a specific phenomenon, and then moving gradually to the most complex top-down approach that includes the full interaction between markets and/or models. Only partial equilibrium models belong to a bottom-up approach in the sense that pay more attention to specific features of the subject being modelled and may include some relations between markets that compete for the use of closely related resources, or analyse the implications of the production of one commodity in other sectors. Even though these models may apparently focus only on a partial view of a whole system, they can also provide some macro-economic insights when the subject is analysed in an aggregated setup such as the country or global level.

Partial Equilibrium (PE) models have their foundations on the explicit establishment of supply and demand that meet to reach market equilibrium through price adjustments. This approach allows to describe in detail the characteristics of the market being analysed and may include one or few different markets that are directly linked through either the use of the same resource, or because there is a vertical integration indicating demands derived indirectly from downstream sectors.

Econometric models are strongly based on empirical evidence and rely on data collection to test hypotheses proposed to verify theoretical propositions. Their analysis can validate and reveal interactions between apparently isolated markets but that have specific channels proposed in a theoretical setup. These theoretical formulations can range from a single detailed market to macro-econometric models that can also use time series and cross panel data analysis.

Input–output (IO) models consider the flow of goods and services in an economy organized in several sectors using national or regional statistics and following the system of national accounts for the construction of a consistent dataset that provides the picture of an economy in a particular year. The advantage of this approach is that it relies on technical input–output coefficients that describe the embedded demand for inputs in all sectors of the economy following the theory laid down by Leontief (1951). This allows tracking direct and indirect demands (and relationships) within the whole economy, which is useful for an extended analysis of the resource nexus that takes into account the consequences of the indirect demands triggered in the rest of the economy.

Computable General Equilibrium (CGE) models go beyond the IO framework extending not only the data but also the conceptual framework, since they provide a microeconomic based theoretical structure to describe the interactions of different representative economic agents within a region. The models are based on social accounting matrices that are built upon the IO model but include also additional data such as household surveys. This allows providing a framework in which, unlike the IO model, the supply and demand of the agents and sectors are all described in a consistent model and are interconnected throughout price feedback mechanisms. While these characteristics are useful for a resource nexus analysis because of the transmission channels present in the CGE structure, there are some limitations derived of the complexity of the approach such as assuming that markets are in equilibrium or the existence of full employment in the economy.

Integrated assessment (IA) models represent the combination of two or more models by soft-linking the results of one model and using them in other models, or by establishing a hard-link between a representative or specific feature of one or several models (usually bottom-up) on another (usually top-down) model. This setting up of several models allow to address the resource nexus in a more complex way taking into account not only the interrelationships among sectors in the regional or global economy, but also respecting the detailed specificities of a particular model that may represent the use of a particular resource.

Nexus elements integrated in macro-economic modelling

There are many alternatives to introduce the analysis of the resource nexus in macro-economic models, and most of the contributions have been made addressing research questions that do not pose explicitly the resource nexus issue but aim to analyse specific policies or developments. In this context, the main elements identified in the literature as part of a nexus analysis correspond with the following elements: water, energy, food, land use, the environment, and resource use in general. There are direct links such as that between energy and water (Gu et al., 2014; Liu et al., 2015) investigating the implication of energy policies on water demand or the use of water in electricity generation. This nexus is then extended to include consequences on food production and food security (Bazilian et al., 2011; Prieler et al., 2013; Foran et al., 2014; Steinbuks and Hertel, 2014; Biggs et al., 2015; Daher and Mohtar, 2015; Mirzabaev et al., 2015; Miralles-Wilhelm, 2016). When food and energy are put in the same analysis, the expected choice is to comprise also land-use changes (Panichelli and Gnansounou, 2015) since the production of bioenergy will create competition with food production and water use, and eventually could put pressure to expand croplands with consequences on deforestation. The use of energy also has consequences on the environment and specific policies designed to address that problematic raise the energy–environment–growth nexus that adds the economic behaviour and Greenhouse Gas (GHG) emissions as other elements in the analysis (Zeshan and Achmed, 2013; Cai et al., 2015; Meyer et al., 2013). While these are only examples of the nexus elements that have been linked in previous studies, they point to a wider challenge that is the aim of a circular economy which should reduce pressure on the environment and provide a solid basis for a sustainable behaviour. There is much potential to address this challenge, yet there are also some limitations that should be overcome to properly address it. In this spirit, this chapter gives a summarized review of the main characteristics of current studies extending the nexus in macro-economic modelling along with further research that should help to bridge existing gaps.

The remainder of the chapter is organized as follows. The second section describes the different macro-economic modelling frameworks found in the literature with specific nexus linkages and/or analysis. This section constitutes the core of the chapter's literature review providing the main criteria and elements to identify not only knowledge gaps but also propose further developments in the nexus macro-economic modelling. The third section summarizes and critically analyses the model categories overviewed in the second section. The fourth section gives a brief overview of studies soft-linking nexus indicators with macro-economic models. This constitutes an alternative to hard-link modelling of the nexus which could open new options for future developments. The fifth section explores the implications of using the nexus in the analysis of mainstream issues such as sustainable development, climate change, poverty and nutrition. The final section concludes with a final discussion about prioritization of knowledge gaps as well as the feasible and best ways to address them.

Extending macro-economic models with the nexus interlinkages

Modelling the resource nexus requires addressing the complex interactions and dynamics of its different components. This implies the synergic confluence of different disciplines and modelling tools. Several conceptual frameworks of the nexus are found in the literature varying in scope, goals and understanding of the nexus. A recent review of the main conceptual frameworks by the UN (2015) notes that organizations have tended to expand the scope of the nexus in order to include elements such as ecosystems, climate, and minerals among others. However, the interactions between water, energy and food/land introduced by the World Economic Forum in 2011 (WEF, 2011; Biggs et al., 2015) still remain at the core of the nexus analyses in the literature.

For instance, two recent reviews shed light on the influence of water, energy and land entanglements within modelling methods. With a focus on water accounting and sustainability, Daniels et al. (2011) introduces an approach that combines process-based and environmental IO methods to support effective regional water policy at a macro level. The review stresses the benefits of coupling detailed and site-specific data from process analysis with consistent economy-wide water account data in order to enhance the capacity of environmental multi-region IO in identifying relevant water flows.

On the energy and land side, Panichelli and Gnansounou (2015) review the modelling approaches for biofuel assessments which take into account the implications on land-use change. This critical analysis of the energy–land use–GHG emissions nexus identifies the modelling choices and concludes that although the research progress in the field is significant there is still room for further developments. New research should aim at improving the supply chain modelling as well as the considering the necessity to advance with land use and forests dynamics, along with the spatial dimension representation. Nevertheless, despite the intrinsic link between water and biofuel and crop production apparently water is not part of those modelling frameworks.

A research agenda considering the main features of the water–energy–food nexus with an integrated approach including interactions that are valid also at the world level is proposed by Miralles-Wilhelm (2016). Issues like regional or spatial scale should be considered as well as creating a community of practice to foster cooperation and knowledge sharing.

Recent applied work uses different modelling tools to assess the resource nexus and support integrated policy analysis at different scales. The modelling tools vary in complexity: from price dynamics analysis using econometric models to combining knowledge from multiple disciplines in IA models. The complexity of the modelling tools goes hand in hand with the scope of the nexus. The remainder of this section describes the latest examples of the modelling efforts following the categories introduced in earlier.

Partial equilibrium

Land-use models have the advantage to bring geographical detail to the nexus assessment and to capture domestic and international market adjustments to changes in the resource supply and demand conditions. In this vein, the Global Biosphere Management Model (GLOBIOM) (Havlík et al., 2011) analyse the competition of land between agriculture, forestry and bioenergy. GLOBIOM is a global recursively dynamic PE model that assesses the demand and supply of water for irrigation and livestock production. The model accounts for the 18 most important crops, a range of livestock production activities, forestry commodities, first and second generation biofuels, and water.

Rosegrant et al. (2013) use the IMPACT model to analyse how the interaction between water, energy and food are affected by the bioeconomy concept. The IMPACT model is a PE, multi-commodity, multi-country model which generates projections of global food supply, demand, trade and prices. Food production is determined by crop and input prices, the rate of productivity growth and water availability. They explore a future where the increased use of natural resources goes hand in hand with higher resource efficiency and sustainability, all these aspects captured by the bioeconomy concept.

Steinbuks and Hertel (2014) make use of a global land-use model focusing on the food–energy–environment nexus to investigate the optimal allocation of world resources considering economic, agronomic and biophysical aspects. The study develops the relationships between forest, agriculture and biofuels in a land-use model with environmental services (FABLE) which is a dynamic, forward-looking, multi-sectoral, PE model. Specifically the nexus is considered by accounting for the food processing industries as well as biofuel production along with forestry activities and biodiversity. The approach puts emphasis in modelling preferences in the sense that by increasing wealth in the society, a greater value is placed on ecosystem services with a decreasing marginal value of additional consumption. Climate mitigation efforts are also considered as part of the analysis in the form of GHG reductions. While the model addresses the nexus with a new development, some limitations are acknowledged by the authors. There are trade-offs between implementing an intertemporal optimization and increasing the regional detail and, at that stage of development, priority was assigned to the intertemporal dimension. The approach follows a PE analysis and although it is considered appropriate for the analysis, it would be useful to extend it to include feedback channels from land use. These are well-known limitations in in the research and modelling community that will eventually be improved in the future.

Based on functional relationships, the water–energy–food nexus modelling tool (WEF Nexus Tool 2.0) introduced by Daher and Nohtar (2015) quantifies the interconnectivity between these three resources. A conceptual framework is developed as a foundation for defining explicit relationships between the three systems. Their modelling tool consists of inputs that reflect national food, water and energy strategic options and allows for the creation and assessment of different scenarios aiming to identify sustainable national resource allocation strategies at the country level.

Probably the most studied nexus is the competition between biofuels and food production for land resources. Several PE models have been used to inform policymakers on the impact of national, regional and global biofuel policies: AgLink-COSIMO (OECD, 2008), ESIM (Deppermann et al., 2016), FAPRI (Fabiosa et al., 2010), CAPRI (Britz and Hertel, 2011), IMPACT (Rosegrant, 2008), FABLE (Hertel et al., 2015) and GLOBIOM (Havlík et al., 2011) are among the most representative.

Econometric

Econometric models are used to infer a causal relationship between selected variables that govern the resource nexus. Abdelrabi and Serra (2015), for example, analyse food-energy price relationships in Europe between November 2008 to June 2012 using a multivariate generalized autoregressive conditional heteroscedasticity model, allowing for both asymmetric volatility spillovers and the influence of exogenous variables in the conditional variance (i.e. euro–dollar exchange rate and international rapeseed oil stock). In addition, they use a semi-parametric approach to capture the remaining information that is not reflected by the parametric estimation.

Natanelov et al. (2013) use a Johansen co-integration methodology to analyse the complex relationship between crude oil, corn and ethanol between 2006 and 2011 in the US. Using futures prices as their data source, they are able to capture potential market linkages based on production or trade. A similar analysis for the Brazilian sugar–ethanol–oil nexus is done by Balcombe and Rapsomanikis (2008).

To study the energy–environment–growth nexus in five South Asian countries, Zeshan and Achmed (2013) use a 30-year panel data for five Bangladesh, India, Pakistan, Sri Lanka and Nepal. The econometric estimation focuses on the causal relationship between energy use, CO_2 emissions and economic activity which should be taken into account at the time of designing and also evaluating energy and climate policies.

Input–output

Both PE and econometric models describe particular markets but do not account for cross-sectoral or economy-wide analysis. IO-based analysis effectively incorporates the different components of the nexus at the economy-wide level. Focused on the water–energy nexus Gu et al. (2014) use an IO framework to evaluate energy-saving policies in China. Basically, the study analyses the relationship between energy and water supply sectors by means of IO tables and calculating the corresponding water savings derived from energy-saving policies. The advantage is that with the IO data it is possible to estimate the direct and indirect water-use coefficients of the primary energy sectors. Even though the paper does not provide an explicit direct link between energy and water, which could be improved through specific modelling choices, it is a good example of how IO models can be used to provide useful insights for the nexus analysis. In addition, it also highlights existing limitations in economic analysis which should be addressed in the future, such as the fact that there are factors like water (or the environment) that are difficult to integrate in economic datasets given that they do not have a market price.

Using a dynamic multi-region IO simulation model (Global INterindustry FORecasting System-GINFORS; Lutz et al., 2010), Meyer et al. (2013) analyse the economic–energy–environment nexus at the global level. The GINFORS model covers 39 world regions, 35 industries and 59 products which are traded bilaterally. GINFORS accounts for emissions of 28 energy carriers and explains the resource extraction for 12 kinds of materials, the agricultural land use for four types of land and the freshwater abstraction for different industries. Although the approach may seem close to a general equilibrium framework, the modelling choices in GINFORS include bounded rationality and imperfect markets which could provide a different perspective when analysing the nexus.

Environmental extended IO models are also combined with life cycle analysis to investigate nexus-related issues. For example, Li et al. (2012) use an input–output-based hybrid life cycle analysis to evaluate water consumption and CO_2 emissions of wind power in China.

Computable general equilibrium

As described in the introduction the advantage of CGE models over the IO framework is the fact that they can account for price feedback mechanisms which allows bringing into the analysis the substitution of goods and factors and therefore choices based on economic evidence. Cai et al. (2015) analyse the energy–carbon–environment nexus using GTEM-C, a combination of a global CGE model with bottom-up engineering details of energy production and consumption. Their modelling framework allows to account for global energy trade flows and combustion-based GHG emissions as well as to analyse economic trade-offs associated with the global environmental and energy issues such as climate change, carbon mitigation, and energy transformation. Climate-economy interactions are considered by linking GTEM-C to a global climate model. The projected GHG emissions are used by the global climate model to estimate changes in surface air temperature that is linked to regional total factor input productivities. Changes in regional temperature and precipitation are linked to land productivity. The heat stress due to change of temperature and humidity is linked to changes in labour productivity.

With a focus on the water–food nexus, Calzadilla et al. (2010) use the GTAP-W model, a global CGE model with water resources as an explicit factor of production in irrigated agriculture, to investigate the role of green and blue water resources in agriculture. They evaluate different scenarios of sustainable water use in the agricultural sector and found a clear trade-off between economic welfare and environmental sustainability.

Using a more developed model that includes bioenergy production and two degrees of competition of crops, one for land resources within each agro-ecological zone and another for irrigation water within each river basin (GTAP-BIO-W), Liu et al. (2014) analyse the potential impact of water scarcity on food production in 2030 considering specific areas where irrigation stress is growing.

The food versus fuel debate raised questions about the land-use implications of biofuel policies. Several CGE model have been used to address the economy-wide and global implications of biofuel expansion: GTAP-BIO (Golub and Hertel, 2012), GTAP-BIO-AEZ (Taheripour et al., 2013), MAGNET (Smeets et al., 2014), EPPA (Gurgel et al., 2007) and FARM (Suttles et al., 2014) are among the most representative.

Integrated assessment

Nexus conceptual frameworks focus on integrated analytical approaches that study the interactions of the different nexus elements. The modelling counterparts to these analytical frameworks are the IA models. With a focus on the water–energy nexus, Liu et al. (2015) use the Global Change Assessment Model-USA (GCAM-USA) to analyse future water demands for electricity generation. They extend the analysis of Davis et al. (2013) to include a more sophisticated description of generation technologies, fuel portfolios and available cooling technologies. The study assesses the interaction between socio-economic factors, electricity demand, mitigation policies, electricity trading and water-saving technologies. Among the gap identified to improve this kind of studies the authors cite the inclusion of water supply constraints (or other sources such as ground or desalinized water) as well as climate change impacts on energy generation.

Prieler et al. (2013) focus on the food versus fuel competition analysing the implications of the food–energy security–environment nexus in the agricultural systems. The study uses a modelling framework encompassing the Global Agro-ecological Zone model (IIASA/FAO, 2012) and the World Food System model (Fischer et al., 2005) which is an applied general equilibrium model. One of the main characteristics of this kind of framework is that it allows combining the results from a CGE model with a spatial grid analysis which provides a better understanding of agricultural activities. In addition, the biofuel production representation has been expanded to consider resource use as well as by-products.

The climate, land-use, energy and water strategies (CLEW) integrated framework (Bazilian et al., 2011; Howells et al., 2013) is a module-based approach that integrates existing well-tested methodologies for each of the three resources with climate change. LEAP (Long-range Energy Alternatives Planning tool) is used to model energy, WEAP (Water Evaluation and Planning tool) to model water and AEZ (Agro-Ecological Zoning) to model land use.

Giampietro et al. (2013) analyse the nexus between food, energy and water security using the Multi-Scale Integrated Analysis of Societal and Ecosystem Metabolism (MuSIASEM) tool, which integrates quantitative information generated by distinct types of conventional models based on different dimensions and scales of analysis. The concepts on which MuSIASEM bases on include the flow-fund model (derived from bioeconomics), and three conceptual tools: the multi-scale accounting, the multipurpose grammar and the impredicative loop analysis (derived from the complexity theory).

Knowledge gaps in the nexus macro-economic modelling

Critical analysis of the nexus macro-economic modelling

Several are the challenges to address the nexus within macro-economic modelling tools and the existing research efforts have proven to effectively address many of them even before the nexus concept was coined. The interlinkages and trade-offs present in the resource nexus have been matter of investigation with many macro-economic models searching for answers to problems focusing on the evaluation of environmental consequences of commodity production and markets integration. Furthermore, the explicit definition of the nexus and its entanglements with production processes, markets and other economic activities has fostered the discussion to put forward a research agenda aiming to improve coordination efforts as well as to provide enhanced tools for integrated assessments that will contribute for a sustainable development.

Table 15.1 presents a summary of these efforts by comparing the model categories reviewed in the second section with the different analysis conducted in the nexus and highlighting the geographical coverage. The main features of the modelling technique along with the limitations or caveats of each approach are also shown. The scope of the different conceptual frameworks on the research agenda concentrates mainly on the water–energy–food nexus. Partial equilibrium assessments have tackled the nexus focusing on water, energy, food and the environment at the country and global level. Despite that PE studies can offer a much detailed description of the relationship between an economic activity and the corresponding nexus elements, there are some limitations derived of the compromises that must be adopted sacrificing other important features of the analysis such as the intertemporal, sectoral and regional dimensions. Another crucial point is related to disregarding possible reverse causalities when designing the model which could overlook specific details of the analysis. However, this problem can be contained by carefully selecting all variables that should be taken into account for the problem being analysed.

Econometric models can provide insightful considerations about energy- and food-related elements of the nexus as well as the natural environment using sophisticated analysis techniques. Paradoxically, while the main strength of these models is based on the causality analysis of existing data, the availability of data may constitute a crucial limitation for the approach. This highlights the fact that data collection and gathering is a fundamental requisite for all modelling efforts.

The advantage of using IO models is that their structure has embedded interlinkages that are suitable for the specific analysis of the nexus. The IO framework takes into account the direct and indirect requirements within a generic production process and this makes it easier to track down the effects on indirectly related markets and the demand for inputs from the nexus. It is possible to extend the accounting framework by including specific coefficients that link additional information of the nexus elements to the demands and outputs of an IO model. While the insights they provide are useful for accounting for changes in the provision of quantities within supply chains, one disadvantage is the absence of feedback effects derived from the interaction between supply and demand.

This last limitation is addressed in CGE models by including price transmission mechanisms that allow for substitution possibilities. In this way, prices constitute the linking element that equalizes supply and demand. This brings out a framework for the nexus analysis that is based on the extension and use of complementary databases in order to improve the representation of trade-offs and cross-sectional effects. However, this approach shares also some limitations of the IO analysis in the sense that there are specific nexus elements such as water, the environment or natural capital that have proven to be difficult to include given the difficulties to put a monetary value on them. Notwithstanding this, there are some initiatives that are starting to address this limitation.

The last category is represented by IA models from which those being based on an ensemble of several models are the better ones suited for studying the intertwined nature of the resource nexus within a macro-economic approach. The sole possibility of combining different models by either linking or coupling them within a harmonized set of assumptions brings forward a consistent way of analysing the implication of the nexus in many dimensions and combinations. The disadvantages of this approach are related to the fact that it is demanding on data and computational resources, and that also shares some of the limitations inherited in each of the single models used, being based either on partial or general equilibrium.

Table 15.1 Overview matrix of the nexus macro-economic modelling

Generic model type	*Generic nexus type*	*Geographic application*	*Modelling technique*	*Limitations/caveats*	*References*
Conceptual framework	Energy related, food related, water–energy–food, energy–LUC–GHG	-	-	• Modelling approach and research agenda	WEF (2011), Rezai et al. (2013), Foran et al. (2014), Biggs et al. (2015), Panichelli and Gnansounou (2015), UN (2015), Miralles-Wilhelm (2016)
Partial equilibrium (PE)	Bioenergy–food, water–energy–food, energy–food–environment	Country and global	Detailed bottom-up analysis	• Introducing the intertemporal dimension could be costly • Aggregated geographic and sectoral coverage • Many crucial variables are set as exogenous (income, population, the rate of technological progress, and variable input prices) disregarding possible reverse causalities	OECD (2008), Rosegrant (2008), Fabiosa et al. (2010), Britz and Hertel (2011), Havlík et al. (2011), Rosegrant et al. (2013), Wise et al. (2013), Steinbuks and Hertel (2014), Daher and Mohtar (2015), Deppermann et al. (2016)
Econometric	Energy related, energy–food, energy–environment–economy	5 South Asian countries, US, EU	Panel data, parametric and semi-parametric approaches	• Analysis focused on a narrow set of markets and indicators • Data availability is a crucial constraint	Zeshan and Achmed (2013), Natanelov et al. (2013), Abdelradi and Serra (2015)
Input–output (IO)	Water–energy, energy–environment–economy	Catalonia Spain, China, multi-region	Accounting framework; extending IO analysis with additional coefficients linking results to nexus elements	• Estimations based on direct water coefficients linked to IO values • There are no feedback effects • Does not recognize the water extracted from the environment without a market price • The use of hybrid techniques has potential, provided there is an availability of bottom-up water consumption data	Gonzales (2011), Daniels et al. (2011), Li et al. (2012), Meyer et al. (2013), Gu et al. (2014)

Computable general equilibrium (CGE)	Water–food, bioenergy–food, water–bioenergy–food, energy–environment–GHG	Country and global (18 regions)	Extending databases; analysis of cross-sectional trade-offs. Hard linking nexus features	• Including specific nexus elements such as water, the environment or natural capital have proven to be difficult given that it is not possible to put a price consistent with the System of National Accounts	Gurgel et al. (2007), Sarkar (2007), Calzadilla et al. (2010), Golub and Hertel (2012), Carraro et al. (2013). Taheripour et al. (2013), Smeets et al. (2014), Suttles et al. (2014), Cai et al. (2015)
Integrated assessment (IA)	Water–energy, water–energy–food, energy–water–LUC, energy–food–environment	50 US states, country and global	Soft linking or coupling different models within a harmonized scenario	• Data resource and computational demanding approach contrasting the potential of the research developed with IAMs • Shares the same limitation of the previous categories since it combines several models in one integrated framework	Bazilian et al. (2011), Prieler et al. (2013), Howells et al. (2013), Liu et al. (2015), Giampietro et al. (2013)

Main knowledge gaps and future developments

The cornerstone of analysing the nexus within a macro-economic setup is the existence of many interrelationships between not only the resources used in a productive process but also the commodities that constitute inputs for a final product. This can be illustrated considering three aspects present as the main dimensions that constitute the basis of the modelling framework as shown in Figure 15.1. The first aspect is the nexus itself at the left side of the figure including resources such as water and land, as well as commodities such as food and energy among others. The second aspect regards the different modelling approaches shown at the right of the figure going from bottom-up with PE models to the more complex top-down approaches under the umbrella of integrated assessment. The third aspect regards the spatial analysis represented in the middle of the figure.

Figure 15.1 shows how modelling has evolved developing each category taking into account one or more elements of the nexus as well as using results from other type of model. This reveals both a vertical and horizontal integration of modelling efforts and despite being the proof of the progress made in research and modelling, there is one dimension in which there is still a gap represented at the centre of the figure highlighting the contrast between the macro-economic and spatial analyses. The development of spatial analysis techniques has been driven by many factors including the advance of geographical information systems in generating and processing data. This has opened the possibility for the enhancement of macro-economic modelling in the future.

On the basis of these considerations and the evidence discussed previously the following actions could be undertaken to exploit the potential of modelling the resource nexus. The collection of complementary or satellite datasets to cover the additional requirements for macro-economic models will not only improve current analysis but will encourage further refinements of modelling the nexus. Using spatial data coupled with economy-wide models will allow for

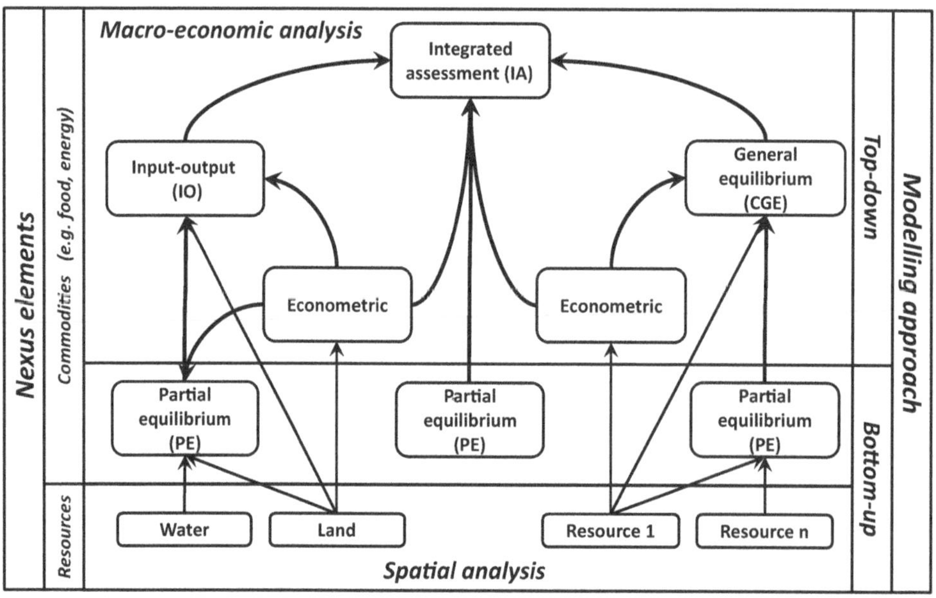

Figure 15.1 Integrating the nexus in macro-economic modelling

increasing the granularity of the analysis. This has been done at a certain point with water and land use but should extend to other resources as well.

The crucial elements in macro-economic modelling of the nexus are the supply chains that describe the different stages involving the transformation of a specific element into a commodity until their final destination. While these chains can be modelled in detail using a PE approach they lack the interactions with the other supply chains present in the economy and therefore overlook the possible trade-offs with other nexus elements.

Macro models mainly IO, CGE and IA provide a consistent umbrella framework that can adopt several features of PE models dealing with specific resources. This offers an opportunity to analyse the nexus considering many elements along with the interactions derived of their exploitation. This calls for an ultimate hybrid top-down-bottom-up approach that should be developed in the following years. This does not mean that IA will constitute the final paradigm to analyse the nexus but that there will be different hybrid models based on IO, CGE and IA, depending on the data availability and on the scope of the analysis.

Linking nexus indicators to macro-economic models

While the studies cited in the second section represent a straightforward approach to analyse the nexus within a modelling context where the nexus elements are explicitly present, there is also a recent stream of literature that has been dealing with linking indicators related to the nexus elements with macro-economic models. Despite that some of these studies do not analyse explicitly the resource nexus, they offer an implicit analysis that is grounded mainly on the sustainable development framework.

For instance, Sarkar (2007) introduces the nexus within the Millennium Development Goals (MDGs) and economic growth, suggesting connotations for resource use (implicitly for the resource nexus) and sectoral development that could have also consequences on poverty reduction and MDG achievement. The paper makes a conceptual analysis taking into account general equilibrium implications of policy analyses in a simple macro-economic model.

With a broader scope based on sustainable development, Carraro et al. (2013) link the output of a multi-region and multi-sector CGE model to a set of indicators based on three sustainability pillars to analyse policy implications on sustainability. Even though the nexus is not addressed directly, three elements of the nexus are present among the selected set of indicators: water and energy use, and the budget share of food consumption over private expenditures.

Another initiative with sustainability connotations is that proposed by Biggs et al. (2015) who identify potential linkages of nexus approaches with sustainable livelihoods approaches. The main focus is on the interrelated dynamics between human population and the natural environment. The study purports an integrated framework with the capacity to measure and monitor environmental livelihood security of whole systems by accounting for water, energy and food requisites within livelihoods at multiple spatial scales and institutional levels. The role of indicators in this integrated framework does not limit to present information about the energy–water–food nexus elements but also includes external factors that may influence the system.

Further analysis and implications considering the resource nexus

The nexus is a concept that provides a more integrated view of the resource use and scarcity with an emphasis towards a sustainable development. The water–energy–food nexus was one of the main approaches considered by the UN in setting its Sustainable Development Goals (Benson et al., 2015). Thus, the role of the nexus is expected to be developed outside the natural resource linkages. Besides the possible extension of nexus indicators within macro-economic models that points towards implications on sustainable development as discussed in the fourth section, there are other areas in which the resource nexus will definitely foster contributions. Following is a summary of the main areas in which it is reasonable to expect further advancements.

Directly related with food production and the impacts of climate change on rural livelihoods is poverty alleviation. Agbola (2011) use meteorological and historical data to highlight the relationship of the climate change and poverty nexus in Nigeria. The paper emphasizes on the importance of climate vulnerability when analysing poverty.

Mirzabaev et al. (2015) review the trade-offs and synergies of bioenergy, food security and poverty reduction within the water–energy–food security nexus. They argue that a transition from traditional to modern bioenergy use is essential for a sustainable development and poverty reduction. Several studies highlight the contribution of bioenergy development to poverty alleviation through mainly higher income for small-scale farmer and increased employment opportunities (Ewing and Msangi, 2009; Arndt et al., 2012).

A report by the High Level Panel of Experts on Food Security and Nutrition (HLPE, 2015) highlights the importance of water as a central element to food security and nutrition by investigating the relationship between water, food security and nutrition from household to global level. The report looks into these multiple linkages in a context of competing demands, rising scarcities and climate change. With a focus on sub-Saharan Africa, Mabhaudhi et al. (2016) stress the need to link water use in agriculture to food and nutrition security, and human health and well-being. They call for adopting the water–food–nutrition–health nexus to ensure that nutrition is made an explicit output in agricultural interventions.

These are some examples of the direction in which the macro-economic modelling of the nexus will progress revealing the intricate nature of poverty alleviation, climate change and sustainable development to which the resource nexus is one of the main, if not the foremost, foundations.

Concluding remarks

Albeit the primordial role that primary resources and input commodities play in production processes has remained invariant, the nexus conceptual frameworks have evolved in time varying in scope and goals. Consequently, the modelling responses have effectively adjusted to these changes and addressed many of them.

Different types of macro-economic models have been used to assess the complex interactions and dynamics of the nexus. While partial equilibrium models give a detailed representation of the corresponding nexus elements, they do not account for impacts on other resource or commodity markets. Econometric models can validate and reveal interactions between different elements of the nexus; however, they rely on data availability and quality. IO models have built-in interlinkages among some nexus components in their core structure, but they do not account for feedback mechanisms between supply and demand. CGE models include feedbacks through price transmission mechanisms enabling substitution possibilities; however, there are some components of the nexus, especially those with no market value, that are not part of their database and require further considerations. As integrated assessment assembles several models from different disciplines, it is best suited to deal with the nature of the nexus, yet IA models are data and computational demanding and share the limitations of each single model used in that approach.

There is mostly a vertical integration of the different methodologies while the elements of the nexus are susceptible of both horizontal (e.g. water–land) and vertical (e.g. water–energy–food) integrations. This reveals a potential gap in the granularity of the top-down analyses. In fact, the incorporation of spatially disaggregated information derived from the horizontal integration of the nexus elements is not common in the reviewed modelling approaches. Although considerable progress has been made towards the spatial representation of water and land resources in top-down models, more effort is required on the wider components of the nexus.

Special attention needs to be paid to supply chains and market linkages at the local, national, regional and global scales. Natural resources are used to produce goods that will be consumed locally, used as an input in other production process and/or sold in foreign markets. Therefore, macro-economic modelling should clearly capture the resource linkages at different stages and scales. In addition, bearing in mind a consistent set of supply chains may facilitate identifying potential synergies and trade-offs when considering a macro-economic modelling effort. Acknowledging the essential role of each nexus element is as fundamental as integrating it properly in the supply chain comprised in any macro-economic analysis.

References

Abdelradi, F., and Serra, T. (2015). Food–energy nexus in Europe: Price volatility approach. *Energy Economics*, 48, 157–167. doi: 10.1016/j.eneco.2014.11.022.

Agbola, B. (2011, Spring) Climate change and poverty in Nigeria. *Regional Development Dialogue*, 32(1), 54–81.

Arndt, C., Pauw, K., and Thurlow, J. (2012). Biofuels and economic development: A computable general equilibrium analysis for Tanzania. *Energy Economics*, 34(6), 1922–1930. doi:10.1016/j.eneco.2012.07.020.

Balcombe, K., and Rapsomanikis, G. (2008). Bayesian estimation and selection of Nonlinear vector error correction models: The case of the sugar-ethanol-oil nexus in brazil. *American Journal of Agricultural Economics*, 90(3), 658–668. doi:10.1111/j.1467-8276.2008.01136.x.

Bazilian, M., Rogner, H., Howells, M., Hermann, S., Arent, D., Gielen, D., Steduto, P., Mueller, A., Komor, P., Tol, R. S. J., and Yumkella, K. K. (2011). Considering the energy, water and food nexus: Towards an integrated modelling approach. *Energy Policy*, 39(12), 7896–7906. doi:10.1016/j.enpol.2011.09.039.

Benson, D., Gain, A. K., and Rouillard, J. J. (2015). Water governance in a comparative perspective: From IWRM to a 'nexus' approach? *Water Alternatives*, 8(1), 756–773.

Biggs, E. M., Bruce, E., Boruff, B., Duncan, J. M. A., Horsley, J., Pauli, N., McNeill, K., Neef, A., Van Ogtrop, F., Curnow, J., Haworth, B., Duce, S., and Imanari, Y. (2015). Sustainable development and the water–energy–food nexus: A perspective on livelihoods. *Environmental Science & Policy*, 54, 389–397. doi:10.1016/j.envsci.2015.08.002.

Britz, W., and Hertel, T. W. (2011). Impacts of EU biofuels directives on global markets and EU environmental quality: An integrated PE, global CGE analysis. *Agriculture, Ecosystems & Environment*, 142(1–2), 102–109. doi:10.1016/j.agee.2009.11.003.

Cai, Y., Newth, D., Finnigan, J., and Gunasekera, D. (2015). A hybrid energy-economy model for global integrated assessment of climate change, carbon mitigation and energy transformation. *Applied Energy*, 148, 381–395. doi:10.1016/j.apenergy.2015.03.106.

Calzadilla, A., Rehdanz, K., and Tol, R. S. J. (2010). The economic impact of more sustainable water use in agriculture: A computable general equilibrium analysis. *Journal of Hydrology*, 384(3–4), 292–305. doi:10.1016/j.jhydrol.2009.12.012.

Carraro, C., Campagnolo, L., Eboli, F., Giove, S., Lanzi, E., Parrado, R., Pinar, M., and Portale, E. (2013). The FEEM sustainability index: An integrated tool for sustainability assessment. In M. G. Erechtchoukova, P. A. Khaiter and P. Golinska, (Eds.), *Sustainability appraisal: Quantitative methods and mathematical techniques for environmental performance evaluation*. Springer Science + Business Media, pp. 9–32.

Daher, B. T., and Mohtar, R. H. (2015). Water–energy–food (WEF) nexus tool 2.0: Guiding integrative resource planning and decision-making. *Water International*, 40(5–6), 748–771. doi:10.1080/02508060.2015.1074148.

Daniels, P. L., Lenzen, M., and Kenway, S. J. (2011). The ins and outs of water use – a review of multi-region input-output analysis and water footprints for regional sustainability analysis and policy. *Economic Systems Research*, 23(4), 353–370. doi:10.1080/09535314.2011.633500.

Davies, E. G. R., Kyle, P., and Edmonds, J. A. (2013). An integrated assessment of global and regional water demands for electricity generation to 2095. *Advances in Water Resources*, 52, 296–313. doi:10.1016/j.advwatres.2012.11.020.

Deppermann, A., Offermann, F., Puttkammer, J., and Grethe, H. (2016). EU biofuel policies: Income effects and lobbying decisions in the German agricultural sector. *Renewable Energy*, 87, 259–265. doi:10.1016/j.renene.2015.10.005.

Ewing, M., and Msangi, S. (2009). Biofuels production in developing countries: Assessing tradeoffs in welfare and food security. *Environmental Science & Policy*, 12(4), 520–528. doi:10.1016/j.envsci.2008.10.002.

Fabiosa, J. F., Beghin, J. C., Dong, F., Elobeid, A., Tokgoz, S., and Yu, T. (2010). Land allocation effects of the global ethanol surge: Predictions from the international FAPRI model. *Land Economics*, 86(4), 687–706. doi:10.3368/le.86.4.687.

Fischer, G., Shah, M., Tubiello, F. N., and van Velhuizen, H. (2005). Socio-economic and climate change impacts on agriculture: An integrated assessment, 1990–2080. *Philosophical Transactions of the Royal Society B: Biological Sciences*, 360(1463), 2067–2083. doi:10.1098/rstb.2005.1744.

Foran, T., Butler, J. R. A., Williams, L. J., Wanjura, W. J., Hall, A., Carter, L., and Carberry, P. S. (2014). Taking complexity in food systems seriously: An Interdisciplinary analysis. *World Development*, 61, 85–101. doi:10.1016/j.worlddev.2014.03.023.

Giampietro, M., Aspinall, R. J., Bukkens, S. G. F., Benalcazar, J. C., Diaz-Maurin, F., Flammini, A., Gomiero, T., Kovacic, Z., Madrid, C., Ramos-Martin, J., and Serrano-Tovar, T. (October 2013). An Innovative Accounting Framework for the Food-Energy-Water Nexus Application of the MuSIASEM approach to three case studies. Food and Agriculture Organization of the United Nations. Retrieved from http://www.fao.org/docrep/019/i3468e/i3468e.pdf

Golub, A. A., and Hertel, T. W. (2012). Modeling land-use change impacts of biofuels in the GTAP-BIO framework. *Climate Change Economics*, 3(3), 1250015. doi:10.1142/s2010007812500157.

González, J. F. (2011). Assessing the macroeconomic impact of water supply restrictions through an input–output analysis. *Water Resources Management*, 25(9), 2335–2347. doi:10.1007/s11269-011-9811-4.

Gu, A., Teng, F., and Wang, Y. (2014). China energy-water nexus: Assessing the water-saving synergy effects of energy-saving policies during the eleventh Five-year plan. *Energy Conversion and Management*, 85, 630–637. doi:10.1016/j.enconman.2014.04.054.

Gurgel, A., Reilly, J. M., and Paltsev, S. (2007). Potential land use implications of a global Biofuels industry. *Journal of Agricultural & Food Industrial Organization*, 5(2), 1–34. doi:10.2202/1542-0485.1202.

Havlík, P., Schneider, U. A., Schmid, E., Böttcher, H., Fritz, S., Skalský, R., Aoki, K., Cara, S. D., Kindermann, G., Kraxner, F., Leduc, S., McCallum, I., Mosnier, A., Sauer, T., and Obersteiner, M. (2011). Global land-use implications of first and second generation biofuel targets. *Energy Policy*, 39(10), 5690–5702. doi:10.1016/j.enpol.2010.03.030.

Hertel, T. W., Steinbuks, J., and Tyner, W. E. (2015). What is the social value of Second generation Biofuels? *Applied Economic Perspectives and Policy*,), ppv027. doi:10.1093/aepp/ppv027.

High Level Panel of Experts (HLPE). (2015). *Water for food security and nutrition*. A report by the High Level Panel of Experts on Food Security and Nutrition of the Committee on World Food Security, Rome.

Howells, M., Hermann, S., Welsch, M., Bazilian, M., Segerström, R., Alfstad, T., Gielen, D., Rogner, H., Fischer, G., van Velthuizen, H., Wiberg, D., Young, C., Roehrl, R. A., Mueller, A., Steduto, P., and Ramma, I. (2013). Integrated analysis of climate change, land-use, energy and water strategies. *Nature Climate Change*, 3(7), 621–626. doi:10.1038/nclimate1789.

IIASA/FAO. (2012). *Global Agro-Ecological zones (GAEZv3.0)*. Laxenburg, Austria and FAO, Rome, Italy: IIASA.

Leontief, W. (1951). *The structure of the American economy, 1919–1939: An empirical application of equilibrium analysis*. New York: Oxford University Press.

Li, X., Feng, K., Siu, Y. L., and Hubacek, K. (2012). Energy-water nexus of wind power in china: The balancing act between CO2 emissions and water consumption. *Energy Policy*, 45, 440–448. doi:10.1016/j.enpol.2012.02.054.

Liu, L., Hejazi, M., Patel, P., Kyle, P., Davies, E., Zhou, Y., Clarke, L., and Edmonds, J. (2015). Water demands for electricity generation in the U. S.: Modeling different scenarios for the water–energy nexus. *Technological Forecasting and Social Change*, 94, 318–334. doi:10.1016/j.techfore.2014.11.004.

Liu, J., Hertel, T. W., Taheripour, F., Zhu, T., and Ringler, C. (2014). International trade buffers the impact of future irrigation shortfalls. *Global Environmental Change*, 29, 22–31. doi:10.1016/j.gloenvcha.2014.07.010.

Lutz, C., Meyer, B., and Wolter, M. I. (2010). The global multisector/multicountry 3-E model GINFORS. A description of the model and a baseline forecast for global energy demand and CO_2 emissions. *International Journal of Global Environmental Issues*, 10(1/2), 25. doi:10.1504/ijgenvi.2010.030567.

Mabhaudhi, T., Chibarabada, T., and Modi, A. (2016). Water-food-nutrition-health nexus: Linking water to improving food, nutrition and health in Sub-Saharan Africa. *International Journal of Environmental Research and Public Health*, 13(1), 107. doi:10.3390/ijerph13010107.

Meyer, M., Distelkamp, M., Ahlert, G., and Meyer, B. (2013). Macroeconomic modelling of the global economy-energy-environment nexus – an overview of recent advancements of the dynamic simulation model GINFORS. GWS Discussion Paper, No. 2013/5

Miralles-Wilhelm, F. (2016). Development and application of integrative modeling tools in support of food-energy-water nexus planning – a research agenda. *Journal of Environmental Studies and Sciences*, 6(1), 3–10. doi:10.1007/s13412-016-0361-1.

Mirzabaev, A., Guta, D., Goedecke, J., Gaur, V., Börner, J., Virchow, D., Denich, M., and von Braun, J. (2015). Bioenergy, food security and poverty reduction: Trade-offs and synergies along the water – energy – food security nexus. *Water International*, 40(5–6), 772–790. doi:10.1080/02508060.2015.1048924.

Natanelov, V., McKenzie, A. M., and Van Huylenbroeck, G. (2013). Crude oil – corn – ethanol – nexus: A contextual approach. *Energy Policy*, 63, 504–513. doi:10.1016/j.enpol.2013.08.026.

OECD. (2008). *Biofuel support policies: An economic assessment*. Paris: OECD Publishing. doi:http://dx.doi.org/10.1787/9789264050112-en.

Panichelli, L., and Gnansounou, E. (2015). Impact of agricultural-based biofuel production on greenhouse gas emissions from land-use change: Key modelling choices. *Renewable and Sustainable Energy Reviews*, 42, 344–360. doi:10.1016/j.rser.2014.10.026.

Prieler, S., Fischer, G., and van Velthuizen, H. (2013). Land and the food-fuel competition: Insights from modeling. *Wiley Interdisciplinary Reviews: Energy and Environment*, 2(2), 199–217. doi:10.1002/wene.55.

Rezai, A., Taylor, L., and Mechler, R. (2013). Ecological macroeconomics: An application to climate change. *Ecological Economics*, 85, 69–76. doi:10.1016/j.ecolecon.2012.10.008.

Rosegrant, M. W., Zhu, T., Msangi, S., and Sulser, T. (2008). Global scenarios for Biofuels: Impacts and implications. *Review of Agricultural Economics*, 30(3), 495–505. doi:10.1111/j.1467-9353.2008.00424.x.

Rosegrant, M., Ringler, C., Msangi, S., Sulser, T. B., Zhu, T., Cline, S. A. (June 2008). International Model for Policy Analysis of Agricultural Commodities and Trade (IMPACT): Model Description. Washingdon, D.C.: International Food Policy Research Institute. Available at https://www.researchgate.net/profile/Siwa_Msangi/publication/267995611_International_Model_for_Policy_Analysis_of_Agricultural_Commodities_and_Trade_IMPACT_Model_Description/links/550d6daa0cf2ac2905a6a384/International-Model-for-Policy-Analysis-of-Agricultural-Commodities-and-Trade-IMPACT-Model-Description.pdf

Rosegrant, M. W., Ringler, C., Zhu, T., Tokgoz, S., and Bhandary, P. (2013). Water and food in the bioeconomy: Challenges and opportunities for development. *Agricultural Economics*, 44(s1), 139–150. doi:10.1111/agec.12058.

Sarkar, H. (2007). The nexus between achieving the millennium development goals and economic growth: The role of policy. *Asia-Pacific Development Journal*, 14(1), 1–24.

Smeets, E., Tabeau, A., van Berkum, S., Moorad, J., van Meijl, H., and Woltjer, G. (2014). The impact of the rebound effect of the use of first generation biofuels in the EU on greenhouse gas emissions: A critical review. *Renewable and Sustainable Energy Reviews*, 38, 393–403. doi:10.1016/j.rser.2014.05.035.

Steinbuks, J., and Hertel, T. W. (2014). Confronting the food – energy – environment trilemma: Global land use in the long run. *Environmental and Resource Economics*, 63(3), 545–570. doi:10.1007/s10640-014-9848-y.

Suttles, S. A., Tyner, W. E., Shively, G., Sands, R. D., and Sohngen, B. (2014). Economic effects of bioenergy policy in the United States and Europe: A general equilibrium approach focusing on forest biomass. *Renewable Energy*, 69, 428–436. doi:10.1016/j.renene.2014.03.067.

Taheripour, F., Hertel, T. W., and Liu, J. (2013). The role of irrigation in determining the global land use impacts of biofuels. *Energy, Sustainability and Society*, 3(1), 4. doi:10.1186/2192-0567-3-4.

United Nations (UN). (2015). *Conceptual frameworks for understanding the water, energy and food security nexus*. United Nations Economic and Social Commission for Western Asia (ESCWA) Working Paper, Beirut, Lebanon.

Wise, M., Dooley, J., Luckow, P., Calvin, K., and Kyle, P. (2014). Agriculture, land use, energy and carbon emission impacts of global biofuel mandates to mid-century. *Applied Energy*, 114, 763–773. doi:10.1016/j.apenergy.2013.08.042.

World Economic Forum (WEF). (2011). *Water security: The water – food – energy – climate nexus. The world economic water initiative* (D. Waughray, Ed.). Washington, DC: Island Press.

Zeshan, M., and Ahmed, V. (2013). Energy, environment and growth nexus in south Asia. *Environment, Development and Sustainability*, 15(6), 1465–1475. doi:10.1007/s10668-013-9459-8.

16

THE FIVE-NODE RESOURCE NEXUS DYNAMICS

An integrated modelling approach

Catalina Spataru

Research needs for integrated modelling approaches of the five-node resource nexus

Integration of modelling tools can support decision-making at all levels to identify local targets that are in line with global Sustainable Development Goals. There are various modelling methods in literature of speciality covering various parts of the resource nexus. Most of them do not capture the interlinkage among the different resources and the dynamics of them. Most are limited to two or three resource components of the nexus (e.g. water–energy, water–land, water–energy–land, water–energy–food, water–energy–land) – which certainly can be considered progress compared to the many modelling attempts focusing on environmental variables such as GHG emissions. The different methods to assess resources range from input–output analysis (IOA) to Computable General Equilibrium (CGE) methods as discussed in the previous chapters in this volume. Each of them has their own flexibility and limitations. Some of these provide a fully integrated approach to represent the nexus relations, while others provide an entry point approach considering the influence of one resource on the others or able to specify impacts on either the environment or the economy.

Some current coupling approaches are: coupling partial equilibrium or CGE models with spatially explicit models including knowledge or biophysical processes (e.g. KLUM@GTAP; Ronneberger et al., 2009 or GCAM and AGLU; Brenkert et al., 2003). These have already demonstrated a multidisciplinary approach, providing an understanding of the socio-biospheric systems. However, these integrated approaches ensure biophysical consistency but not dynamics and high resolution.

The socio-economic models lack integration of institutions and governance, capital and labour or climate variability in the nexus assessment. There are a number of approaches taken by CGE models such as ENV-Linkages (OECD) coupled to the IMAGE model (PBL Netherlands Environmental Assessment Agency) who proved to be suitable for analysing the economy-wide implications of the resource use interlinkages (OECD, 2013). However, they do lack technological details.

The nexus approach argues that even though the subsystems as water, and energy may require independent analysis, such a decision would merely overlook the multiplicity of feedback to be obtained and the interdependencies that may jointly interfere with the sustainability of the wider socio-ecological system (Bazilian et al., 2011).

In general, as it has been discussed in Chapter 1 in this handbook, the interface between bio-physical and socio-economic models could be better represented and developed, since not many frameworks have focused on the role of firms in acting as a mediator between the behavioural and environmental outcomes.

While there is no single best solution of nexus modelling, a first step is a systematic mapping of existing tools and models to assess the resource nexus. There are numerous efforts in the literature showing compilation of modelling tools for specific topic (e.g. hydrology – the Hydrological Modelling Inventory)[1] or domain (e.g. global change – Global Change Master Directory (GCMD)).[2] Mannschatz et al., 2016 introduced the Nexus Tools Platform (NTP), an interactive inter-model comparison of existing modelling tools, comprising 73 models (the alpha version September 2015) covering a wide range of potential applications and processes related to water, soil and waste. This platform allows us to check current trends in model development, such as in environmental sciences or to look for specific topics such as use of wastewater in agriculture by finding in which project has been addressed.

To explore policy formulation the best candidates are management tools, which can be applied to different systems. Such an example is the Water Evaluation and Planning (WEAP)[3] tool, which can evaluate and compare a different scenario for water systems ranging from municipal to complex basin systems. The Long-range Energy Alternatives Planning (LEAP)[4] tool, which provides an integrated analysis of different energy portfolios and their environmental effect, can be used to track energy consumption, production and resource extraction. The WEAP-LEAP framework has started to become integrated with a land-use framework through the Agro-Ecological Zones (AEZ) methodology developed by the International Institute of Applied Systems Analysis and the Food and Agriculture Organization of the UN (IIASA/FAO, 2012). A step forward towards more integrated use of these tools is the climate, land-use, energy and water (CLEW) modelling framework (Howells et al., 2013) demonstrated through a case study for Mauritius (Welsch et al., 2014) showing the importance of dynamics for such a systems approach.

However, in a longer perspective, significant improvements in spatial and temporal representation, and quantification of critical interlinkages in the five-node nexus is needed because resources vary across the globe and the diversity of the spatial representation will help understand the global impacts and other transgressions of planetary and regional boundaries.

Advantages/disadvantages of partial models

Partial models come with their own challenges when it needs to look into both the environment and economy within the sustainable development. Environmental concerns are similar worldwide as everyone must have access to clean water and air. In addition, sectors such as agriculture and industry depend on common environmental resources as soil, forests, land, rivers, mineral deposits, etc.

In order to determine institutional trade-offs and management practices, we need to understand the dynamics of coordination between different mix of factors and relationships between environmental resources. For example, analysing the water management, by considering the nexus approach, means taking into consideration the risks for water quality and agricultural soil salinization, soil processes and waste and socio-economic impacts. At local level some relationships are relatively clear (e.g. local industrial pollution); at global level, because things are so spatially and temporally spread out, the global environmental problems aren't clear. Combinations of cross-level and scale-level interactions at one point in time can sometimes undermine the resilience of a system (Cash et al., 2006). Trajectory analyses can help understand the temporal social-ecological system dynamics and capture the trade-offs and synergies between the different constituents of policies institutions knowledge in time. Since people's behaviours have an impact on resource use at household level, understanding how this shapes and is shaped by a multi-level policies–institutions–knowledge (PIK) nexus operating at larger scales is important to guide socio-ecological systems research (Stringer et al., 2014). Improving resilience in the short term can help improving it over the long term and vice versa (Cabell and Oelofse, 2012), which means developing a research framework by taking into account shortcomings alongside multiple spatial scales. Stringer et al. (2014) presented a novel, integrated framework to guide future social-ecological systems (SES) research, linking the multi-scale water–energy–food security (WEF) nexus with a multi-level policies–institutions–knowledge (PIK) nexus to explore resilience outcomes at different points in time. They use empirical analyses of local economic equity, social justice and environmental equity to identify the factors that inhibit and enable equitable resilience outcomes across a range of social-ecological systems.

Key issues in analysing the five-node nexus resource at different levels

This section illustrates the use of nexus at different spatial levels (regional, country and global) and discusses key issues. A number of initiatives have been taken to date in view of adaptation of nexus approaches. In the 1990s the objective was mitigation of climate risks by using a protective sector-based approach; this shifted in the 2000s to sectorial planning and mitigation of risks and uncertainties for decision support and in the 2010s towards a trans-sectorial and trans-boundary approaches for mitigation and social welfare improvement and climate change adaptation to development.

The nexus approach takes into account finding out about resources that can be defined in the context of water, food and energy within a geopolitical and geographical boundary. Thereafter, it continues to quantify the available resources and some of the ways through which the use of one resource leads to the problem of sustainability of the related socio-ecological systems. For instance, the water–energy nexus in Texas focuses on water that is used for energy and energy that is used to generate water. The energy production in Texas consumes water that can be used to provide for the needs of averagely 3 million people. One the other hand, the energy that is used for water treatment could produce enough power for approximately 100,000 households (Granit et al., 2012). Therefore, the nexus analysis argues that there should be greater attention on the linkages between the energy and water subsystems in order to design and affect the policies that may account for the fundamentally interconnected nature. Many researchers have tended to lay emphasis on the existence of trade-offs. There has been much deliberation between the relationship that exists between the water consumption and crop production in the city of Tuscany in Italy. A number of research studies have shown that this is required to produce biofuel crops that can supersede input used or irrigating maize crops. This creates a severe impact on the availability of water for food production and domestic needs. In other words, production of biofuel in Tuscany may not result in sustainability of water, food and energy nexus thereby contributing to the social-ecological balance. There is an urgent need to assess critical thresholds across resource systems.

Scientific literature includes a number of studies arguing that human activities drive multiple interactive affects that cascade throughout the Earth system. Rockstrom et al., (2009) quantified nine interacting planetary boundaries with possible threshold effects such as climate impacts, global freshwater use, atmospheric aerosol loading, chemical pollution, biodiversity and land-system change. These planetary boundaries estimate the level of uncertainty within defined limits.

Analytical frameworks are predominantly used to understand trade-offs and synergies, improve governance between water, energy and food systems (Hoff, 2011; Smajgl and Ward, 2013). Recent studies on ecological systems have moved towards dynamic non-equilibrium systems with non-linear responses and multiple stable states.

At global level, a number of studies have evaluated the current nexus in countries such as Australia, Spain, The Netherlands, Brazil, India, Sri Lanka, China and South Africa. These studies have suggested different frameworks that are relevant in examining the dynamics of the water–energy–food nexus as well as the life cycle analysis (LCA) with material flows analysis and value chains. Virtually all of them have certain features in common. For instance, they share a process-based approach to resource use, and tend to prefer modelling and system analysis over other empirical strategies. These frameworks have potential applications in the study of the water–food–energy interlinkages. However, they are limited to narrow and more technical range of research questions, hence, disregarding the very role of institutionally mediated human agency. The Institutional Analysis and Development (IAD) framework[5] originally conceived in 1980s by Elinor Ostrom (the Nobel Prize winner in economics in 2009) aim to study commons pool resources and their governance by communities without state intervention. This approach has been widely instrumental in the study of socio-ecological systems and has made it possible for policies to be comprehended as amalgams of interacting institutions that may influence the production and use of energy, water and food resources. If critical thresholds are being transgressed, the combined value chain of the IAD approach may be put to the test in several cases. For example, in the irrigation sector cross-resources and institutional dynamics are interconnected. Water, energy and food share a common interface in that their use in the irrigation sector is crucial.

Connection across the water, food and energy nexus can be comprehended using the input–output analysis (IOA) method in context of some countries that have been engaged in energy production. Examples of India and Spain can be given here that make use of the energy as an input for water distribution and extraction. Spain poses different reservoirs that can store and convey water from various irrigation areas. The Spanish government practices transfer management techniques to ensure sustainability of agricultural practices in effective use of semi-arid lands. A trade-off takes place between the energy utilization and water consumption to ensure conservation of the resources (Hussey and Pittock, 2012). Since the government has invested heavily on the programs that ensure subsidies in irrigation technology, there has been increased production of food and energy in the past few decades. The number of districts that still practice sprinkler irrigation technique have increased from 22 to 44% implying that the government remains committed to ensuring that sufficient water and power production for various activities.

The situation in Germany and Kenya are similar to Spain when it comes to water utilization for food production. However, Germany has gone steps further by adopting four approaches or actor groups to assist with the distribution of water in the country. These include (1) food and bio energy production farmers, (2) urban population and wastewater producers, (3) Braunschweig water association, and (4) municipality and city water authorities. In order to understand the resource use and the dynamics of supply within the water supply context, it is imperative to mention the (1)-(4) actors in the supply and use of water. In contrasts, India has focused on two main players in the water–energy–food nexus: (1) the local farmers and utilities in the distribution of energy, and (2) the linkages that work to regulate resources and maintain the supply dynamics in various situations (Waughray, 2011).

General assessment of existing interlinkages of using other resources over time and trends analysis thus is necessary to understand issues of economics, scale, the dynamic price mechanisms, changes over time and unintended consequences. For example assessing consequences related to the global risk of extreme volatility in energy and agriculture prices and food shortage crises among various risks due to climate change adaptation, governance failure, urbanization, political fluctuations.

Based on the above discussions the following improvements are needed: enable analysis of resources and their interlinkages at different scales over time; assess critical thresholds and risks of what happens if those are being transgressed; incorporate dynamics to capture understanding of volatility driven by either environmental, market-based and regulatory factors; short time which steps will allow the value chains mirrored by different innovation and production cycles.

To address the issues identified and capture the critical interlinkages across different five-node nexus will require connecting databases and developing coefficients and metrics. The purposes of such metrics, which can help in the development of modelling tools, are briefly discussed in the next section.

Purpose of metrics for five-node nexus model approach

To manage the nexus, evidence needs to be collected in the form of reliable and timely data. This could include demand and trade of the system, as well as the interactions between systems. In general, the metrics offer the most effective way of quantifying different objectives. This is essential for effective management of natural resources security. At the moment, a number of metrics have been introduced for this purpose. Examples of such studies is Spang et al., 2014 who listed the water consumption of energy production (WCEP) intensity indicators, each of them providing a view of a particular dimension (e.g. the intensity of water consumption for energy). The per capita WCEP is used to reflect the total water consumption assigned to each individual within a given nation. Normalization means assignment of units of consumption in m^3/person. This type of metric therefore allows us to compare the consumption of different regions or countries. The economic WCEP intensity looks at another view of water consumption versus energy production in that it measures the contribution of water consumption or its impact to the economy. Economic WCEP therefore makes use of a country's economic production levels rather than its population as a basis for normalization. Technological WCEP intensity is another metrics that compares water consumption with energy production within a country based on the nexus resource frame work. The metric measures the contribution of water consumption in technological growth and its application.

The fourth metric is physical WCEP intensity that relates to distribution of water consumption for a country as compared to its energy production based on the available renewable energy sources available. Information obtained through the metric may be critical in knowing about the renewable water source utilization levels globally that reflects consumption levels in different countries or regions. Based on these information models may be developed that point to areas whose renewable water sources have been depleted as well as those that still have potential of energy production mainly due to renewable water source underutilization. The model can be then utilized in making decisions on how to improve sustainability in WCEP for multiple countries.

A proposed dynamic integrated theoretical assessment model framework for five-node resource nexus (IDA5)

Motivation for IDA5 framework

This section discusses motivation behind a framework for an integrated approach to understand and assess the five-node nexus and how interlinkages between natural resources will achieve the SDGs:

- food security, agriculture (SDG2)
- water and security (SDG6)
- energy and energy security (SDG7)
- climate (SDG13)
- land/terrestrial ecosystems (SDG15)

The aims for assessing the five-node nexus are:

- to quantify the interlinkages and feedbacks between natural resources in space and time;
- to analyse trade-offs and practices for five-node resource nexus;
- to assess transitional and future pathways to achieve the SDGs (2, 6, 7, 13 and 15) and implement the 3R's principles (reduce, re-use and recycle) of the circular economy.

The motivation for the development of this framework follows a review of existing integrated natural resource assessment and modelling literature and the key issues identified and discussed in previous sections.

The literature of speciality provides examples of integration of separated tools (such as WEAP-LEAP-AEZ) with additional input from a climate model (GCM) to model together water, energy and land. Welsch et al. (2014) demonstrated through this exercise the additional value of the combined framework by undertaking the same case study in Mauritius. This approach requires knowledge and understanding of each of these models and of the algorithms and assumptions made which sometimes can be a time-consuming exercise. In addition, validation, benchmarking and model intercomparisons are needed for shared learning across multiple models. All these may contribute to larger modelling efforts.

At UCL, our approach was to start with an energy model and build on top of that additional modules for other resource nexus. A first dynamic version of the model called IDA3 (The Integrated Dynamic Assessment 3-nodes nexus model) was completed in 2014 (Spataru, 2014) with the possibility of adapting it for any desired area (e.g. a country, island, region). The model has been developed initially to reveal the mutual dependencies of water, energy and land (WEL) and to show with cross-sectorial integration these can be managed effectively in space and time.

An Integrated Dynamic Assessment 3-nodes (water, energy, land) nexus model (IDA3)

The IDA3 model has been developed as a vehicle for teaching multi-country, multi-regions nexus modelling, capturing the dynamics and trade-offs between resources. Furthermore, it allows for the subdivision of the study area into sub-areas, allowing the representation of regions within a country. The fully encompassed model emphasizes the importance of dynamics between nexus components, providing flexibility both in terms of geographical boundaries and scale. The model has been adapted since then to a number of countries: Brazil, France, Jordan, Egypt and more are underway. These case studies have been chosen and explored so far because each has an interesting history regarding the nexus and implications of water scarce sources on energy production and land use. Figure 16.1 provides a representation of the water–energy–land nexus in the IDA3 model.

The model uses a pyramid-thinking structure, in that a whole country can be subdivided into sub-areas with the possibility of the results to be examined at a micro, meso or macro level. The latter consists of all results of its subordinated units. Entities are modelled as knots or real extent that is only recorded as figures. The model can be easily adapted spatially and temporally for any country in the world. Table 16.1 shows how it was adapted for Brazil, France and Egypt.

The IDA3 model takes into account trades with neighbouring states, which gives the possibility of exploring regional and global trade-offs for the three-node nexus, subject to available input data. A model description can be found on the UCL website.[6]

The model can run in double mode: at country level and regional level. At regional level, the model regions have been clustered as follow: little or no water scarce regions, physical water scarce regions, approaching physical water scarce regions and economic water scarce regions.

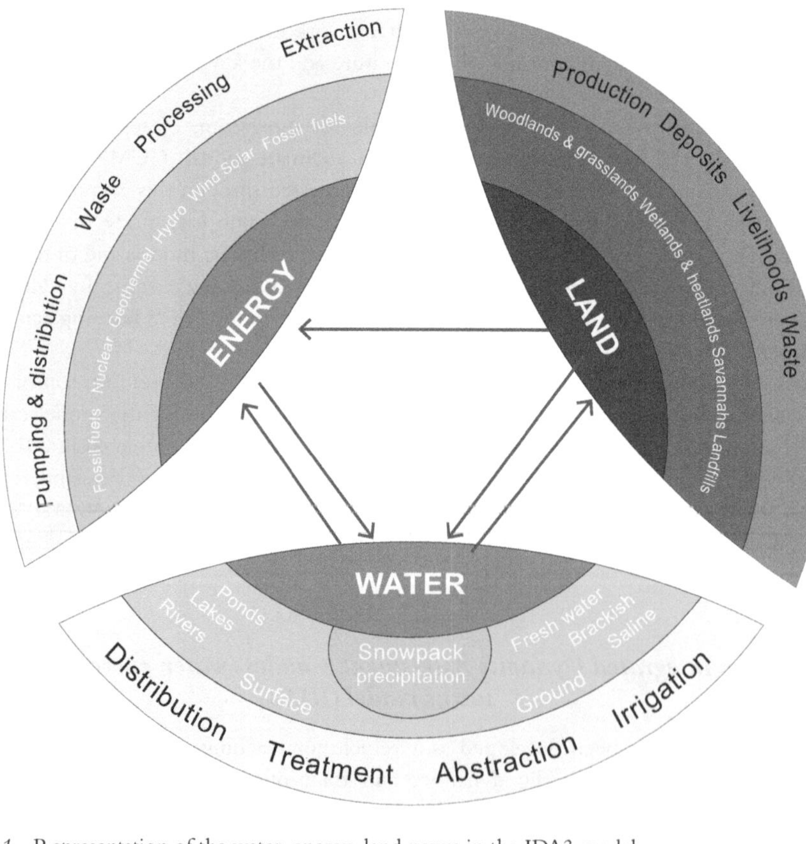

Figure 16.1 Representation of the water–energy–land nexus in the IDA3 model

Table 16.1 Adaptation of IDA3 model main script per country studied

IDA3_ Model_MainScript	*Brazil*	*Egypt*	*France*
	Modifications brought	*Modifications brought*	*Modifications brought*
Regions	**[N;NE;SE;S;CW]** regions	**[C;ED;MD;WD;UE]** regions	**[NPdCP;N;IdF;ACAL;B;PdL;CVdL;BFC;ALPC;ARA;LRMP;PACA;C]** regions
Number of years	nyears = 38;	nyears = 41;	nyears = 36;
% Electricity base demand start value, excluding autoproducers	% Electricity base demand start value **2013**, excluding autoproducers	% Electricity base demand start value 2010, accounting for losses PC_base_2010_GWh = 131236	% Electricity base demand start value **2014**, excluding autoproducers
% Electricity demand growth rates Gr_PC_base	% Electricity demand growth rates **2013–2017, 2018–2023, 2050** Gr_PC_base = [0.042, 0.039, 0.01]	% Electricity demand growth rates 2010–2014 (actual values), 2015–2022, 2023–2050 Gr_PC_base = [0.045, 0.06, 0.0635]	% Electricity demand growth rates **2014–2020, 2021–2030, 2031–2050** Gr_PC_base = [0.02, 0.012, 0.005]
% Electricity demand year from start year, which changes growth rate Gr_PC_y	% Electricity demand year from start year, which changes growth rate Gr_PC_y = [**5, 11**];	% Electricity demand year from start year, which changes growth rate Gr_PC_y = [5, 13]	% Electricity demand year from start year, which changes growth rate Gr_PC_y = [**7, 17**]
% Growth rate **wind** capacity factors each year Gr_CF_wind	% Growth rate **wind** capacity factors each year Gr_CF_wind = 0.008	% Growth rate **wind** capacity factors each year Gr_CF_wind = 0.005	% Growth rate **wind and solar** capacity factors each year Gr_CF_wind = 0.0106; % mean value (2011/2012/2013/2014) Gr_CF_solar = 0.0143; % mean value (2012/2013/2014)
% Water withdrawal start value [domestic, agriculture, industry]	% Water withdrawal start value [domestic, agriculture, industry] **2013**	% Water withdrawal start value [domestic, agriculture, industry] **2010**	% Water withdrawal start value [domestic, agriculture, industry] **2014**
% Water withdrawal year from start year, which changes growth rate Gr_WW_y	% Water withdrawal year from start year, which changes growth rate Gr_WW_y = [**4, 7**]	% Water withdrawal year from start year, which changes growth rate Gr_WW_y = [**4, 7**]	Gr_WW_y = [**7, 17**]

(Continued)

Table 16.1 (Continued)

IDA3_ Model_MainScript	*Brazil*	*Egypt*	*France*
	Modifications brought	*Modifications brought*	*Modifications brought*
% Water consumption start value [domestic, agriculture, industry]	% Water consumption start value [domestic, agriculture, industry] **2013**	% Water consumption start value [domestic, agriculture, industry] 2010 WC_2010_hm3 = [1800 40400 1400]	% Water consumption start value [domestic, agriculture, industry] **2014 (0.219*WW_dom_hm3, 0.675*WW_agri_hm3, 0.198*WW_ind_hm3)**
% Volume of treated municipal wastewater	% Volume of treated municipal wastewater in **2012**	% Volume of treated municipal wastewater 2010 MWWT_2010_hm3 = 3400;	% Volume of treated municipal wastewater in **2007**
% Land-use start values for each region	% Land-use start values **2006 [N;NE;SE;S;CW]**	% Volume of desalinated water 2010 Desal_2010_hm3 = 200	% Land-use start values **2010 [NPdCP;N;IdF;ACAL;B;PdL;CVdL;BFC;ALPC;ARA;LRMP;PACA;C]**
% Land-use growth rates **for each region**	% Land-use growth rates **2006–2050 [N;NE;SE;S;CW]** Gr_LC = **[0.0133; 0.0026; -0.0055; 0.0043; -0.0073]**	% Land-use start values 2015 [C;ED;MD;WD;UE] LC_2006_km2 = [190.4; 35021.30; 13761.18; 12683.06; 13150.35]	% Land-use growth rates **2010–2050** [NPdCP;N;IdF;ACAL;B;PdL;CVdL;BFC;ALPC;ARA;LRMP;PACA;C] Gr_LC = **repmat(0.0033,13,1)**; % Agreste, 2014

In addition to this, geography of international trade is also important. For example when the US experience a supply drop (drought/flooding) or a demand increase (biofuels), production will respond to other parts in the rest of the world. Increased water reduction for irrigation particularly in South Asia and China lead to reductions in irrigated area in key producing regions with shortfalls in supply, which lead to more land conversions and GHGs. Food trade is one of the key adaptation to water shortages. Regions facing the most severe water shortages are most likely to increase net food imports.

Example of scenarios analysis: modelling the impact of climate change on water–energy–land nexus to inform planning and to mitigate environmental impacts

We conducted exploratory analyses on water stress regions. The case study included in this section is Brazil. It has been chosen because of its uneven distribution of resources and unequal distribution of population with unequal access to water and sanitation services. Brazil has been divided spatially in five regions: Northeast, North, Southeast, South and Central West. The following generation scenarios were considered to analyse the impacts on water, energy and land use:

- S1: increase hydro generation at maximum capacity;
- S2: keep hydro at the same capacity, but increase wind and photovoltaic generation;
- S3: keep hydro at the same capacity, but increase thermoelectric generation capacity, biomass and nuclear capacity;
- S4: a generation mix.

Five climate scenarios were simulated to assess possible future developments (Senger and Spataru (2015)):

- Assume business as usual;
- Assume constant capacity factors for hydropower;
- Assume a fast decline of water availability in the Southeast and Central;
- CS_A2 and CS_B2 derived from (Lucena et al., 2010) representing the effects of climate scenarios IPCC SRES A2 and B2, respectively.

A number of datasets were utilized. Water and land-use factors for power generation vary in the literature, so we performed calculations with different datasets. Reduced water availability for hydropower was based on results from Lucena et al., 2010. For water use for power generation the following were considered: lower and upper boundaries of factors (McMahon and Price, 2011); the water footprint of energy from biomass; the operational water consumption and withdrawal factors for electricity-generating technologies (Macknick et al., 2012). For land use for power generation the following sources were used: Fthemakis and Kim, 2009 (land transformations); Gagnon et al., 2002 (land requirements); McDonald et al., 2009 (land-use intensity).

Data for water usage for power plants in Brazil weren't available therefore we considered the water withdrawal and consumption factors from relevant literature. Figure 16.2 shows examples of output (the electricity supply mix) for the four generations scenarios considered under the CS_A2 climate scenario as analysed in Senger and Spataru (2015).

Our case study demonstrates the highest water-use results from scenario S3 at 126 billion cubic meters (bm^3) withdrawn and 193 bm^3 consumed by 2050. The scenario with the second highest water use is S4, whereas S3 peaks at 201 bm^3 withdrawal and 323 bm^3 consumption

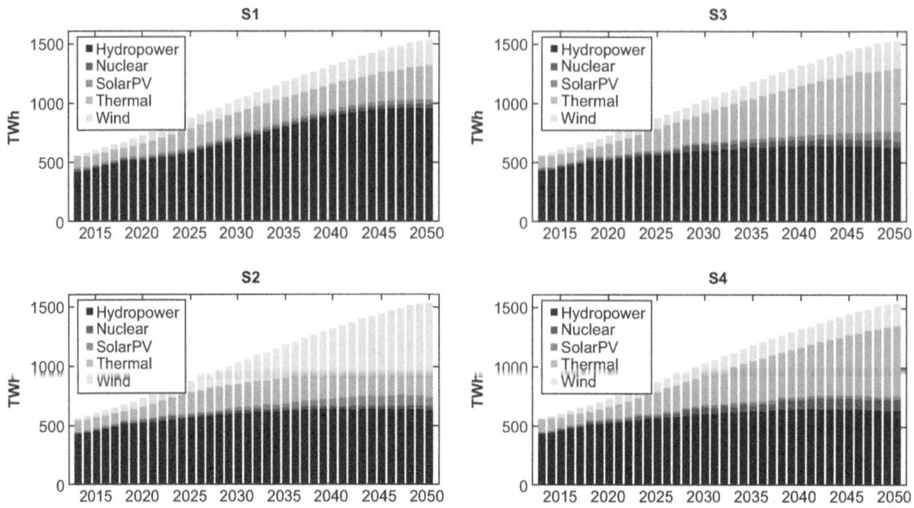

Figure 16.2 Hypothetical generation scenarios for electricity supply mix under climate scenario CS_A2

by 2050. In the case of land use for electricity generation, scenario S3 has the highest land requirements as 315,260 km^2 are projected for 2050 compared to 44,800 km^2 in 2013 – sevenfold mainly due to high land demands for biomass cultivation. Scenario S4 has the second and scenario S1 the third highest land requirements. The most land efficient generation mix is scenario S2 with 119,030 km^2 land requirement. Thermoelectric generation contributes the most to land use in all generation scenarios. The conclusion can be drawn that investments in wind and photovoltaic power can limit additional water and land demands for future power supply in Brazil. The spatial analysis on river basins or watershed level can determine different outcomes.

Other case studies analysed were: Egypt, France, Jordan, etc. In the case of Egypt, the area has been divided into five sub-regions: Greater Cairo, East Delta, Middle Delta, West Delta and Upper Egypt. It was determined the water availability per capita can decrease below the threshold of absolute scarcity limit per year in 2030, while water use per capita, is expected to double by 2050. Thus the gap between supply and demand needs to be covered from reusing by applying circular economy principles or/and considering the trade with neighbouring countries.

Through such case studies we can demonstrate the practical steps involved in modelling the transformation of resource systems, simulating plausible scenarios and indicating changes needed which can be translated into policy recommendations.

As highlighted in Table 16.1 the model can be adapted spatially and temporally for any country/region in the world. Furthermore because the model has the capability to divide into subareas; a country may be modelled as several regions and analysis can be done at different levels.

Defining the nexus components for the five-node resource nexus model

The water systems consume energy, mainly electricity for pumping, drainage and water management, desalination, wastewater treatment, water transport and distribution. On the other hand, water is used for energy generation, cooling, resource extraction and treatment, transportation and bioenergy production, accounting for 15% of the annual global freshwater withdrawal, 11% of which is consumed (IRENA, 2015). Consequently, any disruption to the energy system would directly affect water security, whereas water availability for energy uses is significant for achieving energy security.

The purpose of this framework is to investigate the five-node nexus by integrating and synthesizing across model, data, analytic components and visualization to inform policy, planning and design for improving resource use efficiency and enabling long-term sustainability. We can feed in the model various input data, time-series climate data (of temperature, solar irradiation, precipitation, wind speed), parameters describing type of demands and allowed sources. We can then compute scenarios for the required region.

Resource input coefficients can be derived for main minerals, biomass, land, water and energy with countries or regions being represented according to the country typology. Such approach will allow to assess the future views on climate and sustainability policy, including trade and supply chains. In addition modelling critical interlinkages across the SDGs will require developing coefficients for critical interlinkages. Such an example is the ones for minerals for sustainable and affordable energy, and the nexus thereof with water and energy use.

Two modules (materials and food) are currently developed and plan to be added to the three-node nexus model (IDA3 model): materials and food. From my knowledge, up-to-date, current models do not include physical quantities such as minerals. Including minerals and materials is important because base metals, critical materials and minerals are relevant to water provision, energy production and distribution, and mineral fertilizers for food production; construction materials are relevant for urbanization processes. Furthermore, base metals are environmentally intensive as fossil fuels and agricultural commodities. We will consider in this approach physical amounts from the IEA database by assigning to the extraction sector to which they belong. Unfortunately the flow of physical quantities of materials for resources is not always clear for some sectors and core activities. In this respect including data on metals and mining will be important if the private sector will make them available. We will also plan to specify in detail the production structure for the mineral extraction sectors. One novel element is to include inputs; for example, water in the production of the mineral products as a factor of production. Lecca et al. (2011) provides an explicit production function where energy, materials, capital and labour are recognized separately. Adopting their approach will help us to make the representation at sector and material basis. However, this will require practical analysis.

A first step done was to identify key materials relevant to current consumption patterns (such as cement, aluminium, iron and steel, chemical and petrochemicals, fertilizer, pulp and paper) and demand for services (such as rare earths and precious metals). Each of these came with its own difficulties. For example the iron and steel production sectors has a number of technologies. The ones considered were blast furnace and basic oxygen furnace. For pulp and paper category, it is difficult to separate wood feedstock from fuelwood use.

After defining the materials and food modules, the modules were combined dynamically to illustrate a number of scenarios to address sustainable goals.

A 'meta-database' for the five-node resource nexus model

The choice of initial databases is important for the ability of the model to undertake nexus-based analysis. Relevant data will be extracted from a number of databases in a meta-database. The construction of this is important for accurate output analysis from the model. However, the extent to which such information are available vary significantly across regions worldwide or some data are not available publicly or some data have inconsistencies. Materials industries sector was the most difficult one especially that it contains a wide variety of products and production techniques, but not much information on demand and technical data, which then makes aggregation of data difficult. Databases on materials-energy intensity exist but are often expensive.

There are a number of comprehensive databases. Such as is GTAP database[7] covering 140 regions and 57 sectors, comprising macro-economic data, bilateral merchandise trade data, time-series bilateral trade data, energy data, decomposition of tariff, CO_2 emission and five labour skill categories. Another example is the EXIOBASE dataset,[8] a global multi-regional environmentally extended supply and use database for 43 countries, estimating emissions and resource extractions by each of the 163 types of industry. Extensions can be aggregated to compile indicators such as total material requirement. Comparing to GTAP, which aggregates all activities into 11 different extraction activities, the EXIOBASE dataset does separate out all the mining and quarrying activities. This provides us to undertake detailed analysis on material use.

For the food sector, there are several comprehensive databases. For example the Global Food Prices Database (WFP) includes data on food prices for 76 countries and around 1,500 markets being updated monthly.[9] The World Data Atlas contains different information and datasets for each country profile. The Socioeconomic Data and Applications Center (SEDAC)[10] includes various datasets and maps for population, infrastructure, land use, etc.

An Integrated Dynamic Assessment methodology for the five-node nexus (IDA5) to achieve SDGs by using the R's principles of the circular economy

IDA3 model was developed for its advantages in simplicity and flexibility as shown in Table 16.1. Three modules were set up for energy, water and land sectors. The energy module was based on various energy systems models at UCL (www.ucl.ac.uk/energy-models), while the water and land modules were newly developed. The second version of the model called IDA5 (which is currently under development) (Spataru, 2016) is an extension of the IDA3 model including two new modules, food and materials. The purpose of this model framework is to connect water–energy–land–food–materials together in the so-called resource nexus concept and determine trade-offs between different resources and analyse different paths of achieving the SDGs by using the R's principles of the circular economy. The main interdependencies identified in the five-node nexus components are:

- Changes in water availability, river flow, evaporation in hydropower dams, biofuels crop irrigation, fossil-fuel extraction;
- Increased water demand due to changes in the climatic conditions;
- Water consumption and pollution in refining and production processes;
- Energy for water pumping, pesticides, machinery, transport, food production;
- Land use for biofuels and renewable energy technologies (solar, hydro, wind);
- Crop and other uses of land competing with extraction of resources;
- Land, energy and water needed for food production;
- Materials-embedded energy and high energy intensity of new materials;
- Materials needed for water sector (extraction, desalination, purification, pumping, etc.).

Sustainable development is a pattern of resource use that aims to meet human needs, while preserving the environment (Miller and Spoolman, 2011). Using and exploiting our resources carefully will help future generations. Combining the resource nexus with SDGs and circular economy principles (the practice of the three R's – reduce, re-use and recycle) will help extend

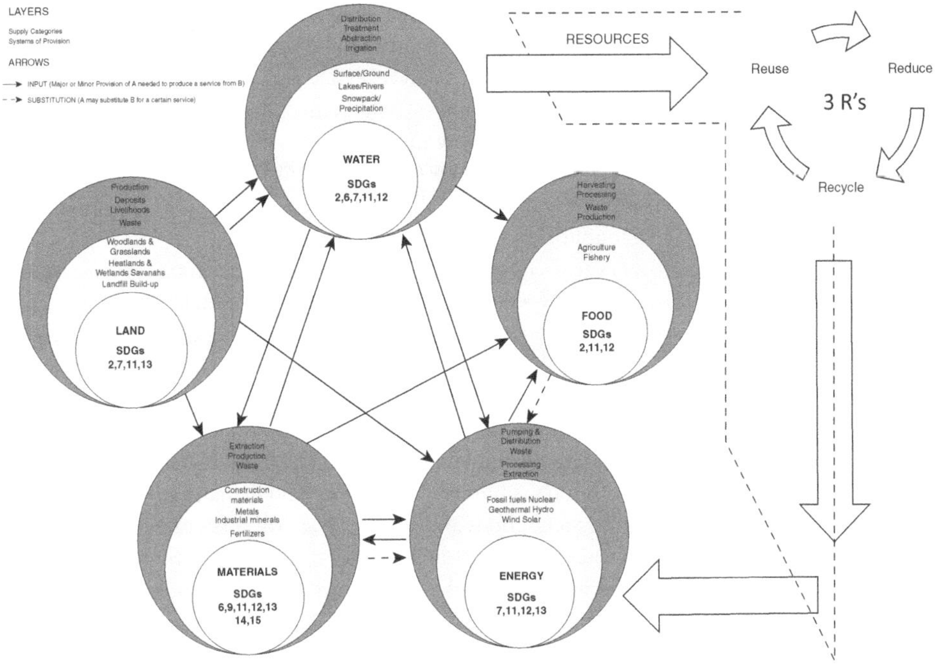

Figure 16.3 Representation of the five-node nexus resource, SDGs and R's interconnections for IDA5 model

the availability of resources and make our health better by reducing GHG emissions. Figure 16.3 shows these principles together.

The future modelling strand in combining existing systems models (energy and water as being the most eminent one), with food models (such as the IMPACT model)[11] and materials to assess SDGs goals. The transition towards the circular economy is the primary objective of research agenda for resource efficiency. Managing waste to increase recycling rates to reduce the pressure on use of raw materials, help re-use materials and reduce energy consumption will help boost competitiveness, ensure security of essential resources and reduce negative impact on environment. By adopting circular economy principles and building a large recycling infrastructure, with a market for repairs, re-use and recycling will boost economic, environmental and health benefits. Research on technology together with economic analyses and consumer behaviour studies will help the transition towards a circular economy to meet the SDGs.

Conclusion

This chapter discusses some general principles of integrated modelling across different natural resources and the importance of developing more integrated approaches across scales to address several key sustainability development challenges. Ideally this integration should be done across spatial and temporal scales, linking several global boundary conditions, and allowing to account for different regions and human behavioural decision-making. In order to do this, metrics could help in defined and structuring the emerging approaches of integrated systems to model the nexus. General methods used in this field are explained and related to applied examples and their specific challenges.

This chapter explored the dynamics of the resource nexus. In order to find the relationship between resource factors it is important to understand the practicality of the cross-sector integration particularly in areas where homogeneous and relatively reduced players are involved.

Examining the interlinkages between resources will help us better understand the path dependencies in different sectors. It will also allow us to evaluate different scenarios that call for compensating asymmetries upstream and downstream of the nexus model. Adopting a resource nexus may help address environment and policymaking issues, enabling a more future-oriented and sustainable trade beyond geopolitics and resource nationalism.

Previous integrated approaches (e.g. ecosystem approaches, CBD); integrated water resources management (GWP); multi-functional systems (IAASTD) e.g. crop-livestock systems; ecological sanitation; etc. have been sectoral and lessons have not been learned – upscaling and transfer is missing. The nexus approach may help assess risk of transgressing critical thresholds, across scales and from local sustainability boundaries to planetary boundaries. Nexus approach can be implemented worldwide to assess numerous issues: MENA region, for assessing soil and water conservation for better green water productivity and climate resilience; renewable energy for seawater desalination; etc. Assessing trade-offs such as how the increase in land demand affect other ecosystems (e.g. grasslands) and their ecosystem goods and services (e.g. livestock, carbon, biodiversity).

That being said, nexus approaches need collaborative efforts and innovation to ensure robust integration such as integration between different water management institutions so as to cope with the changes, such as in the energy and food market prices. The only way to solve the present scarcity of resources is to formulate policies that ensure sustainable and responsible use of limited natural resources. However, attaining this aim will continue to be a challenge mainly due to increase energy resource demand due to rising population and increasing urbanization.

This necessitates analysis of different factors and developing integrated policies on a global scale. The policymakers need to be conscious of different human actions in relation to unsustainable use of human resources. As a result, a sustainability modelling approach that is taken by the policymakers should integrate different parts of the available resources. Accounting of interactions and trade-offs will help understand shocks, such as extreme weather events, which can have cascading impacts across systems.

There are no golden rules for undertaking nexus assessments, but key aspects involve understanding the interactions between systems and people, modelling of systems and stakeholder participation. Understanding the interactions and trade-offs requires an understanding of the social, political and environmental context. So far decisions made relating to water, energy, food and material resources are often politically contested. They have tended to prioritize short-term benefits over the long-term security of systems. The five-node resource nexus model concept can be used to guide the modelling approach. The various spatial and temporal scales considered in the model can help decision-makers.

Notes

1 http://hydrologicmodels.tamu.edu/models.htm

2 http://gcmd.gsfc.nasa.gov/index.html

3 WEAP, the Water Evaluation and Planning system, is developed and maintained by the Stockholm Environment Institute: www.sei-us.org/software

4 LEAP, the Long-range Energy Alternatives Planning system, is developed and maintained by the Stockholm Environment Institute: www.sei-us.org/software

5 Elinor Ostrom. (2011). Background on the Institutional Analysis and Development Framework, *The Policy Studies Journal*, 39(1). Retrieved from http://fpf.forestry.oregonstate.edu/system/files/Ostrom-Background-Institutional%20Analysis-2011.pdf

6 www.ucl.ac.uk/energy-models

7 www.gtap.agecon.purdue.edu/databases/v9/
8 www.exiobase.eu
9 https://data.humdata.org/dataset/wfp-food-prices
10 http://sedac.ciesin.columbia.edu/data/sets/browse
11 https://www.ifpri.org/program/impact-model

References

Bazilian, M., Rogner, H., R., Howells, M., Hermann, S., Arent, D., Gielen, D., Steduto, P., Mueller, A., Komor, P., Tol, R. S. J., and Yumkella, K. K. (2011). Considering the energy, water and food nexus: Towards an integrated modelling approach. *Energy Policy*, 39(12), 7896–7906.

Brenkert, A. L., Smith, A. J., Kim, S. H., and Pitcher, H. M. (2003). *Model documentation for the MiniCAM, PNNL-14337*, Pacific Northwest National Laboratory, Richland, Washington, DC. Retrieved from www.globalchange.umd.edu/models/MiniCAM.pdf

Cabell, J. F., and Oelofse, M. (2012). An indicator framework for assessing agroecosystem resilience. *Ecology & Society*, 17(1), 18.

Cash, D. W., Adger, W. N., Berkes, F., Garden, P., Lebel, L., Olsson, P., Pritchard, L., and Young, O. (2006). Scale and cross-scale dynamics: Governance and information in a multilevel world. Ecology and Society, 11(2), 8. Retrieved from http://www.ecologyandsociety.org/vol11/iss2/art8/

Fthenakis, V., and Kim, H. C. (2009). Land use and electricity generation: A life-cycle analysis. *Renewable and Sustainable Energy Reviews*, 13, 1465–1474.

Gagnon, L., Bélanger, C., and Uchiyama, Y. (2002): Life-cycle assessment of electricity genera- tion options: The status of research in year 2001. *Energy Policy*, 30, 1267–1278.

Gerbens-Leenes, P. W., Hoekstra, A. Y., and van der Meer, T. (2009). The water footprint of energy from biomass: A quantitative assessment and consequences of an increasing share of bio-energy in energy supply. *Ecological Economics*, 68, 1052–1060.

Granit, J., Jägerskog, A., Lindström, A., Björklund, G., Bullock, A., Löfgren, R., de Gooijer, G., and Pettigrew, S. (2012). Regional options for addressing the water, energy and food nexus in Central Asia and the Aral Sea Basin. *International Journal of Water Resources Development*, 28(3), 419–432.

Hoff, H. (2011). *Understanding the nexus, background paper for the Bonn 2011 conference: The water, energy and food security nexus*. Stockholm: Stockholm Environment Institute

Howells, B., and Rogner, R. (2014). Assessing integrated systems. *Nature Climate Change*, 4(7), 246–247.

Howells, M., Hermann, S., Welsch, M., Bazilian, M., Segerström, R., Alfstad, T., Gielen. D., Rogner, H., Fischer, G., van Velthuizen, H., Wilberg, D., Young, C., Röhrl, A., Mueller, A., Steduto, P., and Ramma, I. (2013). Integrated analysis of climate change, land-use, energy and water strategies. *Nature Climate Change*, 3, 621–626. doi:10.1038/nclimate1789

Hussey, K., and Pittock, J. (2012). The energy–water nexus: Managing the links between energy and water for a sustainable future. *Ecology and Society*, 17(1), 31. http://dx.doi.org/10.5751/ES-04641-170131

IIASA/FAO. (2012). *Global Agro-Ecological Zones (GAEZ ver 3.0)*. Laxenburg, Austria and FAO, Rome, Italy: IIASA. Retrieved from www.fao.org/fileadmin/user_upload/gaez/docs/GAEZ_Model_Documentation.pdf

IRENA Report, Authors Ferroukhi, R., Nagpal, D., Lopez- Peña A., Hodges, T. (IRENA), Mohtar, R. H., Daher, B., Mohtar, S. (Texas A&M University); Keulertz, M. (Purdue University) (2015). *Renewable energy in the water, energy and food nexus*. Retrieved fromwww.irena.org/DocumentDownloads/Publications/IRENA_Water_Energy_Food_Nexus_2015.pdf

Lecca, G., Petitdidier, M., Hluchy, L., Ivanovic, M., Kussul, N., Ray, Nicolas, Thieron, V. (2011). Grid computing technology for hydrological applications. *Journal of Hydrology*, 403(1–2), 186–199.

Lucena, A. F.P, Schaeffer, R., and Szklo, A. S. (2010). Least-cost adaptation options for global climate change impacts on the Brazilian electric power system. *Global Environmental Change*, 20, 342–350. doi:10.1016/j.gloenvcha.2010.01.004.

Macknick, J., Newmark, R., Heath, G., and Hallett, K. C. (2012). Operational water consumption and withdrawal factors for electricity generating technologies: A review of existing literature. *Environmental Research Letters*, 7(4). Retrieved from http://iopscience.iop.org/article/10.1088/1748-9326/7/4/045802/meta

Mannschatz, T., Wolf, T., and Hülsmann, S. (2016). Nexus tools platform. *Environmental Modelling & Software Journal*, 76(C), 137–153. Elsevier Science. doi:10.1016/j.envsoft.2015.10.031

McDonald, R. I., Fargione, J., Kiesecker, J., Miller, W. M., and Powell, J. (2009). Energy sprawl or energy efficiency climate policy impacts on natural habitat for the United States of America. *PLoS One*, 4(8). http://dx.doi.org/10.1371/journal.pone.0006802

McMahon, J. E., and Price, S. K. (2011). Water and energy interactions. *Annual Review of Environment and Resources*, 36, 163–191.

Miller, G. T., and Spoolman, S. (2011). *Living in the environment: Principles, connections, and solutions* (17th ed.). Belmont, CA: Brooks-Cole. ISBN:0-538-73534-1.

OECD. (2013). *The OECD ENV- linkages modeling framework, Projecting economy-environment interactions in the coming decades*. Retrieved from www.oecd.org/env/indicators-modelling-outlooks/flyer%20ENV-Linkages%20model%20-%20version%2025%20Sept%202013.pdf

Rockstrom, J., Steffen, W., Noone, K., Persson, A., Chapin, F. S., Lambin, E. F., Lenton, T. M., Scheffer, M., Folke, C., Schellnhuber, H. J., Nykvist, B., de Wit, C. A., Hughes, T., van der Leeuw, S., Rodhe, H., Sörlin, S., Snyder, P. K., Costanza, R., Svedin, U., Falkenmark, M., Karlberg, L., Corell, R. W., Fabry, V. J., Hansen, J., Walker, B., Liverman, D., Richardson, K., Crutzen, P., and Foley, J. A. (2009). A safe operating space for humanity. *Nature*, 461, 472–475.

Ronneberger, K.; Berrittella, M., Bosello, F., and Tol, R. (2009). KLUM@GTAP: Introducing biophysical aspects of land-use decisions into a computable general equilibrium model a coupling experiment *Environment Modeling and Assessment*, 14, 149–169.

Senger, M., and Spataru, C. (2015). *Water-energy-land nexus – modeling long-term scenarios for Brazil*. 2015 IEEE European Modelling Symposium. Retrieved from http://uksim.info/ems2015/data/0206a266.pdf

Smajgl, A., and Ward, J. (Eds.). (2013). *The water-food-energy nexus in the mekong region: Assessing development strategies considering cross-sectoral and transboundary impacts*. New York, Heidelberg, Dordrecht, and London: Springer.

Spang, E. S., Moomaw, W. R., Gallagher, K. S., Kirshen, P. H., and Marks, D. H. (2014). Multiple metrics for quantifying the intensity of water consumption of energy production. *Environmental Research Letters*, 9(10), 105003.

Spataru, C. (2014). *Integrated dynamic assessment methodology for three-node nexus (water, energy, land (WEL)) in space and time (ST) Model Documentation*. Energy Institute, UCL.

Spataru, C. (2016). *Integrated dynamic assessment methodology for five-node nexus (water, energy, land, food, materials) in space and time (IDA5), Model Documentation*. Energy Institute, UCL.

Stringer, L. C., Quinn, C. H., Berman, R. J., Le, H. T. V., Msuya, F. E., Orchard, S.E, and Pezzuti, J. C. B. (2014). *Combining nexus and resilience thinking in a novel framework to enable more equitable and just outcomes*. Sustainable Research Institute, Paper no. 73. Retrieved from www.see.leeds.ac.uk/fileadmin/Documents/research/sri/workingpapers/SRIPs-73.pdf

UN Water. (2014). *World water development report*. Retrieved from www.unwater.org/topics/water-food-and-energy-nexus/en/

Waughray, D. (2011). *Water security: The water-food-energy-climate nexus: The world economic forum water initiative*. Washington, DC: Island Press.

Welsch, M., Hermann, S., Howells, M., Rogner, H. H., Young, C., Ramma, I., Bazilian, M., Fischer, G., Alfstad, T., Gielen, D., Le Blanc, D., Röhrl, A., Steduto, P., and Müller, A. (20140. Adding value with CLEWS – modelling the energy system and its interdependencies for mauritius. *Applied Energy*, 113, 1434–1445. Elsevier.

PART IV

International political economy and the resource nexus

17

THE RESOURCE NEXUS IN AN UNCERTAIN WORLD

A non-equilibrium perspective

Shilpi Srivastava and Jeremy Allouche

Introduction

Nexus 'emerged' at a critical moment against the backdrop of cascading global crises in energy, food and global finance in 2008. These were further amplified by the 'uncertainties' brought about by climate change which were slowly becoming visible to the policymakers (Intergovernmental Panel on Climate Change, 2014). Nexus or interlinked governance across sectors was fronted as a potential solution to the spectre of scarcities in food and energy sectors, and social changes that included population growth, globalisation, economic growth and urbanisation as well as to potentially tackle the anomalies that climate change would bring in future (Hoff, 2011). Some solutions that are proposed in addressing this nexus – i.e. the nexus solutions – tend to gravitate towards a paradigm of control (i.e. stability and durability solutions) in which there is a much higher perception of certainty (Allouche et al., 2014). This proclivity to search for control-oriented solutions for complex systems that are by nature dynamic, ridden with uncertainty and are non-linear may do more harm than good.

We begin with the premise that uncertainty is an intrinsic element of nature itself (Penrod, 2007), and the nexus thinking as well as the nexus solutions should not overlook this element. For instance, climate change-related uncertainties can be approached in two ways. First, would be to speculate on this uncertainty and reduce it to 'controllable risks' via market-based instruments (a pathway that many propose through greening the economy) and the other would be to appreciate this uncertainty as fundamental to the socio-ecological world. We propose the latter. Embracing resilient systems that are open to change(s) is likely to yield more sustainable outcomes than control-oriented, top-down solutions that are amenable to policy lock-ins and may lead to perverse outcomes in terms of resource depletion and deprivation.

This chapter is divided into four sections. First, we will outline the limits to the scarcity narrative which seems to be currently driving wide parts of the nexus discourse, and make a case for viewing water–energy–food–climate systems as complex and dynamic. Second, we discuss the concept of the nexus primarily the market-technical framing of the World Economic Forum (WEF). This is driven by business imperatives and the logic of environmental economics overlooking the inequalities in access to resources. Third, drawing on examples from water and energy systems, we unpack how this particular framing is guided by equilibrium thinking which promotes top-down, control-oriented solutions, which – as we argue – have limited potential in tackling uncertainty and complexity. Finally, we conclude with arguing in favour of dynamic and resilient systems that could provide sustainable pathways for resource use.

Scarcity, uncertainty and complexity

To a large extent, it was the multiple scarcities of 2007–2008 that led to the return of 'limits to growth' thesis, with a pronounced sense of urgency and fear. In the 1970s, resource discourses were highly influenced by the Limits to Growth argument; by showing apocalyptic versions of a future fuelled by Malthusian narratives, critical concerns were raised around scarcity. In the backdrop of the oil crisis, and fuel shocks, this created the presumption, at least among policy-makers, that resources are 'scarce' and they need to be managed and governed in a better way (Mehta, 2010). This very preoccupation with scarcity was further stretched into the commodification argument through the 1990s. In the case of water, for example, to put a premium on the scarcity value of a resource, pricing the resource became the 'best' way to curtail wasteful consumption and use with the understanding that through pricing water will gravitate to high-value uses (Winpenny, 2003). This scarcity-efficiency linearity led to different kinds of policy interventions ranging from water pricing, private sector participation, and decentralised models of service delivery. On the one hand, these reified the economic value of water, on the other they also led to water inequities of various kinds leading to deprivation and marginalisation of the resource poor, especially for the small-scale farmers (water for agriculture use) and urban poor (drinking water supply).

The justification of nexus thinking also rests on similar presumptions of compounding scarcities, and solutions tend to follow a similar line of argument-monetisation of natural resources as natural capital (see next section and Chapter 2 of this handbook). For instance, in its 2011 World Economic Forum report, it is stated that the greatest threat to water scarcity and its risk to economic growth is the under-pricing of water as a resource (World Economic Forum, 2011). These in turn have created regional economic bubbles that sustain on poor tariffs and inefficient trade flows having a corresponding effect on food and energy systems. The food and energy crises also fuelled a growing perception of scarcity that was inevitably reinforced by climate change and environmental degradation narratives (e.g. Hartmann, 2014).

Though nexus evokes an 'integrative imaginary' (Cairns and Krzywoszynska, 2016) as an integration across various sectors, the boundaries of the nexus are still very much disputed; Srivastava and Mehta, 2014). If we understand nexus to be interlinkages across different kinds of resources – water, energy, food, water, climate (and land and minerals as proposed in the editorial chapter) – then we are essentially referring to not only interdependent systems (as emphasised by the nexus thinking) but inherently complex systems. Bazilian et al. (2011) reveal this complexity when they identify analytical and policymaking entry points. They argue that from a water perspective, both food and energy are users of the resource. From a food perspective, energy and water are inputs (Mushtaq et al., 2009), while from an energy perspective, water is a bioresource required as an input and food is an output. Food and water supply as well as wastewater treatment also require significant amounts of energy. Of course, areas such as food-as-fuels (i.e. biofuels) tend to blur these description (Bazilian et al., 2011). Thus it is hard to define the causal links in a neat input–output framework. This becomes more complex as ecological, social, technological, economic and political systems that govern these systems interact within themselves in the context of climate change. For example, the hydrological cycle is a highly dynamic system – socially, economically and technologically – and will become even more dynamic under the conditions of climate change (Mehta et al., 2007).

The Intergovernmental Panel Climate Change (IPCC) predicts that the effects of climate change will manifest themselves in extreme weather conditions – droughts, floods and heat-waves – which will increasingly become more frequent and unpredictable; these will in turn have negative consequences on soil and crop productivity, and displace seasonal hydrological cycles, the implications of which are still clouded with many uncertainties. Though this has reiterated fears of crises thinking and resource scarcity, this resurgence is also dovetailed with an aspect: uncertainty; that is, the inability to foresee changes that will come through changes in the climate. Thus the concerns over human induced climate change lay a considerable focus on establishing the rate, magnitude and patterns of change so that 'scarce resources' can either be 'stocked, conserved or used efficiently' for these unforeseeable changes.

Though uncertainty and risk are often used interchangeably, uncertainty means what we cannot know for certain in terms of outcomes, effects or impacts of a particular event; that is, 'all other forms of ignorance about a system for which we do not have well founded precise probabilities' (Swart et al., 2009, 5). In classical economic models, '[u]ncertainty is understood as the character of situations in which agents cannot anticipate the outcome of a decision and cannot assign probabilities to the outcome' (Beckert, 1996, 804). This certainty, to borrow from Beckert, has an element of fictionality – '[an] inhabitation in the mind of an imagined future state of the world and the beliefs in causal mechanisms leading to this future state. Actors are motivated in their actions by the imagined future and organize their activities based on these mental representations'. Scenarios for future, modelling and predictions are at the heart of such representations.[1] One such powerful representation is the framing that emerged from the World Economic Forum in 2008, and has been reiterated in various international conferences and platforms.[2]

The nexus sectors – food, energy and water – are thus understood to be interdependent and in need of integration (Hoff, 2011). Three guiding principles are proposed: investing to sustain ecosystem services; creating more with less; and accelerating access, integrating the poorest (ibid). These principles are also at the heart of the green economy approach.

The nexus according to the World Economic Forum

It was in the 'perceived' scarcity over food, energy and water resources and the various energy and food crises in 2007 and 2008 that the WEF first proposed the concept of the nexus. A year later, in 2009, UN Secretary-General Ban Ki Moon also addressed the WEF meeting at Davos and underlined the imperative of private sector participation to deal with these crises. This growing momentum led to a proliferation of special bodies within the World Economic Forum to deal with water issues including the creation of a Global Compact CEO Water Mandate and the Water Security Global Agenda Council. The WEF's formulation of the nexus has primarily been driven by international private actors, who see the nexus as an opportunity and a constraint to their business. It strongly links water security to economic growth in the following way:

> Water lies at the heart of a nexus of social, economic and political issues – agriculture, energy, cities, trade, finance, national security and human lives, rich and poor, water is not only an indispensable ingredient for human life, seen by many as a right, but also indisputably an economic and social good unlike any other. It is a commodity in its own right [. . .] but it is also a crucial connector between humans, our environment and all aspects of our economic system.
>
> (World Economic Forum, 2011, 3)

The underlying causality is as follows: to grow, economies should shift their water allocations away from farming and towards uses that deliver higher economic value per litre, especially energy production, industry and manufacturing. Within this logic, governments will therefore have to make choices about the allocation of water between sectors, and, of course, are encouraged to pursue high-value water uses. At the same time these shifts mean that they become more reliant on water use-efficient agriculture alongside food imports. To respond, the world system will need more trade flows in agriculture across more countries (World Economic Forum, 2011).

However universalised portrayals of scarcity that guide the nexus may be simplistic as limits on economic growth due to water is place specific and may not be the general case (Barbier, 2004). Furthermore, it is rainfall variability rather than general scarcity that shapes economic growth (Brown and Lall, 2006). Moreover water scarcity is often mediated through local narratives of conflicts and power differences (Mehta, 2005). So deploying economic rationales and modelling variables may not factor in such differences that are often place and context based (Srivastava, 2015). When does a resource become 'scarce' is contingent on how rules of access and use are set, and how they are enforced. Therefore how scarcity is defined, what is the potential source of resource use and the political economy around the use, extraction and in some cases exploitation of the resource will determine the shape of the nexus.

WEF's resource realism

WEF's perspective has ushered into a new brand of resource realism (Wales and Winston, 2012). It has provoked new public–private collaborations between International Financial Institutions (IFIs) and large transnational corporations such as Coca Cola, Nestlé and SABMiller who want to harness the private sector's 'comprehensive value-chain viewpoint' to tackle nexus governance and also advise governments, other corporates and communities in nexus governance (Wales and Winston, 2012). The WEF stressed four follow-up measures: (1) a task force for data collection; (2) a major program of economic modelling of interdependencies; (3) new models for collaborations to help governments make policy changes; (4) re-imagining institutions to improve governance (World Economic Forum, 2011). The operationalisation of the nexus is already underway in debates on fracking or SABMiller's study on evaluation of malting barley in Rajasthan (India) to determine value trade-offs across sectors (Wales and Winston, 2012; Reig et al., 2014).

This approach to the nexus stresses the business imperative and the need to prepare for investment scenarios in the near future. It underlines that the economics of water is both compelling and challenging and that water security, economic development and Gross Domestic Product (GDP) are interlinked (World Economic Forum, 2009). It thus argues for recognition that future global investments will be significantly driven by the consideration of water and will become a mainstream theme for investors. Global financial regulators, therefore, will have to develop clear-cut rules to manage the flow of innovative water funds.

Scramble for resources

The scramble for resources is also evident in the Global Risk Assessment Reports published by WEF. The 2011 reported that the 'land-grab' phenomenon is a response to a larger structural and correlated global risk to water, energy and food systems. Over the years, climate change and extreme weather events and their impacts on the nexus system have also become an increasing focus of these reports (World Economic Forum, 2016) which suggest diverse forms of public–private collaborations for resource management. The use of market-based instruments for conserving natural capital have also let to enclosures and securitisation of these resources depriving communities of their traditional ownership rights and livelihood practices. Borras, Saturnino and Franco (2012, 37) describe this as a process through which "national governments in 'finance-rich, resource-poor' countries are looking to 'finance poor, resource-rich' countries to help secure their own food and especially energy needs into the future" resulting in many large-scale land deals related to grains and biofuels, very often driven by transnational corporations and often with the active collaboration of the host national governments.

Thus the nexus thinking, in its simplistic form, might lead to the commodification of resources most readily or profitably monetised (perhaps for short-term gain), underplaying other long-term environmental externalities, such as biodiversity protection, pollution or climate change (Dupar and Oates, 2012). The 2011 Bonn paper also underlined this competition over resources against the backdrop of urbanisation, growing population and increasing levels of consumption, globalisation facilitating trade and investment, and resource degradation, which will be amplified through climate change (German Federal Ministry for the Environment Nature Conservation and Nuclear Safety, 2011; Bogardi et al., 2012).

The green Washington consensus

This framing of the nexus solutions is very much aligned with green economy debates. The green economy refers to a range of ideas including that of 'clean energy' and other ecological modernisation concepts, although so-called clean energy options raise other complex socio-environmental issues, in terms of energy use. For example, the rising production of biofuel crops have been linked to deforestation and competing uses of land with agriculture, including smallholder agriculture (Borras et al., 2010).

> The Green Economy, an economy that results in improved human wellbeing and social equity, while significantly reducing environmental risks and ecological scarcities [. . .]. In Green Economy natural capital is valued as a critical economic asset as a provider for benefits for the poor [. . .] it is the nexus approach par excellence.
>
> (Hoff, 2011, 6)

On similar lines, the starting point for the OECD's deliberations on *Green Growth* is the risk of climate change and depletion of natural resources such as unchecked biodiversity loss, overfishing, and the growing scarcity of land and water. The OECD report (2011, 4) states that-"We need green growth because risks to development are rising as growth continues to erode natural capital." The way forward should be characterised by: (1) increased productivity; (2) increased technological innovation; and (3) stimulating new markets.

Though green economy was originally conceptualised as a set of economic tools to operationalise the Sustainable Development Goals (Jacobs, 2012), the concept has been broadened and has become deeply contested over time (Benson and Greenfield, 2012). In the backdrop of the financial crises, it was seen as an entry point to promote green investments (Ocampo et al., 2011) in the wake of impending resource scarcity and climate change. The green economy also pushes for a commodification of natural resources, which provides for a way to deal with resource scarcity linked to the nexus by putting a price on natural resources whereby nature is made subject to financial speculation (Fatheuer, Fuhr and Unmuessig, 2016). It promotes monetisation and financialisation of nature, which is a set of three processes: (1) creation of commodities or marketable goods through market or valuation techniques; (2) introduction of commercial principles such as efficiency, accounting or cost-benefit assessment and profit maximisation; and (3) creation of tradable instruments such as environmental credits and offsets (Huff, 2015). The World Social Forum has called it 'the Green Washington Consensus', as the 'latest phase of capitalist expansion seeks to exploit and profit by putting a price value on the essential life-giving capacities of nature' (Working Group on Green Economy, 2012).

Limits to linear solutions in a non-equilibrium world

In large parts of the current nexus framing, a sense of alarmism and impending scarcity is widely evident which sets into motion supply side responses of augmenting water availability. This – as the nexus thinking suggests – can best be achieved either by making trade-offs or ensuring efficient mechanisms of resource use with maximum avoidance of negative externalities on other resources. As Leese and Meisch (2015, 698) put it, "all nexus conceptions share general perceptions of present and future crises and offer solutions for more efficient resource management within a green economy, thereby specifically calling for integrated solutions with regard to water, energy and food' close with double quotation marks. However, solutions proposed are often about increasing productivity through resource efficiency or availability of resources rather than focusing on supply side challenges such as questioning factors which increase the demand for such resources (Fatheuer, Fuhr and Unmuessig, 2016).

Within the nexus thinking, scarcity is often characterised as a risk – where decisions are made faced with unknown outcomes but known ex-ante probability distributions, rather than uncertainty where the probability distribution of a random outcome is unknown (cf. Knight, 1921). For instance, over the years, the Global Risk Assessment Reports (World Economic Forum) have characterised various elements of the social, economic, environmental and technological systems under the category of 'risks', which are interconnected (to a degree) but are largely framed as non-dynamic systems (for example, how does health insecurity or conflict-related migration speak to the nexus question?).

Uncertainty defies absolute solutions as one can offer only a range of possibilities, and that too with high degree of spatial variation. Nexus, however, operates on stable system thinking, which is usually informed by certain disciplinary trajectories (Leach, Scoones and Stirling, 2010). These tend to offer highly instrumentalist and oversimplified versions of pervasively complex, interlinked, dynamic social and ecological challenges (Stirling, 2015). For example, the investment scenarios and trade flows – one method for value-chain optimisation in the nexus thinking – may account for the patterns of production and consumption of particular resources, and not the dynamic interactions within these systems as well as across these systems. Furthermore, these large-scale and seemingly 'objective' systems are also embedded in social, political, ecological and physical systems that are dynamically interlinked with each other (Stirling, 2015). Compounding this, is also the degree of uncertainty that veils the current climate change scenarios and models (Curry and Webster, 2011). The complexity of decision situations, unforeseen interaction affects between water, food, energy and climate as well as overlapping – and often contradictory regimes – coupled with the choices of other actors limit the predictability and applicability of such scenarios (Dequech, 2011) in a dynamic and non-equilibria world.

Water can be a case in point here especially as 'water system' lies at the heart of the nexus thinking. In January 2014, when the World Bank launched its 'Thirsty Energy' initiative, it argued that the world's energy systems are inextricably linked with water systems' and "with demand rising for both resources, water scarcity can threaten the long term viability of energy projects and hinder development".[3] In another report, the global demand for food and energy was predicted to grow by 50% and for freshwater by 30% by 2030 (Beddington, 2009). The supply side challenges of availability of water (for food and energy) will be affected due to increasing uncertainties caused by climate change and its ecological consequences, in addition to the fast changing socio-economic boundary conditions, including global redistributions of wealth and power, and changing flows of people, resources and knowledge (Schmidhuber and Tubiello, 2007). The thrust of green economy on clean sources of energy such as hydropower also depend critically on water resources, but climate variability may have cascading effects on hydropower generation.

Large water infrastructure (e.g. dams) which have almost become a *sin qua non* of augmenting water availability can offer little to mitigate such conditions. Dams have been a preferred choice of policymakers to augment water storage in contrast to small-scale technologies which play a critical role in ensuring water security for small-scale farmers (FAO, 2008; Hagos et al., 2012). However, alternative options such as natural wetlands, enhanced soil moisture, groundwater aquifers, ponds and tanks, and small/check dams are not equally considered within the planning and assessment process. The inherent characteristics of the storage are also affected by site-specific conditions and the way the storage is planned and managed within the context of its geographic, cultural and political location (McCartney and Smakhtin, 2010). Therefore, for water storage to be more reliable and resilient, in the context of climate change, "storage 'systems' [i.e. a mix of storage types] that combine and build on complementarities of different storage types or other water harvesting measures are likely to be more effective" (Allouche, Middleton and Gyawali, 2014; Rockström and Falkenmark, 2015). Initiatives for harnessing plural water storage options have largely been ad hoc and left to local communities, civil society organisations or local business, though they have proved highly effective in addressing the uncertain effects of climate change.

The case of Maharashtra (India) elucidates this point further. Maharashtra witnessed one of its worst droughts in 2016. The water levels reached abysmally low levels and the government was forced to recommend water rationing for its users. Maharashtra is not an exception but represents another side of the nexus thinking – a nexus between sugarcane cultivation, power generation and growing needs of urbanisation. The growing demands on water in the face of unpredictable monsoons for a period of two years resulted in this situation. At the local level, the effects of this scarcity manifested in crop loss, social conflicts and contested allocation politics – which nexus seems to conveniently bypass (Srivastava, 2015) as with the thinking on plural pathways. For instance, in the drought vulnerable regions of Maharashtra, farmers rely on small check dams to support them during the lean and uncertain monsoon periods and have also migrated to the cultivation of less water-intensive crops. These experiments that are largely localised predate the nexus thinking but have a critical bearing on dealing with uncertain rainfall conditions.

The search for control-oriented, singular and linear solutions is also dynamically interlinked with vested interests in the political economy, which 'close-in' the potential for scaling up alternative pathways. For instance, in their 2010 paper, Smits and Bush argue that despite a high potential for hydropower generation, the Lao government has neglected rural electrification, and promoted the singular model for solar home systems. Other priorities such as providing electricity to neighbouring countries, the government's interests in centralisation of water resources, and "maximising foreign investment and export revenues" have prevailed over ensuring electricity access to rural households (Smits and Bush, 2010, 126).

Complexity and uncertainty is not unique to water or energy systems but has been a way of life for populations who live with this uncertainty on a day-to-day basis. Research on pastoral livelihoods in India (Mehta, 2005) and Africa (Scoones, 2004) also highlights these limits to singular, technology-led policy solutions that are at odds with the unruly realities that people face in their everyday lives and the multiple coping strategies they employ to mitigate these effects; these include planning migration patterns or diversification of livelihoods. Of course, these measures may be opportunistic, short term and unsustainable in the face of large-scale uncertainties but they tend to have better proofing outcomes than solutions guided by standard equilibrium thinking (cf. Scoones, 2004).

To accommodate complexity of ecological, social and political systems would also require a shift in governance from 'stability and durability thinking' towards incorporating 'resilience and robustness thinking' where the limits to control are acknowledged and adaptive solutions that incorporate plural solutions are pursued to address uncertainty and complexity.

Conclusion: living with uncertainty and complexity

In the preceding sections, we have argued that large parts of the current nexus thinking and its solutions are primarily guided by a scarcity narrative, which is often described in absolute and apolitical ways, overlooking the political economy that 'constructs' this scarcity. Correspondingly, the solutions are often framed on the technology-led demand side i.e. increasing availability via hard infrastructure, efficient uses of resources through necessary trade-offs or value-chain optimisation. This framing, as we argue, has problems: First, such notion of the nexus draw on simplistic causal relations between availability and access. For example, the fears around potential food crises portray a simple causality between food and water availability and reduction in hunger or improved access to water (Allouche et al., 2014). The relationship between availability and access is often mediated by monetary (resources available to access) and non-monetary factors (inequitable power relations in the given context, institutional arrangements) that are critical in providing access. Second, the nexus thinking tends to propose technology-led solutions which are closed, top-down, and control oriented rather than questioning the inequalities within the system that mediate access and use of these resources. Instead of questioning the structural inequalities within the nexus, this framing confirms them (Leese and Meisch, 2015). It fails to ask the next critical questions: Who benefits from these improvements and at whose or at what costs?

The logic of optimisation, embedded within the nexus thinking, has clear limits. It treats resource allocation in a perfect equilibrium model (as evidenced through neoclassical economics) (Allouche et al., 2015). However, this does encourage the commodification of resources, and may result in neglecting other issues such as environmental externalities, poverty alleviation and everyday realities at the local level (cf. Dupar and Oates, 2012). For many rural communities, food, water and energy have never been conceptually separated in the way that experts have sought to understand them. Indeed, it may be that the water–energy–food nexus is the (re)discovery by experts working in silos of what practicing farmers and fishers already know. However valuing nature as particular kind of capital and the regimes of financialisation and commodification that accrue with them may not only engender inequalities related to access and control, but could also lead to disrupting these 'local nexuses' which are more attuned to addressing complexity and uncertainty. In light of the myriad physical uncertainties and social vulnerabilities that different societies in varied ecosystems face, one needs to acknowledge pursuing plural pathways that work in bottom-up ways and are more attuned to local systems of resource availability as well as its uses and consumption.

We live in a complex world with a multitude of interactions between the material, social and the biophysical elements that are by nature intractable and difficult to grasp. The nexus thinking may be described as an attempt to render these interactions visible and amenable to policymaking (cf. Dupar and Oates, 2012) and policy solutions, but as we have argued that there are clear limits to such a framing and their corresponding solutions. First, it is difficult to predict and capture the dynamics of this interaction as they are ridden with uncertainties – the unknown unknowns – more so in the context of climate change. Second, the solutions provided by the nexus often prioritises particular causal loops which are over simplistic and neglect the political economy of these interactions especially around the questions of resource access and use; there are inherent dangers that this 'closed' framing may render those local actions and needs invisible or peripheral as they get pushed out of the policy discourse.

In this light, reversing the gaze and thinking of systems as dynamic and open-ended becomes crucial. We need to think of better ways to understanding and working with alternative development choices which privilege local and diverse ways of knowing, as we work across scales to understand these ecological, social and technical interactions. The way forward begins with acknowledging the value of plural frames and solutions, and recognising the deeply political nature of resource use and access.

Notes

1 Drawing on ongoing work on the project 'Uncertainty, Climate Change and Transformation', funded by the Norwegian Research Council (forthcoming IDS Working Paper). www.ids.ac.uk/project/climate-change-uncertainty-and-tranformation

2 Bonn 2011 Nexus Conference: *The Water Energy and Food Security Nexus – Solutions for the Green Economy*; Planet Under Pressure International Conference (March, 2012)-'*Interconnected risks and solutions for a planet under pressure*'; *Rio +20 Conference*; Nexus 2014: *Water, Food, Climate and Energy Conference* (USA); Nexus Declaration delivered to the UN Secretary-General on March 26 2014 (for input to formulation of Sustainable Development Goals; *Nexus Network* launched in the UK – funded by Economic and Social Research Council (ESRC) 2014; *Sustainability in the Water – Energy – Food Nexus International Conference* (2014) (Leck et al., 2015).

3 www.worldbank.org/en/news/feature/2014/01/16/will-water-constrain-our-energy-future (8 March 2014)

References

Allouche, J., Middleton, C., and Gyawali, D. (2014). *Water and the nexus*. Brighton: STEPS Centre.

Allouche, J., Middleton, C., and Gyawali, D. (2015). Technical veil, hidden politics: Interrogating the power linkages behind the nexus. *Water Alternatives*. 8(1), 610–626.

Barbier, E. B. (2004). Water and economic growth. *Economic Record*, 80, 1–16.

Bazilian, M., Rogner, H., Howells, M., Hermann, S., Arent, D., Gielen, D., and Yumkella, K. K. (2011). Considering the energy, water and food nexus: Towards an integrated modelling approach. *Energy Policy*, 39(12), 7896–7906.

Beckert, J. (1996). What is sociological about economic sociology? Uncertainty and the embeddedness of economic action. *Theory and Society*, 25(6), 803–840.

Benson, E., and Greenfield, O. (2012). *Surveying the 'green economy' and 'green growth' landscapes*. Retrieved October 23, 2016, from www.greeneconomycoalition.org/sites/greeneconomycoalition.org/files/GEC%20background%20paper_final%20.pdf.

Beddington, J. (2009) Food, Energy, Water and the Climate: A Perfect Storm of Global Events? London: Government Office for Science, http://webarchive.nationalarchives.gov.uk/20121212135622/http://www.bis.gov.uk/assets/goscien ce/docs/p/perfect-storm-paper.pdf (accessed 24/11/16)

Bogardi, J. J., Dudgeon, D., Lawford, R., Flinkerbusch, E., Meyn, A., Pahl-Wostl, C., Vielhauer, K., and Vörösmarty, C. (2012). Water security for a planet under pressure: Interconnected challenges of a changing world call for sustainable solutions. *Current Opinion in Environmental Sustainability*, 4, 35–43s.

Borras, Jr., S. M., and Franco, J. (2012). Global land grabbing and trajectories of agrarian change: A preliminary analysis. *Journal of Agrarian Change*, 12(1), 34–59.

Borras, S., McMichael, P., and Scoones, I. (2010). The politics of biofuels, land and agrarian change: Editors' introduction. *Journal of Peasant Studies*, 37(4), 575–592.

Brown, C., and Lall, U. (2006). Water and economic development: The role of variability and a framework for resilience. *Natural Resources Forum*, 30, 306–317.

Cairns, R., and Krzywoszynska, A. (2016). Anatomy of a buzzword: The emergence of 'the water-energy-food nexus' in UK natural resource debates. *Environmental Science & Policy*, 64, 164–170.

Curry, J. A., and Webster, P. J. (2011). Climate science and the uncertainty monster. *Bulletin of the American Meteorological Society*, 92(12), 1667–1682.

Dequech, D. (2011). Uncertainty: A typology and refinements of existing concepts. *Journal of Economic Issues*, 45(3), 621–640.

Dupar, M. and Oates, N. (2012). Getting to grips with the water-energy-food "nexus". Climate and Development Knowledge Network (London). Available online: http://cdkn. org/2012/04/getting-to-grips-withthe-water-energy-food-nexus (accessed 10/07/15).

FAO. (2008). *Water for the rural poor: Interventions for improving livelihoods in sub-Saharan Africa*. Rome: Food and Agriculture Organization, Land and Water Division.

Fatheuer T, Fuhr L, Unmuessg B. (2016). Inside the Green Economy: Promises and Pitfalls. UIT: Cambridge.

German Federal Ministry for the Environment Nature Conservation and Nuclear Safety (2011) Thematic Profile, paper presented at the Bonn2011 Conference: 'The Water, Energy and Food Security Nexus Solutions for the Green Economy', Bonn.

Hagos, F., Jayasinghe, G., Awulachew, S. B., Loulseged, M., and Yilma, A. D. (2012). Agricultural water management and poverty in Ethiopia. *Agricultural Economics*, 43, 99–111.

Hartmann, B. (2014). Converging on disaster: Climate security and the Malthusian anticipatory regime for Africa. *Geopolitics*, 19(4), 757–783.

Hoff, H. (2011) 'Understanding the Nexus', background paper for the Bonn 2011 Conference: 'The Water, Energy and Food Security Nexus', Stockholm: Stockholm Environment Institute.

Huff, A. (2015). *Understanding relationships between the green economy, resource financialization and conflict*. IDS Policy Briefing. Brighton: IDS.

Intergovernmental Panel on Climate Change. (2014). Climate Change 2014: A Synthesis Report, Contribution of Working Groups I, II and III to the Fifth Assessment Report of the Intergovernmental Panel on Climate Change. Geneva, Switzerland: IPCC Retrieved from https:// www.ipcc.ch/report/ar5/syr/ (Accessed 10/10/2015).

Jacobs, M. (2012). Green growth: Economic theory and political discourse. London: Centre for Climate Change Economics and Policy Working Paper No, 108.

Knight FH. (1921). Risk, uncertainty and profit. Chicago: Chicago University Press.

Leach M, Scoones I, Stirling A. 2010. Dynamic Sustainabilities: Technology, Environment, Social Justice. Oxon: Earthscan.

Leck, H., Conway, D., Bradshaw, M., and Rees, J. (2015). Tracing the water–energy–food nexus: Description, theory and practice. *Geography Compass*, 9(8), 445–460.

Leese M. and Meisch, S. (2015). Securitising sustainability? Questioning the 'water, energy and food-security nexus'. *Water Alternatives*. 8(1), 695–709.

McCartney, M., and Smakhtin, V. (2010). *Water storage in an era of climate change: Addressing the challenge of increasing rainfall variability*. Colombo: International Water Management Institute.

Mehta, L. (2005). *The politics and poetics of water: The naturalisation of scarcity in Western India*. New Delhi: Orient Longman.

Mehta, L. (2010). Introduction. In L. Mehta (Ed.), *The limits to scarcity: Contesting the politics of allocation*. London: Routledge.

Mehta, L., Marshall, F., Movik, S., Stirling, A., Shah, E., Smith, A., and Thompson, J. (2007). *Liquid dynamics: Challenges for sustainability in water and sanitation*. STEPS Working Paper 6. Brighton: STEPS Centre.

Mushtaq, S., Maraseni, T. N., Maroulis, J., and Hafeez, M. (2009). Energy and water trade-offs in enhancing food security: A selective international assessment. *Energy Policy*, 37, 3635–3644.

Ocampo, J. A. (2011). *The macroeconomics of the green economy: The transition to a green economy: Benefits, challenges and risks from a sustainable development perspective*, 16. Retrieved from www.tni.org/en/article/is-the-green-economy-a-new-washington-consensus

OECD, 2011. Towards Green Growth. Paris: OECD Publishing.Penrod, J. (2007). Living with uncertainty: Concept advancement. *Journal of Advanced Nursing*, 57(6), 658–667.

Reig, P., Luo, T., and Proctor, J. (2014). *Global shale gas development: Water availability and business risks*. World Resources Institute (WRI). [Online]. Retrieved June 5, 2014, from www.wri.org/sites/default/files/wri14_report_shalegas.pdf.

Rockström J. and Falkenmark, M. (2015). Increase water harvesting in Africa. Nature. 519(7543), 283–285.

Schmidhuber, J., and Tubiello, F. N. (2007). Global food security under climate change. *Proceedings of the National Academy of Sciences*, 104(50), 19703–19708.

Scoones, I. (2004). Climate change and the challenge of non-equilibrium thinking. *IDS Bulletin*, 35(3), 114–119.

Smits, M., and Bush, S. (2010). A light left in the dark: The practice and politics of Pico-hydropower in the lao PDR. *Energy Policy*, 38(1), 116–127.

Srivastava, S. (2015). *Rules over regulation: The making of water reforms and regulatory cultures in Maharashtra, India*. PhD Thesis, University of Sussex.

Srivastava, S., and Mehta, L. (2014). *Not another nexus? Critical thinking on the new security convergence in energy, food, climate and water*. Brighton: STEPS Centre.

Stirling, A. (2015, June 29–30) *Developing 'Nexus Capabilities': Towards transdisciplinary methodologies*. Draft discussion paper presented at the Nexus workshop, University of Sussex.

Swart, R., Bernstein, L., Ha-Duong, M., and Petersen, A. (2009). Agreeing to disagree: Uncertainty management in assessing climate change, impacts and responses by the IPCC". *Climatic Change*, 92(1–2), 1–29.

Wales, A., and Winston, A. (2012). *Ecosystem economics: Navigating the water-food-energy nexus*. London: The Guardian Sustainable Business Partner Zone. Retrieved from www.theguardian.co.uk.

Winpenny, J. (2003). *Report of the world panel on financng water infrastructure*. Retrieved from www.worldwatercouncil.org/fileadmin/wwc/Library/Publications_and_reports/CamdessusSummary.pdf

Working Group on Green Economy. (2012). *Is the 'Green Economy' a new Washington consensus?* Montreal: Global Research. Retrieved January 16, 2014, from www.globalresearch.ca/is-the-green-economy-a-new-washingtonconsensus/29462

World Economic Forum (2011). Water security: the water-food-energy-climate nexus, Washington: Island Press.

World Economic Forum (2009). 'The Bubble Is Close to Bursting: A Forecast of the Main Economic and Geopolitical Water Issues Likely to Arise in the World during the Next Two Decades', draft paper for discussion at the World Economic Forum Meeting 2009, Geneva: World Economic Forum.

World Economic Forum. (2016). The Global Competitiveness Report 2016–2017. Geneva: World Economic Forum.

18
MINING AND THE RESOURCE NEXUS

David Humphreys

Mining is the business of extracting useful minerals from the earth's crust. Since minerals, once exacted, cannot be extracted again, a preoccupation of the industry historically has been the continued physical availability of resources in the ground. Would resources be sufficient to meet society's long-term requirements or would the depletion of these resources eventually result in shortages of mineral products and escalating prices?

In recent years, there has been a growing realisation that this might not be the right question. The matter of minerals availability, it turns out, is more complex than previously thought. For one thing, it is evident that improvements in productivity have served to offset the effects of resource depletion with the result that production costs have not risen in the manner that early analysts of the industry had supposed they would. For another, and of greater significance for the future, there is a growing awareness that the recovery of minerals is more heavily dependent than previously thought on developments in other parts of the resource nexus; moreover, that the integration of the various parts of the nexus in relation to mining poses major challenges for corporate policy and for governance in mineral host countries.

The activity of mining is keyed into the resource nexus at several points. Most obviously, mining requires land for the exploitation of ores and for the disposal of wastes. These wastes are substantial, and become more so as ore grades decline, so this remains a challenging area. But mining also requires considerable amounts of water for washing and processing minerals. Large amounts of energy are required to recover minerals and even more to smelt and refine them to finished marketable forms. The connection between minerals and food is more tenuous, although it should be observed that mining is the source of two of the three key plant nutrients: potassium, phosphates and nitrogen. Over time, the linkages between mining and these other resource systems have become progressively more important. Indeed, it is argued here that over the long-term pressure on these other resource systems may turn out to be more critical in conditioning, and possibly constraining, the development of mining than is the physical availability of mineral ores.

Tectonic shift

The first decade of the new millennium was a momentous one for the mining industry. As a consequence of China's rapid industrialisation, demand for mineral products surged and mineral prices surged with it. Caught unprepared, the mining industry moved fast to step up its rate of exploration and investment. Some companies sought to boost their production and capture the benefit of the boom by buying up competitors. There was a wave of mergers and acquisitions (M&A) in the industry between 2005 and 2008.

The growth in demand for minerals from China and from other emerging economies brought with it a decisive shift in the customer base of the mining industry, away from traditional consumers in the advanced economies towards the emerging market and developing economies (henceforth emerging economies).[1] During the mid-1990s, advanced economies accounted for a half to two-thirds of global consumption for the major metals (Figure 18.1). Only 20 years on, the position has been wholly reversed and three-quarters of consumption takes place in emerging economies.

This shift in the locus of mineral consumption gave mining new economic relevance. Whereas in earlier years the fate of the industry had been largely tied to the fortunes of a few hundred million people living in slow-growing and dematerialising advanced economies, suddenly it was being associated with the aspirations of billions of people striving to lift themselves out of poverty and to raise their material standards of living. Investors were quick to grasp the significance of the shift and to take a renewed interest in the mining sector. There developed a narrative of a prolonged commodity 'super-cycle' underpinned by material-intensive, investment-heavy growth in emerging economies, and of commodities as a 'new asset class'. Investment funds poured into the industry.

A similar, albeit rather less dramatic, shift was taking place on the supply side of the industry. Mine production too was moving away from the advanced economies towards emerging

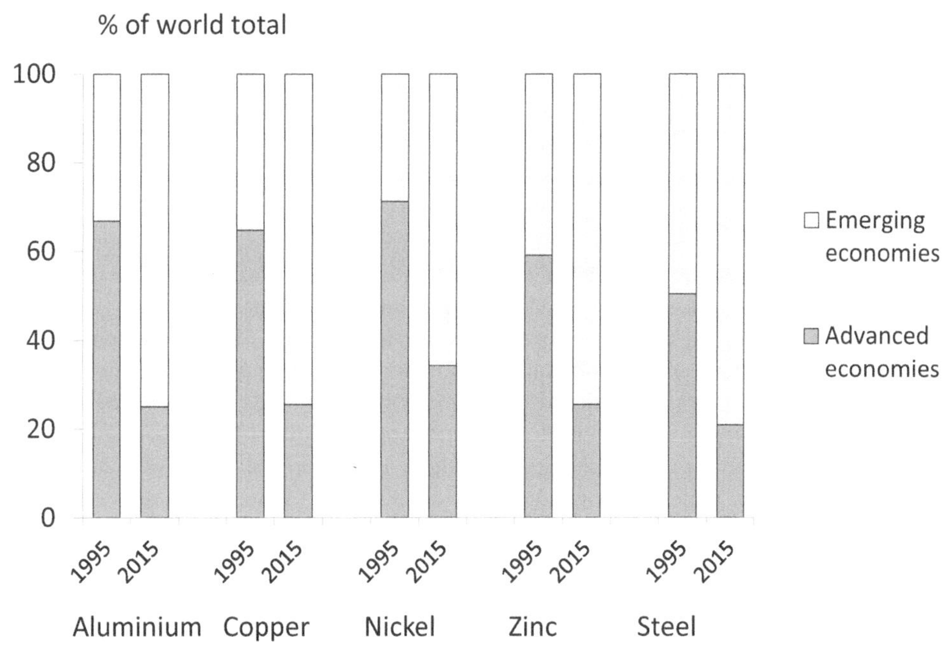

Figure 18.1 Distribution of global mineral consumption, 1995 and 2015

Source: WBMS, WSA

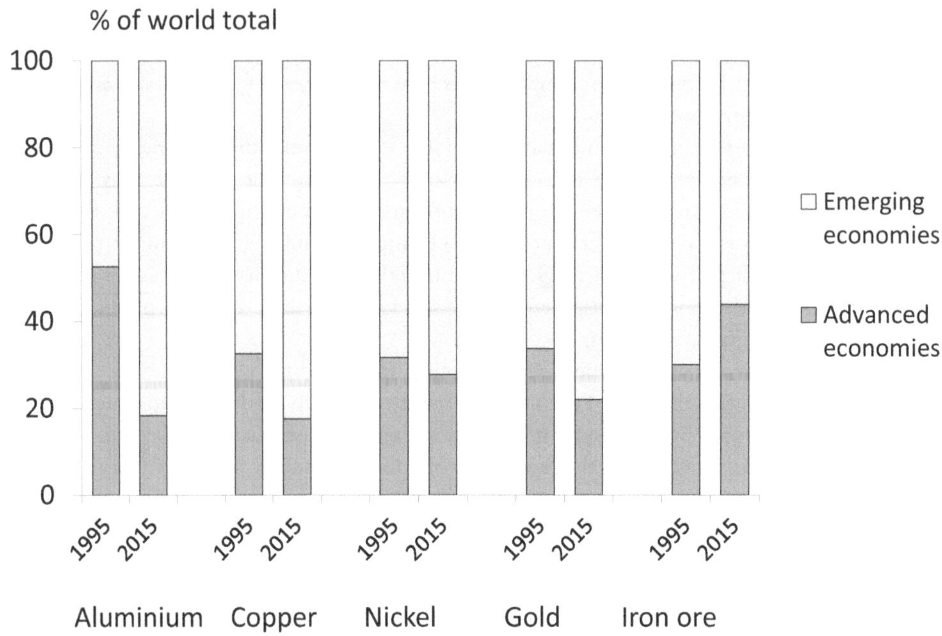

Figure 18.2 Distribution of global mineral production, 1995 and 2015

Source: WBMS, WSA

economies. (See Figure 18.2) In part, this simply reflected the dictates of geology. Emerging economies collectively account for around three-quarters of the world's land area and three-quarters of its resources, so it should not be too surprising that they were gradually moving towards a situation where they accounted also for around three-quarters of global mineral production. (Humphreys, 2015) The exception to this trend was iron ore, where rapid growth in production in Australia led to the advanced economies increasing their share of global production over the period shown in Figure 18.2. Alongside the growth in mine production from emerging economies was growth in the global role played by emerging economy mining companies, including those from the former Soviet Union (Russia, Kazakhstan, Ukraine), from formerly non-aligned countries (India, Mexico, Peru) and from China.

A distinct, but related trend, and one that is relevant to some of the developments discussed later in the chapter, is that more mineral production was taking place in mineral-driven economies; that is, in economies that depend heavily on the mining sector for their growth and exports. Historically, most mineral production took place in mineral-consuming areas. European, US and Soviet industrialisation was largely based on local resources. But, with the opening up of global trade and cross-border investment, more of the world's mineral supplies started to come from mineral-driven economies. Thus, in 1960, 52% of the world's copper and 79% of nickel originated in the major consuming countries. By 2010, the proportion had roughly halved, to 27% and 38% respectively. (Humphreys, 2013b) According to consultants McKinsey, by 2011 the world had 81 resource-driven countries, up from 58 in 1995, although it should be noted that their analysis included oil and gas producers as well as mining countries. (McKinsey, 2013)

The industry was moving into new territory and it was having to change.

Triggers of change

In point of fact, pressures for the industry to adapt to the new conditions had begun to emerge some time earlier. The 1980s and 1990s had been difficult years for mining. De-industrialisation in the US, Western Europe and Japan, coupled with the economic collapse of the Soviet Union, served to depress the demand for minerals and mineral prices. Countries in which government had taken a big stake in the mining industry during the commodities boom of the 1970s, in the hope that this would foster their development, had instead found the sector a burden on their public finances and sought to disengage from it through a series of privatisations. Indeed a view began to take hold that mining, far from bringing benefits to less-developed economies, often distorted them and corrupted them in ways that inhibited rather than promoted growth. Development economists pointed to negative correlation between resource endowment and rates of economic development and began to talk of a 'resource curse' (Auty, 1993; Sachs and Warner, 1995).

Against this background, the mining industry faced a crisis of trust. After years of weak prices and poor profits, investors had lost interest in the industry. Technology provided them with a much more compelling investment story. The public, never much enamoured of the industry anyway because of its conspicuous environmental impacts, was increasingly critical of it. The flames of this disenchantment with the sector were fanned by lurid tales about how revenues from the mining of diamonds and other high-value minerals were being used to fund rebel groups in Africa. New communication technologies were increasing the ability on non-governmental organisations (NGOs) to monitor mining activities around the globe and to draw the spotlight of public opinion onto mining industry transgressions. The legitimacy of mining was under attack along a broad front. This legitimacy had historically been a utilitarian one, one rooted in means and ends. Mining was a dirty activity but society needed the products of mining to support economic growth. However, this was no longer enough. The public was less and less impressed by justifications for the industry based on what it produced. They were increasingly interested in *how* it produced.

By the late 1990s, the mining industry realised it could no longer ignore these pressures. Moreover, it was becoming increasingly apparent that the many problems it confronted were interconnected. To attract money from investors and to fulfil their social obligations, companies in the industry needed to be profitable. In order to be profitable, they needed to manage better the operational risks associated with their environmental performance and their community relations. In order to gain access to the best mineral deposits, they had to persuade host countries that they understood the importance of operating in an environmentally and socially sensitive fashion and that they were ready to collaborate constructively in the promotion of national economic development.

Awakening to the nexus

Accompanying this growing awareness of the interconnectedness of the business challenges facing the industry, there was a growing awareness in the industry of the interconnectedness of the various resource systems with which it interacted. It was, in effect, waking up to the imperative of the resource nexus, albeit that this was not necessarily the term used to describe it.

Historically, the primary resource issue preoccupying the mining industry was geological availability. Were there sufficient mineral resources in the ground to sustain future production and could they be economically recovered? The economist W Stanley Jevons was the first to articulate concerns about mineral depletion in his 1866 book *The Coal Question* – the 'question' being whether Great Britain was going to run out of coal. The Club of Rome returned to the question in its landmark study *The Limits to Growth* in the early 1970s. (Meadows et al., 1974) More recently, concerns about physical availability of minerals have resurfaced in the guise of the theory of 'peak' minerals, the theory that production naturally peaks as the better, more accessible, resources are used up and producers have to move to lower quality or deeper resources. (Bardi, 2014)

However, the industry was gradually being drawn to the realisation that resource availability was no longer the central issue it had to deal with. The resource equation was becoming a whole lot more complex. For sure, miners needed ore in the ground but their activities impact on a whole raft of other resource systems, the relationships with which needed to be managed just as much as did those with their ore resources. Specifically, they had to manage their impacts on land, on water and on energy and emissions.

Mining is an intensive user of land and miners have to compete with other land users for space in which to locate their mines and related infrastructure and on which to dispose of the considerable amount of waste that mining generates. And they have to do this in a way that does not detract from the biodiversity of the areas in which they operate, despoil water resources or pose physical threats such as subsidence. Pressures from this source occur everywhere but are particularly acute in the more advanced economies where pressures from alternative land uses are often greatest and organised opposition most acute.

Some appreciation of the scale of mine waste can be obtained by reference to a typical copper mine. The average ore grade of copper mines around the world is approximately 0.7% copper while the recovery rate of the copper from the ore is around 84%. Assuming an open-pit mine with a strip ratio of 2.5:1 (this is a measure of the amount of waste rock which has to be removed to recover one tonne of copper-containing ore), a miner needs to remove nearly 600 tonnes of waste material to produce one tonne of refined copper metal. For gold, where grades are measured in parts per million (grams per tonne) rather than in percentages, the proportion of waste generated to product recovered is significantly higher. For bulk commodities, like iron ore and coal, where most of what is dug up is useable product, it is significantly lower. Across the mining industry as a whole, it is estimated that some 60 billion tonnes of material are removed from the earth each year. (Ericsson, 2016)

Mining also uses a lot of water, this for concentrating minerals into a marketable form and, in the case of coal, for washing. The range of water use varies enormously from mine to mine, and many mines have highly sophisticated systems for recovering water used in their processes, but to give some idea of what is at issue here, it is estimated on average to take around 172 tonnes (thousand litres) of water to produce one tonne of copper and 107 tonnes to produce one tonne of nickel. For precious metals, the amount required per unit of output is significantly higher, although obviously the amount of metal produced is very much smaller. Some 600–700 tonnes of water are required to produce 1 kilogram of gold. Only 0.3 tonnes of water are required to produce a tonne of black coal, but, then again, the tonnages of coal produced are very large. (Mudd, 2008) The challenges of water to mining are compounded by the fact that mining occurs in some of the most water-stressed places on earth. The Atacama Desert in Northern Chile is at once both one of the world's most important areas for producing copper globally as well as one of driest places on the planet. Other important mining regions facing water stress are southern Africa, the southern states of the US, Australia and northern China.

The third major resource system which mining impacts is energy. Substantial quantities of energy are required to dig out mineral ore, to dispose of mine waste, to crush and to grind the ores to liberate the sought-after minerals and, in the case of metals, to power the metallurgical

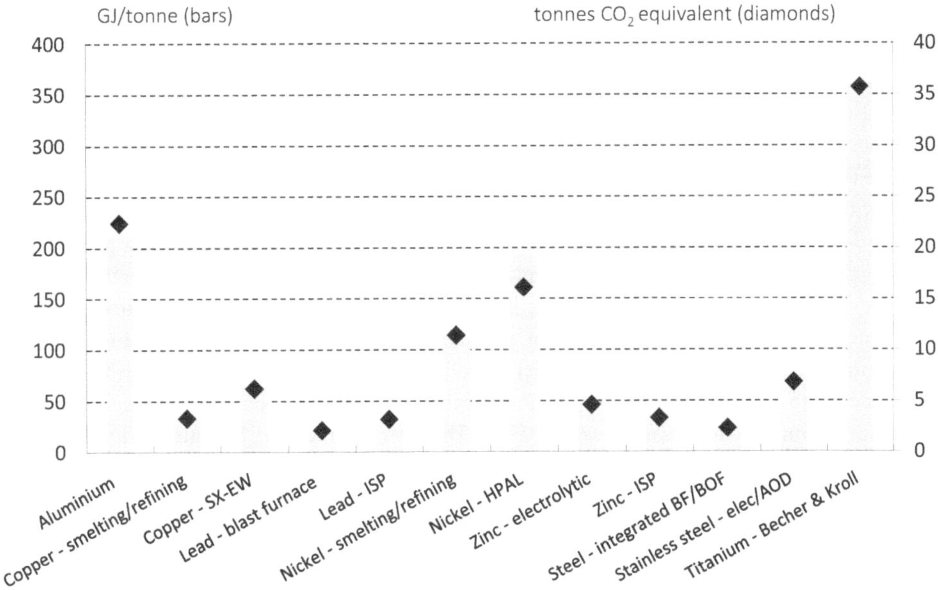

Figure 18.3 Energy use and CO_2 emissions per tonne of metal production

Source: Norgate et al. (2007)

plants that convert ores into finished metal. Very approximately, energy represents a quarter of the total cost of producing the major metals. Associated with this use of energy, the industry is responsible for significant emissions of greenhouses gases to the atmosphere. Figure 18.3 shows the energy required to produce a number of important metals and the associated global warming potential (GWP) of that energy use expressed in tonnes of CO_2 equivalent. The experience varies widely between metals. The production of a tonne of steel results in the generation of around two tonnes of carbon dioxide, while the production of aluminium results in the generation of over 20 tonnes of carbon dioxide. It should be pointed out, however, that this analysis assumes all power is generated using coal. If the aluminium is produced using hydroelectric power, then the associated carbon emissions are reduced by half. (Norgate et al., 2007) The mining of coal, which does not feature in Figure 18.3, is not particularly energy intensive but the burning of the coal is, of course, one of the world's principal sources of greenhouse gases.

These other resource systems, land, water and energy, are, in their own way, as essential to mining as is the availability of mineral ore in the ground. Moreover, problems with any of them have the capacity to constrain investment and production in mining just as surely as does the depletion of mineral ore. In some parts of the world, these other resource systems represent a potentially greater challenge to miners than do the technical demands of mining itself. In short, in conducting its business it was becoming essential that the industry consider all of these resource systems and the connections between them in an integrated fashion.

Industry response

Collectively the issues outlined above represented a serious challenge for the mining industry. It could not simply respond by claiming to be misunderstood and address the issue with a programme of public education. Yes, the industry faced a problem of perception but this was rooted in an underlying problem of performance. The mining industry had simply not adapted to society's changing expectations of it. Some of the issues it faced, notably the negative image projected by artisanal mining (low-technology, pick-and-shovel operations), were somewhat beyond what the large miners could do anything about, but, with respect to what was within their power to manage, it was clear that they needed to raise their game.

The demand for business to give greater focus to the social and environmental dimensions of its activities was not, of course, restricted to the mining industry. The Rio Earth Summit of 1992, which did much to popularise the notion of sustainable development, put strong emphasis on the role that businesses of all kinds could play in helping to deliver growth in an environmentally and socially responsible fashion. Business is an integral part of the economic and social fabric and it needs a good-quality environment and social cohesion if it is to flourish. To build good environmental and social practices into its behaviour was therefore good business. This insight, and a desire to institutionalise and promote further the thinking behind it, led, subsequent to the Rio summit, to the creation of the World Business Council for Sustainable Development (WBCSD), based in Geneva.

The big mining companies were similarly being drawn to the conclusion that they needed to address their many challenges within a framework of sustainable development. However, as a business which involves the depletion of non-renewable resources, this represented a special challenge. It meant, in practice, that the focus of the industry's efforts towards sustainable development would have to be on minimising the negative impacts of its activities on the environment and on the communities in the vicinity of their operations, while ensuring that a large part of the wealth generated by the drawdown of natural capital through mining was invested in other forms of capital which could continue to deliver returns after the mine was exhausted. (This notion of the conservation of the total capital stock is sometimes referred to by economists, somewhat unhelpfully, as 'weak sustainability'.) These 'other' forms of capital include built capital, human capital, social capital and financial capital.

To assist with the development of its thinking on this matter, the industry launched a process of analysis and engagement called the Global Mining Initiative (GMI). The centrepiece of the GMI was a worldwide research and consultation exercise entitled Mining, Minerals and Sustainable Development (MMSD). The work took place over two years and the final report was presented at a major conference in Toronto in 2002. At the meeting, the industry announced (in a move paralleling the creation of the WBCSD) the setting up of a CEO-led organisation, the International Council on Mining and Metals (ICMM) to take the agenda of sustainable development in the mining industry forward on a continuing basis.

The MMSD report was the first ever attempt to examine the mining industry holistically, in all its forms and with all its ramifications. On the basis of this examination it sought to identify paths by which the industry could contribute to the global transition to sustainable development. In line with nexus thinking it urged the industry to better manage its interactions with other resource systems.

> Managing [mining] impacts more effectively requires dealing with unresolved issues of handling immense amounts of waste, developing ways of internalizing the costs of acid drainage, improving both assessment and environmental management systems, and doing effective planning for mine closure.

It also urged companies to look beyond the mine site towards the entire life cycle of the minerals they were producing. "Companies at different stages in the minerals chain can benefit from learning to work together exploring further recycling, re-use, re-manufacture of products and developing integrated programmes of product stewardship and supply chain assurance." (MMSD, 2002).

Another key strand of the report focused on the social and political dimensions of mining. Companies and the communities in which they operated should seek to build partnerships to ensure that the community's voice was heard in the design and development of mining projects and that the communities benefited fully from them. Tax revenues from mining should be transparent and should be used to improve social conditions of host nations as well as being invested in other activities that could generate streams of income for the future. "Minerals have the potential to contribute to poverty alleviation and broader economic development at the national level." However, it also emphasised that this was not a matter wholly in the gift of mining companies. To realise the full economic and social value of mining, the sector required consistent, honest and informed governance, something which had all too frequently been lacking in the past.

The MMSD process, quite apart from its success in opening up a constructive dialogue with all parties affected by the mining industry's activities, also encouraged the industry to address its broader interactions with other resource systems and with the societies in which it operated in a much more thorough and systematic way than it had done hitherto. Subsequent to the MMSD, the monitoring and reporting of companies' social and environmental programmes improved dramatically. The ICMM collaborated in the development of a special supplement to the Global Reporting Initiative's (GRI's) Sustainability Reporting Guidelines, launched in 2010, to ensure that reporting covered such sector-specific matters as biodiversity management, community consultation, indigenous peoples' rights, land disputes, mine closures and materials stewardship, in addition to the more regular reporting of its generation of wastes, its use of water, its emissions to the atmosphere, and its employment practices and health and safety performance. The industry also increasingly sought to be more proactive in its development and implementation of voluntary codes of conduct, as, for example, with the International Cyanide Management Code. (Franks, 2015)

This activity was not, of course, altruistic. It was done because the informed public had come to expect higher standards of mining companies than previously and, also perhaps, because by being proactive in adopting rigorous voluntary standards, the industry hoped to avoid having less well-thought-through legislative measures thrust upon it. Mining companies undoubtedly hoped that, by showing responsibility and demonstrating their commitment to the values of sustainable development, they would make themselves more acceptable as inward investors in mineral host countries and welcome and respected partners with local and regional communities.

Consequences of the commodity boom

The commodities boom referred to in the first section, which took off around 2004, had several important implications for mining and the resource nexus.

First, it deepened the geographic interconnectedness of the industry. Although itself a major producer of mineral raw materials, China's demand far outstripped its capacity to supply the raw materials it required from domestic sources, giving a major boost to international trade in minerals and to the prospects of countries with significant mineral endowments. The principal flows of minerals (including fuels) to China are illustrated in Figure 18.4. Note the particular importance of Australia, Africa and Latin America to the trade in non-fuel minerals. Some 70% of the world's annual production of iron ore is today traded across national boundaries. Thirty years ago the proportion was half this. In short, the boom served to increase further the mutual dependency of mineral-producing and mineral-consuming countries.

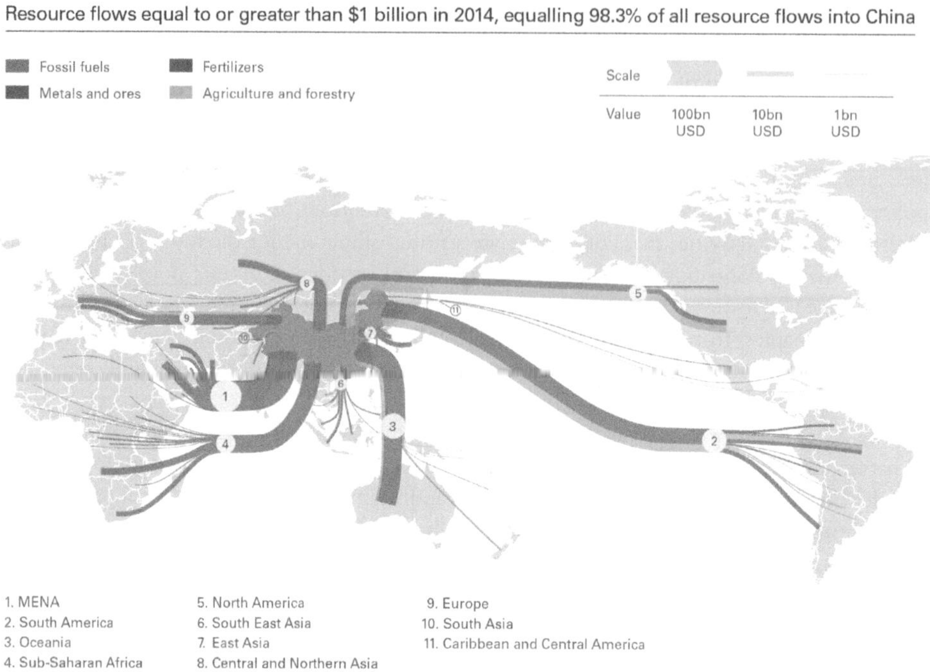

Figure 18.4 China resource flows

Source: figure reproduced with the permission of Chatham House, the Royal Institute of International Affairs.[2]

Second, the boom significantly raised the stakes in dealings between mining companies and mineral host countries and communities, generating tensions with regard to the terms on which the companies operated and the nature and scale of their contributions to local economic and social development. A study by Chatham House found that disputes in the resource sector increased dramatically in the 2000s relative to the previous decade. (Stevens et al., 2013)

While commodity prices were low there was not much to argue about but, with commodity prices soaring, everything about the industry became more important and many mineral host countries determined that they were not getting an appropriate share of the benefits. Taxes and royalties on mining were raised pretty much everywhere. Concerns also emerged in mineral host countries about ownership in the sector and a number of countries, including the Democratic Republic of the Congo, Zimbabwe, Kenya, Guinea, Burkina Faso, Madagascar, Sierra Leone, South Africa, Mongolia and Indonesia, enacted legislation requiring that a certain percentage of new mining projects be reserved for indigenous investors. Elsewhere, mining countries sought to ensure that minerals mined within their borders were processed domestically so as to capture the associated value added. Most dramatically, Indonesia banned the export of unprocessed mineral raw materials such as nickel ore and bauxite from the start of 2014. These developments, collectively referred to as expressions of resource nationalism, served to politicise the relationship between mineral host countries and inward investors in the mining sector. (Humphreys, 2015)

Below the national level, the new, higher profile of the industry and the emergence of plans to build new mines sometimes sparked a response from communities affected by mine developments. On the face of it, the engagement of communities with large mining companies was about ensuring that the companies behaved responsibly towards their neighbours and that the communities secured the best possible deal for their members. In practice, the situation was often more complex, with community opposition to mining being sometimes used to bring pressure to bear on government authorities to grant concessions on wholly other agendas, such as regional or indigenous peoples' rights. Peru, which became a target for significant new investment in copper mining, was the scene of numerous local disputes based around environmental concerns (often water) and the rights of indigenous peoples, as well as the authority of the central government to make decisions about mine development in the regions. There were major outbreaks of violent unrest at the Tia Maria copper mine in the Moquegua Region, at the Conga gold mining project in the northern region of Cajamarca and at the Las Bambas mine in the Apurímac region of southern-central Peru (Economist, 2016). An official government report in 2015 stated that the country was subject to around two hundred incidents of civil unrest a month, of which half were associated with mining (Telesur, 2015).

A third consequence of the boom was that it encouraged the development of small-scale, high-cost, and often badly polluting mine producers. Modern large-scale mine production is highly capital intensive and, accordingly, new mines often take many years to build and bring into production. Because of the speed and extent of the surge in demand and prices in the middle part of the 2000s, the large established producers were simply unable to respond with new supply in a timely fashion. As a result, smaller, more labour-intensive low-technology operations stepped in to plug the gap. This happened across many regions, including Africa, India and Indonesia, but the biggest surge in this type of production was in China, which was determined that shortages of mineral raw materials should not inhibit its industrial development. Between 2002 and 2013, China's mine production of copper tripled from 0.6 million tonnes to 1.7 million tonnes, while its iron ore production rose from 145 million tonnes in 2002 to a peak of 400 million tonnes in 2007 before dropping back. The surge in production propelled China into the front ranks of miners for a number of commodities but it came at a high environmental price, with the upsurge in mining, much of it unregulated or poorly regulated, doing extensive damage to water systems and to the health of employees and those living in the vicinity of mining operations. (Humphreys, 2015)

A fourth consequence was that the big mining companies were finding that the effort they had put into raising their standards of environmental and social performance was not paying the dividends they had hoped for with respect to relations with mineral host countries. Far from making these companies more attractive as inward investors, many mineral host countries viewed their fastidious and cautious behaviour as interfering and bureaucratic, preferring the engagement of investors with a more pragmatic ('no strings') approach to investment. The ability and willingness of Chinese companies to involve themselves in infrastructure projects, often with the support of state-backed development agencies and banks, often made them more attractive investment partners than their Western counterparts. While Western companies would doubtless argue that they were doing what they had to do anyway, since they could not be seen to be adopting different standards in different parts of the world, there was inevitably some disillusion about the fact that pragmatism and politics were frequently trumping the progressive business practices in which they had invested so much effort. In Africa in particular, big companies struggled to find and to commit to major investments, leaving large parts of the field to an assemblage of smaller opportunist companies, oligarchs, traders and middleman. (Burgis, 2015)

A fifth, and final, consequence of the boom was the politicisation of the issue of mineral supply in mineral-consuming countries. Rising commodity prices and the resurgence of resource nationalism triggered concerns in importing countries about the potential implications of these developments for their economic security. A number of studies in the US and European Union

(EU) pointed up their exposure to potential supply disruptions for a number of so-called critical minerals, many of these sourced from China. (Abraham, 2015) Despite widespread expressions of official concern, this did not result in many practical initiatives, save that in 2010, and again in 2012, the US and EU brought cases against China at the World Trade Organisation (WTO) challenging their imposition of export restrictions on a range of mineral products including rare earth minerals, molybdenum, tungsten, manganese and magnesium. These governments were, however, visibly reluctant to address the threat to their supplies through the active encouragement of domestic mining activity, presumably in part because of public antipathy towards the sector in these areas.

Where they were rather more active was with regard to the governance of mining companies listed on their stock markets and to the provenance of minerals used by their manufacturing industries. Thus there were significant efforts made through the Extractive Industries Transparency Initiative (EITI), launched in 2003, the 2010 Dodd-Frank Act in the US and under the EU Accounting Directive of 2013 to improve the transparency of payments made by companies to mineral host nations and to prevent the payment of bribes by companies domiciled in their jurisdictions. The Kimberly Process, launched in 2000, created a certification scheme designed to prevent the import of so-called conflict diamonds from regions where the revenues from diamond sales were believed to be supporting rebel activities. A similar certification scheme was introduced by the 2010 Dodd-Frank Act to restrict the import into the US of minerals such as tin, tungsten, tantalum and gold produced by rebel groups from the eastern provinces of DR Congo.

In short, the commodity boom created a whole new series of challenges with respect to the management of the nexus. The sudden and pressing requirement for mineral supplies which resulted from China's rapid industrialisation produced a response in some regions which harked back to an earlier, more rugged, type of industry practice when ends dominated means in the mining sector, this even as major mining companies and the governments of advanced economy countries pushed ahead with their attempts to encourage sustainable development and raise industry standards. It also injected a new, more political, dimension into the business of resource development.

After the boom

The commodity price boom peaked in 2011, while the peak in industry investment came the following year. China's growth rate began to slacken at the same time that production from new mines, the construction of which had been stimulated by higher prices, began to impact on mineral supply. Markets moved into balance and then into surplus with the result that mineral prices declined across the board, drastically reducing the profitability of the sector. As tends to happen at such times, the focus of the industry shifted from volume growth towards capital conservation and productivity improvement.

However, if prices were displaying their normal cyclical pattern, there was little likelihood that the larger companies in the mining industry could revert to their pre-boom, 1990s, behaviour in respect of the nexus. A line had been crossed. The nexus had become an integral part of the industry's understanding of the world and its role within it; it could not unlearn the insights that the nexus had given it. In time, growth would return and the industry would have to think about expanding again. Moreover, rising population, increased standards of living across the globe and the push for a lower carbon economy, meant that pressures on land use, on water resources and industry emissions would only increase, and these issues become increasingly politicised.

An illustration of how perspectives had changed was provided by the collapse of a tailings dam at the Samarco iron ore mine owned by two of the world's largest miners, BHP Billiton and Vale in the state of Minas Gerais in Brazil in late 2015. (Tailings dams are dams behind which mining companies accumulate liquid wastes arising from the process of concentrating ores so as to allow suspended solids to settle out.) Tailings dams are an undeniable weak point in the mining industry's environmental defences and history has witnessed a number of such failures before in many countries, including Papua New Guinea, the Philippines, Romania, Guyana, Canada and Spain. (Franks, 2015) However, the public reaction to the new disaster suggested that the world had moved a long way from the point where such a collapse could be regarded as 'just one of those things', an unfortunate side effect of a necessary activity. Influenced no doubt by the public outcry in the US to the oil spill at the Deepwater Horizon drill rig in the Gulf of Mexico in April 2010, the public and government reaction was sudden and severe. Although a relatively small part of BHP Billiton's portfolio, the company's share price fell 25% as the government launched a multi-billion dollar case against the companies involved. The eventual settlement for damages came to over $5 billion. The message was clear. Events of this sort are no longer an acceptable part of the cost of our mineral supplies; they simply should not happen.

Although slower market conditions and lower profitability will add challenges to the mining industry, they will also provide opportunities for positive change. They will reduce the pressure to build new mines and encourage the consolidation and rationalisation of existing enterprises. Some of the more marginal and less reputable companies will be driven from the business while in countries such as China and India, which saw uncontrolled mining growth in the boom years, there will be opportunities to close small inefficient and polluting mines and to focus production on larger, better regulated, and cleaner mines.

None of this, of course, will relieve the leading companies in the industry of their obligation to lift their standards with respect to its environmental and social performance still higher. The agenda of the MMSD, while it prompted major improvements in industry practices, still remains highly relevant to the next stage in its development. Events like Samarco notwithstanding, the industry has done much to persuade the public that it is serious in its attempts to minimise the negative impacts of its activities on the environment and on the communities among which it works. But it still has much to do to move beyond the defensive posture of mitigation of impacts towards a focus on the positive contributions the sector can make to sustainable development through the transformation of natural capital into increased wealth and enhanced well-being.

There is also much still to learn about the implications of the shift that has taken place from the industry's traditional preoccupation with below-ground resource issues to above-ground resource issues, and the constraints on its development potentially arising from the growing stresses on these above-ground resources. It is already clear that a significant proportion of the coal resources in the ground will remain there as 'unburnable carbon' since their recovery and use is incompatible with policies designed to combat global warming. In Chile, the world's largest copper producing country, there is simply insufficient freshwater available to support continued growth of production and many future projects will have to factor into the costs of development the procurement of water supplies from elsewhere or the construction of water desalination facilities.

In addressing pressures on these other resource systems with which mining interacts, greater emphasis will doubtless come to be placed on reducing the demand for minerals through increased resource efficiency and the push to a circular economy. (McKinsey, 2011) This still has a long way to run and in some emerging economies has scarcely even been embarked upon but the pressure for it will grow with time and there is significant research being undertaken to expedite its progress. The ultimate objective is to decouple the use of mineral raw materials from economic growth. Following the example of coal, it seems entirely plausible, after decades of worrying about the availability in the ground of minerals such as copper, that a certain amount of the resources we know about may turn out to be unusable long before physical exhaustion becomes an issue. (Humphreys, 2013a)

Perhaps the biggest outstanding challenge for mining and the nexus is that of governance; this at both international and national levels.

China has emerged in a relatively short space of time as much the largest consumer of the world's mineral raw materials. For many of the major base metals it accounts for almost 50% of global consumption; for seaborne iron ore its share is 70%. This is a degree of market dominance unparalleled in recent history (or at least since the US dominance of mineral markets in the 1950s). The speed of China's economic emergence and the urgency of its requirement for raw materials to support its industrialisation resulted in it adopting pragmatic policies towards trade and investment which have sometimes strained established multilateral arrangements and understandings intended to govern these matters. Its trade in minerals has, for example, on several occasions, brought it in conflict with the rules of the World Trade Organisation, while its practices of using large amounts of ethnic Chinese workers in Africa have brought it into conflict with local populations. The fact that non-transparent state organisations have been behind many of these activities has added a further difficulty. There is a need for the institutions governing trade and investment in minerals to be updated to reflect the new economic and political realities and, if they are to stand any chance of working, China and its principal raw materials suppliers need to play a central role in helping to design them. (Preston et al., 2016)

With respect to governance at the national level, while there is much that industry can contribute, here also the primary responsibility for channelling the preferences of society, and signalling to industry what it expects from it, rests with government. Although, as remarked above, mining has become increasingly political in recent times, the involvement of government in the affairs of the industry has often been piecemeal and confused. During the boom years, too much of the effort of mineral host countries was directed towards extracting quick gains from the industry rather than considering how mining could best contribute to long-term national economic development, a tendency which too often favoured less transparent and less scrupulous investors.

To serve a country's long-term interests, mining policy needs to incentivise good investors who are prepared to work closely with governments and communities in constructive and long-lasting partnerships. This implies a need to establish for those investors conditions of stability, transparency, security and predictability. This means good mining law, effectively and honestly enforced. Despite all the information about past events and behaviours which is available, it is noticeable how few countries have broken through to become attractive destinations for mining investment in recent times. The list of countries deemed attractive to investors by research organisations such as the Fraser Institute has changed little over the years and most investment in mining still goes to the same established mining countries with a proven track record as reliable, trustworthy hosts. (Figure 18.5) Among emerging economies, only Chile and Botswana have consistently been deemed to have attractive regimes for inward mining investment. Lower commodity prices will reinforce the tendency for investors to focus on the tried-and-tested, depriving other minerally well-endowed countries of the opportunity for development.

Good governance is also key to the effective management of mining and the resource nexus. Well-thought-through and properly implemented regulations for mining are critical to ensuring the internalisation of environmental costs of mining, acid drainage and use of land for waste. In approaching these matters, countries can usefully draw on the experiences of other countries which have successfully addressed these issues, as well as guidance from schemes such as the

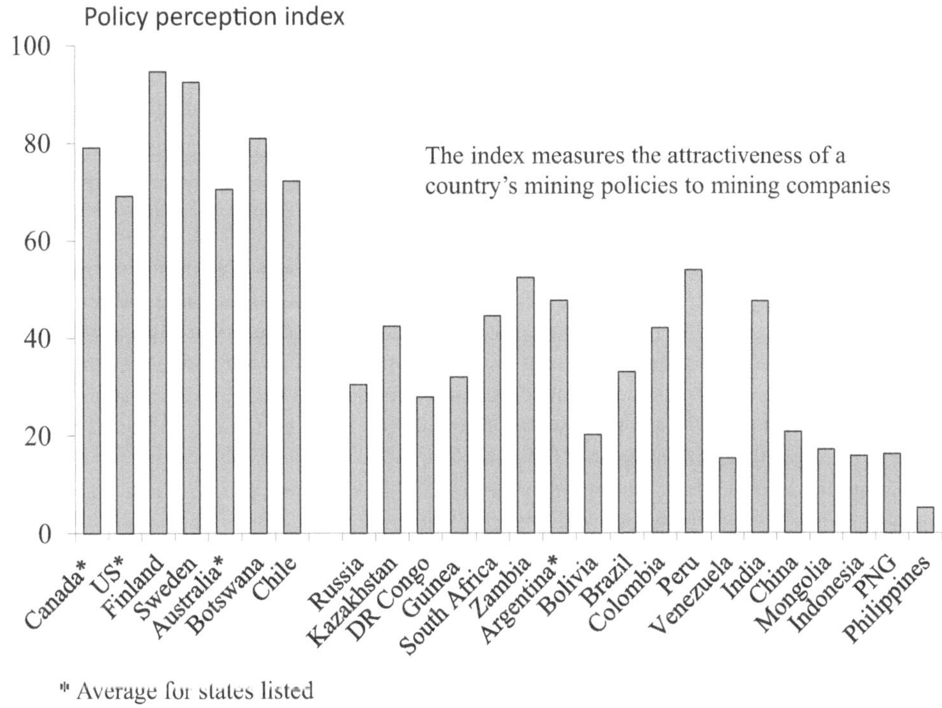

Figure 18.5 Perceived policy attractiveness of world's largest mining countries

Source: Fraser Institute (2015)

World Bank's *Extractive Industries Value Chain*. (World Bank, 2009) Good property law is needed to ensure that a company's entitlement to mine is matched by a responsibility for ensuring that the property is not subject to abuse at the cost to the wider community. Governments need to co-opt the power of the market to ensure that the economic factors used in mining fully reflect their social and environmental costs. Much of world's freshwater is under-priced leading to its inefficient use. Carbon emissions are under-priced everywhere and in some cases, as a result of poor energy policies, their generation continues to be subsidised.

The policies of the major mineral-consuming countries have scarcely been more coherent than those of the mineral-producing countries and too often the policies adopted have become victims of the law of unintended consequences. The initiatives described above to improve the transparency of payments made by companies to mineral host nations may be laudable but the primary impact of these measures all too often falls on the companies which are already the most responsible and the most heavily regulated. Little good is served by tightening rules for listed companies to a point where these companies are incentivised to delist from the stock market and go private. Moreover, as has been pointed out by one of the EITI's early advocates, greater transparency in these payments has sometimes had the perverse effect of putting political pressure on host governments to direct tax revenues from resource projects to higher wages and vanity projects rather than invest for the longer term, precisely the opposite of the effect intended. (Collier, 2015) The same applies to the schemes to suppress the trade in conflict minerals. The Kimberley Process, although it has generally been a success, has seen its credibility badly dented by the recognition of diamonds produced by the government of Zimbabwe as an approved source. Meanwhile, the attempts by the US government under the Dodd-Frank Act to restrict the import into the US of conflict minerals from the eastern provinces of DR Congo have proved so complex and unwieldy as to lead would-be buyers of these minerals to avoid purchases from all producers in that part of the world, both legitimate and illegitimate. (Thomson Reuters, 2014)

Conclusion

Recent years have seen a growing awareness of the extensive and complex manner in which mining interacts with other major resource systems. This has shifted the mining industry's focus away from its traditional concerns about physical mineral depletion towards pressures arising from the resource nexus more broadly, and on those from land use, water and energy in particular. This shift in focus has necessitated the development of business strategies and systems of governance which address the whole network of interactions between mining and other resource systems in a systematic and integrated fashion.

Significant advances have undoubtedly been made by the mining industry in understanding and addressing the challenges posed by this shift in focus. However, the results have been patchy and much remains still to be done. While some of the larger companies have made great strides in placing their activities into the broader framework of the nexus, large parts of the industry, in employment numbers and environmental impact if not necessarily in terms of value added, remain largely untouched by its exigencies. But perhaps the biggest gap to be bridged in advancing the cause of the nexus in mining is the gap in governance. All companies operate within an institutional context established by government, and the commodity boom exposed major deficiencies in this area. China's emergence as the world's dominant mineral consumer has shown that there is a pressing requirement to look at the institutional arrangements which govern international trade and investment in mined products. At the same time, governments in many mineral-rich parts of the world still have much to do to develop effective policies which extract the full economic and social benefits which mining potentially has to offer while at the same time supporting companies' efforts to manage the nexus by way of well-thought-through sector schemes and regulation.

Notes

1 The chapter follows the IMF distinction between advanced economies and emerging market and developing economies, save that in the interests of consistency over time the advanced grouping is held to include only the 'old' OECD countries. For brevity, the emerging market and developing economies are referred to simply as 'emerging economies'.

2 We are grateful to Chatham House, the Royal Institute of International Affairs, for permission to reproduce the figure entitled 'Chatham House and the DRC (2016)' in 'Navigating the New Normal: China and Global Resource Governance, 2016, Felix Preston, Rob Bailey, Siân Bradley, Dr Wei Jigang, Dr Zhao Changwen.'

References

Abraham, D. S. (2015). *The elements of power*. New Haven, CT: Yale University Press.

Auty, R. (1993). *Sustaining development in mineral economies: The resource curse thesis*. London and New York: Routledge.

Bardi, U. (2014). *Extracted*. White River Junction, VT: Chelsea Green.

Burgis, T. (2015). *The looting machine*. London: William Collins.

Collier, P. (2015). *Quoted in 'Transparency as building block, not stumbling block'*. Retrieved from https://eiti/org

Economist. (2016, February 6) From conflict to co-operation. *The Economist*.
EITI. (2003). Retrieved from https://eiti.org/
Ericsson, M. (2016). *Mining's contribution to sustainable development*. Presentation to seminar at UCL Institute of Sustainable Resources, February 24, 2016.
Franks, D. M. (2015). *Mountain movers: Mining, sustainability and the agents of change*. Oxford and New York: Earthscan/Rouledge.
Fraser Institute. (2015). *Survey of mining companies 2014*. Retrieved from https://www.fraserinstitute.org/
Global Reporting Initiative (GRI). (2010). *Mining and metals sector supplement*. Retrieved from www.global-reporting.org
Humphreys, D. (2013a). Long run availability of mineral commodities. *Mineral Economics*, 26, 1–2.
Humphreys, D. (2013b). New mercantilism: A perspective on how politics in shaping world metal supply. *Resources Policy*, 38, 3.
Humphreys, D. (2015). *The remaking of the mining industry*. Basingstoke and London: Palgrave Macmillan.
Kimberley Process. (2000). Retrieved from www.kimberleyprocess.com/
McKinsey Global Institute. (2011). *Resource revolution: Meeting the world's energy, materials, food, and water needs*. Retrieved from http://www.mckinsey.com/mgi/
McKinsey Global Institute. (2013). *Reversing the curse: Maximizing the potential of resource-driven economies*. Retrieved from http://www.mckinsey.com/mgi/
Meadows, D. H., Meadows, D. L., Randers, J., and Behrens, W. W. (1974). *The limits to growth*. London and Sydney: Pan Books.
MMSD. (2002). *Breaking new ground, mining, minerals and sustainable development*. London: Earthscan. Retrieved from https://www.iied.org/
Mudd, G. M. (2008). *Sustainability reporting and water resources: A preliminary assessment of embodied water and sustainable mining*. Melbourne: Monash University.
Norgate, T. E., Jahanshahi, S., and Rankin, W. J. (2007). Assessing the environmental impact of metal production processes. *Journal of Cleaner Production*, 15, 838–848.
Preston, F., Bailey, R., Bradley, S., Jigang, W., and Changwen, Z. (2016). *Navigating the new normal*. Royal Institute of International Affairs/Development Research Council of the State Council. Retrieved from https://www.chathamhouse.org/
Sachs, J. D., and Warner, A. M. (1995). Natural resource abundance and economic growth. *National Bureau of Economics Research Working Paper 5398*.
Stevens, P., Kooroshy, J., Lahn, G., and Lee, B. (2013). *Conflict and coexistence in the extractive industries*. Royal Institute of International Affairs. Retrieved from https://www.chathamhouse.org/
Telesur. (2015). *200 Social conflicts in Peru every month*. Retrieved from www.telesurtv.net
Thomson Reuters. (2014, July 1). Conflict minerals law starts working despite poor industry response. *Inside Metals*.
World Bank. (2009). *Extractive industries value chain*. Extractive Industries for Development Series #3. Retrieved from http://siteresources.worldbank.org/
World Steel Association (WSA). (2016). Retrieved from www.worldsteel.org/
World Bureau of Metal Statistics (WBMS). (2016). *World metal statistics*, various editions.

19

SCARCITIES, SUPPLY, AND NEW RESOURCE CURSES?

Raimund Bleischwitz and Jun Rentschler

Introduction

The supply side of resource provision is often seen from two contrasting views: one narrative is the scarcity debate and 'limits to growth', according to which the availability of resources is limited and uncertain. The other narrative is one of optimism and argues in favour of abundant supply, where any potential scarcity is prevented through technical progress and international trade. Both narratives feature prominently in contemporary media and have resembled in earlier debates (Benes et al., 2015). Yet, the mainstream of the evidence nowadays gravitates towards the middle. A survey would most likely conclude as follows: Basic minerals and fossil fuels are sufficiently available for a foreseeable timeframe (Meinert et al., 2016), the potential of biomass is significant, and concerns primarily relate to short-term issues around food and water and some critical materials.

In line with a cautiously optimistic outlook on the supply side of resources, the governance dimensions of turning natural subsoil assets into well-being for the people are more or less known, and many lessons have been learned towards better resource governance.

This contribution looks at those issues and underlines related nexus challenges. It puts forward a proposition about future constraints on using resources that will result from climate change and other planetary boundaries translating into critical drivers for supply. Looking at different scales this chapter highlights risk of a 'resource curse redux' in resource-rich developing countries originating from stress in water, food, and energy systems, as well as knock-on effects this may have on international supply chains and resource security.

Sustainability and the supply of natural resources

The evidence for physical supply scarcity and 'peak everything'

The debates on 'supply peaks' have focused most prominently on oil: Initiated by M. K. Hubbert in 1956, proponents of the 'Peak Oil' theory argued that the supply of crude oil – a finite resource – will reach a peak (supposedly around 1970), and decline subsequently. Data for the US appeared to confirm the validity of the Peak Oil theory until around 2008. However, thereafter annual oil production increased substantially up to nearly the peak level of 1970 due to the exploration and extraction of vast shale oil reserves.

This illustrates that simple extrapolations on resource availability fail to account for various uncertain factors, and thus yield inadequate predictions of a potential peak: In particular, climate change mitigation policies, the rate of exploration and discoveries, cost effectiveness of resource recovery, and changes in energy usage due to technological advances have all introduced and increased the uncertainties associated with the supply and affordability of oil.

These uncertainties are characteristic, not just for the supply of oil, but of most natural resources. Moreover, they illustrate the complexity of causal links between natural limits such as physical scarcity, the economics of prices, policy measures including licensing processes, political instability, as well as competing and changing interests and development priorities. From a geological perspective, it is crucial to understand the different meaning of reserves and resources. In simple terms, a *reserve* is a known quantity of a resource as established by drilling and sampling and its extraction is economically feasible; it typically is expressed as X tonnes of material with an average grade of Y at a cutoff grade of Z. *Resource* is a broader and more general term than reserve and includes identified material that may be less well characterized, possibly of lower grade and less certain to be economically recoverable. Resources can be converted to reserves by additional drilling or changes in economic factors, such as price or technology.

However, neither reserves nor resources are the same as "all there is" (Meinert et al., 2016). Both are highly dynamic categories. There are permanent efforts of new explorations shifting the frontier as well as new technologies turning previously known resources into economically feasible options. In that regard, one should use indicators such as 'static availability', which is usually expressed in years, with great caution as it suggests more supply limits than actually exist. Analysing exploration expenditures and success in terms of new discoveries is actually a much more valid indicator for future availabilities. It is here where economics matters: explorations are a risky and expensive business, which is why they are usually undertaken when commodity process are high. This also implies that explorations are cyclical and one driver of volatility, since new discoveries then require licensing and active companies for the supply to come on stream.

The debate can be illustrated by the case of phosphorus (Cordell et al., 2009). This is undeniably an essential and life-supporting mineral for food security, and it is often regarded as non-substitutable. There have been new discoveries of reserves since 2010, and the efficiency of supplying and using phosphorus fertilizers has been improving and can be expected to increase further in the future. After all, phosphorus can be recaptured and recycled, suggesting a much wider availability of secondary phosphorus in the future. Thus one should be very cautious to speak of any 'peak phosphorus'. Wellmer and Scholz (2016) conclude that there is a need for monitoring the geopotential in order to avoid bottlenecks in future supply and to enhance exploration technology development, recycling, and other means in a timely manner.

The new scarcities: planetary boundaries, unburnable fossil fuels, and water–energy nexus implications

Overall, it emerges that the traditional concern of scarcity (i.e. diminishing resource stocks) is no longer the central issue. In the case of oil for instance, Hubbert's fear of the depletion of fossil fuels, has reversed into a policy debate – motivated by the objective of climate change mitigation – which focusses on how to reduce the excessive use of fossil fuels. Too much, not too little, oil is the challenge.

Since Johan Rockström et al. (2009) published the *Planetary boundaries: a safe operating space for humanity*, the concept has been widely adopted. Planetary boundaries define and quantify a series of ecological stress thresholds, which must not be surpassed if catastrophic environmental consequences are to be avoided. In particular, thresholds and safe operating spaces surrounding them for nine specific dimensions need to be recognized: climate change, change in biosphere integrity (biodiversity loss, species extinction), stratospheric ozone depletion, ocean acidification, biogeochemical flows (phosphorus and nitrogen cycles), land-system change (e.g. deforestation), freshwater use, atmospheric aerosol loading (microscopic particles in the atmosphere that affect climate and living organisms), and introduction of novel entities (e.g. organic pollutants, radioactive materials, nanomaterials, and micro-plastics).

Going beyond these safe operating spaces can result in irreversible, sudden, and global environmental damages, which are likely to put the human living space at significant risk. Further publications, such as by Steffen et al. (2015), have subsequently improved the robustness of quantitative estimates. The key concept and some of the main conclusions associated with these estimates are reflected in major international policy agendas, including the UN' Sustainable Development Goals, and the Paris Climate Change Agreement (both 2015).

Indeed, 'new scarcities' have emerged: The question of how much oil and coal *can* be supplied, has turned into how much oil and coal *should* be supplied given constraints such as climate change, health, or air quality. For instance, recent studies on 'unburnable fossil fuels' have determined upper limits to energy consumption or economic activity, not based on physical resource supply constraints, but based on limited carbon budgets for the world in line with climate change requirements. Globally, a third of oil reserves, half of gas reserves and over 80% of current coal reserves should remain unused from 2010 to 2050 in order to meet the target of limiting global warming to 2°C (McGlade and Ekins, 2015).

The question arises to what extent similar methods can be applied to determine targets for overall resource consumption. Based on the assumption that all use of abiotic resources puts pressure on ecosystems, Bringezu (2015) proposes a target corridor for tonnes/per capita levels with fairly strong reduction implications. However, many of the assumptions made can be seen as controversial (e.g. construction minerals such as sand and gravel being of comparable environmental relevance like fossil fuels). We conclude here the new scarcities give evidence to using materials in a more responsible manner with likely reduction imperatives for environmentally intensive categories rather than proposing general reduction targets (Nickless, 2016; Akenji et al., 2016).

Ore grades as new sustainability indicator

For long, the sustainability principles for using abiotic resources have been related to their availability and shifting the resource base of economies to a more renewable one. David Pearce and Herman Daly once proposed that any investment into the exploitation of a non-renewable resource must be paired with a compensating investment into a renewable substitute (such as oil extraction paired with reforestation). These principles are well understood and often quoted, but not necessarily in complete alignment with contemporary evidence. First, because of availability being highly dynamic and not a reason of concern compared to environmental constraints. Second, because a worldwide shift to renewable resources is likely to transgress the boundaries of biosphere integrity; ecosystems may be unable to provide the scale of renewable resources, if those are to fully substitute non-renewable ones.

Against this background, ore grade development might be a relevant sustainability indicator for future assessments. By and large, and with the exception of iron, ore grades are declining globally (Giurco et al., 2014). This leads to higher energy intensity in mining, and also generates more mining waste. In addition, the water intensity of mining has been increasing. Similarly, extraction in remote and landlocked areas requires additional efforts for transportation. Thus, future sustainability assessments of resources could assess ore grade developments and other indicators related to environmental pressures. A nexus approach might help assessing the resource input requirements for mining and processing, in terms of energy, water and land, helping to improve overall productivity, reducing capital expenditures, identifying bottlenecks and eco-innovation opportunities across resources adapted to resource-rich countries. Such nexus research can achieve the following:

- Improve integrated knowledge of input-side resource constraints; for example due to decreasing ore grades and increasing energy- and water intensity of extraction of minerals, or increasing land-intensity of some renewable energy forms. Such evidence might inform mineral resource development and/or possible guidelines for planetary mineral consumption.
- Apply research findings on inter-related constraints at the output side ('limited absorptive capacities of ecosystems and environment') to analyse and assess inter-related impacts of resource use (e.g. due to higher energy demand and CO_2 emissions of land- and water-smart agriculture).

The governance of resource supply

The management of natural resources has been an important yet contentious policy issue, in particular in resource-rich developing countries. Resource wealth is associated with a mixed development track record, as there are exceptional development stories (both negative and positive) from both resource-rich and -poor countries. A key observation has been the possibility that resource wealth may be an impediment to economic development and the building of strong institutions, or, worse, be the cause of conflict and instability.

Diversified development: sustainability issues for resource-rich countries

In a prominent report for the World Bank, Gill et al. (2014) focus on the resource-rich countries of Central Asia (in particular former members of the Soviet Union) and analyse the track record of various approaches to industrial policy and resource management. Their report emphasizes that in these countries resource wealth played an important role in dampening and overcoming the severe economic crisis following the fall of the Soviet Union. As a side note one should keep in mind that the oversupply of resources on international markets following the industrial breakdown of the former USSR had cushioned much of the new demand from emerging economies in the nineties and helped keeping commodity prices low in that period. Similarly, the report suggests that in this region resource revenues and their smart reinvestment have contributed to mitigate consequences of the global financial crisis of 2007/2008.

Overall, since the fall of the Soviet Union, countries in Central Asia have benefited significantly from their resource wealth. Nevertheless, continuous adjustment and modernisation of national policies on resource management and industrial and economic development are essential. Lessons from several countries in Africa and South America show that such smart and flexible policymaking are the key to ensuring that resource wealth indeed remain a blessing and do not turn into a curse. In particular, Gill et al. (2014) argue, measures for the continuous improvement of economic efficiency are crucial: this includes increasing productivity, creation of jobs in the private sector, and reduction of economic volatility and instability. What we add here is the notion of resource productivity (i.e. the value added from using natural resources) that should become part of overall multifactor productivity increases.

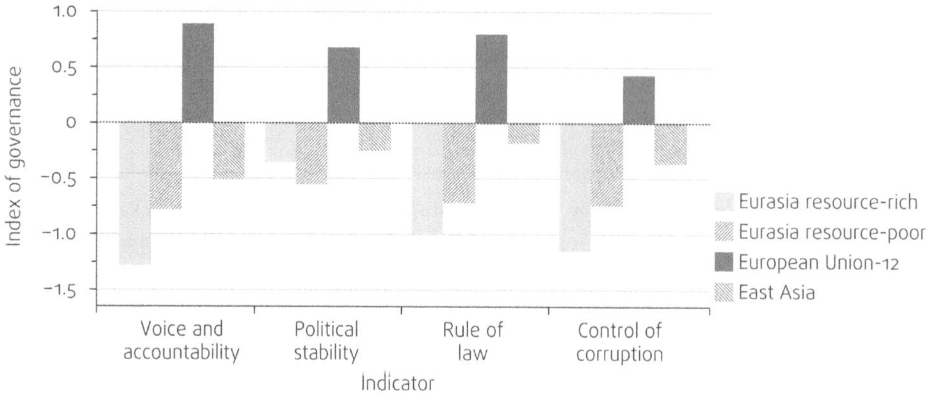

Figure 19.1 Selected indicators of governance in Central Asia. The indicator ranges from -2.5 (weakest governance) to 2.5 (strong)

Source: Gill et al. (2014)

However, it is important to note that economic efficiency is not equivalent to diversification: There is little – if any – evidence that economies which are highly concentrated on particular industries (e.g. natural resources) necessarily end up in a resource trap with high unemployment and economic volatility (Gill et al., 2014). The authors also show that sectoral diversification per se is not a sufficient condition for sustainable development.

On the basis of this evidence, the conclusion and recommendation is that resource-rich countries should not focus excessively on diversifying their export mix. Rather, governments should enhance the versatility and capacity of their country's factor of production. In particular, they should aim to achieve a balance between natural resources, physical assets, and human capital, as well as strengthening economic and administrative institutions.

The report by the World Bank also offers evidence showing that – in international comparison – countries in Central Asia have been remarkably successful in using resource wealth to the advantage of their citizens. The report flags that revenues from resource exports have been used primarily to build and expand physical capital; in particular oil and gas infrastructure, roads, railways, schools, and hospitals. At the same time (too) little attention has been paid to promoting and investing in human capital and effective institutions. Especially areas such as social services, transparency, governance, and unbiased regulation of business leave ample room for further progress (Figure 19.1).

Overall, empirical evidence shows that physical factors of production are not necessarily the key factor to distinguish whether resource wealth is a blessing or a curse. Rather, soft factors, including strong institutions, social services, transparency, as well as prudent and forward-looking policymaking have made the critical contribution to convert resource wealth into economic innovation and sustainable development. Accordingly, this approach also lies at the heart of the work of the Natural Resource Governance Institute.

The resource curse and potential remedies

The paradoxical observation that resource wealth may have devastating effects on economic development prospects and governance has been coined the "resource curse" (Auty, 1994). A large number of studies have since documented empirically that the potential negative side-effects of resource wealth are real and substantial (Sachs and Warner, 2001). In particular, various studies have empirically shown the link between natural resource wealth and numerous economic detriments, including insufficient economic diversification, underinvestment in public infrastructure, revenue volatility, and increased poverty. In a wider socio-economic context, further adverse effects have been documented, such as chronic corruption, a tendency to political instability and violent conflict, as well as a general lack of resilience to external shocks (Collier and Hoeffler, 2005).

The fact that particular oil-exporting countries had poor economic performances compared to some of the top-performing emerging economies, such as South Korea, led to two main observations:

Resource exports may lead to adverse macro-economic effects

In general, booming resource extraction sectors can lead to various macro-economic deficits, including the overvaluing of currency in foreign exchange rate terms, the neglecting of non-extractive sectors, the failure to establish a robust fiscal policy system, and lacking economic diversification of the economy beyond resource extraction. The key reason for these adverse economic effects is an appreciation of the currency due to major resource discoveries and exports. An overvalued currency results in deteriorating terms of trade for other sectors of the economy. This can result in further reductions of exports of other goods and services (known as 'Dutch Disease').

Resource wealth tends to result in weak institutions and governance

The sustainable management of natural resource reserves and revenues depends crucially on institutional capacity and systems of governance. Prudent reinvestment of natural resource revenues in factors of production, infrastructure, and public services are crucial for promoting economic diversification and strengthening alternative revenue sources. On the contrary, the absence of strong institutions is typically associated with symptoms such as high levels of corruption and political interference of interest groups.

The sustainable and prudent management of natural resource revenues has emerged as a key policy issue for hedging against uncertain and diminishing natural resource reserves, and preventing the resource curse. In principle, the basic rationale for such resource revenue management is to use revenues in contribution to the long-term sustainable development of a country. Such development financing can take many forms, and include investments in infrastructure, social safety nets, health and education, or the strengthening of institutions, and factors of production.

Conventional sovereign wealth funds (following Norway's example), or investment funds for enhancing domestic infrastructure, social safety nets, or health and education systems (as in Mozambique, Tanzania, Uganda, Zambia, Kazakhstan, and Malaysia) can be effective vehicles for long-term reinvestment if operated with transparent management and oversight (Halland et al., 2014). Gelb et al. (2014) acknowledge that there is a strong rationale for investing natural resource revenues domestically, but also point out several macro-economic risks (e.g. exacerbating boom-bust spending cycles). However, if done right (e.g. using transparent public tenders and international partnerships), allocating resource revenues to public domestic investments is likely to yield far larger development benefits than wasteful programs such as fossil-fuel subsidies (Halland et al., 2014).

These principles for natural resource management, while valid and of undoubted importance, focus on particular (extracted) natural resources – typically oil or minerals – but do not fully account for the fact that resources are part of a 'resource nexus'. However, as we discuss in this chapter, the management of resource reserves and revenues and the associated challenges cannot be seen in isolation, as they interact with other uncertainties, including environmental, economic, and political pressures. As we outline in the next section, the common commercially extracted natural resources are closely linked to a wide range of other resources, including water and food. It is also important to recognize a potential role of industry in efforts to obtain a social license to operate and help stabilize the fiscal and socio-political condition in extractive regions (ICMM, 2016).

Resource conflicts

Besides the adverse effects on long-term economic development and institutional capacity, natural resource wealth has also been associated with violent conflicts. In fact, both scarcity and abundance of natural resources can be underlying factors in conflicts (Humphreys, 2005; Collier and Hoeffler, 2005; Melvin and De Koning, 2011; Dinar, 2011). Common pool resources especially, such as water and land, can foster cooperation if affected groups manage property rights, limit free access, and establish mechanisms of trust and enforcement, but also results in conflict in fragile states. The legal and regulatory contexts around property rights vary widely across jurisdictions (at the local, national, and international levels) just as the adaptive capacities and resilience also vary widely.

The literature on conflicts has focused on what drives people to become involved in armed conflicts. 'Neo-Malthusian' approaches link demographic changes, especially population growth, with resource scarcity, environmental problems, and increased propensity for violent conflict (e.g. Homer-Dixon, 1991). One may thus derive drivers such as a young population and malnutrition as preconditions for conflicts. A second school of thought, represented by the work of Paul Collier, has been influential because it challenged previously held assumptions that resource scarcity was the primary driver of conflict (Collier and Hoeffler, 2005); in fact, abundance is strongly related to the presence of violent conflict. Abundance creates incentives and opportunities for looting of resources and making profits of transporting minerals from remote landlocked areas to ports.

Large-n studies of conflict have identified a correlation between the proximity to mines, for example, and the level of violence. These first two types of research usually intermingle in terms of causation, and are topped with a third strand of research focused on grievances over ethnic, class, or racial identity as a cause of conflict (Humphreys, 2005; Le Billon, 2001; Korf, 2011). From our perspective, however, usually these studies focus on a limited number of commodities, rather than on the strategic interconnections of the resource nexus.

The transparency agenda for extractive industries and resource-rich countries

Historical experience and countless pieces of anecdotal evidence show that resource revenues – rather than being used for economic diversification and modernisation – have been channelled into prestigious 'white elephant' investments, fuelled corruption, or disproportionately benefited foreign resource firms. According to the NGO Global Witness, more than US$4 billion in public resource revenues have disappeared without a trace in Angola between 1997 and 2002. An evaluation report by an UN expert council has determined in 2008 that a substantial share of resource revenues in the Democratic Republic of Congo is used for the purchase of weapons by regional war lords. In these cases natural resources are often regarded as a direct cause of conflict.

Thus, a number of non-governmental and international organisations have made the transparency of revenue streams a top priority. Their experience shows that publishing the magnitude, recipients, and application of resource revenue streams can be critical for ensuring that funds are used towards sustainable development objectives. These institutions argue that transparency is a pre-condition for accountability and efficient allocation of public expenditures.

At the G8's 2007 summit in Heiligendamm (Germany) heads of state of leading economies committed to increase transparency of flows of revenues and expenditures – particularly in the natural resource sector. The G8 Summit's final statement emphasizes that transparency in the natural resource sector is of decisive importance for ensuring accountability, responsible public decision-making, and sustainable economic development.

Some of the most prominent NGOs active in strengthening transparency of the natural resource sector are Publish What You Pay and the Extractive Industries Transparency Initiative (EITI), initiated by the then-British Prime Minister Blair.

The Publish What You Pay initiative focuses in particular on resource companies and developed economies, and pushes for transparent and complete reporting of monetary and physical flows. EITI emphasizes the role of resource-rich economies in creating transparent systems of resource governance. In particular, EITI member countries commit to:

- Publishing regularly all monetary flows between oil, gas, and mining sectors and public accounts, as well as all government revenues from these industries;
- Conducting regular and independent audits of resource-related public payments and revenues, and clarifying potential inconsistencies;
- Subjecting state-owned resource firms to the same transparency requirements and auditing scrutiny;
- Establishing a multi-stakeholder process, which serves the purpose of reviewing national EITI activities and information disclosure practices.

In 2016, 51 countries were implementing EITI standards, including a large number of sub-Saharan countries, as well as several Central Asian and Western developed economies. A notable absence among member countries are oil-rich countries of the Middle East, as well as large emerging economies such as the BRICS countries. EITI is supported by over 80 extractive companies, and over 90 large financial institutions.

Moreover, the US Dodd-Frank Act of 2010 was initiated following the financial crisis of 2008, with the objective of increasing regulatory oversight in financial markets. Section 1502 of the Dodd-Frank Act contains rules on the conflict minerals by publicly listed companies, while Section 1504 contains rules on transparency in extractive industries. For instance, companies must submit annual reports to the US Securities and Exchange Commission, disclosing payments made to governments at both the country (including sub-national) level and the project level. In August 2012, the commission adopted the long-awaited rules for implementation (Bleischwitz, 2014). The Trump administration in the US has since announced its intention to dismantle this legislation. Alongside, the European Parliament discusses a legal agenda on transparency, integrity, and accountability for extractive industries.

The focus of the above initiatives emphasizes the importance of transparency with respect to financial flows in particular. Thus, recent debates on natural resource revenues and development policy have considered institutionalized 'resource dividends' as a potentially highly effective method for redistributing resource rents directly to citizens; they have a proven track record in Alaska and detailed viability studies have been undertaken for a range of resource-rich developing countries, in particular in sub-Saharan Africa (van der Ploeg et al., 2012). For instance,

Sala-i Martin and Subramanian (2013) offer a discussion of the dimensions and consequences of the resource curse experience in Nigeria. They argue that increasing transparency is critical for overcoming the resource curse, and propose implementing a universal cash transfer scheme. While acknowledging the challenging political economy, they suggest that distributing oil revenues directly to citizens would significantly change the incentives for governance, and thus address the stifling problem of corruption.

Reducing fossil-fuel subsidies towards green fiscal reforms: the annual trillion dollar bailout of the fossil-fuel sector[1]

In 2015, after years of negotiations, governments from 197 countries adopted the Paris Agreement on Climate Change. As part of this treaty, the international community reaffirmed its commitment to mobilize $100 billion of 'climate finance' annually to promote renewable energies and low-carbon development.

However, the International Energy Agency (IEA) estimates that in 2014 countries around the world had spent close to $500 billion to subsidize fossil-fuel consumption. Another $450 billion had been spent in the G20 countries alone to subsidize fossil-fuel production. In other words, the same governments that would commit to the Paris Agreement were spending a multiple of their climate finance commitment on subsidizing fossil fuels.

There is no doubt that the subsidisation of fossil fuel consumption and production directly undermines any effort to mitigate climate change. However, the necessity and urgency of reform can only be fully understood when considering the complete range of adverse environmental, social, and economic side-effects of fossil-fuel subsidies (FFS). Besides driving climate change, FFS dis-incentivize investments, innovation, and efficiency; escalate fiscal burdens; crowd out funds for health, education, and other public infrastructure; incentivize corruption; aggravate air pollution; and reinforce poverty and income inequality. Overall, fossil-fuel subsidies cause severe economic distortions that compromise countries' prospects of achieving equitable and sustainable development.

Yet, the sheer magnitude of fossil-fuel subsidies being paid globally also hints at the size and complexity of the challenge. Although fossil-fuel subsidy reform indeed plays a critical role in climate policy, experience shows that the rationale for such reforms is determined in a complex environment of political economy challenges; macro-economic, fiscal, and social factors; as well as external drivers such as energy prices. The mixed track record of past reform attempts – a number of which resulted in violent street protests – illustrates how difficult reforms are in practice.

Experience also shows that FFS reform is not only about removing subsidies, but requires an integrated strategy comprising a range of carefully designed and sequenced policy measures. In particular, reforming subsidies means understanding and gauging economic and political trade-offs, mitigating adverse effects on vulnerable households, assisting firms with implementing efficiency enhancing measures, and ensuring the long-term contribution of subsidy reform to sustainable development. While the political economy obstacles to reform can be immense, strong government commitment, thorough preparation, careful reform design, effective communication, and smart timing can significantly improve the success rate.

Supply chain management and final consumption

Resources are traded as commodities and subject to business transactions via open markets and supply chain management. The latter can be characterized as design, planning, execution, control, and monitoring of supply chain activities with the objective of creating net value, building a competitive infrastructure, leveraging worldwide logistics, synchronizing supply with demand, and measuring performance globally. Resources figure in as risks, if supply is interrupted or constrained and if prices are volatile and relevant for purchasing costs.

Despite the commodity boom of the 2000s and despite the substantial sustainability impact of the initial stages of minerals, it seems these remain largely outside the scope of sustainable supply chain management so far (Sauer and Seuring, 2017). Thus a relevant future area of activities should address the life cycle of resources being used along supply chains. Such supply chain management should have three dimensions:

- The relevant costs to business and the establishment of a resource costing model with a view to also enable resource cost savings;
- A tracking of the wide resource use as indicated through the various footprints;
- Adding sustainable values, such as efforts for product innovations, systems innovations, and new business models.

Nexus challenges and soft extraction pathways for resource-rich countries

Future stress factors and a fresh view on 'criticality'

The resource nexus has become more pressing in recent years because volatile commodity prices transmit turbulences from local to global markets faster than ever before. Recent evidence on the dangerous conjunction of high prices for food and water and social tensions could be witnessed during the Arab uprisings in 2011. Sternberg (2012) points to the drought that occurred in Northern China as a global trigger mechanism for higher food prices. Johnstone and Mazo (2011) discuss possible impacts of climate change as stress multipliers.

The International Food Policy Research Institute underlines additional domestic factors such as malnutrition and the phasing out of food support programs (Breisinger et al., 2011). In a historical perspective the impact of food prices on the great revolutions in France (1789), Russia (1917), and other civil wars illustrate the security dilemmas of the population and the contingent political outcomes.

However, the relationship between local water and food shortages, resource constraints, and socio-economic struggle and political unrest is a complicated one. In general terms, we observe two new main challenges. First, global drivers can overshadow local drivers in the management of common pool resources such as river basins management and agriculture. Second, increasing connectivity allows local turbulences to spread rapidly, with unintended side-effects on other resources and regions. Thus, fragile countries and regions are likely to become more vulnerable, but other regions including the industrialized countries will have to cope with the indirect impacts too. Adaptive capacities will have to be developed for the directly affected regions and others more indirectly affected.

Figure 19.2 distinguishes three more general stress multipliers: climate change and weather extremes, volatile commodity prices, and population growth from the following drivers at the local and regional level:

- **Malnutrition:** a local and regional food and water crisis can lead to malnutrition for a relevant number of people.
- **Migration:** A local crisis drives people into migration, quite often into neighbouring cities or regions. Migration is also a key social issue arising during and after the time when a

mine is constructed. Rudra and Jensen (2011) and Bearce and Laks Hutnick (2011) provide insights on the nexus between migration and natural resources.

- **Fundamentalism and terrorism:** Some actors may adopt hostile attitudes, in particular if their national government appears to cooperate with foreign powers and offers little support to local people. Natural resources could also provide a way to finance rebellions that have been started for other reasons and may extend the duration of civil wars (see, e.g., Blomberg et al., 2011; Okpanachi, 2011).
- **Secessionism:** This is often seen as a relevant source of violent conflicts and is especially relevant for those extraction activities and reserves that are located in well-defined areas of a country with socio-cultural heterogeneity. Sudan and South Sudan may be seen as a case study.
- **Organized crime:** This is an issue that starts at a lower level with a vengeance-seeking group grievance but could increasingly involve extraction and trade with conflict minerals (but also other natural resources), drugs, and other illicit activities. The cases of Mexico's drug trade, maritime piracy, and illicit trade routes may be illustrative.
- **Civil wars or severe violations of human rights:** This is especially prominent in those regions that have long-lasting civil wars such as Iraq, Afghanistan, and Sudan where natural resource production is hampered and chances for peace are undermined (Mitchell and Thies, 2012; Wennmann, 2011).

The combination of these factors could translate into what we call a redux of the resource curse: triggered by the emergence of a food and/or water crisis – whatever the causes may be – local and national governance mechanisms are vulnerable and may not be able to cope with such a shock. If people's access to water and food is obstructed and if existing institutional resilience is low, fragile states and regions will be put at risk of increased instability, where the mechanisms outlined in this section might escalate (Figure 19.2). Any such escalation may then lead to severe conflicts and may also have knock on effects on supply chains for essential materials with international repercussions.

Many resource-producing countries and regions can be considered fragile. In particular, new resource suppliers in Africa and Asia that started extracting on a large scale during the boom of the years 2000–2008 and in the aftermath of the financial crisis have to struggle with basic governance challenges, including:

- Negotiating fair agreements with extractive industries;
- Managing the switch from the employment-intensive construction period of a mine to the more capital-intensive extraction period, which is often associated with considerable lay-offs;
- Establishing transportation infrastructure that meets the needs of broader development purposes and environmental standards, in particular if the country is landlocked;
- Balancing the needs of the affected local communities with other regions and the general public (quite often the population in the capital);
- Establishing a robust fiscal regime with permission grants, royalties, and rents; and
- Dealing with environmental issues that partly arise through mining processes but usually have other causes.

Primary materials, green economies, and soft extraction pathways

Green economies are likely in demand for more materials in order to meet the SDGs and requirements of low-carbon societies. Many of the new SDGs will lead to an increase in material demand; for instance:

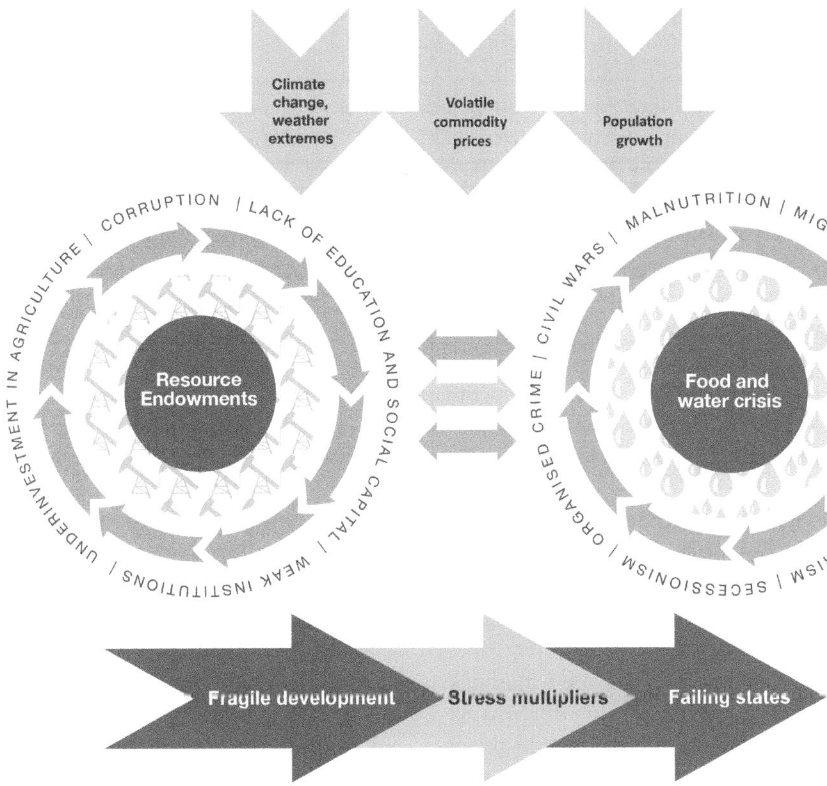

Figure 19.2 Drivers of a new nexus vulnerability

Source: Bleischwitz et al. 2014

- Goal 2 to "End hunger, achieve food security and improved nutrition and promote sustainable agriculture" implies increasing demand for land, mineral fertilizers, water, biomass and food.
- Goal 6 to "Ensure availability and sustainable management of water and sanitation for all" implies investments in water supply and a water distribution infrastructure (i.e. increasing demand for materials).
- Goal 7 to "Ensure access to affordable, reliable, sustainable and modern energy for all" is likely to imply increasing demand for bio-energy and renewable energy, which again implies more demand for land, biomass, water, and materials.
- Goal 9 to "Build resilient infrastructure, promote inclusive and sustainable industrialization" will require more construction materials, metals, and other materials.

Adding the promotion of economic growth to this agenda, the signals are that future material demand stemming from the SDGs is likely to increase. At least for key metals (aluminium, iron ore, copper, and nickel, which altogether make up for more than 80% of world production of metals), for construction minerals, and for biomass, the SDGs are very likely leading to new and additional demand compared to business-as-usual forecasts.

The situation for energy fuels is less straightforward as climate policy will probably lead to restrictions for fossil-fuel usage. On the other hand, the expected increase in unconventional fuels and renewable energies will lead to an additional demand for materials (see Chapter 24), and bio-energy requires land, water, and mineral fertilizers.

The SDGs, on the other hand, also endorse the sustainable production and consumption agenda, and call for global increases in resource efficiency as well as for aims to achieve sustainable and resource-efficient infrastructures by 2030 (SDG9) and sustainable management and efficient use of all resources by 2030 (SDG12).

The balance between such expected demand increases and other goals however is not entirely clear, in particular as key terms (such as sustainable management and efficient use of all resources) are insufficiently defined and will leave space for quite different implementation pathways. Nexus research will thus be pivotal in addressing interlinkages and trade-offs in the implementation of the SDGs.

Soft extraction pathways could become a smart strategy coping with the dilemma of needing primary resources and turning resource endowments into well-being for the people, and accepting limits to an excessive extraction. Future planning for extraction should seek to take nexus issues into account via assessing:

- Availabilities against future risks of water and food stress, energy intensities, and requirements for transportation;
- Opportunities of applying green technologies in extraction processes and early stages of the supply chain, including renewable energies, circular economy options, as well as by-production of critical materials;
- Capacities of institutions to cope with shocks, make fiscal arrangements promoting sustainable growth, and facilitate a social license to operate;[2]
- Long-term implications via foresight processes and biophysical and economic modelling that include the entire value chain, changes in final demand, and expected markets for secondary materials.

Conclusions

The supply side of minerals has long been seen as a subject for geology. Likewise, forestry, agriculture, and hydrology have been their counterparts for other dimensions of the resource nexus. While geology proves a dynamic availability of resources from the earth crust for the future, the broader picture is more complex. With declining ore grades making extraction more energy intensive; stress multipliers from water, food, and climate change; and limited absorptive capacities in a number of countries, we are witnessing new scarcities and supply risks. These risks are especially high in vulnerable countries and may spark international conflicts if agreements on access to resources are either not reached or not honoured. This is why the term 'resource curse redux' seems appropriate. In some circumstances there might even be conditions for a 'perfect storm'. Many of the SDGs will lead to trade-offs and thus nexus challenges for implementation; yet SDG12 also offers an avenue for more integrated approaches. Future nexus research should thus address the supply interlinkages of resources and apply principles of sustainability, resource governance, and a circular economy for extractive industries. Removing subsidies for fossil fuels could yield a number of benefits beyond energy and macro-economics. In the political economy setting more should be done to develop soft extraction pathways for resource-rich countries.

Notes

1 Based on Rentschler and Bazilian (2017, 2016).

2 See on this issue: www.resourcegovernance.org.

References

Akenji, L., Bengtsson, M., Bleischwitz, R., Tukker, A., and Schandl, H. (2016). Ossified materialism: Introduction to the special volume on absolute reductions in materials throughput and emissions. *Journal of Cleaner Production*, 132, 1–12.

Auty, R. M., Jan. (1994). Industrial policy reform in six large newly industrializing countries: The resource curse thesis. *World Development*, 22(1), 11–26.

Bearce, D. H., and Laks Hutnick, J. A. (2011). Toward an alternative explanation for the resource curse: Natural resources, immigration, and democratization. *Comparative Political Studies*, 44(6), 689–718.

Benes, J., Chauvet, M., Kamenik, O., Kumhof, M., Laxton, D., Mursula, S., and Selody, J. (2015). The future of oil: Geology versus technology. *International Journal of Forecasting*, 31(1), 207–221.

Bleischwitz, R. (2014). Transparency in the extractive industries: Time to ask for more. *Global Environmental Politics*, 14(4), 1–9.

Bleischwitz, R., Johnson, C. M., and Dozler, M. G. (2014). Re-assessing resource dependency and criticality. Linking future food and water stress with global resource supply vulnerabilities for foresight analysis. *European Journal of Futures Research*, 2(1), 34.

Blomberg, S. B., Broussard, N. H., and Hess, G. D. (2011). New wine in old wineskins? growth, terrorism and the resource curse in sub-Saharan Africa. *European Journal of Political Economy*, 27, 50–63.

Breisinger, C., Ecker, O., and Al-Riffai, P. (2011). *Economics of the Arab awakening: From revolution to transformation and food security*. IFPRI Policy Brief No. 18. Washington DC: International Food Policy Research Institute.

Bringezu, S. (2015). Possible target corridor for sustainable use of global material resources. *Resources*, 4, 25–54. doi:10.3390/resources4010025

Collier, P., and Hoeffler, A. (2005). Resource rents, governance, and conflict. *Journal of Conflict Resolution*, 49(4), 625–633.

Cordell, D., Drangert, J-O., and White, S. (2009). The story of phosphorus: Global food security and food for thought. *Global Environment Change*, 19(2), 292–305.

Dinar, S. (2011). *Beyond resource wars: Scarcity, environmental degradation, and international cooperation*. Cambridge, MA: MIT Press

Gelb, A., Tordo, S., Halland, H., Arfaa, N., and Smith, G. (2014). Sovereign wealth funds and long-term development finance risks and opportunities. *World Bank Policy Research Working Papers, 6776*.

Gill, I. S., Izvorski, I., van Eeghen, W., and de Rosa, D. (2014). *Diversified development – making the most of natural resources in Eurasia*. Washington, DC: The World Bank.

Giurco et al. (2014) Responsible mineral and energy futures: views at the nexus, *Journal of Cleaner Production* 84, 322–338.

Halland, H., Beardsworth, J., Land, B., and Schmidt, J. (2014). *Resource financed infrastructure – a discussion on a new form on infrastructure financing*. Washington, DC: The World Bank.

Homer-Dixon, T. F. (1991). On the threshold: Environmental changes as causes of acute conflict. *International Security*, 16(20), 76–116.

Humphreys, M. (2005). Natural resources, conflict, and conflict resolution. *Journal of Conflict Resolution*, 49(4), 508–537.

International Council on Mining and Metals (ICMM). (2016). *Role of mining in national economies* (3rd ed.). London: International Council on Mining and Metals.

Johnstone, S., and Mazo, J. (2011). Global warming and the Arab spring. *Survival*, 53(2), 11–17.

Korf, B. (2011). Resources, violence and the telluric geographies of small wars. *Progress in Human Geography*, 35(6), 733–756.

Le Billon, P. (2001). The political ecology of war: Natural resources and armed conflicts. *Political Geography*, 20(5), 561–584.

McGlade, C., and Ekins, P. (2015). The geographical distribution of fossil fuels unused when limiting global warming to 2 °C. *Nature*, 517, 187–190.

Meinert, L. D., Robinson, G. R., and Nassar, N. T. (2016). Mineral resources: Reserves, peak production and the future. *Resources*, 5, 14. doi:10.3390/resources5010014

Melvin, N., and De Koning, R. (2011). *Resources and armed conflict. In: SIPRI yearbook 2011*. Stockholm: Stockholm International Peace Research Institute.

Mitchell, S. M., and Thies, C. G. (2012). Resource curse in reverse: How civil wars influence natural resource production. *International Interactions*, 38(2), 218–242.

Nickless, E. (2016). Resourcing future generations: A global effort to meet the world's future needs head-on. *European Geologist*, 42, 46–50.

Okpanachi, E. (2011). Confronting the governance challenges of developing Nigeria's extractive industry: Policy and performance in the oil and gas sector. *Review of Policy Reserch*, 28(1), 25–47.

Rentschler, J. E., and Bazilian, M. (2016). Reforming fossil fuel subsidies: Drivers, barriers and the state of progress. *Climate Policy*, 891–914.

Rentschler, J. E., and Bazilian, M. (2017). Principles for the effective design of fossil fuel subsidy reforms. *Review of Environmental Economics and Policy*, 11(1), 138–155.

Rockström, J., Steffen, W., Noone, K., Persson, A., Chapin, F. S., Lambin, E. F., Lenton, T. M., Scheffer, M., Folke, C., Schellnhuber, H. J., Nykvist, B., de Wit, C. A., Hughes, T., van der Leeuw, S., Rodhe, H., Sorlin, S., Snyder, P. K., Costanza, R., Svedin, U., Falkenmark, M., Karlberg, L., Corell, R. W., Fabry, V. J., Hansen, J., Walker, B., Liverman, D., Richardson, K., Crutzen, P., and Foley, J. A. (2009). A safe operating space for humanity. *Nature*, 461, 472–475.

Rudra, N., and Jensen, N. M. (2011). Globalization and the politics of natural resources. *Comparative Political Studies*, 44(6), 639–661.

Sachs, J. D., and Warner, A. M. (2001, May). The curse of natural resources. *European Economic Review*, 45(4–6), 827–838.

Sala-i-Martin, X., and Subramanian, A. (2013). Addressing the natural resource curse: An illustration from Nigeria. *Journal of African Economies*, 22, 570–615.

Sauer, P. C., and Seuring, S. (2017). Sustainable supply chain management for minerals. *Journal of Cleaner Production*, 151, 235–249. ISSN: 0959-6526. http://dx.doi.org/10.1016/j.jclepro.2017.03.049.

Steffen, W., Richardson, K., Rockström, J., Cornell, S. E., Fetzer, I., Bennett, E. M., Biggs, R., Carpenter, S. R., de Vries, W., de Wit, C. A., Folke, C., Gerten, D., Heinke, J., Mace, G. M., Persson, L. M., Ramanathan, V., Reyers, B., and Sörlin, S. (2015). Planetary boundaries: Guiding human development on a changing planet. *Science (80-.)*. http://science.sciencemag.org/content/347/6223/1259855/tab-article-info

Sternberg, T. (2012). Chinese drought, bread and the Arab spring. *Applied Geography*, 34, 519–524.

van der Ploeg, F., Stefanski, R., and Wills, S. (2012). *Harnessing oil revenues in Ghana*. London: International Growth Centre London, School of Economics and Political Science.

Wellmer, F. W., and Scholz, R. (2016). Peak minerals: What can we learn from the history of mineral economics and the cases of gold and phosphorus? *Mineral Economics*, 30(2), 73–93.

Wennmann, A. (2011). Breaking the conflict trap? Addressing the resource curse in peace processes. *Global Governance*, 17(2), 265–279.

20
THE INTERNATIONAL COMMODITY TRADE
Stylized facts

Vincenzo De Lipsis, Paolo Agnolucci and Raimund Bleischwitz

Introduction

The fundamental question about what conditions ensure sustainable water security must be understood having in mind the crucial nexus between water, energy and food production. Future water security will depend on the behaviour of several economic actors that interact mainly within the mechanism of international commodity markets. One of the common complaints about markets as allocation mechanism is their inability to correctly assess the sustainability of certain economic activities. The value of the water resources consumed to produce important staples, for instance, is not reflected in market prices, with the consequence of ignoring the risks related with an excessive use of water resources (Allan, 2011). Despite most of the virtual water included in food commodities remains within the national borders of the producers, an increasing portion is now being internationally traded, which has helped provide water security in regions in need but has not guaranteed its sustainability. On the other hand, a crucial role is played by consumer preferences with respect to water consumption, which is a function of demographic trends, food behaviour and diets. For all these reasons it becomes fundamental to understand how the very place of interactions among all these players actually work, and how its features have changed over time. Therefore, to fully understand the implications that the nexus has for current important global issues we cannot transcend from analysing the characteristics of international commodity markets.

The recent dynamics in the price of primary commodities has attracted considerable interest in these markets. After a long period of high price levels, we have witnessed a substantial decline in commodity prices, since 2011 in the case of metals and agricultural products, and since 2014 in the case of crude oil. These changes are related to modifications in the numerous factors that lie behind supply and demand in the market, with some of these aspects being temporary shocks, while others reflecting more permanent transformations of the market structure. In this chapter we shed some light on this latter group of phenomena by looking at important stylized facts of the structure and the evolution of some specific commodity markets. We put the recent events in a broader historical perspective by focusing on the international dimension of the modern commodity industry, and in particular on the recent trade pattern that has emerged among different countries and world regions.

The new role played by Asian emerging economies, and in particular by China, has had a strong impact on the configuration of the trade flows in both manufactures and primary commodities. By considering the price of important commodities some commentators have referred to this as an example of super cycle, drawing a comparison with previous episodes of industrialization involving major countries.[1] We analyse recent data on both the monetary and physical dimensions of the international trade in major commodities to uncover the trade interdependence between different countries in the world, and highlight some important peculiarities in the ongoing development of commodity markets. Our attention is focused on three main dimensions that characterize the commodity sector. First, we use publicly available datasets to understand to which extent are individual markets globally integrated, looking at volumes of exports and production. Second, we examine the direction of commodity trade flows between different geographical regions, in order to identify which countries can be described as world suppliers or consumers with respect to important commodity products. Third, we measure the supply concentration of commodity markets, which represents an important feature of market structure.

We start in the second section by providing a brief taxonomy to illustrate the distinctive features of specific groups of commodities, and assessing their importance in the economic development of a country. In the third section we offer a brief description of the most important historical events that have transformed the structure of commodity markets. In particular, we will describe how technological progress has dramatically reduced transportation costs and facilitated the emergence of global commodity markets. A description of the geography of production and consumption of different commodities is the topic of the fourth section. In that section we examine how the importance of different countries in the international trade has evolved over time, and analyse recent data on real imports of commodities to assess the magnitude of the influence of China in the international trade patterns. In the fifth section we examine how market concentration for some essential commodities has changed in the last few decades, and identify which countries have played the role of dominant supplier of important primary goods, while the growing importance of international trade in water is the subject of the sixth section. Final remarks and a summary of the main findings can be found in the final section.

Taxonomy

A well-known feature of economic development is the declining importance of the primary sector in the national output. Interestingly, most of the reduction in the relative size of the primary sector is due to agriculture. Only a slightly decreasing trend is generally observed for the mining industry, while the energy sector has been responsible for a fairly constant, sometimes increasing, share of the national output. As shown in Table 20.1, a fall in the size of the primary sector as a proportion of GDP can be observed in most countries between 1970 and 2014. A slower decrease, or sometimes an increase, in the importance of the primary sector in the wealth generation process can be explained as the result of large investments in the energy production, e.g. in Norway and Russia, or mining capacity, as for instance in Australia and Argentina.

While defining commodities as the output from the primary sector is useful when analysing economic growth, it is not appropriate when studying commodity trade between countries. A commonly accepted definition of commodities in the international trade literature is the one incorporated in the Standard International Trade Classification (SITC), adopted by the UN.

Table 20.1 Output from the primary sector, and the mining and energy sector, as a share of GDP

		1970	*1978*	*1987*	*1996*	*2005*	*2014*	
Argentina	Primary	10.4	9.5	10.8	9.4	16.0	16.5	↑
	Mining and energy	2.2	3.1	3.9	4.4	7.6	8.3	↑
Australia	Primary	13.6	16.2	14.1	11.2	12.8	14.4	↗↘
	Mining and energy	7.1	9.2	9.5	7.6	9.8	12.0	↑
Brazil	Primary	14.6	13.1	14.3	8.7	12.0	12.0	↗↘
	Mining and energy	3.0	2.8	4.9	3.5	6.5	6.4	↑
Canada	Primary	11.5	13.5	11.4	11.0	13.3	11.8	–
	Mining and energy	7.1	9.3	8.3	8.0	11.5	10.1	↑
Chile	Primary	19.0	16.6	20.5	15.3	22.8	18.2	↗↘
	Mining and energy	13.7	11.0	13.9	10.1	18.7	14.8	↗↘
China	Primary	71.6	72.1	64.5	60.6	21.2	17.0	↓↓
	Mining and energy	36.4	43.9	37.8	41.0	9.1	7.6	↓↓
France	Primary	9.8	7.1	6.6	5.7	4.6	4.2	↓
	Mining and energy	2.3	2.5	3.1	3.0	2.7	2.6	–
Germany	Primary	8.2	7.0	5.8	4.4	3.9	3.8	↓
	Mining and energy	5.1	4.5	4.3	3.3	3.1	3.1	↓
India	Primary	46.5	41.2	37.0	34.6	25.6	21.8	↓↓
	Mining and energy	2.8	4.0	6.2	6.3	6.3	4.8	↗↘
Indonesia	Primary	47.7	43.1	31.8	21.8	24.2	25.0	↓↓
	Mining and energy	5.0	17.1	12.1	8.0	12.0	11.3	↗↘
Israel	Primary	9.4	9.5	6.7	3.9	3.8	3.8	↓
	Mining and energy	3.6	4.3	3.0	2.1	2.1	2.5	↗↘
Italy	Primary	11.5	9.0	7.1	6.3	5.0	5.3	↓
	Mining and energy	2.8	2.4	2.9	3.0	2.7	3.1	↗↘
Japan	Primary	8.8	7.3	6.0	4.5	3.6	3.1	↓
	Mining and energy	2.9	2.9	3.3	2.8	2.4	1.9	–
Mexico	Primary	19.7	19.5	29.2	16.8	16.6	15.8	↗↘
	Mining and energy	9.2	10.4	21.8	11.3	13.4	12.5	↗↘
Nigeria	Primary	29.4	26.7	35.8	47.4	40.5	31.8	↗↘
	Mining and energy	5.3	8.1	10.8	20.1	14.9	11.6	↑
Norway	Primary	9.8	14.8	17.4	22.6	30.3	26.5	↑↑
	Mining and energy	4.0	10.1	13.7	20.1	28.7	24.8	↑↑
South Korea	Primary	36.6	28.2	16.0	8.2	6.0	5.3	↓↓
	Mining and energy	2.8	2.7	4.0	2.7	2.8	3.0	↗↘
Russia	Primary				18.1	19.4	18.4	–
	Mining and energy				10.9	14.4	14.2	↓
South Africa	Primary	18.1	23.8	21.0	14.0	11.9	14.6	↗↘
	Mining and energy	10.9	17.0	15.4	9.8	9.2	12.1	↗↘
United Kingdom	Primary	7.4	9.9	9.3	6.9	5.1	4.8	↓
	Mining and energy	5.0	8.0	7.8	5.5	4.5	4.1	↗↘
United States	Primary	6.0	6.9	5.7	4.9	4.5	5.7	↗↘
	Mining and energy	3.7	4.6	4.2	3.5	3.5	4.5	↗↘

Source: United Nations Statistics Division

Key: upward trend (↑); strong upward trend (↑↑), fluctuating (↗↘), stable (–), downward trend (↓), strong downward trend (↓↓)

Within the approach incorporated in the SITC, we will use the taxonomy of commodities in four major classes proposed by Radetzki (2008):

1. unprocessed and processed food
   a. Food and live animals (SITC 0)
   b. Beverages and tobacco (SITC 1)
   c. Oil seeds, oil nuts and oil kernels (SITC 22)
   d. Animal and vegetable oils and fats (SITC 4)
2. agricultural raw materials
   a. Hides, skins and fur skins, undressed (SITC 21)
   b. Crude rubber including synthetic and reclaimed (SITC 23)
   c. Wood, lumber and cork (SITC 24)
   d. Pulp and paper (SITC 25)
   e. Textile fibres, not manufactured, and waste (SITC 26)
   f. Crude animal and vegetable materials, n.e.s. (SITC 29)
3. minerals and metals
   a. Crude fertilizers and crude minerals, n.e.s. (SITC 27)
   b. Metalliferous ores and metal scrap (SITC 28)
   c. Iron and steel (SITC 67)
   d. Non-ferrous metals (SITC 68)
4. mineral fuels
   a. Mineral fuels, lubricants and related materials (SITC 3).[2]

Globalization of commodity markets

Technological progress and trade policy in the 19th and 20th centuries have been the main drivers of the considerable integration of commodity markets, although this process has not occurred in a linear fashion due to historical events such as wars and recessions (Findlay and O'Rouke, 2003). Before the full completion of the Industrial Revolution, huge transportation costs prevented long-distance shipping of goods other than valuable commodities like coffee, spices and certain metals. Radetzki (2008) identifies two fundamental moments in history when innovation transformed the transportation technology, especially for bulk materials. The first occurred in the second half of the 19th century and consisted in the application of the steam engine to transport and in the introduction of refrigerated ships.[3] The second historical episode is the block of the Suez Canal in the 1950s, which created the incentives for the shipping industry to develop large carriers suitable for long-distance transportation of bulk materials such as iron ore and coal. Both occasions determined a plunge in the incidence of transportation costs on the final price level, and a substantial rise in the long-distance maritime trade.

The most important consequence of these changes was a greater price convergence of many commodities across the regional markets of the time, a clear signal of the emergence of global markets for food, minerals and oil. The globalization of the market for gas came only in the mid-1990s, when innovation in the transportation process and reductions in the costs of liquefaction of natural gas facilitated the connection of a multitude of supply sources, in this way replacing the previous geographical segmentation of three isolated regional markets. The most evident results of these technological innovations is the fact that almost every commodity is currently traded in global markets.

The levels of economic growth and the increasing market globalization in the last 50 years explain the impressive increase in world exports from 107 billion US dollars in 1962 to 17,651 in 2014 (UN COMTRADE, 2016). What happened to commodities in the meantime? The upper panel of Figure 20.1 displays the time series of total world exports, expressed in current

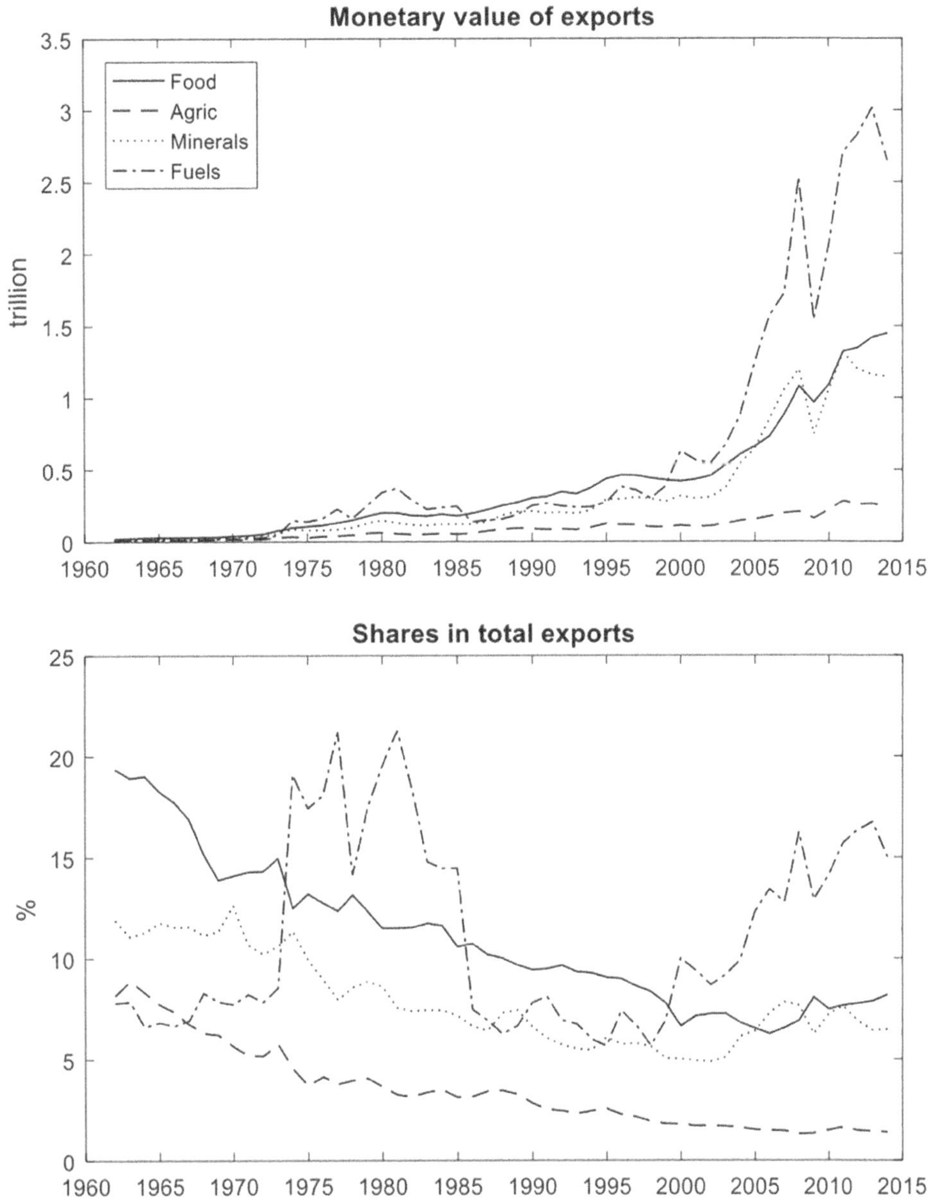

Figure 20.1 World commodity exports. Values are expressed in US dollars at market exchange rate.

Source: authors' calculations on Comtrade data

US dollars, for the four categories of commodities introduced earlier. Although all series are upward trending, the rate of increase is relatively low and stable for agricultural raw materials, which changes from 8.8 to 247 billion. Food and minerals exhibit very similar pattern, with a rise from 20 to 1,445 and from 12 to 1,140 billion, respectively. What clearly emerges from the plot is the considerable increase in the value of fuel exports, from 8 to 2,363 billion, accompanied also by a markedly higher variance compared to the other commodities, partly due to the large changes in the price of oil, especially in the last 10 years of the sample.[4]

Figure 20.1 shows that monetary values of exports for all commodity groups except agricultural raw materials appear to undergo an acceleration at the beginning of the 21st century, strengthened by the commodity price boom started in 2003. In addition, all exports experienced a sudden drop in correspondence with the 2008 financial crisis, but this is especially marked for the fuels and, to a lower degree, for minerals. This is not surprising given their strong sensitivity to the business cycle and the dominance of demand shocks in the price determination process. In the last 50 years, virtually all primary commodities have seen a rise in the traded physical quantity, in addition to the increase in the monetary value discussed above. The share of the value of commodity exports out of all traded goods, however, has decreased considerably from 47% in 1962 to 29% in 2014, reflecting the growing importance of manufactures in international trade. As shown in the lower panel of Figure 20.1, the share of food exports has dropped from 18% to 9%, whereas agricultural raw materials and minerals show a more gradual fall, and fuels exhibit no general trend over the five decades period, but a clear plateau in the 1970s and

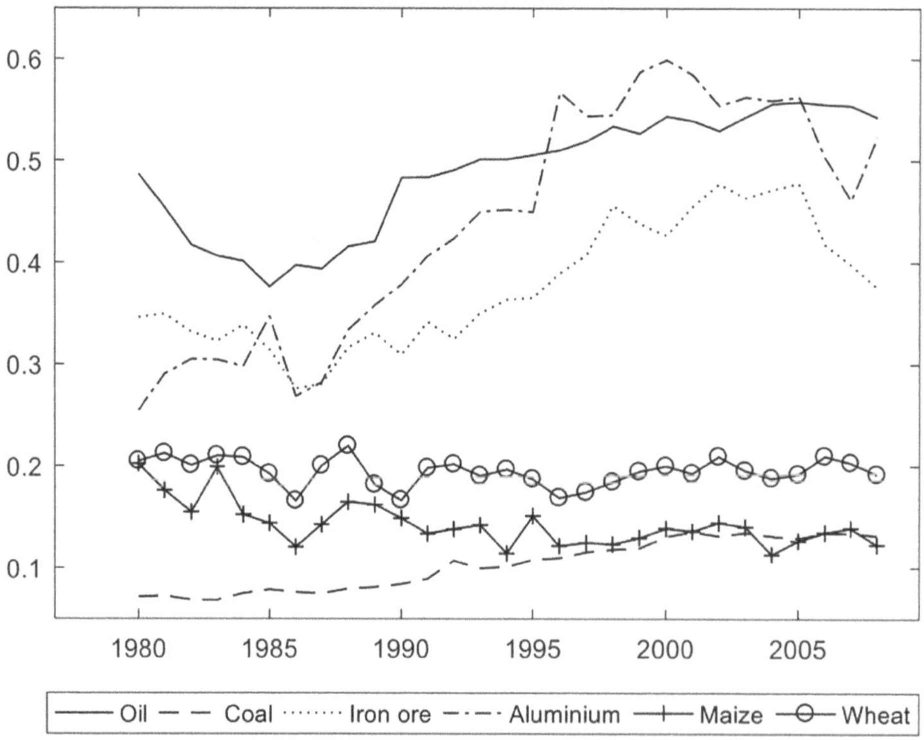

Figure 20.2 World export as share of production. Ratios are obtained from real quantities in tonnes

Source: Comtrade, FAOSTAT, BP, and IEA

Table 20.2 Physical trade balance in 1991. Physical trade balance is defined as volume of imports minus exports, expressed in thousands of tonnes

	Crude oil	*Natural gas*	*Hard coal*	*Iron ore*	*Copper*	*Aluminium*	*Wheat*	*Maize*	*Sugar*	*Cotton*	*Rubber*
Argentina	-668 (0%)	1,453 (0%)	645 (0%)		27 (0%)	-62 (1%)	-5,498 (5%)	-3,896 (9%)	-53 (1%)	-161 (4%)	32 (0%)
Brazil		1,306 (0%)	11,553 (0%)	-113,115 (34%)	405 (1%)	-6,093 (3%)		815 (0%)	-677 (9%)	-18 (3%)	98 (0%)
Chile		217 (0%)	1,518 (0%)	-6,262 (2%)	-2,472 (27%)	8 (0%)	162 (0%)	279 (0%)	202 (0%)		11 (0%)
Mexico		-2,376 (2%)	25 (0%)	-31 (0%)	-246 (3%)	83 (0%)	352 (0%)	1,369 (0%)	784 (0%)	-13 (1%)	81 (0%)
China	-16,265 (2%)	157 (0%)	-18,656 (6%)	18,549 (0%)	105 (0%)	-25 (3%)	12,366 (0%)	-7,781 (17%)	670 (2%)	170 (5%)	318 (0%)
India		168 (0%)	5,141 (0%)	-29,440 (9%)	50 (0%)	-175 (1%)			-92 (1%)	-157 (4%)	12 (0%)
Indonesia	-32,125 (4%)	-25,597 (14%)	-7,367 (2%)	0 (0%)	-551 (7%)	-1,143 (7%)	2,221 (0%)	290 (0%)	309 (0%)	350 (0%)	-1,223 (26%)
Malaysia	-21,569 (2%)	-7,547 (4%)	2,074 (0%)	1,156 (0%)	-30 (1%)	-232 (2%)	1,306 (0%)	1,292 (0%)	657 (1%)	40 (0%)	-1,042 (24%)
Thailand		-813 (1%)	442 (0%)	0 (0%)	69 (0%)	146 (0%)	453 (0%)	-1,227 (3%)	-988 (15%)	345 (0%)	-1,259 (27%)
Saudi Arabia	-309,830 (30%)	-14,342 (8%)	12 (0%)	6 (0%)	50 (0%)	141 (0%)	-829 (1%)	99 (0%)	63 (0%)	3 (0%)	1 (0%)
Russia											
South Africa											
Australia	1,328 (1%)	-3,492 (2%)	-120,936 (37%)	-112,481 (34%)	-405 (5%)	-956 (5%)	-11,571 (11%)	-18 (0%)	-2,370 (13%)	-396 (11%)	35 (0%)
Japan		52,158 (0%)	111,953 (0%)	127,186 (0%)	582 (1%)	4,874 (0%)		16,646 (0%)	1,838 (0%)	637 (0%)	703 (0%)
US	289,324 (0%)	9,960 (2%)	-97,588 (30%)	8,556 (1%)	-385 (9%)	13,588 (5%)	-30,359 (30%)	-24,196 (55%)	1,332 (3%)	-1,528 (41%)	759 (1%)
Canada	-10,536 (4%)	-14,172 (8%)	-20,717 (10%)	-25,031 (9%)	-639 (8%)	1,374 (8%)	-23,241 (22%)	-492 (2%)	911 (0%)		74 (0%)
Norway	-93,658 (9%)	-25,443 (15%)	3 (0%)	-2,128 (1%)	-126 (1%)	-710 (4%)		8 (0%)	162 (0%)		2 (0%)
Germany	104,669 (0%)	41,684 (1%)	11,565 (1%)	43,332 (0%)	1,145 (1%)	3,067 (3%)	-1,319 (2%)	866 (1%)	-1,421 (9%)	189 (1%)	228 (0%)
France	73,362 (0%)	25,875 (1%)	20,921 (0%)	15,053 (1%)	432 (0%)	1,646 (1%)	-17,125 (16%)	-4,444 (11%)	-2,106 (13%)	106 (0%)	174 (0%)
UK	-6,371 (6%)	3,959 (1%)	17,717 (1%)	13,675 (0%)	239 (0%)	354 (1%)	-3,281 (4%)	1,503 (0%)	1,104 (1%)	22 (0%)	124 (1%)

Source: authors' calculations from Comtrade data

early 1980s as a result of the oil price shocks occurred in this period, and a more recent steady increase starting at the end of the 1990s.

A better perspective into how commodity trade has grown in the last few decades can be obtained by assessing the proportion of world production that has been subject to international trade. In Figure 20.2 we plot the share of the exports out of production for six commodities that are representative of three of the groups discussed earlier, namely wheat and maize for unprocessed and processed food (group 1), aluminium and iron ore for minerals and metals (group 3) and oil and coal for mineral fuels (group 4).[5] As shown in the figure the share of the exports either increased or remained stable. A similar pattern can be observed for other commodities belonging to the four groups discussed earlier. Fuels have increased by only a small amount, with coal steadily rising from 7% in 1980 to 12% in 2008 and crude oil reaching 54% in 2008 (slightly higher than the level observed in 1980) after decreasing to 38% in 1985. Among metals, export of aluminium as a share of total production have more than doubled, from 25% in 1980 to 52% in 2008, whereas the level of iron ore exports in 2008 were not markedly different from those in 1980, as a consequence of a fall in the last part of the sample. On the contrary, the share of maize and wheat has remained fairly stable, like other food commodities.

A significant contribution to the development of globalized markets for commodities has come from the emergence of commodity exchanges, and the fact that the list of traded commodities has been steadily increasing over time, especially as a result of improved storage and conservation technology. Although commodity exchanges are dominated by purely financial transactions that are not accompanied by a physical exchange, this type of market delivers greater transparency with regard to price and grade of the commodity, and ensures also higher liquidity, which have facilitated the expansion of the international physical trade in commodities.[6]

Geography of trade

Given the increased international integration of commodity markets, it is important to analyse how the geographic pattern of the corresponding flow of resources has evolved over time. Monetary figures on imports and exports are influenced by a set of factors that transcend the real volume of trade, as the price of commodities can change, exchange rates are subject to their own dynamics and final prices can be affected by tariffs or transfer pricing. For these reasons, if we want to uncover the actual pattern in the real volume of trade and identify the producers and the consumers of world primary resources, we need to look at the physical quantities imported and exported by each country. Dittrich and Bringezu (2010) provide a recent comprehensive analysis of real trade flows within different regions of the world for broad classes of commodities.[7] As available databases show considerable gaps in relation to physical trade flows, these authors propose a simple method to overcome this problem whenever monetary values are available. We follow Dittrich and Bringezu (2010) to calculate physical quantities of imported and exported commodities across countries, but we decide to look only at 11 individual commodities, as this ensures a greater degree of homogeneity when calculating the corresponding weight measures, while covering the most important commodities in the manufacturing process.[8]

In Tables 20.2 and 20.3 we display the physical trade balance, defined as imports minus exports in thousands of tonnes, in 1991 and 2011, for a set of 20 countries, which are chosen for their importance in world trade and their being representative of different geographical regions of the world. We also show in parenthesis the countries' share in world exports so as to highlight the extent of each country's contribution to the world supply.[9] We now analyse one group of countries at a time, underlining any changes in the material trade profile that occurred during those 20 years.

Table 20.3 Physical trade balance in 2011. Physical trade balance is defined as volume of imports minus exports, expressed in thousands of tonnes

	Crude oil	*Natural gas*	*Hard coal*	*Iron ore*	*Copper*	*Aluminium*	*Wheat*	*Maize*	*Sugar*	*Cotton*	*Rubber*
Argentina	-3,024 (0%)	3,771 (0%)	2,047 (0%)	6,623 (0%)	-437 (2%)	-180 (0%)	-8,411 (6%)	-15,802 (14%)	-76 (0%)	-84 (1%)	37 (0%)
Brazil	-14,837 (2%)	9,908 (0%)	19,935 (0%)	-330,830 (29%)	-8 (3%)	-7,108 (8%)	3,390 (2%)	-8,831 (9%)	-25,359 (48%)	-614 (10%)	227 (0%)
Chile		3,647 (0%)	9,154 (0%)	-10,287 (1%)	-5,034 (22%)	13 (0%)	655 (0%)	620 (0%)	476 (0%)	0 (0%)	10 (0%)
Mexico	-66,768 (4%)	9,035 (0%)	7,336 (0%)	-3,446 (0%)	-439 (3%)	495 (0%)	3,212 (1%)	9,434 (0%)	-1,074 (3%)	214 (1%)	74 (0%)
China	251,255 (0%)	22,568 (0%)	167,396 (1%)	685,127 (0%)	9,641 (1%)	-433 (1%)	1,209 (0%)	1,617 (0%)	2,860 (0%)	3,337 (0%)	2,115 (0%)
India	157,820 (0%)	18,139 (0%)	88,888 (0%)	-38,228 (4%)	1,712 (1%)	-366 (1%)	-500 (0%)	-3,940 (4%)	-2,644 (5%)	-1,830 (24%)	119 (1%)
Indonesia	-4,566 (1%)	-32,669 (5%)	-323,547 (9%)	-12,209 (1%)	-1,494 (7%)	-40,404 (46%)	5,605 (0%)	3,195 (0%)	2,502 (0%)	546 (0%)	-2,536 (30%)
Malaysia	-3,148 (1%)	-25,111 (4%)	21,480 (0%)	-2,845 (1%)	215 (0%)	67 (0%)	1,135 (0%)	2,859 (0%)	-259 (1%)	16 (1%)	-280 (11%)
Thailand	37,940 (0%)	11,124 (0%)	16,203 (0%)	-405 (0%)	234 (0%)	437 (0%)	1,426 (0%)	-186 (0%)	-6,508 (12%)	319 (0%)	-2,993 (35%)
Saudi Arabia	-359,242 (22%)	(1%)	69 (0%)		145 (0%)	29 (0%)	1,997 (0%)	1,649 (0%)	890 (1%)		-3 (0%)
Russia	-219,100 (14%)	-190,567 (25%)	-108,801 (10%)	-26,928 (2%)	-197 (1%)	-3,319 (4%)	-15,184 (10%)	-607 (1%)	2,310 (0%)	76 (0%)	58 (0%)
South Africa	17,829 (0%)	2,346 (0%)	-71,195 (6%)	-55,736 (5%)	-222 (1%)	-514 (1%)	1,583 (0%)	-2,866 (3%)	-109 (1%)	22 (0%)	51 (0%)
Australia	10,664 (1%)	-24,667 (4%)	-281,306 (25%)	-460,234 (41%)	-2,137 (9%)	-13,424 (15%)	-19,742 (13%)	-43 (0%)	-68 (0%)	-852 (11%)	7 (0%)
Japan		90,783 (0%)	175,286 (0%)	128,399 (0%)	4,053 (2%)	3,671 (0%)			1,534 (0%)	81 (0%)	795 (0%)
US	458,132 (0%)	14,997 (3%)	-105,488 (10%)	-5,805 (1%)	318 (2%)	11,611 (1%)	-30,822 (22%)	-45,150 (42%)	3,204 (0%)	-2,757 (35%)	1,002 (1%)
Canada	-74,238 (8%)	-30,165 (6%)	-24,254 (3%)	-26,807 (3%)	-498 (2%)	843 (3%)	-16,271 (11%)	7 (1%)	1,105 (0%)	2 (0%)	144 (0%)
Norway	-69,337 (4%)	-80,416 (11%)		-2,308 (0%)	-36 (0%)	-939 (2%)	409 (0%)	53 (0%)	131 (0%)	0 (0%)	0 (0%)
Germany	87,054 (0%)	70,200 (3%)	42,676 (0%)	41,612 (0%)	1,661 (1%)	4,517 (1%)	-1,802 (4%)	1,136 (1%)	-416 (2%)	47 (0%)	287 (2%)
France	62,757 (0%)	44,376 (1%)	15,422 (0%)	13,402 (0%)	179 (0%)	1,911 (0%)	-19,849 (14%)	-5,740 (6%)	-1,880 (4%)	17 (0%)	167 (0%)
UK	19,160 (2%)	23,266 (2%)	31,292 (0%)	8,933 (0%)	11 (0%)	22 (0%)	-1,456 (2%)	944 (0%)	1,168 (1%)	1 (0%)	90 (0%)

Source: authors' calculations from Comtrade data

South America emerges as an important exporter of commodities in 1991 with the exception of energy commodities. In 1991 the main countries in Latin America are dependent on foreign supply of energy, i.e. oil, gas and coal, with the exception of Mexico, which is self-sufficient in terms of coal and is a net exporter of natural gas (2% of world total). From the tables one can see that South America is an important source of metals for the rest of the world. In 1991 Brazil generated 34% of the world exports of iron ore while Chile contributed to 27% of world copper exports. The contribution of South America to world trade of agricultural commodities was also significant, with Argentina exporting 4%, 5% and 9% of the world volume of cotton, wheat and maize respectively, while Brazil represented an important supplier of sugar, with 9% of world exports. In 2011, development of oil fields turned Mexico and Brazil into important oil exporters – the net exports of the whole region being about a quarter of net exports from Saudi Arabia in the same year. It is however worth mentioning that South America is still a net importer of coal and gas, as shown by the positive trade balance in the table. On the other hand, net exports of metals have increased considerably, especially for iron ore and copper, which for Brazil has tripled and doubled between 1991 and 2011 to attain a volume of 330 and 5 million tonnes, respectively. South America as a whole exported in 2011 about a third of the world's exports of iron and copper, although its contribution to the aluminium market was considerably smaller, with Brazil accounting for about 10% of global exports. The fact that the increased South American commodity exports is mainly driven by Brazil emerges also in the agricultural sector. This country's share in world exports indeed is approximately 10% for maize and 50% for sugar. In the case of maize, the combined exports originating from Argentina and Brazil is about a quarter of the world's exports.

Changes between 1991 and 2011 have been staggering with regard to the trade balance of energy and metal commodities in China and India. The former was a moderate net exporter of oil, coal and maize in 1991 which more than compensated its being net importer of iron and wheat. Similarly, India accounted in 1991 for about 10% and 4% of global exports of iron and cotton, respectively. In 2011 China and India were two of the biggest consumers of commodities, especially of energy and metals. As an example, China's and India's net importing position was equal to Saudi Arabia's net-exporting position in the case of oil and almost as big as Australia's net-exporting position in the case of coal. China's net imports of iron ore in 2011 was close to the staggering level of 700 million tonnes, about 200 million higher than Australia's net exports. India, instead, was a net exporter of iron ore, although its exports account only for 4% of world's total, and an important cotton exporter, as it accounted for a quarter of global exports.

Among Southeast Asian countries, Indonesia plays an important role in the trade of energy commodities. In 1991 it accounted for 14% of the world exports of natural gas, although this share has recently decreased. In 2011 it had the highest net-exporting position in the coal market, about 320 million tonnes, even higher than Australia's net exports. Remarkably, Indonesia contributed in 2011 to almost half of the world aluminium exports and about a third of natural rubber. This commodity is characterized by a very high concentration with Indonesia, Malaysia and Thailand holding about 75% of global exports in 1991 and 2011. Thailand was also a strong exporter of sugar, both in 1991 and 2011, with about 15% of global share. On the other hand, Indonesia, Malaysia and Thailand depend on outside sources for food commodities such as wheat and maize.

Between 1991 and 2011 Saudi Arabia retained its dominant share of world oil exports, i.e. 22% in 2011, while Russia confirmed its position of leading world supplier of energy commodities, with shares in global exports equal to 14% for oil, 25% for natural gas and 10% for hard coal. Contribution of South Africa to commodity trade is moderate with the exception of coal and iron ore. In these two markets South Africa hold a 6% and 5% share of 2011 world exports, respectively.

In the last group we include a set of OECD countries that are either significant consumers due to the size of their economy or important exporters due to the abundance of natural resources. Among the latter, Australia arises as a world leading supplier of a wide set of commodities, such as coal (25% of world exports in 2011), iron ore (41%), aluminium (15%), copper (9%), wheat (13%) and cotton (11%). Among the OECD countries in Tables 20.3 and 20.4 that are significant commodity consumers, Japan has a prominent role in the fuel and metal markets, although there has been no substantial increase in volume between 1991 and 2011 in the case of metals. The US emerges as the greatest consumer of imported oil, with more than 450 million tonnes of net imports in 2011, more than the net exports of Saudi Arabia (359 million tonnes). The US is also an important supplier of coal, with 30% of the global exports in 1991 and 10% in 2011, and a dominant supplier of wheat, maize and cotton, with respectively 22%, 42% and 35% of world exports in 2011. The volume of US exports of maize and cotton has almost doubled between 1991 and 2011, whereas those of wheat have remained constant. Canada is a net exporter of energy commodities, with a considerable increase in the net physical trade of oil and gas between 1991 and 2011. Also noticeable are Canadian net exports of wheat, although Canada's share of exports in 2011, 11%, is only half of the share in 1991. Between 1991 and 2011, Norway's importance as an oil-exporting country has decreased both in terms of net physical trade and share of global exports. It however remains an important world supplier of natural gas, with 11% of global exports in 2011 and a considerable increase in net exported volume compared to the levels observed in 1991. Between 1991 and 2011 Germany and France reduced their dependence on imported oil, but have increased gas net imports. The UK changed from becoming a relevant supplier of oil in 1991, with 6% of world exports, to a net consumer in 2011, while increasing its dependence on world supply of gas and coal. The role of France in the world supply of wheat, with 14% share in world exports in 2011, is also noticeable.

Even though Tables 20.2 and 20.3 provide us with an indication about the direction of the trade flows in commodities across different world regions, it remains important to analyse how the trade balance has evolved year by year for broad classes of commodities. In Figure 20.3 we plot the monetary trade balance, defined as the value of imports minus value of exports, of the four groups of commodities defined earlier, as well as of manufactures, for five regions we have selected on the basis of their contribution to international trade. We focus on three industrialized regions, i.e. Europe (the 27 EU countries plus Norway), North America (US and Canada) and OECD Asia Pacific (Australia, New Zealand, Japan, South Korea), and two developing regions, i.e. China and India, and South America (Argentina, Bolivia, Brazil, Chile, Colombia, Ecuador, Falkland Islands, French Guiana, Guyana, Paraguay, Peru, Suriname, Uruguay, Venezuela).[10]

Starting with food, both Europe and OECD Asia Pacific are predominantly net importers throughout the whole period, increasingly so for the Asian countries, while Europe very recently became a net exporter. Wheat and maize from US and Canada are the main causes of this region being a net exporter. South America has undoubtedly become the leading world supplier, reflecting the large share of world exports in wheat, maize and sugar from Brazil, and wheat from Argentina, which was apparent in Tables 20.3 and 20.4. One can notice a considerable acceleration in the size of South America trade balance starting in 2003. Lastly, China and India became net importers in 2008, having been net exporters since 1960. One can notice the speed at which the size of their net import is increasing between 2006 and 2014.

China and India have also become heavily dependent on imported agricultural raw materials, with surging needs from the end of the 1990s, mainly satisfied by North and South America (recall for instance the exports of cotton from these two regions which can be seen in Tables 20.3 and 20.4). Europe and OECD Asia Pacific are also net importers, although their net imports have fallen considerably in recent years.

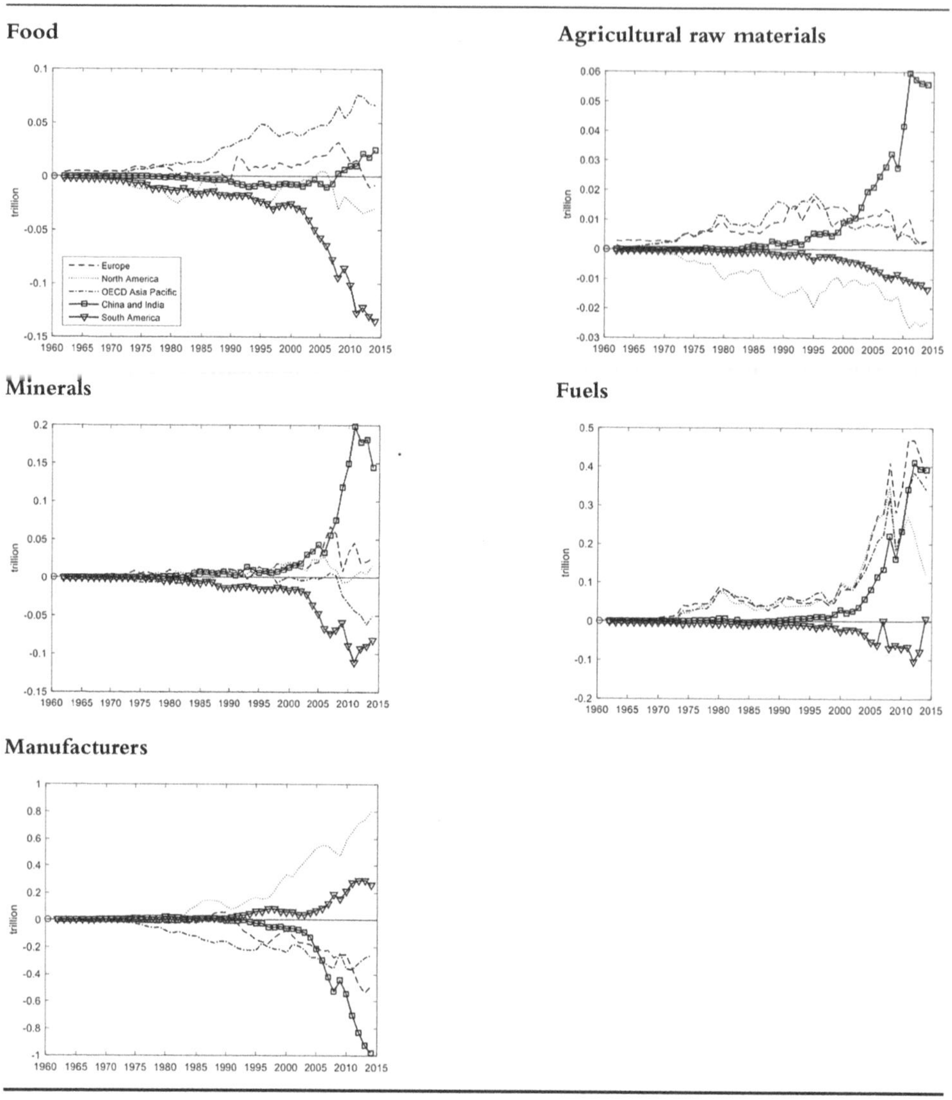

Figure 20.3 Net monetary trade balance for classes of commodities. Net monetary trade balance is defined as value of imports minus value of exports, measured in current US dollars.

Source: authors' calculations using Comtrade data

In the case of minerals, all regions with the exception of South America had fairly balanced trade positions until the end of the 1990s but have since then witnessed an impressive surge. Net imports of China and India have increased dramatically, especially since 2007, to reach a peak of 200 billion dollars in 2011 (recall that only China imported a net of 685 billion tonnes of iron ore as shown in Table 20.4), although it moderately decreased in the following years. European demand for foreign supply of minerals is modest but has increased in the last decade. North America has traditionally been a net importer, although its trade account has recently been more balanced. The graph reveals South America as the main source of minerals, a consequence of strong net exports from Brazil for iron ore and Chile for copper, as shown in Tables 20.3 and 20.4. A considerable acceleration in the size of the region's net exports started in 2004. Increasing net exports from Australia caused a change in the sign of the monetary trade balance in the OECD Asia Pacific group in 2009, with the whole region being net exporters since then.

As some of the most important gas and oil-exporting countries, e.g. Saudi Arabia and Russia, are not included in Figure 20.3, the plot for fuel commodities highlights the net monetary deficit of all four regions in the graph, with the exception of South America. All three industrialized regions have witnessed a considerable rise in the value of fuel net imports since the beginning of the 2000s, which is mainly caused by oil for North America, and a mixture of gas and oil for Europe, although this partially reflects the rise of oil price in this period. The demand from China and India has increased exponentially since the end of the 1990s generating a trade deficit that surpassed that of Europe in 2014, with almost 400 billion dollars of net imports. One can also notice how the recent 2009 financial crisis caused the sudden drop in all series with positive values, a change caused by a pronounced fall in both price and quantity traded. Finally, oil exports of Venezuela, Mexico and Brazil explain the net-exporting position of South America, although it was close to zero in 2014.

In terms of manufactures, North America has increased its net imports since 1982, reaching a value of 800 billion dollars in 2014. South America is also a net importer with increasing values of its deficit since 2003. On the contrary, OECD Asia Pacific countries have always been important world suppliers, with a positive linear time trend, while Europe has generated positive and rising monetary trade balances since the beginning of the 1990s. The main feature of the graph is however the staggering growth in the size of China and India monetary trade balance. Although the two countries have been net exporters since 1992, the size of the balance has experienced an impressive acceleration since 2004 rising from 126 in 2004 to 979 billion US dollars in 2014.

Figure 20.3, jointly with Tables 20.2 and 20.3, allow us to derive a few important observations about structure and evolution of international commodity trade. Industrialized countries, with the exception of Australia, Canada and Norway, depend heavily on commodities imported from developing countries, something which is confirmed also when looking at real traded quantities (Dittrich and Bringezu 2010). This phenomenon can be explained with the significant fall in bulk transportation costs, the industrialization process that Western countries experienced in the 20th century and the more competitive prices of emerging markets (Radetzki, 2008). At the country level, US, Europe and Japan have considerably increased their dependence on foreign supply of commodities in recent years, whereas South America has expanded substantially its exports in all four classes of commodities discussed earlier.

It is worthwhile to highlight that the trade profile of different countries reflects their economic and production structure. While imports of food are strictly related to population density and protective policy measures, minerals demand reveals the stage of economic development that a country is experiencing. This is particularly relevant for India and China that have been important resource consumers only since the mid-1990s, although their contribution to consumption of commodities has shown an exceptional increase in the last decade. Imports of minerals in countries like the US, Japan, and those of the EU have been relatively low and stable since the 1990s as the result of the mature stage of economies, where production inputs are mainly allocated to the tertiary sector. On the contrary, China has witnessed over the same period an accelerated industrialization process that required the establishment of important parts of the physical infrastructure, e.g. transport network, spurring the development of the construction sector and industries characterized by high capital and resource intensity.[11] The country has undergone a similar development process to that experienced decades earlier by the OECD economies, but at a much faster pace. The exponential rise in the dependence on world supply of all four classes of commodities is perfectly matched by an impressive expansion of China's net exports of manufactured goods, especially in the last decade. We also notice a recent reduction in the net imports of minerals and agricultural raw materials, which might signal the inception of a different phase in the Chinese development process.

It is interesting to frame the size and the changes in China's consumption of primary commodities within international consumption and trade. In order to do so, Figure 20.4 compares imported quantities of iron ore, copper, aluminium and oil for China, the US and the eight European countries with the highest levels of GDP (Germany, UK, France, Italy, Spain, Netherlands, Sweden and Denmark). The speed of China's structural transformation is apparent from the sudden changes in the quantity of imported iron ore, aluminium and copper. From 2000 imports of iron ore have increased exponentially, with an average annual growth rate of 12%, whereas they have stayed virtually constant in Europe, and have slightly decreased in the US. In terms of size, China imported 932 billion tonnes of iron ore in 2014, compared to 98 billion of Europe and 5 billion of the US. A similarly increasing trend can be seen in the Chinese imports of copper and aluminium, even if the magnitude of the difference with the US and the European countries is lower. Imports of crude oil have been approximately stable for the European countries, whereas they have decreased in US since 2004, almost reverting the increases

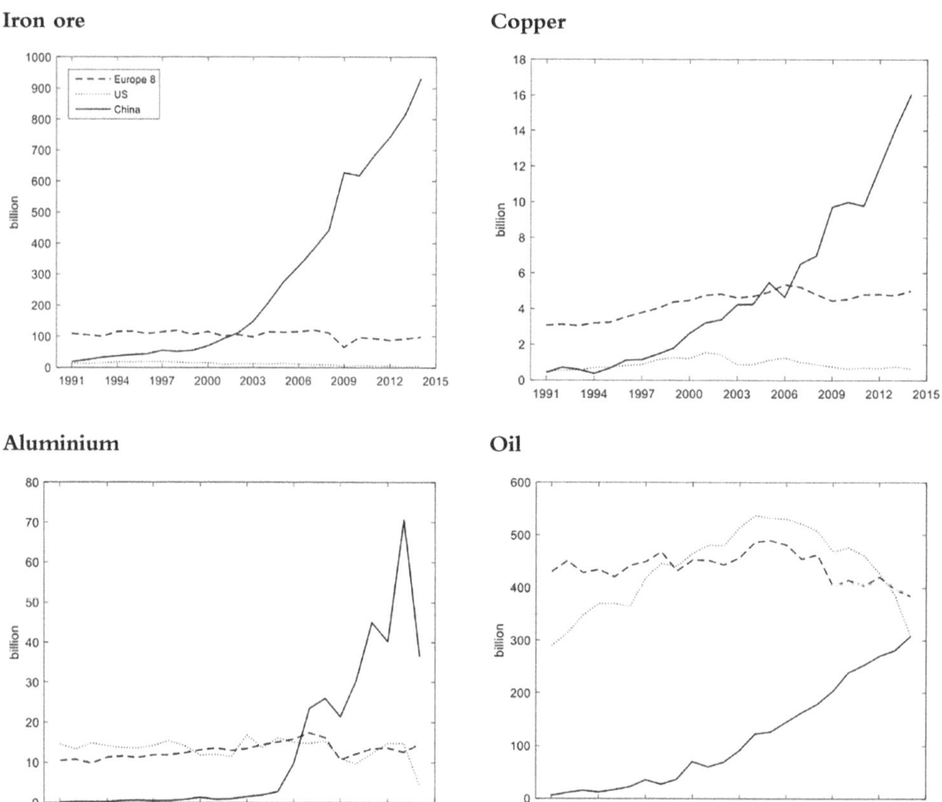

Figure 20.4 Physical imports of industrial commodities. Quantities are expressed in tonnes.

Source: authors' calculations using Comtrade data

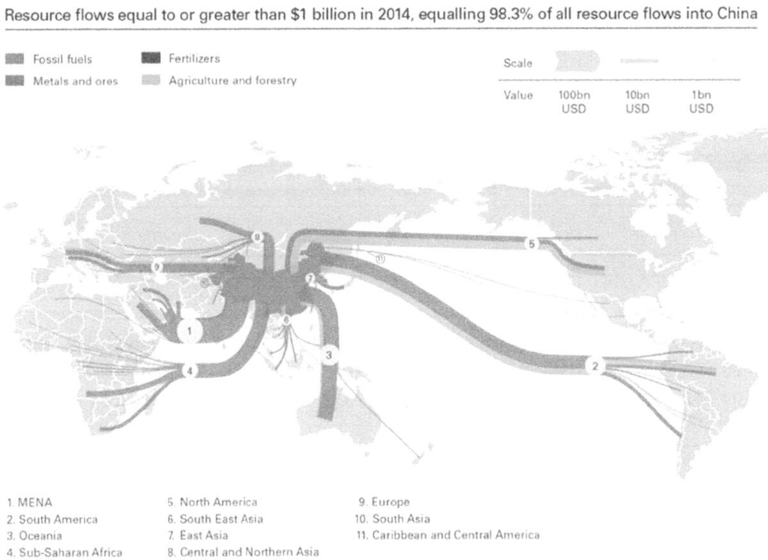

Figure 20.5 Resource flow to China

Source: Chatham House and DRC (2016)

that occurred between 1991 and 2004, presumably as a consequence of the boom in shale oil production stimulated by the advances in extraction technology and high energy prices. Physical imports of oil in China, on the other hand, present a positive trend throughout the sample, with an average annual growth of 15%, which resulted in the amount of oil imports from abroad being 52 times bigger in 2014 than in 1990, i.e. an increase from 6 to 308 billion tonnes, reaching the same level as the US in 2014.

The Sankey diagram in Figure 20.5 helps to get an even more vivid picture of both the size and the origin of Chinese imports in 2014. All these figures confirm an impressively rapid shift of the world production centre of commodity-intensive manufactured goods towards China, which has occurred in the last two decades.[12]

Market concentration

Both economic theory and empirical analysis suggest that market concentration is a key factor in determining the degree to which a market is competitive. Factors influencing the concentration of supply in commodity markets include geographical distribution of natural resources, and the role played by governments in resource extraction. While in theory the increased international integration of a market is likely to be accompanied by increasing competition, it remains interesting to verify whether this is what actually occurred in the last few decades, and in particular how much of world supply is currently distributed among different producing countries. In order to do so we use the Herfindahl index, which is widely used in the trade literature.[13] Figure 20.6 displays the time series of the Herfindahl index for three individual commodities in each of the four broad classes discussed earlier over the period 1974–2014, while Table 20.4 reports the share of the four largest exporters in 1974 and 2014.

In the case of fuel commodities, the Herfindahl index has decreased for oil, gas and coal since 1975, indicating an overall reduction in the degree of market concentration, but at varying speed in different periods and across fuels.[14] This is confirmed by the fact that the sum of shares of the

Table 20.4 Countries' shares in world exports. H indicates Herfindahl; UAE and SACU stand for United Arab Emirates and Southern African Customs Union, respectively

Crude oil				*Natural gas*				*Hard coal*			
1974		*2014*		*1974*		*2014*		*1974*		*2014*	
Exporter	*share*	*Exporter*	*share*	*Exporter*	*share*	*Exporter*	*share*	*Exporter*	*share*	*Exporter*	*share*
Saudi Arabia	0.36	Saudi Arabia	0.22	Netherl.	0.32	Qatar	0.21	US	0.55	Australia	0.36
Iran	0.21	Russia	0.14	Canada	0.24	Russia	0.15	Germany	0.19	Indonesia	0.20
Kuwait	0.10	Canada	0.08	Venezuela	0.09	Norway	0.09	Australia	0.15	Russia	0.12
Libya	0.09	Nigeria	0.07	Kuwait	0.06	Algeria	0.06	Canada	0.06	US	0.09
H index	0.20		0.10	H index	0.18		0.09	H index	0.37		0.20

Iron ore				*Copper*				*Aluminium*			
1974		*2014*		*1974*		*2014*		*1974*		*2014*	
Exporter	*share*	*Exporter*	*share*	*Exporter*	*share*	*Exporter*	*share*	*Exporter*	*share*	*Exporter*	*share*
Australia	0.23	Australia	0.53	Chile	0.20	Chile	0.32	Canada	0.18	Canada	0.12
Brazil	0.15	Brazil	0.23	Zambia	0.14	Peru	0.08	Norway	0.15	Russia	0.11
Canada	0.15	South Africa	0.06	Canada	0.12	Australia	0.07	Netherl.	0.08	Australia	0.08
Sweden	0.12	Canada	0.04	Congo	0.10	Zambia	0.06	Suriname	0.06	UAE	0.08
H index	0.13		0.34	H index	0.10		0.13	H index	0.08		0.05

Wheat				*Maize*				*Soybean*			
1974		*2014*		*1974*		*2014*		*1974*		*2014*	
Exporter	*share*	*Exporter*	*share*	*Exporter*	*share*	*Exporter*	*share*	*Exporter*	*share*	*Exporter*	*share*
US	0.46	US	0.16	US	0.62	US	0.34	US	0.85	US	0.41
Canada	0.22	Canada	0.15	Argentina	0.11	Brazil	0.12	Brazil	0.14	Brazil	0.39
Australia	0.13	France	0.11	France	0.09	Argentina	0.11	Paraguay	0.00	Argentina	0.06
France	0.11	Russia	0.11	SACU	0.05	Ukraine	0.10	Canada	0.00	Paraguay	0.04
H index	0.29		0.10	H index	0.41		0.16	H index	0.75		0.33

Cotton				*Natural rubber*				*Wood*			
1974		*2014*		*1974*		*2014*		*1974*		*2014*	
Exporter	*share*	*Exporter*	*share*	*Exporter*	*share*	*Exporter*	*share*	*Exporter*	*share*	*Exporter*	*share*
US	0.34	US	0.32	Malaysia	0.39	Thailand	0.37	Canada	0.16	Canada	0.15
Egypt	0.18	India	0.21	Singapore	0.27	Indonesia	0.29	US	0.14	US	0.12
Turkey	0.06	Australia	0.13	Indonesia	0.15	Vietnam	0.10	Sweden	0.11	Russia	0.09
Syria	0.05	Brazil	0.10	Thailand	0.08	Malaysia	0.09	Indonesia	0.08	Sweden	0.06
H index	0.16		0.18	H index	0.25		0.24	H index	0.08		0.06

Source: authors' calculations using Comtrade data

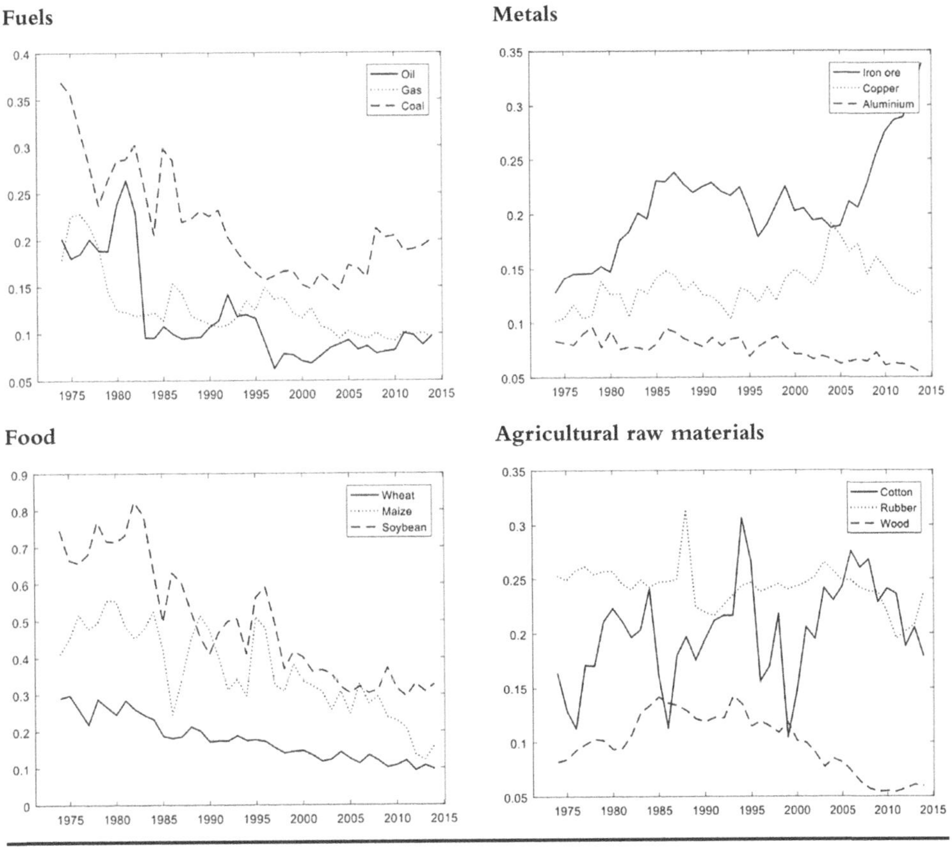

Figure 20.6 Market concentration measured by the Herfindahl index

Source: authors' calculations using Comtrade data

four largest exporters is smaller in 2014 (51%) than in 1974 (76%). Consequently, the Herfindahl index for the fuel market as a whole (i.e. comprising oil, gas and coal) has halved from the levels observed in 1974. It is important to notice however that a considerable degree of concentration remains present in all three fuel markets, but in particular in the case of hard coal, where the sum of the share of the four largest exporters was a very substantial 77% in 2014, although down almost 20 percentage points from 95% in 1974.

Metal commodities, with the exception of aluminium, display a less competitive structure over time. The rise in market concentration is evident from 2008 onwards in the case of the iron ore, for which the Herfindahl index has almost tripled between 1974 and 2014, with the four largest suppliers accounting for 86% of global exports in 2014, and Australia remaining the top contributor with a striking share of 53%. Aluminium, on the other hand, appears to have an increasingly competitive structure, with the four largest suppliers having almost equal shares.

The market for food commodities seems to have become considerably less concentrated over time, at least as far as wheat, maize and soybean are concerned, with the Herfindahl showing a negative trend throughout the last 40 years despite temporary oscillations. US remains the leading supplier of all three staples, but its share has dropped dramatically by half in the case of maize and soybean, and by two-thirds in the case of wheat. Concentration in wheat, as measured by the

Herfindahl index, was in 2014 about one-third of the level observed in 1974. Market concentration is considerably higher in the case of soybean, where US and Brazil alone made up 80% of world exports in 2014, down from 99% in 1974.

In the case of agricultural raw materials, a fairly flat pattern of the Herfindahl index can be seen for cotton and rubber. Concentration in the former, however, is subject to wide fluctuations within the sample, probably due to the fact that new major suppliers have emerged over time (i.e. India, Australia and Brazil). Market dominance of the US is still significant, with its share of exports being unaltered between 1974 and 2014. The market for natural rubber remains fairly concentrated, with the four largest suppliers accounting for 85% of world exports, whereas the market for wood, on the other hand, has taken a more competitive structure over time, with the concentration index reaching 0.06 in 2014.

Trade in water

Most of the water readily available for human consumption is devoted to agriculture, which is accountable for 70% of all global freshwater extraction (reaching 90% in the least developed countries), whereas water withdrawals for energy production are 15% of the total (UN, 2015). A long list of factors – i.e. population growth, industrialization, urbanization and migration – is responsible for generating a surging demand for freshwater in the last 60 years and increasing pressure on existent resources. Bearing in mind environmental degradation, climate change, and increased agricultural production, there is a serious risk of compromising the planet's capacity to provide enough water for the growing population especially when increasing access to standard water sources (UN, 2012).

Water is a commodity that, compared to other primary commodities, has not been subject to significant international trade, despite being distributed unevenly across the globe and being severely scarce in some countries rather than others. One reason for the lack of water trade is

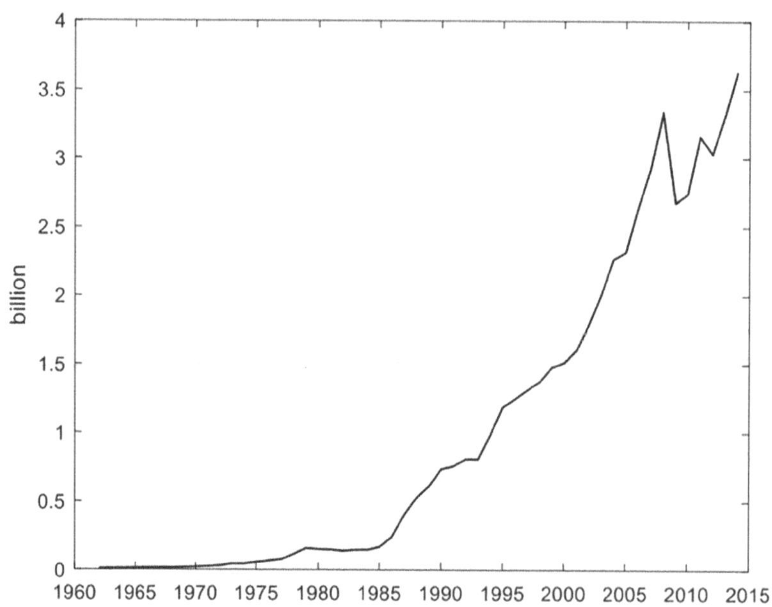

Figure 20.7 World exports of water. Values are expressed in US dollars at market exchange rate.
Source: Comtrade

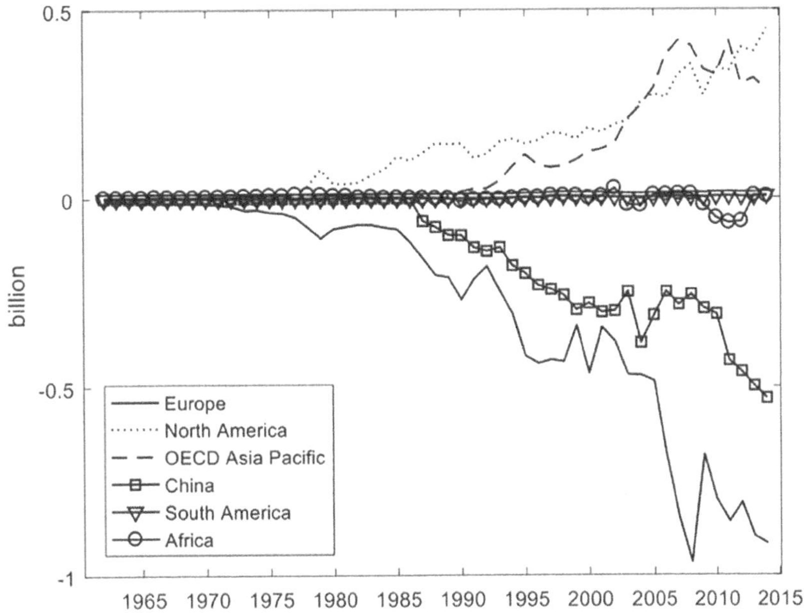

Figure 20.8 Net monetary trade balance for water. Net monetary trade balance is defined as value of imports minus value of exports, measured in current US dollars

Source: authors' calculations on Comtrade data

related to the fact that water is not cheap to transport given its bulky nature, and at the same time countries that are poor in water are geographically close to each other, so transportation needs to occur over long distances (WTO, 2010). On the other hand, international trade in water-intensive, mainly agricultural, commodities is an important way in which water is indirectly traded among countries. In fact, the volume of this traded water has doubled from 1986 to 2007, and has contributed to important savings in world resources (Dalin et al., 2012).[15]

Focusing on the actual, rather than indirect, trade of water, Figure 20.7 shows the time series from 1962 to 2014 of the monetary value of world exports. Any actual rise in the trade volume might be exaggerated due to the Comtrade database being largely incomplete in the early part of the sample, and to the fact that the number of countries reporting their data to UN Comtrade increase over time. Nevertheless, we observe a considerable rise in world exports of water, namely from 1.5 to 3.6 billion dollars between 2000 and 2014, when the same number of countries reported their data to Comtrade.[16] Physical exports have more than doubled, from 8.38 million tonnes in 1992 to 18.41 in 2014.[17] In terms of trade flows across different regions of the world, Europe and China are net exporters of water, whereas North America and the OECD countries of the Pacific are net importers. In all cases this pattern has increased as time went by, as shown in Figure 20.8.[18] It is also worth mentioning that "virtual water" is discussed in Part II of this handbook.

Final remarks

The trade dimension is a crucial part to comprehend the actual way in which the nexus operates in the real world. Because most commodities, including energy and water, are now subject to international trade, it is important to know the core features of the international commodity markets of today. If the recent attention given to primary commodities has been driven mainly by the dynamics of their prices, we should not forget that the observed price level is the result of the interactions among a multitude of factors underlying the market structure. Especially if we are interested in uncovering the long-run trend in future price developments, features like degree of market globalization, geographical structure of supply and demand flows, and degree of market concentration play a crucial role.

We have used data on international trade to highlight how these three dimensions in particular have evolved over time for individual and broad classes of commodities. Our mainly descriptive analysis sheds light on some important stylized facts about relevant characteristics of the commodity markets, which would be interesting to further investigate in dedicated empirical works.

We found that in the last few decades international scope has increased faster for some commodities than others, that the direction of trade has changed as a likely response to different stages of economic development across countries, and that the trend in production concentration differs substantially across commodities. In particular, we observed that international trade has increased more intensively for fuels and minerals than agricultural products. Moreover, while most industrialized countries have remained dependent on foreign supply of important natural resources, we have shown how the extraordinary growth of China has translated into a modification of the international trade balance in both industrial commodities and manufactured goods. As for the degree of market concentration, we found that this has been decreasing for food commodities, which may help explain the observed long-run gradual fall in their prices. On the contrary, we could not detect any clear pattern in market concentration in agricultural raw materials, while mineral fuel supply is less concentrated than 40 years ago, and some widely used metals, like iron ore and copper, have been characterized by rising degree of concentration. Finally, we discovered that, despite being far less traded than other commodities, water is becoming increasingly subject to international exchange, with China and Europe representing important net exporters.

Notes

1 See, for instance, Heap (2005), Radetzki (2006), Tilton (2006), Radetzki et al. (2008), Humphreys (2010) and Clayton (2015).
2 This includes energy fuels such as fossil fuels, both conventional and unconventional.
3 Harley (1988), for instance, finds that the adoption of metal ships and steam propulsion from 1850 was the main reason for the dramatic decline in the freight rates in the years thereafter. See, also Radetzki (2010).
4 A note a caution is in order here since any increase in exports might be exaggerated by the fact that in the Comtrade database not all countries report trade data for each year, and for China, for instance, data are available only from 1984.
5 We limit the graph to the period 1980–2008 because of lack of data for earlier years, and the presence of significant outliers in more recent figures.
6 Some authors have also emphasized how the increasing world market integration has reduced commodity price volatility, facilitating the growth process of poor countries (see Jacks et al., 2011). On the other hand, there are negative consequences from decreasing prices of food commodities for farmers in developing countries (see, for instance, Gilbert and Varangis, 2004).
7 See also the analysis in UNEP (2015).
8 Given the lower reliability of real quantity data, we decide to apply their approach also to recalculate the real quantity whenever the recorded figure implies a price that is an extreme outlier relative to the distribution of values for all countries in a given year. See Dittrich and Bringezu (2010) for more details.
9 Bear in mind that since re-exports are included, a country's share in world trade can be overstated.

10 We skip the Middle East, Russia and North Africa since they are relevant only with respect to fuels trade.
11 The implications for resource-rich Asian economies of the commodity price boom triggered by the rapid growth of China and India is analysed by Coxhead and Jayasuriya (2010), looking in particular at the growth and environmental impacts wherever there is lack of solid institutions and governance.
12 Montalbano and Nenci (2012) examine the recent trade specialization of major developing countries, and find that China appears to emerge as a competitor for the main industrialized economies, while India, Brazil, and South Africa represent dominant players in markets supplied by other developing countries. Qureshi and Wan (2008) assess in detail the competitiveness and complementarities of China and India in their trade with the rest of the world, and they highlight how the export structure of China has been changing over time with an increasing share of high-technology goods and less labour-intensive products.
13 See, for instance, Van Long and Soubeyran (1997), and Oladi and Gilbert (2012). Notice that while 1 is the upper bound of the index, indicating perfect monopoly, the lower bound is determined by the reciprocal of the total number of suppliers in the market.
14 The usual note of caution is necessary here since the number of reporting countries have increased over the years, but, at the same time, it is very likely that each country has reported its exports as far as they reached a non-negligible figure. Thus, we still believe that the index is capable of capturing the overall degree of market concentration.
15 See also Chapagain and Hoekstra (2008).
16 Also, the difference in terms of set of countries is negligible when looking at their share in world exports.
17 These two years allow a balanced comparison since the difference in the set of countries reporting their figure is not significant when we consider their share in world exports.
18 Notice that data for China are available from 1987, and we exclude India since it is a net importer of negligible size.

References

Allan, T. (2011, Spring). The-water-food-trade nexus and global water resource security. *UK Irrigation*, 37.

Chapagain, A. K., and Hoekstra, A. Y. (2008). The global component of freshwater demand and supply: An assessment of virtual water flows between nations as a result of trade in agricultural and industrial products. *Water International*, 33(1), 19–32.

Chatham House and DRC (2016). Navigating the new normal: China and Global Resource Governance. Royal Institute of International Affairs.

Clayton, B. C. (2015). *Commodity markets and the global economy*. Cambridge: Cambridge University Press.

Coxhead, I., and Jayasuriya, S. (2010). China, India and the commodity boom: Economic and environmental implications for low-income countries. *The World Economy*, 33(4), 525–551.

Dalin, C., Konar, M., Hanasaki, N., Rinaldo, A., and Rodriguez-Iturbe, I. (2012). Evolution of the global virtual water trade network. *Proceedings of the National Academy of Sciences*, 109(16), 5989–5994.

Dittrich, M., and Bringezu, S. (2010). The physical dimension of international trade: Part 1: Direct global flows between 1962 and 2005. *Ecological Economics*, 69(9), 1838–1847.

Findlay, R., and O'Rourke, K. H. (2003). Commodity market integration, 1500–2000. In: Bordo, M. D., Taylor, A. M. and Williamson, J. G. (Eds.), *Globalization in historical perspective*. Chicago: University of Chicago Press, pp. 13–64.

Gilbert, C., and Varangis, P. (2004). Globalization and international commodity trade with specific reference to the West African cocoa producers. In: Baldwin, R. E. and Winters, L. A. (Eds.), *Challenges to globalization: analyzing the economics*. Chicago: University of Chicago Press, pp. 131–166.

Harley, C. K. (1988). Ocean freight rates and productivity, 1740–1913: The primacy of mechanical invention reaffirmed. *The Journal of Economic History*, 48(4), 851–876.

Heap, A. (2005). *China-the engine of a commodities super cycle*. New York: Citygroup Global Markets Inc., Smith Barney 67, pp. 8–9.

Humphreys, D. (2010). The great metals boom: A retrospective. *Resources Policy*, 35(1), 1–13.

Jacks, D. S., O'Rourke, K. H., and Williamson, J. G. (2011). Commodity price volatility and world market integration since 1700. *Review of Economics and Statistics*, 93(3), 800–813.

Montalbano, P., and Nenci, S. (2012). The trade specialization of China, India, Brazil, and South Africa: A threat to whom? *The International Trade Journal*, 26(5), 363–384.

Oladi, R., and Gilbert, J. (2012). Buyer and seller concentration in global commodity markets. *Review of Development Economics*, 16(2), 359–367.
Qureshi, M. S., and Wan, G. (2008). Trade expansion of China and India: Threat or opportunity? *The World Economy*, 31(10), 1327–1350.
Radetzki, M. (2006). The anatomy of three commodity booms. *Resources Policy*, 31(1), 56–64.
Radetzki, M. (2008). *A handbook of primary commodities in the global economy*. Cambridge: Cambridge University Press.
Radetzki, M. (2010, August). Primary commodities: Historical perspectives and prospects. In R. Arezki, T. Gylfason and A. Sy (Eds.), *Beyond the curse: Policies to harness the power of natural resources*. Algiers: IMF, 2011. Invited address, High Level Seminar, IMF Institute & Central Bank of Algeria.
Radetzki, M., Eggert, R. G., Lagos, G., Lima, M., and Tilton, J. E. (2008). The boom in mineral markets: How long might it last? *Resources Policy*, 33(3), 125–128.
Tilton, J. E. (2006). *Understanding cyclical and secular trends in metal prices: Mine management handbook*. Carlton, Victoria: Australasian Institute of Mining and Metallurgy.
UN. (2012). *The millennium development goals report*. New York: United Nations.
UN. (2015). *Water for a sustainable world*. The United Nations World Water Development Report.
UN COMTRADE (2016). *UN Comtrade Database*. Accessed online at: https://comtrade.un.org/data.
UNEP. (2015). *International trade in resources: A biophysical assessment*. Report of the International Resource Panel.
Van Long, N., and Soubeyran, A. (1997). Cost heterogeneity, industry concentration and strategic trade policies. *Journal of International Economics*, 43(1), 207–220.
WTO. (2010). *Trade in natural resources*. World Trade Report 2010.

21
RARE EARTH ELEMENTS AND A RESOURCE NEXUS PERSPECTIVE

Eva Barteková

Introduction

Recently, there has been a renewed interest in the security of supply of non-energy minerals, especially in the context of technology metals used for manufacturing of low-carbon energy technologies. One of the most discussed is the group of elements called rare earths. The recent surge in their demand has been driven by the increased deployment of electric vehicles and wind turbines. In particular, the demand for dysprosium and neodymium, the two rare earth elements used for the production of neodymium-iron-boron permanent magnets, has been projected to increase by 2600% and 700% respectively, by 2035 (Alonso et al., 2012). Additionally, the rare earth supply side is concentrated in China, which currently holds a monopoly of 85% of total market share (U.S. Geological Survey, 2015), and is further skewed due to some resource nationalist policies, along with the inability of the other consumer countries to set up domestic supply chains. Taken together, the increasing demand and the institutional inefficiencies of the market, adversely affect the stability of rare earth supply and pricing.

From a broader resource management perspective, besides resource nationalist policies, the imbalance between rare earth supply and demand is driven by the stovepiped approach to climate change policymaking. In fact, increasing the share of low-carbon technologies within the energy mix (low-carbon energy transition) without considering the implications for raw material availability (used in manufacturing of low-carbon technologies), might create bottlenecks along the global supply chain, and also exacerbate environmental and social costs from dirty production processes, and geopolitical uncertainties from import dependence of the consumer countries. In order to reflect the complexity of the interlinked resource flows and systems, a more interconnected thinking in policymaking is required, as advocated by resource nexus approach (Andrews-Speed et al., 2015). Such approach allows examining the interconnections, tensions and trade-offs between different resources in a wider context of climate change (The Nexus Network, 2015).

In this light, the aim of this chapter is to shed light on the rare earths debate from a resource nexus perspective. The chapter goes beyond the traditional supply–demand discussion, which has been extensively put forward by earlier literature (Alonso et al., 2012; Elshkaki and Graedel, 2013; Habib and Wenzel, 2014; Hoenderdaal et al., 2013). Instead, it offers a more systemic perspective on the interconnections between low-carbon technologies and mineral resources from political economy and environmental perspectives. In particular, it provides insights into how the rare earth import dependence for the production of low-carbon technologies comes with high environmental costs and geopolitical uncertainties, when the extraction and production of the former are carried out in a few developing countries, in an often less efficient and environmentally unsound fashion.

The rest of the chapter is structured as follows: the second section offers a brief introduction of rare earths, and their supply and consumption trends. The third section discusses the environmental implications of their mining and processing, and contrasts these with their uses in manufacturing of low-carbon technologies. The fourth section analyses individual countries' supply chains and considers strategies to securing a stable and equitable access to rare earths. The final section offers conclusions on how sourcing minerals through environmentally dirty processes for the manufacturing of low-carbon technologies causes a dilemma in transitioning to a renewable future. It also points out how the lack of equitable and stable access to minerals by manufacturers of low-carbon technologies poses barriers to innovation in these technologies. This chapter therefore argues for better coordination of climate change policies beyond sectoral stovepipes within and across countries.

Rare earths

The "not so rare" metals

Rare earths are a group of 17 chemical elements. As opposed to what their name suggests, they are neither rare nor earths. They rank among a group of transition metals, also called lanthanoids, which can be further split into two sub-groups of heavy and light rare earths. The data on crustal abundance indicates that rare earths as a group are not any less abundant than other industrial metals: the most abundant are in the same range as nickel and zinc, while the least abundant are more than a hundred times more abundant than gold. However, in terms of individual elements, these are found in earth's crust in different quantities, the lighter elements being generally more abundant than the heavy ones (Cotton, 2007). Additionally, despite their occurrence in as many as two hundred minerals, not all deposits can be commercially processed, owing to the costs, their geographical location and the nature of minerals themselves (Walters and Lusty, 2011). In particular, it is the complexity of processing techniques, determined by mineralogy and chemical composition, which influences the competitive advantage of individual deposits. To put this into perspective, heavy rare earths are generally claimed to be more complicated for processing than light rare earths (Cotton, 2007; Walters and Lusty, 2011). Last but not least, some of the minerals also contain radioactive elements, the disposal of which exacerbates the environmental implications of the rare earth production (U.S. Geological Survey, 2013).

International production and trade

Rare earths have often been mined as by-products of other metals, such as iron ore, titanium, or uranium (Walters and Lusty, 2011). However, with their increasing importance in modern technologies, several projects for mining specifically rare earths have been launched in recent years (Hatch, 2015). In terms of reserves, roughly 42% of global rare earth reserves are concentrated in China (as compared to 80% of heavy rare earth deposits contained in Chinese ionic clays) (U.S. Geological Survey, 2013; Walters and Lusty, 2011). Other substantial deposits can be found in Brazil, the Commonwealth of Independent Countries, Australia, India and the US, while more recent discoveries have been made in Sweden, Greenland and Canada, as well as on the seabed of the Pacific Ocean. According to the official production figures,[1] China's stake in global supply of rare earths peaked in 2009 with a market share of 98% (U.S. Geological Survey, 2015). This has been gradually decreasing to 85% by 2015, due to the new supply which gradually developed in other parts of the world. The most important sources of non-Chinese light rare earth supply are Australia and the US, with global shares of 8% and 3%, respectively (U.S. Geological Survey, 2016). Production has also taken off in India, and continues in Russia which is processing rare earths from stockpiles. Nonetheless opening new mines outside China is a rather lengthy process with low rates of success, especially in the current adverse market conditions owing to long lead times, declining prices, high capital intensity, as well as higher environmental standards and lack of capabilities in terms of technically trained personnel in developed countries (Berry, 2016; Clagett, 2013; Gschneidner Jr, 2010). China is therefore foreseen to maintain its monopoly in the short to medium term.

Besides being the main producer, China is also the main consumer of rare earths, with 58% of the estimated global demand (Kingsnorth, 2011). It's consumption has been increasing rapidly during the first decade of the century by almost 300% (Zhanheng, 2010). In order to satisfy the increasing domestic demand and to assure its industrial upgrading to higher value added products (Barteková, 2014), China has been setting up various industrial policies. The rare earth production and export quotas were intended to prevent overconsumption and to keep the environmental damage under control (Information Office of the State Council, 2012). At the same time, however, these have been distorting the international trade, as illustrated in Figure 21.1. Examining the export quotas jointly with the actual exported quantities, it turns out that while during the initial years the tightening of quotas have limited the actual exports of rare earths, the majority of export quotas remained unfilled in the subsequent years. The latter is expected to reverse, driven by the recovery of the post-financial crisis global demand and with the depletion of importing countries' stockpiled material (Kingsnorth, 2013a; Topf, 2013). Looking at the production figures, these have been exceeding the actual production quotas set by the Chinese government. This discrepancy can be explained by the competition among

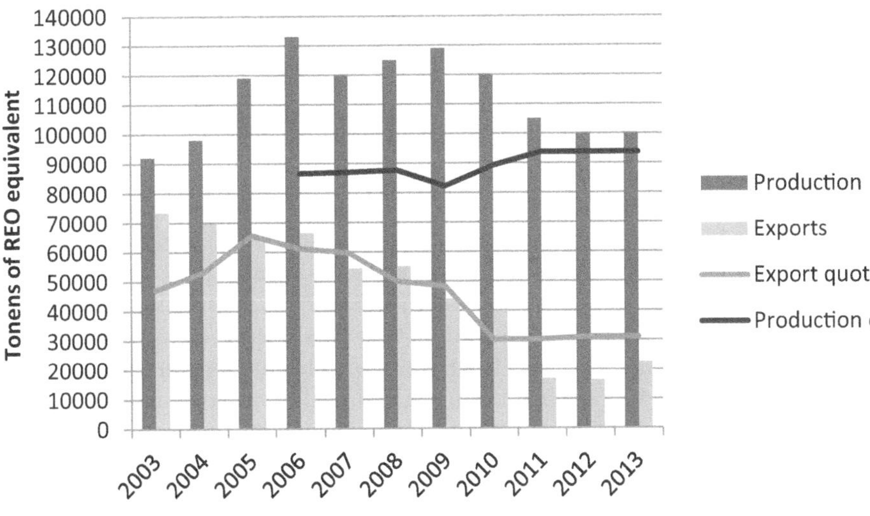

Figure 21.1 Chinese rare earth production and exports, and respective quotas, in tonnes of rare earth oxide equivalent, 2000–2013

Source: based on data collected from MOFCOM (2014); Tse (2011); U.S. Geological Survey (2013, 2014); UN Statistics Division (2013)

local governments, which derive substantial financial resources from mineral production (Tse, 2011). Taken together, the increasing wedge between export and production quotas indicates the increased Chinese domestic consumption. Though export quotas have been recently abolished further to the ruling of the World Trade Organisation, these have been replaced by resource tax and export licensing.

The other main consuming countries of rare earth oxides are Japan and the North East Asian countries with 22%, followed by the US with 12% and rest of the world with 8% of estimated global demand (Kingsnorth, 2011). Due to the rare earth market inefficiencies described above, virtually all countries are import dependent from China (especially on heavy rare earths). Moreover, the rare earth import dependence by developed countries comes with high environmental costs and geopolitical uncertainties. The former comes in form of pollution export by developed countries relocating dirty industrial production to countries with less stringent environmental policies. The latter can be illustrated by the recent diplomatic dispute between Japan and China, which resulted in temporary suspension of the rare earth trade. The environmental and geopolitical perspectives arising from lack of systemic approach to policymaking are discussed within the following two sections, respectively.

From dirty mining and processing to low-carbon technologies manufacturing

Environmental hazards of rare earth production

Historically, rare earths have been mined from monazite, xenotime, bastnasite and ionic clays. While the former two contain impurities of radioactive thorium and uranium, the ionic clays of China contain no radioactive elements and are also easier and more cost efficient to process (U.S. Geological Survey, 2013; Walters and Lusty, 2011). It has been estimated that for each tonne of rare earths produced, there are up to 12'000 cubic meters of waste gas, 75 cubic meters of acidic wastewater, 1 tonne of radioactive residue as well as 2'000 tonnes of mine tailings containing thorium emitted to the environment (Hurst, 2010). Worse even, due to the lax environmental policies in most of the producing countries, these are discharged into rivers without any prior treatment, causing water and soil contamination, with adverse effects on lives of local residents and biodiversity. Besides the emissions of pollutants, surface mining and in-situ leaching techniques used for the extraction of rare earths from ionic clays have led to destruction of forests and severe landslides in mining areas (Packey and Kingsnorth, 2016).

The severity of rare earth production emissions on environment and people's health has now also been recognised by the Chinese government (Information Office of the State Council, 2003, 2012). Yet, the implementation and enforcement of environmental protection and sustainable resource management remains weak (Liu and Maughan, 2012). This is further exacerbated by the rise of illegal mining activity in China, driven by increased global demand which cannot be satisfied by official supply, and facilitated by favourable conditions for extracting rare earths from ionic clays (Packey and Kingsnorth, 2016). Of a particular concern is mine sterilisation, whereby illegal miners extract high-grade ore and leave behind the medium- and low-grade ores. This not only causes environmental damage, it also has adverse implications for the sustainability and economic viability of rare earth deposits over time (Packey and Kingsnorth, 2016). It is also for this reason that China has started to actively combat illegal mining, having closed down 110 mines in southeastern provinces during the past years (Information Office of the State Council, 2012).

Evidence from other parts of the world underpins further the severity of societal and environmental costs of rare earth mining and processing: the La Rochelle plant switched the feed materials from monazite to thorium-free rare earth chloride, after the debate on radioactive thorium has intensified in France in the 1990s (Hedrick, 1994); the only American rare earth mine in Mountain Pass, California was forced to shut down in 2002, after a series of wastewater leakages containing heavy metals and radioactive material (Nystrom, 2003); and the Australian rare earth producer in the Mount Weld deposit outsourced the processing of rare earth concentrates to Malaysia, where also the waste is disposed of (Lynas Corporation, 2016). Regarding the latter, the Malaysian government granted the operating licence for the refinery, despite the large opposition by local civil societies (Stop Lynas, 2015), and the earlier experience of Malaysia with severe negative externalities from rare earth processing on local people's health (Consumers Association of Penang, 1993).[2]

In spite of this evidence, the consumer countries continue neglecting China's environmental motivations in setting up rare earth industrial policies. Undeniably, China has been using some of them to keep back rare earth supply for its domestic uses, so as to fuel the country's industrial upgrading. At the same time though, these also aim at internalising the externalities to people's health and environment. Disregarding these externalities by importing countries further propagates the implications of environmental consequences across the global supply chain. In this respect, rather than devising industrial policies, which ultimately distort international markets, producing countries should be developing eco-innovations in terms of cleaner production processes and more sustainable pathways to resource management. These would not only help minimising the externalities in producing countries, but once adopted internationally they would also allow establishing supply in countries with more stringent environmental policies, and thus contribute to decreasing import dependence and threat of supply disruptions. From a resource nexus perspective, this would significantly contribute to minimising pollution export, and ultimately contribute to greening the entire supply chain.

Rare earths as bottlenecks to low-carbon energy transition

Besides the negative externalities from the production of rare earths, the latter also create bottlenecks on the downstream side of the supply chain, due to their uses in various technologies. In fact, rare earths are used in various forms across an unusually wide spectrum of applications: ranging from mischmetal in metallurgic applications, through phosphors in medical applications (MRIs, X-rays, lasers) and in electronics (energy-efficient fluorescent lamps), to highly purified forms in high-tech products (cell phones, hard disk drives, microphones) and in defence applications (radars, sonars). At present, the largest pressure on the demand for rare earths is driven by low-carbon energy technologies. Their accelerated deployment is driven by industrialisation and energy poverty eradication in developing countries on the one hand (Miroux et al., 2011), and by decarbonising electricity generation and transport on the other (UNFCCC, 2015). The key technologies to achieving these objectives are wind turbines and electric vehicles. Their dependence on rare earths mainly comes from the use of rare earth permanent magnets in generators of some of the wind turbine technologies, as well as in electric motors, batteries and other parts of electric vehicles. For example, permanent magnet excitation generators within offshore wind turbines contain up to 250 kg of rare earths per megawatt (MW) of power range, or 2.5 tonnes of rare earths for a typical 10 MW wind turbine. In case of electric vehicles, a typical mid-size hybrid-electric vehicle with a permanent magnet electric motor and a nickel metal-hydride (NiMH) battery contains up to 4.1 kg of rare earths. An additional amount of up to 0.44 kg has been estimated to be scattered across various parts of vehicles, such as in small magnets in sun roofs and automatic door locks, and in catalytic converters (Barteková, 2015). Note that the latter are contained in both internal combustion engine and advanced technology vehicles (i.e. hybrid-electric and all-electric vehicles).

Considering the tighter supply of some of the rare earths (Kingsnorth, 2013b) and the increased adoption rate of these technologies, concerns prevail that the deployment of the automotive market and its electrification could be severely disrupted by sourcing volatility of rare earths (Barteková, 2015). In terms of pricing volatility, the general increase in dimensions of offshore wind turbines, and the higher relative material cost in production of rare earth-intensive batteries, are the main concerns to keeping the offshore wind and electric vehicles competitive with other technologies on the market.[3] Manufacturers hedge against pricing and sourcing volatilities by optimising the design of technologies towards less rare earth intensive technologies. Currently, several rare earth-free alternatives to permanent magnets in wind turbine drivetrains and electric motor and batteries co-exist on the market. While some of them provide potentially viable future alternatives still waiting to be marketed (e.g. high temperature superconductors which make use of more abundant rare earths), some others have already been deployed yet have been found inferior in terms of their performance, efficiency and reliability (e.g. less efficient induction motors and generators due to excitation losses, or less reliable Li-ion batteries due to security issues and lower drive ranges, as well as geared drivetrains and hybrid drivetrains due to moving parts), as compared to rare earth-intensive technologies (Barteková, 2015).

The above discussion demonstrates how increasing reliance on low-carbon technologies within the energy mix, without considering the availability of rare earths (in terms of capacities to increase the supply in short to medium term), creates bottlenecks to a sustainable and equitable low-carbon energy transition. From a resource nexus perspective, a stovepiped approach to climate change policymaking creates bottlenecks in other parts of the system. Though individual manufacturers strive to develop strategies in response to pricing volatilities and availability risk, they aim at innovating in less rare earth-intensive technologies, which however often tend to be inferior in terms of their performance. Should sourcing of minerals for low-carbon technologies be left to distorted global markets only, these might cause manufacturers to get locked in myopic investment decisions leading to inefficient uses of materials and technologies. Instead, a more systemic approach to policymaking beyond the sector (i.e. considering the interlinkages across supply chains) and a better coordination across supply chains (rare earth strategies within and across rare earth consuming countries) is required. This translates into a need for a more active involvement by governments of consumer countries through comprehensive rare earth strategies and international collaboration.

Rare earth strategies in the international context

Concerns about volatilities in pricing and availability of rare earths for the manufacturing and innovation in rare earth-intensive low-carbon technologies have resulted in countries developing individual mineral strategies. However, despite their common objective, different countries responded differently to the global problem of securing stable supply of rare earths over time. This can be largely attributed to the fact that policy responses have been shaped by individual countries' historical courses of action in terms of their national interests, resource endowment and experience in tackling earlier events of supply risk, as well as their respective policy styles (Barteková and Kemp, 2016).

The Chinese hegemony across the entire supply chain

China started to mine rare earths in the 1950s, and has gradually been increasing its monopoly position ever since (U.S. Geological Survey, 2015). It has often been claimed that this success owes to China's cheap labour force and less stringent environmental regulations, as well as to the rare earth deposits which are easier and more economic to mine (Haxel et al., 2002). While this is generally true, there is more to China becoming the global leader in the rare earth industry. In particular, China's early mineral policy encompassed strengthening of geological work, strategic prospecting and enactment of mineral laws and regulations (Information Office of the State Council, 2003). Moreover, China has also started to build up the rare earth innovation system in the 1960s. It established the first rare earth research institutes and state key laboratories, which were involved in R&D on industrial uses of rare earths, and basic and applied research on rare earth chemistry and physics, respectively (Barteková and Kemp, 2016).

Later on, the strategy of moving up the rare earths value chain has become an explicit part of the Chinese vision for industrial upgrading (Baotou National Rare Earth Hi-Tech Industrial Development Zone, 2015). After the classification of rare earths as "protected and strategic materials" (Information Office of the State Council, 2012), China started actively introducing industrial policies to limit the access of foreign enterprises to its supply chain, to restrict export and production of rare earths, to consolidate the upstream supply chain through merging rare earth producers into six state-owned industrial groups, to eliminate smuggling and to establish strategic stockpiles (Barteková, 2014). Recently, China has also attempted to assure upgrading through technology transfer. An example of such foreign direct investment is the purchase of an American permanent magnet producer, which has been holding patents on bonded permanent magnets (Robison and Ratnam, 2010). This acquisition has clearly helped China to establish production of permanent magnets, and ultimately also contributed to setting up production of higher valued added technologies which make use of these magnets. From its more recent policies, China has introduced the "go global policy" whereby it targets acquisition of resources abroad. Evidence shows successful acquisition of projects in Australia, forming joint ventures in Canada, and considering new untapped sources of rare earths in North Korea (InvestorIntel, 2015; Schearf, 2014; The Sydney Morning Herald, 2009).

Taken together, China has well-integrated activities across the entire supply chain, both in terms of extraction and processing, but also in terms of manufacturing of component parts like magnets, motors and generators, as well as of end-use technologies like wind turbines and alternative technology vehicles.

Weak rare earth strategies in the US, Europe and Japan

As opposed to the well-developed mineral, innovation and industrial policies targeting the rare earth industry in China, policies in other countries involved in the rare earth supply chain are far less advanced. While the downstream parts of supply chains tend to be well established across Europe, Japan and the US, these countries suffer a general lack of well-integrated upstream supply chain activities.

In the US, rare earth extraction was established in first half of the 20th century, the country subsequently becoming the major global producer of rare earths between 1965 and 1995. The early development of rare earth supply chain has been facilitated by a well-established regulatory environment rooted in early national legislations, in materials and minerals policies, as well as in the early establishment of institutions to manage potential supply disruption of minerals

(Barteková and Kemp, 2016). Besides the upstream activities, the US was also known for its permanent magnet industry, which was first established in the 1970s. Over time, however, its strategic perception of rare earths and their security of supply considerations have weakened, owing to the change in geopolitics (Humphreys, 1995). These have in turn been reflected in the decline of the American rare earth industry: the operations of Molycorp were suspended in 2002 (the production has resumed some years later, but the company filed for bankruptcy proceedings in 2015); and the permanent magnet production facilities of Magnequench have been closed down and relocated to China (Robison and Ratnam, 2010). Moreover, the US has also suffered a loss of know-how and expertise in all areas of rare earth mining, processing, chemistry and manufacturing of higher value added products (Gschneidner Jr, 2010). Thus, as opposed to China, which has comprehensive policymaking with long-term vision, the American rare earth supply chain faced disruption due to the discontinuity of political interests.

The case of Europe is different in that it has been historically a resource-poor country in metallic minerals. And while the debate of security of raw material supply dates back to the late 1970s (European Commission, 1975), building largely on raw material diplomacy, an integrated approach to securing minerals has been missing from the European agenda until recently. Moreover, on the country level substantial differences exist, some of them having an established track record in domestic rare earth policymaking (Sweden and Finland), while some others pursuing resource diplomacy to secure resources from abroad (France and Germany) (DEFRA, 2012). Yet, mineral strategies by most other European countries are still in their infancy (EUROMINES, 2016). As a result, extraction and processing activities are practically inexistent across Europe. In terms of mining, only two advanced projects exist in Greenland and Sweden (the actual start of their production depending on environmental permitting and financing), along with two separation plants located in Estonia and France. Some production of polishing powders, compounds and alloys occurs in the UK, Germany and Slovenia (ERECON, 2015). The main hurdles in establishing the European rare earth supply chain have been the missing EU-wide knowledge base on mineral deposits, as well as the lack of access to land, of streamlined permitting processes and of public funding of exploration (Barteková and Kemp, 2016). In terms of downstream activities, while production of electric vehicles and wind turbines is well established in Europe, manufacturing of component parts, such as permanent magnets, is practically inexistent. Taken together, the lack of accomplishment in establishing a European rare earth supply chain has been largely owing to the uncoordinated non-energy mineral policies across countries, and to Europe's reliance on resource diplomacy in sourcing mineral resources from abroad.

Similar to Europe, Japan has also been a historically resource-poor country dependent on foreign imports. However, as opposed to Europe, it has had a more hands-on approach when it comes to stockpiling of minerals (dating back to its experience during the Second World War), and in securing these through long-term supply agreements via public–private partnerships, but also through pioneering seabed exploration within its exclusive economic zones (Deep Ocean Resources Development, 2015; JOGMEC, 2015; Sojitz, 2013). This diversification strategy has been driven mainly by the industrial lobby and the government's active engagement in securing alternative sources of rare earths (Barteková and Kemp, 2016). Japan's economy is significantly dependent on rare earths: the imported oxides are processed domestically into metals and alloys, and used as input materials for manufacturing of component parts and high value added products. In terms of the former, the production of neodymium–iron-boron permanent magnets dates back to the 1980s. Japan is also the market leader in production and export of home appliances, robotics and electric motors of electric vehicles. Taken together, the well-established rare earth industry in Japan is driven by national interests, by its active policy style in securing access to new sources of rare earths, and by the existence of well-functioning institutions and public–private partnerships.

Rare earth supply chain development from a nexus perspective

Following from the above analysis, China appears to be the only country with a well-integrated rare earth supply chain. On the contrary, the upstream rare earth supply chains remain largely underdeveloped across the other countries. The US has seen its mining operations disrupted yet again, while in Europe further development of mines heavily depends on future permitting and public financing. Japan, which does not have mineable deposits of rare earths on land, attempts to establish seabed mining of rare earths, which however is more of a long-term project. In terms of separation and refining, besides China, facilities also exist in Japan, but are rather scarce in Europe and the US. With regard to downstream supply chain, the production of components and of end-use technologies is well established in all the four countries (though to some extent less in Europe, which has been historically lacking capabilities in the magnet industry).

Yet, in order to secure stable and equitable supply of rare earths, a more active involvement by governments of individual countries is required. These have now started formulating their respective rare earth strategies: the Raw Materials Initiative in Europe (European Commission, 2008, 2011), the Critical Materials Strategy in the US (Bauer et al., 2010) and the Strategy for Ensuring Stable Supplies of Rare Metals in Japan (Advisory Committee on Energy and Natural Resources, 2009). In general these strategies focus on: establishing stable access to raw materials on global markets through strategic resource diplomacy, supply diversification, and actions to removing distortionary measures by producing countries; establishing domestic supply by setting up framework conditions to facilitate exploration and extraction of minerals; decreasing primary consumption of minerals through increasing resource efficiency, establishing recycling and promoting use of alternative materials; stockpiling; and financing basic research programs carried out through public–private partnerships on minimisation and replacement of rare earth uses, and improvement of performance of rare earth-free technologies. While some developments have already been accomplished, establishing domestic supply of rare earths requires more active industrial, mineral and innovation policies across individual countries. These include devising mineral policies to streamline permitting, grant land access, and facilitate prospecting and industrial policies to (re-)establish knowledge infrastructure, pioneer mining from new deposits of rare earths and establish secondary sources of rare earths (recycling), and last but not least innovation policies to develop environmentally responsible mining and production techniques.

Yet, given the global nature of supply chains, and the system-wide implications of individual strategies, coordination should not only be pursued across sectoral stovepipes within countries, but also internationally across them. From a resource nexus perspective, coordination of strategies on global level is required for securing a stable access to minerals at fair prices within distorted global markets. Therefore, a need arises for global governance, with the overall aim of regulating the equitable access to supply chains and overseeing sustainability issues related to this, as well as reducing barriers to diffusion of technologies and moving into better coordinated innovation and market development processes on both ends of the supply chain. This in turn calls for establishing an international multi-stakeholder forum or board, composed of representatives from across the entire supply chain. This could examine the pressing nexus issues and could facilitate the international discourse, as well as the strategic short- and medium-term decision-making, for a smooth transition to low-carbon energy systems globally. Moreover, in order to promote an inclusive low-carbon growth, governments, industry associations and civil societies from both developed and developing countries should be involved.

Conclusions

The aim of this chapter is to discuss the challenges of the rare earth market from a broader resource nexus perspective. In particular, this chapter aims at introducing the debate on how ambitious climate change policies devised without considering their impact beyond the sector, transcend into import dependence on rare earths, leading to high environmental costs and geopolitical uncertainties. These occur when extraction and production are carried out in an environmentally unsound fashion, concentrated in few developing countries such as China. The chapter also illustrates how individual investors might take myopic investment decisions leading to inefficient uses of materials and technologies. Therefore, in order to solve the rare earth dilemma, which arises from sourcing minerals gained through environmentally dirty processes for the manufacturing of low-carbon technologies, an active involvement of individual countries' governments is required. Besides diversifying foreign supply, establishing recycling, and minimising the consumption of rare earths through efficiency improvements, a need emerges for setting up respective domestic upstream supply chains. Currently, the lack of rare earth mining and processing in developed countries is driven by environmental considerations and economic viability. This ultimately leads to pollution export to countries with laxer environmental regulation. Therefore, in order to avoid increasing further the externalities to environment and human health from accelerated deployment of low-carbon technologies, rather than imposing industrial policies which distort the internal market, there is an urgent need to develop eco-innovations for sustainable mining and processing techniques. These could on the one hand solve the problem of externalities in producing countries, and thus minimise their exposure to pollution export. On the other hand, they could enable establishing supply in countries with more stringent environmental policies, and thus tackle the issue of import dependence and threat of supply disruptions. A contrasting force here could be the "not in my backyard attitude" (NIMBY) which has been on increase in Europe and in the US, and relates to the citizens' resistance towards development projects which are believed to have negative impacts on environment. Nonetheless, taken together, eco-innovations in mining and processing rare earths could contribute to continued industrialisation and green growth of both producing and import-dependent countries, as well as to greening of the global supply chain, and to a better coordinated climate change policy.

In a more general perspective, a stovepiped and territorial approach to climate change policymaking has been shown to create bottlenecks in other parts of the supply chain. In contrast, the resource nexus approach looks at critical interlinkages within socio-economic systems across scales, and allows thus identifying potential negative externalities and system-wide implications of using of specific resources. Therefore, policymakers should aim at devising policies beyond their sectoral reach, and accounting for the impacts internationally; that is, rather than adopting a silo-thinking to resource management, they should seek to establish international innovation alliances which could contribute to faster deployment of low-carbon technologies. What is more, it is important to involve various actors at multiple scales in the process of policymaking, so as to accommodate their distinct and often competing interests (e.g. the contrasting environmental concerns by civil societies with the profit-oriented industry). Such integrated approach to low-carbon energy transition concludes on the need for a multilevel polycentric governance mechanism, which could facilitate further the coordination in a world of interconnected systems, heterogeneous national policies and priorities, and distorted markets. Ultimately, since rare earths are not so rare, a proper application of nexus thinking might help to deliver the Paris Agreement on climate change for multiple parties, and at the same time pave the way for integrated thinking in other areas where critical raw materials are required for a sustainable resource management.

Notes

1 It has been estimated that as much as 40% of China's total output was coming from the illegal mining. To put it into perspective, the illegal mining in China alone has been larger than the entire non-Chinese supply of rare earths (Packey and Kingsnorth, 2016).
2 An earlier example of pollution export in the Malaysian context is the case of Asia Rare Earth, the main shareholder of which (Mitsubishi Chemicals) despite of its established know-how in handling radioactive materials, has relocated the rare earth processing from Japan to Malaysia. While products were shipped back to Japan, the radioactive waste remained in Malaysia.
3 Assuming the variation of prices in permanent magnets of up to 420 USD/kg, the cost of electricity production by a permanent magnet excited direct drive wind turbine has been estimated to increase by 50% relative to a competing rare earth-free gearbox driven wind turbine with the same parameters (Barteková, 2015).

References

Advisory Committee on Energy and Natural Resources. (2009). *Strategy to secure rare metals* (Draft). Ministry of Economy, Trade and Industry. Retrieved from www.meti.go.jp/committee/materials2/downloadfiles/g90603a07j.pdf (Available in Japanese only)

Alonso, E., Sherman, A. M., Wallington, T. J., Everson, M. P., Field, F. R., Roth, R., and Kirchain, R. E. (2012). Evaluating rare earth element availability: A case with revolutionary demand from clean technologies. *Environmental Science & Technology*, 46(6), 3406–3414.

Andrews-Speed, P., Bleischwitz, R., Boersma, T., Johnson, C., Kemp, G., and VanDeveer, S. D. (2015). *Want, waste or war? The global resource nexus and the struggle for land, energy, food, water and minerals*. Abingdon: Routledge.

Baotou National Rare Earth Hi-Tech Industrial Development Zone. (2015). *Rare earth: An introduction*. Retrieved July, 5, 2015, from www.rev.cn/en/int.htm

Barteková, E. (2014). *An introduction to the economics of rare earths*. Working Paper. MERIT Working Papers 043, United Nations University – Maastricht Economic and Social Research Institute on Innovation and Technology (MERIT). Retrieved from http://digitalarchive.maastrichtuniversity.nl/fedora/get/guid:f9f3e4d5-e094-4266-9c46-4763fd25b427/ASSET1

Barteková, E. (2015). The role of rare earth supply risk in low-carbon technology innovation. In I. B. de Lima and W. L. Filho (Eds.), *Rare earths industry: Technological, economic, and environmental implications*. Amsterdam, Netherlands: Elsevier, pp. 153–169.

Barteková, E., and Kemp, R. (2016). National strategies for securing a stable supply of rare earths in different world regions. *Resources Policy*, 49, 153–164. doi:10.1016/j.resourpol.2016.05.003

Bauer, D., Diamond, D., Li, J., Sandalow, D., Telleen, P., and Wanner, B. (2010). *US Department of energy critical materials strategy*. US Department of Energy. Retrieved from http://energy.gov/sites/prod/files/edg/news/documents/criticalmaterialsstrategy.pdf

Berry, C. (2016). Infrastructure for the 21st century: Building a non-Chinese rare earth supply chain. *The Disruptive Discoveries Journal*. Retrieved from www.discoveryinvesting.com/blog/2016/2/24/infrastructure-for-the-21st-century-building-a-non-chinese-rare-earth-supply-chain

Clagett, N. (2013). A rare opportunity: Streamlining permitting for rare earth materials within the United States. *George Washington Journal of Energy & Environmental Law*, 4, 123.

Consumers Association of Penang. (1993). *Wasted lives: Radioactive poisoning in Bukit Merah*. Penang, Malaysia: Consumers' Association of Penang.

Cotton, S. (2007). *Lanthanide and actinide chemistry* (Vol. 27). Chichester: John Wiley & Sons.

Deep Ocean Resources Development. (2015). *About DORD*. Retrieved August 22, 2015, from www.dord.co.jp/english/corporate/index.html

DEFRA. (2012). *A review of national resource strategies and research*. Department of Environment Food and Rural Affairs – United Kingdom Government. Retrieved from www.gov.uk/government/uploads/system/uploads/attachment_data/file/69526/pb13722-national-resource-strategies-review.pdf

Elshkaki, A., and Graedel, T. E. (2013). Dynamic analysis of the global metals flows and stocks in electricity generation technologies. *Journal of Cleaner Production*, 59, 260–273.

ERECON. (2015). *Strengthening the European rare earths supply-chain: Challenges and policy options*. European Rare Earths Competency Network. Retrieved from http://ec.europa.eu/DocsRoom/documents/10882/attachments/1/translations/en/renditions/native

EUROMINES. (2016). *Update on national implementation of RMI*. Retrieved March 30, 2016, from www.euromines.org/what-we-do/raw-materials-initiative/update-national-implementation-rmi

European Commission. (1975). *The community's supplies of raw materials*. Communication from the Commission to the Council. Retrieved from http://aei.pitt.edu/1481/1/raw_materials_COM_75_50.pdf

European Commission. (2008). *The raw materials initiative – meeting our critical needs for growth and jobs in Europe*. Communication from the Commission to the European Parliament and the Council. Retrieved from http://eur-lex.europa.eu/LexUriServ/LexUriServ.do?uri=COM:2008:0699:FIN:en:PDF

European Commission. (2011). *Tackling the challenges in commodity markets and on raw materials*. Communication from the Commission to the European Parliament, the Council, the European Economic and Social Committee and the Committee of the Regions. Retrieved from http://eur-lex.europa.eu/LexUriServ/LexUriServ.do?uri=COM:2011:0025:FIN:en:PDF

Gschneidner Jr, K. (2010). *The rare earth crisis – the lack of an intellectual infrastructure*. Retrieved March 22, 2014, from www.ameslab.gov/les/RE_Crisis_Intell_Infrastruc_12.9.10.pdf

Habib, K., and Wenzel, H. (2014). Exploring rare earths supply constraints for the emerging clean energy technologies and the role of recycling. *Journal of Cleaner Production*, 84, 348–359. doi:http://dx.doi.org/10.1016/j.jclepro.2014.04.035

Hatch, G. (2015). *TMR advanced rare-earth projects index*. Retrieved July 30, 2015, from www.techmetalsresearch.com/metrics-indices/tmr-advanced-rare-earth-projects-index/

Haxel, G. B., Hedrick, J. B., and Orris, G. J. (2002). *Rare earth elements – critical resources for high technology*. Retrieved from http://pubs.usgs.gov/fs/2002/fs087-02/fs087-02.pdf

Hedrick, J. B. (1994). *Thorium*. Retrieved March 1, 2014, from http://minerals.usgs.gov/minerals/pubs/commodity/thorium/690494.pdf

Hoenderdaal, S., Tercero Espinoza, L., Marscheider-Weidemann, F., and Graus, W. (2013). Can a dysprosium shortage threaten green energy technologies? *Energy*, 49, 344–355.

Humphreys, D. (1995). Whatever happened to security of supply? Minerals policy in the post-cold war world. *Resources Policy*, 21(2), 91–97.

Hurst, C. (2010). *China's rare earth elements industry: What can the west learn?* Institute of the Analysis of Global Security. Retrieved from www.iags.org/rareearth0310hurst.pdf

Information Office of the State Council. (2003). *China's policy on mineral resources*. Retrieved from http://news.xinhuanet.com/zhengfu/2003-12/23/content_1244900.htm

Information Office of the State Council. (2012). *Situation and policies of China's rare earth industry*. Retrieved from http://ycls.miit.gov.cn/n11293472/n11295125/n11299425/n14676844.files/n14675980.pdf

InvestorIntel. (2015). China is looking everywhere for rare earths outside of China?. Retrieved from http://investorintel.com/technology-metals-intel/china-is-looking-everywhere-for-rare-earths-outside-of-china/

JOGMEC. (2015). *Japan oil, gas and metals national corporation – about us – history*. Retrieved August 17, 2015, from www.jogmec.go.jp/english/about/about003.html

Kingsnorth, D. J. (2011). *Meeting the challenges of supply this decade*. Presentation to the Environmental and Energy Study Institute. Washington, DC: USA-Industrial Minerals Company of Australia (IMCOA).

Kingsnorth, D. J. (2013a). *Professor Dudley Kingsnorth's Bullish position on REE markets*. Retrieved March 1, 2014, from www.youtube.com/watch?v=yUlUH5BNvDU

Kingsnorth, D. J. (2013b). *The rare earths industry: Undergoing rejuvenation*. Paper presented at the Rare Earths 2013 Conference, Shanghai, China. Retrieved from http://investorintel.com/wp-content/uploads/2013/08/AusIMM-CMC-2013-DJK-Final-InvestorIntel.pdf

Liu, H-W., and Maughan, J. (2012). China's rare earths export quotas: Out of the China-raw materials gate, but past the WTO's finish line? *Journal of International Economic Law*, 15(4), 971.

Lynas Corporation. (2016). *Locations – the LAMP, Kuantan, Malaysia*. Retrieved April 30, 2016, from www.lynascorp.com/Pages/Kuantan-Lynas-Advanced-Materials-Plant.aspx

Miroux, A., Hamdi, M., Sampath, P. G., Lim, M., and Razo, C. (2011). *Technology and innovation report 2011: Powering development with renewable energy technologies*. UNCTAD. Retrieved from http://unctad.org/en/Docs/tir2011_en.pdf

MOFCOM. (2014). *Industrial exports – non-ferrous metals* (Copper, Lead, Zinc, Tungsten, Antimony, Tin, Rare Earth, Silver, etc.). Retrieved November 20, 2013, from http://wms.mofcom.gov.cn/article/zt_gypck/ysjsh/?

The Nexus Network. (2015). *What is the nexus?* Retrieved March 28, 2016, from www.thenexusnetwork.org/about/

Nystrom, E. C. (2003). *From neglected space to protected place: An administrative history of Mojave national preserve.* Retrieved March 1, 2014, from www.nps.gov/parkhistory/online_books/moja/adhi/adhi.htm

Packey, D. J., and Kingsnorth, D. (2016). The impact of unregulated ionic clay rare earth mining in China. *Resources Policy*, 48, 112–116.

Robison, P., and Ratnam, G. (2010). Pentagon losing control of bombs to China's monopoly. *Bloomberg Business.* Retrieved from www.bloomberg.com/news/articles/2010-09-29/pentagon-losing-control-of-afghanistan-bombs-to-china-s-neodymium-monopoly

Schearf, D. (2014). North Korea's rare earths could be game changer. *Voice of America.* Retrieved July 5, 2015, from www.voanews.com/content/north-korea-rare-earths-game-changer/1832018.html

Sojitz. (2013). *Rare earth development & production in Australia & Malaysia.* Retrieved August 17, 2015, from www.sojitz.com/en/special/project/post-9.php

Stop Lynas. (2015). *Opposition in Malaysia.* Retrieved August 26, 2015, from http://stoplynas.org/opposition-in-malaysia/

The Sydney Morning Herald. (2009). Concerns raised over China's 'Rare Earth' dominance. Retrieved from www.smh.com.au/business/concerns-raised-over-chinas-rare-earth-dominance-20090901-f6xw.html

Topf, A. (2013). *Prices, quotas and smugglers: Making sense of rare earths in China.* Retrieved August 26, 2015 from https://investingnews.com/daily/resource-investing/critical-metals-investing/rare-earth-investing/prices-quotas-and-smugglers-making-sense-of-rare-earths-in-china/

Tse, P-K. (2011). *China's rare-earth Industry.* U. S. Department of Interior – U. S. Geological Survey.

U. S. Geological Survey. (2013). *Minerals yearbooks – rare earths: 1962–2011.* Retrieved November 23, 2013, from http://minerals.usgs.gov/minerals/pubs/commodity/rare_earths/#pubs

U. S. Geological Survey. (2014). *Mineral commodity summaries – rare earths: 1988–2014.* Retrieved November 23, 2013, from http://minerals.usgs.gov/minerals/pubs/commodity/rare_earths/#pubs

U. S. Geological Survey. (2015). *Mineral commodity summaries – rare earths, 2015.* Retrieved July 4, 2015, from http://minerals.usgs.gov/minerals/pubs/commodity/rare_earths/mcs-2015-raree.pdf

U. S. Geological Survey. (2016). *Mineral commodity summaries – rare earths, 2016.* Retrieved April 30, 2016, from http://minerals.usgs.gov/minerals/pubs/commodity/rare_earths/mcs-2016-raree.pdf

UN Statistics Division. (2013). *UN comtrade database.* Retrieved November 18, 2013, from http://comtrade.un.org/db/

UNFCCC. (2015). *What is COP 21?* Retrieved December 3, 2015, from www.cop21.gouv.fr/en/learn/what-is-cop21/

Walters, A., and Lusty, P. (2011). *Rare earth elements.* British Geological Survey.

Zhanheng, C. (2010). *Outline on the development and policies of China rare earth industry.* Beijing: The Chinese Society of Rare Earths.

22
GOVERNING LAND IN THE GLOBAL SOUTH

Julia Tomei and Darshini Ravindranath

Introduction

Land is an essential natural resource and central to effective functioning of the five-node nexus (see Chapter 1). Land is vital for the survival and prosperity of humanity, and for maintenance of all terrestrial ecosystems. The vast majority of human food is derived from the terrestrial environment, which also provides fodder, fibre, fuel, fertiliser, flowers and many other vital products. Land resources are closely tied to water and biomass resources, together providing the supporting, provisioning, regulating and cultural services that underpin sustainable development. Land can hold special significance to populations in the Global South, where a majority continue to reside in rural areas. Rural communities rely deeply on land and its resources for their livelihoods, well-being and existence: crops are grown for subsistence and for sale in local markets; fodder is collected for livestock; fuelwood is used for cooking and heating; biomass is used for constructing homes and stables; and wild species are used for health and well-being. Here, land signifies far more than the productivity of its soil: it is an assertion of socio-cultural identity and heritage.

The World Economic Forum (2011, 5) argues that "a combination of increasing scarcity of some natural resources, climate change and growth in global population to 9 billion by 2050 are creating the conditions for a 'perfect storm'." These combined drivers place pressure on land and terrestrial ecosystems, leading to increased demand for water, food, energy and minerals. However, land resources are finite, while the demands on them are not; indeed, as populations grow and consumption patterns change, demand for food, fibre and fuel will continue to rise, placing additional pressure on increasingly scarce resources. There is therefore an urgent need to conserve and safeguard land as the foundation of the global ecosystem. However, land-use change is a complex, dynamic and physically varied process. Biophysical as well as socio-economic conditions differ considerably from one location to another. The potential production of arable land and its susceptibility to degradation are dependent on inherent soil and other characteristics, as well as the management strategies employed and the wider political economic context. Sustainable, holistic treatment of land, water and biomass is often thwarted by issues of access and single silo regulatory structures. Although the interlinkages between different elements of the resource nexus are increasingly recognised, in the Global South land-use policies and the associated science are often compartmentalised and the interdependencies rarely acknowledged.

land-use policies also continue to rely on 'modern' interpretations and understandings of land and its values, while alternative imaginings of land are often overlooked.

Land governance represents a key challenge for the global community. Its importance is reflected in a number of global initiatives, including the UN Sustainable Development Goals (SDGs). SDG15 aims "to protect, restore and promote sustainable use of terrestrial ecosystems, sustainably manage forests, combat desertification, and halt and reverse land degradation and biodiversity loss". The individual targets are ambitious; for example, Target 15.3 aims to combat desertification, restore degraded land and soil, and strives to achieve a land degradation-neutral world by 2030. Frameworks, such as the UN Food and Agricultural Organisation (FAO) Voluntary Guidelines on the Responsible Governance of Tenure (VGGT), will be important in guiding and governing future land use. However, for some, these guidelines do not go far enough and run the risk of further embedding unsustainable land-use patterns (Borras Jr. and Franco, 2010).

Nexus thinking has been presented as an advance on sector-specific, or silo-based, approaches to natural resource governance. Yet it has tended to downplay the human and livelihood dimensions of resource use. This chapter focuses on one dimension of the resource nexus – land, and specifically land governance in the Global South. Using case studies, this chapter examines how the resource nexus plays out in practice in different settings in the Global South in order to pull out some insights for incorporating the human dimensions into land governance within a nexus framework. The rest of this chapter is structured as follows: the next section outlines key global trends in land use and land-use change. It then draws on four case studies to provide examples of how these trends manifest in different settings, and the consequences for those whose livelihoods depend on land and affiliated resources. The subsequent section identifies four issues to emerge from the case studies which are vital in efforts to govern the resource nexus. The chapter concludes by offering some thoughts on the potential (dis)benefits of the resource nexus in addressing some of the challenges facing land governance in the Global South.

Land use and drivers of land-use change

Land-use changes are influenced by a combination of global and local flows of commodities, information, capital and people. Additionally, political decisions are crucial in changing how land is used. Understanding these interlinked drivers of change, which need to be balanced against a wide array of demands on land resources, represents a key challenge for land governance. An important first step is for policymakers to recognise critical drivers of land-use change, and the subsequent consequences for populations and governance.

Of the 15 billion hectares (ha) of land available globally, around 30% is used for agriculture, including 10% for cropland. As cities have expanded at the expense of cropland, so cropland has expanded at the expense of forests, savannahs and grasslands, particularly in tropical regions. As a consequence of agricultural expansion, biodiversity is lost and ecosystem services are degraded which hinders the ability of societies and species to adapt and evolve.

The International Resource Panel found that between 1961 and 2007, the area used for crops increased by around 11% (Bringezu et al., 2014). Bruinsma (2009) analysed historical growth in world crop production over the same period and found that 77% of growth came from increased yields, 9% was due to increase in cropping intensity, and just 14% was due to expansion in arable land area. However, these global figures mask regional differences. For example, while cropland has decreased in Europe and North America, it has increased in South America, Africa and Asia (Kastner et al., 2014). Further, in Asia, the Green Revolution led to the heavy use of fertiliser and high-yielding crop varieties, along with institutional aid and rural credit programmes, which contributed to the higher productivity of crops such as rice and wheat. By contrast,

Denning et al. (2009) show that in sub-Saharan Africa, per capita growth of agricultural GDP was negative during the 1980s and 1990s. Production of subsistence crops such as maize were based almost entirely on extending cultivated area, with only minor contributions from yield growth. Changing consumption patterns, political decision-making and the emergence of new markets for agricultural products all affect global and national land-use patterns. Understanding the factors that drive land-use change, and how these manifest in different settings, is critical to developing strategies to promote sustainable land use and support rural livelihoods.

Drivers of land-use change

The global population is expected to reach nine billion by 2050 with less-developed regions expected to contribute most to this growth. Not only will there be more people on the planet, but rising incomes, urbanisation and changing diets will also contribute to increased food demand. Bond et al. (2013) estimate that demand for food will have increased by 50–70% by the middle of the century. Increased consumption of animal-based products, which have greater land and water requirements, also have higher associated greenhouse gas (GHG) emissions. This will affect efforts to address climate change, particularly since food production is already estimated to be responsible for between 17–32% of GHG emissions (Bajželj et al., 2013).

Globalisation has led to a dramatic transformation of the food chain, and has resulted in the increased distance between the sites of food production and consumption. International trade between high-yield and low-yield/land-constrained regions has increased, which has evened out differences in resource endowment in terms of land suitable for cropping. However, trade has also been a central driver of deforestation and the loss of ecosystem services. Karstensen et al. (2013), for instance, found that 30% of carbon emissions associated with deforestation were exported from Brazil in the past decade, of which 29% was due to soybean production and 71% cattle ranching. While small farmers continue to provide a large share of food for domestic markets, it has been export-oriented agriculture which has increased most rapidly (Bringezu et al., 2014). For many, the growth in the accompanying large-scale, industrialised agricultural system is a key cause for concern, since it invariably leads to the vertical and horizontal concentration of food supply chains. For example, according to Bringezu et al. (2014), the 10 largest seed companies control 50% of all commercial seed sales, while the top five grain trading companies control 75% of the market. Furthermore, large-scale monocultures have led farmers to adapt their agricultural practices, moving away from crop rotation and abandoning locally adapted plant varieties in favour of uniform, high-yielding varieties. Indeed, the FAO estimates that between 1990 and 2000, 75% of plant genetic diversity was lost, while 17% of livestock breeds are at risk of extinction (FAO, 1999, 2015). Further, three crops – rice, maize and wheat – contribute nearly 60% of calories and proteins obtained by humans from plants.

The emergence of new markets for agricultural products also affects global and national land-use patterns. For example, since the early 2000s, national governments in both the Global North and South have promoted the use of biofuels as a substitute for fossil fuels in the transportation sector. Since biofuels are typically produced from cash crops, such as sugarcane, maize, oil seed rape and African palm, increased demand has raised concerns about the impacts on food production and availability, as well as large-scale land acquisitions. However, the impacts on land use are uncertain as many biofuel feedstocks have multiple end uses; they can be used in direct human food consumption, for animal feed, in industrial applications, for bioplastics, and the generation of electricity. Assessments of the direct and indirect impacts of new products, such as biofuels, therefore need to take into account not just the complex interactions between agricultural markets, but must also recognise the multifunctionality of land and the crops it produces.

The transformation of demand and supply chains brought about by globalisation, and the consequent move towards large-scale intensive agricultural systems, has contributed to increasing duality in agriculture. For example, the Green Revolution of the mid-20th century promoted the use of high-yielding varieties, irrigation and agrochemicals, and resulted in a doubling of food production in some countries. However, it benefitted only those who were able to access these innovations, in many instances deepening the divisions between rich and poor farmers, and has led to environmental degradation (see case study on Rajasthan). Despite productivity gains, food security also remains a critical developmental challenge. In contrast to the resource-intensive agricultural systems that have resulted from agricultural modernisation, traditional agricultural systems have developed over time as humans have adapted to their local environments. These agroecological systems emphasise biodiversity; local knowledge systems; nutrient recycling; synergies among crops, animals and soils; and the regeneration and conservation of resources. Such systems create unique and dynamic farming practices and food cultures, but are rarely given sufficient attention by policymakers when formulating governance strategies. In between these two extremes lies a range of other farming systems (organic, industrial, monoculture, mixed cropping systems, inter-cropping systems, subsistence, commercial, outgrower schemes, family, contract, tenanted farming), which represent different configurations of circumstances, markets, values, and ownership arrangements (Tomei and Helliwell, 2016).

Knowledge gaps between policy and practice persist, contributing to the embedding of unsustainable land-use patterns. Furthermore, as a result of these land-use trends, populations in the Global South are facing mounting challenges related to macroeconomic instability, worsening income inequalities, rising food prices and multiple environmental crises. With their high dependence on land and allied resources for their livelihoods, this could have devastating results on local populations. Effective governance frameworks are therefore essential to support livelihoods, conserve resources and prevent unsustainable practices from becoming ingrained into the management of land.

Challenges of governing land use

A challenge for the governance of land in the Global South is the development of an institutional context that incorporates not only issues of 'conventional' governance (as measured through promotion of economic growth and agricultural expansion), but also the social and environmental aspects. Issues of land tenure, access and security are critical. Thus, in addition to ensuring enforcement of contracts and securing individual property rights, care needs to be given to the extension of protected areas, recognition of indigenous peoples' right of access to resources and management of common pool resources. Many communities lack recognised titles to the land they farm or hold in common; tenure security is important not only for crop productivity, but also to enable farmers to diversify their livelihoods by using land as collateral, renting it out or selling it. It also influences the extent to which farmers are willing and able to invest in the long-term well-being of their land and to adopt newer more intensive farming techniques focussed on short-term benefits.

Over the past decade, there has been an increase in land deals, or land grabs, by public and private corporations for agribusiness investments in Africa, Asia and Latin America. There is no single definition of a land grab; land may be sold outright, acquired through long-term leases or obtained through outgrowing schemes, and typically favour large-scale, export-oriented agriculture. Land deals are difficult to document and as a result estimates of the scale of the phenomenon vary. While the amount of land involved is uncertain, it is clear that this new wave of land acquisitions involves deals that are large scale; in 2010, the median size of land acquisitions was

40,000 ha, while a quarter of all deals involved areas greater than 200,000 ha (Schiedel and Sorman, 2012). Despite a focus on international actors, national actors (including both governments and private sector) are also playing a major role in land acquisitions. For some, they provide an opportunity to attract investment in agricultural and infrastructural development; however, the economic opportunities for countries involved in land acquisitions rarely extend to local communities where poor governance and weak enforcement of customary land rights has caused rapid and widespread changes in land-use patterns and ownership. Women may be particularly affected. While women represent the largest percentage of the workforce in the agricultural sector, many lack formal access to land and productive resource; in Africa, for example, less than 20% of women own land (FAO, 2016a). Although critics have been quick to blame demand for biofuels and transnational actors, research reveals a more complex picture that involves multiple drivers, actors and pressures on land (Anseeuw et al., 2012).

At a global level, initiatives by the African Union, the World Bank and the FAO have developed principles or recommendations for stronger land governance in countries that are the targets of large-scale investments. The most widely accepted has been the FAO's VGGT. The VGGT seek to improve the governance of land tenure with respect to all forms of tenure: public, private communal, indigenous, customary and informal. The overarching goal is to support the right of people to adequate food, with an emphasis on vulnerable and marginalised people. They outline a set of best practices and frameworks that governments can use when developing their own strategies and legislation. However, the VGGT have been criticised for not establishing binding applications and for doing little to challenge the status quo (Borras Jr. and Franco, 2010). Furthermore, the voluntary nature of the VGGT mean that they have not been able to replace existing laws, treaties and agreements and, as a result, the value added of the VGGT in specific contexts may be unclear. A key challenge will therefore be to translate international guidance into real change on the ground.

Efforts to govern the negative local impacts of commodities have focused on standards and certification systems. Such schemes are typically narrowly focused on a single crop (e.g. sugarcane, tea), product (e.g. biofuels, timber) or sector (e.g. fisheries), and often fail to consider the wider landscapes and systems within which production takes place. How boundaries are set determines not only which issues are subject to reporting, but also which impacts are considered and where. There is a risk of leakage, or indirect impacts, wherein regulation in one domain leads to unintended consequences in another. For example, concern about the indirect impacts of increased global demand for biofuels on land use has generated much controversy and debate, eventually leading the EU to set a limit on the amount of biofuels produced from food crops that could be utilised within the bloc. This provides an example of the danger of a narrow focus on one component of complex and multi-dimensional land-use challenges.

At the national level, even those countries which have developed legal frameworks to govern land, a key challenge relates to the capacity of government institutions to implement, monitor and enforce the law. Many countries in the Global South are plagued by inefficiency, corruption, and a lack of accountability and transparency. This means that even where laws and policies exist they may be ineffective, exacerbating poor management practices and deepening inequalities. Strong institutions are recognised to underpin sustainable development, and SGD16 aims to promote the rule of law and develop effective, accountable and transparent institutions at all levels. Mechanisms that promote good governance and strengthen accountability are therefore required to deliver sustainable livelihoods and natural resource management. This will require recognition of the interdependencies between the different elements of the resource nexus, and the impacts of land-use changes on communities and livelihoods. However, developing effective mechanisms will require an understanding of the diverse social, political, economic and environmental contexts within which land is used. Demonstrating this diversity, the next section sets out four case studies to illustrate how some of the challenges discussed in this section play out in different settings in the Global South.

Case studies: the resource nexus in practice

Land degradation in Rajasthan

Land degradation is a "reduction or loss in the capacity of the land to provide ecosystem goods and services, and assure its functions over a period of time" (FAO, 2016b). In arid, semi-arid and sub-humid areas land degradation takes the form of desertification. In drylands, land degradation, population growth, increasingly erratic climatic conditions, and a general lack of investment have put pressure on the livelihoods of dryland inhabitants and the integrity of their ecosystems, leaving little room for error in matters of land-use management and policy development.

The state of Rajasthan, in northwest India covers several agro-climatic zones and accounts for 67% of India's desertification (ISRO, 2007). This is especially significant for Rajasthan where population is dense (approximately 200/km^2), highly stratified, and heavily dependent on the agricultural sector for livelihoods (about 98% of rural population are involved in agriculture) (Swain and Kalamkar, 2012). Agriculture in turn is largely rain-fed and reliant on increasingly erratic monsoons. Farmers have, for centuries, relied on traditional knowledge systems to maintain the integrity of their land, water and biomass resources. Ingenious adaptations in the face of drought include *khadins*, which use natural depressions in the land to preserve soil moisture and develop localised lay-loam soil pockets, and traditional rainwater harvesting through small earthen dams, *johads*. Organisation of society was, until recently, centred on the distribution and protection of natural resources.

Poor land utilisation patterns have, however, exacerbated degradation and desertification in the state. Between 1880 and 2010, Rajasthan experienced significant loss of grasslands, shrublands and forests at the cost of increased cropland (Tian et al., 2014). The government of India has acknowledged that land degradation continues to be a major environmental concern impeding sustainable development. It highlights a list of determinants including, over-cultivation, overgrazing, improper crop management techniques, excessive groundwater extraction, deforestation and population pressure. A large part of the intensity and change in cropping patterns in the state can be attributed to the rise of groundwater-led irrigation. This, together with developments of the Green Revolution, has led to the exploitation of water resources and government-led 'greening of the desert' in an effort to halt desertification in the region. For instance, the use of tractors in the state has increased rapidly. Tractors churn up naturally stabilised sand, and to allow tractors to move freely in a field and stop competition for moisture between crops and shrubs/trees, all natural vegetation is uprooted, with implications for biomass availability. Furthermore, over-irrigation of these sands has led to salinisation and groundwater contamination. This case illustrates the interplay between land use, water and biomass, which has led to degradation. Most significantly, all of these changes have occurred at the cost of deterioration of indigenous knowledge systems that have sustained land for many generations.

Land redistribution and the affiliated legislation have thus far been unable to halt degradation. Market led-reforms in Rajasthan, including reductions in crop subsidies, large-scale electrification programmes, subsidies for tubewells, high-yield varieties of seeds, fertilisers, and investment by international donors are transforming the agricultural sector along free market principles. Less focus has been placed on issues such as conservation, distribution and equality, which has led to contradictions and corruption in the resource use system. There is a need for a governance approach that, by maximising synergies, maintains a balance between traditional and modern agricultural practices. Such an approach must also consider that the resilience of local communities is dependent on multiple silos: land-use policies inevitably have consequences for the use of water and biomass, which if not tackled simultaneously are likely to exacerbate degradation. One of the key principles of the nexus approach proposed in Chapter 1 is the provision of relevant information about critical interlinkages. In Rajasthan, such an approach would enable decision-makers to overcome endemic silo-thinking, thus adding immense value in addressing governance challenges presented by land use and land degradation.

Biofuels in Guatemala

Between 2000 and 2010, the global production of biofuels increased from 16 billion litres to 100 billion litres per year (IEA, 2011). Currently, most biofuels are produced from agricultural commodity crops, such as maize, sugarcane, sorghum and African palm. While the land required for second generation, lignocellulosic crops will be lower, increased global demand for biofuels has major implications for land availability, water use and the use of other agricultural inputs. Estimates of future land-use requirements for biofuels vary widely, and a review by Murphy et al. (2011) found that estimates of additional land required ranged between 100 to 650 Mha. This is a huge requirement given other competing pressures on land. Much of this land is likely to come from the Global South, due to the perceived availability of land, water and favourable climates. Guatemala is one country that has responded to the creation of new global demand for biofuels.

The Guatemalan sugar sector is well established and world leading. It is comprised of 13 sugar mills, all of which are owned by Guatemala's elite landowning families. The mills directly own or manage around 80% of sugarcane plantations; the remaining 20% is accounted for by independent producers, many of whom are themselves large landowners and there is no small-scale production of sugarcane in Guatemala. Large-scale, export-oriented agriculture – such as sugarcane – has historically been located along Guatemala's Pacific Coast due to the region's fertile volcanic soils, stable climate and plentiful water resources. Since the 1980s, the area along the Pacific Coast which is cultivated with sugarcane has increased steadily, reaching 270,000 ha in 2015–2016 or 10% of the total agricultural area (USDA, 2015). The expansion of monocultures, including sugarcane, has led to the concentration of land, and the dispossession and forced eviction of rural and indigenous communities. Communities rarely have legal ownership of the land, rather they rent land from large landowners in order to cultivate maize, beans and other subsistence crops. However, increasing competition from sugarcane and other agribusiness has forced subsistence and small-scale farmers out of the land lease market. The loss of access to land has consequences for food and economic security, and many families, unable to rent land, are forced to seek employment elsewhere, including on the very plantations that displaced them. Sugarcane cultivation also has implications for water use, and the diversion of waterways to irrigate plantations has impacted local communities, causing droughts in the dry season and floods in the wet season.

It is unlikely that biofuels are solely to blame for the expansion of sugarcane in Guatemala, but the opening up of new biofuel markets represents an important opportunity to diversify production. Across the world, the sugar sector has been affected by low sugar prices and adverse climatic conditions and diversification (i.e. moving from sugar to also produce bioenergy, biofuels and biomaterials) can build resilience by providing additional revenue streams and reducing exposure to fluctuating market prices. In so doing, it strengthens the position of the domestic elites who dominate the sugarcane sector to the detriment of local communities and environments. Certification has been proposed as a way of addressing some of the negative issues associated with biofuels. However, schemes focused solely on one feedstock inevitably overlook wider land-use changes, and the complex political economies that shape the winners and losers of those change. The nexus approach aims to generate improved knowledge on cross-resource needs and impacts for decision-making and management. As illustrated in Chapter 1, in a broader sense, all human activities and development pathways would benefit from information about resource use implications and using this approach might help overcome some of the challenges of silo-thinking identified in this case study.

Agroforestry in Tanzania

Agriculture is one of Tanzania's most important sectors, and approximately 80% of the population depend on agriculture for their livelihoods (IFAD, 2014). Forests are also key resources, accounting for approximately 38% of Tanzania's total land area. Deforestation rates are however high; between 1990 and 2005 an estimated 412,000 ha per annum were cleared, largely to meet crop, wood fuel and charcoal demands (Blomley and Iddi, 2009). In Tanzania, depleting land resources and dwindling forest products have hindered poverty reduction, and constrained efforts to meet the set targets of the Millennium Development Goals. In Tanzania, agroforestry has been proposed as a means of promoting a more dynamic and ecologically friendly use of natural resources in Tanzania's agricultural landscape.

Agroforestry is the integration of multiple natural components, such as trees, shrubs, palms, and bamboos, into traditional land-use systems of crops and/or livestock. Incorporating tree-based farming provides multiple co-benefits to farmers, including resilience to climate change, enhancing carbon sequestration, improved crop yields, income diversification, improved water-use efficiency, provision of fuelwood, and addressing food insecurity. Tanzania is home to several traditional agroforestry systems that have been practiced for hundreds of years: the Chagga home gardens, the Mara region home gardens known as *Obohochere*, and the traditional Wasukuma silvo-pastoral system called *Ngitili* are all traditional methods that rely on the use of multi-layered systems with a mixture of annual and perennial plants. On the slopes of Mount Kilimanjaro, for instance, farmers grow multi-layered combinations of banana trees, coffee shrubs and vegetables that imitate a tropical montane forest. The uppermost layer is formed by sparsely spaced trees which provide shade, medicine, fodder, fruits, firewood and timber. Under this layer more than 15 varieties of bananas are grown. Under the bananas there are coffee shrubs and under these, vegetables of variable species including climbers. This system prevents soil erosion, maximises the use of land and water on small home gardens, and provides nutritionally diverse food.

In addition to these traditional practices, Tanzania has more than 30 years of experience of agroforestry research and development activities. Technologies, such as fertiliser trees and biomass transfer for soil fertility improvement, rotational woodlots, and indigenous and exotic trees for food and health security have been introduced for reclamation and enrichment of traditional land-use systems. Aiming to capitalise on this traditional and scientific expertise, a National Agroforestry Strategy was prepared in 2004. The Strategy envisioned four million rural households adopting and benefiting from agroforestry practices by 2025, an uptake of agroforestry technologies, and enhanced livelihoods of 60% of the country's resource-poor by 2020 (ICRAF, 2009). However, despite a body of evidence indicating the benefits of agroforestry, widespread adoption has thus far been constrained by both intrinsic and extrinsic factors. These include knowledge, attitudes, and perceptions of farmers, economic variables, and lack of strong policy patronage (Meijer et al., 2015).

Tanzania faces the challenge of revitalising the agricultural sector by improving the natural resource base: soil, water and biodiversity (ICRAF, 2009). There needs to be stronger policies at both local and national level. Particularly at the local scale, capacity building and institutional support need to take into consideration a more robust approach, which targets not just soil fertility, but also the benefits of agroforestry for water and energy needs. A nexus approach – one which explicitly considers the interlinkages between land, food, water and energy – may help to ensure the adoption of agroforestry practices in the longer term. However, there remains the need for governance frameworks that address the weaknesses of local institutions that lead to problems with implementation.

Agrobiodiversity in the Peruvian Andes

The Central Andes is a globally important area for agricultural biodiversity or agrobiodiversity. Some 45 species of crop plants were domesticated here including the potato and other tubers (such as oca and ollucu), quinoa, and medicinal plants, such as cocaine and quinine. Peasant agriculture in the Peruvian Andes covers a wide range of agroecological zones: from the humid and tropical *ceja de la selva* at 500 metres above sea level to the cold and arid altiplano at more than 3,500 metres above sea level. These zones represent distinct microclimates, which enable communities to cultivate different types of crops. For instance, a family may live in the altiplano where they cultivate potatoes and quinoa, and have fields in the lower valleys on which they grow maize and beans.

In mountainous areas, farming is limited by ecological stresses including soil erosion, poor and marginal soil, and extremes in rainfall and temperature. Since pre-Hispanic times, Peruvian farmers have developed land management strategies to overcome these limitations, including terracing to prevent land degradation, ridged fields known as *waru-waru* which regulate climate through water management, and a defined crop rotation system called *aynoka* or *layme*. Reciprocal labour exchange known as *ayni* enable farmers to overcome labour constraints, and maintain these modifications to the landscape. Farmers also use the region's agrobiodiversity as a strategy to minimise risk; farming communities cultivate many thousands of potato varieties which, over the centuries, have adapted to the growing conditions (Zimmerer, 1991). Both native varieties and high-yielding varieties are cultivated, and while there is concern that the adoption of the latter has led to the loss of native varieties, research has shown that other factors are as, if not more, important than market drivers (Zimmerer, 1991, 2003; de Haan et al., 2010). Native varieties are not only cultivated because of their adaptation to local environments, thus supporting livelihoods, but also because farmers typically prefer cultivar mixtures for home consumption, as these are associated with quality traits and long-established culinary preferences (de Haan et al., 2010). Depending on their quality, native potatoes may be used for consumption, seeds, gifts, sale, animal feed or for barter.

As the influence of 'modern' agriculture has grown, so the traditional agricultural system of the Peruvian Andes has undergone changes. The introduction of non-native crops and livestock, combined with policies that incentivise the use of agricultural inputs and mechanisation, place pressure on traditional practices and knowledge. Demographic changes, including population growth and temporary outmigration, have also eroded cultural identity and affected gender roles (Zimmerer, 2003). Acknowledgement of the importance of agrobiodiversity, and the traditional agricultural systems that underpin it, has led to the establishment of a number of initiatives which aim to support these farming communities. These include mechanisms to help farmers to capture greater value from their products through the creation of value chains and niche markets, rewards for agrobiodiversity conservation services, and the establishment of Indigenous Biocultural Heritage Areas, such as the Potato Park. Located within the Cusco Valley, the Park covers more than 9,000 hectares, and aims to integrate contemporary science and conservation models with rights-based governance approaches. These initiatives offer the prospect of a dynamic and locally appropriate approach to conservation, one which acknowledges the interconnectedness between livelihoods, culture and nature. This demonstrates how an integrated approach, as proposed by the resource nexus, when applied to one resource can be managed for multiple targets such as biodiversity, water and crop production.

Implications for land governance

The resource nexus encourages decision-makers to move beyond a focus on single natural resources in order to acknowledge the interlinkages between land, energy, water and biomass resources. As the case studies illustrate, in practice land management at the local level is an inherently holistic process. For farmers to survive, she or he must take into account the available resources – environmental, social, cultural, physical and financial – and balance these against multiple pressures. The cases of India and Peru, in particular, show how over time farmers have developed land-use strategies that integrate these multiple resources and needs, and which embody the nexus approach that has been advocated. However, in many instances, it has been exogenous factors, such as policy shifts and climate change, which have threatened the sustainability of these traditional systems. Once lost, it may be difficult to regain the traditional knowledge and cultural systems that underpin such systems. In seeking to develop nexus approaches, there is therefore much to be learnt from land managers.

All four cases highlight the importance of the pre-existing political, economic, social, institutional and environmental conditions under which people engage with and benefit from resource use. This contextual equity affects access to and distribution of resources, as well as acknowledging the power relations which affect peoples' capability to make use of resources. A key task will be to design interventions that reflect local conditions, while accounting for global pressures on land and other resources. Solutions need to fit with local realities or they run the risk of not being adopted, or worse, may lead to negative outcomes for people and resources. Reed and Stringer (2016), for instance, discuss some of the outcomes of the Green Revolution. They argue that while it succeeded in increasing crop productivity in some countries such as Mexico and India, the Green Revolution failed to address issues of access and, despite productivity gains, food insecurity remains a key global challenge. The adoption of Western technologies not only led to biodiversity loss and soil erosion, but also ignored traditional agricultural adaptations and local knowledge systems, as evidenced by the case study of Rajasthan.

Greater understanding of the local realities within which technologies and policies are applied will be essential to deliver effective mechanisms to address resource use at different scales. In this regard, there is a need to review past experiences – both positive and negative – in order to understand what worked, what did not, and why. However, it is important to also be aware that conditions change, and that what worked yesterday may not be applicable today or directly replicable in other contexts. Furthermore, policy priorities can and do shift, and progress that has been made can also be unmade. While stable policy environments are crucial for investment, new incentives for sustainable land governance may need to be found as internal and external conditions change. To promote sustainable land governance will require finding policy approaches and incentives that are tailored to meet the needs and motivations of different actors, although ultimately what is possible will be shaped by a country's institutional arrangements, and the relative power of different actors.

The Green Revolution in Rajasthan provides an example of the ways in which pressure to modernise can lead to unintended, often negative, outcomes. In this instance, the loss of traditional knowledge and agricultural systems which have contributed to worsening desertification in the state. Market-led reforms have exacerbated, rather than alleviated, land degradation, while the failure to review the outcomes has led to a deepening crisis which threatens rural livelihoods and the integrity of the land they depend upon. Conversely, the case of agroforestry in Tanzania suggests that there are solutions which are able to combine the traditional and the modern. Agroforestry systems are able to capitalise on the resilience of traditional agricultural systems and the productivity and efficiency of modern systems, potentially leading to more sustainable outcomes and positive impacts for livelihoods and land use. Despite the benefits of agroforestry, policymakers have not yet been able to create the supportive policy environment required to drive its uptake. An approach to governing land, which explicitly acknowledges the interconnectedness between land, water, energy and food resources, and the livelihoods that depend on them, may provide a means of overcoming current limitations by highlighting the benefits across multiple domains.

Conclusions

The challenges facing land use in the Global South are dynamic and complex, encompassing interlinked environmental, social and economic factors, all of which are shaped by the broader political and institutional contexts within which land is situated. While this complexity is theoretically recognised, solutions continue to focus on the technical aspects and ignore the locally adapted farming practices and cultures within which they are implemented. As the case of Rajasthan highlights, this approach has contributed to the erosion of traditional knowledge and practices, and to unsustainable resource-use patterns across the Global South. In addition to the promotion of inappropriate solutions, the very basis of rural livelihoods is under threat as the increase in large-scale land acquisitions demonstrates. The Guatemalan case study highlights the need for land tenure systems that recognise the dependence of rural livelihoods on the natural resources – particularly land – that surround them. With its focus on the interactions between multiple resources, the nexus approach is a step in the right direction. However, the dependence and influence of local communities and their livelihoods upon resources is not yet adequately incorporated into the resource nexus. The Tanzanian and Peruvian cases demonstrate how communities have long practiced a nexus approach to managing their local environments and livelihoods. Additional research is needed to ascertain how these traditional practices can be integrated into the more technical approaches favoured by policymakers. Addressing the challenges facing land will require an understanding of the local contexts within which resource use is embedded, in order to develop locally appropriate solutions that will contribute towards more sustainable and equitable development.

References

Anseeuw, W., Alden Wily, L., Cotula, L., and Taylor, M. (2012). *Land rights and the rush for land: Findings of the global commercial pressures on land research project*. Rome: International Land Coalition.

Bajželj, B., Allwood, J. M., and Cullen, J. M. (2013). Designing climate change mitigation plans that add up. *Environmental Science and Technology*, 47(14), 8062–8069.

Blomley, T., and Iddi, S. (2009). *Participatory forest management in Tanzania: 1993–2009 lessons learned and experiences to-date*. Tanzania: Ministry of Natural Resources and Tourism, Forestry and Beekeeping Division, Dar es Salaam.

Bond, M., Meacham, T., Bhunnoo, R., and Benton, T. (2013). *Food waste within global food systems*. A Global Food Security Report, Swindon. Retrieved June 2016, from www.foodsecurity.ac.uk.

Borras Jr., S. M., and Franco, J. (2010). From threat to opportunity? Problems with the idea of a 'code of conduct' for land grabbing. *Yale Human Rights and Development Manual*, 13, 507–523.

Bringezu, S., Shütz, H., Pengue, W., O'Brien, M., Garcia, F., Sims, R., Howarth, R.W., Kauppi, L., Swilling, M., and Herrick, J. (2014). *Assessing global land use: Balancing consumption with sustainable supply*. International Resource Panel, United Nations Environment Programme. Retrieved May 2016, from www.unep.org/resourcepanel-old/Portals/24102/PDFs/Summary-English.pdf

Bruinsma, J. (2009). *The resource outlook to 2050: By how much do land, water and crop yields need to increase by 2050*. Expert meeting on how to feed the world. Rome: FAO. Retrieved June 2016, from ftp://ftp.fao.org/agl/aglw/docs/ResourceOutlookto2050.pdf

De Haan, S., Núñez, J., Bonierbale, M., and Ghislain, M. (2010). Multilevel agrobiodiversity and conservation of Andean Potatoes in Central Peru. *Mountain Research and Development*, 30(3), 222–231.

Denning, G., Kabambe, P., Sanchez, P., Malik, A., Flor, R., Harawa, R., Nkhoma, P., Zamba, C., Banda, C., Magombo, C., Keating, M., Wanglia, J., and Sachs, J. (2009). Input subsidies to improve smallholder maize productivity in Malawi: Toward an African green revolution. *PLoS Biology*, 7(1), e1000023.

FAO. (1999). *Women users, preservers and managers of agrobiodiversity*. Rome: UN Food and Agricultural Organisation.

FAO. (2015). *The second report on the state of the World's animal genetic resources for food and agriculture*. Rome: UN Food and Agricultural Organisation.

FAO. (2016a). *The female face of farming*. Rome: UN Food and Agricultural Organisation. Retrieved July 2016, from www.fao.org/gender/infographic/en/.

FAO. (2016b). Rome: *Land degradation assessment*. UN Food and Agricultural Organisation. Retrieved July 2016, from www.fao.org/nr/land/degradation/en/.

ICRAF. (2009). *Agroforestry options for Tanzania/*. Policy Brief 3. World Agroforestry Centre, Nairobi. Retrieved August 2016, from www.worldagroforestry.org/downloads/Publications/PDFS/BR09007.pdf

IEA. (2011). *Technology roadmap: Biofuels for transport*. Paris: International Energy Agency. Retrieved July 2013, from www.iea.org/publications/freepublications/publication/biofuels_roadmap.pdf.

IFAD. (2014). *Investing in rural people in the United Republic of Tanzania, May 2014*. Rome: IFAD. Retrieved August 2016, from www.ifad.org/documents/10180/feb514f1-a0d2-4111-8d98-50be8ddd0184

ISRO. (2007). *Desertification & land degradation*. Ahmedabad: Atlas of India, Indian Space Research Organisation.

Karstensen, J., Peters, G. P., and Andrew, R. M. (2013). Attribution of CO2 emissions from Brazilian deforestation to consumers between 1990 and 2010. *Environmental Research Letters*, 8, 1–7.

Kastner, T., Erb, K. H., and Haberl, H. (2014). Rapid growth in agricultural trade: effects on global area efficiency and the role of management. *Environmental Research Letters*, 9(3), 034015.

Meijer, S. S., Catacutan, D., Ajayi, O. C., Sileshi, G.W., and Nieuwenhuis, M. (2015). The role of knowledge, attitudes and perceptions in the uptake of agricultural and agroforestry innovations among smallholder farmers in sub-Saharan Africa. *International Journal of Agricultural Sustainability*, 13(1), 40–54.

Murphy, R., Woods, J., Black, M., and McManus, M. (2011). Global developments in the competition for land from biofuels. *Food Policy*, 36, S52–S61.

Reed, M. S., and Stringer, L. (2016). *Land degradation, desertification and climate change: Anticipating, assessing and adapting to future change*. London: Routledge.

Schiedel, A., and Sorman, A. H. (2012). Energy transitions and the global land rush: Ultimate drivers and persistent consequences. *Global Environmental Change*, 22, 588–595.

Swain, M., and Kalamkar, S. S. (2012). *State of Rajasthan agriculture 2011–2012*. Anand, Guajarat: Agro-Economic Research Centre, Sardar Patel University.

Tian, H., Banger, K., Bo, T., and Dadhwal, V. K. (2014). History of land use in India during 1880–2010: Large-scale land transformations reconstructed from satellite data and historical archives. *Global and Planetary Change*, 121, 78–88.

Tomei, J., and Helliwell, R. (2016). Food versus fuel? Going beyond biofuels. *Land Use Policy*, 56, 320–326.

USDA. (2015). *Biofuels annual: Guatemala*. Washington, DC: U. S. Department of Agriculture.

World Economic Forum. (2011). *The consumption dilemma: Leverage points for accelerating economic growth*. Cologny/Geneva: World Economic Forum.

Zimmerer, K. S. (1991). Labour shortages and crop diversity in the Southern Peruvian Sierra. *Geographical Review*, 81, 414–432.

Zimmerer, K. S. (2003). Geographies of seed networks for food plants (potato, ulluco) and approaches to agrobiodiversity conservation in the Andean Countries. *Society and Natural Resources*, 16, 583–601.

PART V

Applying the resource nexus

Regional and global scale

23
ELEMENTS OF THE WATER–ENERGY–FOOD NEXUS IN CHINA

Philip Andrews-Speed and Carole Dalin

Introduction

While the links between water, energy and food (WEF) have long been recognized, especially within the river-basin management community, the wider ramifications of these links and the widespread application of the term "nexus" was stimulated by a conference on this nexus held in Bonn in 2011 (Hoff, 2011; Leck et al., 2015). Each of the three factors is intimately linked to the other two. Water is an essential input into food production to such an extent that the agricultural sector is by far the largest consumer of water in most countries. Water also plays a significant role in the energy sector. Though much of this use is in the form of withdrawal, consumption may also be significant depending on the nature of the process and the technologies applied (see Box 23.1 for definitions). Not only do the agriculture and energy sectors consume water, but they may also cause significant damage to water quality.

Energy has been a requirement for pumping water for thousands of years. In modern times, urbanization, industrialization and the commercialisation of agriculture have greatly enhanced the role of energy in the water sector, for purification as well as pumping. The main link between food and energy, other than the food needed for human energy, arises from the recent industrialisation of bioenergy of different types which threatens to create competition between food and energy in some locations.

The nexus of these three factors can be displayed at local, regional and sometimes global scales. Subsistence communities living on marginal land may experience in their daily lives the consequences of poor nexus management. Entire regions of a country or continent may be adversely affected by the excessive or inappropriate use of one resource which has knock-on effects for other resources. In some cases, the impacts may be exacerbated by climate change. Finally, the international trade of energy, food and industrial products transforms the nexus to a global scale.

The aim of this chapter is to examine the nexus of water, energy and food in China. The choice of China as an exemplar of the challenges arising from the inadequate governance of the resource nexus derives from the intensity of the resource challenges that are or will likely soon be experienced on account of the convergence of a number of trends, namely: growing wealth, industrialisation, urbanisation and energy use; the large share of coal in the energy mix; the intensive nature of agriculture in many places; large and still growing populations, a significant proportion of which remain rural; growing divide between rich and poor; and intensifying water stress that may be exacerbated by climate change.

The ramifications of China's nexus challenges are amplified by two further aspects of the country's geography and economy. First, the disparities between northern and southern China are profound. The north and northwest of the country is rich in fossil-fuel resources and hosts much of the grain production, yet is desperately short of water and economically poor relative to the coastal regions. In contrast, much of the south and southeast is well endowed with water resources which can be used for the production of food and hydro-electricity, but is deficient in fossil-fuel resources and agricultural land compared to the north. Second, since the beginning of the 21st century, China has progressively become a major global trading nation. Prominent among its imports are the fossil-fuel commodities (namely oil, natural gas and coal) and food (notably soybean).

We have chosen two aspects of human activity to illustrate this nexus: agriculture, and industrialisation and urbanization on account of their pertinence to the case of China. This chapter continues by first presenting the general features of the WEF nexus (the second section) before moving on to examine the WEF nexus in China's agricultural sector (the third section) and the water–energy nexus as it is expressed in the nation's urbanization and industrialization (the fourth section). The fifth section identifies the key policy implications.

The nature of the nexus

Agriculture

Agriculture is at the centre of interactions in the service nexus of water, energy and food (WEF) and in the resource nexus of land, water and energy sources. Indeed, agriculture uses all three resources (land for cropland and pasture; irrigation water and soil moisture; and fuels to produce energy for pumping irrigation water and for producing and transporting inputs and outputs). It also plays a role in providing WEF services or altering their availability by affecting input resources, via e.g. land-use change, water availability and cycle disturbances, and biofuel production.

Agriculture's use of land, water and energy

Agriculture uses significant amounts of land, as cropland and pasture occupy 50% of total arable land. Irrigation is the largest water-consuming activity, responsible for 70% of global freshwater withdrawals and nearly 90% of global freshwater consumption (see Box 23.1 for water use definition). Food production also utilizes energy to produce, transport and distribute food as well as to extract, pump, lift, collect, transport and treat water for irrigation. It is estimated that the food production and supply chain consumes about 30% of the total energy consumed globally (FAO, 2011).

Box 23.1 Definitions relating to water type and use

Water use (BP, 2013):

- Withdrawn water: water removed from surface water or groundwater, at least temporarily, to produce or process energy or for some other purpose.

- Consumed water: the portion of withdrawn water not returned to the surface or groundwater in the same drainage basin from which it was abstracted.

Types of water (Hoff, 2011; Allan, 1998):

- Blue water: fresh surface and groundwater.
- Green water: rainwater held in the soil and available to plants.
- Virtual water: water used during the production of a good.

The residential and industry sectors also increasingly use water, energy and land resources, while facing environmental degradation issues (e.g. water pollution) and in some cases resource scarcity. This situation is expected to worsen in the next decades as 60% more food will need to be produced to feed the world population in 2050. In addition, global energy consumption is projected to increase by up to 50% by 2035 (IEA, 2010), and global water withdrawals for irrigation are projected to increase by 10% by 2050 (Alexandratos, 2009).

Agricultural yield growth since the 1970s has significantly reduced the use of water or land per unit of food produced (i.e. water or land footprint). However, these yield increases have often been associated with increasing use of chemical fertilisers, pesticides and herbicides, as well as a mechanization of the sector, which lead to risks of pollution, water scarcity, soil depletion and climate change. These environmental footprints of food production are highly different across products and regions (Figure 23.1), due to their dependence on the local climate and agricultural practices.

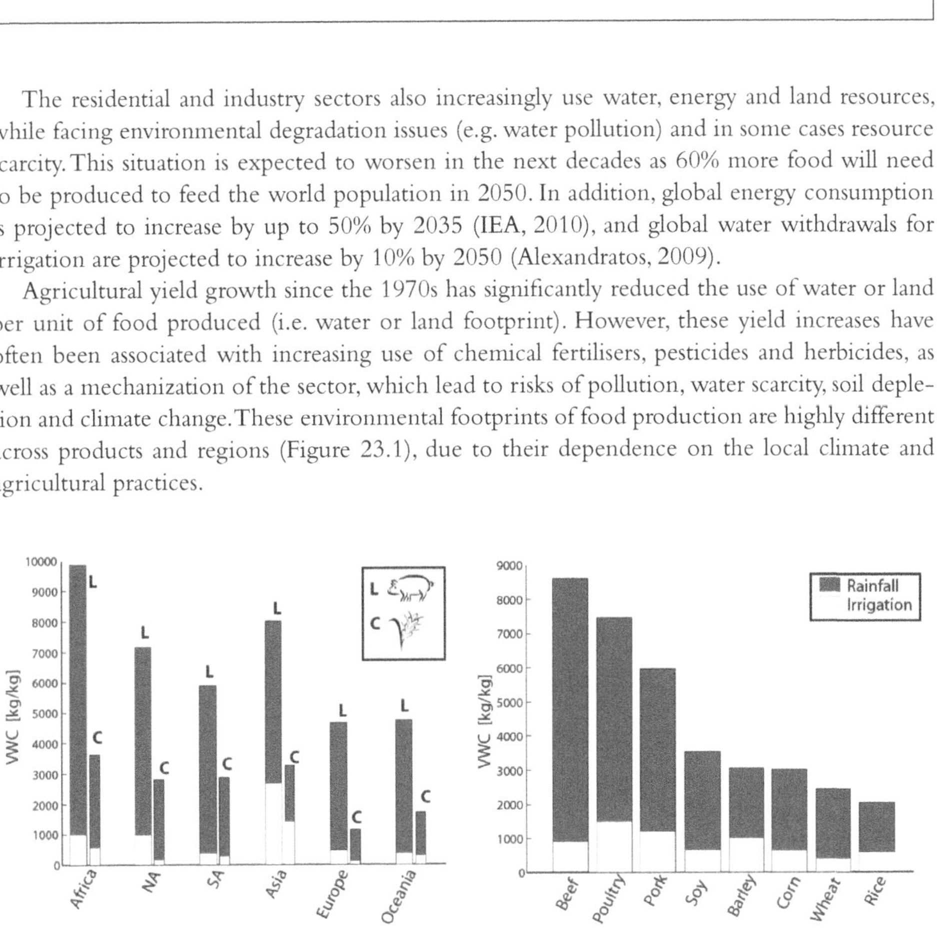

Figure 23.1 Virtual water content (VWC, volume of water consumed per volume of food produced) of major crop (C) and livestock (L) products in 2000, averaged by region (left) and product (right), by water source type (see Box 23.1). High VWC reflects a water-intensive production.

Source: Konar et al. (2011)

Agriculture's effects on provision of water, energy and food services

Agriculture is obviously the direct source of food production. Less obvious are its roles in energy production and water services.

While energy is consumed by the agricultural sector, agricultural outputs – waste and by-products or entire crops – are also used as energy sources called biofuels, mostly biodiesel and bioethanol (Figure 23.2). Biofuels provided 2.7% of the world's fuels for road transport in 2010. The US and Brazil account together for 90% of global ethanol fuel production. The world's largest biodiesel producer is the EU, accounting for 53% of all biodiesel production in 2010 ("Biofuels Make a Comeback Despite Tough Economy". Worldwatch Institute, 2011). The International Energy Agency has a goal for biofuels to meet more than a quarter of world demand for transportation fuels by 2050 to reduce dependence on petroleum and coal (IEA Technology Roadmap, "Biofuels for Transport" 2011).

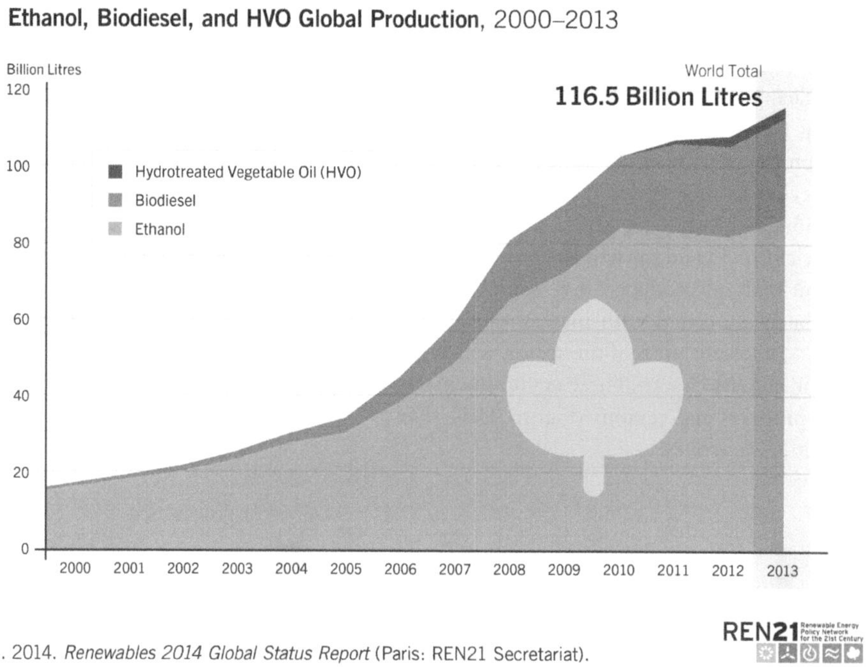

Figure 23.2 Global biofuel production from 2000 to 2013, by fuel type

Source: REN21. 2014. *Renewables 2014 Global Status Report* (Paris: REN21 Secretariat)

The attractiveness of biofuels as a renewable and carbon-neutral energy source has been debated and criticized in the scientific literature. Indeed, growing crops for biofuel will require water and land as well as sometimes chemical fertilisers and pesticides, which can affect ecosystems and resource availability and quality. Delucchi (2010) concluded that biofuels produced from crops using conventional agricultural practices will likely *not* mitigate the impacts of climate change and will exacerbate stresses on water supplies, water quality and land use, compared with petroleum fuels. He recommends the use of marginal land and rain-fed production, with very low inputs of fossil fuels and chemicals.

Besides, as a highly water- and land-intensive sector, agriculture may significantly reduce water availability or quality, potentially hindering the provision of adequate water for industrial and residential uses. Competition – and sometimes conflicts – between cities, farmers and industries over water resources has arisen in multiple regions where water is increasingly scarce, such as the California (Kahrl, 1982) Central Valley (US), Inner Mongolia (China) and South Africa.

Table 23.1 Estimates of water intensity at different stages of the energy supply chain, applying different methodologies and different units

	M^3/TJ	*Variable units*
Extraction and processing		
Conventional oil: offshore	<1	<0.1 bbl/bbl oil
Conventional oil: onshore	<50	<2.0 bb/bbl oil
Unconventional oil	10–250	
Conventional natural gas	<1	
Unconventional natural gas	5–15	<1 bbl/bbloe
Coal: extraction and washing	<60	1.6 m^3/tonne
Uranium: extraction and beneficiation	0.5–3.5	250–1,800 m^3/tonne
Processing		
Crude oil refining	5.5–26	>2.5m^3/m^3
Gas processing	0.7	
Gas-to-liquids	18	
Coal-to-liquids	300	1.5 m^3/m^3
Uranium: conversion, enrichment, fabrication	0.2–3.0	
Power generation		
Coal	60–540	1,600–3,000 l/MWh
Gas	110–220	680–1,000 l/MWh
Nuclear	380–700	<3,200 l/MWh
Renewable energy		
Solar PV		260–640 l/MWh
Wind		40 l/MWh
Hydro-electricity		100,000 l/MWh
Biofuel crops: ethanol and biodiesel	4–60	<2,000 l/l
Biofuel processing: ethanol	130–260	0.36–6.0 l/l
Biofuel processing: biodiesel	15–70	0.5–2.0 l/l
Concentrated solar power	780–900	<5,000 l/MWh
Geothermal	620–1870	>1,000 l/MWh

Data sources: M^3/TJ = BP (2013). Variable units = Lamberton et al. (2010), Meldrum et al. (2013) and Martin and Grossman (2015).

Industrialization and urbanization

Industrialization and urbanization are key components of economic development in most countries. They are also key processes in exacerbating the nexus between water, energy and food (Lawford et al., 2013; Venkatash et al., 2014), not least because the quantity of water and energy consumed per capita and per unit of GDP generally increases with economic development. In this section we focus on the links between energy and water, at the way that modern energy systems use and pollute water, and at the energy needs of water supply.

Water use and pollution by the energy industry

Water is withdrawn and consumed in large quantities along the full length of the energy supply chain, including for renewable energy (Table 23.1). Starting with extraction, the production of oil and gas through conventional techniques requires relatively modest quantities of water for the drilling and production processes. The extraction of unconventional oil and gas uses water in much larger quantities for hydraulic fracturing and, in the case of oil sands, steam injection. The extraction and washing of coal also consumes large amounts of water, notably for washing the coal to remove non-combustible rock materials, but not as much as unconventional oil.

In all these cases there are processes available to significantly reduce the consumption of freshwater, by using re-recycled water, low-quality water, or produced water, for example. Coal washing can be performed with air rather than water. Likewise, liquid nitrogen can be used instead of water to carry out fracturing for the production of shale gas. However, most of these processes add to the costs of the project.

The processing of fossil fuels also consumes significant quantities of water. The largest single process in terms of total global water use is oil refining. The most water inefficient technologies are those which are applied to convert one fossil fuel into another form, for example gas-to-liquids and coal-to-liquids. Where water supplies are short, companies have been developing ways to conserve and re-use water. In addition, new techniques to the reforming stage are being developed to reduce water use.

The energy process which withdraws and consumes the largest quantities of water at a global scale is power generation where the water is required for cooling. It has been estimated that this industry accounts for nearly 50% of global water withdrawals, but only 3% of actual consumption. The quantity of water used per unit of power produced depends on the fuel (coal, oil, gas or nuclear), on the thermal efficiency of the plant and on the cooling process deployed. Nevertheless, this amount of consumption is dwarfed by irrigation which uses an estimated 77% (BP, 2013).

Finally, it is essential to recognize that some forms of renewable energy consume significant quantities of water. While wind power and solar photovoltaic technologies use negligible amounts of water, even including the manufacturing process, the same cannot be said of hydroelectricity, biofuels, concentrated solar power (CSP) and geothermal energy.

In addition to the direct withdrawal and consumption of water, the energy supply chain can cause substantial water pollution if badly regulated and managed. This can occur at all stages from extraction, through transportation and processing. Examples include: accidents in oil fields, mines, power plants, refineries and pipelines; long-term leakage of waste waters into aquifers and freshwater bodies; and deposition of waste materials, dust and chemicals onto soil and water bodies. In addition, the water flows from once-through cooling systems tend to raise the temperature of the water body with detrimental consequences for the ecosystem.

Energy use by the water industry

As in the case of energy, the water industry requires energy along most or all of its supply chain. In developing societies, that energy to take water from where it occurs naturally to where it is needed is supplied by different combinations of gravity and human or animal labour, through the construction of canals and the sinking of wells. In urbanized and industrialized societies, large quantities of water have to be transported from where it naturally occurs to where it is needed and then has to be treated both before and after use.

The energy intensity of water use varies greatly from one place to another, depending on physical geography, and between different sectors. As a result, estimates of energy use at different stages of the supply chain vary greatly and it is much more difficult to produce general estimates of energy intensity (Table 23.2). Large cities in arid climates may require water to be pumped over long distances and possibly over significant topography, as is the case with cities in Arizona (Lamberton et al., 2010). This involves energy consumption that can be much greater than that required to treat the water for household use. In arid climates, the alternative to transporting surface water over long distances is to pump groundwater, a process that also uses significant quantities of energy.

The energy intensity of the treatment of water before first use depends on the quality of the water, generally the salinity, and on the water quality standards for the particular use. One component of the water supply chain which is often ignored in energy calculations is the heating of water for household use (Nair et al., 2014). The final energy-intensive stage of the water supply chain is the treatment of wastewater from industry and households. Once again, the amount of energy required depends on the quality of the wastewater to be treated and on the quality of water required after treatment (BP, 2013).

It has been estimated that the total energy use for global water supply and treatment in the year 2000 amounted to nearly 6% of the world's total electricity production that year. Despite the low level of water treatment, Asia accounted for around 40% of global energy use in water supply. Were high standards to have been applied throughout the world, this would have brought energy consumption for water supply and treatment to nearly 8% of electricity production, with most of this increase occurring in Asia (BP, 2013). Nevertheless, there is increasing recognition that wastewater and sewage sludge are useful sources of biomass energy and some treatment plants are net producers of energy. This is a growing opportunity to draw positive benefit from the energy–water nexus.

Agriculture and the water–energy–food nexus in China

Major connections of the WEF nexus through agriculture

China's economic growth is expected to continue into the next decades, accompanied by persistent urbanization and industrialization – albeit at a slower rate than in the early 2000s. The associated increase in demand for land, water resources and rich foods will add to the challenge of sustainably feeding the population and balancing agricultural and environmental policies (Dalin et al., 2015). Indeed, agricultural policy has incentivized domestic production in order to meet high national food self-sufficiency targets (i.e. share of national food supply produced domestically – set to 95% for grains in 1996, CSC (1996)), sometimes at the expense of environmental conservation, when intensive agriculture is linked with nutrient pollution or water overuse.

Table 23.2 Upper-end estimates of energy intensity of different stages in the water supply chain

		Estimated consumption kWh/m³	*Comments*
Pumping	Water transportation	<2.5	Western US, Australia
	Water transportation	< 4.0	Spain
Treatment	Wastewater treatment: primary	0.0024–0.37	US, Australia
	Wastewater treatment: secondary	0.18–0.82	
	Wastewater treatment: tertiary	0.32–0.88	
	Desalination	1.0–8.5	Membrane technology

Sources: Lamberton et al. (2010); BP (2013); Department of Energy (2014)

Water scarcity is particularly important in the region. Indeed, while China hosts 19–20% of the world's population, it only has 7% of global freshwater resources. Changing diets towards resource-intensive animal products have played an important role in recent changes in China's agricultural production and trade (Dalin et al., 2012), and have further increased the pressure on water resources (Liu et al., 2008) and the wider environment, domestically and abroad. Agriculture, the most water-intensive sector (65% of China's freshwater withdrawal, Food and Agriculture Organization of the UN (FAO), http://fao.org/nr/water/aquastat/countries_regions/china/index.stm), could be a strong lever to reduce the country's rising water use (Figure 23.3).

This section highlights three major connections of the WEF nexus through agriculture in China: water used for crop production (irrigation), resources used for livestock production (irrigation for animal feed and pollution from farms) and crops used for energy production (biofuels)

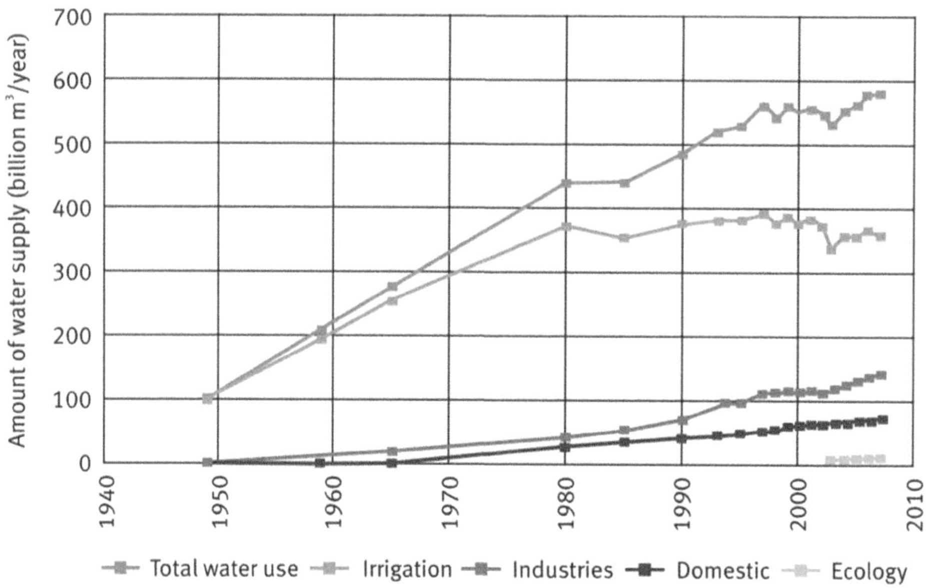

Figure 23.3 Water use by sector in China from 1948–2008

Source: Ge et al. Water. 2017; 9(1):35

Irrigation: water for food

Geographical constraints have forced Chinese farmers to grow crops in the arable lands of the North, even though water resources are much less abundant there than in the South – the northern half of the country has 65% of the arable land and only 17% of national freshwater resources (Zhang and Anadon, 2014). Irrigation has thus quickly developed, covering 45% of the country's croplands (Food and Agriculture Organization of the United Nations (FAO)), http://fao.org/nr/water/aquastat/countries_regions/china/index.stm) and supporting about 70% of domestic grain production (Peng, 2011).

Due to the increasing scarcity of surface water and rainfall – possibly exacerbated by climate change – combined with the growing demand for food, irrigation relies heavily on groundwater reserves in some regions. In the northern provinces of Inner Mongolia and Gansu, respectively

67% and 64% of irrigation originates from groundwater reserves (Wang et al., 2012). Upstream of Beijing, groundwater pumping for irrigation largely contributes to depleting the North China Plain aquifer, which declines at one of the fastest rates in the world (22 mm/y or 8.3 km^3/y, vs. 20.4 mm/y in California Central Valley, US – 3.1 km^3/y, and 40 mm/y in northwestern India – 17.7 km^3/y, Famiglietti (2014)).

Agriculture's share in the water consumption of northern regions remains at about 75% and continues to rise in absolute terms (Jiang, 2015). This already unfavourable situation is exacerbated by two other factors. First, poor environmental management of wastewater discharge has resulted in high and rising levels of water pollution, with 15% of river water being unfit for any use and 28% being unfit for human contact (Jiang, 2015; Shifflet et al., 2015). Second, severe droughts in the southwest of the country appear to be becoming more common, possibly as a result of global climate change, and threaten agriculture, energy supplies, and urban and rural livelihoods (Zhang et al., 2012).

Livestock: more land, more water, more fertilisers for food

China's economic growth has led to a dietary change towards greater consumption of animal and dairy products and thus, an increased demand for animal feed: China's GDP per capita was multiplied by 9.5 from 1986 to 2007 (IMF, 2010) while meat production per capita was multiplied by 3.3 (FAO, 2017). Pig farms have sprung in the rural and peri-urban areas, leading to many environmental issues linked to pollution and greenhouse gases emissions from manure and chemicals.

This increase in pork production has required large amounts of protein-rich soybean cakes for feed, and driven the rapid growth of China's soybean imports since 2001, from Brazil, Argentina and the US (Dalin et al., 2012, Figure 23.4). These imports were facilitated by the lift of soy import restrictions by a domestic trade policy in 2000–2001 (Carter and Rozelle, 2001). Since then, the country has become a prominent importer of virtual water globally. While saving much land and water resources in China, imports of Brazilian soy, sometimes grown on lands cleared from Amazonian forests, can damage the environment there and globally through climate change.

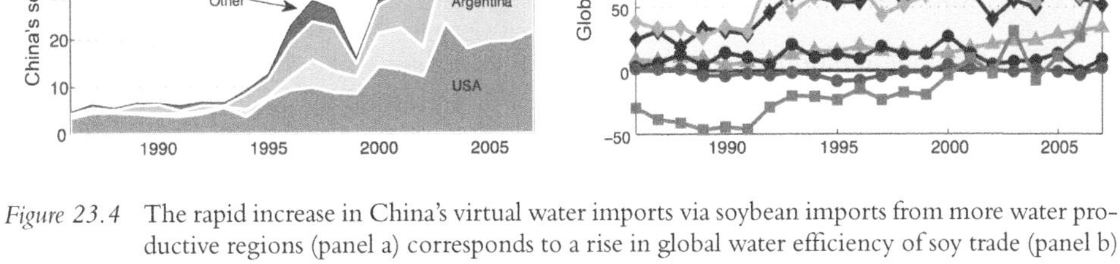

Figure 23.4 The rapid increase in China's virtual water imports via soybean imports from more water productive regions (panel a) corresponds to a rise in global water efficiency of soy trade (panel b)

Source: Dalin et al. (2012)

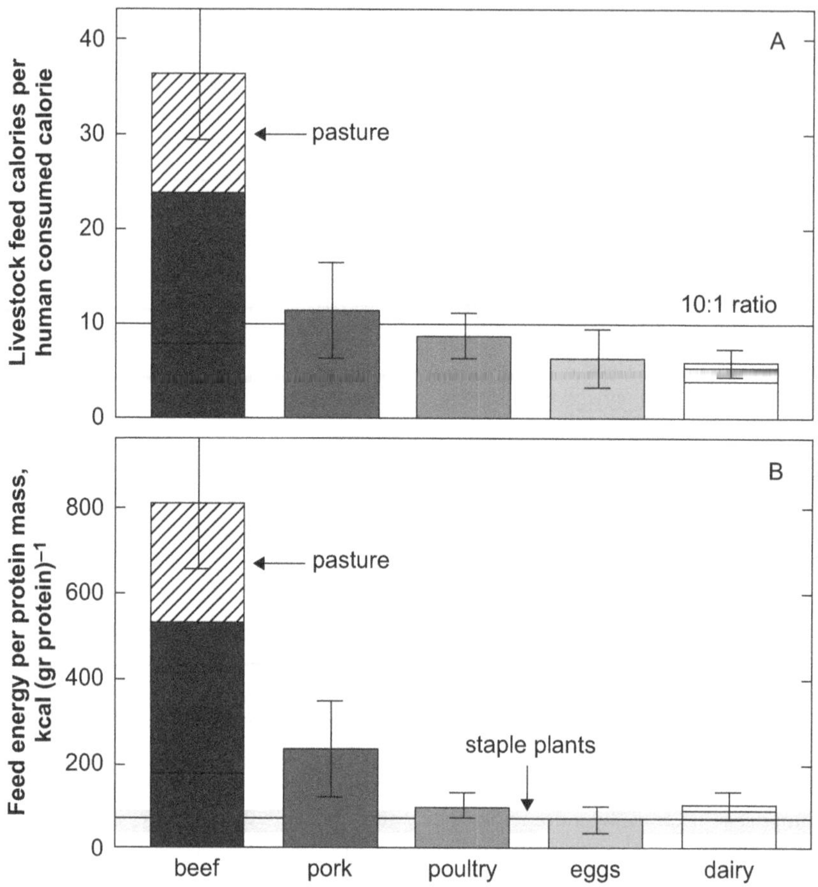

Figure 23.5 Feed-to-food and feed-to-protein conversion factors of different livestock categories. The bar height of each category in A shows the total livestock feed calories used divided by the human-consumed calories they yield. For example, the value of ≈9 for poultry indicates that, on average, nine feed calories fed to poultry yield one calorie consumed by humans. Note that this factor includes the approximately two-fold loss reported by the USDA in the post-farm gate supply chain from primary production through retail to the consumer. The often-quoted 10:1 conversion factor per trophic level arising from studies in ecology is marked as a grey line. B depicts the conversion factors from livestock feed to human-consumed protein mass. For beef and dairy, the contribution of concentrates, processed roughage, and pasture is presented from the bottom to the top, respectively. The grey area marks the upper and lower bounds of the three staple plants. Error (uncertainty) bars indicate SD.

Source: Eshel et al., PNAS (2014) 111:3 11996–12001

Importantly, meat products are much more resource-intensive than crops or vegetables, because of the low efficiency of energy conversion by animals. While the crop–food ratio (the amount of crops needed to obtain one unit of edible food) for grains is close to 1 (~1.2), the feed–food ratio is on the order of 10–40 for livestock products, accounting for transformation of feed calorie by animals and meat processing. This means that, for example, for each one calorie consumed by humans via beef, up to 40 calories of feed need to be produced (Figure 23.5).

Biofuels: food for energy

Agriculture in China has also been providing energy resources for biofuel production. Indeed, China is the third largest producer of biofuel in the world, behind the US and Brazil (Chew, 2006), mostly based on maize feedstock. Even though coal is still the highly dominant source of energy, biofuel development has been strongly supported by the government, motivated in part by energy security concerns (in 2006 about 48% of total oil demand was imported). However, biodiesel has not developed as much, due to lack of political support, limited feedstock and high costs of building large processing plants.

Maize being the main biofuel feedstock (e.g. in large plants built in 2001 in Heilongjiang, Jilin, Henan, and Anhui provinces), bioenergy competes with food production, for the already scarce water and land resources (see Table 23.3 for estimated land and water footprints of biofuel), and for fertiliser and other chemical inputs.

Yang et al. (2009) found that bioethanol production consumes 3.5–4% of total maize production in China, reducing market availability of maize for other uses by about 6%. They project that 5–10% of the total cultivated land in China would need to be devoted to meet the biofuel production target of 12 million metric tons for the year 2020 (Table 23.3). The associated water requirement would amount to 32–72 km^3 per year, approximately equivalent to the annual discharge of the Yellow River. These ambitious targets may thus further tighten the competition between food production, energy production and environmental protection.

Another important resource for food production is fertilisers. Mainly composed of three elements: nitrogen, phosphorus and potassium (N, P and K), fertilisers rely on the limited resource of rock P and increasingly on synthetic N. Importantly, excessive use – which is particularly common in China (Foley et al., 2011) – can lead to these elements running off and infiltrating into surface and subsurface water bodies, with risks of pollution and eutrophication (from N and P). Please refer to Chapter 20 for more details regarding fertiliser use.

Table 23.3 Land and water requirements for the production of China's biofuel targets

Year	*Biofuel target*	*Feedstock crop*	*Feedstock use (million tons)*	*Area for biofuel crops (1000 ha)*	*% of total crop area*	*Total water requirement (km^3)*
2010	2 million tons of ethanol	Maize	6	1112	0.72	5.1
		Cassava	12	735	0.47	6.7
		Sugarcane	30	477	0.31	3.7
		Sugarbeet	28	979	0.63	5.7
	0.2 million tonne Biodiesel	Soybean	1.2	686	0.44	3.8
		Rapeseed	0.66	361	0.23	1.3
2020	10 million tons of ethanol	Maize	30	5562	3.59	25.3
		Cassava	60	3676	2.37	33.3
		Sugarcane	150	2387	1.54	18.6
		Sugarbeet	140	4897	3.16	28.3
	2 million tons of biodiesel	Soybean	12	6857	4.42	38.4
		Rapeseed	6.6	3607	2.33	13.3

Source: Yang et al. (2009)

Water for food and virtual water trade

China's geographic mismatch between its arable land and water availability has led to unsustainable agricultural expansion in dry areas, further supported by food self-sufficiency objectives. In particular, Inner Mongolia (in the Yellow River basin) and the greater Beijing area (Beijing, Tianjin and Hebei provinces, in the Hai River basin) are suffering increasingly severe water scarcity. Major associated environmental issues include soil degradation, water resource over-exploitation and pollution, and land subsidence from groundwater depletion, threatening both ecosystems and human activity (Tilman et al., 2001).

Dalin et al. (2014) estimated the water productivity of various food commodities across Chinese provinces, and the volumes of water embedded in China's interprovincial and international food trade (i.e. virtual water trade, VWT). They found large differences in water productivities across the country, with very low levels in North China, in particular for maize in Inner Mongolia, although it is the largest maize exporting province. Due to such differences, China's domestic food trade is not water-efficient for blue water (i.e. blue water use is larger than in a situation without trade, assuming current consumption could be produced locally). However, much water resources are saved through soybean imports from foreign trade partners (mainly the US, Brazil and Argentina, more water productive for soy than China, Figure 23.4).

China's domestic food trade is efficient in terms of green water resources but inefficient regarding blue water resources. Indeed, domestic wheat trade saves both blue and green water at the national level, but domestic trade of maize loses more blue water (about 8 km^3/y). International imports play a major role in China's VWT network, as well as in the associated savings, especially imports of soy (blue global water savings, 21 km^3/y; total global water savings, 41 km^3/y). Chinese water savings from foreign imports (at the national scale) actually are even greater than the global water savings (at the world scale), because foreign imported soy does not require any use of China's water resources.

Dalin et al. (2015) projected 2030 virtual water trade in China, finding that water-inefficiency of Inner Mongolia's exports will worsen in a business-as-usual scenario. However, they found that blue water resources (water drawn from rivers, lakes and aquifers) could be saved if irrigated areas are reduced in the most water scarce provinces, at the expense of a small reduction in the country's food self-sufficiency. Such trade-offs analysis can help inform policymaking, in combination with decision-makers' choice of the relative importance of environmental and other policy goals.

Policy and outlook

Current solutions to water scarcity are focused on sustaining existing activities; for example, the South to North Water Transfer Project (SNWTP) will increase water supply in the North through physical transfers. However, China's agricultural and water resources strategies could change. The country recently increased its trade openness significantly (e.g., soy imports (Dalin et al., 2012)), and major policy plans, in addition to supporting sustainable agriculture (Xinhua News Agency, 2014), newly emphasize land conservation and reduction of groundwater use as water-saving strategies (CPC, 2013).

The very large SNWTP engineering project has been launched to cope with water scarcity and bring water where demand is high but supply is low: cities of the northeast. But many issues (e.g. soil salinization and groundwater overdraft) in the northwestern region, where land is available and exploited despite scarce water resources, will not be resolved by the SNWTP, as it does not connect to these regions.

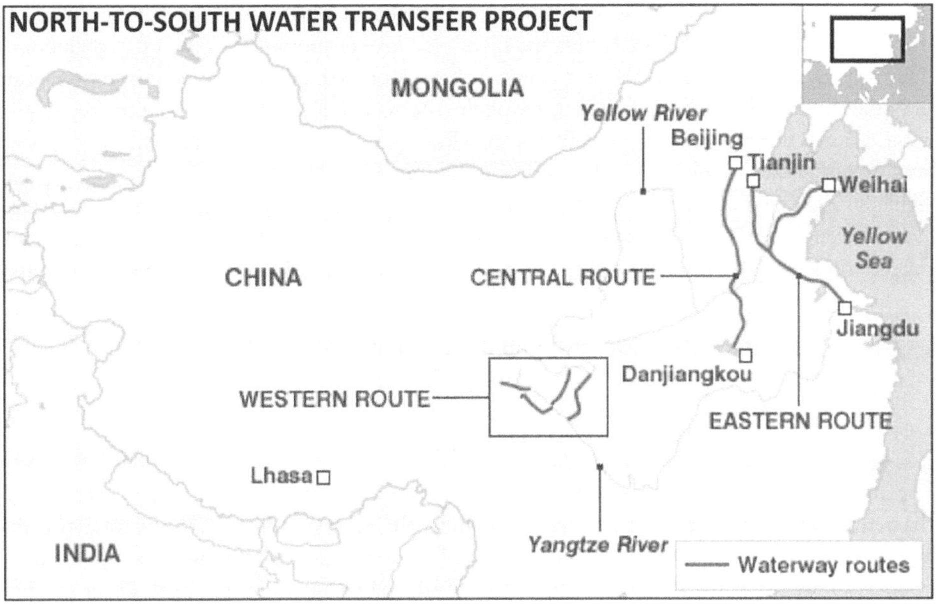

Figure 23.6 Map of the South to North Water Transfer Project in China

Source: © Asia Pacific Memo

The SNWTP ultimately aims to channel nearly 45 billion m^3 of freshwater annually[2] from the Yangtze River in southern China to northern regions through three canal systems (Figure 23.6): the Eastern Route, the Central Route flowing to Beijing and Tianjin and the Western Route, from three tributaries of Yangtze River to provinces such as Qinghai, Gansu, Shaanxi, Shanxi, Inner Mongolia and Ningxia.

The complete project was expected to cost $62 billion – more than twice as much as the Three Gorges Dam (Water Tech, 2008). By 2014, more than $79 billion had been spent, making it one of the most expensive engineering projects in the world (Chang, 2014). The Eastern Route was completed in December 2014, serving Beijing and Tianjin.

One of the main receiving regions of the SNWTP, the Beijing–Tianjin–Hebei region, is a key water scarce area with two large growing cities and a major agricultural province, pressuring surface and groundwater for both drinking water and irrigation needs. The water availability per capita in these three provinces/municipalities is the lowest in the country (less than 100 m^3/cap/y in 2007, Zhang and Anadon).

Alongside the water-supply side option of the transfers, policies supporting water-demand side solution to water scarcity have been developed. The Chinese government's 2011 No. 1 Document outlined a plan to expedite water conservation development and to achieve sustainable management of water resources within this decade (CPC, 2011), planning a four trillion yuan (US$635 billion) investment over this period. While environmental protection has gained weight on the political agenda, implementation, governance and integrated management are still behind (Liu and Yang, 2012).

As China's population is expected to increase until 2030 (World Bank Group, 2016), and as the economy continues to grow, demand for energy and animal products, and in turn water and other resources use, can be expected to further rise. Important developing issues are the conflict between the mining and agricultural sectors for water resources in the dry North, and the risk of exhausting groundwater reserves given current overexploitation in North China, notably for irrigation (Famiglietti, 2014). In addition, policies targeting the highly sensitive environmental issue of air pollution in cities may also enter in conflict with resource conservation goals. Indeed, some coal power plants and other industrial plants are displaced out of cities to reduce local air pollution, but they require water that may already be scarce in their new rural locations. Thus, effective policies need to consider simultaneously the air pollution and water consumption effects of planned industrial activities. This highlights the important potential for nexus approaches to help solve major sustainable development challenges in China. According to Biba (2015), a greater awareness and balance of competing concerns in China's decision and policymaking scene will be needed for this potential to exploited.

Industrialization and urbanization and the water–energy nexus in China

The context

The threats arising from the energy–water nexus in China have their origins in three main characteristics of the country's physical geography: the relatively small scale of total available freshwater, the continuing reliance on coal for a large share of national energy supply, and the co-location of a majority of the coal and other fossil-fuel resources in the north of the country where freshwater is scarce. Despite the steady growth of clean sources of energy, fossil fuels still account for nearly 90% of China's primary energy supply and the share of coal remains above 60%. Despite the growing imports of oil and gas, energy policy seeks to maximize the domestic production of all forms of energy (Andrews-Speed, 2012).

Rapid industrialization and urbanization have driven a fivefold increase in national energy consumption since 1990, and most of this energy has been produced domestically, with the exception of crude oil (BP, 2016). Not only has the production and processing of this energy required large quantities of water, but the demand for water has also risen dramatically. Most of this growth of water consumption since 1990 has been driven by the industrial and household sectors (Global Water Partnership. 2015). As a consequence, greater quantities of energy have been required to extract, transport and treat the water to supply the end users.

Water for energy

There are very few reliable, systematic, quantitative estimates of the role of water in China's energy sector (see, for example, Tables 23.4 and 23.5). Though estimates vary greatly, water withdrawal by the energy sector may account for up to 20% of the national total. Although most of withdrawal and consumption takes place in the generation of electricity, coal has significant impacts on the country's water resources at the stage of mining as well as in the conversion of coal to liquids and gas. This is important because coal continues to account for more than 60% of the nation's primary energy supply.

The main use of water at the mine site is for washing the coal in order to remove non-combustible 'ash'. This is necessary to reduce to amount of rock that has to be transported to the point of use, mainly power stations, and to reduce the dust emitted from these power stations. The proportion of coal washed has risen from 30% in the early years of this century to 55% in 2014, over which period domestic coal production has doubled, requiring an almost fourfold increase in the quantity of water needed for coal washing (Shifflett et al., 2015). Steps are being taken to recycle this water and, in some mines, to apply air washing. The other main threat to water from mining comes from solid waste and wastewater flowing into surface and groundwater (Economy, 2004).

Table 23.4 Estimates of direct use of water per unit of energy output for different forms of energy in China (m^3/TJ)

	Coal	*Crude oil*	*Natural gas*	*Oil products*	*Electricity**
Withdrawal	18.4	37.2	13.6	87.2	4306.5
Consumption	8.9	26.3	9.5	37.9	410.0
Wastewater discharge	28.3	11.0	4.1	49.3	90.6

Source: Zhang and Anadon (2013)

*excluding evaporation related to hydro-electricity

Table 23.5 Estimates of total national direct use of water for different forms of energy in China (million m^3)

	Coal	*Crude oil*	*Natural gas*	*Oil products*	*Electricity**	*Total for energy sector*^
Withdrawal	1025.9	269.9	39.3	1234.2	48,635.9	53,560
Consumption	495.5	210.3	27.5	536.8	4,641.4	6,453
Wastewater discharge	1584.4	87.6	11.8	697.4	1,023.6	3,568

Source: Zhang and Anadon (2013)

*excluding evaporation related to hydro-electricity
^total includes other categories such as coke, heat and other gases

Coal continues to be the feedstock for as much as 70% of the country's electricity supply despite the growth of natural gas, hydro-electricity, wind, solar and nuclear energy. As described above, once-through cooling involves very large water withdrawals; consequently new power stations are increasingly using closed-cycle wet cooling systems. While air cooling is being progressively installed in new power plants in arid regions of the country, it requires informed decisions to be made concerning trade-offs, as this technology requires additional capital investment, reduces energy efficiency and raises carbon emissions (Yu et al., 2011; Zhang et al., 2014). As is the case with coal mining, coal- fired power plants release significant quantities of ash both directly onto the land and into storage ponds, as well as into the air which then contaminates the land and water further away from the plant (Greenpeace, 2010).

A final use of coal that could, if pursued at a large scale, have serious impacts on China's water supply is the conversion of coal to liquids and gas. As indicated in Table 23.1 these chemical processes have large water requirements and this will be no different in China (Yang and Jackson, 2014). Although the current capacity of these plants remains modest, the future rate of expansion is difficult to predict.

These three coal-related uses of water converged in the government's strategy announced in its Twelfth Five-Year Plan (2011–2015) to build a number of major 'coal bases' in Inner Mongolia, Shaanxi, Shanxi and Ningxia, all of which lie in the arid north of the country. These bases bring together coal mining, coal-fired power plants and coal liquefaction and gasification capacity. Great care and the application of the best water management technologies will be needed to avoid an environmental crisis (Greenpeace, 2012).

Although the coal supply chain is clearly the major culprit in China's water–for–energy nexus, other forms of energy also have damaging impacts on freshwater water availability. Hydro-electricity provides about 20% of the country's power supply, from thousands of dams with an aggregate capacity of about 400 GW (Shifflett et al., 2015). Although estimates for water loss through evaporation from hydropower reservoirs are not available, Liu et al. (2015) calculated that evaporation from reservoirs of all types accounts for 22% of the country's total water consumption. This loss diverts water away from cities, factories and farms, especially in times of drought and the changes in water flows damages downstream ecosystems (Shifflett et al., 2015).

Several forms of energy are currently produced or installed on a relatively small scale but are certain or likely to expand in the coming years, exacerbating threats to China's water supply. A massive boom is occurring in the construction of nuclear power plants that will take national capacity from 25 GW at the end of 2015 to as much as 60 GW by 2020. After the Fukushima disaster the government restricted plants to coastal locations in order to protect freshwater bodies, but it is unclear whether this policy will be sustained.

Inspired by the US experience, China has launched a concerted effort to explore for shale gas and tight involve hydraulic fracturing. Some of this activity is taking place in Sichuan and neighbouring areas which, although rich in freshwater resources, is densely populated and agriculturally very productive. Other exploration is taking place in the arid north and northwest of the country. In both regions, great care will need to be taken to constrain the consumption of water and to protect water bodies from pollution (Hu and Xu, 2013).

China's production and consumption of biofuels is at a relatively low level at present, at about 2 million tons per year, or 1% of annual oil consumption (BP, 2016). Given the government's keenness to reduce the nation's dependence on international oil markets, research continues into how best to boost biofuel production in such a way that it does not compete with agriculture and does not consume excessive quantities of water (Chen et al., 2015; Zhao et al., 2015). Another form of renewable energy for which China has big plans is CSP which, as in other countries, will be sited in sunny regions with limited water supplies (Shifflett et al., 2015)

Energy for water

The relatively low abundance of freshwater in China combined with its geographic concentration in the south of the country results in water supply having a relatively high and growing energy intensity (Kahrl and Roland-Holst, 2008; Shifflett et al., 2015). The majority of this energy, an estimated 64% in 2011, is devoted to extracting and transporting water to where it is consumed (Table 23.6; Li et al., 2016).

The growing use of groundwater and the need to transport water over increasing distances results in higher energy intensities of water consumption for two main reasons. The first is the long-standing need to pump water, both vertically and horizontally. The second is the energy which is embodied in the infrastructure built to transport water. In the case of China, this latter factor has been of great importance in recent years after China embarked on a programme to construct three major canal systems to bring water from the south to the north of the country – the 'South-North Water Transfer Project'. This project involves not just vast quantities of energy-intensive concrete, but also, like reservoirs, results in significant water losses through evaporation. For these reasons, surface water and groundwater account for 64% and 33%, respectively, of the total energy use for extraction and transportation. While recycling and, to a much greater extent, desalination are much more energy intensive, the current small scale of these processes in China keep their total energy consumption to a relatively low level (Table 23.6).

The provision of public water supply requires much of the water to be treated before it can be delivered to the end user. This process is relatively energy intensive and accounts for almost 30% of the total energy requirement of the water supply sector (Table 23.6).

A surge in the construction of wastewater treatment plants across China has resulted directly from the progressive industrialization and urbanization and the need to raise living

Table 23.6 Estimates of energy intensity and energy use in China's water sector for the year 2011

	Energy intensity	*Total energy use*	
	KWh/m³	*TWh*	*Share*
Water extraction		124.3	64%
Surface water	< 0.20	79.4	
Groundwater	0.37	41.5	
Desalinated water	5.9	0.7	
Recycled water	0.82	2.7	
Public water supply	0.30	56.0	29%
Wastewater treatment	0.26	13.2	7%
Total		193.5	

Source: Li et al. (2016)

and environmental standards. In their study of wastewater treatment plants in southern China, Gu et al. (2016) estimated that the direct energy intensity of most of the technologies deployed lay in the range 0.4–0.5 kWh/m³, which is comparable with international performances, though the most efficient achieved 0.25 kWh/m³. In their study, Li et al. (2016) used a figure close to this last one of 0.26 kWh/m³ which led them to estimate that wastewater treatment accounts for about 7% of the energy used by the water supply sector (Table 23.6). The government's justifiable determination to rapidly boost the national capacity will inevitably result in increasing energy consumption, especially if tertiary treatment becomes mandatory (Shifflett et al., 2015).

The aggregate energy consumption of China's water sector in 2011 has been estimated as amounting to about 4% of annual electricity demand (Li et al., 2016). If wastewater treatment is excluded, the calculations of Li et al. (2016) showed that the 31% of this energy related to the commercial and industrial sector, 16% to household use, 12% for cooling and 41% for agriculture. Their analysis also revealed significant geographic variations, with the energy use for groundwater extraction being highest in the north and west of the country, and that for wastewater treatment and desalination being greatest in the south and east. The area of China with a high level of energy consumption for all three processes, plus wastewater treatment, was the highly urbanized and industrialized regions around Beijing and Tianjin cities, in Hebei and Shandong Provinces.

Policy and outlook

As is the case in many developing and transition countries, China's rapid economic growth has been achieved through the relatively indiscriminate exploitation of natural resources. The specific climatic, geological, geographic and demographic conditions have resulted in both resource scarcity and environmental damage. Further, the combination of energy intensity, coal-dependence and water scarcity have highlighted the importance of the energy–water nexus.

The inefficient use and poor management of energy and water resources has its origins in a combination of weak or erratic implementation of government policies, low resource prices, a lack of mechanisms to address environmental costs and poorly defined property rights. In addition, as in many countries, national and subnational government lack the institutional structures and mechanisms to manage the energy–water nexus effectively.

In recent years, the Chinese government has been taking steps to improve resource efficiencies in both the energy and water sectors, through both economic and administrative measures as well as massive investment. In the energy sector, the soaring demand for energy led the government to launch a series of programmes to promote energy efficiency and energy conservation starting in 2004 (Andrews-Speed, 2009). These policies achieved a considerable degree of success with national energy intensity declining by as much as 30% between 2005 and 2015. Further, the share of coal in the fuel mix has been diminishing since 2011 through the promotion of renewable and nuclear energy and the reduction of the role of heavy industry. These energy savings have led to significant water savings (Gu et al., 2014). The Thirteenth Five-Year Plan (2016–2020) and its subordinate sector plans emphasize the need to continue these efforts through enhancing the role of renewable energy, constraining the use of coal, and continuing improvements in the efficiency of use of all resources.

Likewise, the effective management of water resources has become an increasingly important priority for China's government, as reflected in successive five-year plans. Over the last 20 years, the government has applied a range of administrative and economic measures, in addition to major infrastructure projects, to improve water use efficiency and reduce pollution (Liu et al., 2013). Nevertheless, energy policy still threatens to undermine water resource management objectives (Qin et al., 2015).

The current leadership of Xi Jinping has reiterated its intention to address the full spectrum of environmental challenges facing the country as part of the goal to create an 'ecological civilisation'. However, a number of obstacles remain to be overcome if recent progress is to be sustained. First, the high cost of the most water-efficient energy technologies and energy-efficient water technologies discourages deployment. Second, the national reluctance to rely on external energy supplies stimulates the ongoing construction of coal bases and search for unconventional hydrocarbons in arid regions. Third, local governments retain their long-standing ability to ignore central government policies unless they are enforced with great vigour. Finally, and relatedly, rivalry between ministries and between neighbouring political jurisdictions (provinces or counties) prevents effective coordination of policies within and between sectors. Underlying these political and economic factors are the geographic realities of the nation, with primary energy resources that require water for extraction and processing being concentrated in the north and west, and water resources in the south and east.

Conclusions

Water, energy and food are all key inputs to human life, and economic development (industrialization and urbanization) raises demand for each of them. The general inability of societies to govern the water–energy–food nexus in a holistic manner arises principally from the silo-like structuring of relevant organizations, in both public and private sectors, and the consequent fragmented thinking and actions of actors in these organizations. International organizations, national and subnational governments, and corporations are increasingly recognizing the need to play a more active role in improving the governance of this nexus. This realization arises, to a great extent, from the growing long-term pressures on many resources resulting from a combination of population growth, rising wealth and climate change. Potential solutions range from better technologies and practices, to improved coordination, capacity building for dealing with the increased complexity that a nexus approach brings, and changes in incentives, including prices.

Across most of Asia, and notably in China, one or more of these resources is scarce or under stress, and policies and actions taken to address issues relating to one resource have knock-on effects for other resources– in the same country or in a neighbouring or more distant country. Climate change is exacerbating these pressures, principally through changing the availability of water for cities, industries and rural populations, but also through increasing variability and uncertainty, thus undermining the path to economic development.

Recognition of the nature and origin of these nexus challenges is growing both within China and across Asia, but policy action is slow on account of a mix of different factors that vary from place to place. These factors include:

- weak governance and state capacity;
- inappropriate governance structures that are poorly suited to coordinate policy across different resources;
- the dependence of political legitimacy of the ruling elites on short-term economic growth;
- the power of economic interest groups;
- the securitization of food, energy and water by national governments.

However, even if these deficiencies can be addressed, the underlying challenge for any government or society is to satisfactorily address and manage the trade-offs that need to be made. These trade-offs include not just between individual resources, but also between human security, sustainable livelihoods and biodiversity, on the one hand, *versus* the imperative of economic growth and 'modern' lifestyles, on the other. The trade-offs may also be geographic, with a single country or between countries. In the case of China, the contrast in resource endowments between the north and south of the country exacerbate the difficulties in managing the WEF nexus. In the context of international trade, China's large imports of soya beans shifts the shifts the nexus problem to the exporting states, such as Brazil. Conversely, the government's preference for maximizing self-sufficiency in fossil-fuel production intensifies the nexus challenges within the country. However, the examples presented in this chapter also show that there are nexus opportunities. For example, efficiency improvements in the use of one resource or shifting the production of a commodity to a more favourable location can yield benefits for other resources or factors.

References

Agricultural Bureau of Hebei (ABH). (2014). *Hebei Action for reducing over-extraction of groundwater by adjusting agricultural production structure and promoting agricultural water saving technology*. Retrieved January 2015, fromwww.gov.cn/xinwen/2014-09/09/content_2747016.htm

Alexandratos, N. (2009). *roceedings of a Technical Meeting of Experts*, How to feed the world in 2050, FAO, Rome, pp. 1–32.

Allan, J. A. (1998). Virtual water: A strategic resource: Global solutions to regional deficits. *Groundwater*, 36(4), 546.

Andrews-Speed, P. (2012). The governance of energy in China. *Transition to a low-carbon economy*. Basingstoke: Palgrave Macmillan.

Andrews-Speed, P. (2009). China's ongoing energy efficiency drive: Origins, progress and prospects. *Energy Policy*, 37, 1331–1344.

Asian Development Bank. (2009). *Status and potential for the development of biofuels and rural renewable energy: The people's Republic of China Mandaluyong City*. Philippines: Asian Development Bank. ISBN: 978-971-561-837-3

Biba, S. (2015). The goals and reality of the water – food – energy security nexus: The case of China and its southern neighbours. *Third World Quarterly*, 37, 51–70.

BP. (2013). *Water in the energy industry: A introduction*. London: BP.

BP. (2016). *BP statistical review of world energy*. London: BP.

Carter, C. A., and Rozelle, S. (2001). Will China's agricultural trade reflect its comparative advantage? China's food and agriculture: Issues for the 21st century. *ERS USDA AIB*, 775, 27–30.

Chang. (2014, August 1). China's water crisis made worse by policy failures. *World Affairs*. Retrieved from www.worldaffairsjournal.org/blog/gordon-g-chang/china%E2%80%99s-water-crisis-made-worse-policy-failures

Chen, H., Xu, M-L., Guo, Q., Yang, L., and Ma, Y. (2015). A review on present situation and development of biofuels in China. *Journal of the Energy Institute*.
Chew, C. (2006). *Current status of new and renewable energies in China: Introduction of fuel ethanol*. Report to the Institute of Energy Economics, Japan.
China State Council. (1996). *The grain issue in China*. Retrieved January 2015, from www.law-lib.com/fzdt/newshtml/24/20050709194656.htm
Communist Party of China (CPC). (2011). *No. 1 Central policy document: Decision from the CPC central committee and the state council on accelerating water conservancy reform and development*. CPC Encyclopedia.
Communist Party of China. (2013). *Major issues concerning comprehensively deepening reforms (Communique of the third plenary session of the 18th central committee of the communist party of China)*. Retrieved November 2014, from www.china.org.cn/chinese/2014-01/17/content_31226494.htm.
Dalin, C., Hanasaki, N., Qiu, H., Mauzerall, D. L., and Rodriguez-Iturbe, I. (2014). Water resources transfers through Chinese interprovincial and foreign food trade. *Proceedings of the National Acadamy of Sciences of United States of America*, 111(27), 9774–9779.
Dalin, C., Konar, M., Hanasaki, N., Rinaldo, A., and Rodriguez-Iturbe, I. (2012). Evolution of the global virtual water trade network. *Proceedings of the National Acadamy of Sciences of United States of America*, 109(16), 5989–5994.
Dalin, C., Qiu, H., Hanasaki, N., Mauzerall, D. L., and Rodriguez-Iturbe, I. (2015). Balancing water resource conservation and food security in China. *Proceedings of the National Academy of Sciences of the United States of America*, 112(15), 4588–4593.
Delucchi, M. A. (2010). Impacts of biofuels on climate change, water use, and land use. *Annals of the New York Academy of Sciences*, 1195, 28–45. doi:10.1111/j.1749-6632.2010.05457
Economy, E. C. (2004). *The river runs black: The environmental challenge to China's future*. Ithaca, NY: Cornell University Press.
Eshel, G., Shepon, A., Makov, T., and Milo, R. (2014). Land, irrigation water, greenhouse gas, and reactive nitrogen burdens of meat, eggs, and dairy production in the United States. *PNAS*, 111(33), 11996–12001.
Famiglietti, J. S. (2014). The global groundwater crisis. *Nature Climate Change*, 4(11), 945–948. doi:http://doi.org/10.1038/nclimate2425
FAO. (2000). *The energy and agriculture nexus environment and natural resources*. Working Paper No. 4. Rome: FAO.
FAO. (2011). *Energy-smart food for people and climate*. Issue Paper. Rome: Food and Agriculture Organization of the United Nations.
FAO. (2017). *Food and Agriculture Organization of the United Nations FAOSTAT database*. Accessible at http://www.fao.org/faostat.
Fishman, R., et al. (2015). *Environment Research Letters*, 10 084022
Jonathan A. Foley, Navin Ramankutty, Kate A. Brauman, Emily S. Cassidy, James S. Gerber, Matt Johnston, Nathaniel D. Mueller, Christine O'Connell, Deepak K. Ray, Paul C. West, Christian Balzer, Elena M. Bennett, Stephen R. Carpenter, Jason Hill, Chad Monfreda, Stephen Polasky, Johan Rockström, John Sheehan, Stefan Siebert, David Tilman & David P. M. Zaks(2011). Solutions for a cultivated planet. *Nature* 478, 337–342.
Global Water Partnership. (2015). *China's water resources management Challenge: The 'Three Red Lines'*. Sweden: Global Water Partnership.
Greenpeace. (2010). *The true cost of coal: An investigation into coal ash in China*. Beijing: Greenpeace.
Greenpeace. (2012). *Thirsty coal: A water crisis exacerbated by China's new mega coal bases*. Beijing: Greenpeace.
Gu, A., Teng, F., and Wang, Y. (2014). China energy-water nexus: Assessing the water-saving synergy effects of energy-saving policies during the eleventh Five-year Plan. *Energy Conversion and Management*, 85, 630–637.
Gu, Y., Dong, Y-N., Wang, H., and others. (2016). Quantification of the water, energy and carbon footprints of wastewater treatment plants in China considering a water-energy nexus perspective. *Ecological Indicators*, 60, 402–409.
Hoff, H. (2011). *Understanding the Nexus*. Background Paper for the Bonn 2011 Conference: The Water, Energy and Food Security Nexus, Stockholm Environment Institute, Stockholm.
Hu, D., and Xu, S. (2013). Opportunity, challenges and policy choices for China on the development of shale gas. *Energy Policy*, 60, 21–26.
International Monetary Fund (IMF). (2010). *World economic outlook*. Retrieved 2011, from www.imf.org/external/pubs/ft/weo/2010/02/weodata/download.aspx.
Jiang, Y. (2015). China's water security: Current status, emerging challenges and future prospects. *Environmental Science and Policy*, 54, 106–125.
Kahrl, F., and Roland-Holst, D. (2008). China's water-energy nexus. *Water Policy*, 10(Supplement 1), 51–65.
Konar, M., Dalin, C., Suweis, S., Hanasaki, N., Rinaldo, A., and Rodriguez-Iturbe, I. (2011). Water for food: The global virtual water trade network. *Water Resources Research*, 47, W05520
Lamberton, M., Newman, D., Eden, S., and Gelt, J. (2010). *The water-energy nexus*. Arroyo: University of Arizona, Water Resources Research Center.

Lawford, R., Bogardi, J., Marx, S., and others. (2013). Basin perspectives on the water-energy-food security nexus. *Current Opinion in Environmental Sustainability*, 5, 607–616.
Leck, H., Conway, D., Bradshaw, M., and Rees, J. (2015). Tracing the water-energy-food nexus: Description, theory and practice. *Geography Compass*, 9(8), 445–460.
Li, X., Liu, J., Zheng, C., Han, G., and Hoff, H. (2016). Energy for water utilization in China and policy implications for integrated planning. *International Journal of Water Resources Development*, 32(3), 477–494.
Liu, J., Yang, W. (2012). Water management. Water sustainability for China and beyond. *Science* 337(6095), 649–650.
Liu, J., Yang, H., and Savenije, H. H. G. (2008, July 24). China's move to higher-meat diet hits water security. *Nature*, 454, 397. doi:10.1038/454397a; Published online July 23, 2008.
Liu, J., Zang, C., Tian, S., Liu, J., Ynag, H., Jia, S., You, L., Liu, B., and Zhang, M. (2013). Water conservancy projects in China: Achievements, challenges and way forward. *Global Environmental Change*, 23, 633–643.
Liu, J., Zhao, D., Gebrens-Leenes, P. W., and Guan, D. (2015). China's rising hydropower demand challenges water sector. *Nature Scientific Reports*, 5, 11446. doi:10.1038/srep11446.
Martin, M., and Grossman, I. E. (2015). Water-energy nexus in biofuels production and renewable based power. *Sustainable Production and Consumption*, 2, 96–108.
Meldrum, J., Nettles-Anderson, S., Heath, G., and Macknick, J. (2013). Life cycle water use for electricity generation: A review and harmonization of literature estimates. *Environmental Research Letters*, 8, 1–18.
Nair, S., George, B., Malano, H.M., Arora, M., and Nawarathna, B. (2014). Water-energy-greenhouse gas nexus of urban water systems: Review of concepts, state-of-art and methods. *Resources, Conservation and Recycling*, 89, 1–10.
National Bureau of Statistics of China (various years). *China statistical yearbook*. Beijing: China Statistics Press.
Peng, S. (2011). Water resources strategy and agricultural development in China. *Journal of Experimental Botany*, 62(6), 1709–1713. doi:10.1093/jxb/err049
Qin, Y., Curmi, E., Kopec, G. M., Allwood, J. M., and Richards, K. S. (2015). China's energy-water nexus – assessment of the energy sector's compliance with the "3 Red Lines" industrial water policy. *Energy Policy*, 82, 131–143.
Shifflett, S. C., Turner, J. L., Dong, L., Mazzocco, I., and Yunwen, B. (2015). *China's water-energy-food rad map*. A Global Choke Point Report. Washington, DC: Woodrow Wilson International Center for Scholars.
Tilman, D., Fargione, J., Wolff, B., D'Antonio, C., Dobson, A., Howarth, R., Schindler, D., Schlesinger, W. H., Simberloff, D., and Swackhamer, D. (2001). Forecasting agriculturally driven global environmental change. *Science*, 292(5515), 281–284.
U. S. Department of Energy. (2014). *The water-energy nexus: Challenges and opportunities*. Washington, DC: Department of Energy.
Venkatash, G., Chan, A., and Brattebo, H. (2014). Understanding the water-energy-carbon nexus in urban water utilities: Comparison of four city case studies and the relevant influencing factors. *Energy*, 75, 153–166.
Wang, J., Rothausen, S. G. S. A., Conway, D., Zhang, L., Xiong, W., Holman, I. P., and Li, Y. (2012). China's water–energy nexus: Greenhouse-gas emissions from groundwater use for agriculture. *Environmental Research Letters*, 7(1), 014035.
Water and power: The conflict over Los Angeles water supply in the Owens Valley by William L. Kahrl (1982).
Water Tech. (2008, September). South-to-North water diversion project, China. *Water-Technology.net*.
World Bank Group. (2016). *World Bank Database*. Accessible at https://data.worldbank.org/.
Xinhua News Agency. (2014). *No. 1 Central document targets rural reform*. Retrieved November 2014, from english.agri.gov.cn/hottopics/cpc/201401/t20140120_21067.htm.
Yang, C-J., and Jackson, R. B. (2014). China's synthetic natural gas revolution. *Nature Climate Change*, 3, 852–854.
Yang, H., Zhou, Y., Liu, J. (2009, May). Land and water requirements of biofuel and implications for food supply and the environment in China. *Energy Policy*, 37(5), 1876–1885. ISSN: 0301-4215. http://dx.doi.org/10.1016/j.enpol.2009.01.035.
Yu, F., Chen, J., Sun, F., Zeng, S., and Wang, C. (2011). Trend of technology innovation in China's coal-fired electricity industry under resources and environmental constraints. *Energy Policy*, 39, 1586–1599.
Zhang, C., and Anadon, L. D. (2013). Life cycle water use of energy production and its environmental impacts in China. *Environmental Science and Technology*, 47, 14459–14467.
Zhang, C., and Anadon, L. D. (2014). A multi-regional input-output analysis of domestic virtual water trade and provincial water footprint in China. *Ecological Economics*, 100, 159–172.
Zhang, C., Anadon, L. D., Mo, H., Zhao, Z., and Liu, Z. (2014). Water-carbon trade-off in China's coal power industry. *Environmental Science and Technology*, 48, 11082–11089.
Zhang, Q., Kobayashi, Y., Alipalo, M. H., and Zheng, Y. (2012). *Drying up: What to do about droughts in the people's Republic of China*. Manila: Asian Development Bank.
Zhao, L., Chang, S., Wang, H., and others. (2015). Long-term projections of liquid biofuels in China: Uncertainties and potential benefits. *Energy*, 83, 37–54.

24
THE ENERGY–MATERIALS NEXUS
The case of metals

Ester van der Voet, René Kleijn and Gavin M. Mudd

Introduction

It takes energy to produce materials. It takes materials to produce energy. To fully appreciate the linkages between materials and energy is a complicated and dynamic issue. So far, the fields of research and policy on energy and on materials are quite separate, and little attention has been paid to their linkages. Some suspect that there may be large consequences if we did pay attention, and that it would significantly change prospects and forecasts of energy and material demand if these linkages were acknowledged. Therefore it is worthwhile to gain a more in-depth understanding of the energy–materials nexus, to better support these respective policies. This chapter contains an exploration of the mutual linkages between energy and materials, their dynamics and their importance. In the second section, we'll briefly describe the developments in the two fields separately. The energy requirements of material production are explored in the third section, both qualitatively and quantitatively. In the fourth section, the focus is on material requirements for energy production. The fifth section touches upon the feedback loops between the energy and the materials system. In the final section, we present a research agenda for the energy–materials link.

Developments in the global energy and materials systems

The most prominent development in the global energy system is its continuous and rapid growth. It has grown from 40 EJ in 1900 to almost 575 EJ in 2014 (IEA, 2016), and still continues to increase. Before the Industrial Revolution, wood and biomass were the prime source. Since then, fossil fuels have become dominant and now constitute over 80% of the world's supply. Recently, not just the absolute amounts but also the share of modern renewable energy (like wind, solar and hydropower) has been increasing a little – from 0 in 1970 to 2% in 2014 (IEA, 2016). Although it presently constitutes a small part of total energy supply, it is now growing and under climate policies might become very significant as a result of a global transition towards a renewable energy system.

Many quantifications of the energy system exist (among others, Smil, 2010). They focus mostly on depletion of fossil resources and exploration of new ones – especially the increasing shift from conventional oil and gas reservoirs to 'unconventional' shale gas (with liquids), coal seam gas and tight gas resources. Another focus of energy related studies is on climate change and CO_2 emissions. Very detailed assessments exist at all scale levels on CO_2 emissions of the various energy production routes. Nexus issues have been outside the scope for a long time. This has led, among others, to a very optimistic view on the sustainability of renewable energy technologies and feedstocks. The most prominent debate has been conducted in the area of bio-energy, especially biofuels. The claimed reduction of CO_2 emissions in essence appears to be less than expected, and to be offset by the environmental impacts (especially the clearing of rainforests and resource inputs) related to biomass production if the complete life cycle is included. Nowadays, nexus issues are on the agenda, especially related to the water requirements of energy in general (IEA, 2012) and the land requirements of biobased energy in particular (UNEP, 2009). IEA stated that presently, 15% of total global water use is for generating energy. Material requirements of the energy system only recently have become a subject of attention. A few scattered publications are available at this point in time (e.g., Graedel & van der Voet (eds.), 2010; Kleijn et al., 2011; Allwood & Cullen, 2015; UNEP, 2016b)

In the realm of materials, a body of literature exists around Material Flow Accounting. Presently, a global database exists of materials extraction and trade, with data per country and time series starting from 1970 (UNEP, 2016a). From such databases we can draw conclusions on the patterns of material use. Globally, materials extraction rises steadily, having roughly doubled since 1970, similar to energy use. Absolutely as well as relatively, the growth is highest in minerals, mainly construction minerals. Metals are a relatively small flow but also increase rapidly. Growth takes place especially in emerging economies that are presently building up their infrastructure. In the more developed parts of the world, the demand for materials has stabilized.

The future use of materials is so far hardly the subject of research. Nevertheless, expectations are that the infrastructure related materials such as construction minerals, steel and copper will continue to rise rapidly in the coming decades under the pressure of population and welfare increase as well as urbanization. As long as the population grows, biomass demand will increase as well. For fossil fuels, this will depend on the development of the global energy system, autonomously as well as under various assumptions of compliance with climate targets such as mentioned in the Paris climate agreements. Drafting scenarios for material use is urgently needed. When these become available, new insights may shape future research.

In the MFA community, nexus issues have had little attention. In the broader industrial ecology field, it is now becoming a topic of interest, though still at a quite general level (see the chapters in Part II). Some first estimates have been made of energy and water requirements of materials, leading to the observation that especially the energy requirement of metals is very high. Even less attention has been paid to the materials requirement of energy production, and so far messages are contradictory (Kleijn et al., 2011; UNEP, 2016b). The reasons for this may be found in different starting points of the analysis rather than in different data on the materials intensity of energy technologies, though. This complex nexus is explored in this chapter.

Energy requirements of material production

An obvious fact is that it takes energy to produce materials. The amount of energy required, and corresponding GHG emissions, differs greatly per material. Gutowski et al. (2013) have investigated this for a number of different materials. Results are shown in Figure 24.1.

From this figure, it appears that there is a wide variability in energy intensity per kg of material, ranging from 1 MJ/kg to one million MJ/kg. Construction minerals, produced in large quantities, can be found at the lower end of the scale, while precious metals appear at the high end. Upscaling that information to include production-level quantities show that cement production is responsible for roughly 5% of total global energy use. Metals are produced on a

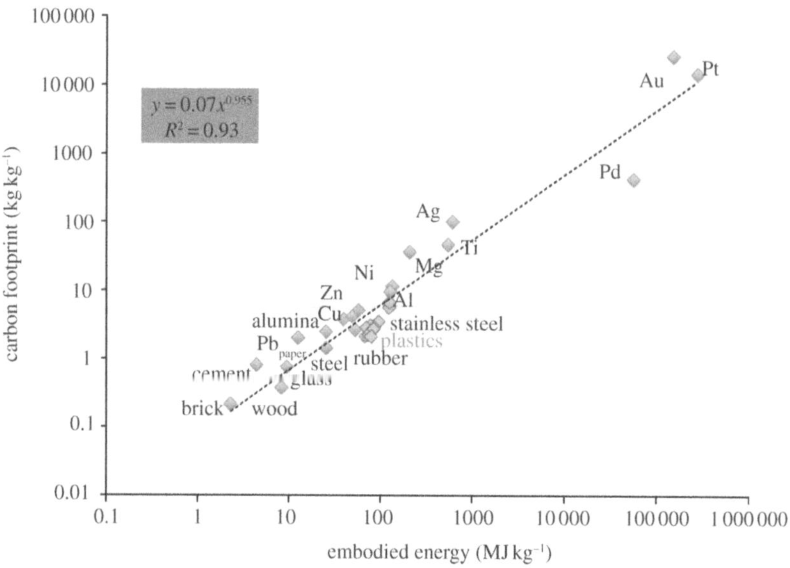

Figure 24.1 Carbon footprint (kg CO_2-eq/kg) and embodied energy (MJ per kg) of materials

Source: Gutowski et al. (2013)

much smaller scale, yet the cradle-to-gate emissions from their extraction and refining contributes considerably to global greenhouse gas emissions as well: 7–8% (UNEP, 2013). Even though the per kg energy use is relatively low, most of that (5%) is caused by iron and steel, aluminium being a good second.

The energy requirements of metals (see Figure 24.2) have been investigated in more detail, also considering the different production routes, and distinguishing between the various processes in the production chain.

Looking forward, we may suspect that the energy requirements for producing metals will change. The demand for metals is presently growing rapidly, therefore all else being equal the energy needed to produce them will also grow. However the energy requirement per kg of metal is also changing, as is explained below.

The demand for major metals is expected to continue increasing well into the 21st century, along with population and GDP growth (Elshkaki et al., 2017). The market for minor metals is much less predictable. New applications may emerge, but may also hit boundaries with regard to availability. Contrary to major metals, prices fluctuate by orders of magnitude, which disturbs not only recycling but even primary production. This is certainly relevant for the other half of the energy–materials link: materials for energy page 373.

With regard to the energy intensity of metal production, several relevant developments occur:

- Ore grade degradation. Throughout the 20th century and into the 21st, the ore grades (i.e. the metal content of the ores being mined) have been decreasing for some metals (Mudd et al., 2013; Mudd et al., 2017; Mudd and Jowitt, 2013). This implies the amount of energy needed to produce these metals from their ores will increase (Norgate and Haque,

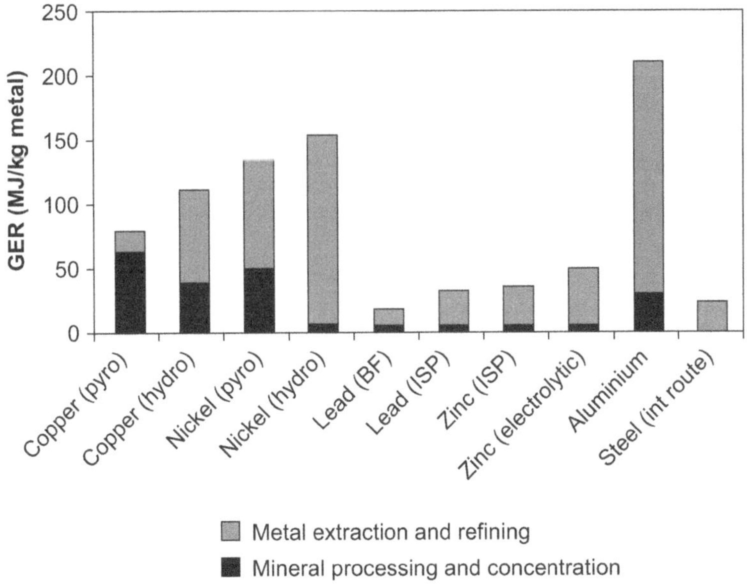

Figure 24.2 Energy requirements for metal production: present situation, gross energy requirement (GER) in MJ/kg refined metal

Source: UNEP, 2013

2010). Figure 4.3 (in Chapter 4) shows this. If this trend continues, energy requirements will go up. This trend is visible for metals such as copper, nickel, lead and zinc, as shown in Figure 24.3. For iron ore and other bulk minerals (e.g. bauxite, manganese), ore grades are often less important than ore type (such as hematite versus magnetite) or impurities present (such as phosphorous, alumina or silica in iron ore), as these factors affect mining costs and metal product quality (e.g. magnetite has much higher energy and water needs than hematite, often related to the high proportion of silica in raw ore; high impurities can affect iron or steel alloy quality). While there is evidence of gradual declines in the grades of raw iron ore declining very gradually in recent decades (Mudd, 2010), the bigger issues faced by mines is input ore types, iron ore product quality and economic conditions (i.e. demand, prices, costs).

- Improvements in the energy efficiency of material production. Over time, the energy efficiency of producing metals can improve (e.g. new technology, moving to larger scales, maximizing recovery, investment in energy efficiency, etc.), although these gains can also be offset by lower ore grades, deeper mines, longer haul roads, aging equipment and so on (see Leveque et al., 2014). Gutowski et al. (2013) have shown that the improvements in material refining processes are leveling off towards the thermodynamic limit. While there is a strong financial incentive for mines to reduce their energy inputs, the reality remains mines are constantly struggling to maintain energy efficiency, as shown in Figure 24.3 for the Olympic Dam and Ok Tedi mines – with Ok Tedi in particular showing a long-term and significant increase in energy intensity.

Therefore, it is not straightforward in what direction the energy intensity will develop given the many complex factors which need to be taken into account.

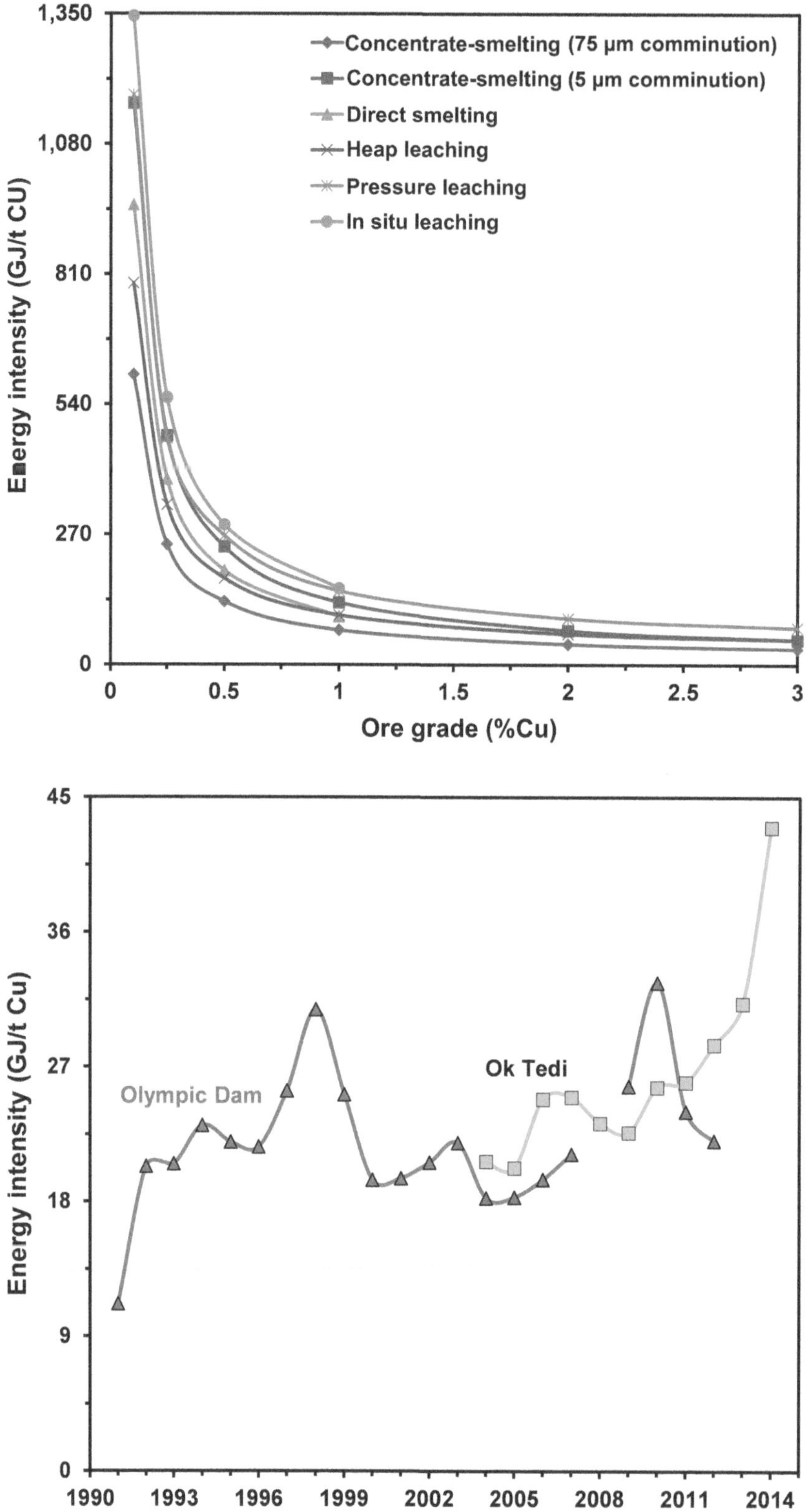

Figure 24.3 Life cycle assessment of copper production showing energy intensity versus ore grade (top) and allocated Cu energy intensity over time for the Olympic Dam Cu-Au-Ag-U mine (Australia) and Ok Tedi Cu-Au-Ag mine (Papua New Guinea) (bottom)

Source: Top – redrawn from Norgate and Jahanshahi (2010); Bottom – data updated from Northey et al. (2013)

Besides these autonomous developments, policies and strategies may be formulated to reduce demand, which will result as well in a reduction of the consequent energy use. To mention a few:

- Substitution. To some extent, metals can be replaced by other materials. However, it takes energy to produce these other materials as well, and there may be other boundaries that come into play, such as water and land that is needed to produce biomass. The relevance of the resource nexus is never more clear than in the case of substitution. Whether or not such a substitution in the end is favourable has to be determined on a case-by-case basis and depends very much on resource nexus issues.
- Increasing the share of secondary production. Secondary production, or recycling, requires (much) less energy than primary production (Allwood & Cullen, 2015). A larger share of secondary production in the total would reduce the demand for primary materials and therefore would reduce energy requirements per average kg produced. There can be economic issues with recycling systems, especially since collection systems can be expensive and not all metal uses lend themselves to recycling due to diffuse uses.
- Stock saturation. This is the notion that not flows, but stocks, are the determining factor for a number of important applications, such as houses, cars or washing machines. When income rises, more of such applications are bought, but at a certain quantity the demand levels off – we don't need more than one washing machine even if we are very rich. At the level of materials, there are indications of stock saturation. Especially for iron and steel the evidence suggests this phenomenon may occur (Müller et al., 2006; Smil, 2010; Allwood and Cullen, 2015). Stock saturation implies that per capita stock will someday stop growing. At that moment, demand will level off as well and will stabilize at the level of stock maintenance. The lifespan of an application is then a very important variable, in fact determining what the replacement rate is. The longer the lifespan, the lower the demand can be, but also the longer it takes before an equilibrium situation will establish itself.

Allwood and Cullen (2015) calculate the possibilities of reducing GHG emissions of materials production by adding up many different options: efficiency improvements, increased recycling rates, substitution and suchlike. They concluded that it would indeed be possible to improve the energy performance of material production, but not sufficient to reach a desired situation. Similar conclusions are reached by Elshkaki et al. (2017) and van der Voet et al. (2017). It suggests a closer look is needed into these dependencies, their dynamics and the feedback loops that may be involved (see the fifth section on page 375).

Material requirements of energy production

As stated above, it is obvious that it takes energy to produce materials. It is equally obvious that it takes materials to produce energy. Even the production of straightforward fossil energy carriers takes materials: drills, mines, an infrastructure, means of transport to get the coal, oil and gas where it is needed, and power plants or refineries to get it in a form directly usable as energy. For renewable energy sources such as wind and solar, material requirements are mentioned and on the agenda, especially related to metals with potential or perceived supply issues (e.g. tellurium) (Dawkins et al. 2012). The critical metal debate is linked to renewable energy technologies. However the issue is broader than that. Wind turbines may require neodymium, but also large amounts of concrete, steel and copper. Solar technologies have minor and rare metals as an essential part of solar cells themselves, but also need module parts (e.g. inverters) and mounting systems using steel, aluminium, plastics and glass (Ashby et al., 2012; Graedel et al., 2015).

Biomass is often excluded from such concerns, since the production of the feedstock is low tech and does not require specialty metals. However, the use of major metals such as copper and steel is actually very high. Agriculture and harvesting require machinery and equipment and are actually relatively materials intensive compared to industrial production. According to data from Ashby et al. (2012) and Kleijn et al.(2011), their material requirement is in the same order of magnitude as that of solar and wind technologies.

All in all, there are large differences in material requirement between the different energy technologies. Figure 24.4 shows this.

From Figure 24.4, it appears that, in general, renewable energy technologies require much more metals than fossil technologies, in some cases several orders of magnitude. Other, similar inventories confirm this (Falconer, 2009; Kleijn & van der Voet, 2010; Elshkaki & Graedel, 2013), although when upscaled the additional materials needed for a renewable energy system do not appear not so large (UNEP, 2016b; De Koning et al., 2015). The differences between the inventories shows that this line of research is still in its infancy – it depends on many assumptions, not just on energy technologies but also on the grid and storage needed, on the fraction of renewables in the mix, and on the composition of the alternative renewable energy system as a whole. But all sources agree on the fact that renewable energy technologies require more metals.

To all probability, the already started energy transition will continue and we will have a larger share of renewables in the energy system. In view of the magnitude of the transition this will be a wide variety of technologies, including bio-energy, solar energy and wind energy. The larger material requirement may lead to some scarcity and criticality issues (Moss et al., 2011). This applies especially to minor and specialty metals used for these technologies. For base metals, copper is associated with potential supply problems (Hertwich et al., 2015).

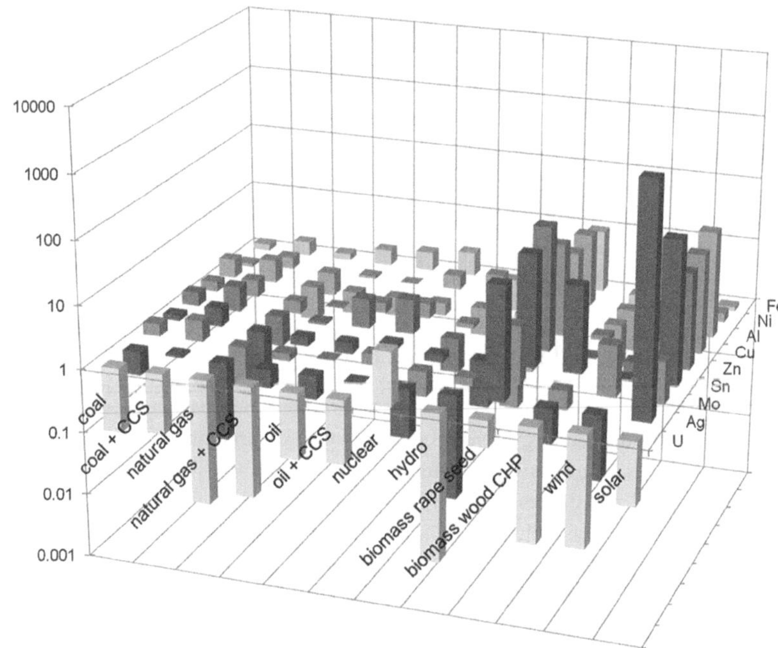

Figure 24.4 Material requirement of energy technologies relative to the 2010 global energy mix

Source: Kleijn et al. (2011)

The transition also implies an increase in iron and steel requirements. For the foreseeable future, this is not expected to have any criticality consequences. It will, however, lead to an increase in energy use, and therefore of GHG emissions as long as the energy transition is not complete. The same is true of other metals: they are relatively energy-intensive materials (van der Voet et al., 2017). Hence it is clear that there are important nexus issues attached to this transition. There is a clear trade-off: reducing GHG emissions goes at the expense of increasing material intensity of energy production, and this in turn also will lead to some increase of GHG emissions during the transition when the renewable energy system must be built up. In a more mature state this investment can be reduced to the level of maintenance.

Feedback loops: EROI and CROI

From the above, it appears that there is not just a mutual dependency of materials and energy, but that there are complex feedback loops. For energy, this is already recognized, for example in the concept of energy pay-back times. It takes energy to produce energy, and it also takes an energy investment to build up an energy system. The energy pay-back time then calculates the time it takes until the energy technology has produced the amount of energy needed for its own production and installation.

A related, though slightly different concept, is the Energy Return on Investment (EROI). This is different from the energy pay-back time: it calculates the amount of energy actually needed to produce energy – it is not about the original investment, but about the operation of energy technologies. A general treatment of the subject is given by Murphy and Hall (2010). The EROI concept has been important, among others, in the debate on biofuels. Pimentel and Patzek (2005) calculated that the EROI for biofuels is close to 1, making it an effort hardly worth the while to begin with. That assessment has been modified since then, but the value of his work was that he, for the first time, included the life cycle and recognized that not just the production of the fuel itself but also the growing of the energy crop needs an energy input. Including the life cycle in the EROI also allows for assessing the energy-for-materials-for-energy feedback loop, a relevant and complex issue.

Since then, a large amount of LCA studies into biofuels has appeared, with widely conflicting outcomes, dependent on system boundaries and choice for allocation (van der Voet et al., 2010) but at least acknowledging the many linkages of this resource nexus. EROI is also used for fossil energy, showing a decrease for oil resources, sometimes even approaching 1 (Guilford et al., 2011). For solar and wind, the energy pay-back time is an important variable, but EROI less so: once in place, the energy input of operating the systems can be low.

Kleijn et al. (2011) took the EROI concept and applied it to materials (i.e. copper) to calculate the CROI, the copper return on investment, in the feedback loop materials-for-energy-for-materials. He showed that in the extreme we may face a situation wherein it actually could take more than 1 kg of copper to produce 1 kg of copper. While these calculations are indicative only, it does show that the approach is applicable and may add value. These feedback loops may in themselves lead to nexus constraints.

The energy–materials nexus: research agenda

The energy–materials nexus is a two-way street: the requirements as discussed in the previous sections change over time, also as a result of the linkages themselves. It is shown that they may be important: the forecasts of energy demand may be underestimated because the nexus is not taken into account, and material requirements of the energy transition is an underrepresented topic in research. The first step, therefore, is recognition of the issue. Once this point is reached, it is important to gain understanding in the complex and dynamic nature of the energy–materials linkages to avoid nasty surprises in the future. This recognition of the energy–materials nexus is especially important in a situation wherein both energy and materials systems may have to undergo major global level changes to meet international agreements on climate change and sustainable development.

Points for the energy–materials nexus research agenda are the following:

1 Expanding the database.

To be able to quantify the materials–energy nexus information is needed, both technological and statistical information. Examples are:

- Data on the past and present energy production system at different scale levels;
- Data on the past and present materials production system at different scale levels, both primary and secondary;
- Data on ore grades of metals;
- Data on material requirements of energy technologies;
- Process data on energy and material production;
- Supply chain and trade data;
- . . . and probably much more.

Such information is partly available, but there are still large gaps and wide differences between individual estimates. Harmonization and standardization are important processes these databases have to go through.

2 The development of energy–materials nexus indicators.

- A good start is the development and standardization of one-dimensional indicators: indicators related to intensity and "footprints". This can go both ways: the energy intensity of materials production, and the material intensity of energy production. Such indicators can be provided at the aggregate level, but also on a detailed level of individual materials, energy resources, countries or even production sites or processes. For such indicators, databases and methods are available. There may still be gaps, but a start has been made already in providing this information (see Part II). Footprint indicators can be broken down, using trade data, to indicate the location of the pressure. In that way, they can be used as indicators of burden shifting.
- Two-dimensional indicators regarding mutual intensity or footprint dependency can provide a good addition, and may provide unique information on nexus boundaries. An already existing example is the Energy Return on Investment, the EROI. When this indicator approaches 1, it is a clear indication that the energy production pathway is becoming meaningless. The EROI idea can be transplanted in the realm of materials to calculate MROIs, for example the Copper Return on Investment, to spot other nexus boundaries. This line of research has hardly started yet, but could be worthwhile to explore.
- A very interesting line of research could be to further explore these mutual dependencies. How do they influence each other? How does the energy system change as a result of changes in the material system, and the other way around? How do the two systems co-evolve, and how could that change as a result of external pressure? How will differences in local or regional situations affect these global level systems, for

example via burden shifting or knock-on effects? This type of information would be very useful for an integrated management of resources that goes beyond the disciplinary boundaries as well as the single-resource boundaries.

3 Providing time series of the indicators.

Time series of the indicators would show how they change over time, and whether feedback loops may be intensifying over time. It is a demanding task, as all data required to calculate these linkages also have to be available in time series. Changing energy systems, changing ore grades, improving efficiencies, changing recycling rates all lead to changed intensities and footprints. A second reason to obtain time-series information is a desire to explore future developments (see point 4). Past time series can be used to derive driving forces and other important variables determining aspects of energy and materials demand and supply.

4 Developing integrated scenarios for resource use and linking those to climate and energy scenarios.

A real challenge is the development of resource scenarios. So far, credible comprehensive global level scenarios exist only for energy and climate change. For other types of resources, future explorations lack comprehensiveness. It would be a large effort, since resource scenarios should not only be driven by socio-economic developments and price dynamics, but also by physical laws, most importantly resource constraints, stock dynamics and, last but not least, resource nexus issues. This is a huge task, and would involve combining models and databases as well. Linking these resource scenarios to energy scenarios could clarify the future challenges originating from the materials–energy linkages, in addition to those of the demand increase itself.

5 Studying the dynamics of the resource nexus.

The dynamics of materials–energy systems, or in general of linked resource systems, are quite complex in nature. There are the developments related to the individual resources. There are the linkages between the resources and how the intensities and footprints may change over time. There are the even more complex dynamic feedback mechanisms captured in concepts like the EROI and MROI. There are the driving forces behind all these, embedded in nature, in technology, in demographics, and in the economy. There are also regional and local aspects that may be relevant: either in the impacts or in the driving forces, regional nexus issues may be different from the global level and may, therefore, have consequences for the global level. Dynamics may be even more complicated when it is not two, but many linkages between different resources that are present. For example, the production of resources can be severely impeded as a result of water stress, which may partly be due to the production of the same resource. Or, what would happen if supply chains for certain technology metals are being interrupted. Or, national initiatives, such as energy plans, could have international consequences, with repercussions that might also backfire to the country in question. For more examples, see Bleischwitz et al. (2013). Using system dynamics models may offer valuable insights in these broader aspects of the materials–energy nexus. This is a huge field of research, involving experts of different disciplines and different topical areas.

Most of the identified challenges are general in nature, not confined to any specific resource nexus. However it can be a good approach to tackle complex issues from one particular point of view. The materials–energy nexus, with all the richness of its dynamics and with an impressive knowledge base already available, would seem as good a place to start as any.

References

Allwood, J., and Cullen, J. M. (2015). *Sustainable materials, without the hot air. UIT Cambridge Ltd, series Without the hot air.* ISBN: 9781906860301. Earlier published as: Allwood, J., and Cullen, J. M. (2012). *Sustainable materials, with both eyes open.* UIT Cambridge Ltd. Retrieved from www.withbotheyesopen.com/index.html

Ashby, M., Attwood, J., and Lord, F. (2012). *Materials for low-carbon power.* A White Paper, Granta Design, 2nd ed.

Bleischwitz, R., Johnson, C. M., and Dozler, M. G. (2013). Re-assessing resource depletion and criticality. Linking future food and water stress with global resource supply vulnerabilities for foresight analysis. *European Journal of Futures Research*, 15–34. doi:10.1007/s40309-013-0034-1.

Dawkins, E., Chadwick, M., Roelich, K., and Falk, R. (2012). *Metals in a low-carbon economy: Resource scarcity, climate change and business in a finite world.* Stockholm: Stockholm Environmental Institute.

De Koning, A., Huppes, G., Deetman, S., and Tukker, A. (2015). Scenarios for a 2 °C world: A trade-linked input – output model with high sector detail. *Climate Policy*, 16(3), 301–317. doi:10.1080/14693062.2014.999224

Elshkaki, A., and Graedel, T. E. (2013). Dynamic analysis of the global metals flows and stocks in electricity generation technologies. *Journal of Cleaner Production*, 59, 260–273.

Elshkaki, A., Graedel, T. E., Ciacci, L., and Reck, B. K. (2017). A potential future mismatch between metal supply and demand.

Falconer, I. (2009). *Metals required for the UK's low carbon energy system: The case of copper usage in wind farms.* MSc thesis, University of Exeter.

Graedel, T., Elshkaki, A., and van der Voet, E. (2015, December). Entangled circles – energy and its resource connections. In G. J. Kramer and B. Vermeer (Eds.), *The colours of energy, essays on the future of our energy system.* Amsterdam, the Netherlands: Shell International BV. ISBN:9789082398403.

Graedel, T. E., and van der Voet, E. (Eds.). (2010). *Linkages of sustainability: Strüngmann forum reports.* Cambridge, MA: MIT Press.

Guilford, M. C., Hall, C. A. S., O'Connor, P., and Cleveland, C. J. (2011). A new long term assessment of Energy Return on Investment (EROI) for U. S. oil and gas discovery and production. *Sustainability*, 3(10), 1866–1887. doi:10.3390/su3101866.

Gutowski, T. G., Sahni, S., Allwood, J. M., Ashby, M. F., and Worrell, E. (2013). The energy required to produce materials: Constraints on energy-intensity improvements, parameters of demand. *Philosophical Transactions of the Royal Society A*, 371, 20120003. doi:http://dx.doi.org/10.1098/rsta.2012.0003

Hertwich, E. G., Gibon, T., Bouman, E. A., Arvesen, A., Suh, S., Heath, G. A., Bergesen, J. D., Ramirez, A., Vega, M. I., and Shi, L. (2015). Integrated life-cycle assessment of electricity-supply scenarios confirms global environmental benefit of low-carbon technologies. *PNAS*, 112(20), 6277–6282.

International Energy Agency. (2011). *Cogeneration and renewables: Solutions for a low-carbon energy future.* Paris: OECD/IEA.

International Energy Agency. (2012). *World energy outlook 2012*, Chapter 17 Water for Energy. Retrieved March 2017, from www.worldenergyoutlook.org/media/weowebsite/2012/WEO_2012_Water_Excerpt.pdf.

International Energy Agency. (2016). *Key world energy trends.* Retrieved March 2017, from www.iea.org/publications/freepublications/publication/KeyWorldEnergyTrends.pdf

Kleijn, R., and van der Voet, E. (2010). Resource constraints in a hydrogen economy based on renewable energy sources: An exploration. *Renewable and Sustainable Energy Reviews*, 14, 2784–2795.

Kleijn, R., van der Voet, E., Kramer, G. J., van Oers, L., and van der Giesen, C. (2011). Metal requirements of low-carbon power generation. *Energy*, 36, 5640–5648.

Leveque, M., Millar, D., and Paraszczak, J. (20140. Energy and mining – the home truths. *Journal of Cleaner Production*, 84, 233–255.

Moss, R. L., Tzimas, E., Kara, H., Willis, P., and Kooroshy, J. (2011). *Critical metals in strategic energy technologies. Assessing rare metals as supply-chain bottlenecks in low-carbon energy technologies.* JRC report no JRC-65592 EUR 24884 EN-C. Luxembourg: Publications Office of the European Union. ISBN 978-92-79-20698-6, ISSN 1018-5593, doi: 10.2790/35600.

Mudd, G. M. (2010). The environmental sustainability of mining in Australia: Key mega-trends and looming constraints. *Resources Policy*, 35(2), 98–115.

Mudd, G. M., and Jowitt, S. M. (2013). A detailed assessment of global nickel resource trends and endowments. *Economic Geology*, 108, 1163–1183.

Mudd, G. M., Jowitt, S. M., and Werner, T. T. (2017, January). The world's lead-zinc mineral resources: Scarcity, data, issues and opportunities. *Ore Geology Reviews*, 80, 1160–1190. ISSN:0169-1368. doi:http://dx.doi.org/10.1016/j.oregeorev.2016.08.010.

Mudd, G., Weng, Z., and Jowitt, S. (2013). A detailed assessment of global Cu resource trends and endowments. *Economic Geology*, 108, 1163–1183.

Müller, D. B., Wang, T., Duval, B., and Graedel, T. E. (2006). Exploring the engine of anthropogenic iron cycles. *Proceedings of the National Academy of Sciences*, 103(44), 16111–16116.

Murphy, D. J., and Hall, C. A. S. (2010). Year in review – EROI or energy return on (energy) invested. *Annals of the New York Acadamy of Sciences*, 1185, 102–118.

Norgate, T., and Haque, N. (2010). Energy and greenhouse gas impacts of mining and mineral processing operations. *Journal of Cleaner Production*, 18(3), 226–274.

Norgate, T. E., and Jahanshahi, S. (2010). Low grade ores – smelt, leach or concentrate? *Minerals Engineering*, 23(2), 65–73.

Northey, S. A., Haque, N., and Mudd, G. M. (2013). Using sustainability reporting to assess the environmental footprint of copper mining. *Journal of Cleaner Production*, 40, 118–128.

Pimentel, D., and Patzek, T. W. (2005). Ethanol production using corn, switchgrass and wood; biodiesel production using soybean and sunflower. *Natural Resources Research*, 14, 65–76.

Smil, V. (2010). *Energy transitions: History, requirements, prospects*. Santa Barbara, CA: Praeger. ISBN: 978-0-313-38177-5, 178 pp.

UNEP. (2009). *Towards sustainable production and use of resources: Assessing biofuels*. Report of the International Resource Panel (S. Bringezu, H. Schütz, M. O'Brien, L. Kauppi, R., W. Howarth and J. MacNeely). ISBN 978-92-807-3052-4.

UNEP. (2013). *Environmental risks and challenges of anthropogenic metals flows and cycles, a report of the working group on the global metal flows to the international resource panel* (E. van der Voet, R. Salminen, M. Eckelman, G. Mudd, T. Norgate and R. Hischier, Eds.). Paris: United Nations Environment Programme. ISBN: 978-92-807-3266-5.

UNEP. (2016a). Global material flows and resource productivity. An assessment study of the UNEP international resource panel (H. Schandl, M. Fischer-Kowalski, J. West, S. Giljum, M. Dittrich, N. Eisenmenger, A. Geschke, M. Lieber, H. P. Wieland, A. Schaffartzik, F. Krausmann, S. Gierlinger, K. Hosking, M. Lenzen, H. Tanikawa, A. Miatto, and T. Fishman). Paris: United Nations Environment Programme.

UNEP. (2016b). *Green energy choices, the benefits, risks and trade-offs of low-carbon technologies for electricity production. Report of the International Resource Panel* (E. G. Hertwich, J. Aloisi de Larderel, A. Arvesen, P. Bayer, J. Bergesen, E. Bouman, T. Gibon, G. Heath, C. Peña, P. Purohit, A. Ramirez and S. Suh, Eds.). Paris: United Nations Environment Programme.

Voet, E. van der and Graedel, T. E. (2010). The emerging importance of linkages. In T. E. Graedel and E. van der Voet (Eds.), *Linkages of Sustainability* (Strüngmann Forum Reports). Cambridge, MT: MIT press, pp. 461–470.

Voet, E. van der, van Oers, L., Verboon, M., and Kuipers, K. J. (2017). Environmental implications of future demand scenarios for seven major metals. *Journal of Industrial Ecology*.

25

UNCONVENTIONAL OIL AND GAS PRODUCTION MEETS THE RESOURCE NEXUS

Tim Boersma and Philip Andrews-Speed

Introduction

Producing and consuming energy resources often come with consequences for other resources. Oil and natural gas are no different. For example, once an oil well has been drilled, water is generally injected into the reservoir (water flooding), to build up pressure in the reservoir and to stimulate the production of oil. To transport natural gas by pipeline, land has to be cleared, with consequences for local flora and fauna. Finally, as widely documented and acknowledged by close to 200 national governments in the Paris Agreement that was signed in December 2015, the consumption (burning) of fossil fuels, including oil and natural gas, results in greenhouse gas emissions, including carbon, which is widely acknowledged as a key contributor to climate change.

One of the most, and possibly the most, impactful developments in global energy markets in the last two decades has been the advance of so-called unconventional technologies to extract oil and natural gas from different types of rock with low permeability. Commencing in the early 2000s for natural gas, and just before 2010 for oil, analysts are still grappling with the long-term economic, geopolitical, and environmental consequences of this phenomenon. In addition, it seems only a matter of time before unconventional production will further spread globally. Next to the US, which has largely served as the global unconventional oil and gas laboratory, commercial production of natural gas from shale now takes place in China, Canada, and Argentina, with more countries to follow.

Unconventional oil and gas production has not come without controversy. Next to more general but increasingly vocal opposition to fossil-fuel production and consumption, several environmental concerns continue to be linked explicitly to so-called hydraulic fracturing, or 'fracking', of shale and other tight reservoirs. These include increased seismic activity, methane leakage, water contamination and land degradation. While initial reports were alarming, by now, a broad body of literature has emerged, which allows us to present a more nuanced and fact-based view of these concerns. Oil sands, previously known as 'tar sands', involve techniques such as surface mining and steam-assisted drainage which involve substantial water consumption, risk of water pollution, significant greenhouse gas emissions and, in the case of mining, land disturbance.

This chapter focuses on the linkages between oil and natural gas production on the one hand, and water, land and air quality on the other. It also contains a brief discussion of other debates surrounding unconventional oil and gas production, but in line with this book, the emphasis is on the nexus between various resources. Our analysis is based on primary and secondary data. We both spent a substantial amount of time researching unconventional oil and gas in various parts of the world, predominantly in North America, Europe, and Asia. This chapter does not include new data, but provides an overview of existing research, following our own publications, and available literature. We focus on places where unconventional oil and gas extraction actually take place, or where at least attempts have been made, but draw mainly on shale gas as this has recently become the most widely practiced, and contentious form. We provide several broad conclusions about unconventional oil and gas and the resource nexus, and initial suggestions for future synergies.

Unconventional oil and gas production and the resource nexus

The term "unconventional" as applied to oil and gas production is rather misleading because what has been classified as unconventional has now become or is becoming conventional. In broad terms, the oil or gas in a conventional field has been created by heat and pressure in an organic-rich source rock, and then migrated to a porous and permeable reservoir rock where it has become trapped in a geological structure. Conventional oil and gas production involves drilling a well into the reservoir and allowing the oil or gas to flow to the surface under natural geological pressure. The oil industry has developed several techniques over the last few decades that can enhance flow rates or improve recovery of the hydrocarbons. These include pumping water, chemicals or gas into the reservoir, hydraulic fracturing of the reservoir to improve permeability and horizontal drilling. By themselves, these techniques are today regarded as conventional, though sometimes referred to as 'enhanced oil recovery (EOR)'. The key characteristic that is common for all unconventional oil and gas accumulations is that the rock which hosts the hydrocarbon has very low permeability. Therefore, what today are called 'unconventional' oil and gas accumulations cannot be easily or economically produced by the techniques described above, at least not when such techniques are deployed in isolation. Instead, commercial production requires either a combination of these techniques or the deployment of new techniques.

Several different types of unconventional oil and gas accumulations can be recognized (Table 25.1), yet fall into two distinct categories. In one category lie those accumulations in which the hydrocarbon has remained in its original source rock (indicated by "S" in Table 25.1). The production of shale gas and tight oil shale generally requires a sophisticated combination of horizontal drilling and hydraulic fracturing with the use of chemical or gas stimulants, as well as other techniques such as measurement-while-drilling and down-hole seismic. It is this approach that has led to the unconventional oil and gas revolutions in North America. Coal-bed methane is another case where natural gas remains in its original source rock, but the use of horizontal drilling and hydraulic fracturing are optional techniques. The distinctive feature of coal-bed methane production is the need to drain massive amounts of water from the coal in order to release the methane. In the case of oil shale, the organic material remains in the form of kerogen and requires heating to convert the kerogen into oil. The normal procedure is to mine the shale and then process it, though in situ processes are being tested.

Tight gas and extra heavy oil are examples of unconventional hydrocarbons that have migrated from their source rock to a reservoir rock. However, production through conventional techniques is not viable because flow rates are too low. In the case of tight gas this is due to the low permeability of the reservoir rock, a problem that is overcome through a combination of horizontal drilling and hydraulic fracturing, as with shale gas and tight shale oil. In contrast, the low flow rates from oil sands arise from the heavy and viscous nature of the oil that may even occur as solid bitumen. Production has normally involved surface mining of the oil sands followed by heating to extract the oil and upgrading by adding lighter oils. More recently, companies have been deploying in situ heating by steam-assisted drainage.

Table 25.1 Key technical attributes of main types of unconventional gas and oil

	Reservoir	*Techniques*				
		Horizontal drilling	*Hydraulic fracturing*	*Chemical injection*	*Heat*	*Mining*
Gas						
Coal-bed methane	Coal (^{S})	O	O	-	-	-
Shale gas	Shale (^{S})	N	N	N/O	-	-
Tight gas	Sandstone, limestone	N	N	N/O		
Gas hydrates	Deep ocean sediment, permafrost	-	-	-	O	-
Oil						
Tight shale oil	Shale (^{S})	N	N	N/O	O	O
Extra heavy oil	Sandstone	O	-	-	N	O
Oil shale	Shale (^{S})	O	-	-	N	O

^{S}: Source rock
^{N}: Technique normally deployed
^{O}: Technique optional
–: Technique not normally deployed

Table 25.2 Key alleged potential negative nexus impacts arising from the extraction of the main types of unconventional gas and oil

	Water		*Land*		*Atmosphere*		
	Consumption	*Contamination*	*Disturbance*	*Contamination*	*Methane*	*CO_2*	*Seismicity*
Gas							
Coal-bed methane	-	X	X	X	X	X	-
Shale gas	X	X	X	X	X	X	X
Tight gas	X	X	X	X	X	X	X
Gas hydrates	-	-	X		X	-	
Oil							
Tight shale oil	X	X	X	X	X	X	X
Extra heavy oil	X	XX	XX	XX	X	XX	
Oil shale	X	XX	XX	XX	X	XX	

X: Moderate probability of measurable impact if best practice is not followed
XX: Significant probability of measurable impact if best practice is not followed

A final form of unconventional hydrocarbon takes the form of methane hydrate that occurs in vast quantities on the deep ocean floor as well as in permafrost. The methane originates from deep geological sources as well as from shallow biogenic processes. Production is still at the experimental phase. Potential techniques to release methane from the hydrate include depressurizing the reservoir, and injecting chemicals, inert gases, hot water or steam. The main nexus risks are associated with the catastrophic release of large volumes of methane into the atmosphere.

The deployment of these unconventional techniques raises the possibility of measurable or significant negative impacts on other resources in the nexus in comparison with conventional techniques, especially if best practices are not followed and the most advanced techniques are not deployed. Table 25.2 summarizes, in a simplified form, the potential negative impacts arising from the production of each form of unconventional hydrocarbon, and shows the connections to water, land, and the atmosphere, as well as seismicity.

Prevailing debates anno 2017

Taking a birds-eye perspective, academic discussions related to unconventional oil and gas extraction have greatly evolved, but are still similar compared to those made a number of years ago. New literature continues to emerge on several environmental concerns that have been linked to (aspects of) hydraulic fracturing. We discuss contemporary debates on these issues in this chapter. It is important to note that unconventional oil and gas extraction is still a relatively new phenomenon, and therefore many questions have not yet been answered in full. We emphasize this because all too often analysts and policymakers are tempted to take firm stances on what the future will bring, albeit regarding future commercial production, or the spread of this technology beyond North America. The truth is that all too often we do not really know and time, and practice, must tell. However, we do observe that in the US the tone of the debate about the benefits and costs of hydraulic fracturing is changing. This has most markedly been demonstrated by the decision of the NY state legislature to ban hydraulic fracturing by law (which is significant as it prohibits industrial activity in a part of the most productive shale gas play in the country, the Marcellus Shale). In addition, in the period prior to the November 8, 2016, elections, several presidential candidates took a tough stance against unconventional oil and gas extraction, the most outspoken one being Senator Sanders, who proposed to ban fracking altogether. We did not expect that to happen, nor do we think it will happen going forward, given that an estimated 60% of all natural gas produced in the country now comes from shale. The Senator did force his opponent, Secretary Clinton, to shift gears as well. Furthermore, several events over the course of recent years, such as the Aliso Canyon methane leak in California and the final confirmation that fracking indeed caused water pollution in Pavillion, Wyoming, and continued resistance against expansion of natural gas infrastructure in New England, have not improved natural gas's image. After November 8, President Trump has, in line with the occasional broad statement made during his campaign, reiterated his ambition to further expand domestic fossil-fuel production, including natural gas and oil. The president is widely believed to incentivize incremental production by lowering regulatory burdens where possible. There is reason to be skeptical whether regulation has under the previous administration in fact curtailed domestic oil and gas production. Instead, it is more likely that prices, albeit domestic or international, incentivize US oil and gas production. In addition, major uncertainties remain whether possible trade restrictions could harm domestic energy producers, rather than support them. At the time of writing, it is too early to proclaim anything about future energy and climate policy in the US with an amount of certainty.

In contrast, efforts to exploit shale gas continue unabated in China, despite the more challenging geological conditions. Companies are also pressing ahead in Argentina. In the UK, initial enthusiasm for shale gas exploration was constrained by public opposition and consequent inability of the companies to obtain permission from local governments. This situation was relieved in May 2016 when the county council of north Yorkshire gave permission for the hydraulic fracturing of a well drilled in 2013, yet public opposition continues to be fierce, making it difficult to predict with certainty whether commercial shale gas production will occur in the UK.

Next to the scholarly literature, and a significant increase of available data, we also observe clear learning curves regarding (at least some) regulatory approaches and/or industry practices to address environmental concerns. To give an example, the amount of wastewater that is recycled has increased substantially, and in most states, discharging wastewater to surface water is not allowed. Jackson et al. (2014) report that prior to 2011 only 13% of wastewater in the Marcellus Shale was recycled, but that by 2011 this was the case for 56% of the wastewater. In 2013 this number had grown to 67% of all the water used in oil and gas extraction (Harkness et al., 2014).

The other form of unconventional hydrocarbon that has attracted a lot of criticism is the extra heavy oil that is exploited in Alberta, western Canada. In the past this has involved open cast mining, and extensive use of water and heat. Although production of these oil sands dates back to the 1960s, a combination of high oil prices and technological advances led to a surge in both production and development plans from the early 2000s. These heavy oils now dominate Canada's crude oil production, and output continued to rise through 2015 despite declining oil prices.[1] Alleged and actual negative consequences include physical land disturbance, pollution of land, water and air, greenhouse gas emissions, as well as consequences for human health and biodiversity (Gosselin et al., 2010; Gordon, 2012; Paskey et al., 2013).

Other prevailing debates regarding unconventional oil and gas extraction focus on economic impacts, albeit local impacts or broader energy market impacts, and geopolitical ripple effects. Although longer term consequences remain the topic of substantial debate, the US has become the world's largest producer of natural gas, and consequently is set to become one of the world's major exporters of natural gas (as a result of expanded pipeline capacity to Mexico, and increasingly significant liquefaction capacity that will impact the market in coming years). Increased natural gas production in 2008/2009 contributed to a broader pattern of global oversupply, which had substantial impacts on pricing structures in the more liquid parts of the European gas market (e.g. Boersma, 2015). In the coming years because of a wave of new liquefaction capacity, particularly in Australia and the US, there will be knock-on effects on global markets, and contribute to what can be labelled the coming of age of global LNG markets (Boersma and Losz, 2017). In the US, very competitive natural gas has made major market inroads. As we discuss in more detail later, in electricity generation natural gas has trumped coal in the merit order, leading to significant carbon and other GHG emissions reductions. We have also witnessed some increased demand for major industrial processes that use natural gas as a feedstock. As investments in gas-fired electricity generation capacity continue, and natural gas is increasingly becoming an interesting feedstock for parts of the transportation sector, US policymakers are facing questions how to square large-scale investments in natural gas with long-term scenarios of deep decarbonization (Boersma, 2016).

In the following sections, we return to the resource nexus, and discuss the most profound environmental concerns, the state of the academic literature, industry practices and regulatory lessons, before we draw conclusions.

Water

Risks to water resources in the literature broadly fall into four categories: (1) contamination of shallow aquifers with fugitive hydrocarbon gases; (2) contamination of surface water and shallow groundwater from spills, leaks, and/or the disposal of inadequately treated wastewater; (3) accumulation of toxic elements in soil or stream sediments near disposal or spill sites; and (4) the over usage of water in areas where water is scarce (Vengosh et al., 2014).

Stray gas contamination mostly happens due to improperly constructed or failing gas wells, but the extent to which this is problematic continues to be debated. Jackson et al. (2014) give detailed descriptions of the importance of wellbore integrity and cite several studies where degraded wellbore integrity contributed to groundwater contamination through fugitive methane. Other studies however found no statistically significant relationship between dissolved methane concentrations in groundwater from domestic water wells and proximity to pre-existing oil and gas wells (Siegel et al., 2015). Earlier analyses have found that casing failures in Pennsylvania between 2010 and 2013 comprised 3% to 6% of total production in the first three years of well life (Ingraffea et al., 2014). Other studies report that the failure rate may be substantially higher, and up to 12% within the first year of operation (Hildenbrand et al., 2015). Although their analysis of 550 groundwater samples from the Barnett Shale in Texas detected numerous volatile organic compounds in aquifers above the Barnett Shale, this did not necessarily implicate unconventional oil and gas extraction as the source of contamination. Notably, Rahm et al. (2015) found that shale gas development entails more risk related to spills and solid waste management, while conventional gas development entails more risk associated with cementing and casing issues, and site restoration. They observe that conventional and typically older gas wells in their case study in Pennsylvania, were perhaps less environmentally protected, thus drawing less regulatory scrutiny.

Wastewater produced during hydraulic fracturing operations includes flowback water, which returns to the surface immediately after fracturing the well, and produced water, which is brought to the surface during oil and gas production. The balance of flowback and produced water across the Marcellus in Pennsylvania in 2011 was 43% flowback and 45% produced water (Jackson et al., 2014). Rough estimates show that each unconventional shale gas well produces between 3500 and 7200 m^3 of wastewater during its lifetime (Harkness et al., 2014). The total annual volume of oil and gas wastewater for Pennsylvania in 2011 and 2012 was estimated between 3000 and 5000 million liters of water (Ibid.). In most states, wastewater is injected in deep saline aquifers after treatment. In states where injection in aquifers is not possible, like Pennsylvania, wastewater is increasingly recycled and then reused, or in some instances treated in municipal treatment plants, and then discharged. Even though in most states wastewater management is regulated, risks remain, particularly from spills, and inadequate treatment. Harkness et al. (2014) in their study of three treatment plants in Pennsylvania found that wastewater release to the environment is associated with significant levels of iodide and ammonium. They recommend that the issue of adequate treatment of wastewater, and tight controls on discharging and leak prevention, is particularly important in countries where deep-well injection is not allowed, such as China and the EU. Jackson et al. (2014) report a number of studies which mention surface leaks and spills from well pads and wastewater holding ponds, as well as inadequate treatment before wastewater discharge.

In parts of the US, and many other parts of the world, water availability can be a (growing) concern. As Jackson et al. (2014) illustrate, context is helpful in interpreting water usage by the oil and gas industry. In Texas, for example, a state with rather intensive drilling, the total amount of water used for hydraulic fracturing on an annual basis is less than 1% of total usage. However, on the local level these numbers may change dramatically, with several counties in the Barnett Shale region in Texas for instance reporting that up to 30% of their annual water use goes to the oil and gas industry. In some cases, this has resulted in competition between the energy industry and the agricultural sector (which in the case of Texas historically holds most of the water rights). The flipside is that as the costs of water rights in some instances increased, so did the incentive for the industry to develop alternative fluids for use in hydraulic fracturing, for instance liquids, or brackish groundwater (which is too saline to use for agricultural activity, or human consumption). As these technologies are increasingly refined and applied, their costs decrease, making them potentially more attractive for application in other countries in the world like China, South Africa, and Algeria (see for instance Vengosh et al., 2014; Boersma et al., 2015). This is to date one of the best examples where nexus challenges have incentivized industry to develop alternatives and create synergies, moving beyond trade-offs for resource usage. It remains to be seen whether, absent financial incentives, industry will be incentivized to apply these practices on a larger scale.

China's quest for shale gas faces both of these challenges, and in an accentuated form (Enoe et al., 2012; Gunningham, 2014). The most favorable geological conditions appear to lie in Sichuan province and the neighboring municipality of Chongqing. This region is home to more than 100 million people, about 50% of whom live in rural areas with intensive agriculture. Any significant contamination of surface or near-surface waters could undermine agricultural productivity and local livelihoods. Further, although Sichuan has historically been blessed with abundant rainfall, recent years have seen occasional dramatic droughts (Krupnick et al., 2014). Many of the other prospective areas for shale gas in China lie in the far north and northwest of the country, regions already suffering from drought.

Aforementioned risks related to water resources can be substantial, but it is important to note that all can be dealt with using existing technologies under proper regulations. Avoiding stray gas contamination for instance could be enforced by installing safe zones between new shale gas sites and already existing drinking water wells. In addition, to increase transparency regarding naturally occurring methane in water resources vis-à-vis contamination, baseline monitoring could be considered. Disclosure of the content of fracking fluids and chemicals used in operations is also preferable. Regarding wastewater, not allowing discharge without proper treatment is advisable, but discharge cannot always be prevented, surely in cases where injection of wastewater into saline aquifers may induce seismicity (see the subsection 'Seismicity').

We now turn to the risks of greenhouse gas emissions that have been linked to unconventional oil and gas extraction.

Greenhouse gas emissions

The effects of unconventional oil and gas in terms of greenhouse gas emissions is decidedly a mixed bag. On the one hand, energy resource extraction is an industrial process, which comes with air pollutants throughout the production and transportation process. At the same time, replacing coal with natural gas for electricity generation has had a downward effect on emissions of several pollutants.

As Jackson et al. (2014) describe, GHG emissions typically occur throughout the shale gas supply chain, from preparing a well pad, installing processing facilities and compressor stations. Then once drilling starts, wastewater is stored in open-pit tanks on site, from which other GHG emissions can flow. Then during the processing of natural gas, vapors may flow into the air, and in the case of oil production in some instances methane is flared, or even vented. The EPA estimates that 58% of methane leakage occurs during field production (leakage from the wells, gathering pipelines, or gas treatment facilities).

Howarth et al. (2011) published a study suggesting that fugitive methane leakage from shale gas wells (between 3.6 and 7.9%) was substantially higher than conventional gas wells (between 1.6 and 7.0%). As a result, they concluded that the carbon footprint of shale was potentially comparable to that of coal. Since then several studies have emerged that have questioned the data used in this paper, and by now it is safe to conclude that shale gas has not fundamentally altered the carbon footprint of the gas industry (Cathless et al., 2012; O'Sullivan and Paltsev, 2012; Heath et al., 2014; Levi, 2015). One reason for this is the increased application of green completions technologies, and increased availability of infrastructure in places where the primary focus is on crude oil production (but a lot of associated gas is produced). In addition, the EPA started regulating emissions from new wells starting January 2015, and rules for existing wells are expected in the future.

Addressing midstream and downstream emissions has proven to be more difficult. One reason may be that leaks are complicated to measure, and data can therefore be incomplete (Boersma and Ebinger, 2014; Larsen et al., 2015). However, detailed studies at the US city level provide worrisome data, with methane emissions often substantially higher than local authorities believe (McKain et al., 2014). There can be several reasons for this, such as inadequate access to emissions, inadequate measuring, and higher-than-expected leak levels from aging pipeline systems, in particular in major urban areas.[2] None of these arguments however should prevent serious thinking and action about how to reduce methane emissions from midstream and downstream gas systems. In many cases that would likely mean major investment in new pipeline systems. Logically, it would in turn mean that household rates are going up, but it is worth keeping in mind that someone is paying for that fugitive methane as well, and that is the customer. Cost arguments come on top of arguments over air quality (ozone formation), climate change, jobs creation, and overall safety.

On the upside, the US has experienced significant overall carbon emissions reductions, and these were partly the result of fuel switching in the electricity sector. It is unrealistic to point to a singular reason for this drop in emissions. Instead, a mix of factors is more likely, including the economic downturn, increased energy efficiency standards and continued investments in renewable energy, particularly solar and wind (due to federal tax breaks and renewable portfolio standards at the state level). Yet natural gas undeniably played an important role in this drop in carbon emissions as well, which according to the US Energy Information Administration in 2015 reached the lowest level since 1993, and was 21% below the 2005 level (the baseline used in the US pledge for Paris).[3] Next to carbon emissions, burning natural gas in comparison to coal comes with significantly less emissions of other GHGs, such as SO_2, H_g, and NO_x and particulates, and thus in some places improves local air quality (Jackson et al., 2014).

The central argument that continues to fuel lively debate in the US is whether natural gas can indeed play the role of bridge fuel, as proponents like to claim. Though some have disputed this, as touched upon before, the reality seems to be that natural gas already is a bridge fuel, based on the very significant investments that have been made, and the fact that 60% of current domestic natural gas production comes from shale rock layers. The more relevant question seems to be what actions are required now to reconcile the expansion of the share of natural gas with the goal of deep decarbonization in 2030 and beyond (Boersma, 2016). Initial studies suggest that natural gas helps to reduce carbon emissions, but that additional targeted climate policies are required to drive down emissions substantially after that initial bump provided by fuel switching (Newell and Raimi, 2014; McJeon et al., 2014). Continued uncertainties about actual methane emissions throughout the supply chain (see e.g. Omara et al., 2016) further underline the need for continued empirical analysis of GHG emissions related to shale gas extraction, and the need to further curtail emissions where possible.

In the medium to long term, technologies like carbon capture and sequestration for gas-fired power plants seem a necessity if countries are to reach levels of what is often called deep decarbonization. Although the focus of this chapter is on unconventional oil and natural gas, it is important to note that the continued cost reduction for some renewable technologies, in particular solar photovoltaics and onshore wind, make these options increasingly attractive for investments in new electricity generation capacity. This does erode the possible market share for fossil fuels in electricity generation, even though it will take decades to make a transition in which renewables comprise the majority of the global fuel mix, especially as long as fossil fuel prices remain relatively low, and negative externalities are not adequately priced. In addition, it is worth considering that in most discussions about renewable energy, little if any time is spent contemplating that resources like oil and natural gas are used in a wide variety of applications, including industrial processes, cooking and heating, and transportation.

Two other forms of unconventional hydrocarbon deserve mention in the context of greenhouse gas emissions. The energy needed to exploit oil sands necessarily involves incremental production of greenhouse gas emissions, notably carbon dioxide, and these emissions are generally greater for in suit recovery than for surface mining. However, technological progress has led to a steady decline in emissions per unit of oil produced (Gosselin et al., 2010). In the case of methane hydrates, the potential emissions arise not from the extraction process itself but from the accidental and catastrophic release of methane from these giant accumulations during production. This could cause catastrophic instability in overlying marine sediments as well as the release of a powerful greenhouse gas into the atmosphere (Khameneh et al., 2012).

Land degradation and societal disturbance

Another environmental aspect of shale gas extraction is land-use changes from production processes. Cumulative risks of clearing existing land cover can be significant, most notably during the site identification and preparation phase and during production. In addition to direct effects on land usage, indirect effects (such as clearing for gas infrastructure construction, intensive truck usage for water transportation, and noise pollution) can be substantial as well.

Jenner and Lamadrid (2013) concluded that in Pennsylvania a range between 34,000 and 82,000 acres of land could be cleared for shale gas production (see also Jantz et al., 2014). They suggest that shale gas extraction turns large areas into industrial zones, creating brownfields that presumably will be left to localities to deal with when producers leave. Pierre et al. (2015) in their study of landscape impacts of oil and gas extraction in La Salle County in Texas, concluded that the disturbance from infrastructure development (3% of the county area) decreased the available core areas by almost three times the area of disturbance, with effects for habitats and local wildlife. Similarly, Moran et al. (2015) in their measurement of land-use changes within the Fayetteville Shale gas region in Arkansas between 2001/2002 and 2012 conclude that considering the large number of wells drilled in other parts of the country, shale gas development will likely have substantial negative effects on forested habitats and organisms that depend upon them.

Land disturbance is much greater when the hydrocarbon is extracted through surface mining, as has been the case for the oil sands in Alberta for decades. Although land reclamation has taken place and continue, the companies struggle to keep up with the pace of disturbance (Gosselin et al., 2010) and, in some places, there are serious conflicts of interest with indigenous Canadian communities (Paskey et al., 2013).

Seismicity

In recent years, scientists have learned more about possible linkages between unconventional oil and gas production and seismic activity, or induced seismicity. Various anthropogenic applications (such as dam impoundment, mining, and wastewater disposal) have long been known to, in some circumstances, potentially accelerate seismic activity (Schultz et al., 2015). USGS data demonstrate that in the period between 1973 and 2008, on average 21 earthquakes with a magnitude of 3 and larger took place annually in the central and eastern US. In the period from 2009 until 2013 this number jumped to 99 earthquakes of the same magnitude or larger and this number reportedly continues to rise.[4]

Studies have focused predominantly not on hydraulic fracturing itself, but the reinjection of flowback water, a very common practice in unconventional oil and gas producing regions in the US. The most important conclusion, however, is that there are several major unanswered questions. For instance, Keranen et al. (2013) studied the impacts of wastewater injection and their links to induced seismic activity in Oklahoma. Their research suggests that time delays can exist between the start of injection and induced seismicity, as it may take time for pressure at the fault to build up. Rosen (2015) reports that the rate of fluid injection into the well, rather than other factors such as volumes of fluids or depth of injection, may play a more prominent role. Next to injection of flowback water, hydraulic fracturing itself in rare circumstances has also been linked to induced seismicity, though mostly these tremors cannot be felt (e.g. Jackson et al., 2014; Schultz et al., 2015). A widely reported case is that of Blackpool in the UK, where in 2011 two relatively small earth tremors halted the emerging unconventional gas industry in its tracks, and operations were suspended for over one year. Hereafter a traffic light system was introduced which the state of Oklahoma also adopted. Essentially, it means that earth tremors are monitored and the regulator can summon operators to stop their activities if considered necessary. It is important to keep in mind though that examples of seismic activity that are attributable to hydraulic fracturing and/wastewater injection are small in comparison to other anthropogenic triggers such as mining and dam impoundment (Jackson et al., 2014, p. 345).

Conclusions

From this analysis, we derive the following lessons. Similar with more conventional extraction processes for oil and natural gas, relatively new technologies or a combination of existing technologies (often labelled unconventional oil and gas extraction) can have major consequences for other resources. This is particularly the case when best practices are not applied. As we have shown, price incentives often stimulate companies to further refine extraction techniques. In parts of Texas where (access to) water resources are relatively scarce, and water rights therefore increasingly expensive, the industry developed alternatives to freshwater to be used in fracking fluids, such as brackish water, or gels. We also observe that at least a substantial number of oil and gas companies are concerned with their corporate image in an age where fossil-fuel production is under public scrutiny, adopting best practices as a result. Other behavior has been enforced by regulations, such as the traffic light system monitoring seismic activity in Oklahoma. Thus, we observe that over the course of the last decade, both industry and regulatory authorities cross the board have gone through a steep learning curve, and many important lessons were learned.

However, that is not the full story. As described, several unconventional techniques, such as oil sands production, come with substantial trade-offs. Extracting oil through a process that involves surface mining comes with significant land disturbance, and GHG emissions. To counter negative externalities like these, production processes must be regulated, and more accurately priced than we often do today. Another solution that a part of the environmental lobby proposes is to leave the resource in the ground. The reality seems to be that it will likely take decades before we phase out fossil fuels, and therefore a more pragmatic approach in the short and medium term appears useful. As we described, unconventional oil and gas has only just begun to expand globally, and we expect more countries to follow suit. Applying best practices in both technology and governance are therefore the way forward.

Although unconventional hydrocarbons are seen as playing a negative role in the resources nexus, an important geographic feature distinguishes them from conventional hydrocarbon accumulations; namely, that many accumulations of unconventional hydrocarbons occur in areas with limited reserves of conventional hydrocarbons. Consequently, countries that have long relied on coal for the bulk of their energy supply may have an opportunity to replace a proportion of the coal supply with unconventional gas. If managed well, this would in the short and medium term reduce carbon emissions, land degradation, and pollution of water and air. In some cases, unconventional gas could also increase access to modern energy. China, India, South Africa and Poland are examples of countries with potentially large unconventional gas resources, but all of them face geological, technological and regulatory obstacles to realizing this opportunity. Each case requires coordinated policy analysis across the relevant resources to decide if the short-term benefits outweigh the long-term costs.

Notes

1 National Energy Board of Canada. www.neb-one.gc.ca/nrg/sttstc/crdlndptrlmprdct/stt/stmtdprdctn-eng.html
2 Robert Jackson. www.brookings.edu/blogs/planetpolicy/posts/2015/02/methane-emissions-boston-white-house-climate-action-plan-jackson
3 Energy Information Administration. www.eia.gov/todayinenergy/detail.cfm?id=26232
4 USGS. http://earthquake.usgs.gov/research/induced/

References

Boersma, T. (2015). *Energy security and natural gas markets in Europe – lessons from the EU and the United States.* London: Routledge Studies in Energy Policy.

Boersma, T. (2016). *Natural gas in the United States in 2016 – problem child and poster child.* Policy brief 16–02, The Brookings Institution, Washington, DC.

Boersma, T., and Ebinger, C. K. (2014). *Prevailing debates related to natural gas infrastructure: Investments and emissions.* Natural Gas Briefing Document #3, The Brookings Institution, Washington, DC.

Boersma, T., and Losz, A. (forthcoming). In Goldthau (Ed.), *The new international political economy of natural gas.* Palgrave MacMillan.

Boersma, T., Vandendriessche, M., and Leber, A. (2015). *Shale gas in Algeria: No quick fix.* Washington, DC: The Brookings Institution.

Cathles, L., Brown, L., Taam, M., and Hunter, A. (2012). A commentary on "The greenhouse-gas footprint of natural gas in shale formations" by R. W. Howarth, R. Santoro, and Anthony Ingraffea. *Climate Change*, 113(2), 525–535.

Enoe, M., He, Y., and Pohnan, E. (2012). *Lessons learned: A path toward responsible development of China's Shale gas resources.* New York: Natural Resources Defense Council.

Gordon, D. (2012). *Understanding unconventional oil.* Washington, D.C.: Carnegie Endowment for International Peace.

Gosselin, P., Hrudey, S. E., Naeth, M. A., Plourde, A., Therrien, R., Van Der Kraak, G., and Xu, Z. (2010). *Environmental and health impacts of Canada's oil sands industry.* Ottawa: Royal Society of Canada.

Gunningham, N. (2014). A Shale gas revolution for China? *Climate Policy*, 14(2), 302–320.

Harkness, J. S., Dwyer, G. S., Warner, N. R., Parker, K. M., Mitch, W. A., and Vengosh, A. (2014). Iodide, bromide, and ammonium in hydraulic fracturing and oil and gas wastewaters: Environmental implications. *Environmental Science & Technology*, 49(3), 1955–1963.

Heath, G. A., O'Donoughue, P. Arent, D. J., and Bazilian, M. (2014). Harmonization of initial estimates of shale gas life cycle greenhouse gas emissions for electric power generation. *Proceedings of the National Academy of Sciences USA*, 111(31), 1–10.

Hildenbrand, Z. L., Carlton, D. D., Fontenot, B. E., Meik, J. M., Walton, J. L., Taylor, J. T., Thacker, J. B., Korlie, S., Shelor, C. P., Henderson, D., Kadjo, A. F., Roelke, C. E., Hudak, P. F., Burton, T., Rifai, H. S., and Schlug, K. A. (2015). A comprehensive analysis of groundwater quality in the Barnett Shale region. *Environmental Science & Technology*, 49(13), 8254–8262.

Howarth, R., Santoro, R., and Ingraffea, A. (2011). Methane and the greenhouse gas footprint of natural gas from shale formations. *Climate Change*, 106(4), 679–690.

Ingraffea, A. R., Wells, M. T., Santoro, R. L., and Shonkoff, S. B. C. (2014). Assessment and risk analysis of casing and cement impairment in oil and gas wells in Pennsylvania, 2000–2012. *Proceedings of the National Academy of Sciences, USA*, 111(30), 10955–10960.

Jackson, R. B., Vengosh, A., Carey, J. W., Davies, R. J., Darrah, T. H., O'Sullivan, F., and Petron, G. (2014). The environmental costs and benefits of fracking. *The Annual Review of Environment and Resources*, 39, 327–362.

Jantz, C. A., Kubach, H. K., Ward, J. R., Wiley, S., and Heston, D. (2014). Assessing land use changes due to natural gas drilling operations in the Marcellus Shale in Bradford County, PA. *The Geographical Bulletin*, 55, 18–35.

Jenner. S., and Lamadrid, A. J. (2013). Shale gas vs. coal: Policy implications from environmental impact comparisons of shale gas, conventional gas, and coal on air, water, and land in the United States. *Energy Policy*, 53, 442–453.

Keranen, K. M., Savage, H. M., Abers, G. A., and Cochran, E. S. (2013). Potentially induced earthquakes in Oklahoma, USA: Links between wastewater injection and the 2011 M_w 5.7 earthquake sequence. *Geology*, 41(6), 699–702.

Khameneh, N. A., Bahmani, S., and Mushtaq, W. (2012). *Methane hydrates as potential energy resource*. Project Report – Natural Gas TPG 4140. Department of Petroleum and Applied Geophysics, Norwegian University of Science and Technology.

Krupnick, A., Wang, Z., and Wang, Y. (2014). Environmental risks of Shale gas development in China. *Energy Policy*, 75, 117–125.

Larsen, K., Delgado, M., and Marsters, P. (2015). *Untapped potential – reducing global methane emissions from oil and natural gas systems*. New York: Rhodium Group.

Levi, M. (2015). Fracking and the climate debate. *Democracy: A Journal of Ideas*. Available at: http://democracyjournal.org/magazine/37/fracking-and-the-climate-debate/

McJeon, H., Edmonds, J., Bauer, N., Clarke, L., Fisher, B., Flannery, B. P., Hilaire, J., Krey, V., Marangoni, G., Mi, R., Riahi, K., Rogner, H., and Tavoni, M. (2014). Limited impact on decadal-scale climate change from increased use of natural gas. *Nature*, 514, 482–485.

McKain, K., Down, A., Raciti, S. M., Budney, J., Hutyra, L. R., Floerchinger, C., Herndon, S. C., Nehrkorn, T., Zahniser, M. S., Jackson, R. B., Phillips, N., and Wofsy, S. C. (2014). Methane emissions from natural gas infrastructure and use in the urban region of Boston. *Proceedings of the National Academy of Sciences USA*, 112(7), 1941–1946.

Moran, M. D., Cox, A. B., Wells, R. L., Benichou, C. C., and McClung, M. R. (2015). Habitat loss and modification due to gas development in the Fayetteville Shale. *Environmental Management*, 55, 1276–1284.

Newell, R. G., and Raimi, D. (2014). Implications of shale gas development for climate change. *Environmental Science & Technology*, 48(15), 8360–8368.

Omara, M., Sullivan, M. R., Li, X., Subramanian, R., Robinson, A. L., and Presto, A. A. (2016). Methane emissions from conventional and unconventional natural gas production sites in the Marcellus Shale Basin. *Environmental Science & Technology*, 50(4), 2099–2107.

O'Sullivan, F., and Paltsev, S. (2012). Shale gas production: Potential versus actual greenhouse gas emissions. *Environmental Research Letters*, 7, 044030.

Paskey, J., Steward, G., and Williams, A. (2013). *The Alberta oil sands then and now: An investigation of the economic, environmental and social discourses across four decades*.Oil Sands Research and Information Network, University of Alberta, School of Energy and the Environment, Edmonton, Alberta. OSRIN Report No. TR-38. 108 pp.

Pierre, J. P., Abolt, C. J., and Young, M. H. (2015). Impacts from above-ground activities in the eagle Ford Shale play on landscapes and hydrologic flows, La Salle County, Texas. *Environmental Management*, 55, 1262–1275.

Rahm, B. G., Vedachalam, S., Bertoia, L. R., Mehta, D., Vanka, V. S., and Riha, S. J. (2015). Shale gas operator violations in the Marcellus and what they tell us about water resource risks. *Energy Policy*, 82, 1–11.

Rosen, J. (2015). Pumped up to rumble. *Geoscience*, 348(6241), 1299.

Schultz, R., Stern, V., Novakovic, M., Atkinson, G., and Gu, Y. J. (2015). Hydraulic fracturing and the Crooked Lake Sequences: Insights gleaned from regional seismic networks. *Geophysical Research Letters*, 42, 2750–2758.

Siegel, D. I., Azzolina, N. A., Smith, B. J., Perry, A. E., and Bothun, R. L. (2015). Methane concentration in water wells unrelated to proximity to existing oil and gas wells in Northeastern Pennsylvania. *Environmental Science & Technology*, 49(7), 4106–4112.

Vengosh, A., Jackson, R. B., Warner, N., Darrah, T. H., and Kondash, A. (2014). A critical review of the risks to water resources from unconventional Shale Gas development and hydraulic fracturing in the United States. *Environmental Science & Technology*, 48(15), 8334–8348.

26

FEEDING AFRICA

Nexus-related opportunities, challenges and policy options

Timothy O. Williams, Fred Kizito and Marloes L. Mul

Introduction

The challenge of meeting the food and nutrition requirements of a growing population through sustainable and climate-resilient farming systems is one of the key issues facing African agriculture over the coming decades. This challenge is well recognized by African political leaders. Africa's Agenda 2063 endorsed by the African Union (AU) Summit of January 2015 and the AU Malabo Summit Declaration of June 2014 on Accelerated Agricultural Growth and Transformation both affirmed African governments' commitment to eliminate hunger and food insecurity by 2025; and to enable resilience of livelihoods and production systems to climate change and other shocks (AU, 2014, 2015). Meeting these commitments will critically depend, among other things, on how land, soils, water, energy and agroecosystems are managed and sustained in the production of food and other basic human needs. As essential as these resources are, each one is coming under pressure due to demographic, economic and climatic changes.

Africa's population is projected to increase from 1.1 billion in 2013 to 1.6 billion people by 2030 (UNDESA, 2013). This increase will translate into a need to produce at least 50% more food (AUC-AMCOW, 2016). Urbanization is also growing apace. In 2015, 40% of Africa's population lived in urban areas and this proportion is projected to rise to 56% by 2050 (UNDESA, 2014), leading not only to an increase in the quantity but also in the variety of food demanded. Water use in the agricultural sector is about 184 km^3/year, accounting for about 81% of total water withdrawal (FAO, 2016). The projected increase in population by 2030 is expected to lead to, at least, a tenfold increase in water needs for energy production to support industrial, social and economic growth (AUC-AMCOW, 2016). This points to a greater competition for available water resources in future. At the same time, water and energy resources are at the centre of the recent wave of foreign investments in agricultural land in Africa (Williams et al., 2012; Sidibé and Williams, 2016). Climate variability and climate change also come into play. The projected increase in temperature of up 1.4°C by 2020 in Africa is predicted to result in increased rainfall variability and incidences of extreme weather events.

The mounting pressure on natural resources from various angles and the risks posed to the attainment of the SDGs in Africa point to the need to reexamine, within a nexus perspective, the linkages between natural resources. This reexamination is needed to identify technical, policy and institutional arrangements that will promote resource use efficiency and integrated resource management approaches to meet future food, human and environmental security objectives in Africa. However, in order to understand how linkages between resources can be better managed, it is necessary to first review the status and characteristics of resources and how they impact on food production and food insecurity on the continent.

Characteristics of the natural resource base, food production and food insecurity in Africa

Land and water

Africa is relatively well endowed in land and water resources. Over the three-year period 2012–2014, average cultivated area was about 271 million hectares (24% of cultivable area and 9% of total land area) compared with 238 million hectares in South Asia (75% of cultivable area and 37% total land area) (FAOSTAT, 2017). Long-term average internal renewable freshwater resources (IRFWR) per capita is about 3,319m^3 in Africa and 1,131m^3 in South Asia. Freshwater withdrawals as a percentage of IRFWR is around 6% in Africa compared with 46% in South Asia (FAO, 2016). The overall land and water endowment statistics for Africa, however, mask significant sub-regional differences between land abundant and land constrained countries (Jayne et al., 2014) and between countries experiencing physical and economic water scarcity (CA, 2007). Furthermore, Chamberlain et al. (2014) and Jayne et al. (2014) have pointed out that Africa's underutilized arable land is concentrated in a few countries, many of which are fragile states, with the additional problem that most of this land is located far from input and output markets, thus limiting their economic attractiveness in the short term. Irrigation is not yet widespread in Africa. The total area equipped for irrigation in Africa in 2013 was 15.6 million ha (5.8% of cultivated area) and 98.0 million ha in South Asia (45.7% of cultivated area) (FAO, 2016).

Over most of the cultivated and rangeland areas, land degradation is pervasive. A recent study by ELD Initiative and UNEP (2015) reckoned that desertification affects about 45% of Africa's land area, with 55% of this area at high or very high risk of further degradation. The same study pointed out that an additional 280 million tonnes of cereals per year could be produced if soil erosion is controlled.

Soils

Although Africa has a wide variety of soil types (Figure 26.1), the soils over much of the continent are old and highly weathered, often lacking in essential nutrients and organic matter and thus requiring careful management when used for agricultural purposes.

Arenosols are the most dominant soil type in sub-Saharan Africa (SSA) occupying about 22% of the land area. They are easily erodible sandy soils with low water- and nutrient-holding capacity. Leptosols are the second dominant soils, covering about 17% of the land area. They are shallow soils over hard rock with a weak soil structure and little fine earth. They have low organic carbon and low fertility status which create severe limitations for rooting depth. Next in order of relative share are cambisols covering about 11% of SSA land area. They can have varying characteristics depending on the nature of the parent material, climate and terrain. They have low to medium agricultural potential. In essence, going by the classification in Figure 26.1, up to 50% of SSA soils are of low to medium quality.

On top of the inherent quality problem, continuous cultivation of arable land with little or no nutrient input has led to declining soil fertility. For the period 2010–2012, fertilizer use

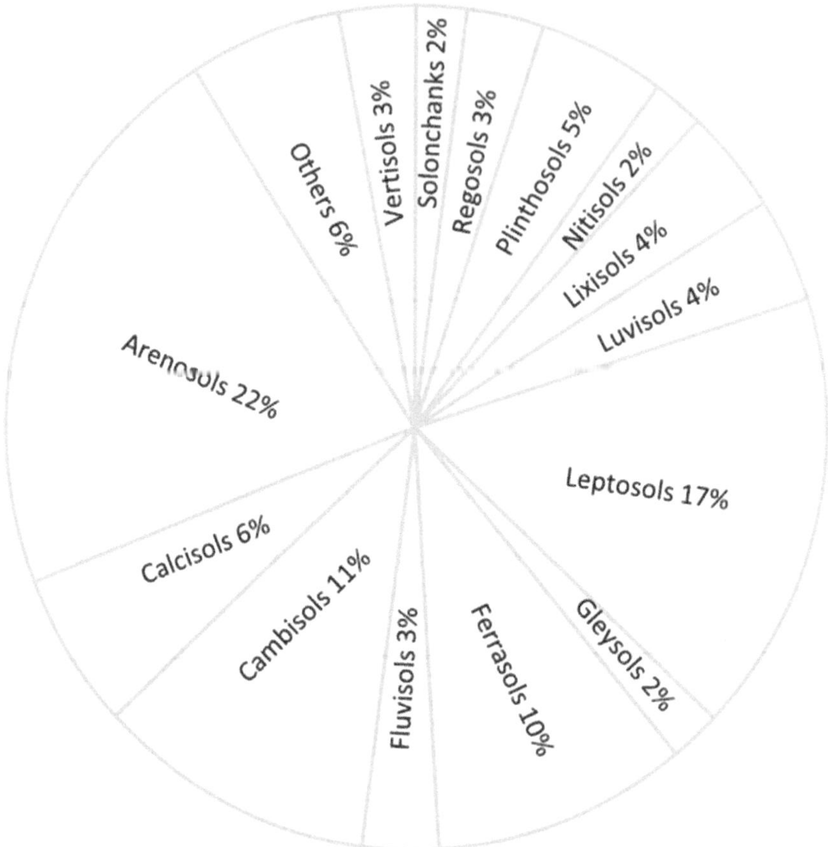

Figure 26.1 Percentage distribution of sub-Saharan Africa's major soil types

Source: adapted from FAO (2015)

intensity, measured as average kilograms of fertilizer nutrients (nitrogen, phosphorus and potassium) per hectare, was 14.9 for SSA, 157.9 for South Asia and 124.0 for the world (FAO, 2015). Other negative consequences include soil acidification and soil erosion.

The conundrum of limited arable land expansion possibility in many African countries coupled with widespread soil and land degradation and the need for irrigation expansion bring to the fore the synergies and trade-offs underlying the resource nexus approach. Under this situation, sustainable agricultural intensification, involving increased fertilizer, organic manure and water use together with high-yielding crop varieties, would seem an appropriate response. Although renewable water resources in Africa are not anywhere near being fully exploited currently, managing trade-offs among competing needs and using water efficiently is critical given climate change and population growth projections.

Energy

Primary agricultural production in Africa is still largely driven by human and animal energy. The use of mechanical energy for land preparation, cultivation and harvesting remains low. Energy input for water lifting and pumping in large-scale irrigation schemes has traditionally been based on electricity generated by hydropower schemes. In 2013, Africa's installed hydropower capacity was reported at 45,936 MW out of an estimated hydropower potential of 304,335 MW (AU-AMCOW, 2016), indicating room for expansion of hydropower generation in new environmentally and socially acceptable ways. Small-scale irrigation schemes rely mostly on small- to medium-size pumps powered by electricity or fossil fuel (diesel and gasoline). In many countries across Africa where such pumps are commonly used by smallholder farmers, the removal of government subsidy on electricity, diesel and gasoline has sharply increased the cost of running these pumps, prompting a gradual switch to solar-powered pumps. Data on mechanization and energy input into agro-processing, storage and transportation are not routinely collected on a national or regional basis. Indirect energy input into food production in the form of sequestered energy in fertilizers is low.

The recent increase in bioenergy crop production in Africa raises the spectre of a trade-off when biofuel and food production competes for land and water (Harvey and Pilgrim, 2011; Bogardi et al., 2012). In a review of 148 large-scale agricultural land investments across sub-Saharan Africa with a total of 3.4 million ha, 68% of the investments were directed towards bioenergy crops, including *Jatropha* and sugarcane (Williams et al., 2015). Apart from the direct competition, large-scale biofuel plantations have often led to the displacement of poor existing land users and this, in turn, leads to loss of access to land and associated water rights (Williams et al., 2012; Kizito et al., 2013).

Food production

With low population pressure in the 1960s, per capita food production in Africa was higher than in Southern Asia and the world (Figure 26.2, based on FAOSTAT database). With increasing population and poor performance of the agricultural sector in Africa in the 1970s and early 1980s, food production failed to keep pace with population growth. Per capita food production for most of the 1980s and 1990s was below the level achieved in the early 1960s in Africa. It is only in the last decade that Africa has witnessed a rapid growth in per capita production but the continent still lags behind the growth achieved in Southern Asia and globally over the same period.

Cereal yield

Only modest gains have been recorded over the past 50 years in increasing the yield per hectare of the most important cereals consumed in Africa (Figure 26.3, based on FAOSTAT database). For the period 2010–2014, rice yield per hectare in Africa was only about 70% and 57% of yield levels in Southern Asia and the world. For the same period, maize yield per hectare was about 70% and 38% of Southern Asia and World yield level. The cereal yield gap indicates the potential for improvement in Africa and show that more needs to be done to close the gap. Nonetheless, a recent study by van Ittersum et al. (2016) provided evidence which showed that it will not be feasible to meet future SSA cereal demand on current production area by yield gap closure alone. Additional measures including intensification (i.e. increasing cropping intensity) and sustainable expansion of irrigated production area will be needed, if SSA is to avoid greater dependence on imports of cereals than it does today. For much of the 1990s and up to 2007, cereal import-dependency ratio hovered around 22–30% per year. It jumped to an average of 35% during the three-year period 2010–2012 and 42% over the period 2011–2013 (FAOSTAT database).

Figure 26.2 Net per capita food production index in Africa, South Asia and the world, 1961–2013

Source: FAO. FAOSTAT database www.fao.org/faostat/en/#data/QI

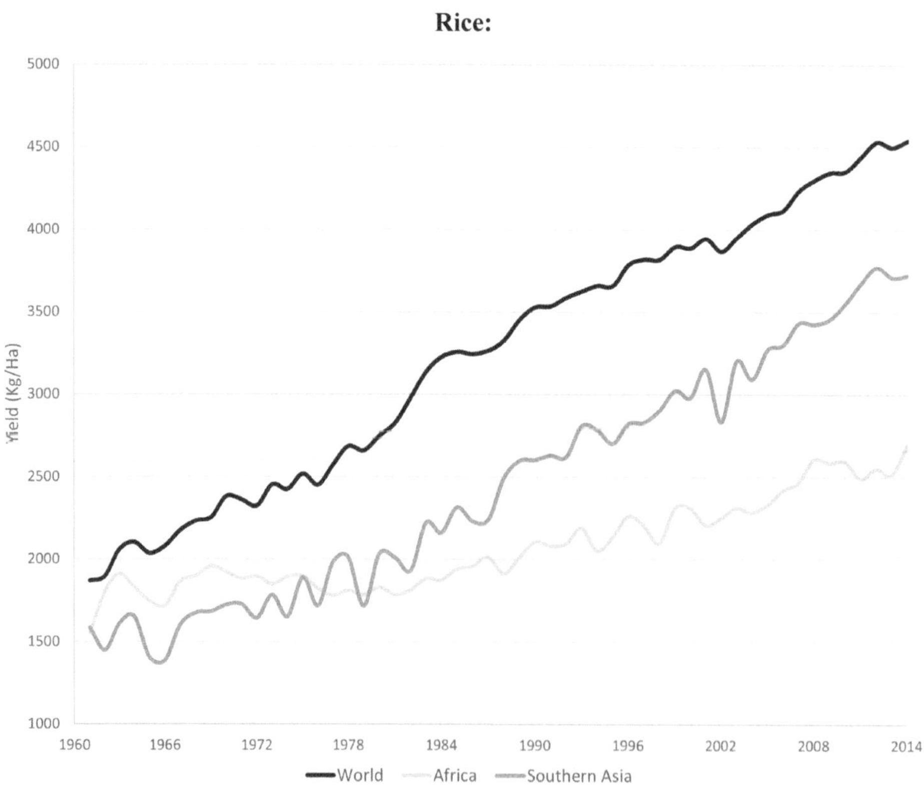

Figure 26.3 Cereals yield (kg/ha) in Africa, Southern Asia and the world, 1961–2014

Source: FAOSTAT database www.fao.org/faostat/en/#data/QC

Maize:

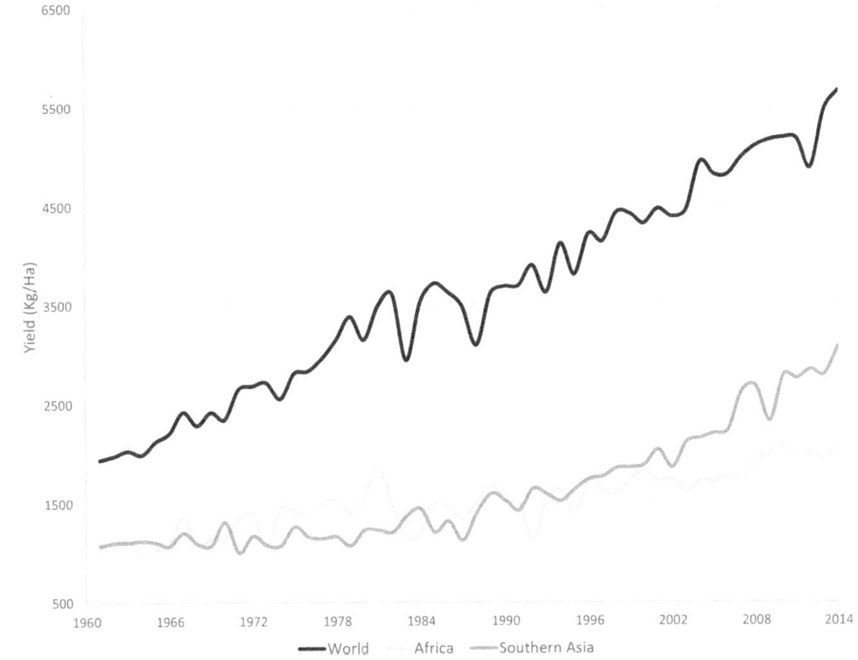

Sorghum:

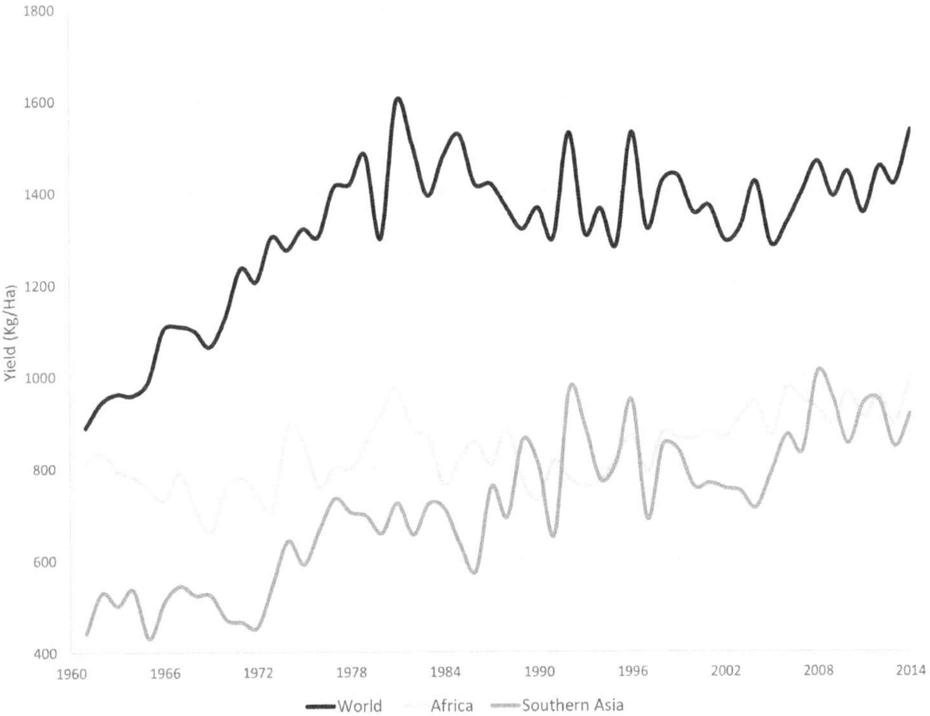

Figure 26.3 Continued

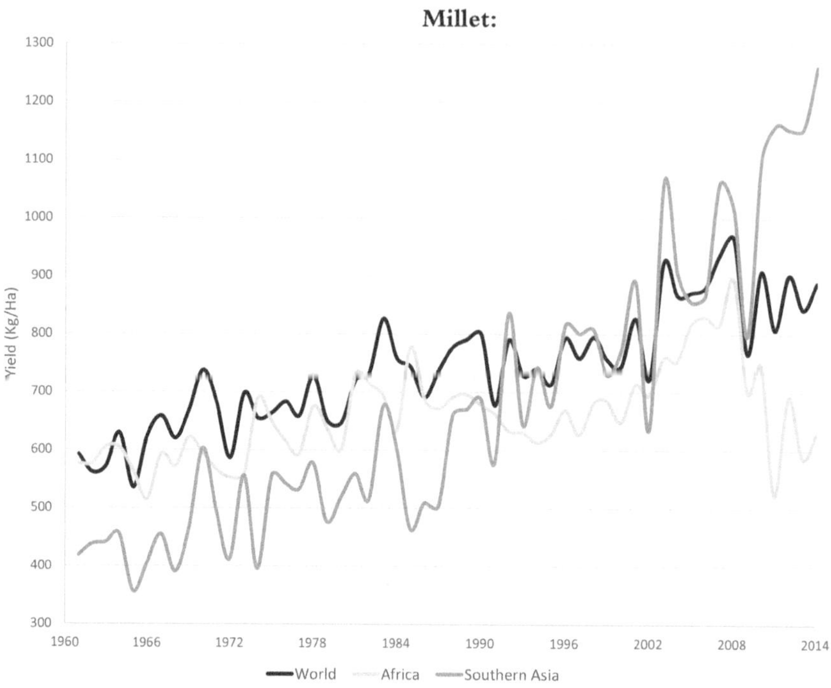

Figure 26.3 Continued

Table 26.1 Undernourishment in Africa, 1990–1992 to 2012–2014

Region	*Number of undernourished (millions) and prevalence (%) of undernourishment*									
	1990–1992		*2000–2002*		*2005–2007*		*2008–2010*		*2012–2014*	
	No.	*%*	*No.*	*%*	*No.*	*%*	*No.*	*%*	*No.*	*%*
Africa	182.1	27.7	209.0	25.2	211.8	22.6	216.8	20.9	226.7	20.5
Northern Africa	6.0	<5	6.5	<5	6.4	<5	5.6	<5	12.6	6.0
Sub-Saharan Africa	176.0	33.3	202.5	29.8	205.3	26.5	211.2	24.4	214.1	23.8

Source: FAO, IFAD, and WFP (2015)

Undernourishment

The slow progress in increasing yield per hectare and overall production has, in turn, led to unsatisfactory progress in eliminating hunger in Africa. Despite progress achieved over the past 20 years in reducing hunger and undernourishment, one in five people remained undernourished in Africa in 2014 (Table 26.1).

Opportunities and challenges of a resource nexus approach

The current status of natural resources, the magnitude of yield gaps and the level of undernourishment depicted above partly provide a justification for an alternative approach to the utilization and management of natural resources for food production. Figure 26.4 illustrates a conceptual framework that can be used to examine the potential opportunities and challenges entailed in a

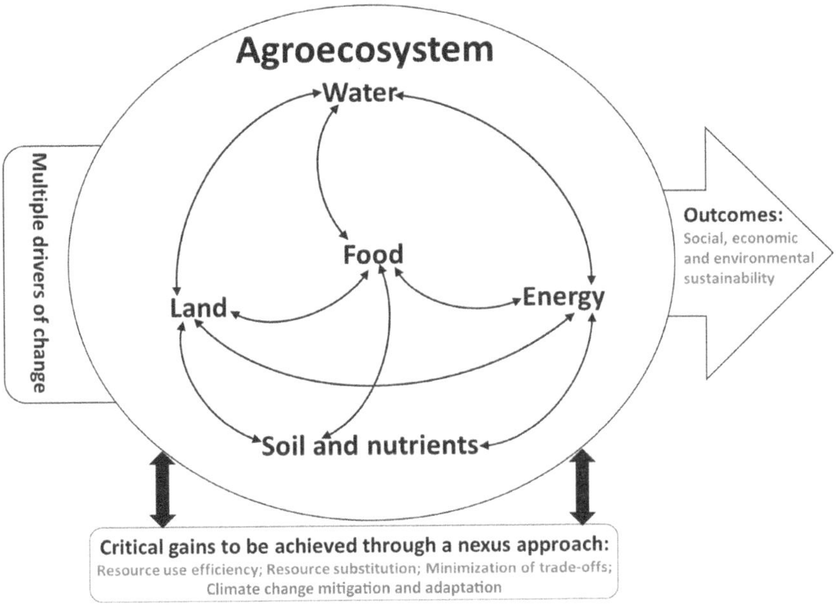

Figure 26.4 A conceptual framework of a resource nexus approach to food production

resource nexus pathway to feed Africa. Drivers of change such as population growth, urbanization, climate change, science and technology, income growth and policies will affect the interlinkages and the way land, soil, water and energy resources within an agroecosystem is managed to produce food and meet development and environmental objectives. Understanding the linkages, trade-offs and synergies between the various resources in an agroecosystem is key to managing them in such a way that they can mutually support food production and allow the critical gains of a nexus approach to be realized while contributing to desirable sustainability outcomes.

There is evidence that the interlocking and reinforcing positive benefits of integrated land, soil, nutrients, water, energy and ecosystem management can bring about increased food production alongside resource use efficiency, resource substitution, climate change mitigation and adaptation. Achieving these objectives will, in turn, lead to positive sustainability outcomes. A few examples will suffice.

Soil and water conservation techniques, including zaïs and rock bunds, in conjunction with tree planting are being used in many countries in West Africa to rehabilitate degraded land and contribute to food security and poverty alleviation (Sawadogo, 2011). Zaïs are traditional constructions of holes about 20 cm across and 15 cm deep used to increase soil moisture and organic matter content. Farmers place organic matter, such as composted manure, in these holes to create a micro environment that allows crops such as sorghum and millet to grow well despite poor soils and dry conditions. Tests of the impact on soil fertility and yields show that zaï can produce significant increase in soil moisture and organic content and can, under certain conditions, double or even quadruple crop yields. However, manual construction of zaï is labour intensive, requiring about 300 man hours per hectare (Barro et al., 2005). Mechanized zaï construction using animal traction has been shown to reduce labour time to about 22–36 hours per hectare and simultaneously increased sorghum yield by 34% compared with manual zaï construction (Fox et al., 2005).

System of Rice Intensification (SRI) is an agroecology-based rice planting method that was developed in Madagascar in the 1980s. It incorporates four synergistic agronomic practices of improving plant establishment, reducing plant population, improving soil conditions and reducing irrigation water application. In practice, single young rice seedlings are planted at wide spacing and compost, manure and crop residues are applied, supplemented as needed by chemical fertilizer. Weeds are controlled using a mechanical hand-pushed or motorized multi-row weeder and irrigation water is applied using alternate wetting and drying (AWD) method. The technology has been successfully applied in several sub-Saharan Africa countries including Madagascar, Mali, Gambia, Ghana, Kenya and Tanzania. Benefits of SRI per hectare reported in several studies can be summarized as follows: rice yields increased by 25–70%; 30–60% less irrigation water utilized; plots used 85–90% less seeds and 30% less chemical fertilizer and herbicides (Styger et al., 2011; Nyamai et al., 2012; Omwenga et al., 2014; Kahimba et al., 2014). These benefits translated to lower costs and improved revenue for farmers, while simultaneously allowing them to optimize the use of natural resources. In addition to rice, farmers have begun to apply SRI agronomic principles to the cultivation of other crops such as wheat, sugarcane and teff with good yield increases and environmental benefits (SRI-Rice, 2014).

Kizito et al. (2013) provided evidence which showed that when large-scale biofuel plantations on marginal lands are properly managed, they can serve as a positive ecosystem buffer to reduce runoff and pollution of water bodies. Model results indicated that runoff under *Jatropha* was 30% lower than runoff under fallow land. The reduced soil erosion arising from reduced runoff can also help in the reclamation and restoration of such marginal lands for future productive agricultural use. By using marginal lands that are not used for food production, the trade-off between biofuel and food production on prime agricultural land is avoided.

Solar-powered irrigation systems (SPIS) are emerging in Africa as an antidote to the challenges faced in using electricity and fuel-powered pumps for irrigation. Many rural communities are located in out-of-grid connection areas. Even in areas with electricity connection, high tariff and erratic power supply make electric pumps extremely unreliable and expensive to use by small-scale farmers (Giordano and de Fraiture, 2014). Use of fuel-powered pumps (diesel and gasoline) is equally expensive for smallholders due to the increasing cost of fuel, as governments gradually withdraw subsidies, and maintenance costs (Namara et al., 2014). In this situation, solar pumps coupled with drip irrigation kits are seen as a cost-effective and environmentally benign long-term solution to expand irrigation and increase agricultural productivity in Africa. Private sector companies and NGOs are at the forefront of promoting SPIS in East, West and Southern Africa while state agencies are prominent in North Africa. Some of the more advanced private sector companies (e.g. SunCulture in Kenya) provide farmers with soil analysis and on-site training using technicians and agronomists certified by the company, thereby creating service-oriented jobs for other workers. The various benefits associated with SPIS, for example, reliance on renewable energy and emissions-free pumping power, water savings, reduction in fossil-fuel demand and agricultural labour costs are all in consonance with the principles of resource substitution, resource use efficiency and environmental sustainability that are at the heart of the nexus approach. Burney et al. (2010) in a study conducted in Sudano-Sahel zone of West Africa showed that solar-powered drip irrigation can provide significant economic, nutritional and environmental benefits. Income earned by women irrigators from the production of market vegetables enabled them to purchase staples and protein during the dry season and their consumption of vegetables increased to the recommended daily allowance of 500–750 g per person per day. Also, the authors estimated that by using a solar-powered pump instead of a fuel pump, each vegetable garden of about 0.5 ha avoided a minimum of 0.86 t of carbon emissions per year (12.9 t over a 15-year lifetime). Nevertheless, there are cost-related challenges and environmental safeguard measures that need to be put in place to promote adoption and sustainable out-scaling of SPIS in Africa. These measures will be discussed in the following section.

Another promising area in the application of the resource nexus approach that also provides a rural–urban linkage is the work that IWMI and its partners are doing to turn fecal sludge and solid waste that end up polluting water sources and the environment in rapidly urbanizing cities across Africa into safe, pelletized organic manure for agricultural use (Nikiema et al., 2014; Cofie et al., 2016). The objective is not only to improve sanitation in urban areas, but also to recover nutrient and organic matter from waste in a business-oriented manner that is also beneficial to soil fertility improvement and agriculture. The fertilizer produced through the co-composting of solid waste and fecal sludge, called *Fortifer*, provides organic manure in a form (i.e. pellets) that is cost effective (lower transport cost) and easier to handle for on-farm distribution. In Ghana, the Environmental Protection Agency and the Ministry of Food and Agriculture in 2016 certified *Fortifer* as an organic fertilizer for commercialization and distribution. As part of this work, IWMI facilitated the establishment of two public–private partnerships (PPPs) for commercialization of *Fortifer*. A co-composting plant, established through one of these PPPs, that will annually transform 5,000 m^3 of fecal sludge and 300 t of organic solid waste into 200 t of *Fortifer* is due to start production early in 2017. Another strand of this work is investigating energy recovery from waste to provide reliable cooking energy in the form of briquettes to households in Uganda (Gebrezgabher et al., 2016).

Policy and institutional measures for a resource nexus approach to future food production in Africa

The interlinked and interdependent nature of the resource nexus makes it imperative to coordinate and integrate policies across sectors. Yet to date in Africa there is a dearth of policy frameworks to explicitly link and coordinate the management, planning and implementation of the nexus. To be sure in some countries, for example, Morocco a single ministry is responsible for energy, mining, water and the environment. Also, several African countries that have signed the Comprehensive African Agriculture Development Program (CAADP) compact have developed a National Agriculture Investment Plan (NAIP) and, with respect to climate change, National Adaptation Plan of Action (NAPA) that include some elements of the resource nexus. A recent review of the NAIPs of countries in West Africa that make up the Economic Community of West African States (ECOWAS) pointed to a missed opportunity to sufficiently integrate energy dimensions into the NAIPs and rural development programmes where agricultural and water management policies and income generation are jointly considered (Bhattacharyya et al., 2015). At the continental level, the African Climate-Smart Agriculture Alliance was launched following the 2014 Malabo Declaration to mainstream climate change into agriculture. Through this alliance, countries and regional economic bodies across the continent are being supported to initiate inter-agency dialogues and cross-sectoral integration of climate-smart agriculture into NAIPs and water management policies. While these efforts are in the right direction, much more still needs to be done to put in place coordination and integration structures across sectoral boundaries. In doing this, thoughtful consideration should be given to what needs to be coordinated at what level and which policy instruments would be appropriate to bring about this coordination (Pahl-Wostl et al., 2014).

The growing importance of SPIS in Africa exemplifies the need for joined-up policies to ensure full realization of the benefits of the resource nexus approach. With increasing cost of fuel and electricity due to withdrawal of government subsidies, farmers are willing to ditch fuel and electric power pumps for solar pumps. But a majority of smallholder farmers cannot afford the initial capital outlay on solar pumps, which are still expensive in many African countries compared to fuel pumps of the same power supply capacity. In this situation, innovative financing arrangements, including credit risk guarantee schemes to lenders by governments, can help to provide loans to farmers to defray the upfront capital outlay on solar pumps and drip irrigation kits. In addition, complementary energy policies to allow the sale of any excess power generated by farmers using SPIS to public and/or private sector utility companies will prevent farmers from over irrigating and drawing down the water table through over pumping of groundwater. IWMI through the TATA Foundation supported Solar Power as Remunerative Crop (SPaRC) project is exploring coordinated policy options not only to ensure sustainable SPIS, but also to improve future solar economy of India (Shah et al., 2015). Similar projects appropriately adapted would be equally useful in African countries.

Policy measures are also needed to effectively bring together various actors (representatives of farmers, public and private sectors, NGOs and civil society) in the resource nexus value chains. In many African countries, private sector firms, NGOs and civil society organizations working together with individual farmers and farmer groups are already actively promoting various aspects of the nexus approach. In addition to the involvement of some of these actors in the SPIS described above, Syngenta Foundation for Sustainable Agriculture (SFSA), in collaboration with AfricaRice (a CGIAR centre), has been working since 2010 to support SRI projects in Senegal, Mali and Burkina Faso. SFSA provides land preparation and grain harvesting and threshing services to farmers through its Farmer Equipment Service Centre model.Yields have increased from 6 to 8 t/ha in the river delta of Senegal and from 4 to 6 t/ha in parts of Burkina Faso and Mali (SFSA, 2016). In addition, farmers are linked to the rice supply chain to improve timely access to inputs and increase productivity and income. Governments can support these developments by providing an enabling environment and regulatory framework to provide appropriate incentives.

Governance of natural resources is at the heart of the resource nexus. The range of scales and institutional settings in which governance decisions are made suggest that flexible but complementary governance arrangements are needed to ensure resource use efficiency and desirable livelihood and environmental outcomes. Such flexibility will allow some decision-making powers to be devolved to local government level, while others are coordinated at the national level (Bhaduri et al., 2015). In countries where such devolution of power has been undertaken (e.g. Burkina Faso, Ghana, Ethiopia, etc.) the problem often is the disconnection between natural resource management decisions taken at the two levels. In other cases, the decision-making authority at the lower tier level lacks the power and resources to put into effect the decisions taken. Establishment of multi-stakeholder dialogue platforms involving representatives from the different tiers of government, resource user groups and civil society is of paramount importance to facilitate the crafting of new governance arrangements that are in tune with the nexus approach (Karlberg et al., 2015).

At the same time, the political issues inherent in the nexus approach need to be recognized and addressed. As pointed out by Woertz (2015), 'affordable food, water and energy are crucial for political legitimacy and so is their nexus'. The trade-offs and co-benefits innate in the nexus will throw up winners and losers and other issues that are best resolved through the political process. On a different but related note, checks and balances in the political system will be needed to ensure that nexus narratives are not used to justify rent seeking or disadvantage poor smallholder farmers while favouring big domestic and foreign investors.

Effective and well-functioning intra-Africa regional trade will be an important adjunct in the quest to promote resource use efficiency and to ensure that food is produced only in areas with the requisite comparative advantage. While the volume of food trade within the continent has been increasing in recent years, removing the remaining bottlenecks that still impede trade will go a long way to promote future improvement in trade and allow more efficient resource use and food production on a regional basis. Furthermore, opportunities to mainstream the basic tenets of the resource nexus approach into future revisions of the Voluntary Guidelines on Responsible Governance of Tenure and Principles for Responsible Investment in Agriculture and Food Systems should be explored.

One criticism that has been leveled against the nexus approach is that it is long on theory but short on translating theory into real-world solutions and outcomes at multiple scales and on demonstrating practical steps of how the nexus approach can be implemented (Leck et al., 2015; Woertz, 2015). Lack of a consensus on what constitutes a successful nexus approach outcome and how it can be achieved, monitored and evaluated poses significant challenge to policymakers (Leck et al., 2015) and puts the onus on the proponents of the approach. This points to the need for investment in research to expand the pool of nexus-related technologies and foster innovation that will allow the processes, solutions and outcomes of the nexus approach to be clearly and practically demonstrated. Equally needed is investment in the development of indicators and methods to monitor and evaluate the outcomes of nexus implementation and provide the evidence base to scale-out successful solutions. In this respect, the SDGs and associated targets and indicators that are now being developed provide a framework to organize and implement the resource nexus approach in many African countries.

Practical utilization of new nexus-related technologies and management practices will necessitate strengthening the capacity of farmers and/or assisting them to develop new competencies to implement approaches they may not yet be familiar with. Yet agricultural extension systems in many African countries are poorly resourced in human-, physical- and financial-capital terms. Non-governmental organizations and international aid agencies have stepped in to provide advice directly to farmers without much coordination among such organizations and with government departments. Polices to foster coordinated public–private sector partnership in advisory services delivery to farmers will need to be implemented to support the resource nexus approach.

Conclusions

Land, water, energy and agroecosystems are the essential pillars of primary food production all over the world. The availability, quality, policies and institutions governing the use and management of these resources impose opportunities and challenges on food security, human well-being and the environment. Linkages between these resources have always been recognized, but often considered in a two- or three-dimensional context that links land and food production; soil fertility and food production; land and water; land, water and ecosystems; or water, energy and food. Recent evolution in nexus thinking, as espoused in the opening chapter of this book and the SDGs, have provided an impetus to look at the linkages through a multidimensional lens in order to identify solutions to improve resource use efficiency and address trade-offs without losing sight of human well-being and environmental outcomes. This task is of special relevance to Africa where increasing population, resource degradation and unsatisfactory progress in reducing malnutrition makes it particularly important to utilize available resources to improve food and nutrition security and enhance livelihoods while simultaneously protecting valuable ecosystems.

This chapter analyzed the characteristics of the natural resource base and the ensuing food insecurity and resource degradation to show that business as usual is no longer tenable. Five promising agricultural initiatives that are in line with the principles of the resource nexus approach were presented, with a discussion of the policy and institutional measures that are needed to further improve them and to ensure that additional similar innovations can be developed and successfully scaled out.

The evidence pieced together in this chapter demonstrates that there is a strong basis for the adoption and implementation of a resource nexus approach to feed Africa over the coming decades. It is both an imperative and a necessity that should be embraced by African governments on the basis of enlightened self-interest. With implementation of appropriate policies and institutional measures progress can be made in the planning and implementation of such an approach to deliver tangible outcomes that will ensure improved resource use efficiency and livelihoods, as well as long-term sustainability of valuable ecosystems.

References

AU. (2015). *Agenda 2063: The Africa we want.* Addis Ababa: AU.

AU, NEPAD, and CAADP. (2014). *Synthesis of the Malabo declaration on African agriculture and CAADP.* Addis Ababa and Midrand: AU and NEPAD.

AUC-AMCOW. (2016). *African water resources management priority action programme 2016–2025 (WRM – PAP).* Addis Ababa and Abuja: AUC-AMCOW.

Barro, A., Zougmoré, R., and Taonda, S. J. B. (2005). Mécanisation de la technique du zaï manuel en zone semi-aride. *Cahiers Agricultures*, 14, 549–559.

Bhaduri, A., Ringler, C., Dombrowski, I., Mohtar, R., and Scheumann, W. (2015). Sustainability in the water–energy–food nexus. *Water International*, 40(5–6), 723–732. doi:10.1080/02508060.2015.1096110.

Bhattacharyya, S., Bugatti, N., and Bauer, H. (2015). *A bottom-up approach to the nexus of energy, food and water security in the Economic Community of West African States (ECOWAS) region.* ESRC (Economic and Social Research Council) Nexus Network thinkpiece Series.

Bogardi, J. J., Dudgeon, D., Lawford, R., Flinkerbusch, E., Meyn, A., Pahl-Wostl, C., Vielhauer, K., and Vörösmarty, C. (2012). Water security for a planet under pressure: Interconnected challenges of a changing world call for sustainable solutions. *Current Opinion in Environmental Sustainability*, 4(1), 35–43. doi:10.1016/j.cosust.2011.12.002.

Burney, J., Woltering, L., Burke, M., Naylor, R., and Pasternak, D. (2010). Solar-powered drip irrigation enhances food security in the Sudano – Sahel. *PNAS*, 107(5), 1848–1853. Retrieved from www.pnas.org/cgi/doi/10.1073/pnas.0909678107.

Chamberlin, J., Jayne, T. S., and Headey, D. (2014). Scarcity amidst abundance? Reassessing the potential for cropland expansion in Africa. *Food Policy*, 48, 51–65.

Cofie, O., Nikiema, J., Impraim, R., Adamtey, N., Paul, J., and Koné, D. (2016). *Co-composting of solid waste and fecal sludge for nutrient and organic matter recovery.* Colombo, Sri Lanka: International Water Management Institute (IWMI). CGIAR Research Program on Water, Land and Ecosystems (WLE). 47p. (Resource Recovery and Reuse Series 3). doi:10.5337/2016.204.

Comprehensive Assessment of Water Management in Agriculture (CA). (2007). *Water for food, water for life: A comprehensive assessment of water management in agriculture.* London: Earthscan and Colombo: International Water Management Institute.

ELD Initiative and UNEP. (2015). *The economics of land degradation in Africa: Benefits of action outweigh the costs.* Retrieved from www.eld-initiative.org

FAO. (2015). *The state of food and agriculture 2015. Social protection and agriculture: Breaking the cycle of rural poverty.*

FAO. (2016). *AQUASTAT database.* Retrieved December 23, 2016, from www.fao.org/nr/water/aquastat/data/query/index.html?lang=en accessed.

FAO. (2017). *FAOSTAT database.* Retrieved January 17, 2017, from www.fao.org/faostat/en/#data/RL.

FAO, IFAD, and WFP. (2015). *The state of food insecurity in the World 2015.* Meeting the 2015 international hunger targets: Taking stock of uneven progress. Rome: FAO.

Fox, P., Rockström, J., and Barron, J. (2005). Risk analysis and economic viability of water harvesting for supplementary irrigation in semi-arid Burkina Faso and Kenya. *Agricultural Systems*, 83(3), 231–250.

Gebrezgabher, S. Amewu, S., Taron, A., and Otoo, M. (2016). *Energy recovery from domestic and agro-waste streams in Uganda: A socioeconomic assessment.* Colombo, Sri Lanka: International Water Management Institute (IWMI). CGIAR Research Program on Water, Land and Ecosystems (WLE). 52p. (Resource Recovery and Reuse Series 9). doi:10.5337/2016.207.

Giordano, M., and de Fraiture, C. (2014). Small private irrigation: Enhancing benefits and managing trade-offs. *Agricultural Water Management*, 131, 175–182.

Harvey, M., and Pilgrim, S. (2011). The new competition for land: Food, energy, and climate change. *Food Policy*, 36, S40–S51. doi:10.1016/j.foodpol.2010.11.009.

Jayne, T. S., Chamberlain, J., and Headey, D. D. (2014). Land pressures, the evolution of farming systems, and development strategies in Africa: A synthesis. *Food Policy*, 48, 1–17.

Kahimba, F. C., Kombe, E. E., and Mahoo, H. F. (2014). The potential of System of Rice Intensification (SRI) to increase rice water productivity: A case of Mkindo Irrigation Scheme in Morogoro Region, Tanzania. *Tanzania Journal of Agricultural Sciences*, 12(2), 10–19.

Karlberg, L., Hoff, H., Amsalu, T., Andersson, K., Binnington, T., Flores-López, F., de Bruin, A., Gebrehiwot, S. G., Gedif, B., zur Heide, F., Johnson, O., Osbeck, M., and Young, C. (2015). Tackling complexity: Understanding the food-energy-environment nexus in Ethiopia's Lake Tana Sub-basin. *Water Alternatives*, 8(1), 710–734.

Kizito, F., Williams, T. O., McCartney, M., and Erkossa, T. (2013). Green and blue water dimensions of foreign direct investment in biofuel and food production in West Africa: The case of Ghana and Mali. In T. Allan, M. Keulertz, S. Sojamo, and J. Warner (Eds.), *Handbook of land and water grabs in Africa: Foreign direct investment and food and water security*. London: Routledge, pp. 337–358.

Leck, H., Conway, D., Bradshaw, M., and Rees, J. (2015). Tracing the water-energy-food nexus: Description, theory and practice. *Geography Compass*, 9/8, 445–460. doi:10.1111/gec3.12222.

Namara, R. E., Hope, L., Owusu Sarpong, E., De Fraiture, C., and Owusu, D. (2014). The adoption patterns and constraints pertaining to small-scale water lifting technologies in Ghana. *Agricultural Water Management*, 131, 194–203.

Nikiema, J., Cofie, O., and Impraim, R. (2014). *Technological options for safe resource recovery from fecal sludge*. Colombo, Sri Lanka: International Water Management Institute (IWMI). CGIAR Research Program on Water, Land and Ecosystems (WLE), 47p. (Resource Recovery and Reuse Series 2). doi:10.5337/2014.228.

Nyamai, M., Mati, B. M., Home, P. G., Odongo, B., Wanjogu, R., and Thuranira, E. G. (2012). Improving land and water productivity in basin rice cultivation in Kenya through System of Rice Intensification (SRI). *Agricultural Engineering International: CIGR Journal*, 14(2), Manuscript No. 2094.

Omwenga, K. G., Banti, B. M., and Home, P. G. (2014). Determination of the effect of the System of Rice Intensification (SRI) on rice yields and water saving in Mwea Irrigation Scheme, Kenya. *Journal of Water Resource and Protection*, 6, 895–901. doi:http://dx.doi.org/10.4236/jwarp.2014.610084.

Pahl-Wostl, C., Gupta, J., and Marx, S. (2014). Governance of the water-energy-food nexus. GWSP. *Global Water News*, 14, 4–5.

Sawadogo, H. (2011). Using soil and water conservation techniques to rehabilitate degraded lands in northwestern Burkina Faso. *International Journal of Agricultural Sustainability*, 9(1), 120–128.

Shah, T., Verma, S., and Durga, N. (2015). *Solar Power as Remunerative Crop (SPaRC): Exploring alternative architectures for India's future solar economy*. International Workshop on Prospects for Solar-Powered Irrigation Systems (SPIS) in developing countries FAO HQ, Rome, Italy, May 27–29. Retrieved December 22, 2016 from https://www.slideshare.net/IWMI_Media/solar-power-as-a-remunerative-crop-sparc-exploring-alternative-architectures-for-indias-future-solar-economy

Sidibé, Y., and Williams, T. O. (2016). Agricultural land investments and water management in the Office du Niger, Mali: Options for improved water pricing. *Water International*, 41(5), 738–755. doi:http://dx.doi.org/10.1080/02508060.2016.1178900

SRI-Rice (2014). The System of Crop Intensification: Agroecological Innovations for Improving Agricultural Production, Food Security, and Resilience to Climate Change. SRI International Network and Resources Center (SRI-Rice), Cornell University, Ithaca, New York, and the Technical Centre for Agricultural and Rural Cooperation (CTA), Wageningen, Netherlands. Retrieved September 27, 2017 from sri.ciifad.cornell.edu/aboutsri/othercrops/SCImonograph_SRIRice2014.pdf

Styger, E., Aboubacrine, G., Attaher, M. G., and Uphoff, N. (2011). The system of rice intensification as a sustainable agricultural innovation: Introducing, adapting and scaling up a system of rice intensification practices in the Timbuktu region of Mali. *International Journal of Agricultural Sustainability*, 9(1), 67–75.

Syngenta Foundation for Sustainable Agriculture (SFSA). (2016). *Improving the livelihood of smallholder farmers.* System of Rice Intensification. Retrieved from www.syngentafoundation.org/index.cfm?pageID=310

United Nations Department of Economic and Social Affairs. (2013). *World population prospects. The 2012 revision. Volume 1: Comprehensive tables*. New York: United Nations.

United Nations Department of Economic and Social Affairs. (2014). *Urbanization prospects. The 2014 revision. Highlights*. New York: United Nations.

van Ittersum, M. K., van Bussel, L. G. J., Wolf, J., Grassini, P., van Wart, J., Guilpart, N., Claessens, L., de Groot, H., Wiebe, K., Mason-D'Croz, D., Yang, H., Boogaard, H., van Oort, P.A.J., van Loon, M.P., Saito, K., Adimo, O., Adjei-Nsiah, S., Agali, A., Bala, A., Chikowo, R., Kaizzi, K., Kouressy, M., Makoi, J. H. J. R., Ouattara, K., Tesfaye, K., and Cassman, K. G. (2016). *Can sub-Saharan Africa feed itself?* Retrieved from www.pnas.org/cgi/doi/10.1073/pnas.1610359113

Williams, T. O., Gyampoh, B., Kizito, F., and Namara, R. (2012). Water implications of large-scale land acquisitions in Ghana. *Water Alternatives*, 5(2),243–265.

Williams, T. O., Sidibe, Y., Baker, T., Windham-Wright, T., and Gemo, H. (2015). Analysis of impacts of large-scale investments in agriculture on water resources, ecosystems and livelihoods and development of policy options for decision makers. Final project report to UNEP, GRID-Arendal and FAO. Accra: International Water Management Institute (IWMI).

Woertz, E. (2015, November). The water-energy-food nexus in MENA. *Oxford Energy Forum*, 102, 45–48.

27
THE FIVE-NODE RESOURCE NEXUS AT SEA

Tundi Agardy

Introduction

Human dependence on oceans for water, energy, food security, and other resources

The world is increasingly looking to the sea for drinking water, energy, food, and strategic minerals. How this bodes for humanity's – and the planet's – future is open to question, as the global population is expected to reach about 9 billion by 2050 and attendant demands for food and energy will be nearly double what they are today. Intense conflicts over scarce surface water and aquifer water resources have already started, and the spectacular growth of energy-demanding desalination in response to water scarcity threatens access to and availability of seafood resources. At the same time, offshore energy development displaces fishers from productive fishing grounds, and catastrophic oil spills harm or even shut down fisheries. To date, siloed thinking about how to manage marine resources has resulted in uncoordinated fisheries, energy, mining, and marine use policies that do not allow the consideration of trade-offs and do not capitalize on the synergies that taking a nexus approach would provide.

Marine management regimes will undoubtedly be put to the test in the near future, given challenges peculiar to the ocean environment, including a lack of property rights, the difficulty and expense of monitoring and surveillance, and lingering ecological uncertainties. Against the backdrop of global climate change, and with the cumulative pressures from unregulated marine use and insufficient watershed management, new demands for ocean space and resources will push marine governance to its limits. This chapter explores the trends in food, water, energy, minerals, and use of space in the ocean realm and reveals the ways that agricultural and energy production on land has and will continue to impinge on future options for obtaining resources from the sea. Current and projected future uses of the ocean environment, including desalination, renewable and non-renewable energy extraction, minerals mining, and both wild-capture fisheries and aquaculture are assessed, with both the positive and the negative interactions between these competing uses highlighted.

As on land, the water, energy, biomass, minerals, and land (herein meant to pertain to both substrate and water column, or comprehensive ocean space) nexus or five-node nexus is one of on the one hand interdependence, and on the other, co-mingled environmental impacts and inevitable trade-offs. To meet rising demands for freshwater, seawater is filtered of its salts, minerals, and other compounds. This desalination requires energy, as does the piping of water to end users, and the treatment of freshwater for drinking water use and in waste management. Conversely, electrical energy generation requires large and sustained quantities of water. For example, in the US in 2005 the production of thermoelectric power accounted for almost half of the nation's total water consumption (ACEEE, 2011). And energy is also required to provide food resources – whether through agriculture on land, farming the sea, or wild capture of fisheries resources. Energy powers fishing vessels, fuels processing operations, provides refrigeration for seafood storage, allows for the movement of goods in local and international trade, and fuels cooking for seafood consumption. Energy is required for fossil-fuel extraction and refining, as well as minerals extraction and getting these resources to market and to consumers. Finally, there are dependencies even within one element of the nexus: wild capture of forage fish is needed to produce the fishmeal commonly used as feed for livestock and farmed seafood and to a smaller extent still also as fertilizer in agriculture. The intense harvest of forage fish to feed the things we humans eat threatens the ecology of the very same marine areas that provide seafood that currently accounts for 20% of per capita protein intake for more than 1.5 billion people worldwide (UNCTAD, 2016).

Understanding these impacts and interdependencies is crucial for ensuring that use of ocean space and resources is sustainable. With ever-greater reliance on the sea, marine managers must ensure that ocean ecosystems remain healthy and productive. Marine and coastal ecosystems provide valuable resources and support livelihoods the world over. The market value alone of marine resources is estimated at US$3 trillion annually, and additional significant values reside in the role that marine ecosystems play in water quality maintenance, coastal flood control, shoreline stabilization and hazard mitigation, climate change mitigation, tourism, and recreational values.

Food from the sea will be in greater demand as the human population continues its growth trajectory and as competing uses for diminishing space on land push resource extraction offshore. The economic value of many fisheries is growing steadily, and both marine wild-capture fisheries and aquaculture are critically important sectors for most coastal countries. Marine fisheries generate revenues, increase GDP, expand trade opportunities, provide livelihoods for coastal residents, and are a strong part of the cultural identity of many coastal societies. But there are many other important goods and services that originate from the very same bodies of water and coasts that have been supplying seafood product to humans for millennia. New uses of ocean space, alongside expanding conventional uses, have brought resource extraction sectors into conflict, and have confounded attempts at marine management that keeps use at sustainable limits.

The ecosystem services that support these water, energy, and fisheries values are particularly concentrated in ecologically critical natural habitats, and these are discussed in detail on page 417. Within these important habitat types, multiple goods and services are generated. For instance, the specific ecosystem services that are linked to fisheries and aquaculture include not only the production of fisheries resources, but also the supporting services such as providing food to fisheries species through nutrient cycling and prey production, providing habitat to escape predation, providing unfettered access from one critical habitat to another, providing access to spawning grounds (especially spawning grounds undisturbed by fishing pressure), and – for the special cases of anadromous species – providing access to good-quality freshwater habitat for reproduction.

There are also additional ecosystem services provided by coastal and marine habitats that influence the quality and viability of fisheries and aquaculture more indirectly, such as shoreline stabilization, flood control, and water quality maintenance. Importantly, the profitability of many marine fisheries is linked to high-quality, safe products – all made possible by ecosystems that maintain water quality, suppress disease, and keep food webs in balance. And with the growing movement towards certification, wherein fisheries products can be sold at a premium if they can certify that harvest practices and levels are sustainable, there are now incentives for protection of habitats that support fisheries and enhance product value. Unfortunately, these are the very habitats that are also under pressure from the impacts of other sorts of resource use, including energy extraction and desalination, as well as being on the receiving end of the cumulative impacts from other sorts of direct use and indirect degradation.

Thus mangrove forests, coastal wetlands, estuaries, rocky shorelines, coral reefs, beaches, seagrass beds, upwelling areas, and other coastal and marine habitats all contribute to providing food resources, maintaining water quality, and stabilizing shorelines that accommodate not only housing but also energy infrastructure and ports that make the transshipment of all these important resources possible. Collectively, these ecosystem services benefit humanity and make sustainable development possible. Many of these habitats provide the foundation for one of the most important drivers of the economy of the world's approximately 150 coastal nations: tourism. Additionally, these coastal and marine habitats contribute ecosystem services that benefit local and national economies and contribute to culture and identity: maintaining environmental quality and reducing public health risks, stabilizing land and protecting coastal property and infrastructure, supporting agriculture and aquaculture, mitigating the effects of climate change, providing educational opportunities, and preserving sites of cultural, historical, and spiritual interest (see Figure 27.1).

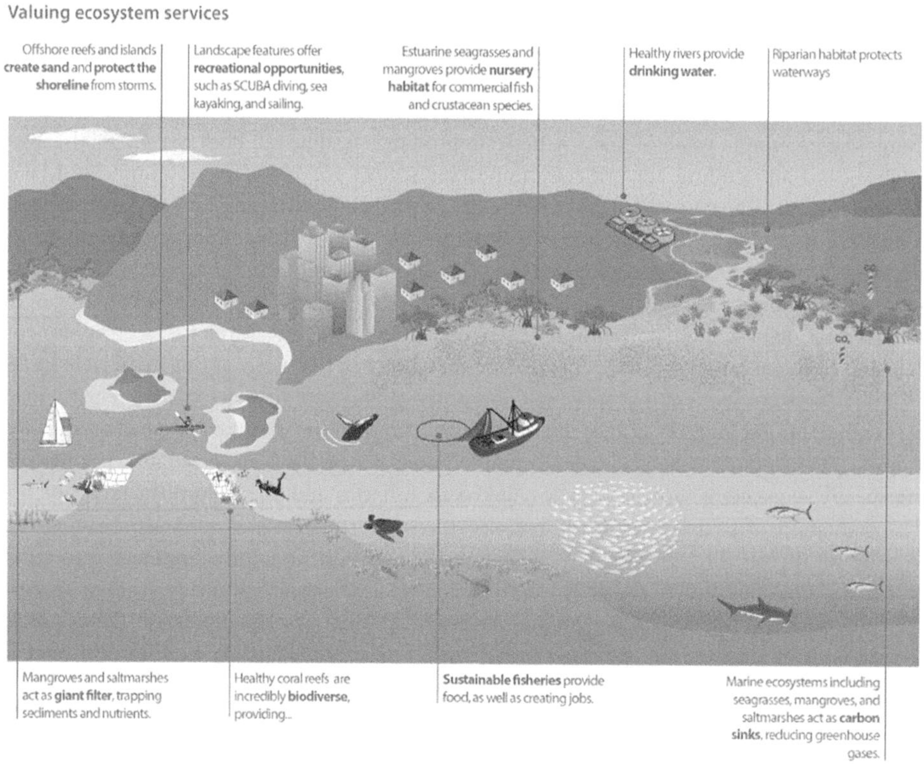

Figure 27.1 Coastal and marine ecosystem services schematic

Source: Agardy et al. (2011) EBM Manual

Collectively, marine and coastal habitats support the well-being of communities and the economic vitality of coastal nations, and the various opportunities for resource use that they provide are interrelated (Hinrichsen, 2013). Take for example how coastal habitats support nature-based tourism, which creates demand for local fisheries products and allows for a more diverse portfolio of livelihood opportunities, thereby reducing competition and conflict within the fisheries sector. Some of these very same coastal habitats also protect ports and harbors, stabilize shipping channels, and buffer land from storms, and these services contribute to the economic health of the shipping sector and the export of fisheries resources. Shoreline stabilization and erosion control is a natural service that also reduces the costs that would be incurred by municipal, state, or regional governments to maintain fishing port and fish processing plant infrastructure. And although difficult to value economically, the coasts and oceans also offer recreational opportunities, spiritual rejuvenation, and inspiration to coastal communities – all of which contribute to human well-being – well-being that can become compromised when trade-offs made under the water–energy–biomass–minerals–land nexus are not considered carefully. How these uses of marine resources support or preempt one another, and how the five-node nexus on the terrestrial side affects the very same in the seas, is the subject of this chapter.

The water–energy–biomass–minerals–land nexus in the ocean realm

Our planet is ocean-dominated. The vast majority of the world's 195 sovereign countries and the bulk of their populations are coastal, with economies inextricably tied to the health and productivity of marine ecosystems. Around the world seas and coasts are filled with valuable assets that generate substantial revenues for economic development, support livelihoods, improve the well-being of local communities and visitors, and have a key role in climate change mitigation. Healthy marine and coastal ecosystems are vital for providing water and energy resources, as well as overall food supply and food security, not only through marine fisheries and aquaculture but also in support of the agricultural sector on land. Ocean environments are likely to be increasingly important as global populations grow, land becomes scarce, the climate changes, and new markets for seafood and marine products emerge.

Water

Desalination of seawater for use as drinking water, watering livestock, or irrigation of crops, gardens, golf courses, and landscape plantings seems on the surface like the perfect solution to the planet's water scarcity problem. After all, the bulk of humanity lives near the coasts, and the volume of ocean water ready to be filtered of salts seems almost limitless. However, desalination takes a toll on the marine environment when the brines that are by-products of water purification are disposed at sea (the least cost option). The degradation caused by desalination can affect the other elements of the five-node nexus, especially fisheries production. Some countries that are highly dependent on desalination, such as many in the Arabian Gulf region, face an uncertain future if pollution from brines continues, as it will at some point no longer be economically feasible to practice large-scale desalination to meet water needs. Moreover desalination is highly energy intensive, with fossil fuels still providing the only major source of energy, making desalination also a growing greenhouse gas emitter. Renewables are only very slowly entering the desalination market, being limited at the moment to pilot plant scale.

Nonetheless, desalination of seawater for drinking water and for irrigation is increasing exponentially. Currently over 120 countries rely on desalination to meet their water needs, and many new plants are coming online. New technologies allow for economical filtration of salts, heavy metals, and other compounds from seawater, and waste processing of the brines has also seen technological advance. However, desalination is still not without environmental costs, and regulators have been slow to deal with pollution and access issues arising from wide scale desalination.

There is another link between seawater and potable freshwater. Seawater intrusion and the water quality of nearshore coastal waters can affect the purity of water in aquifers, as has been shown in the Middle East. Paradoxically, this lowering of water quality happens in many cases at the hand of desalination, when toxic and highly saline brines are released back to the marine environment.

Desalination infrastructure can preempt other uses of the coast, and reduce the value and thus profitability of the marine system, seascape and coastal land areas. Paradoxically, the need for larger and larger desalination plants is occurring in response to booming tourism populations in many places – tourists who have been drawn to these coastal and island locations due to the aesthetic and other values the coasts provide. Coastal nations will have to plan the siting of desalination plants carefully, so as to minimize the risk to tourism-related businesses, the health of the coastal ecosystems and the natural resources that flow from it, and public health.

As well as being a valuable resource in its own right, water is an input into the other elements of the nexus, especially energy. The vast majority of electricity-generating plants utilize thermally driven, water-cooled energy conversion cycle. Thermoelectric power plants withdraw a tremendous amount of water. Depending on the type of cooling, the percentage of water that is evaporated and therefore lost to other inputs/uses varies significantly (Torcellini et al., 2003). Locating power plants at the coast and using seawater for cooling can reduce the pressure on scarce inland water resources.

Energy

Oil and gas

Hydrocarbons have fueled the growth and economic development of the industrialized world for over a century. In recent decades, the source of this oil and natural gas has increasingly been the sea, with major oil and gas operations in the North Sea, Persian Gulf, western Pacific (especially Sakhalin oil field) and Southeast Asia, eastern Pacific off California, the Caspian Sea, the Atlantic Ocean off Brazil, northwest Atlantic, West Africa, the Indian Ocean (especially off the coast of India), and the Gulf of Mexico. In the US, approximately one quarter of all oil currently comes from the Gulf of Mexico deposits. And while early offshore oil and gas exploration and recovery was in the nearshore domain, technological advances have allowed drilling in areas far from shore, in deep waters. Oil and gas operations in these extreme environments were brought into the limelight after the spectacular 2010 Deepwater Horizon oil spill in the Gulf of Mexico, in which a blowout at the wellhead some 1600 meters below the surface resulted in an explosion, the death of 11 rig workers, and the release of an estimated 4 million barrels (approximately 795 million liters) of oil over the course of 84 days (NOAA, 2013). The spill, as well as the use of dispersants in emergency response, caused US officials to temporarily close over 200,000 sq. km. of formerly productive fishing grounds to finfish and shellfish fisheries. Research into the long-term environmental impacts of the spill continue to this day.

Hydrocarbons have thus been sourced from both land and sea. As easily accessible and economical terrestrial sources reach their limits, oil and gas companies moved offshore. The first extraction of offshore oil and gas occurred in the late 19th century in California, near to shore and close to ports. Today the demand for energy has driven oil and gas operations from continental slope to the far reaches of the continental shelves (Smith and Thrupp. 2015). Currently approximately a third of globally recovered oil and gas derives from offshore sources (see Figure 27.2). Given the need for cheap, reliable energy to fuel industry, agriculture, and housing around the world, marine hydrocarbons are likely to continue to fuel economic development for some time (Odell, 2004).

Although offshore oil and gas operations are carried out with stringent safety, quality control, and environmental restrictions, the industry has caused both chronic and catastrophic impacts on the marine environment and its other users. Ecological damage has occurred during all phases of oil and gas development, including during exploration for deposits, drilling and extraction, delivery of oil to ports, and processing. The catastrophic oil spill incidents certainly get the public's attention – not only the aforementioned 2010 Deepwater Horizon spill in the Gulf of Mexico, but also the Exxon Valdez spill in Alaska two decades before, and the Ixtoc 1 spill in Campeche a decade before that. Yet it is likely that the low level but chronic and cumulative effects of offshore non-renewable energy development exact an even greater toll on marine life and on the communities that rely on intact and thriving coastal ecosystems. The use of sonar arrays to reveal pockets of oil or gas under the seafloor stresses marine life (especially cetaceans), while the maintenance traffic from shore to rigs is thought to contribute to the high rate of loss of wetlands along the shores nearby. These and other chronic impacts, and the risk of catastrophic accidents, are continually being weighed by agencies with jurisdiction for permitting offshore oil and gas development, and figure into the economic development plans of coastal countries.

Oil and gas impacts coastal ecosystems and coastal communities in other ways as well. Port development to service vessels traveling to and from rigs, as well as to create oil storage and refining facilities, can displace fishing boats and recreational vessels. Oil and gas rigs and shore facilities can damage prospects for coastal tourism, and can deprive coastal communities from profiting from tourism-related livelihoods. Finally, oilrigs at sea can interfere with navigation, fishing, and marine recreation, and the disposal of defunct rigs creates challenges and costs for coastal management authorities.

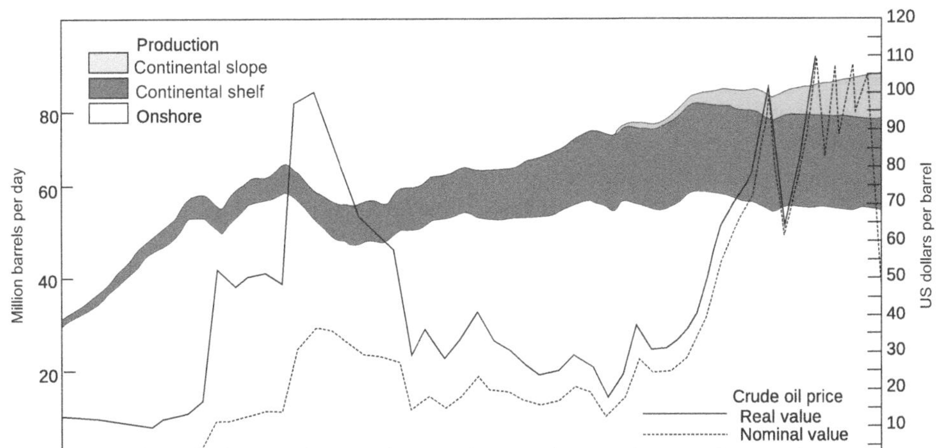

Figure 27.2 Offshore production: continental shelf and slope

Source: Smith and Thrupp (2015)

Renewables

As oil and gas become increasingly difficult to recover or as existing supplies in accessible areas diminish, and as geopolitics complicates the access to both the resources and the markets for such energy products, interest in renewables is on the increase. Additionally, recent attention on carbon emissions from hydrocarbon use and the role of fossil fuels in driving global warming and other climate change means that renewable energy resources are fast becoming a preferred alternative. However, renewable energy resources are still at this point in time generally more expensive than conventional non-renewables, and production tends to be small scale.

Ocean renewables have great potential as a sustainable source of energy for numerous reasons. For one, water is approximately 800 times denser than air providing greater power density and more accurate predictability than most other renewable energy sources. Additionally, ocean renewable energy is easily accessed from the nation's most densely populated areas – on major rivers and near coasts – minimizing the need to construct thousands of miles of new transmission lines. The most common form of renewable energy is offshore wind, though other technologies include offshore solar, kinetic energy obtained from harnessing waves, tides, or currents, and algal biofuels (O'Neill et al., 2015)

The need to mitigate the effects of climate change, coupled with increasing energy demands, has driven governments to develop bioenergy production. However, the large-scale use of first generation biofuel feedstocks have spurred speculation that biofuel production may result in 'food displacement.' Even when biofuel production is moved to marginal land that are agriculturally less productive, if profits from biomass plantations exceed those from food production, farmers will grow bioenergy crops on prime agricultural land in order to maximize profit (Littlejohn et al., 2015). This conundrum is typical in the siloed approaches governments have to date exhibited when it comes to energy production and resource use. A nexus approach would help balance competing demands and ensure that biofuel production does not result in trade-offs with food production, other forms of energy generation, or utilization of ocean space for water, minerals, or other uses. Alternative approaches include producing biodiesel from micro-algae at sea, and in so doing producing co-benefits like carbon sequestration/climate change mitigation, as well as water purification.

Marine biofuel production and conversion for energy is still in its infancy, but shows substantial promise. Many industrialized countries with large demands for power and less and less available space on land for siting renewable facilities like solar arrays or wind turbines are heavily invested in biofuel research and development. In the US, the National Aeronautical and Space Agency is developing floating bags of algae for placement in nutrient rich waters, requiring no additional inputs for biofuel generation. Three-dimensional ocean farming, as demonstrated by groups like GreenWave, allow for farming of seaweeds and shellfish that utilizes ocean space for the co-generation of energy from biofuels, agricultural fertilizers as an alternative to conventional phosphate minerals extraction, and food simultaneously. Additional emerging uses of marine algae in these integrated initiatives include developing their use as biofilms. The global market for marine biotechnology is projected to reach $4.8 billion by 2020, growing at a rate of 5.4% annually. Marine biofuel production can thus not only provide an energy source, but can power desalination for water and support aquaculture, and agriculture on land by providing fertilizers. When such biofuel production is co-located with other renewable energy production and aquaculture, for instance at offshore wind farms, the five-node nexus at sea is achieved, and the costs of taking a siloed approach are avoided.

Marine biomass for food

Marine fisheries products are essential to support human populations, not only in coastal but in inland areas as well (Ye, 2015). Fish products provide essential proteins for human consumption globally, with regional variations (see Figure 27.3). While the proportion of food protein provided by fish is significantly smaller than that provided by meat and dairy products, its share is increasing rapidly around the world as global populations grow from 7.4 billion to an expected 9 billion by 2050.

Throughout the world, marine and coastal fisheries are engines of economic growth. Fish is one of the most highly traded of commodities, and in the record-setting year of 2013 (the most current year for which comprehensive global fisheries data are available), global exports reached US$136 billion (UNCTAD and the Commonwealth Secretariat, 2015). The seafood value chain is long and lucrative; with additional earnings generated by value added industries that process both domestic and imported seafood products (Agardy et al., 2016). Both fisheries and aquaculture are expanding rapidly throughout the world, and even in the most developed countries, fish consumption is increasing both per capita and in absolute terms, with implications for food security, trade and social stability.

Fisheries play a key role in ensuring food security, perhaps even more important than their export value or direct economic output to GDP. Seafood, whether procured through capture fisheries or via aquaculture, is a major food source for coastal populations and in these areas per capita fish consumption is significantly higher than the global average. In addition, food for subsistence and much-needed cash provide social benefits in areas where coastal communities are marginalized or in rural locations (FAO, 1996). In the lesser developed countries, and particularly in remote coastal areas, fish is not only the major source of animal protein, it is also a critical source of micronutrients essential to people with otherwise deficient nutrition (Pauly and Zeller, 2016). For these reasons, it will be increasingly important to bring together governments, companies and local communities to engage in sustainable and innovative fisheries exploration practices, where ecosystem restoration and sustainable fisheries harvesting go hand in hand (Agardy et al., 2016).

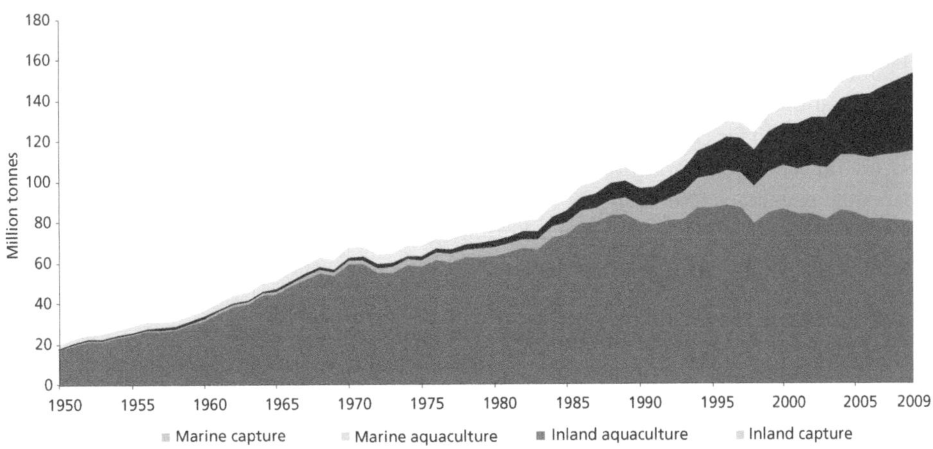

Figure 27.3 World fish production from different sectors of fisheries

Source: Ye (2015)

Seafood and fishing are also culturally important, with millions of people engaged in artisanal fishing as part of traditional use and alternatives to urban occupations. From a cultural perspective, seafood has played a central role in the development of traditional gastronomy, which has become a fundamental part of cultural pride and identity. This supports not only cultural identity, but also the growing marine and cultural tourism trade (Agardy et al., 2016). As a mainstay of many coastal communities, small-scale fisheries and aquaculture play an important role in the social fabric of society (FAO, 2014).

Against this backdrop of increasing local reliance on seafood is the ever-expanding industrial fisheries sector, providing low-cost seafood for transshipment around the world, as well as fishmeal to support agriculture and aquaculture (Ye, 2015). In many parts of the world, competition over access to resources between large-scale commercial fisheries and small-scale artisanal or subsistence fisheries has created conflict. This trend will be exacerbated as stocks become overexploited and perverse subsidies drive overcapitalization that can result in even more overexploitation. This is likely to lead to the degradation of fisheries habitats and the ecosystems associated with them, unless planners mainstream nexus thinking and consider integrated use of oceans across the five-node nexus.

Capture fisheries

Wild-capture fisheries are conventional fisheries – having been practiced for millennia. The harvest of seafood runs the gamut from small-scale subsistence fishing practiced with handlines, handmade traps or weirs, or spearfishing to large-scale industrial fishing utilizing satellite technology to locate fish, industrial machinery to harvest fish, and on-vessel processing facilities. Marine fisheries provide food security and are the backbone of international trade, and include not only fish but also mollusks (shellfish, squid, octopus), crustaceans (crab, lobster, shrimp), and even seaweeds (kelp, macroalgae for agars, etc.).

An even bigger driver of marine fisheries in the future will be the growing global demand for fishmeal and fish oil for agricultural feeds, fertilizer, and human dietary supplements. While use of low-trophic level marine species for agricultural fertilizer eases the pressure to mine minerals for phosphates and nitrogen, this use of ocean space and resources has consequences for other elements in the nexus. Of the 10 largest fisheries by volume, seven are focused on forage fish: the small, low-trophic level species that occur in great volume, such as herring, anchovy, anchoveta, and sardines. In these seven major fisheries across the globe, 90% of product is processed into fishmeal and fish oil (WWF, 2012). Much of this meal makes its way into large-scale terrestrial agriculture, as food for cows, pigs, and other livestock, and as fertilizer to increase crop yields. At the same time, the growing aquaculture industry is adding to demand, since the most lucrative aquaculture operations farm carnivorous species like salmon and shrimp and these seafood species require large volumes of protein meal for fast grow out.

In order for these forage fisheries to be profitable, vast quantities of the low-value small-forage fish need to be harvested. The consequence of this is an undermining of marine food webs – which affects not only other marine species but humans as well – since this base of marine food supports the growth of other marine fish that are targeted in fisheries across the globe. And marine food webs are being hit hard from both sides of the trophic levels: the bottom of the food web is being undermined by large-scale removal of forage fish, while the top is being decimated by both targeted fisheries for top predators like swordfish, marlins, tuna, and shark, and by the accidental catch of these and other species high on the food chain. This wasted by-catch accounts for surprisingly large removals of marine biomass.

Aquaculture

Aquaculture is both a lucrative business enterprise and a means for assuring local food security in underdeveloped regions. By far the most profitable aquaculture industries are those that produce high-value product, including shrimp and salmon. The production of this seafood is input intensive, requiring vast quantities of fishmeal to feed the carnivorous target species growing in tanks or ponds. This pressure on wild stocks of forage fish to be made into fishmeal is discussed elsewhere in this chapter. However, high-value product is also grown in non-conventional ways, using net cages in coastal areas or large floating enclosures far offshore. These methods utilize food in the ambient environment, thereby reducing the need for fishmeal. And recent research suggests that algae can be grown as an alternative to fishmeal (Kiron et al., 2016) – a method which shows promise not only because it relieves pressure on wild forage fish, but because the growth of algae for meal has added co-benefits like carbon sequestration to mitigate climate change, and water filtration to improve water quality.

In recent decades, the worldwide aquaculture industry has expanded in response to new market demand and a spate of new investors. In Asia, aquaculture provides the bulk of protein, though the vast majority of production is freshwater species. Farming of marine species takes on strategic importance in some regions. In Latin America, six countries – Chile, Ecuador, Mexico, Brazil, Colombia and Cuba – currently account for the bulk of production (Agardy et al., 2016). Shrimp and salmon aquaculture targeting markets in the US, Japan and Europe account for more than 80% of aquaculture production in the Latin American region (FAO, 1996, 2014). In Ecuador, shrimp exports generate some US$2.6 billion in 2014 annually, largely driven by strong and increasing US demand for shrimp, combined with a drop in Asian shrimp production due to the early mortality syndrome (Agardy et al., 2016). In addition to Ecuador, other major shrimp producers in the region include Mexico, Colombia, Honduras, and Panama. In contrast, Chile is the sole large-scale developer of salmon farming, accounting for more than 10% of the world salmon supply.

As in other parts of the world where industrial aquaculture is practiced, farming operations are vulnerable to disease outbreaks. These operations have been the source of large-scale habitat destruction (especially the destruction of mangrove forests for shrimp ponds, see UNEP, 2014) and degradation tied to the release of fishery waste products, antibodies and other medicines as well as nutrients into local waters. However, the situation has improved considerably through the application of international certification, emerging regulation, and rising government interest in the protection and management of coastal and marine ecosystem services in more environmentally sound ways (Gunther, 2012).

Aquaculture operations have faced challenges, but many are improving, due in part to the 2009 establishment of the Aquaculture Stewardship Council (ASC) and its standards for the 12 most commonly farmed fish and shellfish species. Many aquaculture operations have already moved to get ASC or Best Aquaculture Practices (BAP) certification. Schemes such as these provide measurable standards and third-party verification in order to ensure that their aquaculture operations adhere to best practices and are attractive to the industry due to the marketing opportunities that certification can provide. Several nations have already developed strategies for sustainable seafood and aquaculture. Nonetheless, there is scope for improvement, both to increase efficiency and net benefits to society, and to ensure that economic development in the fisheries sector does not constrain other maritime industries and benefits in the long run.

One of the practices holding most promise for sustainability and avoiding devastating trade-offs is three-dimensional ocean farming. Typically these aquaculture operations grow seaweed like kelp, culture mussels, oysters, and clams, and support fish populations in relatively small areas of the sea. The plants trap carbon dioxide, produce biomass for biofuel and food, and support the growth of fish and shellfish. Though these farms are small, they are remarkably productive, with a single acre of ocean 'land' producing 250,000 pounds of shellfish and 10–30 tons of kelp per year. The kelp can soak up 5X the amount of carbon as land-based plants, mitigating climate change in the process. Adding kelp to livestock diets could reduce outputs of methane – a dangerous greenhouse gas – by 90%. These 3-D farms can be co-located in areas used for recreation, or renewable energy. Groups like GreenWave in the US have demonstrated how ocean space can be optimized to produce energy, mitigate climate change, deliver high-quality and profitable seafood product, and perform ecosystem services like improving water quality, thereby affecting water production through desalination use as well. Such 3-D farming practices beautifully demonstrate the 5-part node at sea.

In conclusion, fisheries and aquaculture are especially important to the livelihoods of the poor in coastal regions, not just in terms of opportunities for employment in fishing, but also with respect to associated employment, such as jobs in the seafood processing, retailing, and restaurant sectors. Yet as important as the fisheries sector is to countries across the globe, perverse subsidies have meant enormous losses and wasted investments in the fisheries sector (World Bank, 2009). While fisheries are the backbone of the local economy of many nations, and while seafood exports are for some a significant source of revenue, expanding fisheries and growing aquaculture operations have placed fisheries at odds with other extractive and non-extractive industries, including desalinization of seawater for drinking water and irrigation, use of oceans and coasts for tourism and recreation, minerals mining, and development of both renewable and non-renewable energy.

Minerals

The oceans provide significant minerals resources, the extraction of which has become less costly with new technologies. Base metals, strategic minerals, and sands and other minerals for us in construction can all be found in marine substrates. These minerals are all important to the five-node resource nexus.

As with marine fisheries exploitation, the mining of minerals from oceans and coasts runs the gamut from small-scale, subsistence use (shoveling sand or gathering coral rubble from the nearshore for use as home building material, for instance) to very-large scale, high-tech, and profitable operations mining precious metals and minerals like gold and diamonds, strategic minerals like manganese and chromium, and minerals for use as fertilizers like phosphates. While all these uses of the sea carry with them environmental and ecological impacts, the use of mined minerals for fertilizer can have the positive environmental effect of reducing the expansion of agriculture-related land and forest conversion. Utilizing a nexus approach can help take into account both positive and negative externalities, and allow for a strategic and rationalized set of policies for ocean use across the entire five-node nexus.

Land

Unlike terrestrial portions of the globe, land itself is not utilized in the sea, except as part of the base of operations for obtaining food, energy, minerals, or siting infrastructure related to water supply through desalination. However, submerged land leases, for instance Outer Continental Shelf (OCS) leases for oil and gas, as well as marine leases for aquaculture operations further inshore, can confer exclusive property rights to a particular sector. The synergies and conflicts with other sectors that are thus created by such 'land' rights are discussed in the other portions of the water–energy–biomass–minerals–land nexus.

One important application of the nexus though is the integrated management and governance of the land–coasts–oceans continuum, with critical interlinkages (e.g. in terms of water, sediment, nutrient and pollution transports and displacements), as well as opportunities for cross-resource substitution and efficiency improvements.

Critical marine habitats that support the nexus

Coastal and marine ecosystems directly support the resource nexus in two ways: they provide resources or inputs, of course, and they provide the ecological processes that ensure that good-quality water will be available for desalination and high-quality, healthy seafood products will be available to fishers. Fisheries products originating from coastal areas with good water quality and high biodiversity will be of higher quality and will fetch a higher price than products sourced from polluted waters. What correlates to this higher water quality and more profitable fisheries is not necessarily human absence, however – it is instead the presence of coastal habitats that provide ecosystem services like water filtration, hydrological balance, biodiversity maintenance, and disease control.

Some habitats are more crucial to sustaining the nexus in this way than others. Mangrove forests, coastal wetlands, estuaries, coral reefs, seagrass beds, macroalgae assemblages, and upwelling areas all contribute to fisheries production, providing generous opportunities for social cohesion, leisure, and economic activities in the region. Additionally, these coastal and marine habitats contribute ecosystem services that benefit local and national economies and contribute to coastal culture and identity: supporting agriculture, mitigating the effects of climate change, providing educational opportunities and sacred sites. Fisheries and aquaculture also support and indeed increase the profitability of other drivers of economic growth, such as marine-oriented tourism and agriculture. Fisheries and to some extent aquaculture thus provide livelihoods, food security, and cultural identity in thousands of coastal communities on islands and coasts across the world (Agardy et al., 2016). Estuaries in particular are proving grounds that demonstrate that the nexus can be sustainable if managed correctly.

Coastal habitats provide ecosystem services beyond the cornucopia of biomass, energy, and water inputs in the nexus. Some habitat types are particularly valuable in protecting infrastructure investments, worker housing, navigation channels, and other environmental features that support the marine resources nexus. Of particular note are estuaries and deltas (Tessler et al., 2015), coastal salt marshes (Smardon, 2009), mangrove forests, coral reefs, seagrass meadows, and barrier islands (MEA, 2005). Many of these habitats directly support fisheries. Mangrove, for instance, is an extremely important fishery nursery habitat – meaning that the habitat supplies an environment for fisheries species to spend the early stages of their life safe in the shelter of mangrove prop roots and channels (FAO, 2007; Linneweber, 2013). These links can be quantified; mangrove forests of Pacific Mexico support the vast bulk of lucrative fisheries in the Gulf of California (Aburto-Oropeza et al., 2008). Despite these values, the world is losing mangrove habitats at an alarming rate. Worldwide, 3.6 million hectares of mangrove were lost over the last 25 years (UNEP, 2014) – an area larger than the total coverage of all mangrove forest in South America.

Seagrass is an underwater flowering plant that grows in meadows very similar to the fertile meadows on land. Like mangrove, seagrasses perform a multitude of ecological functions and provide a wide variety of regulating and supporting services (Green and Short, 2003). These include water quality maintenance, stabilization of substrates and erosion prevention, providing nursery and breeding habitat for fishery species, and support to a wide array of biodiversity. In fact, higher numbers of species are directly reliant on seagrass than on mangroves, since seagrasses range from cold temperate waters to the warm waters of the tropics – and are thus foundational in virtually all shallow-water ecosystems.

Beaches and barrier islands provide other ecosystem services. Though beaches are formed by physical oceanographic processes involving sediment transport and sand deposition, biota that colonize barrier islands stabilize the land forms. The presence of barrier islands buffers land from the daily erosive power of tides and waves, and are even more important in natural hazard mitigation in that they buffer land and infrastructure from cyclones and hurricanes as well as tsunamis. Gradual loss of beaches is occurring in many places due to environmental degradation and interference with coastal processes (as happens the world over with coastal constructions that interfere with currents, violation of setbacks on shorelines that attenuate beach erosion, and alterations to river channels and outlets to improve navigation). Recent attempts to model the effects of barrier beach loss in terms of compromises to infrastructure and property damage (see Ulm et al., 2016, for example) shows that the benefits provided by barrier beaches far outweigh the costs of effective management that preserves barrier beach integrity.

Finally, healthy coral reef systems in which biodiversity is intact and environmental quality is maintained have supported humankind for millennia. Today approximately 500 million people live within 10 kilometres of a coral reef and tens of millions of people in impoverished communities depend on coral reef fisheries for their protein, as well as coastal protection against storm surges (Agardy et al., 2011). According to The Economics of Ecosystems and Biodiversity (TEEB) initiative, which synthesized over 80 studies performing economic valuations on coral reefs, the ecological services coral reefs provide can reach US$375 billion per year (TEEB, 2010). This represents values of US$130,000 to US$1.2 million per hectare per year, including for the provisioning of food and materials (up to US$6000 per hectare per year), climate regulation, waste treatment, shoreline stabilization and storm buffering (up to US$35,000 per hectare per year), cultural services (up to US$1.1 million per hectare per year), and maintenance of genetic diversity (up to US$57,000 per hectare per year) (TEEB, 2010).

Yet despite their beauty and value, reefs continue to be at risk from direct destruction and indirect degradation (Burke and Maidens, 2004). Desalination in the tropics threatens coral reefs by direct impacts of industrial infrastructure footprints, and by the release of brines that can kill corals if discharge pipes are not placed far enough offshore. Energy production offshore can harm coral reefs similarly through infrastructure construction, as well as operational and catastrophic oil spill pollution. At the same time, a dependency on coral reef ecosystems as a food source, which is common to many parts of the tropical world (Sale et al., 2014), can destroy reef ecosystems if fisheries are not carefully regulated and cause them to shift to alternative stable states in which algae dominate and fisheries production declines. In the Caribbean region, for example, 30% of coral reefs have either been wiped out or are at serious risk (Burke and Maidens, 2004). Almost two-thirds of remaining Caribbean coral reefs are threatened by coastal urbanization, sedimentation, contamination by toxic substances, water acidification, and overfishing (UNEP, 2010). Coral reefs thus provide virtually all the ecosystem services nature offers, and demonstrate the trade-offs that need to be made in the marine five-node nexus.

Water, energy, and food security conflicts

Coastal areas continue to experience growing population and exploitation pressures; estimates vary but 40–60% of the world population is thought to live in this thin fringe of land. Demographic trends suggest coastal populations are rapidly increasingly, mostly through migration, increased fertility and tourist visitation to these areas. Population densities on the coasts are nearly three times that of inland areas and communities and industries increasingly exploit fisheries, timber, fuelwood, construction material, oil, natural gas, sand and strategic minerals, and genetic resources. Additionally, demands on ocean space for shipping, waste disposal, military and security uses, recreation, aquaculture, and even habitation are increasing (MEA, 2005).

The condition of the coastal and semi-enclosed seas and their ability to supply inputs to the resource nexus is not solely a function of the condition of coastal areas and the pressures exerted in situ, however. Much of what we do on land affects watersheds indirectly, and this in turn affects coasts and oceans indirectly. Approximately 77% of all marine pollution derives from land-based sources (MEA, 2005). Agricultural fertilizers pollute coastal waters and cause eutrophication, which can then lead to fish kills, fisheries collapse, lowered water quality that make shorelines unsuitable for recreational use, and human health hazards. At the same time, agricultural pesticides and herbicides can find their way into coastal waters, affecting not only wildlife and ecosystem functioning, but also contaminating seafood and in extreme cases making it unfit for human consumption.

Infrastructure to support humanity and in particular the industries revolving around the five-node water–energy–food nexus is ever on the increase in coastal areas. The buildings, industrial facilities, and roads all increase runoff and contribute in a major way to coastal pollution. The amount of non-imperious surfaces in an area is directly related to water quality conditions on coast. And while we are increasing the inputs of pollutants, we are at the same time reducing the roles that coastal habitats can play in mitigating those impacts. The loss of ecosystem services like water filtration done by mangroves, seagrass, oyster reefs, etc. – which occurs through both habitat conversion to accommodate infrastructure, and also through overfishing and maintenance dredging – can exacerbate the water-quality problems (Sale et al., 2014). The end result is that across the globe, coastal waters are the most chemically altered environments on earth (MEA, 2005). Reduced water quality means lessened fisheries production and food availability and also can have human health impacts. Climate change is driving further degradation, increasing ocean acidification and pathogen spread, and stressing the few critical coastal habitats that remain.

While pollution of nearshore waters is on the increase, activities on land also cause other pressures. Hydroelectric energy production even high up in watersheds can have severe negative impacts on ocean ecology, by reducing sediment loading (30% decrease worldwide, with regional variations) and freshwater inputs to estuaries, destroying nursery areas and fishing grounds (MEA, 2005). Reduced sediment delivery to coasts also exacerbates coastal erosion, affecting resource availability and livelihoods, and costing coastal municipalities and local governments billions of dollars annually in shoreline stabilization and beach renourishment.

What we do at sea matters as well. Poorly managed marine resource use fisheries threaten marine values, and limit the opportunities for a diversified portfolio of livelihood options (Bovarnick et al., 2010). Many countries worldwide are reliant on fisheries that run the risk of being overexploited, or are at maximum capacity. In fact, close to 30% of the world's well-studied commercial marine fisheries are exploited beyond what scientists have determined is the maximum sustainable yield – in other words, almost a third of the world's fisheries are overfished and thus unsustainable (however it must be noted that the FAO estimates that more than three-quarters of the world's marine fish stocks are currently unassessed) (IISD, 2016). But the quality of fisheries management is not the only factor behind unsustainable use of food from the sea – in many cases, the very habitats that support fisheries production – providing ecosystem services such as nutrient loading or nursery habitats – are being lost as trade-offs between sustainable fisheries and unsustainable coastal development are being made.

Against this backdrop of challenging marine management and competing demands for ocean space to secure food, water, and energy resources is the larger environmental condition: global climate change and its attendant ocean warming, sea level rise, acidification, and marine pathogen/disease spread (Verner, 2010). Ocean acidification is particularly threatening to global food security, as projections of acidification-related impacts suggest that many lucrative fisheries will have decreased production or may cease altogether.

With the adoption of more ethical and sustainable resource use practices, there are good possibilities that the conflicts between land use and marine use, and among marine uses, will be minimized. For instance, direct fisheries management (controlling catch) must go hand in hand with the protection of habitats that maintain this resource. Fisheries management organizations are increasingly utilizing tools such as Marine Protected Areas, green-financing mechanisms (for instance, payments for ecosystem services) and certification schemes to ensure that management is more committed to efficiency and a holistic ecosystem-based approach (Potts et al., 2016). More importantly, systematic and large-scale marine spatial planning integrates the management of many marine sectors, and the management of land, freshwater, and oceans. These and other options for finding harmony in the five-node nexus are discussed in the next section.

Opportunities for enhancing water, energy, and food security in the blue sphere

There is great potential for developing conservation interventions and policy to maintain the value of ecosystem services at stake in the marine water–energy–biomass–minerals–land nexus. Clear trade-offs are made when converting coastal habitat for resort development or when marine and coastal ecosystems become degraded from inadequate control of pollutants, ineffective fisheries management, or unregulated use. Decision-makers need to understand these trade-offs – and characterizing the ecosystem services and option values that will be lost when natural areas are impacted may help nations make more informed decisions about the future. And because marine resources and ocean space are shared, the room that an ecosystem services perspective gives for shared and transboundary management is significant (López-Hoffman et al., 2010).

Along the coast and in marine areas of the world, a mix of pressures acts cumulatively to erode the health of those natural habitats that remain in the wake of large-scale urbanization, tourism infrastructure, energy development, and industrial farming. Only a small percentage of the exclusive economic zone of coastal countries is under some sort of protection and most marine reserves and marine parks are not properly managed (Agardy, 2015). What marine and coastal protected areas do operate well are generally focused on high visibility habitats like coral reefs, and on maintaining one service.

Yet two developments hold great promise in ensuring that water, energy, and food use from the marine environment can be sustained. One is the drive to increase the efficiencies of water use, energy generation, and fisheries production, while at the same time investing in natural capital that supports these industries. To achieve this, the first step is understanding and pinpointing inefficiencies. Increasing the reliability and decreasing the energy demand of pumps and ensuring that pipes do not leak is critical in the water arena. Similarly, improving the technology and bringing down the cost of alternative energy production at sea will reduce the environmental and social costs associated with oil and gas production. And increasing the efficiency of fishing, while also increasing the controls on fisheries that prevent overfishing, can be coupled to mechanisms to increase efficiencies along entire supply chains. This will allow fishers to reap more benefits from marine fisheries and aquaculture and exert fewer negative impacts on marine systems. Another example of efficiency increase is provided by polyculture marine farming. These innovative aquaculture enterprises provide resources and ecosystem services simultaneously, and are typically surrounded by a buffer area that promotes wider marine conservation. Disruptive technologies/approaches like GreenWave's 3-D ocean farming couple food production to biofuel energy generation, requiring no water for use, and are changing the nature of marine resource nexus.

The second major development in the marine resource nexus is the expansion of holistic and integrated approaches to management of uses at sea, made possible by marine spatial planning (Agardy, 2015). Through marine spatial planning, countries have great opportunities to integrate management of existing and emerging uses while at the same time ensuring that development does not undermine ecosystem health and productivity (Agardy, 2015). This is especially important as nations commit to the "blue economy", and invest in industries that unlock the economic potential that rests in the myriad ocean uses. These are not only environmental concerns, but also social ones, for as countries ramp up their industrialized use of marine resources and ocean space, they may be denying access to coastal communities that depend on the very same marine resources and ocean space.

National and international governance will need to consider the interactions between elements of the nexus, and manage natural resources accordingly. The move towards integrated coastal management that took place in coastal nations in the 1980s ushered in a movement towards cross-sectoral policy and interagency cooperation. This began in Europe, the US, and Australia, and has spread to other places as this trend continued and as development interests focused on areas farther offshore. Thus we have seen the formation of councils, commissions, and para-statal organizations to oversee management of large marine areas for multiple objectives. At the same time, UN agencies and development banks have been helping developing country governments restructure for cross-sectoral management. The international community and the UN Conference on Trade and Development (UNCTAD) have been pushing the World Trade Organization to address the issue of perverse subsidies that cause nexus disconnects and conflicts (such as fuel subsidies for industrial fishing of forage fish for animal feed that drives overexploitation of food fisheries, for instance). There has been vertical integration as well, such that not only are marine and coastal areas being managed across different sectors, that are also increasingly being managed by systems of co-management that couple national government ambitions to local community realities. All this activity has created opportunities for multi-sectoral governance that could help insure that the development of any node in the five-node nexus does not compromise the other nodes – at least in theory.

In practice, we are entering uncertain times. The recent emergence of populism and nationalism in Europe and the US (as well as other countries) not only diminishes prospects for international cooperation in marine science and management, but could also fuel a competitive rush to deplete resources quickly, before another country or another company does so. The dissolution of the World Order that witnessed the formation of many of the UN agencies, the World Bank, and other multilateral institutions may be imminent, and the future, especially vis-à-vis cooperation in environmental matters, is unclear.

Yet the world has little choice but to address and reduce these conflicts, and carefully manage direct and indirect impacts on the marine environment. The community of nations has signaled that it is serious about sustainability: the recent launch of the UN 2030 Agenda for Sustainable Development includes a specific goal (Goal 14) for sustainably using oceans, seas, and marine resources. The Sustainable Development Goals outline specific resource management targets, and suggest the strong aspirations of the global community at the highest political level to prioritize attention on restoring the health and resilience of oceans and marine resources over the next 15 years (UNCTAD, 2016).

We have seen that the water–energy–biomass–minerals–land nexus is not confined to the Earth's land areas. Demands for resources to support human life that come from the marine environment are on the increase, and the interactions between fisheries, energy extraction, and other marine uses are correspondingly on the rise. Not all these interactions are negative, with one use displacing another. In the promising new world of marine renewables, for instance, consideration is given to co-location of offshore wind platforms and aquaculture. These marine uses not only co-exist, they can enhance one another, strengthening the five-node nexus and providing a solid foundation for humanity into the future.

References

Aburto-Oropeza, O., Ezcurra, E., Danemann, G., Valdez, V., Murray, J., and Sala, E. (2008). Mangroves in the gulf of California increase fishery yields. *PNAS*, 105(30), 10456–10459.

ACEEE. (2011, May). *Addressing the energy-water nexus: A blueprint for action and policy agenda*. Alliance for Water Efficiency and American Council for an Energy-Efficient Economy. Retrieved from www.aceee.org/files/pdf/white-paper/Water-Energy%20Blueprint.pdf

Agardy, T. (2015). Marine protected areas and marine spatial planning. Ch 31 In H. D. Smith, J. L. Suarez de Vivero, and T. S. Agardy (Eds.), *Routledge handbook of ocean resources and management*. Abingdon: Routledge.

Agardy, T., Davis, J., Sherwood, K., and Vestergaard, O. (2011). *Taking steps toward marine and coastal ecosystem-based management*. Nairobi: UNEP.

Agardy, T., Vignati, F., and Gomez-Garcia, R. (2016). *Nature's benefits: Latin America's valuable marine fisheries and aquaculture*. Geneva: UNCTAD.

Bovarnick, A., Alpizar, F., and Schnell, C. (Eds.) (2010). *The importance of biodiversity and ecosystems in economic growth and equity in Latin America and the Caribbean: An economic valuation of ecosystems*. New York: United Nations Development Programme.

Burke, L., and Maidens, J. (2004). *Reefs at risk in the Caribbean*. Washington, DC: WRI

The Economics of Ecosystems and Biodiversity (TEEB). (2010). *The economics of ecosystems and biodiversity: Ecological and economic foundations*. Abington: P Kumar Editor Routledge.

FAO. (1996). *Fisheries and aquaculture in Latin America and the Caribbean: Situation and Outlook in 1996*. FAO Fisheries Circular No 921 FIPP/C921.

FAO. (2007). *The world's mangroves: 1980–2005. A thematic study in the framework of the Global Forest Resources Assessment*. Rome: FAO

FAO. (2014). *The State of World Fisheries and Aquaculture (SOFIA)*. Rome: Food and Agriculture Organization.

Green, E. P., and Short, F. T. (2003). *World atlas of seagrass*. Cambridge: UNEP WCMC

Gunther, M. (2012, August 6). *Shrimp farms' tainted legacy is target of certification drive. Environment 360*New Haven, CT: Yale University

Hinrichsen, D. (2013). *Coastal waters of the world: Trends, threats, and strategies*. Washington, DC: Island Press.

IISD. (2016). *State of sustainability initiatives review: Standards and the blue economy*. London: IISD.

Kiron, V., Sorenson, M., Huntley, M., Vasanth, G. K., Gong, Y., Dahle, D., and Palihawadana, A. M. (2016, May 17). Defatted biomass of the microalga, *Desmodesmus* sp. can replace fishmeal in the feeds for Atlantic salmon. *Frontiers in Marine Science*,.

Linneweber, V. (2013). *Mangrove ecosystems: Functions and management*. Berlin, Germany: Springer Science and Business Media.

Littlejohn, C., Curran, T. R., and Wratten, S. (2015). Farmland, food, and bioenergy crops need not compete for land. *Solutions*, 6(3), 34–48.

López-Hoffman, L., Varady, R. G., Flessa, K. W., and Balvanera, P. (2010, March). Ecosystem services across borders: A framework for transboundary conservation policy. *Frontiers in Ecology and the Environment*, 8(2), 84–91.

Millennium Ecosystem Assessment (MEA). (2005). *Ecosystems and human wellbeing*. Washington, DC: Island Press

NOAA. (2013). *Oil spill education resources*. Retrieved fromwww.education.noaa.gov/Ocean_and_Coasts/Oil_Spill.html

Odell, P. R. (2004). *Why carbon fuels will dominate the 21st century's global energy economy*. Brentwood, IL: Multi-Science Publishing

O'Neill, S., Elefant, C., and Agardy, T. (2015). Renewables: An ocean of energy. In H. Smith, J. L. Suarez de Vivero, and T. S. Agardy (Eds.), *Routledge handbook of ocean resources and management*. Abingdon: Routledge, pp. 283–295.

Pauly, D., and Zeller, D. (2016). Catch reconstructions reveal that global marine fisheries catches are higher than reported and declining. *Nature Communications, 7, 10244. doi:10.1038/ncomms10244.*

Potts, J., Wilkings, A., Lynch, M., and McFatridge, S. (2016). *The state of sustainability initiatives: Standards and the blue economy*. IISD/IIED/FAST.

Sale, P. F., Agardy, T., Ainsworth, C. H., Feist, B. E., Bell, J. D., Christie, P., et al. (2014). Transforming management of tropical coastal seas to cope with challenges of the 21st century. *Marine Pollution Bulletin*, 85, 8–23.

Smardon, R. (2009). *Sustaining the world's wetlands: Setting policy and resolving conflicts*. Berlin, Germany: Springer Science and Business Media.

Smith, H., and Thrupp, T. (2015). Oil and gas. Chapter 18 In H. D. Smith, J. L. Suarez de Vivero, and T. S. Agardy (Eds.), *Routledge handbook of ocean resources and management*. Abingdon: Routledge, pp. 269–282.

Tessler, Z. D., Vörösmarty, C. J., Grossberg, M., Gladkova, I., Aizenman, H., Syvitski, J. P. M., et al. (2015). Profiling risk and sustainability in coastal deltas of the world. *Science*, 349, 638–643.

Torcellini, P., Long, N., and Judkoff, R. (2003). *Consumptive water use for U. S. power production*. National Renewable Energy Lab. Retrieved from www.osti.gov/bridge

Ulm, M., Arns, A., Wahl, T., Meyers, S. D., Luther, M. E., and Jensen, J. (2016). The impact of barrier island loss on extreme events in the Tampa Bay. *Frontiers in Marine Science*, 3(56). Retrieved from http://dx.doi.org/10.3389/fmars.2016.00056

UNCTAD. (2016). *Sustainable fisheries: International trade, trade policy, and regulatory issues*. UNCTAD/WEB/DITC/TED/2015/5

UNCTAD and the Commonwealth Secretariat. (2015, September 25). *Sustainable fisheries: International trade, trade policy and regulatory issues*. Background Note.

UNEP-WCMC, WorldFish Centre, WRI, TNC (2010). Global distribution of coral reefs, compiled from multiple sources including the Millennium Coral Reef Mapping Project. Version 1.3. Includes contributions from IMaRS-USF and IRD (2005), IMaRS- USF (2005) and Spalding et al. (2001). Cambridge (UK): UNEP World Conservation Monitoring Centre. URL: http://data.unep-wcmc.org/datasets/1.

UNEP. (2014). The importance of mangroves to people: A call to action. In J. van Bochove, E. Sullivan, and T. Nakamura (Eds.), *United Nations environment programme world conservation monitoring centre*. Cambridge, 128 pp.

Verner, D. (2010). *Reducing poverty, protecting livelihoods, and building assets in a changing climate: Social implications of climate change in Latin America and the Caribbean*. Washington, DC: World Bank.

World Bank. (2009). *The sunken billions: The economic justification for fisheries reform*. Washington, DC: World Bank.

World Wildlife Fund. (2012). *The 2050 Criteria: Guide to responsible investment in agriculture, forest, and seafood commodities*. Washington, DC: WWF US

Ye, Y. (2015). Global fisheries. Chapter 14 in H. D. Smith, J. L. Suarez de Vivero, and T. S. Agardy (Eds.), *Routledge handbook of Ocean resources and management*. Abingdon: Routledge, pp. 215–231.

PART VI

Governing the resource nexus

Emerging responses

28

URBAN METABOLISM AND NEW URBAN GOVERNANCE

Corey Johnson

Introduction

> I think our governments will remain virtuous for many centuries; as long as they are chiefly agricultural; and this will be as long as there shall be vacant lands in any part of America. When they get piled upon one another in large cities, as in Europe, they will become corrupt as in Europe.
>
> – Thomas Jefferson in a letter to James Madison (Jefferson, 1955, 438)

> Towns are like electric transformers. They increase tension, accelerate the rhythm of exchange and constantly recharge human life.
>
> – Fernand Braudel, *The structures of everyday life* (1981, 479)

These quotes have little in common with each other except that they reflect humankind's long-standing love-hate affair with cities. Cities are the basis of civilization, yet they have always combined the characteristics of a blessing and a curse. They provide a focus for wealth creation, scientific and artistic creativity, and social interaction. But they also display profound income inequality, pollution, and crime. By virtue of their densities – of population and built environment – cities tend to magnify and accelerate many of the challenges of the nexus. Yet they are also sites of innovative solutions to the challenges we face. Urbanity at its very core represents intensification: of interactions, infrastructure, communication, and, of course of relevance to this volume, resource use. Massive worldwide urbanization commenced roughly with the Industrial Revolution in Europe and has continued, nearly unabated if in a geographically uneven manner, until the present. Human existence on the planet is now better than 50% urban. More people live in cities, variably defined, than do not, and as a result, urban lifestyles, which are in general more resource intensive than traditional rural lifestyles, are having an increasingly outsized imprint on the flows, transformations, and wastage of most natural resources globally. Much like the tension reflected in the two quotes above, there is debate in scholarly literature about precisely what cities mean for resource nexus issues. While urban lifestyles engender certain types of consumption that increase demand for energy, water, land, and minerals over time, once a society has reached a certain level of urbanization and development, living in cities is on balance *less* resource intensive than living outside the city largely owing to size of housing, distances traveled, and modes of transportation used. To put it more provocatively, one might ask if the key to sustainably managing resource consumption in the future lies in "piling" an ever larger share of the world's population in cities, as some have suggested (Owen, 2009).

This chapter provides an overview of the urban resource nexus and draws on a range of literatures from geography, political ecology, and planning. At the outset it is important to note that there is no well-defined area of urban nexus research, with the term connoting in various contexts the urban–rural interface, the water–energy–food nexus as manifested in urban areas, integrated approaches to urban metabolism (Moss, 2016), but also issues of social equity and provision of basic resources (Roy, 2016). This chapter acknowledges this lack of conceptual clarity in current nexus research, while adopting the basic understanding of the nexus referring to the interlinkages of demand, consumption, and waste cycles between different categories of resources, and puts this into dialogue with renewed interest in "urban metabolism" in scholarly literature.

Why cities?

According to the UN Environmental Program, cities occupy 2% of land surface, produce 50% of global waste, account for between 60 and 80% of global greenhouse gas emissions, consume 75% of natural resources, and produce 80% of global GDP (UNEP, 2013). There are 800,000 new urban residents added every week, and if this mass movement of humans continues, by 2050 the rate or global urbanization will be 70%, up from the current 50+%, in what is being described as the "second wave" of urbanization. Most of the 3 billion or so people added to the global population will be in Asian and African cities (ibid.). Accompanying this wave will be the creation of massive infrastructural works, housing, industrial, and commercial spaces. Public and private efforts to house new city dwellers have historically not been able to keep up with massive growth, especially in poorer parts of the world, which means that informal settlements (slums) will continue to play an important role in the urban landscape. While the percentage of people living in slums decreased over the last decade, the actual number of slum dwellers increased due to urban population growth.

The oft-quoted UN statistic of the percentage of humans living in cities is illuminating, but it raises the definitional question of what precisely "urban" means. There is no standard definition, but there are general characteristics that governments and scholars look to in order to devise their definitions, such as population size and density, economic functions, and characteristics of the built environment (dense settlement, presence of paved streets, water supply and sewerage systems, etc.). Urban geographers, planners, and sociologists have through time also resorted to a range of metaphorical representations to facilitate understanding urbanity. The MIT planners Julian Beinart and Kevin Lynch (Beinart, 2013) describe three normative modes of urban form: *supernatural* or *cosmic*, which considers the non-material drivers of people being drawn to cities (if Biblical heaven is a "city upon a hill," this metaphor sees cities as beacons representing some sort of spiritual release or realization to non-urban populations); *city as machine*, wherein urban spaces emerged as multiple systems of moving parts and "grids of expediency" to accommodate the waves of people moving pushed and drawn to cities by material realities of economic development, land enclosures, etc. (think of Fritz Lang's dystopic depiction of the city in his film *Metropolis*); and *city as organism*, which has origins in the Enlightenment unification of nature and reason (Gandy, 2002) but then re-emerged from the scientism of the 19th century. As a metaphorical organism, cities are born, grow, mature, and exist within a surrounding habitat.

The concept of urban metabolism (UM) emerged from this last category, and it has seen renewed interest over the last decade by scholars seeking to destabilize the neat categories of urban and rural that tend to conceal the multiple layers of interactions and mutual interconnections between cities and their immediate and more distant surroundings (Figure 28.1). Urban metabolism conceives of cities as akin to organisms that take in food and other resources and release waste into the environment.

Abel Wolman first developed the concept in the 1960s as a means of quantifying inputs and outputs of cities, and he identified the major metabolic challenges as water supply management, sewage disposal, and air pollution (Holmes and Pincetl, 2012; Wolman, 1965). Systems ecologists such as Odum further developed the concept during the 1970s by conducting numerous studies on the solar energy equivalents ("emergy") that were well received by the deep sustainability and ecology movements but found little traction in urban governance approaches (Kennedy et al., 2011). Meanwhile, more engineering-focused contributions have looked at material flows using conventional units for measurement for various resources.

Urban metabolism helps in analyzing the technical and socio-economic processes of a city (Kennedy et al., 2007). This can be quantified using material flow analysis (MFA), which seeks a system-based understanding of how urban areas transform materials using readily available data (but typically neglects the role of energy) (Holmes and Pincetl, 2012). The material flows in an urban area can be conceptualized as:

- Inputs: domestic extraction of resources, and imports of raw materials and products;
- Outputs: emissions and wastes, and exports of raw materials and products;
- Internal processes: intermediate and final consumption; and
- Addition to stock: share of the consumption that is accumulated in the system.

Life cycle assessment (LCA), a technique widely used in industrial ecology, is another technique for examining the cradle-to-grave use of resources in urban areas (Holmes and Pincetl, 2012).

UM has experienced a resurgence in recent years in the development community, by scholars in urban studies and urban political ecology, and in the context of increasing concern over climate change (Heynen et al., 2006). Some urban nexus studies that might benefit from an UM

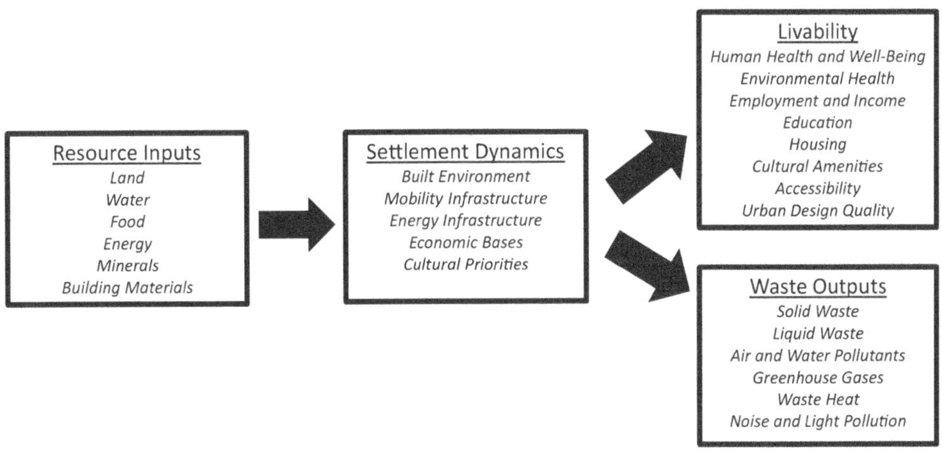

Figure 28.1 Extended metabolism model of human settlements

Source: adapted from Newman (1999)

approach include work on "patch fragmentation," heat island effects, reduced biodiversity, water supply and sewage, etc.

Given the wide-ranging work on UM spanning five decades, it is useful to put the urban nexus into conversation with this longer standing body of work that has similar conceptual and practical aims. The idea of the "urban nexus" approach was conceptualized at the Bonn 2011 conference on Water, Energy, and Food Security. It is aimed at describing the optimization of urban system and processes as a means to the opportunities for resource sustainability and meeting the basic human requirements. It is therefore important to consider a nexus approach for the cities that includes integrated planning of infrastructure and management of resources such as water and energy. The urban nexus is complicated by governance structures of urban areas that often have overlapping, highly fragmented jurisdictional boundaries. A major city such as Mumbai, India, must negotiate across multiple jurisdictional boundaries in order to secure enough freshwater to supply the population, and of course moving precious supplies of water across boundaries creates almost by definition political frictions, as any student of water-scarce regions throughout the world well knows. In effect, the metabolism of the city is in competition for many of the resources required to sustain contemporary urban lifestyles, while the waste and pollution produced by producing electricity, moving people through space, etc, produce effects felt well beyond the city's confines. The fragmented nature of urban local bodies and their management system are not able to respond to cities' complex, fast-growing, and interdependent systems. The "silo-thinking" and focusing on development of only one sector does not resolve the issue of urban problems.

One iteration of the urban nexus that has gained traction in certain circles comes from the German development community:

> The Urban NEXUS is an approach to the design of sustainable urban development solutions. The approach guides stakeholders to identify and pursue possible synergies between sectors, jurisdictions, and technical domains, so as to increase institutional performance, optimize resource management, and service quality. It counters traditional sectoral thinking, trade-offs, and divided responsibilities that often result in poorly coordinated investments, increased costs, and underutilized infrastructures and facilities. The ultimate goal of the Urban NEXUS approach is to accelerate access to services, and to increase service quality and the quality of life within our planetary boundaries.
>
> (GIZ and ICLEI, 2014, 6)

This approach places a great deal of emphasis on integration across scales, systems, etc. The report includes nearly 40 case studies of public and private projects in cities spanning the globe that exemplify this cross-sectoral frame of mind around questions of resource access, use, and disposal in urban environments.

The urban resource nexus can be useful in this context by considering not just how and in what quantities different resources are used *within* cities, but rather by destabilizing the supposed binary of urban–rural and thinking through the life cycles of different resources as they move through space, interact with one another, and are fixed in place at various moments in buildings, roads, and a host of other material objects that have come to define what we think of as a city.

The added value of urban metabolism is precisely how it can break down nature–society oppositions; we, humans, and the lives we lead in cities, the technologies we deploy, the structures we erect are all entirely products of the natural world, and in turn human activities impact the natural environment in myriad ways. Cities exist within nature's economy, and

Table 28.1 Urban problems

Problems arising from badly designed and managed cities:

- Land, air and water pollution
- Destruction of agricultural land, biodiversity and of historical heritage
- Waste of resources (land, water, energy and mineral)
- Great social and economic inequality
- Poor health and physical security, especially for marginalized groups
- Sectors of many cities become 'single use'(work/leisure), and the preserve of the rich for living
- Poor transportation, traffic congestion
- Unreliable, nonexistent utility services
- Lack of physical security – crime, breakdown in social order, violent/inhuman police and security services

Threats to cities:

- Inadequate water and energy supplies
- Flood and storm damage in low-lying coastal or riverine locations
- Also, in some places, earthquakes and landslides

Sources of these problems:

- Old cities designed with aesthetic priorities
- Urban sprawl – dormitory suburbs and towns, low density (in some countries)
- Energy inefficient buildings and factories
- Transport systems based on the road/motor vehicles
- Inadequate household, commercial and industrial waste management
- Inefficient and dirty sources of power generation and transport

Source: Andrews-Speed et al. (2015)

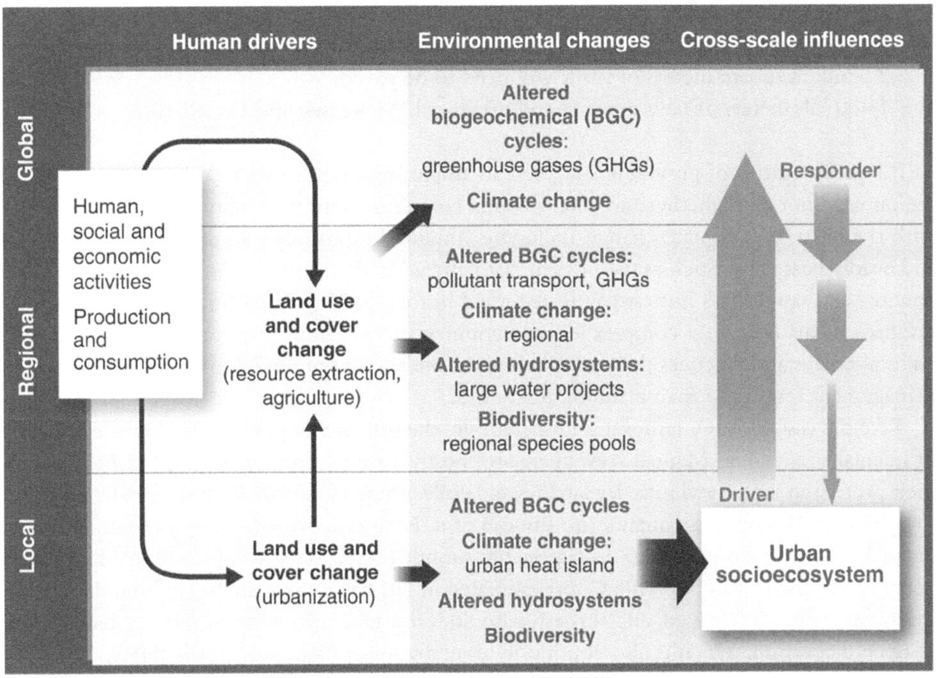

Figure 28.2 Cross-scale ecological cycles in cities

Source: Grimm et al. (2008)

they are part of a "hybridized and historically contingent interaction between social and bio-physical systems" (Gandy, 2004, 364), in what Gandy describes as a dialectical set of feedback loops. Therefore, it is important not to overly fetishize the distinction between urban and rural space, since cities are thoroughly integrated into their immediate as well as global hinterlands by means of infrastructure networks that make available water, energy, food, minerals, building materials, and of course bits of information in our increasingly digital world (Brenner, 2014; Gandy, 2002).

Cities are interesting from a resource nexus perspective because of the ways in which they magnify and concentrate nature's economy in particular locations, and it goes without saying that cities provide endlessly fascinating examples of the transformation of resource endowments to facilitate human endeavors (Figure 28.2). The examples are virtually limitless, but standing on the Bund in Shanghai with a view of the massive, asparagus-like proliferation of a new skyline in the span of just a few decades, or looking at the expanse of concrete runways and steel-and-glass terminal buildings at Dubai's soon-to-be busiest airport in the world, or wandering through an informal settlement in some South Asian megacity, with all the challenges in provisioning basic needs that entails, provide at least a taste of where to look for the resource nexus in the global urban context.

Case study of sanitation in New Delhi, India

By 2025, the UN projects that South Asia will be home to five of the 10 most-populous cities in the world (Delhi, Mumbai, Dhaka, Kolkata, and Karachi). By 2030, 590 million people will live in cities in India (twice the US population today), and that staggering growth will require:

- 700–900 million square meters of commercial and residential space needs to be built yearly (a new Chicago every year);
- 2.5 billion square meters of roads will need to be paved;
- 7,400 kilometers of new subways and urban rail (McKinsey and Co., 2010).

If basic questions of providing suitable and ample basic necessities to urban dwellers in this region were not enough, there are also growing concerns about the ability of megacities to deal with the impacts of climate change, including amplified monsoons and droughts, sea level rise, and deadly heat waves such as the ones in 2015 and 2016 that caused thousands of deaths in the country and saw India's hottest day in recorded history in May 2016. Strategies to govern the resource nexus issue in a complex, diverse country such as India also must take into account cultural norms and practices that are seemingly at odds with the models of urban infrastructural management found in urban planning textbooks.

Take the case of waste removal, and specifically the still widely practiced caste-based practice of manual scavenging. Manual scavengers are mostly women, mostly from the *dalit* caste, and their profession is to crawl into dry latrines and collect human excreta with their bare hands and carry it as head-load in a container to dispose of it. Each toilet of a higher caste family earns a manual scavenger around 20 to 30 rupees per month (US$0.30–0.40). In spite of the practice being illegal, the Indian Supreme Court estimated in 2014 that there are 9.6 million dry latrines still being manually cleaned, and this estimate does not take into account the manual cleaning of open defecation, which is also commonly done by lower caste individuals (HRW, 2014). Of the nearly 8,000 towns and cities in India, fewer than a thousand have a sewer system, and even where a sewer system exists it usually only partially covers the city. Moreover, only 160 cities and towns have a sewage treatment plant, meaning that the remaining communes with sewer systems release untreated waste into the environment, either into waterways or cesspools.

Manual scavenging is a filthy, often dangerous profession. It also comes with a host of accompanying discrimination, including exclusion from places of worship and, perversely, from communal water sources (HRW, 2014). Prime Minister Narendra Modi, prior to taking office in 2014, expressed a commitment to building more modern sanitation systems ("build toilets before you build temples"), but there have been only limited efforts to improve the situation for manual scavengers (ibid.). Manual scavenging is an issue of social justice, but there are also serious environmental implications of the sanitation crisis in South Asian cities.

Efforts by civil society actors have complemented government attempts to eliminate manual scavenging. The work of the Sulabh International Social Service Organization in New Delhi is an NGO that works towards improving the lives of manual scavengers, and it has also designed a toilet to be used in buildings that are not connected to sewage systems. The Sulabh toilet uses an on-site composting system in which collected waste is transformed into an odorless dry sludge and used as manure and biogas. The toilet requires 1–1.5 liters of water to flush, in contrast to the 12–14 liters required by conventional toilets. This alternate source of energy is used for cooking and electricity purposes. This approach provided cost-effective solutions, use of indigenous technologies while expanding sanitation infrastructure and aiming towards achieving the goal of 100% sanitation coverage in the country. Using the 'Sulabh Model' design, public toilet facilities have been constructed across the country and linked to around 200 biogas plants of 35 to 60 cubic meters capacity. Human excreta-based biogas technology remained unnoticed for long due to the fact that the available technology was not socially acceptable, as it required manual handling of human excreta, which contains a full spectrum of pathogens. The design developed by Sulabh does not require manual handling of human excreta and there is complete recycling and resource recovery from the wastes.

Creative solutions nexus issues in cities such as this one are ones that can be encouraged by appropriate governance and encouragement of local actors to solve intractable environmental and resource issues.

Case study of cement in China

Cement is the world's most popular building material, and the industry is the largest user of natural resources in the world, in particular minerals, water, and energy. Cement is composed largely of clinker, which is sintered limestone and clay that must be kilned and is therefore a major producer of GHG. There is a large body of literature mostly focusing on how cement production affects air quality and energy use, such as the carbon footprint of the industry and the dispersal of mercury, acidic gases, and particulate matter. There has been far less attention paid to the entire life cycle of the material, as well as what the concretization of cities globally, but especially in places such as China, means for water (runoff, infiltration, water pollution), land (soil sealing), and other resources.

China produces more than half of the world's cement, and between 2011 and 2013 the country used more cement than the US used during the entire 20th century (6.4 vs. 4.4 gigatons). Cement production accounts for 7% of total energy consumption in China, 15% of CO_2 emissions, 21% of particulate matter, 4% SO2 and 10% of NO_x; around 99% of energy for cement production comes from coal. While a recent paper in *Nature* suggests emissions from cement production in China are lower than earlier thought and demand for the material has recently slowed considerably, the new, lower estimates still make cement a major environmental factor locally and globally (Liu et al., 2015). Additionally, very little cement is recycled in China, which means that old material is landfilled.

The reason for the concrete boom in China is the urbanization boom of the last three decades. Most of the construction of infrastructure and buildings has been to accommodate the rural-to-urban migration that has occurred, constituting the largest scale mass movement of humans in history. Cement is valued for its versatility and strength, but when applied in such concentrated and mass quantities, there are numerous ripple effects. Paving over soils, "soil sealing," profoundly inhibits the natural role that soils play in absorbing heat, water, exchanging gases, hosting plant cover, and controlling erosion of fine particles (Scalenghe and Marsan, 2009). The horrendous air-quality issues faced by cities such as Beijing are a very visible manifestation of the types of issues urban China can expect to face, but there are other impacts as well.

One of the largest infrastructure projects in China in recent years has been the South-to-North Water Diversion, a $60 billion megaproject that when fully operational will divert 45 billion cubic meters of water from the relatively water-rich Yangtze River system to the relatively water-poor northern provinces. The per capita water availability in Beijing is now below 240 cubic meters, below the 300 cubic meters per capita standard set by UNESCO for the absolute minimum supply required to support a "modern and productive social life" (cited in Crow-Miller,

Table 28.2 Component affected, effects, timing and consequences of the sealing of soils

Symbols indicate: short-term (■); medium-term (■ ■); and long-term (■ ■ ■) effects

	Effect	*Time*	*Consequence*
Heat	Decreased radiation absorption	■	More reflective surfaces
		■■	Heat island (HUI)
Water	Less infiltration	■■	Reduced chemical reactivity
		■■■	Less filtering action
		■■	Cracking
		■	Loss of biomass
		■■■	Diminishes the natural recharge of aquifers
	More runoff	■	Increased water through adjacent areas
		■■	Increased ponding time
		■■	Probability of anaerobiosis
		■	Transfer of contaminants
		■■■	Increased risk of flash floods
	Barrier for perched water table	■	Increased risk of anaerobiosis
		■■	Release of contaminants
Gas	Reduced/interrupted exchanges	■■■	Risk of anaerobiosis
		■	Partial trapping
Biota	Loss of plant cover/biomass	■■	Reduced biodiversity
		■■■	Reduced carbon sink
	HUI	■■	Thermal specialization
Landscape	Increased wind erosion	■■	Increased air-borne particulate
	Increased water erosion	■■	Increased erosion of adjacent areas
	Uniformity	■	Reduced aesthetic appeal
		■	Reduced visual appearance
		■■	Reduced attractiveness

Source: Scalenghe and Marsan (2009)

2015). In short, Beijing is the most water-stressed megacity in a world increasingly full of water-stressed megacities. These concrete arteries, the largest interbasin transfer project in history, has been framed as essential to sustain the population of the North China Plain and surrounding cities, but by one estimate the project could have been avoided if a third of buildings in Beijing collected rainwater (Barnett et al., 2015). Thus, while cities such as Beijing lack abundant water resources to sustain the types of growth they are experiencing, the transfusion of distant water in this case results in a clear case of the urban nexus: massive quantities of materials transformed into concrete, steel, and other materials, to construct a series of canals that disrupt natural water cycles in order to sustain a city that if other governance decisions had been made likely could have avoided the need for the project in the first place. In and around the city itself, of course, the paving over of vast tracts of land further disrupts the replenishment of groundwater stocks while runoff carries the pollutants one would expect in any urban area, such as petroleum distillates, fertilizers, heavy metals, and other road-deposited sediment (Zhao et al., 2010).

As urbanization continues to expand its geographic and consumption footprints, understanding the material flows and resource requirements of cement will need to be a high priority. As the case of cement in China illustrates, it is not just the resource requirements required to produce cement that are relevant, but also the longer term shaping of consumption patterns that environments built with cement create. Canals have long-term implications for water flows and energy consumption, just as building roads and buildings involves a host of effects on resource consumption. Infrastructures being built today create significant path dependency for future resource requirements in cities; planners, other policymakers, and private sector actors must take this into account and steer and shape investments with a well-informed sense of these long-term implications (UNEP, 2013). In the case of cement, promising new technologies such as eco-cements that reduce the reliance on energy-intensive clinker and sequester CO_2 emissions from production into the actual product through synthetic carbonate aggregate are certainly part of a more sensible future for cement (Mathews, 2012). But reconsidering the design of cities – and questioning the paving over of vast swathes of land – must also be part of any urban governance strategy.

Governing the urban nexus and conclusion

Governance of the urban resource nexus involves questions of urban design and ecological management. The fragmented nature of urban local bodies and their management system are not able to respond to cities' complex, fast-growing, and interdependent systems. "Silo-thinking" and focusing on development of only one sector does not resolve the issue of urban problems. There is a long tradition of cities learning from each other, and the urban nexus presents an ideal set of challenges for cities to share experiences around how best to integrate nexus thinking into planning and design processes. The Urban International movement at the start of 20th century witnessed the exchange of town planning practices (Saunier, 2001), as did the Urban Internationale des Villes (UIV), International Garden Cities and Town Planning Association.

Urban nexus has received strong institutional backing from institutions such as ICLEI – Local Governments for Sustainability, the German Federal Ministry for Economic Cooperation and Development (BMZ), the German Society for International Cooperation (GIZ GmbH) (GIZ and ICLEI, 2014), and implicitly from a major recent report the German Advisory Council on Global Change (WBGU, 2016) on the "transformative power of cities." GIZ and ICLEI (2014, p6) define the urban nexus as:

> The Urban NEXUS is an approach to the design of sustainable urban development solutions. The approach guides stakeholders to identify and pursue possible synergies

> between sectors, jurisdictions, and technical domains, so as to increase institutional performance, optimize resource management, and service quality. It counters traditional sectoral thinking, trade-offs, and divided responsibilities that often result in poorly coordinated investments, increased costs, and underutilized infrastructures and facilities. The ultimate goal of the Urban NEXUS approach is to accelerate access to services, and to increase service quality and the quality of life within our planetary boundaries.

ICLEI has a toolbox for best practices related to the urban nexus and has initiated pilot projects seeking at to get key stakeholders to talk with one another across boundaries of institutional silos (www.iclei.org/urbannexus.html).

The German Advisory Council on Global Change, meanwhile, sees as its overarching goal for cities:

> moderate densification, resource conservation, and conversion to a circular economy, for cities that are energy-efficient and emissions-neutral in the long term, and marked by socially mixed districts in which the urban societies are substantively involved in urban development. In addition, more polycentric spatial-development concepts and decentralized settlement structures should be promoted in order to avoid spatial socio-economic disparities, and to counteract potential agglomeration disadvantages in growth regions.
>
> (WBGU, 2016, 414)

These ambitious goals for sustainable cities require buy-in from multiple stakeholders in urban, national, and global scales. However, it is in cities themselves, as key sites for economic and environmental transformation and with a commitment to shaping and planning a sustainable urban future where innovations are most critical. WBGU advocates a "social contract for urban transformation" that emphasizes the importance of polycentric responsibility in bringing about transformative action in the fields of decarbonization, mobility, urban form, climate change adaptation, and material flows. They have core recommendations for what self-assured, forward-thinking cities should have as their goals in these various areas in line with development and environmental goals established by, for example, the UN that are more squarely targeted at the national level. Similarly, UNEP and the International Resource Panel have developed a specific set of recommendations and best practices for cities to "decouple" their growth from ever-increasing resource consumption (UNEP, 2013). "Eco-urbanism" that refers to new developments following sustainable design principles is one important facet of this, and already figured in developments in various parts of the world, including the Lingang "Carbon Valley" development on the outskirts of Shanghai.

Cities in many parts of the world are already realizing their potential to effect positive change. In recent years, climate change has been an important motivator for urban platforms for information and best-practice sharing. The cities for climate change movement includes **C40**, now comprising 80-some megacities, 550 million people. It was started in 2005 as a G8 initiative by Ken Livingston, then mayor of London. Movements such as this reflect "growing influence of a range of non-state actors in shaping urban climate governance and an ever more complex political economy of climate change, woven between notions of carbon control, resource scarcity, resilience and security" (Bulkeley and Betsill, 2013). Municipal voluntarism such as practiced in the realm of climate change politics could serve as a useful model for the urban resource nexus, and conversely, already established transnational municipal networks could benefit from incorporating nexus-like thinking into their forums, analysis, etc.

One of the major challenges from a social justice perspective is how to avoid a situation in which transnational policy networks arrive at a consensus about "best practices" that are not in the interests of the poorest populations in cities (Clarke, 2012). For example, market-oriented water and sewerage reform and infrastructure investments designed to benefit private, Global North firms may seem in the best interest of expanding access to these key infrastructures, but evidence suggests this is not always the case (Goldman, 2007). Sustainable development goals that consider social equity and environmental health as mutually interrelated are therefore key to addressing the urban nexus moving forward.

References

Andrews-Speed, P., Bleischwitz, R., Boersma, T., Johnson, C., Kemp, G., and VanDeveer, S. D. (2015). *Want, waste or war? The global resource nexus and the struggle for land, energy, food, water and minerals*. London and New York: Routledge.

Barnett, J., Rogers, S., Webber, M., Finlayson, B., and Wang, M. (2015). Sustainability: Transfer project cannot meet China's water needs (Comment). *Nature*, 527(7578), 295–297.

Beinart, J. (2013). *Video lectures from open courseware course 'Theory of City Form'*. Retrieved June 22, 2016, from http://ocw.mit.edu/courses/architecture/4-241j-theory-of-city-form-spring-2013/this-course-at-mit/

Braudel, F. (1981). *The structures of everyday life: The limits of the possible* (1st U. S. ed.). New York: Harper & Row.

Brenner, N. (Ed.). (2014). *Implosions/explosions: Towards a study of planetary urbanization*. Berlin: Jovis.

Bulkeley, H., and Betsill, M. M. (2013). Revisiting the urban politics of climate change. *Environmental Politics*, 22(1), 136 151.

Clarke, N. (2012). Urban policy mobility, anti-politics, and histories of the transnational municipal movement. *Progress in Human Geography*, 36(1), 25–43.

Crow-Miller, B. (2015). Discourses of deflection: The politics of framing China's South-North water transfer project. *Water Alternatives*, 8(2), 173–191.

Gandy, M. (2002). *Concrete and clay: Reworking nature in New York City*. Cambridge, MA: The MIT Press

Gandy, M. (2004). Rethinking urban metabolism: Water, space and the modern city. *City*, 8(3), 363–379.

GIZ and ICLEI. (2014). *Operationalizing the Urban NEXUS: Towards resource-efficient and integrated cities in metropolitan regions*. Berlin, Germany: Deutsche Gesellschaft für Internationale Zusammenarbeit (GIZ), ICLEI (Local Governments for Sustainability).

Goldman, M. (2007). How "Water for All!" policy became hegemonic: The power of the World Bank and its transnational policy networks. *Geoforum*, 38(5), 786–800

Grimm, N. B., Faeth, S. H., Golubiewski, N. E., Redman, C. L., Wu, J., Bai, X., and Briggs, J. M. (2008). Global change and the ecology of cities. *Science*, 319(5864), 756–760.

Heynen, N. C., Kaika, M., and Swyngedouw, E. (2006). *In the nature of cities: Urban political ecology and the politics of urban metabolism*. London: Routledge.

Holmes, T., and Pincetl, S. (2012). *Urban metabolism literature review*. Los Angeles, CA: California Center for Sustainable Communities at UCLA.

HRW. (2014). *Cleaning human waste: "Manual Scavenging," caste, and discrimination in India*. New York: Human Rights Watch. Retrieved June 19, 2016, from www.hrw.org/sites/default/files/reports/india0814_ForUpload_0.pdf

Jefferson, T. (1955). Letter to James Madison, 20 December 1787. In J. P. Boyd (Ed.), *The papers of Thomas Jefferson*. Princeton, NJ: Princeton University Press.

Kennedy, C., Cuddihy, J., and Engel-Yan, J. (2007). The changing metabolism of cities. *Journal of Industrial Ecology*, 11(2), 43–59.

Kennedy, C., Pincetl, S., and Bunje, P. (2011). The study of urban metabolism and its applications to urban planning and design. *Environmental Pollution*, 159(8–9), 1965–1973

Liu, Z., Guan, D., Wei, W., Davis, S. J., Ciais, P., Bai, J., Peng, S., Zhang, Q., Hubacek, K., Marland, G., Andres, R. J., Crawford-Brown, D., Lin, J., Zhao, H., Hong, C., Boden, T. A., Feng, K., Peters, G. P., Xi, F., Liu, J., Li, Y., Zhao, Y., Zeng, N., and He, K. (2015). Reduced carbon emission estimates from fossil fuel combustion and cement production in China. *Nature*, 524(7565), 335–338

Mathews, J. (2012). *Eco-cement, the cheapest carbon sequestration on the planet*. Retrieved December 9, 2016, from https://theconversation.com/eco-cement-the-cheapest-carbon-sequestration-on-the-planet-10978

McKinsey & Co. (2010). *India's urban awakening: Building inclusive cities, sustaining economic growth*. New Delhi: McKinsey Global Institute.

Moss, T. (2016). *Unpacking and re-assembling the 'urban nexus': A socio-technical perspective on urban infrastructures*. Paper presented at Nexuses of the Urban Workshop, May 12, Brighton.

Newman, P. W. G. (1999). Sustainability and cities: Extending the metabolism model. *Landscape and Urban Planning*, 44(4), 219–226.

Owen, D. (2009). *Green metropolis: Why living smaller, living closer, and driving less are keys to sustainability*. New York: Riverhead Books.

Roy, D. (2016). *The urban nexuses of social injustice*. Paper presented at Nexuses of the Urban workshop, May 12–13, Brighton.

Saunier, P.-Y. (2001). Sketches from the Urban Internationale, 1910–50: Voluntary associations, international Institutions and US philanthropic foundations. *International Journal of Urban and Regional Research*, 25(2), 380–403.

Scalenghe, R., and Marsan, F. A. (2009). The anthropogenic sealing of soils in urban areas. *Landscape and Urban Planning*, 90(1–2), 1–10.

UNEP. (2013). *City-level decoupling: Urban resource flows and the governance of infrastructure transitions*, A Report of the Working Group on Cities of the International Resource Panel'. Nairobi: UNEP. Retrieved June 18, 2016, from www.unep.org/resourcepanel/portals/24102/pdfs/Cities-Full_Report.pdf

WBGU. (2016). *Humanity on the move: Unlocking the transformative power of cities*. Berlin: German Advisory Council on Global Change.

Wolman, A. (1965). The metabolism of cities. *Scientific American*, 213(3), 179–190.

Zhao, H., Li, X., Wang, X., and Tian, D. (2010). Grain size distribution of road-deposited sediment and its contribution to heavy metal pollution in urban runoff in Beijing, China. *Journal of Hazardous Materials*, 183(1–3), 203–210.

29
ECO-INNOVATION AND RESOURCE NEXUS CHALLENGES
Ambitions and evidence

Michal Miedziński, Will McDowall and Raimund Bleischwitz

Introduction

Resource nexus is a concept coined to describe complex interdependencies across the life cycle use of natural resources applied to deliver fundamental societal and economic functions, notably food, water and energy. The nexus approach becomes particularly relevant in situations of competing claims over natural resources for different applications. These situations can become acute and even lead to conflicts when access to resources is limited or when their extensive use is associated with negative socio-economic and environmental impacts. Preventing and resolving problems associated with resource nexus is, therefore, among the key challenges of sustainable development.

One of the major over-arching strategies for resolving resource-efficiency and nexus challenges relates to decoupling economic activities from resource consumption. If households and economic agents develop capabilities to use less resources to ensure well-being the struggle for access may decrease over time. It is here where, in addition to the WEF nexus, all abiotic resources become important as the vast majority of resources used are construction minerals, metals and fossil fuels. Going beyond the WEF nexus and including minerals (usually referred to as materials) has been a key within the debate about decoupling. Most discussions identify a difference between *absolute decoupling*, in which resource use or environmental damages decline while economic activity continues to expand, and *relative decoupling*, in which resource use continues to grow alongside expanding economic activity, but at a relatively slower rate than economic growth. In either case, it is widely agreed that innovation, in particular eco-innovation, is critical for achieving decoupling.

A reflection is needed, however, about what types of eco-innovation – and what 'eco-innovation mixes' – can be *both* feasible and effective in resolving resource nexus problems, and deliver on the promise of relative or even absolute decoupling. Nexus challenges are complex and multidimensional, and so they need innovation strategy that addresses both specific technical and technological shortcomings as well as the system-level deficiencies in institutional and regulatory settings. Importantly, at the heart of nexus often lie fundamental dilemma and social choices that require innovation going well beyond technological domain, and beg novel modes of governance and collective deliberation. In line with global challenges such as climate change, system innovations of transformative nature appear as the most desired strategy to tackle the most persistent nexus problems.

We will propose that the nexus approach, decoupling and eco-innovation can be mutually reinforcing and relevant in both developed and developing countries' context. This book contribution critically reviews the current literature and evidence on eco-innovation and decoupling, addresses issues of measurement, and gives an outlook on research and policy challenges of eco-innovation.

Decoupling and the promise of eco-innovation: evidence and debates

Policy discussions about the environment have often been informed by one of two conflicting perspectives: the view that that economic growth is fundamentally incompatible with the environmental sustainability of a finite planet; and the opposing belief that growth can be decoupled from resource use and environmental pressures, so that the economy can grow while environmental limits are respected. While the latter view has long been the dominant framework through which environmental policies are debated and pursued, the evidence on decoupling is mixed.

On several measures of resource consumption, a range of studies have presented evidence that decoupling has occurred. For CO_2, it is well known that – at a global scale – relative decoupling has occurred, but that absolute decoupling remains elusive. For flows of materials, food and water, the evidence is less well developed, but a growing number of studies in recent years have applied material flow analysis to examine this issue. These studies suggest a simple story: as economies have developed, they have, in relative terms, dematerialised – perhaps through shifts from goods to services, and certainly through higher levels of efficiency with which resources are used to produce both goods and services. The higher levels of decoupling seen in industrialised countries suggest the tantalising hypothesis – embraced by UNEP's International Resource Panel report on resource efficiency (UNEP, 2016a, 2016b) – that at high levels of per capita income, resource consumption can be increasingly decoupled from economic growth. Figure 29.1 shows evidence of apparent relative decoupling of resource consumption from GDP growth in OECD countries.

However, this picture is complicated by the difficulties of accurately assessing the full resource consequences of consumption in a national economy. The measures illustrated in the figure above makes use of domestic material consumption (DMC), which (as explained in Chapters 5 and 6) neglects the upstream resource consequences of imported goods. When these are included, as with indicators such as material footprint or raw material consumption, the strong decoupling of economic growth from material consumption in rich countries disappears (Wiedmann et al., 2015; UNEP, 2016a). The accounts of other countries, notably exporters of goods and energy, however, may look slightly better compared to a DMC-based view. It has been argued that this reveals a fundamental tension between economic growth and resource use, and that evidence for decoupling, may be a misconception or even a smokescreen to distract from the need for absolute limits on resource consumption (Akenji et al., 2016; Monbiot, 2015). The optimistic narrative that rising per capita incomes will, sooner or later, translate into absolute decoupling from resource use, looks less plausible when material footprint data is used. In particular, it seems clear that there is not necessarily a natural transition to a dematerialised growth path beyond a certain threshold of per capita income.

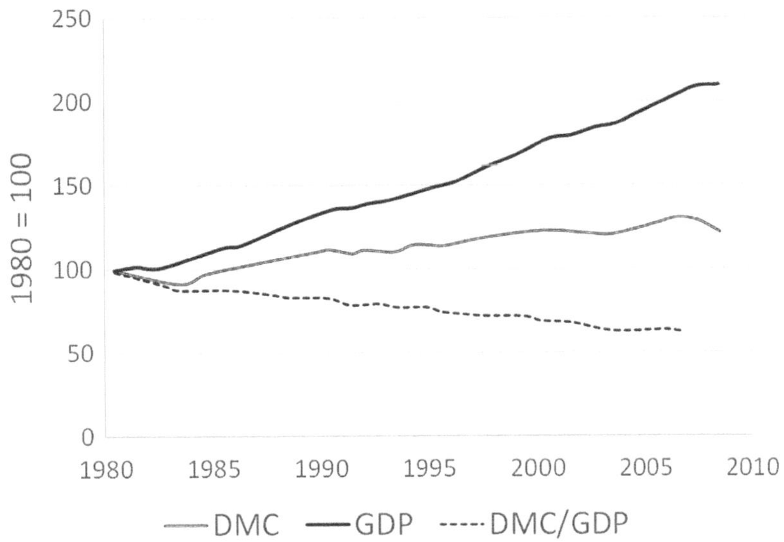

Figure 29.1 GDP and domestic material consumption in OECD countries

Source: adapted from OECD (2011)

Data at a global level suggest that there has been a relative decoupling of resource consumption from GDP during the latter half of the 20th century (Krausmann et al., 2009; UNEP, 2011), though this trend has been reversed since 2000 with the global material intensity of GDP rising since then (UNEP, 2016a). A relative decoupling may result from a combination of processes, including saturation (once the early industrialisation has taken place and physical infrastructures have been established), substitution (towards dematerialised consumption) and improved material productivity (improving the efficiency with which resources are used to produce goods and services) has been at play. These processes are underpinned by various forms of innovation and technological changes that in turn changes the global production and consumption patterns. A systemic outlook on these processes is key for a better understanding of how eco-innovation contributes to economic and environmental impacts, including the notion of decoupling.

What is eco-innovation?

Definitions and classifications of eco-innovation

The term 'eco-innovation' entered the public debate in the second half of the 1990s on the wave of the sustainable development debates preceding and following the Rio Earth Summit in 1992 (Fussler and James, 1996; Rennings, 1998; Rennings, 2000). The concept has climbed the policy agenda over the last decade, notably in Europe and OECD. The debate was further reinforced by the explicit recognition of the role of innovation in meeting the Sustainable Development Goals (UN, 2015). The concept of eco-innovation itself emerged from the earlier debates in the 1980s that started pointing at a possible alignment of economy and the environment ('ecological modernisation', 'green economy', 'cleaner production systems') to tackle the emerging environmental challenges and to move towards sustainable development. Claude Fussler and Peter

James (1996) called for "super innovation" or "innovation for sustainability" and linked it to the WBCSD's definition of eco-efficiency:

> the delivery of competitively priced goods and services that satisfy human needs and bring quality of life while progressively reducing ecological impacts and resource intensity, through the life cycle, to a level at least in line with the earth's estimated carrier capacity.

In academia, eco-innovation has been first underlined in ecological economics (Rennings, 1998; Kemp. 1999). Rennings (2000) articulated a notion of "double externality": private firms struggle to appropriate the full value of investments in eco-innovation, because others can copy good ideas (a problem known as positive externality or knowledge spillovers) and because damages to the environment are not captured in market prices. Firms adopting or selling environmentally benign goods and services need to compete with others who sell at prices that do not incorporate the costs associated with environmental degradation (the negative externality). Public intervention is, therefore, required to give a direction to innovation and counteract the negative environmental externalities. The definitions of innovation that underpin most public innovation policies are neutral in the sense that they do not determine the direction of change resulting from the implemented novelty. The OECD and European Commission, for example, define innovation as "the implementation of a new or significantly improved product (good or service), or process, a new marketing method, or a new organisational method in business practices, workplace organisation or external relations" (OECD, 2005). Rennings (2000) noted that this neutral approach is not sufficient to address major environmental challenges and enable the shift towards sustainable development.

All definitions of eco-innovation proposed since mid-1990s combine the notion of innovation *with* the orientation towards an expected sustainable outcome of the implemented change. There is a strong alignment between the definitions in relation to the core attributes of eco-innovation: eco-innovation needs *a degree of novelty* and must contribute to *environmental sustainability*. The Eco-Innovation Observatory (EIO, 2011, 2012a), for example, defined eco-innovation as

> the introduction of any new or significantly improved product (good or service), process, organisational change or marketing solution that reduces the use of natural resources (including materials, energy, water and land) and decreases the release of harmful substances across the whole life-cycle.

Most definitions include the whole life cycle perspective to the notion of environmental impacts, and implicitly for resource intensity too. The definitions differ in emphasising importance of the resource efficiency and various environmental impacts, and in their openness to consider other impacts, such as economic competitiveness or well-being. The definitions are generic and implicitly optimistic regarding environmental impacts by integrating them into one large category. They do not recognise links, including trade-offs, between changes in environmental performance in different dimensions of environmental impact. For example. reducing carbon emissions may not come hand in hand with gains in material efficiency.

In terms of types of innovation, most definitions are very open, including innovations ranging from new products to novel systems. Some definitions, however, choose to emphasise selected types of innovation. For example, UNIDO (2015) focuses on products and while UNEP (2014) emphasises business model innovation. Table 29.1 presents a classification of eco-innovation

Table 29.1 List of indicators in the Eco-Innovation Scoreboard

Name of indicator	*Source*	*Year*	*Updated for 2016 version*
1 Eco-innovation inputs			
1.1 Governments environmental and energy R&D appropriations and outlays (% of GDP)	EUROSTAT	2015	Yes
1.2 Total R&D personnel and researchers (% of total employment)	EUROSTAT	2015	Yes
1.3 Total value of green early stage investments (USD/capita)	Cleantech	2013-2016	Yes
2 Eco-innovation activities			
2.1 Enterprises that introduced an innovation with environmental benefits obtained within the enterprise (% of total firms)	EUROSTAT / CIS questionnaire	2014	Yes
2.2 Enterprises that introduced an innovation with environmental benefits obtained by the end user (% of total firms)	EUROSTAT / CIS questionnaire	2014	Yes
2.3 ISO 14001 registered organisations (per mln population)	ISO Survey of Certifications	2015	Yes
3 Eco-innovation outputs			
3.1 Eco-innovation related patents (per mln population)	Patstat	2013	Yes
3.2 Eco-innovation related academic publications (per mln population)	Scopus	2015	Yes
3.3 Eco-innovation related media coverage (per numbers of electronic media)	Meltwater	2016	Yes
4 Resource efficiency outcomes			
4.1 Material productivity (GDP/Domestic Material Consumption)	EUROSTAT	2014	Yes
4.2 Water productivity (GDP/total fresh water abstraction)	EUROSTAT	2010	Yes
4.3 Energy productivity (GDP/gross inland energy consumption)	EUROSTAT	2014	Yes
4.4 GHG emissions intensity (CO2e/GDP)	EEA	2014	Yes
5 Socio-economic outcomes			
5.1 Exports of products from eco-industries (% of total exports)	EUROSTAT	2015	Yes
5.2 Employment in eco-industries and circular economy (% of total employment across all companies)	Orbis	2015	Yes
5.3 Revenue in eco-industries and circular economy (% of total revenue across all companies)	Orbis	2015	Yes

Source: EIO (2017)

EIO (2017) EU Eco-Innovation Index: 2016 version. Technical note. Eco-Innovation Observatory.

based on the selected classifications in the literature (Kemp and Pearson, 2007; Huppes et al., 2008; Reid and Miedziński, 2008; EIO, 2012a, 2013).

The focus of the eco-innovation debate has, over the years, extended to encompass the notion of system innovation (Geels, 2005; EIO, 2013) and 'transformative innovation' (Kemp, 2011). The focus on a systemic perspective emerged because it became clear that attention to single products, services or firms is not sufficient (Arundel and Kemp. 2009; Huppes et al., 2008; EIO, 2012a, 2013). Some argue that the emphasis has to shift to value chains or a product-service systems or even to the functional socio-technical systems (EIO, 2013). The current approaches to system eco-innovation rest on several interconnected analytical perspectives. Three main approaches can be distinguished:

- Innovation system perspective;
- Transitions perspective; and
- Innovation management perspective.

The innovation system perspective emerged in the 1980s (Lundvall, 1985; Freeman, 1995) arguing for a system-level approach to understanding innovation processes that evolve in networks of various actors and institutions. The approach drew heavily from evolutionary and political economy as well as from the general systems theory. The system of innovation perspective provides a crucial perspective for understanding of determinants of knowledge creation and innovation diffusion. It is a conceptual framework underlying a majority of the current thinking and policies supporting innovation. Among criticisms of innovation system approach has been that it lacked the explanatory power about the process of structural change or socio-economic transformation. In 2000s a technological innovation system (TIS) perspective emerged that focused on specific dynamic functions that innovation systems should have to meet societal challenges (Hekkert et al., 2007; Bergek et al., 2008). The TIS attempts to bring the notion of change to the core of the concept.

Box 29.1 Classification of eco-innovation

Process innovation

- Pollution control and pollution treatment technologies:
  - Pollution control technologies;
  - Cleaning technologies that treat pollution released into the environment;
  - Noise and vibration control; and
  - Environmental monitoring technologies (meters).
- Waste prevention and waste management:
  - Waste management processes and equipment; and
  - Integration of secondary materials in the production.
- Resource-efficient processes:
  - Material-, energy- and water-efficient production processes; and
  - Renewable energy uses in manufacturing.

Product and service innovation

- Innovative products:
  - Novel competitive products with a reduced life cycle-wide environmental impact (e.g. urability, reparability, re-usability, recyclability, biodegradability); and
  - Novel eco-innovative materials with a reduced life cycle-wide environmental impact and capacity to substitute existing alternatives.
- Innovative services:
  - B2B: provision of services aimed at improving processes of clients (e.g. waste management, water and wastewater management, environmental consulting, eco-design); and
  - B2C: provision of services that are less resource intensive and reduce emissions (e.g. extended warranties and repair services).

Organisational innovation

- Environmental management and auditing systems (such as EMAS, ISO 14001).
- Introduction of Total Quality Management to the organisation.
- Introduction of Extended Producer's Responsibility solutions (CSR).

Marketing innovation

- Informing customer choices (e.g. independently verified eco-labels).
- Engagement in science-based educational campaigns raising awareness of importance of customer choices for sustainability.

Business model innovation (predominantly single-actor models)

- Radical changes in value proposition and product-service systems of companies (e.g. product sharing, functional sales).

Social innovation

- Eco-innovation that meets the social needs and contributes to civil society (e.g. non-for-profit green business models).

System innovation (multi-actor models)

- Radical changes in value proposition and product-service systems engaging a group of functionally connected organisations:
  - Multi-actor product-service systems (e.g. product sharing platforms and infrastructure);
  - Industrial and urban ecology; and
  - Integrated mobility systems.

Source: Kemp and Pearson (2007); Huppes et al. (2008); Reid and Miedziński (2008); EIO (2012a, 2013)

The transitions perspective emerged in the literature in early 2000s, initially driven mainly by the Dutch scholars (Rotmans et al., 2001; Geels, 2005) drawing from evolutionary economics and socio-technical systems perspective on technological and social change (Bijker et al., 1987; Bijker and Law, 1992). The transitions perspective provides a critical perspective on system shifts, pointing to both successful and failed *technological niches* from the historical perspective and the challenges of extending or upscaling those niches into larger transitions (regime changes, landscape alterations).

The last perspective was brought about more recently as a response to the challenge of *enabling* system-level transitions with a particular orientation. System eco-innovation in this context is a system change designed to deliver on both economic and environmental objectives. System innovation from this perspective is *an instrumental* system-level change designed to alter an entire value chain or a functional system (e.g. mobility, health, shelter). System innovation includes a range of functionally connected changes of both technological and non-technological nature (including policy framework) designed to contribute to both economic and environmental sustainability. This approach rests on the premise that design and process of system innovation can in fact *be managed*. This opens up questions on business models and modes of governance suitable for tackling complexities, uncertainties and risks inherent in innovation process.

Figure 29.2 positions system eco-innovation in the context of other types and levels of innovation. Importantly, the system innovation embraces changes of *both* incremental and radical

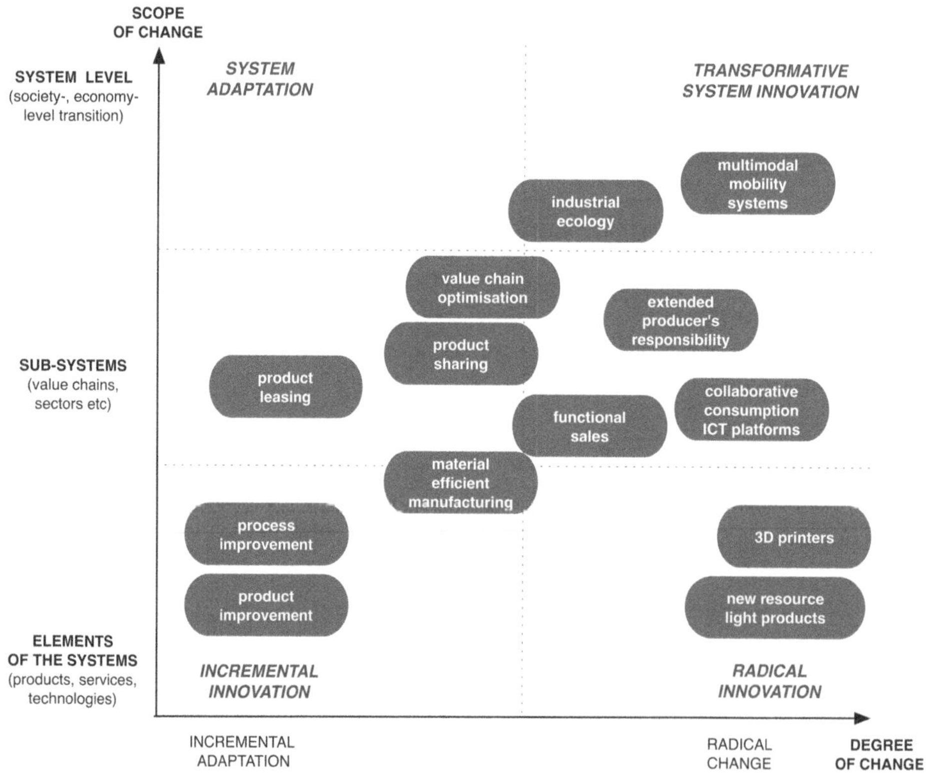

Figure 29.2 Mapping eco-innovation

Source: based on EIO (2013)

nature, and exploits functional connections between them. Their impact of system innovation depends on the strength of systemic synergies between its elements, and on its capacity to radically reconfigure or substitute existing systems. Industrial ecology model, for example, relies on the combination of organisational and process adaptations as well as infrastructural investments. These individual changes may not be radical on their own; however, their combination can be considered radical in specific contexts. This illustrates that the relevance of system innovation should be perceived from the perspective of its single component.

Eco-innovation and challenges of resource nexus and decoupling

To what extent is the concept of eco-innovation fit for addressing the challenges of resource nexus and decoupling? In principle, all kinds of eco-innovation have a potential to contribute to resolving the resource nexus challenges and contribute to relative or even absolute decoupling. Most definitions of eco-innovation, and much of the application of the concept in analysis and practice, have adopted generic and inclusive boundaries, considering its role for all key environmental pressures and impacts in a similar way to a five-node nexus on water–energy–food–materials–land as used in this book. On the other hand, the eco-innovation debate has not sufficiently addressed the issue of interdependences and possible synergies between the use of different resources, and often paid attention to selected resource flows or environmental pressures (notably materials, water and harmful emission). Land issues have not been at the centre of the debate.

The concept of eco-innovation catered for the resource nexus challenges needs broad analytical boundaries allowing to construct an explicit theory of change behind eco-innovation and its impact. Simply put, it needs to be explicit about the systemic nature of the nexus challenge and about the expected impact on the identified problem. First, the nexus challenges are systemic in nature and, therefore, require understanding of the system-level dynamic and innovations designed with a system perspective in mind. A micro-level perspective focused on specific eco-innovative technologies or products is necessary, but will not be sufficient to offer innovation activity a desired systemic direction. The complexity of systemic feedbacks, such as diffusion, substitutions or rebounds following the introduction of new products or services, implies that even disruptive technological eco-innovations may not translate into wider social or environmental benefits. A new product may trigger wider changes in the system, but their scale and impacts will depend on the structures and dynamics beyond a single product or value chain. The extent to which product or service eco-innovation can contribute to resolving nexus challenges cannot be confidently answered *a priori*. The explanatory power of micro-level definitions of eco-innovation is, therefore, limited to the system boundaries that the novelty directly affects (e.g. product footprint), but it remains speculative regarding its wider impacts on sustainability.

Nexus eco-innovation needs to be a system eco-innovation based on mutually reinforcing mix of changes to address systemic problems. The boundary of this 'eco-innovation mix' may encompass combinations of new technologies, products or services, as well as changes in surrounding innovation system, including policy and regulatory framework. Nexus eco-innovation strategy needs to internalise changes in the innovation system that usually reside outside realm of influence of any single actor. It will require new collaborations, often involving public–private partnerships, within *and across* sectors, value chains and supply chains, and may require adapting regulatory frameworks.

Nexus eco-innovation has to differentiate and compare key resource interlinkages, and likely environmental pressures and impacts associated with alternative eco-innovative solutions. Due to its broad boundaries, it needs to include both economic and social rationales behind the choices to support one or another innovation pathways. This requires a deep reflection, collective deliberation and different type of evidence on how trade-offs between different uses of different resources can be addressed. Together, the resource nexus and eco-innovation offer an opportunity to realise co-benefits and address a number of challenges more synergistically:

- Deliver the Sustainable Development Goals 2 (food), 6 (water), 7 (energy), 9 (infrastructure and industrialisation), 12 (sustainable consumption and production) in a more integrated manner;
- Develop business niches with local people towards eco-innovations with a potential to grow and become interconnected;
- Enable new alliances for collaborations with international companies seeking for community involvement and eco-innovation across borders with local benefits; and
- Engage with investors, large companies and international organisations that are under pressure to serve long-term goals with more short-term returns.

Analysing and measuring eco-innovation

Measuring eco-innovation is as much about direct measurements of innovation inputs, outputs and impacts, as it is about understanding what drives and hampers innovation process. There is a wealth of studies focusing on drivers and barriers to eco-innovation and innovation diffusion. The studies focused on drivers of environmental innovation emerged in the 1990s, notably in relation to the so-called Porter's hypothesis (Porter and van der Linde, 1995) suggesting an overall positive relationship between environmental regulation and company's innovation performance and productivity. The literature has since focused on various dimensions of the hypothesis and developed a rich body of evidence supporting, questioning and refining the original claim. This stream of literature has had dominant emphasis on regulatory drivers of innovation.

Innovation researchers looked at the issue of determinants in a more systemic way, incorporating many types of drivers and barriers to eco-innovation emerging in the innovations system (Horbach, 2016; Reid and Miedziński, 2008; EIO, 2012a). A systemic approach incorporates determinants that are both internal and external to the firm. The latter include framework conditions, such as regulatory framework, social capital, entrepreneurial culture and institutional setting. Importantly, the barriers and drivers to eco-innovation have to be analysed taking into account the capacity of innovator, the type of innovation as well as the specific context in which innovation emerges or diffuses. What is considered a driver by one company may be seen as a barrier by other players.

Micro-level approaches

There is considerable evidence linking specific eco-innovations to decoupling at the micro-level. The evidence relates to innovations or organisations (notably firms). Most abatement (or 'end-of-pipe') technologies result in relatively straightforward decoupling: they come at a small economic cost, but result in considerable improvements in environmental performance (examples range from catalytic converters to sewage treatment facilities and industrial flue gas scrubbers). Technologies that are more efficient can also usually be straightforwardly identified as improving environmental performance at the micro-level. Combined cycle gas turbines for example, which diffused widely throughout global power generation markets since the 1990s, generate power with considerably higher efficiencies than either steam turbines or gas turbines, resulting in decoupling between power generation output and associated environmental pressures. Process innovations that lower the amount of resources used compared to value added are a low risk strategy as all benefits remain within the company while payback times of investments are often less than one year. A McKinsey report estimates savings of US$2.9 trillion worldwide in 2030 if all such resource productivity potential are exploited, rising to US$3.7 trillion if carbon is priced and subsidies on water, energy and agriculture are removed (McKinsey Global Institute, 2011).

There are well-established methods allowing for measurement of environmental pressures and impacts on the micro-level, notably LCA and various approaches to measuring product and service footprints. Evidence generated with the use of these methods is now widely used by policymakers, businesses and researchers. The nexus perspective requires a combination of such methods in order to assess critical resource interlinkages (see Part II, especially Chapter 9), with a life cycle perspective on consumption in order to address the upstream implications of using resources. The micro-level approaches, however, have a limited explanatory power to explain and anticipate macro-level trends.

Macro-level approaches

Macro-level approaches often rely on composite indicators comprising country-level proxy measurements relevant for eco-innovation performance. Eco-Innovation Observatory (EIO, 2012a, 2017), for example, developed an index-based country-level indictors of eco-innovation performance (including innovation inputs, activities, outputs) and wider economic and resource-efficiency impacts on a country level. Table 29.1 overviews indicators in the EU eco-innovation scoreboard.

Macro-level analyses of eco-innovation effects is constructed by linking country-level innovation potential and performance with indicators of environmental pressures, such as GHG emissions or resource use, using material flow or footprint indicators. The macro-level analyses point to a correlation between eco-innovation performance, competitiveness and resource productivity (Reid and Miedziński, 2008; Flachenecker, 2017). The considerable growth in resource productivity and the reduction of emissions intensity, notably in the OECD countries, is therefore often linked with an improved innovation performance that captures market diffusion of low-carbon and resource-efficient processes and products. The analyses of decoupling cautiously point to several structural reasons driving resource productivity, such as technological change, de-industrialisation, shift to services in the OECD economies and economic crisis.

Despite the limitations, the research allows one to conclude that so far the eco-innovation activities have not had a disruptive effect on the production and consumption patterns towards an absolute decoupling. In order to better understand the mechanisms of decoupling, there is a need to embrace complexities of different types of innovation and their diffusion processes, and how they influence the use of various resources. In doing so, the nexus perspective suggests to go beyond materials and energy, and analyse consumption of water, food and land.

The explicit research focus on eco-innovation as a driver of decoupling on the macro-level has been limited, partly because of the methodological complexity and data availability challenges. Importantly, however, the macro-level analysis alone has a limited explanatory power to explain causal links between eco-innovation and macro-level decoupling, and needs to be combined with micro- and meso-level enquires.

Linking the micro and macro levels

The most troubled question concerns impact pathways linking the micro-level changes of products or innovative processes in companies to macro-scale decoupling. There are a number of methods that seek to understand the linkages between innovation and decoupling. Decomposition analysis provides a powerful set of tools for examining the driving forces behind changes in resource- or eco-efficiency over time. Two major variants exist: index and structural decomposition analysis. Index Decomposition Analysis (IDA) involves decomposing changes in an aggregate environmental indicator (such as water footprint or CO_2) for a set of sectors, and identifying changes that can be attributed to changes in overall economic activity, structural changes in the shares of activities undertaken by each sector, and the eco-efficiency with which each sector produces its output. This third component typically corresponds well with innovation, though it should be noted that changes in efficiency can also arise from price-mediated shifts in supply. A large number of IDA studies exist, applied to a range of resources and associated pollutants, and they have identified the contribution of eco-innovation to decoupling and found it to be substantial in many cases (e.g. Prawatky and McDowall, 2016). And policy organisations such as the European Environment Agency have used decomposition analysis to monitor the driving forces of change in key environmental indicators.

Structural Decomposition Analysis (SDA) involves decomposition of an environmentally extended input–output (EEIO) table, and identifies two different forms of technological change. In addition to changes in demand, SDA identifies the changes in environmental intensities of production of both intermediate and final outputs throughout the supply chain (i.e. direct eco-efficiency effects) as well as the changes in intermediate inputs for each sector (which can be understood as representing technological change understood as changes in the process of conversion of inputs into outputs by each sector). Structural decomposition has shown significant contributions of technological change to decoupling in a range of sectors and for a range of environmental pressures (e.g. Andreoni and Galmarini, 2012; Mattila, 2012).

Statistical and econometric methods are also able to provide evidence as to whether observed decoupling can be attributed to technological change or structural factors. Statistical analysis can suggest that particular innovations have driven environmental improvements, independently of other factors such as structural changes. However, such techniques cannot capture the economic feedbacks that could undermine the absolute environmental consequences of decoupling. As Van den Bergh (2013) have noted, where eco-efficiency also generates rebound effects, the magnitude of these can substantially undermine net environmental improvements. Even for ex-post analysis, it is not straightforward to identify the extent to which macro-level trends in key environmental indicators are being mediated by identifiable micro-level processes of innovation. Font Vivanco et al. (2014) recently developed an approach to incorporating various sources of rebound effects into efforts to scale up micro-level assessments of environmental savings. The method that they develop links micro-level LCA approaches to economic analysis using input–output models to examine the environmental implications of economy-wide indirect rebound effects. Methodological developments of this kind are increasingly necessary to understand the way in which macro-level environmental outcomes are influenced by micro-level innovation and diffusion processes.

Some scholars have sought to study the associations between eco-efficiency, innovation and environmental policy at a macro-level. Given the very large literature on the Porter Hypothesis (Cohen and Tubb, 2017; Ambec et al., 2013), which posits a causal relationship in which environmental regulation can lead to both environmental and economic benefits via innovation, it is perhaps surprising that the evidence here on environmental outcomes is relatively scarce:

most of the studies assessing Porter-type effects at the macro-level (and, for that matter, at the micro-level) either assume that the environmental improvements are being achieved, or proxy environmental innovation with input-indicators, such as 'green' patents, 'green' R&D (Kesidou and Demirel, 2012), or the stated adoption of environmentally beneficial innovations based on the European Community Innovation Survey (e.g. Horbach et al, 2012, Horbach 2016). In any case, a persistent challenge for this area is that the causal structures are not straightforward. Eco-innovation begets environmental regulation, just as environmental regulation drives eco-innovation. This makes formal causal inference challenging. There seems little doubt that, on a cross-country comparison, higher levels of resource productivity are also found in countries with higher levels of innovation, but untangling potentially confounding factors and making a causal claim is difficult (see e.g. Kemp and Pontoglio, 2011).

The resource nexus challenges for measuring eco-innovation

Considering complex nexus and decoupling dynamics, measuring impacts of eco-innovation requires further efforts on identifying, attributing and measuring economic and environmental impacts and verifying causal pathways explaining how the reduction of environmental pressures (e.g. material and energy use) translates into environmental impacts. The field of eco-innovation research, notably impact assessment, needs innovative transdisciplinary approaches that meaningfully bring together methodologies and metrics from social sciences, engineering and environmental sciences. Analysing resource nexus dimension requires methodological innovation, including improvements in:

- Identifying and verifying causal assumptions in alternative impact pathways of eco-innovation and critical resource interlinkages (sequential attributions and contribution analysis);
- Identifying and assessing relevance of internal and systemic determinants of innovation process (innovation system analysis and contribution analysis);
- Using both production and consumption-based (footprint) indicators on various analytical levels (countries, sectors, technology and product groups);
- Experimenting with viable ways of integrating diverse datasets and metrics from different disciplines and sources, as well as exploring innovative data sources, to investigate environmental, social and economic impacts of processes and products in the nexus perspective (e.g. focus on interlinkages between various uses of resources);
- Analysing and assessing ex-ante and ex-post meso- and macro-level economic and environmental impacts of eco-innovation on nexus (impact assessment); and
- Experimenting with practical ways of incorporating expert and stakeholder insights on risks and uncertainties involved in choosing particular eco-innovative solution.

Another challenge is to relate eco-innovation and resource nexus to the challenge of sustainability transitions. The nexus perspective provides a useful approach to assess resilience of emerging eco-innovation niches. By looking at life cycle-wide uses of resources, and their critical interlinkages, it could contribute to integrated innovation assessments that identify risks and systemic resilience of emerging technologies (e.g. risk of droughts or floods for renewable energy systems). The nexus can provide a perspective to assess value creation potential of innovation that optimises access to key resources with minimal footprints, as well as creates possible synergies and co-benefits of connected innovations (e.g. solar PV supporting desalination in water-stressed areas). The nexus perspective can also help to design viable governance of these solutions by suggesting new alliances of actors.

Aligning eco-innovation with the resource nexus offers a reframed narrative that incorporates multiple perspectives including the socio-economic and environmental dimensions, as well as incommensurable aspects such as value systems, ethics and social justice. The latter are key in framing problems involving conflicting agendas and social choices (e.g. short-term choices between human health versus environmentally preferable options). Therefore, the conceptualisation of nexus eco-innovation requires a systemic approach from the outset and, given their nature, may require innovative participatory research approaches. Resource nexus challenges need to be considered in a close connection with their local context and *at the same time* have to be regarded taking into account planetary boundaries (see Chapter 3). Nexus eco-innovation needs to include a multi-level assessment to identifying and assessing possible impact pathways.

Decoupling and resource nexus as an innovation policy challenge

Science, technology and innovation (STI) policy has been called on to provide vital contributions to resolving societal and environmental challenges (Rennings, 2000; Reid and Miedziński, 2008; Kemp, 2011). There is a strong rationale for public intervention in the case of decoupling and the resource nexus challenges: they are often both created and exacerbated by market and systemic failures. Bleischwitz (2012) has developed the following guidelines for a policy mix: market order by getting the framework conditions right; provide data and information; facilitate learning processes; promote market development towards regime changes; provide a long-term orientation. This section outlines how resource nexus and decoupling can be addressed in by STI policy in a systemic manner.

Improving evidence base and policy learning

The resource nexus requires a robust evidence based on interdisciplinary knowledge and diverse expertise actively brought in along the whole policy process. The STI policies tackling complex challenges of nexus and decoupling cannot rely only on 'research as usual' approaches, and need to encourage collaborative transdisciplinary research engaging various stakeholders in generating and validating context-specific evidence for policy intervention. The dynamic nature of challenge requires that evidence base needs to be continuously adapted based on an ongoing policy learning process that interprets and prioritises validated evidence in a transparent way. The complex and emerging nature of nexus requires that the evidence base provides data on critical resource interlinkages as well as other key data on footprints, and incorporates the long-term prospective view by using both formal modelling approaches as well as inclusive foresight methods. There is a need to integrate the notion of risk and uncertainty in the analysis, while integrating the pre-cautionary principle at the heart of the process.

Shared understanding of the nexus challenges and key projects

Nexus challenges involve many stakeholders who bring with them diverse perceptions, understandings and interests that explain how they frame the problem. Various stakeholders will perceive the same problem through different lenses focusing on aspects of technology and infrastructure, environment, economic and business models, policy and regulations, as well as culture and values. All these perspectives are valuable for the STI policy in the context of the nexus and decoupling. Policy process should directly include the process in which these various frames are transparently presented, explained and supported with the use of evidence. It should help to facilitate key innovation projects on such nexus eco-innovations by comparing successful niches and scaling them up regionally and internationally. This process enriches the evidence base by revealing the motivations and positions of stakeholders as well as prepares the foundations for the vision and strategy on how STI can tackle the nexus issues.

Leadership, participation and shared vision of the future

The nexus challenges require a long-term perspective as well as a strong buy-in from both stakeholders affected by the challenge as well as those who may be instrumental in solving it. The latter may not be directly affected by the challenges, but they may consider their contribution beneficial, if evidence on short-term benefits and co-benefits can be established. The STI policy need to provide a platform for developing future scenarios of dealing with nexus and a shared vision of the future. The main value added of the process is creating a shared vision based on various perspectives, and combining potentially conflicting interests into a 'future project' by creating short- and long-term incentives for key actors.

Short- and long-term scenarios and transition roadmaps

The vision needs to be translated into a more concrete strategy on how to both kick-start the process, identify potential asset losses and sunk investments, and follow up on it in the medium to long term. Planning will need to explicitly recognise the interdependencies between uses of various resources and seek flexible solutions to overcome the current and avoid future lock-ins in the resource- and capital-intensive functional systems (e.g. energy and water supply) that contribute to unsustainable use patterns. Scenarios and roadmaps are examples of strategic tools well suited for dealing with complex challenges of winners and losers requiring both short- and long-term outlooks. They provide a practical framework for developing shared understanding of alternative innovation pathways (scenarios) and contribute a foundation for implementation of complex multi-actor innovation projects (roadmaps). It will be essential to bring new valuation perspectives in such scenarios and roadmaps in order to identify risk avoidance and gains of eradicating poverty and enabling access to key resources for the world's poor.

Systemic policy for system innovation

Policymakers are an important stakeholder in making system innovation possible. Nexus challenges require an innovative, coordinated and possibly coherent policy response based on a policy mix that both directly supports eco-innovation and ensures that wider regulatory and policy frameworks favour the sustainability transition. The direct support can be delivered by dedicated market and financial instruments, allowing a degree of risk in the case of particularly promising investment, whereas the enabling environment can be based on a strong regulatory framework as well as removing regulatory barriers, subsidies and other support that favours actors or technologies that contradict the direction agreed in the vision.

Governance for resource nexus

The resource nexus requires revisiting existing governance structures and mechanisms. This concerns all dimensions of STI governance, including leadership, strategic deliberation, participation, responsibility and accountability, implementation, and monitoring and evaluation. Resolving nexus challenges require envisaging alternative forms of governance that exist in parallel to, or even substitute, established institutions and organisations. On the one hand, the challenge is to design and implement viable governance approaches that make optimal use of existing capacities and power structures by stimulating collective action on the ground. On the other hand, the challenge may be to innovate and create new governance structures when existing settings do not suffice or are mobilised against the desired change. The latter suggests that design of policy and governance needs to take into account the dimension of power and leadership as well as organisational capacity, technical competences and budgets. In the global perspective, the governance catered for the nexus challenges is likely to rely on the regionalised polycentric coordination of collective action with a degree of the global coordination. This regionalised perspective is a relevant complement to other planetary governance approaches, such as 'earth system governance' (Biermann et al., 2012).

References

Akenji, L., Bengtsson, M., Bleischwitz, R., Tukker, A., and Schandl, H. (2016). Ossified materialism: Introduction to the special volume on absolute reductions in materials throughput and emissions. *Journal of Cleaner Production*, 132, 1–12.

Ambec, S., M. A. Cohen, S. Elgie and P. Lanoie. (2013). The Porter hypothesis at 20: can environmental regulation enhance innovation and competitiveness? *Review of environmental economics and policy*, 7(1), 2–22.

Andreoni, V., and Galmarini, S. (2012). Decoupling economic growth from carbon dioxide emissions: A decomposition analysis of Italian energy consumption. *Energy*, 44(1), 682–691.

Arundel, A., and Kemp, R. (2009). *Measuring eco-innovation*. MEI project.

Bergek, A., et al. (2008). Analyzing the functional dynamics of technological innovation systems: A scheme of analysis. *Research Policy*, 37(3), 407–429.

Biermann et al. (2012). Navigating the anthroposcene: Improving earth systems governance. *Science*, 335, 1306–1307.

Bijker, W. E., Hughes, T. P., and Pinch, T. J. (1987). *The social construction of technological systems: New directions in the sociology and history of technology*. Cambridge, MA: MIT Press, x, 405 p.

Bijker, W. E., and Law, J. (1992). *Shaping technology/building society: Studies in sociotechnical change*. Cambridge, MA: Inside Technology, MIT Press, vii, 341 p.

Bleischwitz, R. (2012). Towards a resource policy – unleashing productivity dynamics and balancing international distortions. *Mineral Economics*, 24, 135–144.

Cohen, M.A. and Tubb, A. (2017). The Impact of Environmental Regulation on Firm and Country Competitiveness: A Meta-Analysis of the Porter Hypothesis (August 14, 2017). *Vanderbilt Owen Graduate School of Management Research Paper No. 2692919*. Available at SSRN: https://ssrn.com/abstract=2692919.

EIO. (2011). *The eco-innovation challenge: Pathways to a resource-efficient Europe*. Eco-Innovation Observatory.

EIO. (2012a). *Methodological report*. Eco-Innovation Observatory.

EIO. (2012b). *Closing the eco-innovation gap: An economic opportunity for business*. Eco-Innovation Observatory

EIO. (2013). *Europe in transition: Paving the way to a green economy through eco-innovation*. Eco-Innovation Observatory.

EIO (2017) *EU Eco-Innovation Index: 2016 version*. Technical note. Eco-Innovation Observatory.

Flachenecker, F. (2017). The causal impact of material productivity on macroeconomic competitiveness in the European Union. *Environmental Economics and Policy Studies*. doi:10.1007/s10018-016-0180-3. https://link.springer.com/article/10.1007/s10018-016-0180-3#citeas

Font Vivanco, D., Freire-González, J., Kemp, R., and Van Der Voet, E. (2014). The remarkable environmental rebound effect of electric cars: A microeconomic approach. *Environmental Science & Technology*, 48(20), 12063–12072.

Freeman, C. (1995). The 'National System of Innovation' in historical perspective. *Cambridge Journal of Economy*, 19.

Fussler, C., and James, P. (1996). *Driving eco-innovation: A breakthrough discipline for innovation and sustainability*. London and Washington, DC: Pitman Pub., xx, 364.

Geels, F. (2005). Processes and patterns in transitions and system innovations: Refining the co-evolutionary multi-level perspective. *Technology Forecasting and Social Change*, 72(6), 681–696.

Hekkert, M.P., R. A. A. Suurs, S. O. Negro, S. Kuhlmann and R. E. H. M. Smits (2007). Functions of innovation systems: A new approach for analysing technological change. *Technological Forecasting and Social Change*, 74(4), 413–432.

Horbach, J. (2016). Empirical determinants of eco-innovation in European countries using the CIS. *Environmental Innovation and Societal Transitions*, 19, 1–14.

Horbach, J., Rammer, C., and Rennings, K. (2012). Determinants of eco-innovations by type of environmental impact – the role of regulatory push/pull, technology push and market pull. *Ecological Economics*, 78, 112–122.

Huppes, G., et al. (2008). *Measuring eco-innovation: Framework and typology of indicators based on causal chains*. Eco-Drive Project.

Kemp, R. (1999). *Evolutionary theorising on technological change and sustainable development*, OCFEB research memorandum, no.9912; Rotterdam: Ocfeb

Kemp, R. (2011). Ten themes for eco-innovation policies in Europe. *SAPIENS*, 4(2).

Kemp, R., and Pearson, P. (2007). *Final report MEI project about measuring eco-innovation*. MEI-Measuring eco-innovation.

Kemp, R., and Pontoglio, S. (2011). The innovation effects of environmental policy instruments – A typical case of the blind men and the elephant? *Ecological Economics*, 72, 28–36.

Kesidou, E., and Demirel, P. (2012). On the drivers of eco-innovations: Empirical evidence from the UK. *Research Policy*, 41(5), 862–870.

Krausmann, F., Gingrich, S., Eisenmenger, N., Erb, K. H., Haberl, H., and Fischer-Kowalski, M. (2009). Growth in global materials use, GDP and population during the 20th century. *Ecological Economics*, 68, 2696–2705.

Lundvall, B-Å. (1985). *Product innovation and user-producer interaction*. Aalborg: Aalborg Universitetsforlag.

Mattila, T. (2012). Any sustainable decoupling in the Finnish economy? A comparison of the pathways and sensitivities of GDP and ecological footprint 2002–2005. *Ecological Indicators*, 16, 128–134.

McKinsey Global Institute. (2011). *Resource revolution: Meeting the world's energy, materials, food, and water needs*. November 2011. McKinsey & Company.

Monbiot, G. (2015, November 25) False promise. *The Guardian*. Retrieved from www.monbiot.com/2015/11/24/false-promise/

OECD. (2005). *Oslo manual-guidelines for collecting and interpreting innovation data*. Luxembourg: Eurostat.

Porter, A. L., and van der Linde, C. (1995). Toward a new conception of the environment-competitiveness relationship. *The Journal of Economic Perspectives*, 9(4), 97–118.

Prawatky, L., and McDowall, W. (2016). *SO2 emissions in the EU28 coal power sector-an LMDI decomposition*. Paper presented at the International Association for Energy Economics (IAEE), Bergen, Norway.

Reid, A., and Miedziński, M. (2008). *Eco-innovation*. Final Report for Sectoral Innovation Watch, Report for the European Commission.

Rennings, K. (1998). *Towards a theory and policy of eco-innovation – neoclassical and (co-)evolutionary perspectives*. ZEW Discussion Paper 98–24.

Rennings, K. (2000). Redefining innovation – eco-innovation research and the contribution from ecological economics. *Ecological Economics*, 32(2), 319–332.

Rotmans, J., Kemp, R., and van Asselt, M. (2001). More evolution than revolution: Transition management in public policy. *Foresight*, 3(1).

Steward, F. (2008). *Breaking the boundaries: Transformative innovation for the global good*. London: NESTA.

UN. (2015). *Transforming our world: The 2030 agenda for sustainable development*, A/RES/70/1.

UNEP. (2011). Decoupling natural resource use and environmental impacts from economic growth, a report of the working group on decoupling to the international resource panel (M. Fischer-Kowalski, M. Swilling, E. U. von Weizsäcker, Y. Ren, Y. Moriguchi, W. Crane, F. Krausmann, N. Eisenmenger, S. Giljum, P. Hennicke, P. Romero Lankao, A. Siriban Manalang, S. Sewerin). Paris: United Nations Environment Programme.

UNEP. (2014). *The business case for eco-innovation*. Paris: United Nations Environment Programme.

UNEP. (2016a). Global material flows and resource productivity: An assessment study of the UNEP international resource panel (H. Schandl, M. Fischer-Kowalski, et al.). Paris; United Nations Environment Programme.

UNEP. (2016b). *Resource efficiency: Potential and economic implications. A report of the International Resource Panel* (P. Ekins, N. Hughes, et al.).

UNIDO. (2015). Industrial development report 2016: *The role of technology and innovation in inclusive and sustainable industrial development*.Vienna: United Nations Industrial Development Organization.
van den Bergh, J. C. J. M. (2013). Environmental and climate innovation: Limitations, policies and prices. *Technological Forecasting and Social Change*, 80(1), 11–23.
Wiedmann T. O., Schandl H., Lenzen M., Moran D., Suh S., West J, Kanemoto K. (2015). The material footprint of nations. *Proceedings of the National Academy of Sciences*, 112(20), 6271–6276.

30

GREEN CHEMISTRY

Opportunities, waste and food supply chains

Avtar Matharu, Eduardo Melo and Joseph A. Houghton

Introduction

Human activity, at all stages and levels, relies on the use of natural resources as feedstock for technology and development. This gives rise to a highly symbiotic phenomenon involving the use of these common goods, which is often described as the three-point *water–food–energy nexus* (Garcia and You, 2016). However, a more integrative definition for this phenomenon is suggested, i.e., the five-node *water–energy–food–land–minerals resource nexus*, which is discussed in Chapter 1. For simplicity, herein it is referred to as the resource nexus (RN). Although this concept is crucial for maintaining the demands and lifestyle of our society, the existing RN is unsustainable due to several interlinked global problems: increasing population, limited availability of food, natural resource depletion, increasing pollution and adverse climate change, social inequality, etc. Global population is expected to increase 40% by 2050, yet today, around 1 billion people struggle to have their daily sustenance and more than 1.2 billion still have no access to clean drinking water (Garcia and You, 2016). Ironically, 1.3 billion tonnes of food is wasted every year (FAO, 2011a). With an increasing population, global food production will have to increase by at least 60% (OECD, 2013) while natural resources will become even more scarce. These facts illustrate the urgency in developing a well-distributed and sustainable RN management which allows the next generation to access water, energy and food accordingly to specific local needs.

A proximal quantitative distribution of specific resources (water, energy and food) within the nexus and supply network is illustrated in Figure 30.1. As per this diagram, our global food supply chain is highly water and energy intensive, consuming 30% of all produced energy and 70% of all distributed water. From the globally produced food directed for human and animal consumption, at least one-third (1.3 billion tonnes) is lost or wasted every year in the several stages of the food supply chain (production, transportation, processing, distribution and consumption) corresponding to a carbon footprint of 3.3 billion tonnes of CO_2 equivalent. The combination of food production and consumption by humans has been estimated to account for 20–30% of anthropogenic greenhouse gas (GHG) emissions, and food was found to account for 48–70% of a household's impact on land and water consumption (Golden and Handfield, 2014). Food waste represents an economic loss of $750 billion (excluding fish and seafood) (FAO 2011a, 2013). In the UK alone, avoidable food waste (60% of total food waste) accounts for an annual economic loss of £11.8 billion (Bond et al., 2013).

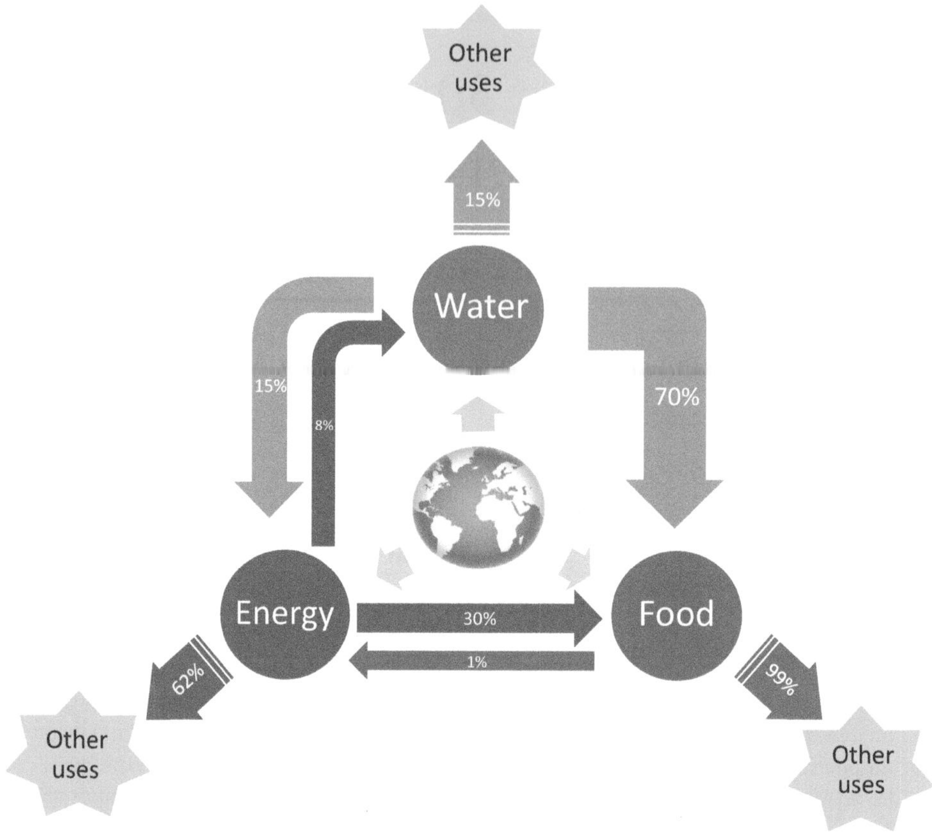

Figure 30.1 Distribution of water, energy and food into the global supply system and its nexus
Source: adapted from Martin and Danielsson (2016)

While the traditional 3-point nexus is a good method for predicting the effect of actions within the increasingly complex area of water, food and energy, there is scope for it to be improved upon with the inclusion of minerals and land as discussed in Chapter 1. This change from a 3-point to a 5-point nexus allows for a much more complete and accurate depiction of the total effect actions taken on one part of the nexus will have on the whole nexus. Within the scope of food management and waste, mineral and land use are intrinsically important. The land used, not only in the growing of food crops, but also devoted to the processing and storage/treatment of the waste produced, influences the nexus as a whole. Plants uptake valuable minerals from the soil they are grown in, so accounting for these minerals throughout the supply chain is of paramount important to the continuation of good crop yield and quality from that plot of land. Commonly the inedible parts of the plant are left or tilled back in the soil they were grown in to allow for re-enrichment of the field with the minerals the plants absorbed for their growth.

Food waste can derive from all stages of the food supply chain. Although the contributions to food waste from developed and developing nations are very similar (630 and 670 million tonnes, respectively), their sources are very different. In developing countries food waste is mainly due to poor storage post-harvest. Moreover, economic and political issues and perfect environmental conditions (warm and humid) for food spoilage contributes to the loss of edible food (FAO, 2013). On the other hand, in developed countries most of the wasted food comes from the post-consumer stage (considered avoidable and suitable for consumption), which reflects the consumerism culture present in these societies, along with the over restrictive food quality control and miscommunication between producers, retailers and consumers (FAO, 2013).

Where possible, policies and strategies for prevention and reduction of avoidable food waste should be encouraged coupled with redistribution of surplus food (FAO, 2013; Vandamme, Anthonis, and Dobbelaere, 2011). However, where unavoidable food waste is present as a result of primary and secondary processing (from farm to fork), then it should be considered as a rich source of biobased chemicals and materials in addition to bioenergy (composting, anaerobic digestion (AD), biomass burning and biofuels).

Utilisation of green chemistry practices and principles (also known as the 12 principles as listed below) as developed by Anastas and Warner (1998) coupled with the latest UN Sustainable Development Goals (United Nations, 2015) set in 2015 to protect the people and planet offers valorisation (value added) opportunities for unavoidable food supply waste other than landfill and/or composting/AD while sustaining the water–energy–food nexus (Figure 30.2). Green

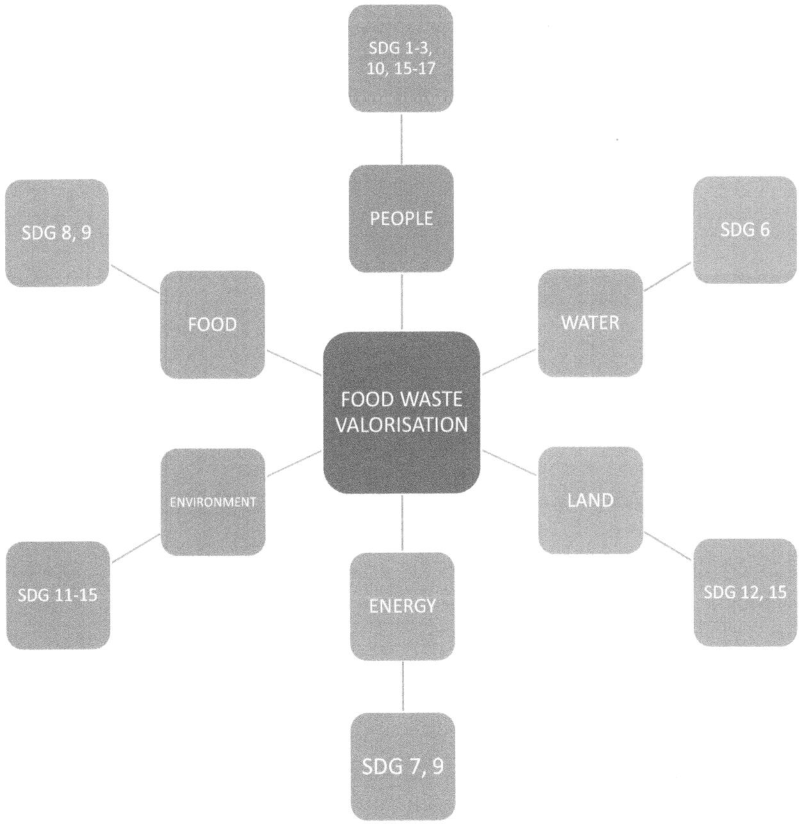

Figure 30.2 Linkage and contribution of food waste valorisation to the expanded nexus and Sustainable Development Goals (SDGs)

chemistry is an attitude that affords policy changes to improve chemical products and processes for the purpose of reducing their damage on human health as well as on the environment.

1 *Waste prevention*
2 *Atom economy*
3 *Less hazardous chemical synthesis*
4 *Designing safer chemicals*
5 *Safer solvent and auxiliaries*
6 *Design for energy efficiency*
7 *Use of renewable feedstocks*
8 *Reduce Derivatives*
9 *Catalysis*
10 *Design for degradation*
11 *Real-time pollution prevention*
12 *Safer chemistry for accident prevention*

According to principle 7, using unavoidable food waste as a renewable feedstock helps our society to move away from an oil-based to a bio-based economy, reducing environmental burden and turning what was once overlooked waste into an economically valuable renewable resource. Obviously, the complete success of a bioeconomy-driven chemical industry will also depend on the application of the other green chemistry principles.

Unavoidable food waste is widely distributed around the world in very high volumes (Figure 30.3), which gives it potential for exploitation at an industrial scale. In developed nations like in Europe and North America, the easier access to cutting-edge green technologies and high volumes of food waste (280–300 kg per capita yearly) (FAO, 2011a) opens up new opportunities for food waste valorisation. For instance, in the UK, where 15 million tonnes of avoidable food

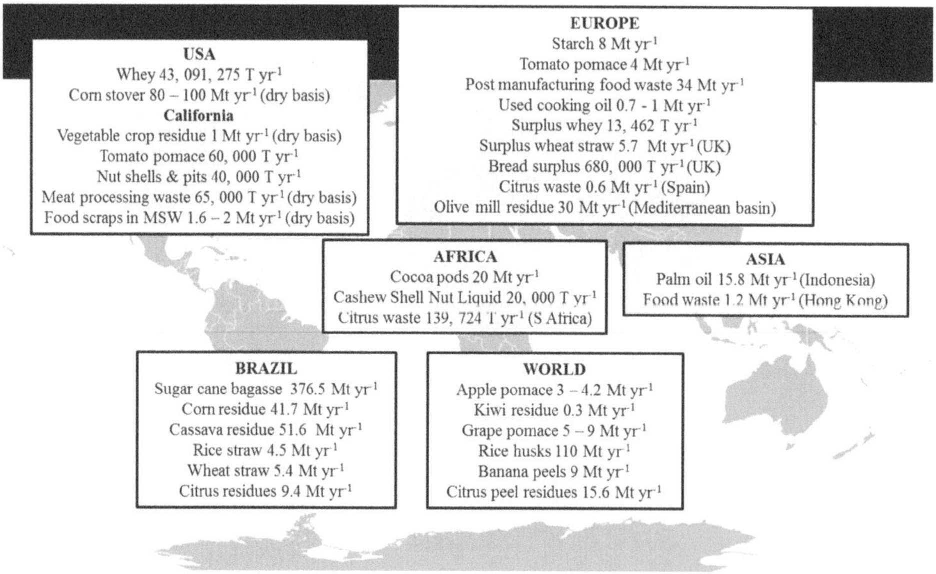

Figure 30.3 Potential global food waste volumes

Source: Matharu et al. (2016)

and drink waste is produced every year (Priestley, 2016), the outcome of biomass/biorefinery research centres allows for the development of new strategies and technologies for the use of food waste as a feedstock for biobased products. On the other hand, in developing nations like Brazil, India and South Africa, although the volume on unavoidable food waste is huge and diverse (see Figure 30.3), scientific research and development receives low support from the public sector and industry does not communicate with research institutions and universities, which then struggle to provide feasible technologies and opportunities to transform biomass into valuable products, instead of environment burden. It is estimated that 18% of anthropogenic CO_2 emissions come from the burning of biomass in the agricultural fields in developing countries (which are the main suppliers of food for the global market). The value of this burned biomass is equivalent to US$120 billion/year if a valorisation approach was applied instead of open air burning (Internet, 2016b). Networking, collaboration and technology/knowledge transfer can have significant impact in helping biomass-rich countries to overcome political and economic barriers and engage academia and industry in the development of biorefineries and suitable technologies for production of bioenergy, biomaterials, and high-value chemicals from food waste.

Unavoidable food waste biorefineries

Like the concept of a traditional petroleum or oil refinery, the "biorefinery" concept is the fractionation of a natural resource into a variety of products including platform chemicals, energy and fuel, but using a renewable resource (biomass) instead of a non-renewable one (petroleum). It can be designed to be versatile regarding the variety or seasonality of the biomass used as input and to maximise outputs, not just in the procurement of product but also potential fuels. The following sections will illustrate three case studies of potential biorefineries for unavoidable food waste.

A case for bioenergy from food waste

The biorefinery concept can be applied to different sets of integrated technologies aiming to convert biomass into highly energetic molecules suitable for use as biogas or biofuels, i.e., *bioenergy*. As shown in Figure 30.1, 30% of all energy produced is directed to the production of food, hence from a circular bioeconomy point of view, it sounds reasonably sustainable that food waste can also be used to produce energy. Recently, new biotechnological approaches have applied renewable feedstock such as food waste to "feed" microorganisms used in the bioprocesses to produce *bioenergy*. For example:

Biohydrogen and biohythane

According to a review on biorefinery models using biotechnologies (Venkata Mohan et al., 2016), biogenic waste biomass (waste such as municipal solid waste or food waste that can produce "life", i.e., microorganisms) could be used as feedstock for the production of a new generation of biogases, namely *Biohydrogen* (molecular hydrogen, H_2, produced by microorganisms) and *Biohythane* (a mixture of H_2 and CH_4). These biogases are produced during the acidogenesis stage within anaerobic digestion, whereby microorganisms "ferment" the biomass by means of various enzymatic reactions producing biohydrogen and/or biohythane, along with some minor volatiles compounds (such as fatty acids). In this process, minimal energy is lost usually as heat (only 3–5%). H_2 has a very high energy content (143 GJ/tonne), surpassing that of any other known fuel and is claimed to be 50% more efficient than gasoline (Waldron, 2007). Moreover, hydrogen is considered the most environmentally friendly biofuel since it does not contain any carbon, hence, does not produce any greenhouse gases when burned.

To date, production of hydrogen has been heavily based on exploitation of non-renewable resources (natural gas, coal and oil), and electrolysis (which accounts for only 4% of total production due to the high costs involved). Besides being produced at low cost by cheap "bacterial cocktails", biohydrogen produced by anaerobic digestion is probably the most cost-efficient method for production because of a high conversion rate, and waste biomass can be used as feedstock (Waldron, 2007). Currently, the EU has funded many research projects on developing green and economically feasible production of biohydrogen and, according to its market development, promising applications for novel hydrogen/hybrid vehicles and chemical storage of hydrogen (hydrogen fuel cells) have already started to appear (Internet, 2016a). Along with the EU, Japan, Taiwan and the US have already targeted biohydrogen as a high priority for providing hydrogen energy in the near future (Lai et al., 2011).

In our view, new biotechnological approaches that rely on the use of renewable biomass, such as the ones described in this case study, can potentially become a competitive alternative to the existing conventional methods of hydrogen production which are neither economically nor environmentally sustainable.

Second generation bioethanol

Although first generation bioethanol has been widely produced from maize and sugarcane (see Chapter 27 for more details), its production from food crops has been highly criticised due to ethical and socio-economic implications, for example, the food *vs* feed *vs* fuel debate. These are also influenced by other contemporary social phenomena, such as the increasing population, higher standards of living, changes in diet (higher consumption of animal derived-products requires more "food for feed"), fluctuant harvest yields and the dependence of agricultural systems on fossil energy (Vandamme, Anthonis, and Dobbelaere, 2011). As an alternative, second generation biorefineries that use residual lignocellulosic biomass as a feedstock (including unavoidable food waste), do not compete for food to produce biofuels. On the contrary, it complements the food supply chain by adding value to its co-products (Vandamme, Anthonis, and Dobbelaere, 2011).

The production of bioethanol from food waste biomass can take different forms and include different feedstocks. The "cascade" biorefinery approach (multiple process steps producing multiple products) has been applied to different types of food processing waste. For example, Tsukamoto *et al.* used orange peel waste (widely available in Brazil since it is a major producer of citrus waste, see Figure 30.3) as feedstock for the simultaneous production of nanocellulose and bioethanol (Tsukamoto, Durán, and Tasic, 2013) and cassava waste for producing bioethanol, biobutanol, and biogas, among other valuable chemicals (Zhang et al., 2016). Rice straw is a major unavoidable food waste in Asia, and Abraham *et al.* have reviewed the potential of rice straw-based biorefineries discussing the availability and feasibility of these models for the production of bioethanol in Asian countries (Abraham et al., 2016). Third generation biofuels are now being developed which use non-land feedstocks such as algae. (Ansari et al., 2015). It is expected that future biorefineries for biofuel production, especially bioethanol, will rely on the use of non-food grade biomass such as unavoidable food waste, making better use of the resource nexus.

A case for vegetable proteins from food waste

The increasing demand for non-animal, vegetable derived proteins within the food supply chain has seen a marked effect on the industry. This demand is not only caused by the growing popularity of vegetarian/vegan diets, but also the general increase in global population. European legislation prohibits the use of animal protein in animal feed intended for human consumption, leading to an increased demand for vegetable protein in the meat/dairy industry as well. An increase of 50% with regard to the vegetable protein needed in animal feed has been observed within the last 10 years, and this value will continue to increase in future years (WRAP and French-Brooks, 2012).

Martin and Danielsson (2016) modelled the effect of reducing animal protein consumption from 51% (2010) to 25% by 2050 while replacing with proteins sourced from vegetables, on greenhouse gas emissions; a reduction of 42% in emissions by 2050 is predicted. Following the same protein-reduction plan an estimated 30% reduction in land usage and 20% in water usage by 2050 is proposed. Furthermore, Martin and Danielsson explored the effect of reduction in food waste on GHG emissions, land usage and water usage, modelling food waste reduction in retail and consumer sectors by 60% (2030) and 85% (2050) with a corresponding reduction in food production needed (as there would be less food going to waste) with a 12% and 16% reduction in 2030 and 2050, respectively. Their study assumed that all food waste at the retail and consumer sector was avoidable and on this basis estimated a 5% reduction in GHG emissions along with a 7% reduction in land use and a 10% reduction in water consumption would be achievable by 2050. However, their study did not consider the potential for reduction and re-utilisation of unavoidable food waste during industrial processing. This is, by far, the easiest stage at which to implement food waste valorisation methods as the waste is highly controlled, categorised, and homogeneous unlike household food waste. If the food waste stream was valorised into ingredients intended for re-use in food then this would aid reduction in GHG emissions, land, and water usage.

Potatoes are a source of protein which have attracted interest in recent years, partly because of their large global production (385 million tonnes in 2014) (FAO, 2014) as well as their relatively large percentage waste (across the entire supply chain roughly 53–55% for fresh potatoes and 41–46% for processed potatoes) (WRAP and French-Brooks, 2012). The reasons for the surprisingly high percentage waste for potatoes is a matter of several different factors including health and safety regulations as well as consumer demand for aesthetically acceptable products (Willersinn et al., 2015). For example, Table 30.1 shows broad ranging amounts of waste for agricultural production, wholesalers, retailers and household in Switzerland (Willersinn et al., 2015), while Table 30.2 focusses on food supply chain wastes from farm to fork (WRAP and French-Brooks, 2012).

Both Tables 30.1 and 30.2 do not consider over-production. Some years the amount of potatoes produced exceeds need. It has been stated that upwards of 9% of potatoes produced are fed to animals due to the production exceeding the needs of the county (Willersinn et al., 2015).

Table 30.1 Potato wastage in Switzerland

Agricultural production (%)	*Wholesalers (%)*	*Retailers (%)*	*Household (%)*
15–24	12–24	1–3	15

Table 30.2 Potato wastage from farm to fork

Agricultural production		*Wholesalers*		*Retailers*
Field loss (%)	Grading loss (%)	Storage loss (%)	Packing loss (%)	Retail waste (%)
1–2	3–13	3–5	20–25	1.5–3

Potatoes are a rich source of starch. Roughly 19% of the potato by weight is starch and starch is used for a variety of industrial purposes including, paper manufacture, as a food additive and in pet and animal food. Potato processing for starch is water intensive and while the starch-extraction methodology has been optimised, there is still potential for improvements. For example, the "waste" water produced during starch extraction is rich in proteins. Approximately, 2% by weight of potato tubers are protein and during the rasping/washing process 85% of that protein is extracted into the potato fruit juice (PFJ) meaning that protein extraction is being carried out incidentally within the starch refining process.

Potato protein can be relatively easily recovered using a range of precipitation methods depending on the desired properties of the final protein, these precipitation methods range from simple heat, through to acid-assisted and use of ammonium sulfate which is commonly used as a fertiliser allowing for the re-use of the water in irrigation of the fields used to grow potatoes. Potato protein is viewed to be of very high quality when compared to proteins sourced from other vegetable sources due to its relatively high concentration of the essential amino acid lysine which is usually in low concentration in other vegetable proteins (Waglay, Karboune, and Alli, 2014).

The market for vegetable proteins is already well established and experiencing a rapid growth in recent years. Potato protein specifically is used in many areas of food production usually as an additive in products ranging from salad dressing, mayonnaise, ice cream, yogurt to high end uses such as a processing agent in wine manufacture. Potato protein has been estimated to have a market value of €0.8/kg if utilised for animal feed and >>€3/kg if used within the human food market, which is comparable if not above that for proteins sourced from other vegetables/cereals, and is approaching the price for animal-based proteins, as to be expected with the rapid growth in need of vegetable proteins within the last few years (Mulder et al., 2016).

A case for tropical fruit waste valorisation

The UK Foresight Project on Global Food and Farming Futures (Parfitt and Barthel, 2011) has stated that:

> The four BRIC (Brazil, Russia, India, China) countries combined account for over a quarter of the world's land area and more than 40% of the world's population. The economic emergence of the BRICs will have unpredictable consequences on the global environment and patterns for food supply to 2050. Over recent years BRIC countries have been the major influence on changing food production and consumption patterns in the world economy.

Interestingly, foodstuffs from the four BRIC countries are becoming increasingly popular as exported crops as well as increasing in production due to the population growth within these areas. The three tropical countries from BRIC (Brazil, India and China) are major global producers of fresh and processed fruits (FAO, 2011b), and tropical fruits are experiencing a sharp increase in popularity in the western world, and a lot of the fruit is processed before reaching consumers, so along with the rise in exported exotic foods there is a corresponding increase in waste streams associated with these foods (van der Goot et al., 2016). Mango is a prime example, with production seeing a fourfold increase in the last 50 years to a global production of just under 44 million tonnes in 2013, assuming all mango production is processed, with a waste of 16–19% from peel only this would represent a potential waste of roughly 7–8 million tonnes yearly (Parfitt and Barthel, 2011).

Banana is the most consumed tropical fruit and the fourth most produced crop in the world (Ayala-Zavala et al., 2011). Production of bananas reached over 107 million tonnes in 2013 and has been experiencing a steady increase in production volumes over the last 50 years (FAO, 2014). Wastage as a result of processing exotic fruits can create as much as 60% waste by weight (Silva et al., 2014), representing a high volume resource of biobased chemicals and materials, for example, dietary fibres, pectin, colourants, pigments and anti-oxidants (Ayala-Zavala et al., 2011).

Dietary fibre, essentially a mixture of the polysaccharides present in the cell walls, is commonly used as a low-calorie bulking agent in foods, as a rheology (texture) modifier, a clouding agent as well as a binding agent. Dietary fibre is commonly found in relatively large amounts within fruit, but certain exotic fruit such as mango and guava, have bi-products (peel and seeds) that have a very high quantity of dietary fibre (78% and 61.4% respectively) (Ayala-Zavala et al., 2011).

Pectin is commonly used industrially as a thickener for preserves, and it is increasing in demand year by year. The global market for pectin reached US $ 850 million in 2013 and has an average price of roughly US $ 14/kg (Ciriminna et al., 2015). Many studies have shown that pectin can be extracted successfully from tropical fruit waste, with yields up to 20% being reported (Sirisakulwat et al., 2010). Given the large volumes of waste present in exotic fruit production and the relatively high-value bulk products that can be isolated from them, it is easy to see the attraction of valorising these waste streams into commercial products such as pectin and dietary fibres.

Betalains, carotenoids, anthocyanins and flavonoids are all present in abundance within tropical fruit waste streams (Parfitt and Barthel, 2011; Ayala-Zavala et al., 2011). Along with being natural pigments, these compounds also have anti-oxidant properties making them well suited for use in the food industry (Houlton, 2016), this, along with a growing trend of consumers demanding natural additive in their food, make flavour, fragrance and colour specific compounds from natural sources a rapidly expanding industrial endeavour.

Conclusions and perspectives

Intelligent resource recovery from unavoidable food supply chain wastes to yield biobased chemicals, materials and bioenergy is paramount and commensurate with the needs of a sustainable 21st century. As a global nation, we have unsustainable production and consumption patterns as exemplified by the 1.3 billion tonnes of food waste being generated annually, which have a negative impact on the water–energy–food nexus. Making better use of resources (*doing more with less*) destined for landfill and/or incineration mitigates climate change and improves efficiency of both land and water usage.

The adoption of a 5-point nexus approach comprising water, energy, food, minerals and land, is a positive step forward within sustainable food production and consumption, allowing for the exploration of how changing one part of the nexus will affect the rest as a whole. This is invaluable from a resource saving and waste reducing standpoint due to the fact that it takes into account the non-linearity of the problem as a whole. For example, re-utilisation of edible food back into the supply chain not only reduces waste, but also alleviates the pressure to produce as much food initially which will have a knock-on effect in the land, water and energy sectors.

Application of efficient, green and sustainable chemical technologies for the processing of unavoidable food supply chains into high-value molecules for (re)-nutrition (for example proteins, anti-oxidants, pectins, etc.) will maximise the inherent chemical potential, structure and function provided by Nature. For example, modern green technologies such as pressurised fluid extraction, supercritical fluid extraction, ultrasound assisted extraction and microwave assisted extraction can be used to extract valuable chemicals from food waste.

Moreover, the concept of responsible and convergent innovation (which considers the environmental, economic and health impacts of technological innovation) is becoming a must for the industrial sector, especially for the foodstuff processing industry, which will inevitably have to adapt their processes to more efficient and intensive ones, as well as move from the production of global-driven pure and refined ingredients to more complex and specialised functional food fractions containing mixed composition with high bio-functional value (van der Goot et al., 2016)

Additionally, policy targets to reduce consumption of animal protein would decrease environmental burden of food waste, however, there might still not be enough to balance out greenhouse gases emissions, and land and water usage in Europe (Martin and Danielsson, 2016). Furthermore, the growing of agroecological production of food, which promotes the creation of short food supply chains, biodiversity care and local agriculture, has questioned the conventional food production model, considering it unsustainable and incapable of achieving the Sustainable Development Goals. Therefore, policymakers need to address food supply chains problems by acknowledging the lack of communication among food producers, retailers, and consumers as a political and economic challenge to be overcome.

Acknowledgements

Eduardo Melo acknowledges the Brazilian Research Council–CNPq for scholarly postgraduate funding to research at University of York, England.

References

Abraham, A., Mathew, A. K, Sindhu, R., Pandey, A., and Binod, P. (2016). "Potential of Rice Straw for Bio-Refining: An Overview." *Bioresource Technology*, 215(September), 29–36. doi:10.1016/j.biortech.2016.04.011.

Ansari, F. A., Shriwastav, A., Gupta, S. K., Rawat, I., Guldhe, A., and Bux, F. (2015). "Lipid Extracted Algae as a Source for Protein and Reduced Sugar: A Step Closer to the Biorefinery." *Bioresource Technology*, 179, 559–64. doi:10.1016/j.biortech.2014.12.047.

Ayala-Zavala, J. F., Vega-Vega, V., Rosas-Domínguez, C., Palafox-Carlos, H., Villa-Rodriguez, J. A., Siddiqui, W., Dávila-Aviña, J. E., and González-Aguilar, G. A. (2011). "Agro-Industrial Potential of Exotic Fruit Byproducts as a Source of Food Additives." *Food Research International*, 44(7), 1866–74. doi:10.1016/j.foodres.2011.02.021.

Bond, M., Meacham, T., Bhunnoo, R., and Benton, T. G. (2013). "Food Waste within Global Food Systems." Report available at: www.foodsecurity.ac.uk

Ciriminna, R., Chavarría-Hernández, N., Rodríguez Hernández, A. I., and Pagliaro, M. (2015). "Pectin: A New Perspective from the Biorefinery Standpoint." *Biofuels, Bioproducts and Biorefining*, 9(4), 368–77. doi:10.1002/bbb.1551.

FAO. (2011a). "Global Food Losses and Food Waste: Extent, Causes and Prevention." Save Food! Rome. doi:10.1098/rstb.2010.0126.

———. (2011b). "Intergovernmental Group on Bananas and Tropical Fruits." Commitee on Commodity Problems. Available at: https://goo.gl/Xp8VGf

———. (2013). "Food Wastage Footprint. Impacts on Natural Resources. Summary Report." doi:ISBN 978-92-5-107752-8.

———. (2014). "FAO Statistics." The Food and Agriculture Organization of the United Nations. http://faostat3.fao.org/home/E.

Garcia, Daniel J., and You, F. (2016). "The Water-Energy-Food Nexus and Process Systems Engineering: A New Focus." *Computers & Chemical Engineering*, 91(August), 49–67. doi:10.1016/j.compchemeng.2016.03.003.

Golden, J. S. and Handfield, R. B. (2014). "Why Biobased? Opportunities in the Emerging Bioeconomy." http://www.biopreferred.gov/files/WhyBiobased.pdf.
Houlton, S. (2016). "The Natural Food Dye Revolution." *Chemistry World*.
Internet. (2016a). "Production and Use of Biohydrogen in Europe." Retrieved from http://www.biofuel-stp.eu/biohydrogen.html. Accessed on 26/09/2017
———. (2016b). "The Dirty Secret of Biomass Waste." Retrieved from https://tatacenter.mit.edu/the-dirty-secret-of-biomass-waste/. Accessed on 26/09/2017
Lai, W-H., Chen, H-Y., Chang, F-Y., Wu, C-C., Lin, C-Y., and Huang, S-R. (2011). "Market and Patent Analysis of Commercializing Biohydrogen Technology." *International Journal of Hydrogen Energy*, 36(21), 14049–14058. doi:10.1016/j.ijhydene.2011.03.155.
Martin, M. and Danielsson, L. (2016). "Environmental Implications of Dynamic Policies on Food Consumption and Waste Handling in the European Union." *Sustainability*, 8(3), 282. doi:10.3390/su8030282.
Morais Ribeiro da Silva, L., Teixeira de Figueiredo, E. A., Pontes Silva Ricardo, N. M., Gusmao Pinto Vieira, I., Wilane de Figueiredo, R., Montenegro Brasil, I., and Gomes, C. L. (2014). "Quantification of Bioactive Compounds in Pulps and by-Products of Tropical Fruits from Brazil." *Food Chemistry*, 143(January), 398–404. doi:10.1016/j.foodchem.2013.08.001.
Mulder, W., van der Peet-Schwering, C., Hua, N-P., and van Ree, R. (2016). "Proteins for Food, Feed and Biobased Applications." Retrieved from http://www.iea-bioenergy.task42-biorefineries.com
OECD. (2013). "Material Resources, Productivity and the Environment: Key Findings." *OECD Green Growth Studies*. OECD Publishing. doi:10.1787/9789264190504-en.
Parfitt, J. and Barthel, M. (2011). "Global Food Waste Reduction: Priorities for a World in Transition." *Science Review*. Bristol: Foresight, Government Ofice for Science. Retrieved from http://www.bis.gov.uk/Foresight
Priestley, S. (2016). "Food Waste Briefing Paper." House of Commons Library.
Sirisakulwat, S., Sruamsiri, P., Carle, R., and Neidhart, S. (2010). "Resistance of Industrial Mango Peel Waste to Pectin Degradation prior to by-Product Drying." *International Journal of Food Science and Technology*, 45(8), 1647–1658. doi:10.1111/j.1365-2621.2010.02314.x.
Tsukamoto, J., Durán, N., and Tasic, L. (2013). "Nanocellulose and Bioethanol Production from Orange Waste Using Isolated Microorganisms." *Journal of the Brazilian Chemical Society*, 24(9), 1537–1543. doi:10.5935/0103-5053.20130195.
United Nations. (2015). "Transforming Our World: The 2030 Agenda for Sustainable Development." Vol. 64. New York.
Vandamme, E., Anthonis, T., and Dobbelaere, S. (2011). "Industrial Biomass : Source of Chemicals, Materials, and Energy !" Brussels. Retrieved from http://hdl.handle.net/1854/LU-2084500
van der Goot, A. J., Pelgrom, P. J. M., Berghout, J. A. M., Geerts, M. E. J., Jankowiak, L., Hardt, N. A., Keijer, J., Schutyser, M. A. I., Nikiforidis, C.V., and Boom, R. M. (2016). "Concepts for Further Sustainable Production of Foods." *Journal of Food Engineering*, 168(January): 42–51. doi:10.1016/j.jfoodeng.2015.07.010.
Venkata Mohan, S., Nikhil, G. N., Chiranjeevi, P., Nagendranatha Reddy, C., Rohit, M.V., Naresh Kumar, A., and Sarkar, O. (2016). "Waste Biorefinery Models towards Sustainable Circular Bioeconomy: Critical Review and Future Perspectives." *Bioresource Technology*, 215(September), 2–12. doi:10.1016/j.biortech.2016.03.130.
Waglay, A., Karboune, S., and Alli, I. (2014). "Potato Protein Isolates: Recovery and Characterization of Their Properties." *Food Chemistry*, 142, 373–382. doi:10.1016/j.foodchem.2013.07.060.
Waldron, K. (2007). *Handbook of Waste Management and Co-Product Recovery in Food Processing, Vol. 1*. Cambridge, England: Woodhead Publishing Limited.
Willersinn, C., Mack, G., Mouron, P., Keiser, A., and Siegrist, M. (2015). "Quantity and Quality of Food Losses along the Swiss Potato Supply Chain: Stepwise Investigation and the Influence of Quality Standards on Losses." *Waste Management (New York, N.Y.)*, 46(December), 120–132. doi:10.1016/j.wasman.2015.08.033.
WRAP and French-Brooks, J. (2012). "Reducing Supply Chain and Consumer Potato Waste." WRAP. Retrieved from http://www.wrap.org.uk/content/reducing-supply-chain-and-consumer-potato-waste
Zhang, M., Xie, L., Yin, Z., Kumar Khanal, S., and Zhou, Q. (2016). "Biorefinery Approach for Cassava-Based Industrial Wastes: Current Status and Opportunities." *Bioresource Technology*, 215(September), 50–62. doi:10.1016/j.biortech.2016.04.026.

31
CALIFORNIA INNOVATIONS @ WEN

Blas L. Pérez Henríquez

Introduction

In the US, the interdependent relationship between water and energy, known as the water–energy nexus (WEN), has become a salient policy issue at the federal, local and regional level. Water demand from the energy sector is growing at a faster rate than competing uses such as potable water and agricultural production (U.S. Department of Energy (DOE)/U.S. Energy Information Agency (EIA), 2016). Moreover, recent technological advances enabling unconventional oil and gas (O&G) development "ha[ve] added complexity to the national dialogue on the relationship between energy and water resources" (DOE, 2014).

The American West is highly vulnerable to shifts in climate patterns due to both limited water availability and high spatial and temporal variability in what water is available. Extreme climatic events aggravate the risks faced under the WEN to secure reliable and affordable energy production and freshwater supply. California's WEN management, defined as such, has been at the forefront of policymaking at the state, regional, local and city level since the start of this century. It is state policy to "promote all feasible means of energy and water conservation and all feasible uses of alternative energy and water supply sources" (15 California P.R.C. § 25008). Since then, the challenge of how to best manage the trade-offs at the WEN finds a rich source of experiential and experimental policy know-how in a constantly evolving physical, technological and political environment. This chapter presents a review of key institutional and technological developments that effect California's WEN management.

A formidable challenge is presented by the tasks of developing innovative governance structures that best integrate interdisciplinary approaches for WEN policy action in a multi-dimensional, multi-stakeholder context. Moreover, an intergenerational or long-view planning perspective should also be part of this organizational design challenge for effective decision-making. An underlying issue is how to overcome the "silo mentality," as it is commonly referred to in the business and public administration literature (Diamond and Allcorn, 2009). The "silo effect" limits the performance of individual agents within an organization to efficiently and effectively advance a common vision. Despite deep knowledge and expertise of those working in discrete tasks or areas of managerial responsibility, a silo management style becomes a hurdle to produce well-integrated solutions that maximize the objectives of the agency or corporation

(Tett, 2015). This consideration is central to avoid such pitfalls and instead provide smart policy and business solutions in the context of a shared resource systems issue such as WEN.

A high degree of interagency cooperation and public consultation is needed to develop well-integrated policies and programs to best balance the competing needs of powerful stakeholders, such as urban, industrial and agricultural sectors, in the WEN context. Emerging innovative technologies and smart information systems can assist to inform policy and business decisions. We need to take advantage of advanced information technologies (IT), remote sensing and data analytics to better understand WEN interactions and develop efficient shared resource management strategies. Such processes so far have lacked momentum given that WEN as a coherent policy focus is a relatively new concept in policy formulation. It has taken a back seat to the ambitious clean energy and climate goals of the state.

However, the prolonged drought that began in 2012 prompted some institutional and policy realignment towards a more holistic, integrated approach to this issue, including investment in research, development and demonstration (R&DD) to support the emergence of smart, high-tech solutions. The state government declared that California was in "state of emergency" because of severe drought conditions at the start of 2014 (Office of the Governor, 2014). This was followed by the introduction of California's first-ever mandatory water consumption cut backs for urban centers on April 2015. Based on water usage from 2013, urban water (i.e., business and residential) users were forced to reduce by 25% their consumption against that baseline. Some of these restrictions were eased by the summer of 2016, given healthier levels of precipitation that season. In April 2017, Governor Brown lifted the drought state of emergency, but retained the water-use reporting requirements and restrictions on water waste (Office of the Governor, 2017). The dangers of crisis-driven policymaking, however, is precisely that the longer-view, integrated policy choices are ignored in favor of quick fixes to the crisis conditions. In order to ensure this long-view, holistic and integrated policy design for WEN, then, there is a need to maintain political will beyond the windows of public concern.

Californians historically, however, have shown willingness to support forward-looking environmental regulation. Since 2006, state residents have strongly favored a trailblazing legislated climate action plan for the California economy despite the lack of federal legislation on this issue (PPIC, 2015). As the state government moves forward on implementing an integrated clean economy model to address climate change, careful analysis and attention on how to best manage the WEN in this context is becoming an increasingly urgent issue. For instance, water supply uncertainties to cities and industry in general, and in this case the electric power sector, provide a renewed sense of urgency on how to best address WEN in the state. Thus, interest is on the rise from researchers, innovators, investors and businesses in general to provide long-term, effective and commercially viable solutions to the WEN challenge. Therefore, both policy and technological innovations are needed in this area.

There are some common-sense solutions to increases sustainability of the WEN, like eliminating wasteful practices in general, and more specifically, uptake of water-efficient energy alternatives and energy demand management (Carter, 2011). In California, examples of the solutions being advanced include the promotion of new technologies supporting the phasing out of water-based cooling processes in power plants despite delays in federal programs due to litigation; mitigating water-intensive agricultural practices through smart sensing and pressurization; and modernizing heating systems in the urban built environment. However, alternative conservation solutions represent trade-offs. For instance, installing pressure-based irrigation systems can increase water efficiency, but will require more energy than standard irrigation technologies. Air-based cooling of thermal power instead of once-through water cooling reduces water use in the electricity generation sector, but can reduce power production efficiency. (Dale et al., 2008, 8)

Another logical step in the evolution of the implementation of smart water conservation strategies is expanding re-use of water. For instance, closed loop water systems are a common practice in cities with major water resource constraints such as Singapore, as seen with its NEWater program. (Xie, 2006). In the US, however, development of unconventional water source augmentation programs would require socialization and education to eliminate the "yuck"[1] effect of "toilet to cup" processes, along with the expansion of indirect systems such as injection of highly treated wastewater into the aquifer as in the case of Orange County, CA and under certain circumstances around Los Angeles City and County (EPA Region IX, 1998, 4). A growing number of companies, entrepreneurs and investors in California are already taking the lead in developing innovative, smart and profitable answers to best manage this nexus. Communities in California's innovation hubs, such as iconic Silicon Valley and emerging ones in Los Angeles and San Diego, are working closely with industry to provide new solutions.

These are particularly important emerging strategies in the context of a stressed WEN system in California. Citizen support for a long-view, holistic systems policy perspective on these issues is key given the necessary investments, societal change and uptake of new technological solutions that the transition to a resource-efficient, decarbonized economy requires. Such vision calls for enhanced policy integration and organizational alignment in the management of water and energy resources. Government to government (G2G) interagency cooperation, and government and business (G2B) collaboration is key to best integrate technological advances and new business models to support cost-effective WEN management in California.

Risks and vulnerabilities at the WEN in California

Droughts produce serious impacts on California given the water dependency and growing energy demand from its dynamic economic sectors and urban centers. At the WEN, climate and environmental risks are most clear in shared resource systems interaction at the regional level. California recently implemented a collaboration with NASA to use satellite observations from its Gravity Recovery and Climate Experiment (GRACE). Such observations allow policymakers and stakeholders to better assess the impact of droughts and the risks to California that these climatic episodes represent to its freshwater supply, while developing more granular data to facilitate more geographically targeted problem-solving. For instance, before the intense and damaging storms of early 2017, data from this program showed that, "since 2011, the Sacramento and San Joaquin river basins decreased in volume by four trillion gallons of water each year (15 cubic kilometers)" (NASA, 2014a). This volume of water represents more "than California's 38 million residents use each year for domestic and municipal purposes." This first of its kind analysis of satellite data shows that "about two-thirds of the loss is due to depletion of groundwater beneath California's Central Valley" (NASA, 2014b). In a contract with the California Public Utilities Commission to develop a model of the WEN relationship in California, GEI Consultants and Navigant Consulting Inc. estimated that groundwater pumping also consumed more electricity in the state than both the State Water Project and Central Valley Project during spring and summer (2010) (Figure 31.1). Clearly groundwater represents a significant governance gap in the current WEN context in California, as it is largely unregulated.

Contributing to the vulnerability of water supplies in California is the high spatial variability of this supply. The watershed region of the Sierra Nevada, which includes a portion of the southern Cascades, has been essential to California's overall water supply. Per the California Department of Water Resources (CDWR), historically, "60 percent of the state's total annual

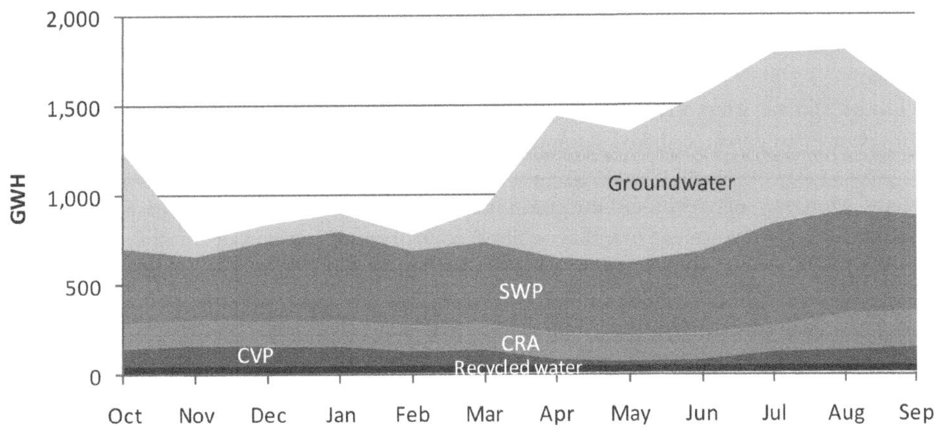

Figure 31.1 Estimated monthly electricity consumption in California by water supply (2010)

Source: GEI/Navigant Consulting, Inc. (2010)

precipitation falls in the form of rain and snow" in the winter, which in turns melts during warmer seasons to flow into natural lakes and man-made reservoirs downstream (Water Education Foundation, 2017). This watershed provides drinking water supply to more than 23 million Californians. In the case of the City and County of San Francisco, 85% of its water comes from the Sierra Nevada. Additionally, half of the water that flows into the Sacramento-San Joaquin Delta originates in that region as well (Ibid.). New governance and innovative approaches in forest management are needed to enhance the sustainability and reliability of the water supply coming from this region.

New research in other areas points to the effects of climate change on water availability in California by focusing on the link between high temperatures and dry years as a cause for increased drought risks (Diffenbaugh, Swain and Tourma, 2015; Mann and Gleick, 2015). If the trends as suggested in these findings hold, drought episodes will continue, exacerbating the duration and severity in the lack of freshwater for human uses and ecosystem services in the state for the foreseeable future. However, there is not yet scientific consensus on such connection in meteorological terms (Neelin et al., 2013; Funk, Hoell and Stone, 2014; Seager et al., 2014, 2015; Simpson et al., 2015).

As stated in Williams et al. (2015), these recent studies consider the analysis of individual sets of data from individual climate variables "without addressing the structural uncertainty" across a diverse set of statistical analyses. Nevertheless, while "natural variability dominates" for instance in terms of precipitation levels, it is now more certain that human activity has indeed "substantially increased the overall likelihood of extreme California droughts." (Williams et al., 2015) Moreover, Mann and Gleick (2015) point to California's ongoing extreme drought episode that began in 2012, showing that "[m]easured both by precipitation and by runoff in the Sacramento and San Joaquin River basins, 10 of the past 14 years have been below normal, and the past 3 year have been the driest and hottest" since instrumental data has been available.

In addition to the high spatial variability of water supply in California, climate change may be increasing the temporal variability of this supply as well. The California Climate Change Adaptation Strategy operates on the premise that temperature increase in tandem with recent changes in precipitation patterns producing extended dry periods are producing more rain rather than snow during the winter. Another effect of climate variability is early snow runoffs in the spring. These changes are increasing the temporal variability of water supply to communities in the state as the snowpack is less effective as freshwater reservoir to smooth supply throughout the year.

Under climate stress, critical infrastructure at the WEN becomes vulnerable and in need of investment to enhance shared systems resilience. For instance, uncertain water supply limits the capacity to reliably supply energy from hydropower and power plants using water cooling systems. Moreover, dry spells and uncertain natural systems water supply require adjustments in management and investment to enhance flood control and water conservation systems capacities. The strong storm events of 2017, while providing needed relief to most regions of the state from extreme drought conditions, pinpointed to the need to develop innovative systems to also reduce extreme water releases, minimize flood risk, and capture excess runoff from extremely wet seasons for future use.

Forest and land-use policy to restore mature forest cover can delay snowpack melt and increase snow storage capacity in the Sierras which hit a 500-year low in 2015 (see Figure 31.2) (Varhola et al., 2010). In 2012, California passed Assembly Bill 1750 allowing the development of innovative, decentralized rain harvesting systems to address temporal variability. Temperature increases will also require adapting existing reservoir operations to best capture earlier runoff water in the spring, as well as greater volumes of water during intense storm episodes such as those due to climate variations, which may negatively impact electricity generation for dual use hydropower reservoirs in the state.

In relation to environmental risks, research shows that hot weather brings significant reductions in hydropower generation (Rheinheimer, Ligare and Viers, 2012). Around 7.1% of California's electricity comes from hydropower (CEC, 2017a). Therefore, power switching to conventional dirtier electricity generation technologies increases air pollution levels in the state.

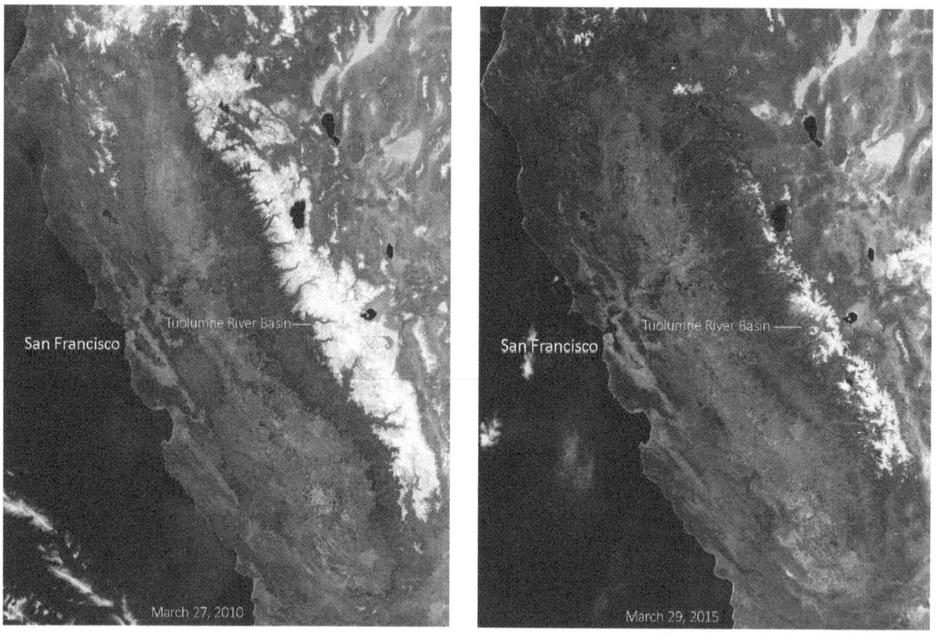

Figure 31.2 Sierras snowpack extreme variations

Source: NASA/MODIS

California's ambitious clean energy deployment program mitigates in part this negative effect through source diversification and decarbonization of the portfolio of electricity suppliers. Still, compared to a similar period (2009–11), the California drought of 2012 has increased the level of GHG emissions in the state by approximately 22 million metric tons of CO_2 emissions (Hardin et al., 2017). Natural gas use for electric power generation increased to make up the losses in hydropower. California's overall CO_2 emissions between 2012–2014 show "a 33% increase in annual CO2 emissions compared to 2011, emissions equivalent to the annual energy use of 1 million homes or annual emissions of 1.5 million passenger vehicles" (Hardin et al., 2017; EPA, 2016).

A major vulnerability in the WEN context is regional water supply needs and the impact of the recent drought to groundwater sources. Water governance is being challenged by thirsty urban centers, inefficient agricultural practices and climate variability in the southern region of the state. Here a major source of water supply is the Colorado River. Its water is shared by seven states along the Colorado River basin and shares estuaries at the US–Mexico border, while supplying water to Mexico's gulf of Baja California ecosystems. An iconic infrastructure feature in this system is the Hoover Dam, and its water reservoir Lake Mead, that serves approximately 20 million in Nevada, Arizona and California. The entire basin serves now 30 million people per day from an original population of 12 million five decades ago. In 2014, the US Bureau of Reclamation reported the lowest levels of water elevation in Lake Mead's history at 1,074 feet. The Colorado River flow itself has declined from approximately 15.5 million acre feet per year in 1965 to 12 million in 2015. Fortunately, at the beginning of 2017, intense storms improved the situation in most of California. However, "the longer-term impacts of the drought are still being observed in relation to groundwater supplies in various California locations" (United States Drought Monitor, 2017).

According to the California Department of Water Resources (CDWR), approximately 30% of California's urban and agricultural water supplies come from groundwater per year under normal conditions. During periods with intense drought, groundwater consumption can rise to 60% or more due to reduced availability of surface water (Carle, 2004; CDWR, 2017). Despite the highest capacity of water stored in groundwater reservoirs (with the largest reservoirs located in the Central Valley), not all the water is usable. More than half is unavailable due to poor quality and high cost of pumping water from the ground.

After the winter and early spring rain and snow storms of 2017, Northern California was technically out of the drought by restoring two of the three main sources of freshwater in the state – surface water and snowpack. However, the lack of water resources in Southern California as a region, and the need to replenish groundwater remain as longer term issues (see Figure 31.3).

Experts argue that the long-term over-exploitation of groundwater can result in permanent loss of groundwater storage capacity and quality due to subsidence and seawater intrusion. Replenishment of California groundwater supplies depleted due to the drought could take many years (USGS, 2017). California can't rely on less frequent wet episodes or intense storms to alleviate chronic droughts effect; a long-term perspective to provide solutions in all areas is needed.

These debates over the risks, infrastructure investments and technical solutions at the WEN call for more integrated, interdisciplinary and multidimensional research efforts to strengthen the evidence of metrological, hydrological, as well as the political economy issues future natural water deficits for California present. California policymakers are assessing how best to proceed with predictions given the extremities of the recent drought of 2012–2016 and the severe storms of 2017 that technically ended it, at least in Northern California. Accurate, timely, diverse and large-scale and localized data and data analysis is required to mobilize a prescient analytic framework.

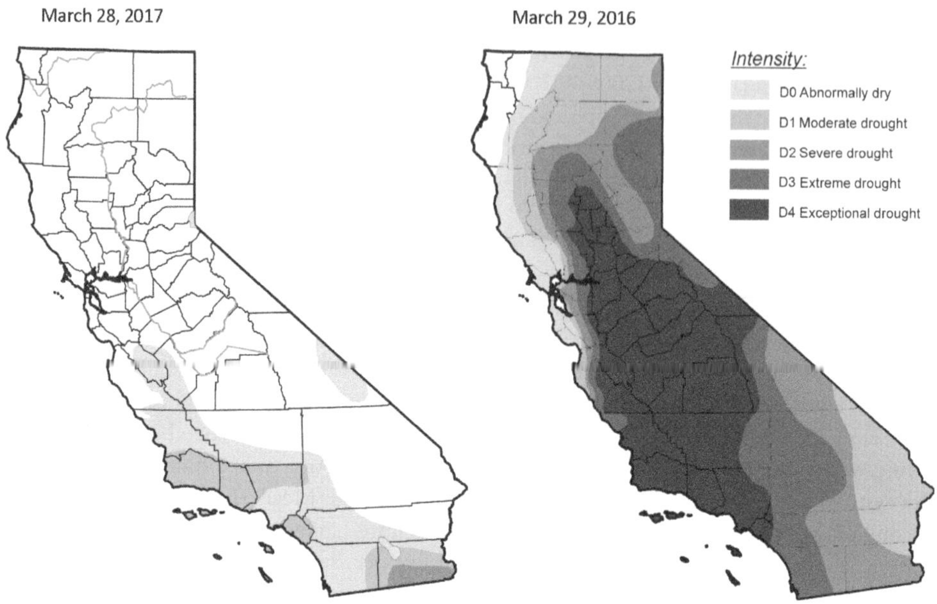

Figure 31.3 Drought relief from winter/spring 2017 storms

Understanding the water–energy relation: smart data for policy action

Water needs power to ensure its conveyance, delivery, purification and availability in homes, businesses and agricultural fields. Users in general expect water to automatically flow from their faucets, cooling or irrigation systems, and lights to turn as we enter a home or business facility. Likewise, fossil-fuel, nuclear and hydropower plants require substantial water resources to produce power. All this is expected without much thought on the amount of effort, energy and investment it takes for that service to be delivered. These are two valuable and interdependent resources, not always well managed or appropriately priced. Therefore, investments in R&DD and smart infrastructure systems to avoid energy and water-intensive models of the past are required. Moreover, economic incentives for behavioral change by users to prompt more resource-efficient utilization are being explored. Ultimately, reliable and equitable access in WEN is essential in the context of smart, prosperous and resilient communities.

As mentioned above, while water governance has been central to California for over a century, the WEN is a relatively new concept in the California policy agenda thus requires an upgrade in institutional design for holistic approaches to problem-solving and to foster collaboration with interested stakeholders. In 2005, the State of California began a formal assessment of the tools available to better understand the water–energy relationship. With limited data, the California Energy Commission (CEC) published a report on integrating available databases and developing methodologies to estimate "the magnitude and intensity of water-related energy consumption by segment of the water-use cycle." (Navigant Consulting, 2006) This report launched the development of appropriate data frameworks, management and collection systems, and data validation systems. From the start, it was clear that the CEC should also invest in a continued refinement effort to improve data and information to "more effectively target research and development decisions at the technological level" (Navigant Consulting, 2006).

To begin data collection on the WEN, metrics had to be developed to represent the amount of energy required to provide various water resources, and the amount of water to provide various energy sources. For water delivery, the CEC developed metrics of upstream and downstream embedded energy, representing the amount of energy required to collect, extract, convey, treat and distribute, for the former metric, and treat and dispose for the latter. (Navigant Consulting, 2006: 14) This enhanced set of parameters included recommendations to take into consideration regional aspects, for instance establishing preliminary relative values for Northern and Southern California water–energy relationships respectively. However, there is not yet an equivalent metric of upstream and downstream embedded water for the energy sector.

Another important recommendation for the design of pertinent databases was to take a sectoral perspective on water-related energy use by assessing the type of water to determine if treatment and disposal is needed. For instance, irrigation and storm water generally would not require treatment. On the other hand, indoor water does consume important amounts of energy for its appropriate treatment and disposal processing. An important preliminary finding was to show that water in agriculture and urban end users accounted for more than 50% of the water-related energy consumption in California. (Navigant Consulting, 2006: 14) Groundwater pumping energy use was still largely left out of these early studies, however, but were subsequently estimated (see Figure 31.4).

Moreover, an area of opportunity for enhanced conservation was highlighted if data regarding energy and water intensity in industry based in California could be made more readily available. However, perhaps the most important preliminary finding of such assessments was that "the water-use cycle outside of the retail water meter with highest variability in energy intensity and magnitude of energy use" was supply and conveyance (Navigant Consulting, 2006: 14). For instance, the most energy-intensive water supplies include large interbasin water transfers through the Colorado River Aqueduct and State Water Project and ocean water desalination (see Figure 31.5). With an emphasis on identifying energy savings potential in water-related usage, the CEC continues to explore this new area of research – the energy–water nexus – through

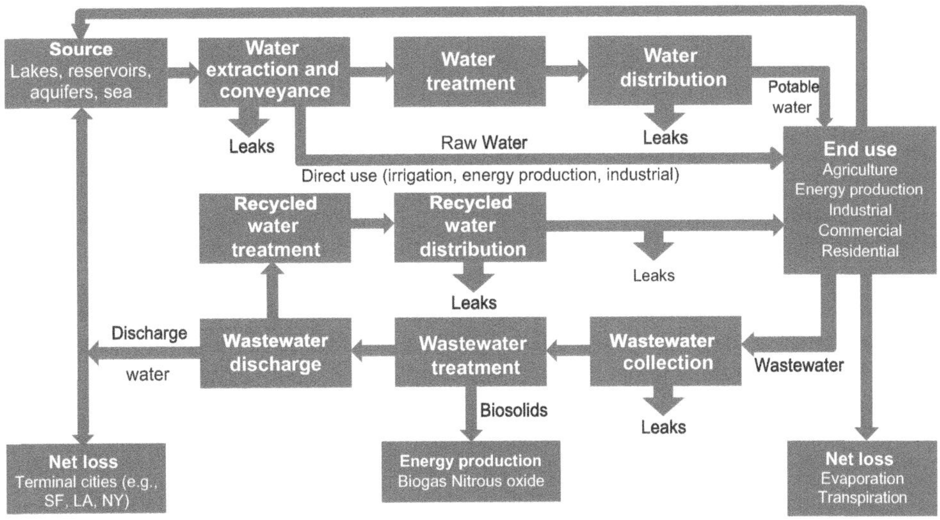

Figure 31.4 Water-use system

Source: Water in the West (2013), adapted from Wilkinson (2000)

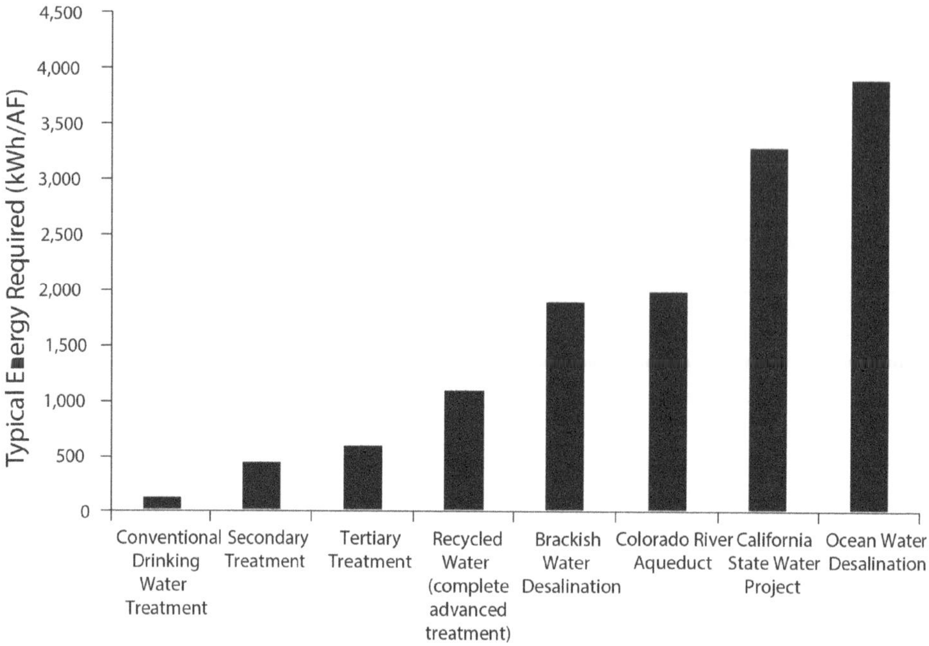

Figure 31.5 Electricity requirements for different water treatment processes

Source: Christian-Smith and Wisland (2015)

investments from the EPIC program (Electric Program Investment Charge) aimed at promoting innovative water–energy efficiency programs and projects across California.

Understanding better the key drivers in the process of water-related energy use was prioritized. Therefore, consultants and policy analysts recommended further disaggregation of data, from energy magnitudes and intensities, to more sub-segments capturing idiosyncrasies of different sectors in the California economy. Filling data gaps and work to improve analytical frameworks and databases began, and a continuous improvement process to refine such information was recommended. Advocacy groups suggest that "Improved information from water and wastewater utilities regarding their electricity use and sourcing practices" can make more salient its role in clean energy development and services provision to a smart grid (Christian-Smith and Wisland, 2015, 1). Moreover, they argue that by engaging more directly the clean energy transition, the water sector can create new revenue sources. For instance, the "sale of excess renewable electricity or receipt of renewable energy credits, cap- and-trade proceeds, payments for providing grid services, and fees for accepting organic wastes for renewable energy generation" can generate substantial revenues for utilities. (Christian-Smith and Wisland, 2015, 2) In a state where information technology and smart applications are a dominant industry, experts have argued in favor of "robust use of water-information systems that enable accurate, timely, and transparent accounting through the water supply and use cycle. This system must extend from mountain headwaters through valley groundwater." (Bales et al., 2016)

Finally, these initial efforts by the CEC to formalize policy analytic work on the energy–water nexus suggested that beyond energy magnitude and intensity, ingenuity to provide innovative solutions in water and energy systems, either through changes to operations or the introduction of novel technological approaches, could improve performance on these two key dimensions. Therefore, improving data collection, developing pertinent baselines, systems and analytics capacity is an important aspect to consider for state's R&DD investments addressing this emerging policy issue.

Policy evolution: interagency coordination and institutional alignment

WEN is a multi-stakeholder, multidimensional, interdisciplinary and highly technical issue. Therefore, new governance structures for integrated policymaking and innovative systems for better interconnectivity for smart management in the WEN shared resource systems context is required. In 2013, Stanford University conducted an assessment of the state-of-the-art in the research, technological and governance advances in California regarding the issue of "conjoined management" of water and energy resources (Water in the West, 2013). The Water in the West review found a policy analysis and action deficit, stating that the demand for "regulatory and operational innovation, and for more interdisciplinary research to capture the full benefits of integrated water–energy resource management, largely remains unmet." Policymakers are currently working towards introducing a long-view perspective in the search for addressing the WEN policy challenge in California. They are pursuing an intergenerational perspective in terms of policy planning, governance and investment under WEN (Interview with Chair of the CEC, 2016). For instance, the increasing role of seawater desalination in California's water resource portfolio serves as a reminder of the importance of an integrated approach to optimizing the water and energy decisions. While emergency measures have been taken in the past during dry episodes, from systems that introduce higher rates for consumers that break pre-established consumption thresholds, to discreet solutions such as water-saving toilets or water-efficient shower heads as in the case of the 1970s, a long-view perspective in the WEN policy formulation process would lead to a more cost-effective, equitable integrated policy framework (Interview with Chair of the CEC, 2016).

Historically, water conservation has been a top policy priority for California. Article 10, Section 2 of the state Constitution states that: "the water resources of the State be put to beneficial use to the fullest extent of which they are capable, and that the waste or unreasonable use or unreasonable method of use of water be prevented, and that the conservation of such waters is to be exercised". Under a WEN integrated, long-view policy perspective, the state should also seek to achieve to deliver energy reliability.

California climate and energy policy currently aims to enhance resilience and energy efficiency to mitigate the risks and dangers that more extreme climate variability and severe drought episodes will present to future generations. Therefore, institutional alignment and policy integration for effective WEN action plans should aim to develop more complete pricing policy mechanisms that can foster efficient use in the interconnected energy and water systems. Moreover, an integrated policy approach to foster development and adoption of commercially viable technological innovations is also key to achieve cost-effective and efficient WEN management in California. The state of the institutional governance structure for water in California is the following:

Standard water management operations are conducted by the agencies above (Figure 31.6). However, given the central role climate and energy policy have in advancing the concept of a clean, prosperous and resilient California, other agencies are now involved at the WEN level. The California Environmental Protection Agency (Cal-EPA) mission is "to restore, protect and enhance the environment, to ensure public health, environmental quality and economic vitality." (2017) However, most of the implementation mandate from climate legislation and rules

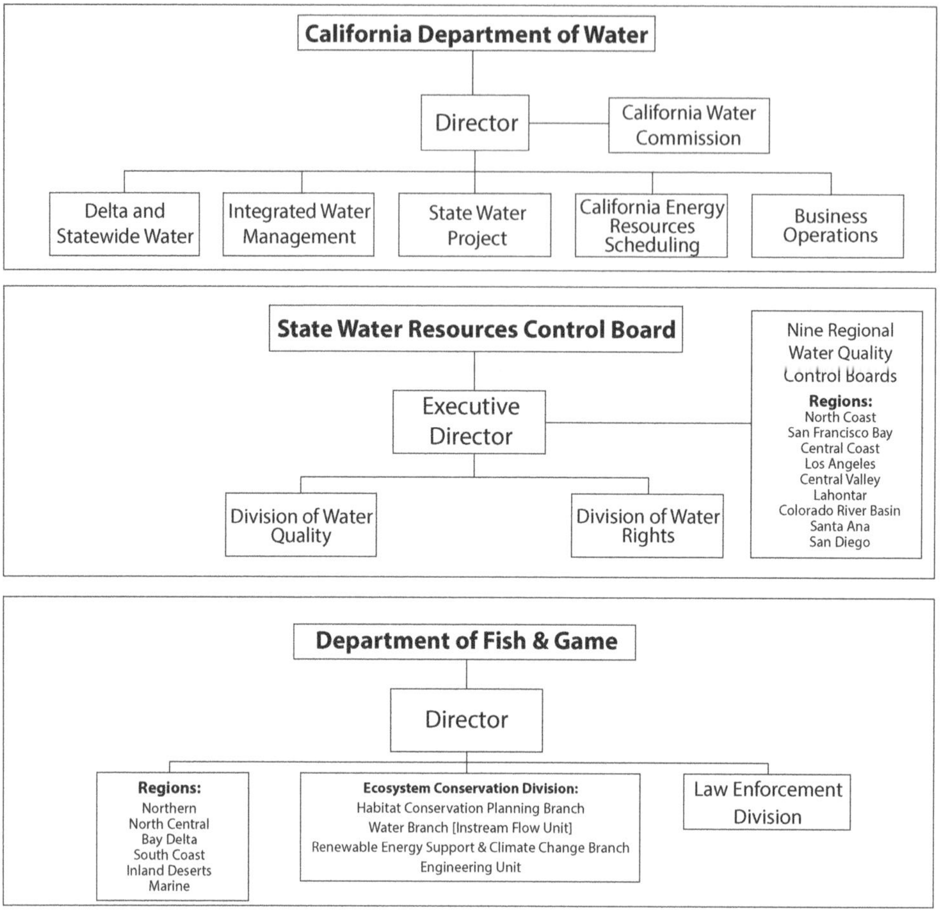

Figure 31.6 Water governance in California

Source: Little Hoover Commission (2010)

falls on the California Air Resources Board (CARB) portfolio of responsibilities, which is a sub agency under the California EPA.[2] CARB is primarily involved with the water–energy nexus as a coordinating institution for the California Climate Adaptation Plan. Through this, they work with the other institutional partners (see list later in this section) to develop the water chapter of CARB's Scoping Plan and Adaptation Strategy.

The emergence of an institutional focus on the WEN in California was initially driven by concern over GHG associated with water use, and climate change impacts on water availability. The California Global Warming Solutions Act (AB32) required the development of a Climate Adaptation Plan (CAP), and the Climate Action Team (CAT) responsible for this included a Water Energy Team (WET) focused on the energy intensity of water provision and the potential climate change risks to water availability in the State. Additionally, the recent series of extended droughts in California raised the specter of the water intensity of the energy industry. The two sides of the WEN have different institutional bases in California, but work together within the WET-CAT for the Climate Adaptation Plan, and thus have individual institutional goals derived from this collaborative work around the WEN. The WET is one of 10 working groups that comprise the CAT. The institutional partners of WET include the:

- California Air Resources Board (CARB);
- California Environmental Protection Agency (Cal-EPA);
- California Department of Food and Agriculture (CDFA);
- California Department of Public Health (CDPH);
- California Department of Water Resources (CDWR);
- California Public Utilities Commission (CPUC);
- California Energy Commission (CEC);
- California Natural Resources Agency (CNRA);
- Governor's Office of Planning and Research (GOPR);
- State Water Resources Control Board (SWRCB); and
- Strategic Growth Council (SGC).

The WET is responsible for developing adaptation strategies for the water sector aimed at decreasing its associated energy use, but within the context of increasing drought in California has also focused on increasing water efficiency within energy systems. The implementation leads for most of these WEN strategies are the DWR, SWRCB, CEC and CPUC.

Gaps and plans on WEN institutional integration and policy action

The State of California policies have begun to promote managing water and energy in tandem, providing funding allocated for water and energy efficiency programs. The California Public Utilities Commission (CPUC) together with utilities analyze ways for quantifying energy savings from water conservation using an evolving water–energy calculator (2017). However, a number of gaps have been identified which require action.

The lack of lead institution and benefit cost sharing

Multijurisdictional issues in the public sector may not get enough policy action given the lack of clarity about which agency leads, has the legal mandate and, more importantly, the funding. For instance, an initial R&DD public investment program mentioned above, the Water Energy Technology Program (WET) was not included in 2014–15 legislative session and failed to secure a budget allocation. The expectation was that given the nature of the issue, it could secure some of the Greenhouse Gas Reduction Fund from which the Legislature appropriates proceeds from the state's cap-and-trade (i.e., carbon market) a key financing mechanism for implementing California's climate action plan. The WET program was intended to be a joint program between the CEC, the Department of Water Resources and the Water Resources Control Board focusing on "water and energy savings at the nexus." Instead, a CEC standalone funding selection and administration program was preferred, the Electric Program Investment Charge (EPIC).

There are, therefore, some institutional barriers to genuine interagency cooperation around WEN. This weakens the case of the WEN in the budgetary policy process, a highly contentious political process given the competing interests and limited resources available for the overall implementation of the state's climate plan led by CARB. Under the current interagency coordination mechanisms to implement California's climate action plan, WEN is a subsidiary issue to CARB's regulatory mandates. There is no WEN institutional lead agency that can introduce a holistic and long-view perspective in the policy formulation processes around this issue. California WEN policies, if lead by one agency with budgetary authority, could go beyond agency administrative responsibilities, thus breaking management silos to provide well-integrated solutions under a WEN horizontal management framework across state agencies.

With regard to stakeholder management and consultation processes at the WEN, distributional resource inequality should also be incorporated as part of a future institutional alignment to improve regulatory quality in this area. This could be done by developing an environmental justice framework that informs the policymaking process on marginalized community challenges within the WEN context. Finally, the next frontier in water management and institutional development seems to be on groundwater governance development, as well as innovating introduction of new behavioral and utility incentive mechanisms using water-use efficiency rate decoupling (Kammen et al., 2015).

Innovations outlook: policy, technology and investment

A pragmatic policy vision that integrates the idiosyncrasies of the shared systems at the California WEN is required to address the needs of all interested parties in a cost-effective and equitable manner. A holistic perspective in policy design and public and private investment plans to foster the advancement and discovery of clean, hi-tech solutions for the deployment of WEN smart infrastructure will facilitate balancing the competing needs of powerful stakeholders, such as urban, industrial and agricultural communities. The following are some examples of investments in innovations and demonstration projects with a vision of achieving commercial viability in technology, planning and finance at the WEN in California.

Next Generation Agribusiness

There are projects funded by EPIC with a direct link to economic sectors that epitomize the shared resource systems challenge. For example, in the agribusiness sector. California plans to fund innovation to minimize water and energy usage in the agriculture and food processing industry. There are three R&DD for commercialization in the pipeline: 1) infrared dry-blanching and drying system for fruits or vegetables that results in high-quality products with potential energy use savings of 40% compared to traditional freeze-drying methods by eliminating the water or steam used in traditional blanching; 2) Infrared heating technology for peeling tomatoes promises to reduce energy consumption and minimize wastewater while preserving product quality; and 3) New drying and nutrients extraction technology allows a higher yield extraction of lipids and nutrients from fruits and vegetables as compared to conventional techniques. The research will also quantify the new technology's energy savings for the extraction process. (CEC, 2017b)

Desalination Renaissance

Some other well-known and somewhat controversial technological approaches are reemerging thanks to both more environmentally friendly and less energy-intensive technologies as in the case of the renaissance of desalination plants in the state. For instance, in December of 2015, after three decades of planning and addressing environmental concerns, the largest and most technologically advanced seawater desalination plant in the Western hemisphere began operations in Carlsbad, southern California. One year later, the plant was already supplying approximately 10% (i.e., 15 billion gallons) of the freshwater consumed in the San Diego region, thus mitigating the effects of one of most severe droughts in state history while reducing dependence from northern water resources (Poseidon Water, 2016). Almost two dozen plants are now planned to be in operation in the next decade or so in the state. In some cases, these plants may be less energy intensive than deliveries from the State Water Project.

Living Laboratories

Finally, universities can apply and experiment in the implementation of new concepts to address policy challenges. Two major R&DD projects at Stanford showcase the gains of integrated shared resource systems through innovative approaches and planning focusing on resource recovery in the WEN context, as well as novel financing approaches focusing on local communities' needs.

a **Energy Systems Innovation:** Stanford University has been developing contracts to source most its campus electricity from renewable sources within California. A Stanford Solar Generating Station, to designed and built by SunPower off site, provides half of all campus electricity. Combined with planned solar power from installations on campus rooftops and the purchase of further renewable power from the grid, renewable energy will supply 65% of all campus electricity. The renewable energy power is now integrated into its WEN management systems with a first-of-its-kind campus heat recovery system. Stanford has converted to a state-of-the-art energy system that relies on renewable electricity and creating a new transformational energy supply model for large organizations, utilities and governments. "The combined new system – Stanford Energy System Innovations (SESI) – makes Stanford one of the most energy-efficient research universities in the world. It far exceeds the aggressive goals of California's climate legislation." (SESI, 2017)

b **Water and Resource Recovery Innovation:** the William and Cloy Codiga Resource Recovery Center is a research project focused on recovery of resources from various campus waste streams, prioritizing wastewater recovery innovation so far. The Codiga water treatment plant uses an anaerobic process instead of the traditional aerobic wastewater treatment to eliminate more trace contaminants (e.g. pharmaceuticals) from wastewater while also generating methane to power the plant from biogas produced by the leftover organic matter (Scherson and Criddle, 2014). This resource recovery process has implications for both public health planning and for WEN.

Water-focused Impact Investment

Innovative socially and environmentally responsible investment practices to address WEN issues are emerging as a possible way forward. Based on expert opinions, opportunities for such type of investments in agriculture and livestock contexts are being developed for instance by "using land purchases or joint ventures to improve profitability and generate marketable water." For instance, innovative municipal financing vehicles such as green bonds and environmental impact bonds are being used to finance alternative municipal water solutions. Community water trusts are also being developed to deploy investor capital at the river-basin scale in a financially self-sustaining fashion. Finally, forest investment in resilience bonds are being used to reduce fire risks in watersheds through pay-for-performance vehicles. (Bowman, 2016)

Silicon Valley, California's knowledge-based regional economic powerhouse will continue to contribute to the emergence of new technological, policy and economic paths that, through scientific discovery, innovation and entrepreneurship, are advancing the vision of clean, smarter and prosperous communities in California and elsewhere. This transformational effort is providing a first-mover advantage for its business and entrepreneurial community to maximize participation in the huge global market for green, smart technological solutions that clean energy, water, transport, food and urban infrastructure require.

Concluding remarks

California's climate vulnerabilities are mitigated only by its capacity to innovate and the support of citizens to act on issues such as energy efficiency, environmental quality and more recently on developing regulatory frameworks and institutions to support the transition to a smart, prosperous clean, resource-efficient economy. California presents an important case study to illustrate some of the challenges and opportunities that WEN currently presents in the context of advancing an ambitious clean economy model. In the case of WEN, new organizational and managerial approaches that can take advantage of decentralized technological solutions and smart data management systems will produce metrics and models that can advance our capacity to best manage shared resource systems.

A holistic policy perspective and an integrated institutional systems approach is necessary as part of the planning and execution of California's clean energy transition to provide smart governance, technological and business solutions that can effectively minimize resource demand tensions as part of this interdependence. Important interagency efforts are being implemented to best support a balanced WEN, but there is room for improvement. An integrated systems management strategy and policy framework for WEN as part of the clean economy model of California can maximize the return from investment in innovation and other advancements to the entire web of systems.

To date, water consumption and energy generation interact in an intricate manner. WEN is no doubt a more salient policy issue in California during and after severe droughts. New governance structures from multi-level, multi-stakeholder dialogue and collaboration are needed to support a high degree of policy and stakeholder coordination. These nascent but now ongoing processes should result in more water–energy efficient operations as policy and technological innovation in WEN solidifies. Integrated WEN management systems in California ultimately aim to maximize the benefits to all communities (e.g., urban and rural), minimize regulatory and infrastructure costs to competing users from industry (e.g., electricity power generators and the agricultural sector), while at the same time protecting our ecosystems.

Since 2006, continuous policy support by citizens and government to the clean economy model has provided a stable regulatory framework that provides certainty for the business and economy-wide transformation of California. Central to this process is an informed and committed citizenry that provides the foundation for societal transformation in support of this vision. Decarbonization and resource-efficient transformation is occurring in the context of an emerging socio-techno-economic paradigm influenced by new exchange and communication frameworks such as highly efficient online connectivity through multiple media and devices. Moreover, the exponential increases in the capacity to collect and analyze big data in real time through IT, artificial intelligence (AI) and remote sensing allow for precision measuring, verification and reporting for enhanced decision-making in business and government. WEN no doubt will benefit from these new technological capacities and smart management systems.

Finally, emerging concepts such as the sharing economy and the internet of things (IoT) are transforming business models and services for our homes, education, transport, energy, water and urban systems. These are the foundations of a set of seamless infrastructure systems needed to support the emergence of smarter, cleaner and prosperous connected communities. More education on the economic, environmental and energy security benefits of a balanced WEN is therefore needed to spark more interagency and multi-stakeholder cooperation. In other words, institutional realignment for effective collaborative action across government agencies, along with close consultation and interaction with communities and the business sector, will bring about innovative and smart resource management approaches and technological solutions to address the WEN challenge.

Notes

1 The "yuck" effect refers to the negative public perception of using treated wastewater for potable water reuse. Groups resistant to such water recycling plans in Los Angeles coined the phrase "toilet to cup" as a way to convince the public of the "yuck effect". (Morin, 2015; Ching, 2010)

2 The California Air Resources Board is a part of the California Environmental Protection Agency, an organization which reports directly to the Governor's Office in the Executive Branch of California State Government. Its stated mission is "To promote and protect public health, welfare and ecological resources through the effective and efficient reduction of air pollutants while recognizing and considering the effects on the economy of the state."

References

Bales, R., Conklin, M., Viers, J., Fisher, A., Fogg, G., and Kiparsky, M. (2016). Foundations for California's water security in a changing climate. *A policy position paper prepared for presentation at the conference on Water and Fire: Impacts of Climate Change*, convened by the Institute on Science for Global Policy (ISGP), April 10–11, 2016, at California State University, Sacramento.

Bowman, M. (2016). *Mobilizing impact-focused capital for water sustainability: A water policy expert's perspective on impact investing*. Talk given to the Stanford Woods Institute of the Environment, May 9. Retrieved from https://woods.stanford.edu/news-events/event/mobilizing-impact-focused-capital-water-sustainability-water-policy-expert%E2%80%99s

Bureau of Economic Analysis. (2016). *The 2016 annual update to the national income and product accounts*. United States Bureau of Economic Analysis

California EPA. (2017). Retrieved from www.calepa.ca.gov/About/

Carle, D. (2004). *Introduction to water in California*. Berkeley, CA: University of California Press.

Carter, N. T. (2010). Retrieved from https://fas.org/sgp/crs/misc/R41507.pdf

Carter, N. T. (2011). *Energy's water demand: Trends, vulnerabilities, and management*. DIANE Publishing.

CDWR. (2017). Retrieved from www.water.ca.gov/waterconditions/background.cfm

CEC. (2003). *Integrated energy policy report*. Retrieved from www.energy.ca.gov/reports/100-03-019F.PDF

CEC. (2016). *California energy demand 2016–2026, revised electricity forecast, Volume 1: Statewide electricity demand and energy efficiency*. California Energy Commission.

CEC. (2017a). *California hydroelectric statistics & data*. Retrieved from www.energy.ca.gov/almanac/renewables_data/hydro/

CEC. (2017b). Retrieved from www.energy.ca.gov/research/iaw/projects.html#water

Ching, L. (2010). Eliminating 'Yuck': A simple exposition of media and social change in water reuse policies. *International Journal of Water Resources Development*, 26(1), 111–124.

Christian-Smith, J., and Wisland, L. (2015). *Clean energy opportunities in California's water sector*. Cambridge, MA: Union of Concerned Scientists. Retrieved from www.ucsusa.org/clean-energy/california-andwestern-states/clean-energy-opportunities-california-water-sector.pdf

CPUC. (2017). *Water energy calculator*. Retrieved from www.cpuc.ca.gov/uploadedFiles/CPUC_Public_Website/Content/Utilities_and_Industries/Energy/Energy_Programs/Demand_Side_Management/EE_and_Energy_Savings_Assist/Copy%20of%20CPUCWaterEnergyCalculatorVersion105.xlsm

Dale, L., Fujita, S. K., O'Hagan, J., and Hanemann, M. (2008). *The interaction of water and energy in California: Climate change and price impacts*. Berkeley, CA: Energy Analysis and Environmental Impacts Division. Berkeley Lab. University of California.

Diamond, M., and Allcorn, S. (2009). *Chapter 3: Private selves in public organizations: The psychodynamics of organizational diagnosis and change*. New York: Springer.

Diffenbaugh, N. S., Swain, D. L., and Touma, D. (2015). Anthropogenic warming has increased drought risk in California. *Proceedings of the National Academy of Sciences*, 112(13), 3931–3936.

Funk, C., Hoell, A., and Stone, D. (2014). Examining the contribution of the observed global warming trend to the California droughts of 2012/13 and 2013/14. *Bulletin of the American Meteorological Society*, 95(9), S11.

GEI Consultants, Navigant Consulting Inc. (2010). *Embedded energy in water studies, study 1: Statewide and regional water-energy relationship, draft final report*. California Public Utilities Comission.

Hardin, E., AghaKouchak, A., Qomi, M. J. A., Madani, K., Tarroja, B., Zhou, Y., and Samuelsen, S. (2017). California drought increases CO 2 footprint of energy. *Sustainable Cities and Society*, 28, 450–452.

Kammen, D. M., Pérez Henrıquez, B. L., and Johnston, J. (2015). California's climate policy and the development of clean energy systems institutional foundations. In A. Bumpus, J. Tansey, B. L. Pérez Henríquez, and C. Okereke (Eds.), *Carbon governance, climate change and business transformation*. Routledge pp. 175–187.

Little Hoover Commission. (2010, August). *Managing for change: Modernizing California's water governance*. Sacramento, CA: Little Hoover Commission.

Mann, M. E., and Gleick, P. H. (2015). Climate change and California drought in the 21st century. *Proceedings of the National Academy of Sciences*, 112(13), 3858–3859.

Morin, M. (2015, May 24). Turning sewage into drinking water gains appeal as drought lingers. *Los Angeles Times*.

NASA. (2014a). *NASA data underscore severity of California drought*. Retrieved from www.jpl.nasa.gov/news/news.php?feature=4412

NASA. (2014b). *Needed: 11 trillion gallons to replenish California drought*. Retrieved from https://science.nasa.gov/science-news/science-at-nasa/2014/16dec_drought/

Navigant Consulting, Inc. (2006). *Refining estimates of water-related energy use in California*. California Energy Commission, PIER Industrial/Agricultural/Water End Use Energy Efficiency Program. CEC-500-2006-118.

Neelin, J. D., Langenbrunner, B., Meyerson, J. E., Hall, A., and Berg, N. (2013). California winter precipitation change under global warming in the Coupled Model Intercomparison Project phase 5 ensemble. *Journal of Climate*, 26(17), 6238–6256.

Office of the Governor. (2014). Retrieved from www.gov.ca.gov/news.php?id=18368

Office of the Governor. (2017). Retrieved from www.gov.ca.gov/news.php?id=19747

Poseidon Water. (2016). *Nation's largest seawater desalination plant marks one-year anniversary*. Retrieved from www.carlsbaddesal.com/news/nations-largest-seawater-desalination-plant-marks-one-year-anniversary

Public Policy Institute of California. (2015). *PPIC statewide survey: Californians and the environment*. Retrieved from www.ppic.org/main/publication.asp?i=1159

Rheinheimer, D. R., Ligare, S. T., Viers, J. H., and Center for Watershed Sciences, University of California, Davis. (2012). *Water-energy sector vulnerability to climate warming in the Sierra Nevada: Simulating the regulated rivers of California's West Slope Sierra Nevada*. California Energy Commission. Publication number: CEC-500-2012-016.

Scherson, Y. D., and Criddle, C. S. (2014). Recovery of freshwater from wastewater: Upgrading process configurations to maximize energy recovery and minimize residuals. *Environmental Science & Technology*, 48(15), 8420–8432.

Seager, R., Hoerling, M., Schubert, S., Wang, H., Lyon, B., Kumar, A., and Henderson, N. (2014). *Causes and predictability of the 2011–14 California drought*. Retrieved from http://cpo.noaa.gov/ClimatePrograms/ModelingAnalysisPredictionsandProjections/MAPPTaskForces.

Seager, R., Hoerling, M., Schubert, S., Wang, H., Lyon, B., Kumar, A., and Henderson, N. (2015). Causes of the 2011–14 California drought. *Journal of Climate*, 28(18), 6997–7024.

SESI. (2017). Retrieved from http://sustainable.stanford.edu/campus-action/stanford-energy-system-innovations-sesi.

Simpson, I. R., Seager, R., Ting, M., and Shaw, T. A. (2015). Causes of change in Northern Hemisphere winter meridional winds and regional hydroclimate. *Nature Climate Change*.

State of California. (2010). *20x2020 water conservation plan*. State of California.

Tett, G. (2015). *The silo effect: The peril of expertise and the promise of breaking down barriers*. New York: Simon and Schuster.

U. S. DOE. (2014). *The water-energy nexus: Challenges and opportunities, overview and summary*. Retrieved from https://energy.gov/sites/prod/files/2014/07/f17/Water%20Energy%20Nexus%20Executive%20Summary%20July%202014.pdf

U. S. DOE/U. S. EIA. (2016, May). *International energy outlook 2016*. Retrieved from www.eia.gov/outlooks/ieo/pdf/0484(2016).pdf

U. S. EPA. (1998). *Water recycling and reuse: The environmental benefit*. Retrieved from https://www3.epa.gov/region9/water/recycling/pdf/brochure.pdf

United States Drought Monitor. (2017). Retrieved from http://droughtmonitor.unl.edu/Home/Narrative.aspx

USGS. (2017). Retrieved from https://ca.water.usgs.gov/data/drought/

Varhola, A., Coops, N. C., Weiler, M., and Moore, R. D. (2010). Forest canopy effects on snow accumulation and ablation: An integrative review of empirical results. *Journal of Hydrology*, 392(3), 219–233.

Water Education Foundation. (2017). *Aquapedia*. Retrieved from www.watereducation.org/aquapedia/sierra-nevada

Water in the West. (2013). *Water and energy nexus: A literature review*. Stanford, CA: Water in the West Program, Stanford University.

Williams, A. P., Seager, R., Abatzoglou, J. T., Cook, B. I., Smerdon, J. E., and Cook, E. R. (2015). Contribution of anthropogenic warming to California drought during 2012–2014. *Geophysical Research Letters*, 42(16), 6819–6828.

Wilkinson, R. (2000). *Methodology for analysis of the energy intensity of California's water systems and an assessment of multiple potential benefits through integrated water-energy efficiency measures. University of California Santa Barbara.*

Xie, J. (2006). *Dealing with water scarcity in Singapore: Institutions, strategies, and enforcement, China*. Addressing Water Scarcity Background Paper No. 4, The World Bank – Environment and Social Development Department – East Asia and Pacific Region, Washington, DC.

32
THE UN, GLOBAL GOVERNANCE AND THE SDGS

Maria Ivanova and Natalia Escobar-Pemberthy

Introduction

Over several millennia, human societies have changed the world around them in search for sufficient food, water, shelter, and fuel. Their natural environment – the land, water, air, and living species – has changed under the increasing pressure of a growing population, economy, and desire to acquire wealth. In the last 50 years, however, humans have altered the environment more rapidly and extensively than ever and created international institutions to guide and regulate collective action. The profound transformation of the planet has led to significant advances in human welfare (although for some more than others) while also putting a significant strain on the ability of the planet to sustain human activity.

Water, food, energy, and land have traditionally been viewed as local concerns since they appear to be confined within a certain physical space and a result of local practices and policies. Addressing them is then expected to lie within a limited set of actors often directed by a national government with the requisite authority. The ability of national actors to provide clean water, adequate land and food, and sustainable energy has, however, been impacted by global concerns such as transboundary pollution, drought, desertification, and climate change. Moreover, in a globalized economy, divergent responses to localized issues create competitiveness tensions that add a further international dimension to their effects. Traditional divisions of the world into sectors and into local, national, and international domains therefore no longer provide an accurate map of the terrain we tread. Most contemporary problems are multidimensional, complex, and interconnected.

Water, energy, and food as well as land and minerals are linked in a nexus of interactions, which in turn links to policy frameworks and is embedded in a broader governance system. Conceptually, the resource nexus approach allows us to connect multiple issues in causal relationships, to map out and understand the consequences, and to begin creating integrated solutions at multiple levels of governance. A nexus approach forces us to connect decisions in one area to impacts in the other areas and to anticipate possible trade-offs and synergies. It allows us to design and prioritize response options viable across the different sectors and thus to use and manage these resources sustainably.

Numerous national and international institutions are active in the nexus domains identifying problems, articulating goals, and implementing them. The Food and Agriculture

Organization is the anchor institution for food issues. The UN Convention on Combatting Desertification is a central body in the land domain. The International Energy Agency, the International Atomic Energy Agency, and the International Renewable Energy Agency[1] manage the energy domain. There is no one agency, however, responsible for water or for minerals. Multiple organizations operate in those domains. In all of the nexus domains, the work of the agencies has taken place largely in silos, with a focus on the specific issue and often missing the interactions. Uncoordinated approaches in policy development and implementation within each domain and across the sectors have led to inconsistencies, inefficiencies, additional cost, and imperfect allocation of human and financial resources. The multiplicity of institutions and the fragmented goals and targets place a burden on all countries, and especially on developing countries, in terms of participation in multilateral processes, compliance with and effective implementation of legal instruments, reporting requirements, and, ultimately, attaining the desired results.

The core instrument for promoting integration across issues, across scales, and across agencies has been the setting of global goals. The Millennium Development Goals articulated common aspirations that focused attention and prioritized assistance and investment but remained focused on specific issues. Recognizing the need to meet growing social and environmental challenges in an integrated manner, in 2015, governments adopted a set of universal and ambitious global goals, the Sustainable Development Goals. The 17 goals present a new integrated approach to development for both industrialized and developing countries. They offer a set of aspirations to attain as a package and will require policy coherence across sectors and across levels of governance.

As governments embark on a new venture to meet growing economic, social and environmental challenges under the integrated approach of sustainable development, the resource nexus demands systematic review and response. In this chapter, we explore how the global governance mechanisms – existing and emerging – address the resource nexus and offer insights on the challenges of integrated approaches, common policy instruments, and institutional architectures. Ultimately, we seek to contribute to improving policy outcomes and generate research agendas for engaged scholarship.

The UN and the resource nexus

As states recognize their inability to address problems on a national basis, they have turned to international organizations for the facilitation of a collective response. In 2015, the UN marked 70 years of its existence as the main international organization with the goal to

> save succeeding generations from the scourge of war . . . to regain faith in fundamental human rights, in the dignity and worth of the human person, in the equal rights of men and women and of nations large and small, and to establish conditions under which justice and respect for the obligations arising from treaties and other sources of international law can be maintained, and to promote social progress and better standards of life in larger freedom.
>
> (United Nations, 1945, Preamble)

The UN has therefore been the core global institution with responsibility to provide for the betterment of humankind.

The UN, however, is a constellation of agencies, programmes, funds, commissions, and other bodies that have a mandate in one or more of the issues that constitute the resource nexus agenda. Each of the resource areas therefore has a number of international institutions operating in the space and responsibilities are spread across multiple organizations. A coherent strategy and response within an issue area have been difficult and connecting across the resource nexus even more so. In a 2008 evaluation of the institutional arrangements for global environmental governance, the Joint Inspection Unit of the UN articulated a dynamic that also holds across the resource nexus: there is a lack of "an overarching authority for global environmental governance within the UN system," and as a result, "[r]esponses . . . have become sector-specific, specialized and fragmented, despite some improvement through the formation of clusters" (Inomata, 2008, 30). An integrated approach would therefore require significant change across two levels – coordination and coherence *within* the resource nexus areas, and coordination and coherence *across* them.

Water

Water is fundamental to life. It ensures the sustainability of ecosystems and of human life. Water security supports livelihoods, health and welfare, production and food security, and reduces vulnerability to disasters (United Nations, 2012). Institutional arrangements for the governance of water exist at all levels – from the local to the national, regional, and global. The basic components of an institutional framework for water management include regulatory and legislative bodies that create rules or laws about water access, use, and quality; regulations and legislation governing the access, use, and quality of water; and bodies or entities that implement programs enabling water access, regulating water use, or improving water quality. Such arrangements vary among countries depending on the availability of water resources, the extent to which water quality is at risk, and the general structure of government (federal or unitary, parliamentary or congressional). It is even more challenging to create international governance mechanisms for water management.

UNESCO is a key institution working to improve scientific knowledge for water resources management through the International Hydrological Programme and the UNESCO-IHE (Institute for Water Education) in the Netherlands and over 20 affiliated research centers around the world. Recognizing the need to bring together the various UN entities working on water, in 2003, governments created UN-Water, a coordination mechanism that includes 31 UN bodies and 38 partner organizations outside of the UN system. Its purpose is to promote synergies and facilitate joint efforts in information, monitoring and reporting on water and sanitation, building knowledge, and catalyzing action (UN-Water, 2016). UN-Water's work deals with freshwater resources, including quality, quantity, development, assessment, management, monitoring, and use in agriculture, domestically, and as ecosystem requirements. Sanitation and water-related disasters are also important areas of focus.

Water was the first of the resource nexus issues to be recognized as an area where multilateral cooperation was necessary and critical. The UN established the international drinking water supply and sanitation decade from 1981 to 1990, the first water decade (Figure 32.1). The goal was to ensure access to clean drinking water and to sanitation across the world by making these issues a priority among countries, by supporting initiatives at the national level to facilitate the achievement of these goals, and to mobilize financing, training and technology development. The second international water decade, Water for Life, was launched from 2005 to 2015 and provided further impetus for delivering on the global water goals. It followed the establishment, in 2004, of the UN Secretary-General Advisory Board for Water and Sanitation to "solve the planet's foremost water and sanitation troubles" (UNSGAB, 2015, 4). The Board comprised 20 individuals with a diverse set of backgrounds chosen by the Secretary-General's office to act in their personal capacity and deploy their networks and influence to positively impact the global water agenda. The Board concluded its work in 2015 as the UN concluded the second water decade and launched the new Sustainable Development Goals.

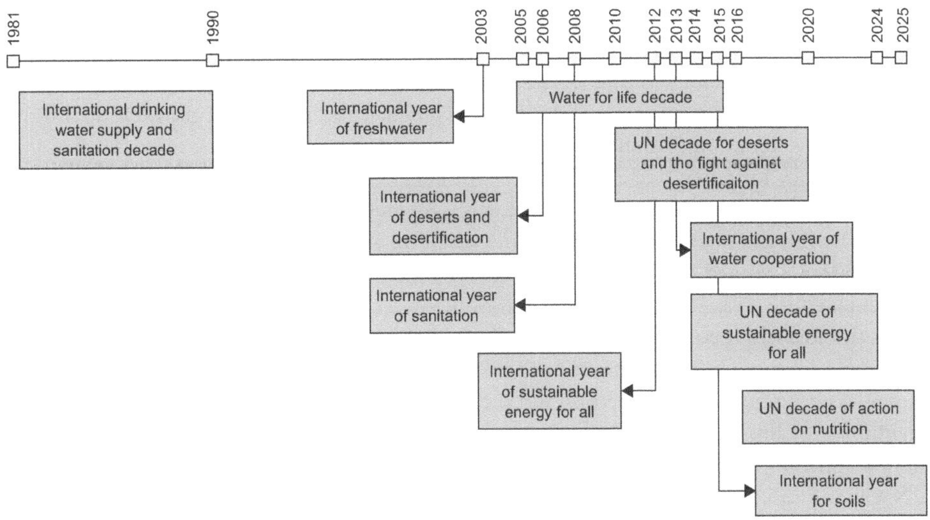

Figure 32.1 Focus areas for international cooperation, as recognized by the UN

Energy

Adequate, secure, and sustainable energy at affordable prices is critical to human well being. Over one billion people, however, have no access to electricity and 2.9 billion cook with inefficient and polluting fuels (United Nations and World Bank, 2016). With growing population and increasing standards of living, global energy demand is rising steadily and rapidly. The World Energy Council projects a doubling of global energy demand by 2050 (World Energy Council, 2013). Energy production and consumption is also the main contributor to climate change. Energy security therefore will be defined by energy transitions to renewable energy sources as countries seek to tackle the challenge of making energy available to all while reducing its carbon intensity. The commitments that countries and companies made as a result of the Paris Agreement have advanced the demand for and supply of low-carbon energy options and shifted investment towards such low-carbon sources. Coupled with a fall in fossil-fuel subsidies from $500 billion in 2014 to $325 billion in 2015, these dynamics point to a transition in the global energy systems (International Energy Agency (IEA), 2016). Importantly, while recent years have seen a number of calls for energy independence, the future of the energy sector will continue to be characterized by interdependence and demand an even greater role of international organizations to produce international energy standards.

Much like in the water domain, national interests, processes, and policies dominate the governance of energy. As Dubash and Florini note:

> Global energy governance consists of inadequate and uncoordinated mechanisms attempting to achieve fragmented and unprioritized objectives which pose as yet unresolved structural trade-offs. We do not currently have the institutional infrastructure needed to address the significant and urgent challenges we face.
>
> (Dubash and Florini, 2011, 15)

Until the 1990s, two intergovernmental energy organizations dominated the field – the International Atomic Energy Agency (IAEA) created in 1957, and the International Energy Agency

(IEA) created in 1974. While the IEA is regarded as "the "world's foremost multilateral energy organization" (Van de Graaf, 2015, 79) and "the closest we have to a World Energy Organization" (Van de Graaf and Colgan, 2016, p. 5), its limited membership (only OECD member countries can be members of the IEA) does not grant it the requisite legitimacy or authority. IEA's exclusive focus on fossil fuels prompted states, in 1981, to call for the creation of an international agency for renewable energy.

IRENA, the International Renewable Energy Agency, was created over 20 years later, in the late 2000s with the objective to promote the adoption and sustainable use of renewable energy sources. Headquartered in Abu Dhabi, the agency is not formally a UN body but is an intergovernmental entity with 150 member states as of the end of 2016 and an additional 27 states have begun the formal process for membership. Two other institutional responses – UN-Energy and Sustainable Energy for All – addressed the increasing global concerns about the sustainability of energy consumption and production, and the impact of the increasing demand on human and environmental health (Van de Graaf, 2013). UN-Energy is an interagency mechanism to promote a coherent approach to achieving sustainable energy. Created as a subsidiary body to the UN Chief Executives Board (UN CEB) in 2004, it engages the multiple UN bodies active in energy. Sustainable Energy for All is a unique partnership between the UN and the World Bank with the UN Secretary-General and the World Bank President co-chairing a high-level group of governments, private sector, and civil society representatives. Its charge was to develop sustainable energy – energy, which, in its production or consumption, has minimal negative impacts on the health of humans and of ecosystems and can be supplied continuously to future generations.

Recognizing the importance of the linkages between energy, poverty eradication, and sustainable development as well as the role of energy as a contributor to pollution, emissions, and climate change, the UN announced 2012 as the International Year of Sustainable Energy for All and 2014–2024 as the UN Decade of Sustainable Energy for All (Figure 32.1). Indeed, the Sustainable Energy for All initiative was launched in response to the 2012 International Year of Sustainable Energy for All. The core concerns for the decade currently underway include the diversification of energy supply and the development of advanced, cleaner, more efficient, affordable and cost-effective technologies, in order to transform the mix of energy in ways that respond to the rising global demand.Virtually all UN agencies are engaged in the decade leading activities related to their particular mandates.[2]

Food

Food is the issue area with a long-standing lead agency in the UN. The Food and Agriculture Organization was created in 1945 with the mission to achieve food security – ensure that "people have regular access to enough high quality food to lead active, healthy lives." Food security is therefore about the availability, access, use, and stability/sustainability of food (FAO, 2014). To this end, three core goals define the work of the organization:

> the eradication of hunger, food insecurity and malnutrition; the elimination of poverty and the driving forward of economic and social progress for all; and, the sustainable management and utilization of natural resources, including land, water, air, climate and genetic resources for the benefit of present and future generations.
>
> (FAO, 2015)

As demand for food grows, there is increasing pressure on water and energy and the linkages among them demand more knowledge and better governance. Large-scale water infrastructure projects, for example, may have positive effects across several dimensions, producing energy and offering water storage. They might, however, negatively impact downstream ecosystems, agricultural systems, and societies as they change land-use patterns and might entail resettlements. Similarly, biofuels production through an irrigated agriculture scheme may help improve energy supply and create jobs, but may also result in increased competition for land and water resources with impacts on local food security. Food security efforts therefore need to be water, land, and energy smart. Yet, agricultural intensification – the core food security strategy – is also water, land, and energy intensive. Increasing energy access and ensuring climate neutrality has to also respect water realities and land use. Improving water availability may demand greater energy use as in the process of desalination and attaining a balance between water, energy, and food demands would require unprecedented integration of information, planning, and prioritization.

In 1974, the first World Food Conference reaffirmed the right to be free of hunger and malnutrition and set the goal of eradicating hunger in the world. Even though progress was made, this goal remained unattained. Agenda 21 and the first World Food Summit held in 1996 reinstated the global commitment to eliminate hunger and malnutrition and to achieve food security for all (FAO, 1996; UNCED, 1992). The summit also raised awareness about the importance of agriculture capacity for food security, and established specific goals towards the reduction of hunger and undernourishment by 2015. This was the baseline for the definition of a specific target for hunger reduction in the Millennium Development Goals (United Nations, 2000). The work of the Committee on World Food Security and the 2014 Rome Declaration on Nutrition and Framework for Action have served as platforms for countries and stakeholders to work together to ensure food security and nutrition for all.

Initially, hunger reduction showed rapid progress but was followed by a decline in the first five years of the millennium, and then a rebound starting around 2008 (United Nations, 2015). Additional efforts then focused on multiple fronts. In 2002, an initial review of the outcomes of the World Food Summit called for additional guidelines and financial support for developing countries to support their food security programs (FAO, 2002). In 2009, the UN Secretary-General established a High-Level Task Force on Food and Nutrition Security (United Nations, 2009) that promoted a comprehensive unified response to achieve the goals established in five areas: eradication of children malnutrition, food security, food systems sustainability, small farming productivity and income, and food waste elimination. This led to the establishment at Rio+20 of the Zero Hunger Challenge with member states reaffirming the right "to have access to safe, sufficient and nutritious food, consistent with the right to adequate food and the fundamental right of everyone to be free from hunger" (United Nations, 2015) and calling for all the actors involved to unite around those five policy areas to end hunger and eliminate malnutrition.

In 2015, the UN General Assembly announced the International Year for Soils, and a year later proclaimed 2016–2025 as the nutrition decade (Figure 32.1). The goal was to mobilize action to eradicate hunger and improve nutrition around the world. The decade will provide platform for a broad group of actors to collaborate on a range of issue related to food and nutrition. FAO and WHO will lead the efforts in collaboration with the World Food Programme (WFP), the International Fund for Agricultural Development (IFAD) and the UN Children's Fund (UNICEF).

Land

Land management has been a critical issue on the UN agenda because of the importance of soil for agriculture and resilience and because of land's vulnerability as an ecosystem. The

Food and Agriculture Organization has promoted sustainable land management and engaged in land degradation assessment, land policies and planning working with local communities. Land degradation, desertification, and drought have been important topics for decades. In 1977, the UN Environment Programme hosted the first UN Conference on Desertification in Nairobi and put the issue on the international political agenda. The conference also triggered greater scientific interest and catalyzed the development of national policies and, subsequently, an international legal instrument – the UN Convention to Combat Desertification, which would only be adopted in 1994 and enter into force in 1996.

Through the UN Convention to Combat Desertification governments acknowledged the connection between environment and sustainable land management and established a partnership to reverse and prevent desertification and land degradation, and to mitigate the effects of drought in affected areas (United Nations, 1994). Under this framework, the UN proclaimed the year 2006 as the International Year of Deserts and Desertification to highlight the threat of desertification and its implications for biodiversity loss. The problems persist and the 2010–2020 decade was named the UN Decade for Deserts and the Fight Against Desertification (Figure 32.1). The UN entities leading the work on land during the decade include UNCCD, IFAD, UNEP, UNDP and the Department of Public Information (DPI). In 2015, the convention established a Global Mechanism to support countries in achieving specific targets, defining national baselines, and making progress in land degradation neutrality.

Minerals

Minerals are essential for economic activities, health, and the provision of multiple materials used in daily activities. Their exploitation is also the source of negative economic, social, and environmental externalities. Importantly, mining has increased in developing countries where the infrastructure and institutions to deal with these externalities are limited. The UN has been focusing its work related to minerals on three aspects: production and consumption, environmental, social and economic impacts, and the maximization of benefits and sustainability. In 1992, Agenda 21 pointed to the need for sustainable management and conservation of areas degraded by extractive industries (UNCED, 1992) and controlling the effects of the industry on environmental and human health and societal dynamics was a key focus of the discussions at the Commission on Sustainable Development in 2010 and 2011. There are no specific UN institutions, however, with an explicit mandate to work exclusively on minerals and mining. The International Seabed Authority created in 1982 under the UN Convention on the Law of the Sea holds such mandate for the seabed and ocean floor beyond areas of national jurisdiction. On land, a number of existing agencies have created initiatives to improve the management of minerals development efforts. For example, UNEP launched the International Resource Panel in 2007 to create knowledge for the improvement of resource use. The Panel includes scientists, government representatives and other stakeholders and produces reports on key global resource use issues. UNDP implements the ACP (African, Caribbean and Pacific)-EU (European Union) Development Minerals Programme, a three-year, €13.1 million capacity-building initiative for industrial minerals, construction materials, dimension stones, and semi-precious stones.

During the Rio+20 summit, mining industries were acknowledged as catalyzers of economic development and poverty reduction. The need for more effective regulatory frameworks, policies, and practices was also acknowledged, however, recognizing that they need to include transparency and accountability measures that avoid the connection between mining activities and illicit financial flows. Stakeholders strongly advocate for ensuring human and environmental rights as related to extractive industries, transparency, standards, and the payment of benefits to local communities. Important international networks such as the International Council on Mining and Metals, also work on the connection between foreign direct investment and mining. To date, the UN has not used the international year and the international decade designation to focus work on minerals.

Global goals as a governance tool

The institutional landscapes for each of the nexus areas will likely remain intact, or may even expand as additional entities begin operating in this space. Efforts at integration within each – integrated water resources management, integrated energy development, and integrated rural development – will continue. Governments and many UN agencies, however, have recognized the need to break the silo approach and work across issues and across institutions to ensure that global resources are appropriately and efficiently managed. The Sustainable Development Goals – regarded as indivisible – offer an opportunity to develop an integrated vision and a range of new collaborative approaches and initiatives to instigate the necessary institutional transformation and action.

Global goals have served to conceptualize and define a common international agenda since the end of the Second World War when governments committed to reduce poverty as a way of ensuring peace, security and stability in the international system (Ivanova and Escobar-Pemberthy, 2016; Kanie and Biermann, 2017). Entering the 21st century, the international community took up once again the idea of a global set of goals to frame international development with the adoption of the Millennium Development Goals (MDGs), a set of eight ambitious goals that offered a multidimensional perspective on poverty (see Figure 32.2). Applicable exclusively to developing countries, the MDGs incorporated different policy areas ranging from education to gender equality and health, environmental sustainability, and foreign aid. They were to provide a mechanism to evaluate countries' progress in comparison to a baseline. The goals, and their 21 targets, established intermediary benchmarks to plan, monitor and evaluate progress towards international development.

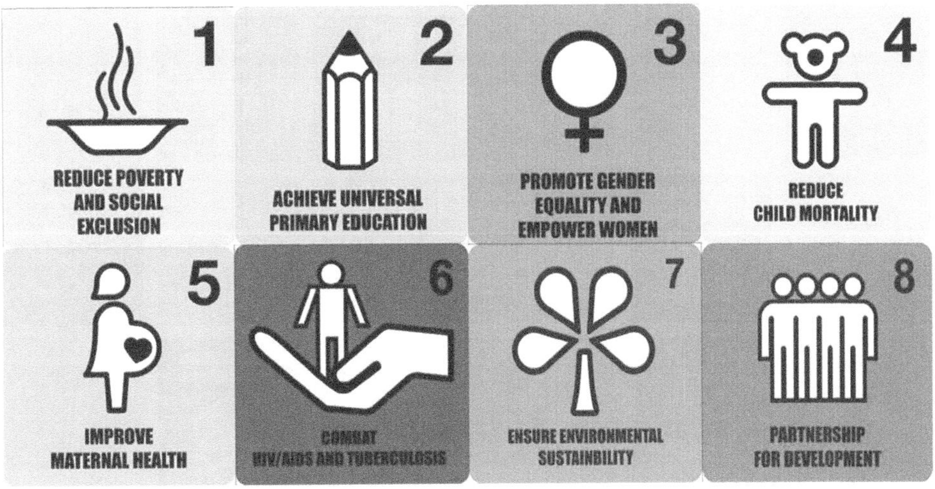

Figure 32.2 Millennium Development Goals

Source: United Nations

Undoubtedly a key policy instrument for the international community, the MDGs have shown both promise to advance the international agenda on human development and challenges to attain human security. Goals and targets have proved useful instruments for monitoring progress and prompting political mobilization, focusing advocacy and targeting financial aid at the global and national level (Manning, 2009; Waage et al., 2010). As instruments of public policy, the MDGs articulated a global political consensus that engaged both political and social actors, and created political structures to focus the efforts, advocacy and motivation of countries and stakeholders around international development (McArthur, 2013). The MDGs also created a culture of data collection, monitoring and assessment central to the evolution and implementation of public policies. Ultimately, the MDGs focused on a systemic contribution: improving human development contributes to promoting and ensuring peace and security as well as to the stability of the international system by meeting the needs that cause conflict, war, and instability, such as poverty, hunger and lack of education and health services (Manning, 2009; Melamed and Scott, 2011; Sachs and McArthur, 2005; Waage et al., 2010).

Implementing the MDGs, however, has not been an easy task. Lack of capacity and financing affected countries' trajectories towards the attainment of the different targets. Importantly, the MDGs focused primarily on poverty reduction and were based on a traditional approach to development. They did not offer an integrated and holistic vision but a rather piecemeal approach that did not account for trade-offs and synergies. At Rio+20, governments agreed to articulate a new set of global goals, the Sustainable Development Goals, to "build on the foundation laid by the MDGs, seek to complete the unfinished business of the MDGs, and respond to new challenges" (UN General Assembly, 2015). While maintaining the same core as the MDGs in policy areas such as poverty reduction, education, health, and environmental sustainability, the SDGs present a more integrated approach to poverty reduction and sustainable development and, importantly, are universal (see Figure 32.3).

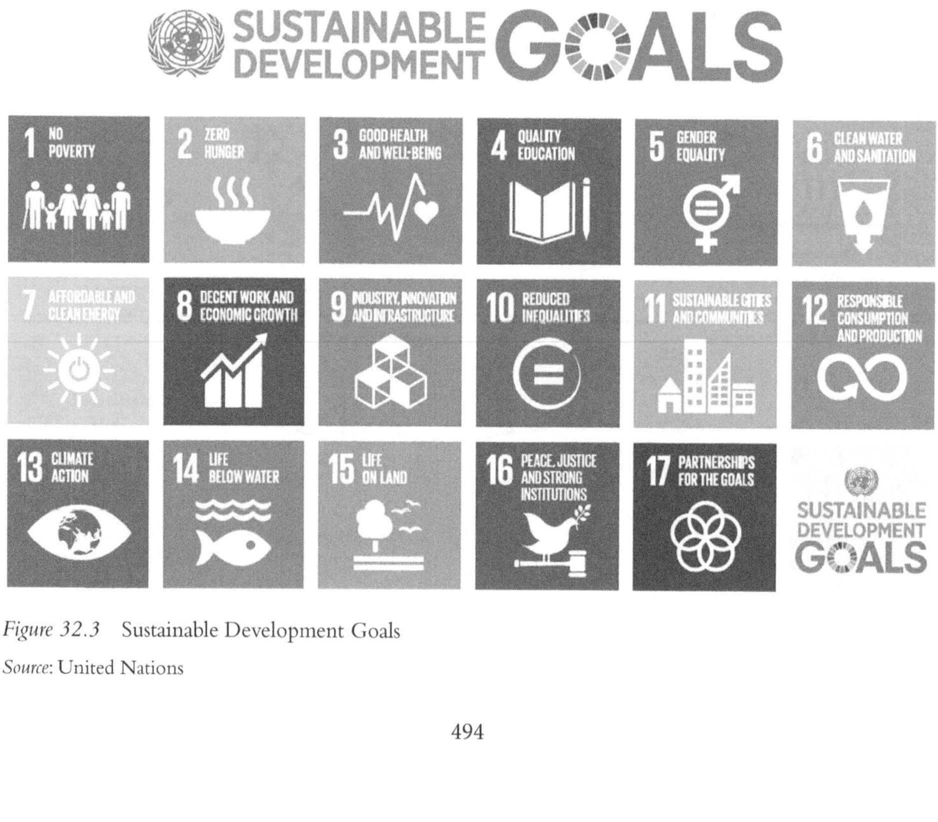

Figure 32.3 Sustainable Development Goals

Source: United Nations

Perhaps the greatest evolution from the MDGs is in the environmental sustainability category. There are now seven discrete goals that recognize the importance of various ecosystems and the services they provide, emphasize the need for "urgent action to combat climate change and its impacts," add a focus on water, energy, food security, sustainable production and consumption, and cities. In the area of global partnership, the SDGs also expand by articulating two ambitious goals, "promote peaceful and inclusive societies for sustainable development, provide access to justice for all and build effective, accountable and inclusive institutions at all levels," and "strengthen the means of implementation and revitalize the global partnership for sustainable development." The SDGs, therefore, establish specific targets regarding partnerships and governance across different levels. As defined, the Sustainable Development Goals constitute an instrument for the international community to prioritize the issues required for improving socio-economic development and responsible environmental stewardship. The key challenge is how to enable the integration among the various goals not only at the conceptual level but also with other initiatives and institutional architectures within the UN system.

Nilsson, Griggs, and Visbeck (2016) suggest a simple and powerful scoring methodology that policymakers can use to think systematically about the many interactions among the 17 SDGs. They identify seven types of interactions among the goals – from indivisible (the most positive and scoring +3) to cancelling (the most negative and scoring -3). Policymakers can focus on a specific SDG and map out, score, and explain potential interactions with the other goals. Taking into account the specific geography, governance, and technologies within which they operate, they can empirically identify ways to create the integration necessary and avoid potential conflicts.

The resource nexus in the global goals

Policy and scholarly discussions of the resource nexus have emphasized the interrelationships among its dimensions and policy responses as well as the need for information, cooperation and partnerships to achieve integrated management of these resources (Hellegers et al., 2008; Scott et al., 2015). The Millennium Development Goals articulated aspirations for only two of the resource nexus dimensions – water and food. The Sustainable Development Goals, on the other hand, offer goals and targets for most of the resources. Table 32.1 illustrates the evolution of the global goals and targets for water, food, energy, land, and minerals from the MDGs through the SDGs as well as their interconnected nature across the sustainable development agenda.

Food has always been present in global development goals from the perspective of hunger eradication (FAO, 2014). In the MDGs, hunger and poverty were the focus of MDG1 – eradicate extreme hunger and poverty. Target 1C required the halving, by 2015, the proportion of people who suffered from hunger. Overall, the 2000–2015 period witnessed an important reduction of almost half in the proportion of undernourished people in developing countries, from 23.3% in 1990–1992, to 12.9% in 2014–2016 (United Nations, 2015). This means that nearly one in nine individuals do not have enough to eat.

The SDGs separated poverty and hunger into two goals: SDG1 – No Poverty and SDG2 – Zero Hunger. The 2030 Sustainable Development Agenda committed "to end hunger and to achieve food security as a matter of priority and to end all forms of malnutrition" (UN General Assembly, 2015). The focus on food expanded to include food security, nutrition, and sustainable agriculture. Food production, consumption and waste management are also part of SDG12 – Responsible Consumption and Production. Four of the eight targets of SDG2 are connected to food security as part of the resource nexus, while SDG12 has specific targets for sustainable production and consumption patterns and the reduction of the per capita food waste and the food production losses.

Table 32.1 The resource nexus in global goals

	MDGs	*SDGs*
Water	**Goal 7** Ensure environmental sustainability • **Target 7C** Halve, by 2015, the proportion of people without sustainable access to safe drinking water and basic sanitation	**Goal 6** Ensure availability and sustainable management of water and sanitation for all • **Target 6.1** By 2030, achieve universal and equitable access to safe and affordable drinking water for all • **Target 6.3** By 2030, improve water quality by reducing pollution, eliminating dumping and minimizing release of hazardous chemicals and materials, halving the proportion of untreated wastewater and substantially increasing recycling and safe reuse globally • **Target 6.4** By 2030, substantially increase water-use efficiency across all sectors and ensure sustainable withdrawals and supply of freshwater to address water scarcity and substantially reduce the number of people suffering from water scarcity • **Target 6.5** By 2030, implement integrated water resources management at all levels, including through transboundary cooperation as appropriate • **Target 6.6** By 2020, protect and restore water-related ecosystems, including mountains, forests, wetlands, rivers, aquifers and lakes • **Target 6.a** By 2030, expand international cooperation and capacity-building support to developing countries in water- and sanitation-related activities and programmes, including water harvesting, desalination, water efficiency, wastewater treatment, recycling and reuse technologies • **Target 6.b** Support and strengthen the participation of local communities in improving water and sanitation management
Energy		**Goal 7** Ensure access to affordable, reliable, sustainable and modern energy for all • **Target 7.1** By 2030, ensure universal access to affordable, reliable and modern energy services • **Target 7.2** By 2030, increase substantially the share of renewable energy in the global energy mix • **Target 7.3** By 2030, double the global rate of improvement in energy efficiency • **Target 7.a** By 2030, enhance international cooperation to facilitate access to clean energy research and technology, including renewable energy, energy efficiency and advanced and cleaner fossil-fuel technology and promote investment in energy infrastructure and clean energy technology • **Target 7.b** By 2030, expand infrastructure and upgrade technology for supplying modern and sustainable energy services for all in developing countries, in particular least developed countries, small island developing States, and landlocked developing countries, in accordance with their respective programmes of support **Goal 12** Ensure sustainable consumption and production patterns • **Target 12.c** Rationalize inefficient fossil-fuel subsidies that encourage wasteful consumption by removing market distortions, in accordance with national circumstances, including by restructuring taxation and phasing out those harmful subsidies, where they exist, to reflect their environmental impacts, taking fully into account the specific needs and conditions of developing countries and minimizing the possible adverse impacts on their development in a manner that protects the poor and the affected communities

Food	**Goal 2** Eradicate extreme hunger and poverty **Target 1C** Halve, between 1990 and 2015, the proportion of people who suffer from hunger	**Goal 2** End hunger, achieve food security and improved nutrition and promote sustainable agriculture • **Target 2.1** By 2030, end hunger and ensure access by all people, in particular the poor and people in vulnerable situations, including infants, to safe, nutritious and sufficient food all year round • **Target 2.3** By 2030, double the agricultural productivity and incomes of small-scale food producers, in particular women, indigenous peoples, family farmers, pastoralists and fishers, including through secure and equal access to land, other productive resources and inputs, knowledge, financial services, markets and opportunities for value addition and non-farm employment • **Target 2.4** By 2030, ensure sustainable food production systems and implement resilient agricultural practices that increase productivity and production, that help maintain ecosystems, that strengthen capacity for adaptation to climate change, extreme weather, drought, flooding and other disasters and that progressively improve land and soil quality • **Target 2.a** Increase investment, including through enhanced international cooperation, in rural infrastructure, agricultural research and extension services, technology development and plant and livestock gene banks in order to enhance agricultural productive capacity in developing countries, in particular least developed countries **Goal 12** Ensure sustainable consumption and production patterns • **Target 12.2** Ensure sustainable consumption and production patterns • **Target 12.3** By 2030, halve per capita global food waste at the retail and consumer levels and reduce food losses along production and supply chains, including post-harvest losses
Land		**Goal 15** Protect, restore and promote sustainable use of terrestrial ecosystems, sustainably manage forests, combat desertification, and halt and reverse land degradation and halt biodiversity loss • **Target 15.3** By 2030, combat desertification, restore degraded land and soil, including land affected by desertification, drought and floods, and strive to achieve a land degradation-neutral world
Minerals		Even though there is no specific goal or target related to minerals and mining activities, this economic activity connects to the following goals, among others: **Goal 1** End poverty in all its forms everywhere **Goal 6** Ensure access to water and sanitation for all **Goal 7** Ensure access to affordable, reliable, sustainable and modern energy for all **Goal 8** Promote sustained, inclusive and sustainable economic growth, full and productive employment and decent work for all **Goal 9** Build resilient infrastructure, promote inclusive and sustainable industrialization and foster innovation **Goal 12** Ensure sustainable consumption and production patterns **Goal 15** Protect, restore and promote sustainable use of terrestrial ecosystems, sustainably manage forests, combat desertification, and halt and reverse land degradation and halt biodiversity loss

Sources: Columbia Center of Sustainable Investment, UNDP, UN Sustainable Development Solutions Network, and World Economic Forum (2016); UN General Assembly (2015); United Nations (2000)

SDG2 recognizes the interlinkages between agriculture, small farming, rural poverty, health, and climate change, and the need to boost food production, agricultural productivity and rural incomes as means to eradicate not only hunger but also poverty. Sustainable agriculture and the management of food supply chains also connects to productivity, waste reduction, and the conservation of multiple ecosystems. Land, soils, water and biodiversity are critical inputs into food production, and their scarcity makes it imperative to promote their sustainable management with other goals for water, sustainable consumption and production, and ecosystems (Milder et al., 2012). Scientific knowledge and integrated decision-making processes are central to achieving SDG2. Progress on SDG2 was reviewed at the High-Level Political Forum in 2017. Governments agreed that despite some progress on attaining the goal, the necessary transformation would require much greater investment of effort and resources or large populations would remain malnourished.

Unlike food, water did not have a standalone goal in the MDGs. It was, however, one of the targets in MDG7 – ensure environmental sustainability. Target 7C required the halving, by 2015, the proportion of people without sustainable access to safe drinking water and basic sanitation. Two efforts shaped the process of implementing target 7C. First, in 2004 the UN Secretary-General established a panel of eminent people to advise on the solution of global challenges in water management and sanitation. The UN Secretary-General Advisory Board on Water and Sanitation provided important input on access to safe water, mainstreaming sanitation, finances, integrated water resources management, pollution, and the prevention of water-related disasters, including the interrelationships between water, energy and food systems and the need to address these synergies (UNSGAB, 2015). Second, the 2005–2015 "Water for Life" decade was critical to promoting efforts to fulfill international commitments made on water and water-related issues, and achieve not only target 7C under MDG7 but the objectives established by Agenda 21 and the 2002 World Summit for Sustainable Development. Its key challenge was to guarantee that governments and stakeholders activities and policies were oriented to long-term sustainable management of water resources and included measures to improve sanitation. Furthermore, in 2010 the UN General Assembly adopted Resolution 64/292 in which governments affirmed the human right to water and sanitation, and acknowledged that clean drinking water and sanitation are essential to the achievement of all human rights. Significant progress was made to halve the proportion of people without access to improved sources of water, and between 1990 and 2015, 2.6 billion people gained access to water, while 2.1 billion people gained access to improved sanitation (United Nations, 2015).

The end of the "Water for Life" decade and of the MDGs timeline facilitated the creation of even more specific and ambitious goals regarding water resources. In the 2030 Agenda for Sustainable Development, member states reaffirmed their "commitments regarding the human right to safe drinking water and sanitation" (UN General Assembly, 2015) and created SDG6 – Clean Water and Sanitation – to ensure the availability and sustainability of water and sanitation for all. Seven out of the eight targets for this goal refer to water management, including the management of water-based ecosystems, and the development of strategies for international cooperation and local communities' participation. Increased emphasis on quality, efficiency, integrated water resources management programs, and the environmental health of water systems reflects on the efforts of UN-Water and the UN Secretary-General Advisory Board for Water and Sanitation (Muller, 2015).

SDG6 has clear linkages to other goals relating to health, food security, climate change, ecosystems, and resilience among, others. Since the definition of the goal, countries and stakeholders have worked to identify these specific connections and to set targets and indicators that establish an ambitious vision for the next 15 years. In this process, the facilitative role of UN-Water will be important. SDG6 is slated for review in 2018.

Energy was not part of the Millennium Development Goals either as a goal or a target. It was only after the UN efforts around the Sustainable Energy for All initiative that discussions about a specific SDG on energy began. The 2030 Agenda for Sustainable Development includes a specific goal on energy. SDG7 – Affordable and Clean Energy – that seeks to ensure universal access to modern energy services, increase the share of renewable energy in the global energy mix, improve energy efficiency, infrastructure and technology, and enhance international cooperation for access to research and technology. SDG6 includes targets for all the key issues research, technology, efficiency, cooperation, and infrastructure to ensure energy access for all.

Energy links directly to the other nexus dimensions. For example, water and energy are interrelated and the linkages will intensify with increased demand for both. The energy sector is a major water user (Van de Graaf and Colgan, 2016). Power generation requires large amounts of water as does the production of fossil fuels and biofuels. The water sector is, in turn, a major user of energy for water supply and treatment and demand for energy will only increase, as desalination becomes a more prevalent water supply option. Understanding and managing these connections will be critical to ensure sustainable development (Dubash and Florini, 2011). Indeed, achieving SDG7 will be critical to attaining most of the other goals, including poverty eradication (SDG1), hunger elimination (SDG2), good health and well-being (SDG3), quality education (SDG4), gender equality (SDG5), clean water and sanitation (SDG6), economic growth (SDG8), industry, innovation and infrastructure (SDG9), reduced inequalities (SDG10), sustainable cities and communities (SDG11), responsible production and consumption (SDG12), and climate action (SDG13).

Desertification and land degradation are addressed in SDG15 Life on Land. Specifically, target 15.3 aims at combating desertification, restoring degraded land, and achieving a land-degradation-neutral world (UN General Assembly, 2015). This target has become the instrument to drive the implementation of UNCCD, and its attainment will also contribute to other SDGs, including climate, biodiversity, food security, water management, disaster risk reduction, and poverty elimination.

There is no specific SDG or target connected to extractive industries. However, it is clear that the industry has connections with all the goals, and should definitely contribute through its engagement, partnerships and industry dialogues that bring together all actors involved. Specific connections have been identified with each of the three dimensions of sustainable development, in issues such as water and sanitation, energy, poverty, inequalities, peace, economic growth, and infrastructure and innovation (Columbia Center of Sustainable Investment, UNDP, UN Sustainable Development Solutions Network, and World Economic Forum, 2016). The mining industry has an opportunity to contribute to the attainment of SDGs, with both improvements in the targets and the prevention and mitigation of negative consequences.

Nexus integration across sectors

As the international community begins implementing the Sustainable Development Goals, the trade-offs and synergies among the goals will become pressing (UN General Assembly, 2015; United Nations, 2000). It will be critical to analyze the role of policies across the nexus areas to enable a more integrated approach that understands each issue not as a separate pillar but as dimensions of the same goal – the integrated management of critical natural resources. Achieving goals in water, food, energy, land, and minerals policies, and contributing to the fulfillment of the SDGs requires instruments and mechanisms that recognize the multidimensional nature of the resource nexus and the interconnections among the different dimensions.

Importantly, there is a clear deficit of information in the management of the resource nexus. Authoritative and comprehensive reports exist in each of the nexus areas, but there is little integration among them. The UN Environment Programme produces the Global Environment

Outlook (GEO) and the Global Chemicals Outlook as well as various reports through the International Resources Panel; UNESCO publishes the UN-Water World Water Development Report, FAO the Global Food Security Update, and the IEA the World Energy Outlook. At most, two of the five nexus domains have been discussed in connection to each other as pairs – water and energy (UN-Water, 2014), water and food (Roberts and Barton, 2015), energy and agriculture. The interconnectedness of all dimensions has been difficult to map out and analyze systematically.

UNEP has begun work in this area through an ecosystem approach. It has launched water–energy–food nexus dialogues to facilitate knowledge acquisition and policy development with a focus at the regional level. In collaboration with the African Ministers' Council on Water (AMCOW) and river-basin organizations, UNEP works to engage a range of sectors in a nexus dialogue, to support environmental ministries in creating the appropriate governance structures, build capacity, and serve as a convener of nexus deliberations. The overall goals of such dialogues are to ensure sustainable livelihoods and reduce environmental risk. The core nexus initiatives underway currently include:

- Broadening the Nexus dialogue to include ecosystems
- Using science to improve understanding of sustainability
- Developing tools for integrated approaches and promoting a common nexus language
- Supporting policy development support and capacity building

The information dimension could be partially addressed by closer collaboration of the key UN institutions producing the flagship reports in each nexus area. They will, however, need to coordinate efforts not only to understand the integration among the various dimensions but also to respond to the new data collection and reporting demands. Common rules and methodologies for monitoring, early warning mechanisms, and scientific expertise are required to effectively respond to the multiple threats and risks associated to the management of natural resources, including in some cases the lack of transparency (Ivanova and Escobar-Pemberthy, 2016).

Facilitating linkages and promoting integration lie at the core of a successful strategy for implementing the SDGs. Solid institutional arrangements and predictable, stable, and coherent financing will be critical both for the resource nexus and the 2030 Sustainable Development Agenda. This does not imply the creation of new global governance institutions but calls for more robust and functional inter-agency and inter-program efforts that recognize the integrated nature of the contemporary sustainable development challenges. The resource nexus is not only relevant to the SDGs directly related to each of its domains, but also to other economic, social, and environmental goals that require a comprehensive and holistic approach for their implementation.

Designating an International Year of the Resource Nexus and potentially a Resource Nexus Decade could promote the integration of the various dimensions of sustainable development. It would bring together the various UN agencies already working on the different aspects of natural resources management and integrate the efforts already underway in water management, agricultural management, food security, and energy and minerals development. It will likely enable increased coherence, improved access, and greater visibility through the generation and dissemination of comprehensive information, the creation of implementation and monitoring mechanisms to support global and national policies, the efficient use of resources, the assessment of progress, and the effective engagement with multiple stakeholders.

Notes

1 IRENA is an intergovernmental agency with more than 170 countries and the EU engaged but is not formally a UN agency.
2 For a list of the agencies and their activities, see www.se4all.org/decade_un-energy-decade

References

Columbia Center of Sustainable Investment, UNDP, UN Sustainable Development Solutions Network, and World Economic Forum. (2016). *Mapping mining to the sustainable development goals: An atlas*. New York: World Economic Forum.

Dubash, N. K., and Florini, A. (2011). Mapping global energy governance. *Global Policy*, 2(s1), 6–18.

FAO. (1996). *Rome declaration on world food security and world food summit plan of action*. Rome, Italy: United Nations.

FAO. (2002). *Declaration of the world food summit: Five years later*. Rome, Italy: United Nations.

FAO. (2014). *The water-energy-food nexus: A new approach in support of food security and sustainable agriculture*. Rome, Italy: United Nations.

FAO. (2015). *Food and agriculture organization – what we do?* Retrieved from www.fao.org/about/what-we-do/en/

Hellegers, P., Zilberman, D., Steduto, P., and McCornick, P. (2008). Interactions between water, energy, food and environment: Evolving perspectives and policy issues. *Water Policy*, 10(S1), 1–10.

Inomata, T. (2008). *Management review of environmental governance within the United Nations system*. Geneva: Joint Inspection Unit. JIU/REP/2008/3.

International Energy Agency (IEA). (2016). *World Energy Outlook*. Retrieved from http://www.worldenergyoutlook.org/publications/weo-2016/

Ivanova, M., and Escobar Pemberthy, N. (2016). The quest for sustainable development: The power and perils of global development goals. In T. Pogge, G. Köhler, and A. D. Cimadamore (Eds.), *Poverty & the millennium development goals (MDGs): A critical assessment and a look forward*. London: CROP/Zed Books, pp. 83–111.

Kanie, N., and Biermann, F. (2017). *Governing through goals: Sustainable development goals as governance innovation*. Cambridge, MA: MIT Press.

Manning, R. (2009). *Using indicators to encourage development: Lessons from the millennium development goals*. DIIS Reports/Danish Institute for International Studies.

McArthur, J. W. (2013). *The declaration of the millennium development goals*. Washington, DC: Brookings Institution.

Melamed, C., and Scott, L. (2011). *After 2015: Progress and challenges for development*. London: Overseas Development Institute.

Milder, J. C., Buck, L. E., DeClerck, F., and Scherr, S. J. (2012). Landscape approaches to achieving food production, natural resource conservation, and the millennium development goals. In *Integrating ecology and poverty reduction*. New York: Springer, pp. 77–108.

Muller, M. (2015). The "nexus" as a step back towards a more coherent water resource management paradigm. *Water Altern*, 8(1), 675–694.

Nilsson, M., Griggs, D., and Visbeck, M. (2016). Policy: Map the interactions between Sustainable Development Goals. *Nature*, 534, 320–322.

Roberts, E., and Barton, B. (2015). *Feeding ourselves thirsty: How the food sector is managing global water risks*. Boston, MA: CERES.

Sachs, J. D., and McArthur, J. W. (2005). The millennium project: A plan for meeting the millennium development goals. *Lancet*, 365(9456), 347–353.

Scott, C. A., Kurian, M., and Wescoat Jr, J. L. (2015). The water-energy-food nexus: Enhancing adaptive capacity to complex global challenges. In *Governing the Nexus*. Cham: Springer, pp. 15–38.

UN General Assembly. (2015). *A/RES/70/1 transforming our world: The 2030 Agenda for Sustainable Development*. New York: United Nations.

UNCED. (1992). *Agenda 21*. New York: Retrieved from www.un.org/esa/sustdev/documents/agenda21/english/Agenda21.pdf

United Nations. (1945). *UNTS XVI charter of the United Nations*. San Francisco, CA: United Nations.

United Nations. (1994). *Convention to combat desertification*. Retrieved from www.unccd.int/en/about-the-convention/Pages/About-the-Convention.aspx.

United Nations. (2000). *A/RES/55/2 United Nations millennium declaration*. New York: UN.
United Nations. (2009). *High-level task force on global food and nutrition security*. Retrieved from www.un.org/en/issues/food/taskforce/index.shtml
United Nations. (2012). *A/RES/66/288 the future we want – outcome document from Rio+20*. United Nations Conference on Sustainable Development, Rio de Janeiro.
United Nations. (2015). *Millennium development goals report 2015*. Retrieved from www.un.org/millenniumgoals/2015_MDG_Report/pdf/MDG 2015 rev (July 1).pdf
United Nations, and World Bank. (2016). *Sustainable energy for all: Our mission/our work/our partners*. Retrieved from www.se4all.org/
UNSGAB. (2015). *The UNSGAB journey water and sanitation*. New York: United Nations.
UN-Water. (2014). *United Nations world water development report water and energy* (Vol. 1). Geneva, Switzerland: United Nations.
UN-Water. (2016). *UN-water work programme 2016–2017*. Geneva, Switzerland: United Nations.
Van de Graaf, T. (2013). Fragmentation in global energy governance: Explaining the creation of IRENA. *Global Environmental Politics*, 13(3), 14–33.
Van de Graaf, T. (2015). The IEA, the new energy order and the future of global energy governance. In D. Lesage and T. Van de Graaf (Eds.), *Rising powers and multilateral institutions*. Palgrave McMillan: London, pp. 79–95.
Van de Graaf, T., and Colgan, J. (2016). Global energy governance: A review and research agenda. *Palgrave Communications*, 2, 15047. doi:10.1057/palcomms.2015.47
Waage, J., Banerji, R., Campbell, O., Chirwa, E., Collender, G., Dieltiens, V., . . . Unterhalter, E. (2010). The millennium development goals: A cross-sectoral analysis and principles for goal setting after 2015. *The Lancet*, 376(9745), 991–1023.
World Energy Council. (2013). *World energy scenarios: Composing energy futures to 2050*. London: World Energy Council.

INDEX

Note: Page numbers in italics indicate figures. Page numbers in bold indicate tables.

For Product Safety Concerns and Information please contact our EU representative GPSR@taylorandfrancis.com Taylor & Francis Verlag GmbH, Kaufingerstraße 24, 80331 München, Germany